3D Radiative Transfer in Cloudy Atmospheres

The Noble Cumulus
oil and acrylic on canvas 22×28 in.

A. Marshak   A.B. Davis

Editors

3D Radiative Transfer in Cloudy Atmospheres

With 227 Figures

 Springer

Alexander Marshak
NASA/Goddard Space Flight Center
Climate and Radiation Branch
Mail Code 913
Greenbelt, Maryland 20771
USA

Anthony Davis
Los Alamos National Laboratory
Space and Remote Sensing Sciences Group
(ISR-2)
Los Alamos, New Mexico 87545
USA

About the painting

It was Ruskin, a noted art critic of the 19th century, who commented on the appearance of clouds in art and recollected that 'There has been so much . . . fog and artificial gloom, besides, that I find it is actually some two years since I last saw a noble cumulus under full light'. The painting 'The Noble Cumulus' is inspired by the nobility of illuminated clouds and their effects on radiation, the principal topic of this book. This painting is one of a series that is inspired by the artist's interest in the science of clouds and the role the study of clouds has played in the development of meteorology as a science over the years. This topic and others are described in the article by the artist: The useful pursuit of shadows. *Am. Scientist*, 91, 442–449 (2003). Other art and a pdf of this article can be found at http://cloudsat.atmos.colostate.edu/CSMart.php.

About the artist

Graeme L. Stephens is a Professor in the Department of Atmospheric Sciences at Colorado State University in Ft. Collins, Colorado. He has spent much of his career studying clouds and their effects on atmospheric radiation, has numerous publications, author of a text book on remote sensing and is a contributing author of this book. Professor Stephens has received a number of major awards for his scientific research on clouds and radiation.

Cover image: 3D Monte Carlo simulation of the solar radiance field from a broken cumulus field itself simulated using a high-resolution dynamical cloud model (Fig. 4.6 in the text).

ISSN: 1610-1677

ISBN-13   978-3-642-06303-9   e-ISBN-13   978-3-540-28519-9

Springer is a part of Springer Science+Business Media
springeronline.com
© Springer-Verlag Berlin Heidelberg 2005
Softcover reprint of the hardcover 1st edition 2005

Cover design: Erich Kirchner

Dedication

To the memory of

Georgii A. Titov

(March 5, 1948 – July 25, 1998)

and

Gerald C. Pomraning

(February 25, 1936 – February 6, 1999)

George Titov and Jerry Pomraning (courtesy of Lucia Levermore).

"[Preliminary] Monte Carlo results are generally modulo a factor of π."

Georgii Titov

"Life is an integral!"

Jerry Pomraning

Physics of Earth and Space Environments

springeronline.com/series/5117/

The series *Physics of Earth and Space Environments* is devoted to monograph texts dealing with all aspects of atmospheric, hydrospheric and space science research and advanced teaching. The presentations will be both qualitative as well as quantitative, with strong emphasis on the underlying (geo)physical sciences.
Of particular interest are

- contributions which relate fundamental research in the aforementioned fields to present and developing environmental issues viewed broadly

- concise accounts of newly emerging important topics that are embedded in a broader framework in order to provide quick but readable access of new material to a larger audience

The books forming this collection will be of importance for graduate students and active researchers alike.

Series Editors:

Rodolfo Guzzi
Responsabile di Scienze della Terra
Head of Earth Sciences
Via di Villa Grazioli, 23
00198 Roma, Italy

Dieter Imboden
ETH Zürich
ETH Zentrum
8092 Zürich, Switzerland

Louis J. Lanzerotti
Bell Laboratories, Lucent Technologies
700 Mountain Avenue
Murray Hill, NJ 07974, USA

Ulrich Platt
Ruprecht-Karls-Universität Heidelberg
Institut für Umweltphysik
Im Neuenheimer Feld 366
69120 Heidelberg, Germany

Preface

A few years ago one of us (AM) was giving a series of lectures on three-dimensional (3D) radiative transfer in cloudy atmospheres at the Summer 1999 School "Exploring the Atmosphere by Remote Sensing Techniques" hosted by the Abdus Salam International Centre for Theoretical Physics in Trieste (Italy). By the end of the series, the instructor was asked by students for an available book on the subject. It turned out that in spite of multiple decades of research, relative maturity of the field, the involvement of dozens of scientists worldwide, and hundreds of journal papers, there was in fact no tutorial book in existence. So there was nowhere for students and young researchers to start or to use as a reference. One of the directors of the school, Rodolfo Guzzi, and the editor of the physics section of Springer-Verlag, Christian Caron, who was also there, suggested that we fill this gap by writing a monograph on the subject.

We enthusiastically accepted the Springer-Verlag commission and attracted many leading 3D radiative transfer scientists as co-authors: H. Barker, N. Byrne, R. Cahalan, E. Clothiaux, R. Davies, R. Ellingson, F. Evans, P. Gabriel, A. Heidinger, Y. Knyazikhin, A. Korolev, R. Myneni, I. Polonsky, G. Stephens, E. Takara, and W. Wiscombe. More than half of them are on the science team of the Atmospheric Radiation Measurements (ARM) program sponsored by the U.S. Department of Energy (DOE). A major goal of ARM is to further our understanding of radiative transfer in the atmosphere – especially the role of clouds – and at the Earth's surface. The DOE's ARM program has therefore provided generous funding for this book project. We also greatly appreciate the ongoing support we receive from our home institutions, Los Alamos National Laboratory and NASA's Goddard Space Flight Center, and the support received from the Joint Center for Earth Systems Technology of UMBC, where one of us worked at the beginning of the project. Technical expertise in Springer-Verlag's LaTeX desktop publishing environment was ensured by Lisa LeBlanc, now with the Canadian CLIVAR Network at McGill University; without her help, we would not have been able to prepare this manuscript.

The title of this book is "Three-Dimensional Radiative Transfer in Cloudy Atmospheres." At one point, we were tempted to use the more provocative title "*Real* Radiative Transfer in Cloudy Atmospheres." Indeed, it is the 3D radiative transfer equation that determines the radiation processes in real cloudy atmospheres. By contrast, the standard 1D model, which can be traced back at least 100 years, is an approximation that should prove useful under certain circumstances. In other words, it is time to think of 3D theory as the golden standard in atmospheric radiative transfer rather than as a perturbation of standard 1D theory.

The book captures and preserves much of the best 3D cloud radiation work done in the last couple of decades, and brings it to better maturity as authors took special care to explain their discoveries and advances to a larger audience. Our primary readership will be made of graduate students and researchers who specialize in atmospheric radiation and cloud remote sensing. However, we hope that remote sensing scientists in other application areas (biosphere, hydrosphere, cryosphere, etc.) will find many portions of the volume stimulating.

Beyond the two introductory chapters, the volume naturally divides into three parts: Fundamentals, Climate, and Remote Sensing. The two last topics are indeed the main concerns in atmospheric radiation science. The chapters are essentially independent but cross-reference each other. We tried our best to avoid overlap; in several places, however, we found it more effective to repeat some material rather than pointing to other portions of the book. Most chapters end with Notes and/or a Suggested Reading list because they open more questions then they answer; these contain input from the authors, the reviewers, and the editors. As much as possible, we tried to use the same notation throughout the whole book. A list of notations and a subject index can be found at the end of the volume. Each chapter has been peer-reviewed by at least one reviewer internal to the author collective and one external reviewer. We wish to thank all reviewers, especially the external ones: Larry Di Girolamo, Qiang Fu, Jeff Haferman, Harshvardhan, Alexei Lyapustin, Andreas Macke, John Martonchik, Lazaros Oreopoulos, Klaus Pfeilsticker, Bill Ridgway, Tamas Várnai, and Tatyana Zhuravleva.

This project took us much longer than we initially anticipated. Being committed to other projects during the daytime, we mostly worked on the book during the evenings and weekends at home, taking time from our families. We are very grateful for their support and understanding. It was rewarding to work on this book, writing our own chapters, reading and editing other chapters. We personally learned a lot and we hope that the readers will enjoy it too.

Finally, we dedicate this book to the memory of two great radiative transfer scientists, G. Pomraning and G. Titov. We consider ourselves lucky to have met them and to have learned so much from them.

Greenbelt, Maryland   *Alexander Marshak*
December, 2004   *Anthony Davis*

Contents

Part I   Preliminaries

1 Scales, Tools and Reminiscences
W.J. Wiscombe   3

2 Observing Clouds and Their Optical Properties
E.E. Clothiaux, H.W. Barker and A.V. Korolev 93

Part II   Fundamentals

3 A Primer in 3D Radiative Transfer
A.B. Davis and Y. Knyazikhin 153

4 Numerical Methods
K.F. Evans and A. Marshak 243

5 Approximation Methods in Atmospheric 3D Radiative Transfer,
Part 1: Resolved Variability and Phenomenology
A.B. Davis and I.N. Polonsky 283

Part III   Climate

6 Approximation Methods in Atmospheric 3D Radiative Transfer,
Part 2: Unresolved Variability and Climate Applications
H.W. Barker and A.B. Davis 343

7 3D Radiative Transfer in Stochastic Media
N. Byrne 385

8 Effective Cloud Properties for Large-Scale Models
R.F. Cahalan 425

9 Broadband Irradiances and Heating Rates for Cloudy Atmospheres
H.W. Barker 449

10 Longwave Radiative Transfer in Inhomogeneous Cloud Layers
R.G. Ellingson and E.E. Takara 487

Part IV   Remote Sensing

---

11 3D Radiative Transfer in Satellite Remote Sensing of Cloud Properties
R. Davies... 523

12 Horizontal Fluxes and Radiative Smoothing
A. Marshak and A.B. Davis 543

13 Photon Paths and Cloud Heterogeneity: An Observational Strategy to Assess Effects of 3D Geometry on Radiative Transfer
G.L. Stephens, A.K. Heidinger and P.M. Gabriel 587

14 3D Radiative Transfer in Vegetation Canopies and Cloud-Vegetation Interaction
Y. Knyazikhin, A. Marshak and R.B. Myneni 617

Appendix: Scale-by-Scale Analysis and Fractal Cloud Models
A. Marshak and A.B. Davis 653

Epilogue: What Happens Next?
.... 665

Notations 671

Index 683

List of Contributors

Howard W. Barker
Meteorological Service of Canada
Downsview, Ontario M3H 5T4
Canada
Howard.Barker@ec.gc.ca

Nelson Byrne
Science Applications International
Corporation
San Diego, California 92121
USA
nbyrne@pacbell.net

Robert F. Cahalan
NASA-Goddard Space Flight Center
Climate and Radiation Branch
Code 613.2
Greenbelt, Maryland 20771
USA
Robert.F.Cahalan@nasa.gov

Eugene E. Clothiaux
Department of Meteorology
Penn State University
University Park, Pennsylvania 16802
USA
cloth@essc.psu.edu

Roger Davies
Jet Propulsion Laboratory
Pasadena, California 91109
USA
Roger.Davies@jpl.nasa.gov

Anthony B. Davis
Los Alamos National Laboratory
Space and Remote Sensing Sciences
Group (ISR-2)
Los Alamos, New Mexico 87545
USA
adavis@lanl.gov

Robert G. Ellingson
Department of Meteorology
Florida State University
Tallahassee, Florida 32306
USA
bobe@met.fsu.edu

K. Franklin Evans
Program in Atmospheric and Oceanic
Sciences
University of Colorado
Boulder, Colorado 8030
USA
evans@nit.colorado.edu

Philip M. Gabriel
Department of Atmospheric Science
Colorado State University
Fort Collins, Colorado 80523
USA
gabriel@atmos.colostate.edu

Andrew K. Heidinger
UW/CIMSS/NOAA
1225 West Dayton St.
Madison, Wisconsin 53706
USA
Andrew.Heidinger@noaa.gov

Yuri Knyazikhin
Department of Geography
Boston University
Boston, Massachusetts 02215
USA
jknjazi@bu.edu

Alexei V. Korolev
Meteorological Service of Canada
Downsview, Ontario M3H 5T4
Canada
Alexei.Korolev@rogers.com

Alexander Marshak
NASA-Goddard Space Flight Center
Climate and Radiation Branch
Code 613.2
Greenbelt, Maryland 20771
USA
Alexander.Marshak@nasa.gov

Ranga B. Myneni
Department of Geography
Boston University
Boston, Massachusetts 02215
USA
rmyneni@bu.edu

Igor N. Polonsky
Los Alamos National Laboratory
Space and Remote Sensing Sciences
Group (ISR-2)
Los Alamos, New Mexico 87545
USA
polonsky@lanl.gov

Graeme L. Stephens
Department of Atmospheric Science
Colorado State University
Fort Collins, Colorado 80523
USA
stephens@atmos.colostate.edu

Ezra E. Takara
Department of Meteorology
Florida State University
Tallahassee, Florida 32306
USA
etakara@met.fsu.edu

Warren J. Wiscombe
NASA-Goddard Space Flight Center
Climate and Radiation Branch
Code 613.2
Greenbelt, Maryland 20771
USA
Warren.J.Wiscombe@nasa.gov

Part I

Preliminaries

1

Scales, Tools and Reminiscences

W.J. Wiscombe

1.1	Why Should We Care About Clouds?	7
1.2	My Life in Cloud Radiation	18
1.3	The Estrangement of Cloud Radiation, Cloud Physics, and Rain Remote Sensing	28
1.4	The "Science as Tool-Driven" Viewpoint	32
1.5	Tools for 3D Clouds, 1970s through 1990s	34
1.6	Current Cloud Observational Tools	39
1.7	Future Cloud Observational Tools	47
1.8	Tomography: The Ultimate Solution?	52
1.9	Cloud Structure Modeling: Introduction	55
1.10	Cloud Structure Modeling: Luke Howard	55
1.11	Cloud Structure Modeling: After Luke Howard	57
1.12	Scale	62
1.13	3D Radiative Transfer	64
1.14	Turbulent Radiative Transfer	70
1.15	Laws of Clouds	70
1.16	Why Cloud Radiation is So Hard	75
1.17	Cloud Challenge Questions for the 21st Century	78
1.18	Epilogue	80
	References	81
	Suggested Further Reading	89

A 3D cloud scientist's mental model of Earth, concocted from real AVHRR and GOES data by Washington University, but with vertically exaggerated clouds to indicate their true importance for global climate.[1]

> *"If the clouds could be rationally and convincingly explained, without recourse to superstition and prejudice, then so could anything else in nature, for they represented the most supreme manifestation of the ungraspable."*

René Descartes (1596–1650)

> *"Clouds themselves, by their very nature, are self-ruining and fragmentary. They flee in haste over the visible horizons to their quickly forgotten denouements. Every cloud is a small catastrophe, a world of vapor that dies before our eyes. [...] And as long as clouds, for the poetic imagination, stood as ciphers of a desolate beauty, gathering in apparently random patterns only to disperse with the wind, how could they ever be imagined as part of Nature's continuous scheme? What could there be to a cloud, beyond a vague metaphorical allure?"*

Richard Hamblyn, The Invention of Clouds (2001)

---

[1] http://capita.wustl.edu/CAPITA/DataSets/MODIS/GlobFused/glob3d.html

Fig. 1.1. The cloud theoretician's worst nightmare: real 3D clouds as viewed from an aircraft window ill-advisedly left unshuttered

A running joke among my colleagues, one that particularly amused 3D cloud radiation pioneer (and author herein) Roger Davies, was my assertion in the 1980s that all good cloud radiation modelers should close their airplane windowshades so as not to be corrupted by the spectacle of real 3D clouds such as that in Fig. 1.1. With a determined effort to dismiss real clouds as an evanescent illusion (a sentiment expressed much more poetically by Richard Hamblyn above), I was able to rationalize my own simple models of clouds, both mental and computational.

My windowshade joke had a serious side, however, for at the time I believed – on faith alone – that 3D cloud effects could be cleverly mimicked by 1D models, at least with sufficient time- and space-averaging. In this belief, I was partly bending to the realities of climate models, which even now cannot afford to calculate full 3D radiation; and partly espousing the time-tested strategy of simplifying a phenomenon to the greatest extreme possible – the same strategy that occasionally leads us to model ice crystals as spheres. However, Einstein's dictum "simplify as far as possible, but no farther" reminds us that scientists must always ascend a knife-edge ridge with the chasms of oversimplification and overcomplexification yawning on either side. This difficult ascent is nowhere more evident, nor are the falls from grace so great, than in this field of 3D clouds, where firehoses of computer power seduce us toward overly complex models on one side, and climate modeling presses us toward overly simple models on the other.

This book is a testament to all those who refused to close their windowshades or take a smeared-out, statistical view of clouds yet who also refused to succumb to the allure of kitchen-sink modeling of every little detail. They stepped up to one of the hardest problems in all of Earth science – fully 3D clouds – with courage and perseverance, and their contributions in this book represent the cream of much of the 3D cloud radiation work of the past 15 years. Now is a good time to look back and take stock of what has been accomplished in this incredible burst of creation.

I attribute this burst mainly to two factors: (a) real progress on understanding cloud structure as a function of scale, and (b) the availability of new tools, both theoretical and experimental. Those will be my main themes in this chapter – scale and new tools. As prelude, however, I shall address the question "Why should we care about clouds?" and then describe my personal odyssey which profoundly shaped my admittedly unique view of this field.

Note that this book is not about all possible wavelengths of cloud radiation. Infrared, microwave and radar wavelengths receive comparatively short shrift. The tilt toward solar wavelengths (0.3 to 4 microns) accurately reflects the preponderance of research in the 1990s. The infrared attracted less interest mainly because clouds act primarily as near-blackbody blobs there and thus provide less theoretical challenge. Takara and Ellingson (2000) and Ellingson (1982) showed that the errors from neglecting finite clouds are typically no more than 20% in the longwave, whereas in the shortwave they can easily be 100% and more. Longwave radiation responds differently to cloud 3D-ness than shortwave – it is much more affected by the actual shape of the cloud, since it tends to come from the outermost 50 m of a cloud. Microwave and millimeter-wave radar observations of clouds remain solidly in their infancy, in spite of several decades of research, and it will probably be another decade before a book like this one could be produced for that wavelength range.

This chapter will not treat cirrus clouds.[2] The reasons are several. First, cirrus cloud scientists tend to think that single scattering by ice crystals, which is not the focus of this book, is at least if not more important than the 3D radiative effects of cirrus. Second, most cirrus are optically thin to sunlight and this simple limiting

---

[2] Cirrus are largely ignored in this book, except for Chaps. 2 and 10, and in much of the 3D radiative transfer literature. A rare exception is by Gu and Liou (2001) who investigated the dynamical feedback of 3D radiative fluxes in cirrus evolution using a Large-Eddy Simulation model.

case is not of much interest for 3D cloud radiation theorists.[3] Third, many of the measurement systems I discuss don't work well in ice clouds, and I don't want to make continual caveats about that. So let us agree that when I talk about clouds, I mean liquid water clouds.

Acronyms are new language and can be useful as such, in spite of my occasional outbursts against "encroaching acrobabble." But their continual redefinition in long, interruptive parenthetical expressions has become a plague upon scientific writing. Most readers know what the usual acronyms mean in a specialized-topic book such as this one. So, in an effort to recapture a free flow of narrative, I will take the shocking step of omitting most of the parenthetical definitions. For the benefit of newcomers to the field, all but the most familiar acronyms are listed with their meanings at the end of the chapter.[4]

This chapter will look at some cloud issues that go beyond just cloud radiation, but only eclectically. Even the treatment of cloud radiation is eclectic. For a comprehensive overview of cloud radiation from a climate viewpoint, I recommend the paper by Wielicki et al. (1995). For a good overview of the entire cloud problem by two titans in the field, I recommend the recent article by Randall et al. (2004) and the older one by Hobbs (1991). Both articles are highly readable and comprehensive, and their delightful styles reveal that scientific writing need not have every ounce of humanity and passion ruthlessly expunged.

1.1 Why Should We Care About Clouds?

According to the remarkable historical book by Hamblyn (2001), before 1800 only poets cared much about clouds. Shakespeare wrote more about them than all scientists put together. Hard as it is to believe, until Luke Howard's revolutionary classification of clouds based on their dynamical processes rather than their form, clouds were barely studied at all. The 1800s then became somewhat of a Golden Age for cloud physics; most of the laws used today were discovered then, and hilarious theories of clouds as bubbles and such like faded into a well-deserved obscurity. In the 1900s clouds passed from a concern of physics and chemistry to a concern of meteorology. The flight of Nobel-Prize-winner C.T.R. Wilson, of cloud chamber fame, from cloud physics to quantum mechanics around 1900 symbolized the end of an era of physicist interest. Once a province of meteorology, cloud physics became primarily involved with precipitation and thus with the rather limited class of strongly convective clouds. Arrhenius, the inventor of simple climate modeling, knew in 1896

---

[3] Optical thinness does allow cirrus to be probed by lidar, however, sometimes from top to bottom, and thus their remarkably turbulent internal structure is better measured than that of liquid water clouds.

[4] Those who find undefined acronyms disturbing are invited to notch the page and return for comfort as often as needed to this footnote: the sight of BNAs (Bare Naked Acronyms) is anathema to the AGU (Acronym Generation Unit) of the AMS (Acronym Manufacturing Society) and the growing dearth of FDAs (Fully Dressed Acronyms) has caused rapid growth of the Bare-Acronym-Angst Anonymous Association (BAAAA).

that clouds also affect the Earth's energy budget, but, for many decades thereafter, no one had the slightest idea by how much; there were guesstimations, but no solid data until the first thorough satellite analyses were published by Vonder Haar and Suomi (1971).

It is easy to understand the past neglect of clouds. Except when they rain on us, they seem little more than insubstantial evanescences in the sky. Like atoms, clouds are mostly empty space. Their solidity is an optical illusion. The authors in this book know how that illusion is created, but that takes nothing away from its wonder. A typical cloud droplet is about 10 microns in radius. A typical marine stratocumulus cloud has about 50 such droplets per cubic centimeter, which therefore fill only about one ten-millionth of the volume in which they reside. That's pretty empty! But what about the big rain clouds? Even they rarely have 1 g/m^3 of liquid water everywhere, but let's suppose that this extreme value fills a cubic cloud 1 km on a side. That's 10^9 g or 10^9 cm^3 of liquid, which would fill a cube 10 m on a side and occupy only one millionth of the cloud volume. Yet a few hundred meters of such tenuous cloudstuff can blot out the Sun and turn bright daylight to gray dusk. Clouds are nothing if not a testament to the almost incredible extinction power of fractionating a mass of material into micron-sized particles.

Now, let us try to sharpen the question in the section title: why have we created a scholarly tome on such a seemingly esoteric subject as 3D cloud radiation, and why now?

1.1.1 Climate

The short answer to the "Why?" question is that clouds are the greatest unknown in all of physical climate modeling; they radically alter the distribution of radiant energy and latent heating in ways that have proven devilishly hard to capture in climate models.

The powerful visual effects of clouds translate into equally powerful energetic effects. But amazingly, we couldn't quantify these effects until the advent of the ERBE (Earth Radiation Budget Experiment) three-satellite constellation in the mid-1980s. We were in fact doubly ignorant. First, beyond crude visual observations by weather observers (at most twice a day), and uncalibrated satellite observations that could not be made sufficiently quantitative, we didn't know how much cloud there was at what altitude, nor how thick it was optically or geometrically.[5] Second, since clouds reflect sunlight but preserve infrared radiation to the Earth (by radiating to space at a colder temperature than the surface they overlie), their solar and infrared radiation effects work against each other, and we had only theoretical calculations of which prevailed. So, as of the mid-1980s, we had only a crude idea of how much cloud there was, and how it affected the total (solar plus infrared) radiation. Things had not really advanced much since the simple models of Manabe and Wetherald

---

[5] This lack was partly remedied beginning in the mid-1980s by the International Satellite Cloud Climatology Program which now has a 20-year data record on certain cloud parameters (Rossow and Schiffer, 1999).

(1967) and Schneider (1972) first elucidated the many ways clouds could modify the climate.

The first numbers from ERBE were announced by Ramanathan et al. (1989). Ramanathan was a strong advocate of processing the ERBE data in such a way that the effect of clear sky could be subtracted out in order to manifest the effect of cloud in stark relief. While the paper had complete results only for April 1985, tentative results for three other months were reported, and all months indicated clouds had a net cooling effect on the Earth. For the first time, a number could be given: clouds' sunlight-reflection effect wins. It wins by about 20 W/m^2, or five times the energy effect of doubling CO_2.

A clear understanding of the importance of cloud radiation had solidified by 1990 (e.g., Arking, 1991), partly as a result of the Ramanathan et al. paper and partly due to Global Climate Model (GCM) intercomparisons (Cess et al., 1989) that showed large disagreements caused mainly by GCM cloud treatments. Spurred by these developments, the rest of the 1990s were a period of intense activity in the subject, not least the remarkable ARM (Atmospheric Radiation Measurements) Program of the U.S. Dept. of Energy whose focus was entirely clouds and radiation (Ackerman and Stokes, 2003). ARM and NASA in the U.S., and the Japanese, Canadian and several European governments abroad, steadily and reliably supported cloud radiation research, as well as its sibling, dynamical modeling for clouds beyond the traditional towering rainclouds that had previously gotten most of the attention. We entered the 1990s at a relatively low level of theoretical and observational capability but as a consequence of this steady support exited with an astoundingly better capability – ranging from how we designed cloud field programs, to the sophistication of cloud instruments, to understanding of how clouds scale, to the quality of cloud parameterizations in GCMs. Thus, in answer to the question "Why now?" there was a feeling in the community that, after this intense burst of activity, it had reached somewhat of a plateau of new knowledge, and that this would be a good time to collect what we have learned in one place and survey the extent of our conquest.

Perhaps the most obvious application of this new knowledge is to the issue of global warming (although it would be equally relevant to global cooling). Clouds, or more precisely lack of knowledge of how clouds interact with the climate system, impede useful forecasts of future global warming. All GCMs predict warming in response to CO_2 increase, but the warming ranges from moderate to severe depending on how they treat clouds. The right part of Fig. 1.2 shows predictions of global-average surface temperature warming for doubled CO_2 as various feedbacks are added to a single GCM, one by one. The range due to adding cloud feedback is 2 to 5°C, depending on what treatment is used. 2°C would already be a serious concern, but perhaps manageable with wisdom and foresight. 5°C would be equivalent to the warming since the last glacial retreat 10,000 years ago and, coupled with the end of the fossil fuel era by the end of this century (Goodstein, 2004), would certainly require unprecedented adaptations. 2°C or 5°C? Clouds hold the fate of the Earth in their hands, and we don't know which number they will pick, if either. They are indeed the lever sought by Archimedes, with which one could move the Earth – or at least the Earth's climate.

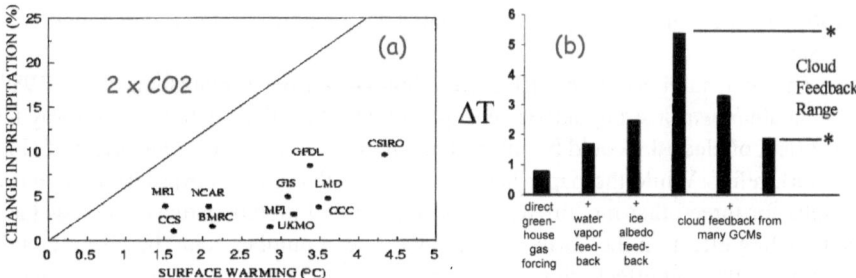

Fig. 1.2. (**a**) The temperature and precipitation responses of the leading coupled ocean-atmosphere GCMs from many countries to an imposed doubling of CO_2. No clustering is evident in the range of responses. (Adapted from Stephens et al. (2002).) (**b**) The global-average surface temperature response of a well-known British climate model to an imposed doubling of CO_2 with feedbacks added one by one: first no feedbacks; then water vapor feedback; then snow and ice albedo feedback; and finally different cloud feedbacks. (Adapted from Senior and Mitchell (1993).)

The left part of Fig. 1.2 shows that this 2–5°C range is not caused by a few outlying models; the range is uniformly populated. Furthermore, a recent 19-GCM comparison (Potter and Cess, 2004) reveals that this 2–5°C range has not diminished significantly in 14 years. This is surprising considering that most GCMs have worked hard to improve their cloud parameterizations over those 14 years, especially by introducing predictive equations for cloud liquid water. The "cloud radiative forcing" in all the GCMs, a measure of the warming or cooling effect of clouds, differs widely from the best satellite measurements. Thus, the cloud problem is proving a tougher nut to crack than anyone suspected!

All the climate models represented in Fig. 1.2 use 1D radiation. Would using 3D radiation make any difference? Figure 1.3 show how big the effect of 3D radiation can be, although for a cubic cloud case which is admittedly extreme. The difference between the Plane-Parallel Approximation commonly used in climate models and the Independent Pixel Approximation is so large that it would change the results of every climate model as well as the range of predicted temperature changes in Fig. 1.2. The point labeled 3D is correct for this particular sun angle, but for other sun angles and other situations could lie above the IPA point or even below the PPA point. In this case IPA does not seem to be an improvement, but in more realistic cases we find that IPA is a decided improvement for spatial averaged radiation. Indeed, for marine stratocumulus, we find that the PPA differs from the IPA by only about 10% and the IPA agrees with 3D to a few percent. Thus, in terms of the size of their 3D radiative effect, marine stratocumulus lie at one extreme and cubic clouds (or popcorn cumulus) lie at the other.

We have assumed that as clouds in GCMs are better calculated, the 2–5°C range will narrow. Some even hope that the range will narrow to a single number. That it has not, after over a decade of "improvements" in cloud treatments, raises another specter: some range (hopefully not 2–5°C!) may be intrinsic – a limit of predictability

Fig. 1.3. (*Left*) A regular array of cloud cubes, each with optical depth 50, asymmetry factor 0.85 and single-scattering albedo 0.999, embedded in a vacuum. Solar zenith angle is 50 degrees and cloud fraction is 50%. (*Right*) Transmittance versus optical depth: *solid curve* is for a 1D slab cloud and the three labeled points refer to the cloud array on the left, infinitely repeated. The point labeled PPA (Plane Parallel Approximation) simply uses the mean optical depth of the array (25) in 1D slab theory. The point labeled IPA (Independent Pixel Approximation) averages the transmissions of each column separately. The ordering shown, IPA above PPA, always holds because the curve of transmission versus optical depth is concave. (Adapted from a presentation by Bernhard Mayer at the 2004 International Radiation Symposium in Korea.)

if you will – and thus not "fixable" by improving the cloud treatments. After all, clouds are a fast random component of the system, loosely analogous to the stochastic "weather" term that was used in simple climate models of the 1970s (Hasselmann, 1976), and thus clouds may prevent a perfectly deterministic solution to the climatic consequences of rising CO_2. That is, the solution may never settle down to a predictable value that all models can agree upon because of the random jiggling of the clouds. Clouds may indeed be completely deterministic in an ideal Laplacian universe, but in any conceivable modeling framework they will always have unknown aspects which will have to be drawn from a probability distribution. This is already true now for any GCM that uses "random overlap" of clouds or any of its variants. Only time (and perhaps the super-parameterizations of Randall et al. (2004)) will tell how far the 2–5°C range can be narrowed, and how much it will resist narrowing no matter how much resolution and how many new parameterizations we throw at it.

The 1990 version of the IPCC Report, in which the world's climate scientists first summarized the state of their knowledge, ranked cloud feedback on temperature as the highest priority issue to resolve, just because of the 2–5°C problem. But the keystone importance of cloud radiation was increasingly obscured as the 1990s wore on. During that decade, there was an increased clamor for attention and resources by climate subfields which are, if truth be told, less important than clouds, even if more loudly advocated. In the face of this clamor, the IPCC gave up on prioritizing and fell back on mere list-making. But the cloud problem did not get solved in spite of its loss of the IPCC's number-one spot, nor has it gone away. As Randall et al. (2004) said, it is the problem that refuses to die.

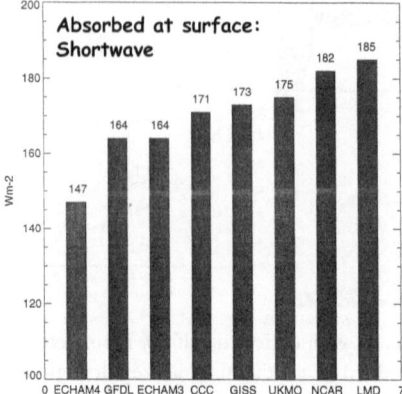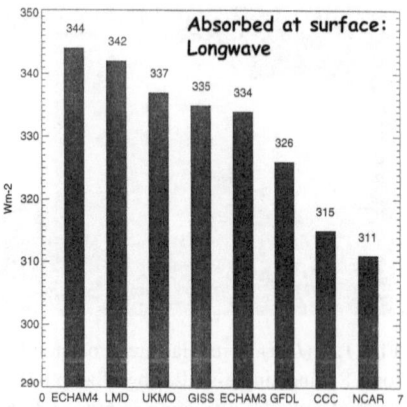

Fig. 1.4. Global-mean-annual shortwave (*left*) and longwave (*right*) radiation absorbed at the Earth's surface calculated by a variety of Global Climate Models (ECHAM: Germany, LMD: France, UKMO: England, CCC: Canada, GISS: NASA, GFDL: NOAA, and NCAR: NSF). The most likely values from observations – 345–353 W/m^2 for longwave, 143–147 W/m^2 for shortwave (Gilgen and Ohmura, 1999) – lie outside the range of GCM values and indicate a major defect in GCM cloud treatment. (Plots taken from Ohmura et al. (1998).)

1.1.2 Surface Radiation Budget

The second important application of cloud radiation is to surface radiation budget. Clouds are by far the greatest modulators of sunlight and infrared radiation reaching the surface. I remember Richard Somerville once telling me at NCAR in the 1970s, at a time when radiation was not as popular as it is now, that his weather prediction colleagues would be hard pressed to explain the land surface temperature at night without cloud radiation.

Figure 1.4 shows how poorly climate models agree among themselves about surface radiation; they agree on only one thing – they are all far from the best observations. The disagreements are mainly because of clouds, although even climate-model-predicted clear-sky shortwave and longwave radiation has apparently not fully benefited from the ICRCCM activity of the 1980s (Ellingson and Fouquart, 1991).

Not getting surface radiation right can cause extreme downstream effects in climate models. Among many roles, surface radiation is the ultimate source of energy for vegetation, for convection, and for ocean mixed layer warming. Up to the mid-1990s, the climate of many coupled atmosphere-ocean models wandered further and further off track because of getting surface radiation wrong. Some of this error was due to the neglect of 3D cloud effects. These models felt compelled to make so-called "flux corrections" whereby a tuned amount of energy, on the same order as the untuned amount predicted by the models, was added to or subtracted from the surface energy budget. While current climate models have largely gotten rid of flux corrections, the range of results in Fig. 1.4 indicates that such models still have a problem with surface radiation. What Wielicki et al. (1995) said a decade ago still holds today: "... present-day GCMs produce unrealistic simulations of the surface energy fluxes associated with solar and terrestrial radiation, *and especially the modulations of surface radiation by clouds*" (italics mine).

Let's think about the nature of those "modulations" for a moment. Without clouds, curves of solar and longwave surface radiation as a function of time look smooth and slowly-varying. Clouds destroy that utterly. They make surface radiation "turbulent," mirroring turbulence in the clouds themselves. This turbulence manifests itself as large vacillations in minute-to-minute measured surface fluxes and even second-to-second variations in the direct beam of the Sun. Among many worthy examples, I cite two. First, Ockert-Bell and Hartmann (1992) showed that, throughout vast areas of the tropics and subtropics, thick cumuliform clouds with small cloud fractions explain most of the variance in solar radiation budgets. Second, Nunez et al. (2005) found fractal behavior typical of turbulence in measured time series of downward shortwave flux under stratocumulus, the most benign of all clouds! They challenge the time resolution of current methods of retrieving surface radiation fluxes from space and show that one-hour sampling gives unacceptably large rms errors of 20–36% in daily averages. Ten-minute sampling reduced the error to 5%, an acceptable value considering uncertainties in other energy fluxes, but no extant satellite system capable of retrieving surface radiation can provide such rapid sampling globally. We have barely begun to study the limitations caused by radiative turbulence. I will briefly revisit the subject in Sect. 1.14.

"Global dimming", the occasion of a session and much press coverage at the Spring 2004 AGU Meeting, refers to a general multi-decade decline of surface solar radiation (Liepert et al., 2003). The evidence is compelling and comes from a 50-year record of pan evaporation as well as some (but not all) direct radiation measurements. After eliminating other possible suspects, changes in cloud seem the most likely cause. This seems to contradict ISCCP results showing cloud fraction has steadily decreased since 1987 while cloud optical depth has not appreciably changed (Rossow and Duenas, 2004; also Fig. 1.17 below). Current climate models with aerosol and cloud parameterizations included cannot even come close to predicting the putative dimming. While this issue is far from resolved, it perfectly illustrates the large and poorly understood role of clouds in surface energy budget.

1.1.3 Radiative Heating Rates

A third application is to radiative heating rates in the atmosphere. Not everyone may appreciate what was at stake in the recent "enhanced shortwave cloud absorption" brouhaha that nearly ripped the radiation community apart. There isn't much solar heating of the atmosphere. Most sunlight gets absorbed at the surface or reflected back to space. Therefore small changes in atmospheric heating, say 10 to 20 W/m^2, can have huge impacts. Dave Randall likes to bring home the relevance of radiational heating by saying that when it changes by 1 W/m^2, that causes a significant change in rainfall somewhere else; radiative heating in the atmosphere and rainfall roughly have to balance each other. Water vapor of course modulates this heating, especially in the infrared and especially in the upper troposphere and stratosphere, but clouds are the dominant creators of variance in the lower troposphere where the main water vapor absorption bands are saturated.

I cite one example among many worthy ones. Barker et al. (1999, 2003) have shown the importance of cloud structure for radiative heating rate profiles in GCMs and studied the effectiveness of the various bandaids (like cloud fraction and cloud overlap assumptions) applied to 1D radiation schemes to mimic 3D effects. They find major errors from the bandaids and a worse prognosis for a future in which GCM voxel[6] sizes shrink. They argue that these bandaids are a dead end and advocate that "unresolved horizontal variability and overlap be treated together within 1D algorithms." Clouds make calculating surface radiation hard, as we have seen, but this is as nothing compared to what they do to heating rates in the atmosphere.

1.1.4 Data Assimilation

A fourth application is to "data assimilation," which is a way to improve weather forecasts by inserting observational data (surface, radiosonde, satellite) at regular intervals during a forecast model run. The data must be inserted carefully so as not to shock the model or unbalance its conservation laws. Currently, assimilation is done every six hours. Future plans call for one hour. Any data can be useful, even from a single place (ECMWF assimilates ARM Oklahoma site data, for example), but satellite data provide the biggest potential help. I say "potential" because throughout the 1980s the forecast models basically rejected satellite retrievals of various quantities like temperature and moisture profiles, often in a non-obvious way. You could remove satellite data from the data mix and not perceptibly worsen the forecasts.

Gradually, it was realized that satellite retrievals lose some of the information in the original radiance data, and that it would be better to assimilate the satellite radiances directly. But the retrieval industry is jealous of its role as middleman and has a lot of inertia, so it took time for this new idea to take hold. Now weather assimilation systems routinely compute IR radiances so they can assimilate satellite IR radiances directly (shortwave radiances are much harder to compute because of clouds). Martin Miller of ECMWF says that much of the improvement in forecast skill over the last decade comes from adding new satellite data to the mix, rather than from model improvements. In the future, especially as model voxels shrink, an understanding of 3D cloud radiation will be necessary in order to better assimilate satellite radiances from clouds.

1.1.5 Cloud Shadows

Cloud shadows, so obvious in Fig. 1.1, are a distinctly 3D cloud radiation phenomenon that were impossible to even contemplate in a 1D mental and modeling framework. I am convinced they will prove of use in the future, once the advances documented in this book become a routine part of the remote sensing arsenal. It is easy to

---

[6] "Voxel" is a small volume, usually a rectangular parallelpiped; it is useful to distinguish voxels from "pixels", a dimensionally ambiguous term which may refer to a 2D polygon or a 3D column.

forget that cloud shadows are a distinct phenomenon with their own special behavior. This was brought home to me when we were planning the Triana satellite for the L-1 Lagrange point, which lies along the Earth-Sun line. Triana would see almost no cloud shadows! That would have been a most unusual view of Earth. Cloud shadows have been ignored in remote sensing because they are intrinsically 3D effects. They occur in the infrared too, as slight changes in surface temperature which may tell us something about surface heat conductivity and/or moisture. One possibility is to observe the edges of cloud shadows, the counterpart of the fabled "cloud silver lining." Transmission through the cloud edges followed by surface reflection might give qualitatively new information about clouds.

1.1.6 Multi-Angle Remote Sensing

Traditional satellite imagers can give no more than a 2D+ view of clouds,[7] they see a cloud pixel at only a single angle. But a new class of multi-angle instruments, pioneered by the European Along-Track Scanning Radiometer (Prata and Turner, 1997) and JPL's ground-breaking nine-camera Multi-Angle Imaging Spectroradiometer instrument (Diner et al., 1998)[8] are providing a much more 3D-ish view of clouds. These instruments are a largely-untapped gold mine of information and ideas for 3D cloud radiation applications. It will take time for the radiation community to really exploit these multi-angle data, however, and they will need the tools and methods in this book to do so. Of course, the clouds will inevitably evolve in the few minutes that the satellite takes to fly over and image the cloud from several angles, and so radiation people will need to learn more about the time evolution of clouds than has been their wont.

1.1.7 CloudSat

CloudSat[9] (Stephens et al., 2002) is the centerpiece of the so-called "A-train" constellation of satellites (Fig. 1.5) that will provide an unprecedented "X-ray view" of clouds. CloudSat's 3-mm-wavelength cloud radar shoots pulses vertically to create cloud cross-sections of the kind illustrated in Fig. 1.5. CloudSat will reveal not just the exterior of clouds – all that we can see easily with shortwave imagers or lidars – but their interior structure as well. Admittedly radar backscatter measures the 6th moment of the drop distribution while interest resides mainly in the zero-th through third moments, but much work has gone into making this leap with some confidence. Still, a single cloud drop 10 times bigger in radius than its brethren will give a radar total return a million times bigger than one of them, and may even dominate the return. Thus, cloud radar works better when there are no drizzle drops (100–300 microns) and, for that matter, no insects, spider webs, spores, and other objects about

---

[7] "2D+" indicates that some vertical information is available: first, from cloud shadows; and second, from infrared radiances that show cloud-top altitudes.

[8] http://www-misr.jpl.nasa.gov/introduction/goals3.html/

[9] http://cloudsat.atmos.colostate.edu/

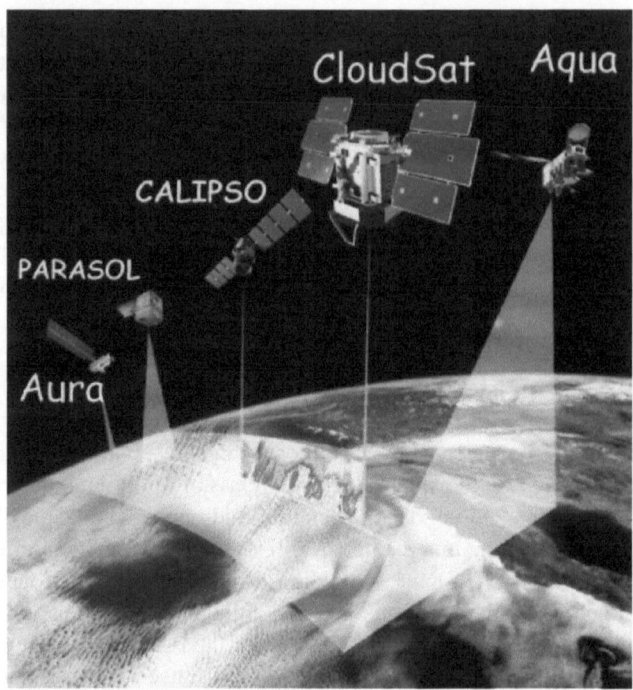

Fig. 1.5. CloudSat, carrying a 95 GHz cloud radar, in orbit with other satellites to form a constellation-of-opportunity called the "A-train." CloudSat is slated for a 2005 launch. The satellites are not drawn to scale. Each satellite has capabilities which will complement Cloud-Sat's radar ex post facto, but the A-train was not designed ab initio as a real constellation and thus does not have all the requisite coordination of time, space, and angle viewing

the size of a drizzle drop.[10] Quibbles aside, however, CloudSat will provide an unprecedented view of Earth's 3D clouds, including the first global picture of cloud geometric thicknesses.

The other members of the A-train can provide important complementary data about clouds. CALIPSO is a lidar with a vertical-shot measurement strategy like CloudSat. Sadly, NASA dropped CloudSat's and CALIPSO's accompanying oxygen A-band spectrometers, which would have provided vital complementary information on photon pathlengths. Aqua is a multi-purpose satellite carrying six disparate instruments; the most significant for clouds are the MODIS imager and the CERES radiation budget radiometers. The French Parasol satellite measures polarization in reflected sunlight, a little-studied new variable for clouds, while Aura measures atmospheric chemical species. Because of cloud evolution, the data from other

---

[10] The cm-wavelength rain radars developed just after World War II are basically blind to the cloud particles that affect solar and infrared radiation; they measure only the precipitation-sized drops of 500 microns and larger. So they see drizzle poorly, while cloud radar sees it too well!

satellites of the A-train diminish in value with every minute that elapses between their overpass and CloudSat's. There will be plenty of need for 3D cloud radiation experts to make sense of this welter of data!

1.1.8 3D Clouds at Microwave Wavelengths

The rain community has its own cadre of 3D microwave cloud radiative transfer experts. Their radiation problem is similar to the shortwave one in many respects, although the high absorption of liquid water at microwave and radar wavelengths keeps the amount of scattering low – six scatters at most. The rain community is not represented in this book, however; there is an estrangement between the rain and cloud radiation communities (cf. Sect. 1.3). But the two communities certainly share a common interest in 3D cloud structure modeling. Rain heterogeneity is much greater than cloud drop heterogeneity, but the two communities could probably develop common multifractal cloud structure models, differing only in choice of parameters. There is, after all, a continuum of states between a raining and non-raining cloud, although from a casual perspective outside a cloud, there seems to be a quantum leap from non-rain to rain. Both communities deal with sub-resolution variability, which the rain community call "the beam-filling problem" since rain only partially fills the field of view of their satellite microwave and radar instruments. Both communities deal with the same drop size distribution, just different subsets of it (cf. Fig. 1.8). Knyazikhin et al. (2002, 2005) and Marshak et al. (2005) have reached a hand across the barrier in studying the clustering of the rare larger cloud drops, the ones that have a chance to become rain, and how this affects radiative transfer and remote sensing. This is an exemplar of how the two communities could cooperate more than they do now.

1.1.9 Clouds and Detection of Ultra-Energetic Cosmic Rays

Recently our ARM research group received a delegation of Goddard space scientists who represented the Orbiting Wide-angle Light-collectors, or OWL, mission. OWL is a pair of stereo-viewing satellites designed to measure the spectacular cascade in the Earth's atmosphere caused by single ultra-high-energy cosmic rays (Sigl, 2001). Satellites are needed because surface stations capture too few events—only about one per year. Ultra-high-energy cosmic rays are theoretically impossible and could shake the foundations of physics, making them of great interest. Each ray creates a glowing pancake of particles a meter thick and 15 meters wide. Secondary collisions with N_2 molecules in air release bursts of faint UV fluorescence, and there is also strongly forward-directed Cerenkov radiation. The OWL delegation would in truth prefer a cloudless Earth, but they came to us for help with the 3D cloud problem. They want to know what they can learn from their satellites as an event propagates through a cloud. This is the most complex 3D cloud radiation problem I have seen yet. It illustrates the kind of new and unexpected applications that the tools and methods in this book will be called upon to deal with.

1.1.10 Earth-Like Planets

My personal favorite future application is to Earth-like planets. We will directly image such planets, if they exist, sometime in this century. We already know that many planets have clouds, mostly made of noxious chemicals not water. We will need to be prepared to understand how 3D structure affects the observed spectra from these planets! Any interpretation of the imagery and spectroscopy from such planets will require the full arsenal of 3D cloud radiation methods.

1.1.11 Cloud Rendering

Just to see what people were doing with 3D clouds out in the wide world, I used Google to search on all reasonable variants of the phrase "3D cloud" and got up to 3000 hits. My sampling indicated that many or even most are not, alas, about applications mentioned above but about computer-graphics renderings of cloudy skies. Most of us have moved beyond our early concern with making our cloud structure models look realistic to the eye, but a whole community of artists definitely worries about it. Their clouds do not bear close inspection by those of us trained to look for structure at all scales, however. Perhaps someday the work in this book will inform cloud renderings in movies and make them more realistic.

1.1.12 A Personal Note

I will close this brief review of applications and future opportunities on a personal note. From 1990 onwards, my ARM project, ably manned by Alexander Marshak, Anthony Davis, Frank Evans, and Robert Pincus, proposed every three years a perfectly logical research plan of 3D cloud radiation studies. Some of the things we actually accomplished, but others fell by the wayside because the field was developing so rapidly under the impetus of ARM that there were always new surprises and new opportunities which we grabbed and ran with. It would have been impossible to anticipate many of the things we wound up discovering. ARM furnished a milieu of tool creation that led naturally to new discoveries, a theme I shall return to later. In our 2001 proposal, as we analyzed the way we had operated, we concluded that our zigzagging research path was mainly due to the fact that ARM frequently had to deal with the fallout from 3D cloud radiation effects, whether they liked it or not. Ofttimes people would have preferred to use 1D models, but ARM data just couldn't be stuffed into a 1D framework very often – it was just too procrustean. So we became like emergency workers who rushed to every new problem and tried to repair or at least minimize the damage. Much was learned in that process. This book tells some of that story.

1.2 My Life in Cloud Radiation

I have had a 33-year odyssey in cloud radiation, first as a theoretician and scientific software developer, later as a field program participant and organizer, and supporter of new instrument development. I have tried to straddle theory, observation, computation, modeling, and new platform and instrument development. This was an exciting challenge, but probably my reach exceeded my grasp. Others have probably succeeded better at Dido's Problem – stretching limited resources to encompass a vast territory. However, during my odyssey, I have seen almost everything that has occurred for three decades and met most of the players. I hope that by sharing the parts of my odyssey relevant to and parallel to the growth of 3D cloud radiation, I can illuminate the larger milieu in which the fine contributions in this book are set.

In the early 1970s, I was seduced from a budding career in nuclear weapons radiative transfer (yes, 3D) into the emergent field of climate. I had not been keen on being part of the weapons establishment, but jobs for physicists were scarce when I got my PhD in 1970. Luckily, my company won a grant from the ARPA Climate Dynamics Program, the first major GCM-centered climate program and the first major effort to lift climate from "the province of the halt and the lame," as Ken Hare once put it. The ARPA program embraced many of the pieces that we call "global change" or "Earth System Science" today, including seminal work by John Imbrie on ice ages and the Milankovitch theory. Our company's proposal was to improve the parameterizations of atmospheric radiation and mountain lee wave drag in the Mintz-Arakawa two-layer GCM. I grabbed the radiation part when the scientist originally committed to it left the company. With a PhD under Gerald Whitham in applied math (applied to nonlinear waves in water and plasmas), plus one course in the Boltzmann transport equation with an emphasis on neutrons from Noel Corngold at Caltech, I was obviously ideally suited to the task! Perhaps having done a senior thesis at MIT with Hans Mueller of Mueller matrix fame predisposed me to take up these radiation cudgels.

Fortunately, I had the benefit of a remarkable mentor, Burt Freeman, who taught me a great deal about radiation and about the Los Alamos-Livermore-General Atomics axis from whose bowels our company and a lot of good but classified radiation work had sprung. I learned Fortran in a trailer while waiting for my top-secret security clearance, and my company offered a course in numerical methods at lunchtime where I discovered not just a talent but a passion for numerical modeling.[11] Later, the mix of a classical analytic training and a talent for numerics enabled me to bring a unique capability to my work.

1D plane-parallel radiative transfer was the norm in atmospheric science in those days, and compared to what we had been doing for nuclear blasts, it at first seemed almost too easy. (The reduced complexity in the spatial dimension was partly counterbalanced by the extra complexity in the wavelength dimension and in the scattering phase functions, though.) My detour into 1D plane parallel radiative transfer lasted almost two decades, but when I finally got back to 3D radiative transfer as a result of the birth of the Dept. of Energy ARM program in 1990, I was under no illusion that we would need to invent the field from scratch.

---

[11] Computer solutions were frowned upon at Caltech in favor of classical analysis, so what little numerical analysis I learned was at night away from the prying eyes of my thesis adviser.

When I entered the atmospheric radiation field in 1970, it was so small that it hadn't even had its first AMS conference yet (Tom Vonder Haar organized one in 1972). The subject was primarily viewed as providing computer codes for GCMs – and GCMs then were General Circulation Models, long before they made their clever segue to become Global Climate Models. I was told upon entering the field that an essential skill was reading a "radiation chart," a kind of graphical parameterization invented in the 1940s, but alas, mastery of the chart always eluded me. More creative radiation work was going on among a small cadre of planetary scientists – William Irvine, Jim Hansen, Carl Sagan, Richard Goody, and Jim Pollack come to mind – but they spent little time worrying about the Earth since that was, with Voyager and other probes, the heyday of planetary science. Radiation theory also guided remote sensing instrument development and data interpretation, but remote sensing furnished little data in return that was incisive and accurate enough to challenge, or cause the improvement of, radiation theory and models. Mainly, radiation played a service role and was not regarded as a subject for real research and discovery. However, I saw it as a subject ripe for renewed attention as a result of the incipient focus on climate in 1970.

In the 1970s I was able to master all that was known of atmospheric radiation, which was very little, especially on the observational side. All the papers worth xeroxing fit in one drawer of my file cabinet. Being a plodding reader who can take a good part of a day to absorb a single paper, radiation, with its thin literature, was a good match for me. While the field was not popular, it offered plenty of untilled ground, unlike older areas of atmospheric science. Even some of the simple things hadn't been done, and adapting and extending ideas and methods from other parts of physics and mathematics could be a real contribution. Thus, I charged in with gusto and made radiation my adoptive field, even to the point of becoming a rather notorious radiation patriot in my earlier days (continuing a long tradition of zealous converts).

While I ranged rather broadly, the energetic role of clouds has been my home base in radiation. Clear sky radiation remains a fascination for some, but to me the challenge was always clouds. Aerosols were a big topic in 1970, due mainly to Reid Bryson and his "human volcano" slogan, but to me it seemed a tempest in a teapot. Clouds are the primary atmospheric modulators of the flow of radiant energy from the Sun back to space. They are more important even than water vapor because they operate at all wavelengths and the Earth is 67% cloud-covered according to the latest ISCCP results. Clouds' role in energetics was known by the time Arrhenius published his amazing energy-balance climate model calculations for doubled CO_2 in 1896, although he assumed a cloud albedo of 78% at all wavelengths and thus assigned them a bizarrely low emissivity of 22% in the infrared.

Early in my career, when people still knew my background, they used to ask me, "How does the state of radiation in the nuclear weapons field compare with that in atmospheric cloud radiation?" The answer is easy, but not encouraging. In the nuclear weapons field, the radiation models, exotic though they were, were tested, refined against measurements, then tested again, in an iterative loop that eventually led to robust models that worked in all cases. Note that I said "all" not "most."

Model failure was not an option, since the consequences of failure were profound. Unfortunately, this kind of urgency and demand for model verisimilitude was lacking in cloud radiation for the first 15 years of my career. Until the advent of NASA's FIRE field program (1986-2000), DOE's ARM (1990-present) and CloudSat (to be launched soon), the resources put into the problem were underwhelming. Perhaps clouds should have wrapped themselves in more weirdness and mystery, or had a better public relations flak, to attract attention proportional to their importance. They are probably more ubiquitous in the galaxy than short-lived elementary particles or supernovae, which receive far greater funding. To my mind, this lack of attention remains inexplicable, since the consequences of not knowing the radiative feedback of clouds on climate are enormous; this uncertainty has brought long-term climate prediction to a virtual standstill. Absent an effort of size appropriate to the importance of the problem, cloud radiation will, sadly, remain somewhat of a wild card and a tuning knob enabling an uncomfortably large range of climate change scenarios.

For the ARPA program, I wound up building the first atmospheric radiation model that worked identically across the solar and IR spectrums – amazingly, solar and IR radiation were separate communities with almost no communication at that time! You had to add solar model results from one community to IR model results from the other to get total radiation (what climate cares about), and worry about the no-man's land between 2.5 and 5 microns wavelength which both communities disowned, and other disharmonies. I called my model ATRAD in an effort to create new terminology without descending into acrobabble. ATRAD took advantage of the then-recent LOWTRAN model for atmospheric absorption from Bob McClatchey and his group at Air Force Cambridge Research Labs (Pierluissi et al., 1987; Dutton, 1993), and also of the elegant Grant-Hunt version of adding-doubling for treating scattering (Hunt and Grant, 1969). Burt Freeman and I developed some improvements for the doubling part in order to do thermal emission and specular reflection correctly. I probably disappointed my ARPA employers by not producing any GCM parameterizations, but that seemed to me startlingly premature considering the primitive state of atmospheric radiative transfer (especially in testing against observations). Hopefully I have done other things which, in the fullness of time, may compensate for that failure (Sect. 1.4–1.8).

At least some of the tools which I helped create, like the delta-Eddington approximation with Joachim Joseph and Jim Weinman, and various snow albedo approximations with Steve Warren, eventually wound up in some GCMs. It is in creating such tools that I have found the most satisfaction, more so than in publishing papers, and this theme runs right through to the present day with my efforts to help build ARM and thereby a whole new paradigm in field observing; to provide useful scientific software like my Mie code and DISORT; and recently to catalyze the development of in situ multiple-scattering lidars (Davis et al., 1999; Evans et al., 2003) and miniaturized cloud physics instruments for small UAVs with Paul Lawson. Since I have often favored tool creation over paper publishing, but never discussed this choice publicly before, I have taken the opportunity to do so as one major theme of this chapter, below.

In 1974, I took a 20% pay cut to join the fledging NCAR Climate Project in Colorado at a time when immigration of bright scientists and engineers from other fields was being actively encouraged by NCAR and NSF. Steve Schneider and Will Kellogg were instrumental in bringing me to NCAR, after admiring some of my work on Arctic stratus cloud radiation (Wiscombe, 1975) for the AIDJEX Program.[12] I was to become the first full-time cloud radiation modeler at NCAR. NCAR had many cloud resources that brought me out my isolation from mainstream cloud physics, including: (1) aircraft data from various field campaigns (on punched cards or paper tables!); (2) Doug Lilly, whose seminal work on cloud-topped boundary layer infrared cooling had put radiation on the map of cloud dynamicists; and (3) two pioneering and somewhat rival cloud dynamics modeling efforts, one by Deardorff and Lilly (without microphysics) and one by Clark and Hall (with microphysics).

Compatriots at my company in La Jolla thought I was crazy to leave our idyllic existence and retire my surfboard for a mile-high life in Colorado, but I was captivated by visions of Earth System Science (before it was called that) resulting from my experiences in the ARPA program. ARPA had introduced me to an amazingly broad range of Earth scientists – Jacob Bjerknes, Akio Arakawa, Larry Gates, Mike Schlesinger, Murray Mitchell, Barry Saltzman, Bill Sellers, Norbert Untersteiner, Diran Deirmendjian, John Kutzbach and John Imbrie, among others – and their influence, plus the seminal 1970 book "Study of Man's Impact on Climate," was heady and seductive stuff for a young scientist. Thus NCAR, with its fledging climate program under Warren Washington, Bob Dickinson, Steve Schneider, and Will Kellogg, seemed like a potential paradise. My fellow immigrants to NCAR included Gerry North, Joe Klemp, Ramanathan, Jim Coakley, Steve Warren, Bob Cahalan, and many others who became leading lights in climate and meteorology. NCAR was a life-changing opportunity for me, as it was for them.

At NCAR, I continued several lines of theoretical investigation prompted by my building of ATRAD. In truth, I never made much use of ATRAD itself, except as a source of ideas for more fundamental work and algorithm improvements. Sometimes I have chastised myself about this, since lying concealed within ATRAD were many discoveries later made by others. Not least of these was a capability to study radiative feedbacks in toto, in a single model, rather than just the shortwave or longwave parts separately.

In spite of being looked upon with suspicion and occasional hostility by many of the NCAR weather prediction crowd, I was given lots of freedom and little pressure of the kind that bedevils young assistant professors today. I made good use of the time to master not only the whole of the radiation field, small as it was then, but to broaden my learning into paleoclimate and other subjects which now go to make up Earth System Science.[13]

---

[12] a sea-ice modeling and observational study including an ice camp in the early 1970s

[13] and I even learned some traditional meteorology, starting with Forrester's wonderful book "1001 Questions Answered about the Weather" – although I found subfields of low population like atmospheric electricity and micrometeorology more interesting than crowded fields like midlatitude stormology

NCAR, with its own aircraft and a wealth of observational scientists visiting and in residence, provided an ideal opportunity to learn all about cloud field programs of that day. Most were single-aircraft, single-PI efforts, aimed at dynamics, turbulence, and cloud microphysics. Radiation was, at best, an afterthought. Drops were collected on sticky tape (Formvar replicators to be precise) and sized manually; thus a single drop distribution, with hours of graduate student labor behind it, seemed of much greater value than today! For a while, I indulged an obsession with gathering drop distributions, calculating their various moments, and comparing them to the distributions popular among cloud radiative transfer scientists. Rarely was there a good match, and indeed for some distributions the famous effective variance parameter of Hansen and Travis (1974), a measure of the width of the drop distribution, was far outside its theoretical range, indicating a much greater prevalence of larger drops than could be accounted for by any exponential-tailed drop size distribution. For years I carried around a set of 2000 punched cards containing manually-sized drop distributions, until card readers went the way of the giant ground sloth.

At NCAR I also met Bill Hall (cf. Fig. 1.6), who with Terry Clark had built one of the first dynamical cloud models complete with microphysics. Bill not only patiently taught me as much as my poor brain could hold about cloud physics, but also willingly collaborated with Ron Welch and me in our later work (Wiscombe et al., 1984; Wiscombe and Welch, 1986) to furnish rising-parcel-model drop distributions for our radiation modeling, rather than the radiation-blessed drop distributions which my own analyses of measured cloud data had shown to be woefully incomplete. A rising-parcel cloud model is 1D, which matched our 1D radiation model. I saw nothing wrong in this; shaped by my experience in the ARPA Program, I always

Fig. 1.6. With Bill Hall in one of his infamous Hawaiian shirts. Bill was one of the earliest cloud modelers to reach "hands across the water" and collaborate with cloud radiation scientists, namely myself and Ron Welch

approached clouds and cloud radiation from the point of view of a climate modeler.[14] I wanted to extract the essence of clouds and avoid their details. In any case, using a cloud dynamics model for input, Hall, Welch and I can claim paternity for much of the current 3D cloud radiation work that draws input from 3D cloud models.

We also horizontally-averaged the drop distributions at various time steps in the Clark-Hall 2D cloud model and made many radiation calculations and plots with them, which were destined for Part II of our work, but Wiscombe et al. (1984) remained a dangling Part I. I still have fat folders of Part II results in my file cabinet. This taught me never to use "Part I" in a paper title! Viewing our abandonment of Part II from this vantage point in time, I suspect that we felt overwhelmed by the quantity of results that could pour forth from a radiation model taking input from a time-stepping cloud model, and didn't know quite how to extract the needle from this vast haystack. This feeling must be magnified a hundredfold among the current generation of 3D radiation modelers.

Until the late 1980s I avoided the 3D cloud problem, in spite of having a background in 3D radiative transfer. There were, after all, lots of other pressing problems, for example, cleaning up numerical problems in 1D radiative transfer, moving beyond simple gas-absorption band models (eventually to k-distributions), studying nonspherical particle scattering for ice and aerosols, and developing useful simple approximations for all these things. In the 1970s, atmospheric radiative transfer was a vast playground with many such things to be done. Even importing advanced numerical techniques constituted progress. There were enough things to do with gaseous absorption to fill a whole career, but they did not attract me as much as the hard problems posed by clouds. I realized by the mid-1980s that 3D clouds were the greatest remaining frontier. While there had been pioneering early work by McKee and Cox (1974) and others (cf. Chap. 3), it wasn't until the 1980s that the geometric, computational, and observational tools really came together to make progress possible.

But here I saw a problem with the radiation community. Entering students were cutting their teeth by writing 1D radiative transfer models, which then naturally came to dominate their most productive early careers. Even many of my colleagues were still fiddling with 1D methods. So I decided that something had to be done: provide a nearly perfect 1D radiative transfer tool, one far superior to anything a graduate

---

[14] In Wiscombe (1983), an invited review of the atmospheric radiation field, I mounted a spirited defense of the simple 1D approach to clouds: "Most cloud-radiation models are 1D. This is the natural milieu in which to test many hypotheses about cloud radiation. However, there has been an explosion of papers in 3D clouds, mostly cubical in shape [...] The actual or implied denunciation of 1D cloud modeling in some finite cloud papers requires comment. First, measurements are the acid test of any model; it is not enough that a model simply 'looks' better. Perhaps weighting 1D albedos by the proper measure of cloud fraction will correctly predict the albedo of patchy cloud fields. But more importantly, our job is not to make our models as complicated as Nature herself; it is to simplify and idealize, in order to gain understanding. 1D cloud modeling is an entirely acceptable way to do this [...] Our job is to learn to make simple adjustments to 1D predictions to mimic patchiness, not to reject this very valuable modeling approach out of hand." ("1D" has been substituted for the increasingly archaic term 'plane-parallel' in this quote.)

student could write. I think I can admit here, for the first time, that this was one of my main motivations for lavishing so much of my time and energy on DISORT: to leverage the field out of its reinvention of the 1D wheel and thereby hasten the movement into the world of 3D clouds. This, I am happy to say, is exactly what occurred within a few years after the release of DISORT (Stamnes et al., 1988).

I also recall a kind of despair at ever completely specifying (for input to radiation models) the 3D clouds I saw out my airplane windows. The customary cloud classifications provided no useful input to a radiation model. During his team's 15-year odyssey to create the most complete surface-based cloud climatology ever, I even occasionally chided my colleague Steve Warren that those old cloud classifications were like a strait jacket preventing us from finding more quantitative and mathematical ways of classifying clouds. I was at the time thinking that clouds should be classified by, for example, a set of multifractal parameters rather than by names. But now I see that a rich tradition dating back to the seminal lecture by Luke Howard in 1802 would be lost if we threw out the old descriptions entirely. They contain fundamental process information that needs to be supplemented rather than supplanted.

So how did I get involved with 3D clouds? That is a strange, winding road indeed. It began with the earliest application I made of ATRAD outside the ARPA Program – to Arctic stratus clouds. I compared the output of ATRAD to a few radiative flux measurements by Gunter Weller in Wiscombe (1975), now somewhat a classic in the polar community. One review of that paper stung me to the core, though, and shaped my future in 3D clouds in ways I could never have envisioned at the time. It simply remarked that ATRAD seemed an excessive load of machinery to throw at a few pathetic broadband flux measurements. I spent the rest of the 1970s doing very theoretical things, but this criticism nucleated a growing conviction that, by merely fiddling with models, I was increasingly part of the problem rather than part of the solution. This discomfort finally led to action when our community undertook the groundbreaking ICRCCM (Intercomparison of Radiation Codes in Climate Models), led by Fred Luther, in the early 1980s. Many bizarre anomalies were discovered, including models that couldn't calculate the Planck function right, and models that differed by $100\,W/m^2$, but our main discontent was a lack of observations with sufficient spectral sophistication to incisively test the models.

The ICRCCM community asked Bob Ellingson and myself to do something about this. Bob had done field work in BOMEX but, like me, had primarily been engaged in theoretical pursuits before this happened. Some might have viewed us as an unlikely pair to nucleate any kind of observational program, given our backgrounds. How wrong any such judgment would prove to be! We both threw ourselves wholeheartedly into learning about atmospheric instruments and what could be done spectrally. We decided on the simplest possible problem for our first outing, the clear-sky longwave measured with IR spectrometers, and created, after the usual setbacks and false starts, the SPECTRE field program (Ellingson and Wiscombe, 1996). While clouds had always been my main focus, clear sky was more appropriate for an effort with the hidden agenda of demonstrating a quantum leap in the sophistication of field radiation measurements. SPECTRE attached itself to FIRE's second field campaign in the Fall of 1991.[15]

SPECTRE became the blueprint for ARM. Most of our SPECTRE instruments survive to the present (in much improved form) in ARM, and some of our SPECTRE documentation became part of the earliest ARM Science Plan. Little did the FIRE participants, who generally viewed us as a rather odd and perhaps spurious appendix, realize what an amazing and revolutionary observational program SPECTRE would lead to.[16] ARM quickly become the leading 3D cloud-radiation program of our time, soon growing larger than the FIRE program within whose bowels it was uneasily born. Of course, as ARM grew beyond its birth in SPECTRE, it adopted many instruments whose development paths began in the FIRE Program. And while FIRE and ARM remained separate and distinct at the top level, there were too few people involved in clouds to prevent a considerable merging at the working level.

My very first proposal to ARM in 1990 concerned 3D cloud radiation, with which I have been more or less involved to the present day. So, my tortuous road to the 3D cloud problem was: a critical review of a 1975 paper sparking a concern over lack of good radiation observations, which grew to a point where I shelved my theoretical work and co-led a clear-sky IR field program, which nucleated the cloud program ARM, which funded me to work on 3D cloud radiation problems for 14 years.

While SPECTRE was hatching over a 6-year period, I made a determined effort to transform myself into someone knowledgeable about observations, if not a true observationalist. I even became somewhat of a pest to my fellow theoreticians, urging them to "get out into the field" and away from their computers.[17]

As part of this makeover, I went in 1985 on my first cloud field program flight as a guest of Vernon Derr, then Director of the NOAA Environmental Research Labs in Boulder. Figure 1.7 shows me about to board a research aircraft in San Diego to fly over stratocumulus clouds, which are abundant offshore and which attracted the first FIRE field campaign two years later. This was the day one of the flight engineers told me they shined a flashlight on the Eppley flux radiometers to make sure they were working! Vernon told me radiation was not a high priority on these research flights. By contrast, the turbulence probes on the nose barber pole, and the radar inside the nose cone, were state of the art. The radiation people who had grown up inside atmospheric science were incapable of seeing the gross incongruity of $1000 commercial radiometers not adapted for aircraft use flying alongside much more expensive instruments designed specifically for aircraft. It was an eye-opening experience, and as seminal in shaping my future drive for better radiation instruments and experiments as the review of my 1975 paper had been.

---

[15] just in time to witness the brilliant sunsets resulting from the Mt. Pinatubo eruption, whose particles were fortunately too small to affect our IR measurements

[16] One revolutionary aspect of ARM, little marked today because it is so well accepted, is its insistence on a permanent presence in the field in order to gather long climatic datasets, as opposed to the evanescence of typical field campaigns which took several years to organize and only lasted several weeks.

[17] I am happy to report that this has become more the norm in atmospheric radiation, rather than the exception as in those days.

Fig. 1.7. On the tarmac at North Island Naval Air Station in San Diego in 1985, ready to board for my first cloud research flight, with the NOAA P-3 research aircraft in the background. North Island was also the base for the first FIRE stratocumulus experiment in 1987, which I attended in order to learn more about cloud measurements

While on the 1985 flight, we dived 3 km to perform what we nowadays call a "tip calibration" on a prototype microwave radiometer, in order to point the (untiltable) radiometer at the horizon. While not bungee jumping, this was certainly a thrill ride! Months later, to my surprise, I got a hazardous duty citation in the mail. It taught me a lesson I never forgot: aircraft are not vertical profiling platforms, and if we ever want to know vertical profiles in clouds, we need other methods. Many of the so-called vertical profiles published from aircraft field programs are in fact nothing of the kind unless you accept an assumption of cloud horizontal homogeneity, since the aircraft glides 50–100 m horizontally for every 1 m it travels vertically.[18] Radar is the method of choice for vertical profiling, of course, but at present we are frustrated by ambiguities in the retrieval of cloud variables like droplet number concentration and liquid water content from radar. These ambiguities are due not merely to the 6th-moment problem mentioned earlier, but to the presence of objects in clouds which give a larger radar return signal than typical cloud drops. Such objects include drizzle drops, insects, spider webs, spores, and various other detritus small and light enough to remain airborne for days.[19] Thus I believe we will always need other methods to supplement radar. This accounts for my advocacy of tethered balloons, small UAVs,

---

[18] In a world obsessed with safety, dives like mine have become a thing of the past.

[19] High quality polarized Dopper-radar spectra may someday enable us to disentangle the return by these objects from the return by cloud drops.

and cloud-physics dropsondes, all carrying miniaturized cloud physics instrumentation.

The last decade of my career has been spent on a deep involvement with ARM and with trying to push new instruments, tools, observing strategies, and platforms, and new ways of doing field science. I am very proud of the work of my ARM group – Alexander Marshak, Anthony Davis, Frank Evans, Robert Pincus, and now Yuri Knyazikhin and Christine Chiu – and am privileged to have been able to point such talented people in promising directions and then give them their heads to run like the wind. Out of this laissez-faire approach have come deep studies of cloud scale, radiation scale, aircraft measurement strategy, innovative cloud lidar techniques, and most recently the spatial distribution of larger drops – and other discoveries which can be found in this book now and which are destined for the textbooks of ten years hence.

1.3 The Estrangement of Cloud Radiation, Cloud Physics, and Rain Remote Sensing

3D cloud radiation scientists would ideally like to know where every drop is located and the size of each one. Then they could average as they please, but always starting from correct information rather than being forced to make uninformed and often incorrect assumptions about cloud drops. In the best of all possible worlds, they would have turned to cloud physicists and cloud dynamicists for this information. This did not happen – hence the title of this section. It is a strange tale indeed.

Knowing the spatial distribution of cloud drops by size would enable the radiation scientists to concentrate on what they do best. They would not have to know the details of cloud physics and dynamics. This happened anyway – that is, radiation people learned almost nothing about cloud physics – but not because they were taking their input variables from cloud physics. In fact, they were relying on other radiation people to invent mythological "radiation clouds" – pancakes with no spatial variation and no variation in drop size distribution. This was convenient for getting on with the radiation business, but so far from reality that communication between the two communities remained virtually nonexistent until the early 1990s. As one of the few who tried to learn something about cloud physics and cloud models and interact with the cloud physics community, I witnessed this firsthand.

If we knew where every drop was located, and its size, we could do something called a "first-principles Monte Carlo" in which photons interact with real drops, not with fictitious "elementary volumes" as in standard radiative transfer theory. However, first-principles Monte Carlo is a far-off goal and in any case overkill for routine work. Mostly, radiation people just want the probability distribution of drop sizes ("drop distribution" for short) and from this they calculate all needed cloud optical properties, mainly through the intermediary of Mie theory.[20] "Drop distribution" is

---

[20] Some didn't want even that much contact with cloud physics, preferring instead to specify cloud optical properties directly. Especially popular was the Henyey-Greenstein phase function, an angular scattering pattern that has never been realized in any known universe.

devilishly difficult to define precisely, both theoretically and observationally, and is the weakest link in standard radiative transfer theory, but its convenience made it a mainstay of cloud radiative transfer modeling. I used it without much introspection well into the 1990s even though I knew the functional forms used by my colleagues to fit drop distribution tails were grossly inadequate.

From the 1970s to the present day, many cloud radiation people got their drop distributions not from a cloud physicist but from radiation scientist Diran Deirmendjian (1969), the middleman who specified a few analytic functions based loosely on drops collected on sticky tape by an aircraft. Deirmendjian's distributions had single humps with exponential tails toward large sizes.[21] The exponential tails were not based on any theory or analysis of observations but mere convenience, and had the intentional side effect of saving radiation scientists from endless hours of Mie computations for scattering by the larger drops. It's hard to believe now, but in the late 1970s Mie computations could still drag down the world's fastest computers. In 1979 I once used 8 hours of time on what was then the world's fastest supercomputer, a Cray-1 (now in the National Air & Space Museum), for Mie computations, and was chastised for it by the head of the division.

Exponential-tailed drop distributions offered the further convenience of making all moments of the distribution converge, since $r^n \exp(-br) \to 0$ (for any n) as $r \to \infty$ where r is drop radius. In fact, the higher moments of many drop distributions that I processed diverged: that is, the contribution to higher moments from the largest drops in the distribution were still increasing strongly. But common practice developed into a mindset that exponential-tailed distributions represented reality, a devilishly difficult misconception to dislodge after 30 years.

Later, Deirmendjian (1975) tried to atone for his neglect of larger drops by prescribing rain distributions, for use by microwave radiation people. This time, however, his distributions had no small drops! This symbolized how the rain radiation people (microwave and radar) and the cloud-climate radiation people (solar and IR) lived out their lives on one side or the other of the "sub-millimeter divide." And as far as radiation people were concerned, drop distributions remained either cloud drops below 15–20 microns, or rain drops above a few hundred microns. Drizzle drops in the 100 micron range fell between the cracks – until millimeter-wave cloud radars were invented and drizzle began, when present, to dominate their reflectivities. I call all the analytic distributions invented by radiation people, with their arbitrary cutoffs, "fantasy drop distributions" because their inventors didn't care what the real functional form was[22] and they never bothered to analyze boatloads of cloud aircraft data to learn the truth. The attitude was pretty much "any drop distribution that smoothes out the awful Mie resonances, makes my Mie computations manageable, and isn't obviously silly is just fine with me." This was not an attitude likely to

---

[21] Some daring souls switched to the Hansen and Travis (1974) analytic drop distributions, again invented by radiation, not cloud, physicists – but in fact these were just repackagings of the same old one-hump exponential-tailed distributions.

[22] in particular they never bothered to determine if the tails were actually slow power laws rather than fast exponentials

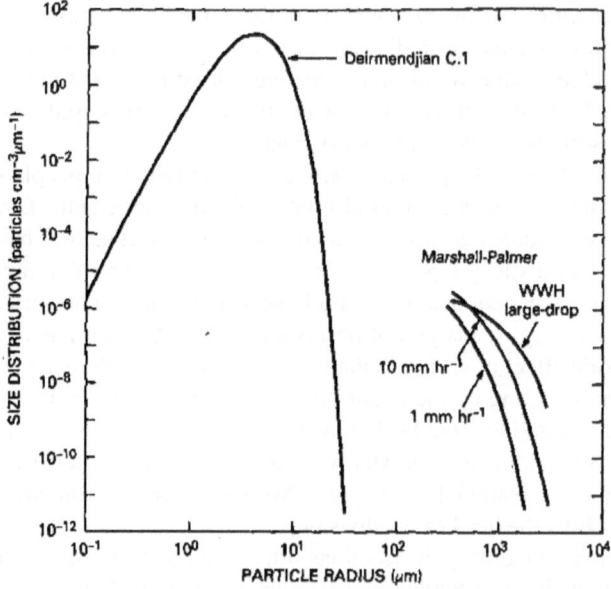

Fig. 1.8. Past estrangement between cloud and rain scientists is mirrored in the assumptions they make about drop size distributions. On the left is a typical radiation size distribution of the 1980s, falling off exponentially with no possibility of larger drops, much less rain. On the right are several traditional raindrop size distributions along with a distribution (WWH = Wiscombe/Welch/Hall) taken from a rising-parcel cloud model; these distributions are typically truncated with no extension to smaller sizes. There is a chasm between the two types of distribution, which Nature fills with drizzle drops largely ignored by both groups, at least until Bruce Albrecht called attention to their ubiquity and copious drizzle was observed in field campaigns like DYCOMS (Stevens et al., 2003) and EPIC (Bretherton et al., 2004). (From Wiscombe and Welch (1986).)

promote collaborations with cloud physicists, nor even with radiation brethren across the "sub-millimeter divide."

Figure 1.8 is taken from a paper (Wiscombe and Welch, 1986) which explored the subject of real vs. fantasy drop distributions. It shows the gulf between typical fantasy drop distributions used in solar and IR radiative transfer and typical raindrop distributions based largely on surface measurements. This gulf is created by rather arbitrary truncation decisions in each field. Figure 1.8 shows that the exponential-tailed distributions of solar and IR radiative transfer fall off far too rapidly to ever link up with reasonable raindrop distributions. Little has changed since that figure was published. Our suggestion in 1986 of a steady state cloud in which larger drops are continually being created, and in which the gulf shown in Fig. 1.8 does not exist, met with a resounding silence from the radiation community, and so it was with pleasure that I read of the discovery in the DYCOMS-II field program of stratocumulus clouds which continually drizzled yet not only maintained themselves but were more robust than their non-drizzling counterparts (Stevens et al., 1999).

Just as there is a chasm between cloud physicists and cloud radiation scientists, so also is there a chasm between scientists who study the interaction of microwave radiation and radar with clouds and those who study the interaction of solar and IR radiation with clouds (the two groups did interact in the ARM Program). The two communities have separate literatures, separate favored journals, and separate meetings. A leader of the microwave-precipitation field once came to me and a few other cloud radiation people and practically begged us to join the TRMM team in order to represent the small drops which his community completely ignores. It seems bizarre that two communities which share a focus on radiation and clouds remain almost completely isolated, like two religious sects shunning each other over small doctrinal differences. Perhaps the budding theoretical and observational efforts in the sub-millimeter wavelength region, lying between the IR and microwave, will help to connect the two alienated communities and bridge the sub-millimeter divide.

The progress of cloud physics is leaving the fantasy drop distributions used by radiation people increasingly in the dust. Those fantasy distributions were picked hastily, for convenience rather than correctness. They may work for some applications, but they are just plain wrong in general. And the sad spinoff is that cloud radiation scientists seem to have lost interest in doing better, or in accounting for the complex way drops are spatially distributed in a cloud, differently for each drop size. In reality, drops have a joint probability distribution in both space and size, and that joint distribution cannot be separated into a function of size times a function of space because of size-dependent spatial clustering.

Lest I let cloud physicists off the hook too lightly, or imply that they are all heroes in this drama, let us be clear that 90% of cloud physicists have remained as militantly ignorant of radiation as cloud radiation people have about cloud physics. The cloud physics-dynamics field historically focused only on storm clouds and precipitation. Anything that wasn't evolving rapidly toward precipitation was not of interest to them. Stratiform cloud – boundary layer, alto, cumulus outflow, or cirrus – remained largely a niche field. Conventional wisdom holds that radiation can gain little grip on a storm that lasts only an hour or two. Besides, the 1D radiation of the 1970s was of little relevance to a patently 3D storm cloud whose sides quickly grow to be larger than its top or bottom. The cubic-cloud ventures into 3D radiation in the 1970s and 1980s were too primitive to interest cloud physicists or seduce them to learn more about radiation. And, since they were unable to solve the seemingly simple problems like warm rain and the initial broadening of the drop distribution, and since they suspected turbulence as a root cause in both cases, they were much more motivated to learn about turbulence than radiation. The few cloud physicists who learned about and used radiation as a natural and normal part of their work, like Lilly, Betts, Randall, and Cotton, were oddly little imitated.

Only in the latter half of the 1990s did cloud physicists and cloud radiation people finally begin to link up, spurred in part by the increasing sophistication on each side. ARM and FIRE actively facilitated that linking. Radiation people were finally ready to handle realistic 3D clouds and actively sought input from cloud-resolving models. For instance, such input is now being used in intercomparisons of 3D radiation codes (the I3RC Project spearheaded by Bob Cahalan). Cloud-resolving models, although started in the 1970s, apparently weren't ready to consider radiation until the 1990s. And, it must be admitted, many cloud physicists still retain the inherited prejudice that radiation is unimportant. But the work of Betts, Randall, and others has slowly chipped away at this prejudice. In the new world of super-parameterization (Randall et al., 2003), not only radiation, but 3D radiation, will be essential. The fates of cloud physicists, rain radiation scientists, and cloud radiation scientists will be forever linked, and our professional grandchildren will look upon our past estrangement as senseless and inexplicable. They will ask, how could three communities, all basically working on clouds, not have talked to each other? I would be hard-pressed to explain it to them.

1.4 The "Science as Tool-Driven" Viewpoint

Looking back over my career, I see that I have always unconsciously subscribed to the idea that science is mainly tool-driven. Perhaps my applied mathematics background inclined me that way. I was furnishing software tools to the community long before that was commonplace and long before I could have articulated any philosophy justifying it as the best use of my time. I intuitively understood that if the tools are not there, no amount of wishful thinking – and often, not even great ideas – will lead to progress. In the case of 3D cloud radiation, I sniffed around the subject in the 1970s and early 1980s, but turned away mainly because the tool situation was so hopeless. Cubic clouds, the paradigm of those days, seemed too unrealistic geometrically, and the paltry 50,000 Monte Carlo photons that we could throw at them seemed too small to learn much other than that holes between clouds really do matter. The deep ideas of scaling necessary to properly model cloud structure were in their infancy then. Starting in the late 1980s, through the ARM Program, I was able to participate in creating new observational and theoretical tools for 3D clouds that made the subject more tractable.

Some may say that the idea that science is tool-driven is too obvious to be worth stating. Not so. The argument is not so much about whether tools are important – everyone agrees on that – but whether tool creation in and of itself is a high priority activity, worth supporting for its own sake. Some famous scientists have argued vigorously for this idea; they would hardly have bothered if it was universally accepted. Even at NASA, where I have worked for 20 years, I have seen a devaluation of simply creating excellent tools – platforms, instruments, datasets, software – and an increasing valuation placed on "science justification." It's a nice buzzphrase, but nobody bothers to ask what actually moves science forward most effectively. And in the view of many historians of science, developing science tools *for their own sake* is the most demonstrably effective way to move science forward.

One of the most persuasive writers on this subject is Freeman Dyson, a legendary physicist who became a big supporter of ARM.[23] I never had a strong philosophical

---

[23] Dyson once told me "you are smart to be involved with a great program like ARM" and I think I glowed for a while afterwards.

underpinning for my tendency toward tool-driven science until I read his recent book (Dyson, 1999). I'll briefly summarize his main points from that book.

Dyson says that new tools, much more than new concepts, power scientific revolutions. Examples abound, beginning with Galileo's adaptation of a Dutch telescope design and ending with the double-helix revolution in biology and the big-bang revolution in astronomy. Thomas Kuhn's influential book *The Structure of Scientific Revolutions* overemphasized the role of new ideas and almost ignored the role of new tools in creating scientific revolutions. The most strongly growing areas of science (e.g., astronomy and biology today) tend to be those which have new tools. Scientific software is also a tool. The first extra-solar planets were found by a software program analyzing a pulsar time series, not by a big telescope. Only software made possible the Sloan Sky Survey and the Human Genome Project. Even plate tectonics, often cited as the quintessential example of concept-driven science, in fact languished in the dustbin until new tools like deep ocean drilling and sea-floor magnetometers revived it.

Dyson's opinion is strongly seconded by Harwit (2003), who wrote in *Physics Today* that "progress [in physics] came primarily from the introduction of new observational and theoretical tools" which were created for their obvious general utility. These tools were able to explore the time, wavelength, and angle dimensions much better than before – something which is also vital in ARM. People tend to overemphasize the role of great ideas. Almost all the important discoveries in astronomy, Harwit says, came as huge surprises; theoretical anticipation had little to do with it. He points out that "at critical junctures [...] there is generally an overabundance of ideas on how to move ahead [...] Resolution is usually attained only with the arrival of new theoretical tools that can cut through to new understanding and set a stagnating field in motion again."

Of course, Dyson and Harwit tend to focus on revolutions or crises, since those are the most dramatic events in science. 3D cloud radiation has not undergone a classic revolution, where a pre-existing theory is overthrown. No such dominant theory existed – certainly not that old bugaboo 1D radiation. 1D radiation was merely a waystation at which to bide time, and do what useful work could be done, until the proper tools for 3D were available. No one seriously contended that 1D would suffice for the 3D problem except in very circumscribed situations like GCMs. There was a small scuffle in the 1980s between the Euclideanists and the fractalists over how best to model 3D structure, but the Euclidean cloud shape model, like the 1D model, was merely a waystation. It required only a slight push, not a revolution, to topple it.

So, there was really nothing to revolt against. The field of 3D cloud radiation has been created ab initio in the past 30 years, and by a mere handful of people including many authors in this book. These pioneers spent a goodly portion of their time creating general tools. I would like to make a brief survey of some of those new tools, in particular: new instruments and observing strategies; new models of cloud structure based on fractal and other scaling concepts; and new radiative transfer methods. My second major theme of scale will arise naturally in this survey. Let us begin by first looking back at where we were in the 1970s.

1.5 Tools for 3D Clouds, 1970s through 1990s

I would really prefer to skip the past and just read my crystal ball about future tools. But as in so many endeavors, the past is prologue, and many of the issues we first faced in the 1970s still haunt us today. The defects of the past are the spur to tools of the future, and without some understanding of those defects, it is hard to appreciate exactly what problems the new tools are fixing.

In the 1970s, we didn't have any observational tools to characterize 3D clouds in a way that would enable 3D radiation modeling. Precipitation or "weather" radars, used since their accidental discovery in World War II, had centimeter wavelengths too long to get a significant return from the 1–30 micron drops important for solar and IR radiation, although they gave us our first look inside the bowels of a raincloud. And the look was only semi-quantitative according to a weather radar pioneer I spoke to, due to calibration and problems of data interpretation.

Microwave remote sensing was just getting started in the 1970s, both from the ground and from satellites, leveraging off detector advances in astronomy and other fields,[24] but it would be a decade before any credible quantitative results for clouds emerged. The microwave absorption line parameters and continuum were still being debated and regular bake-offs were held, usually as a spinoff of ICRCCM, with the dozen people participating usually winding up at loggerheads. The spectroscopic input to microwave radiative transfer models was sufficiently uncertain when ARM fielded its first microwave radiometers in 1991 that the routine ARM microwave retrieval algorithm for cloud liquid water path had to be empirical, with tunable parameters set for each location, and only recently has a "physically-based retrieval" (one based on microwave radiative transfer theory) without tunable parameters been used operationally.

For in situ measurements in clouds, we had only aircraft measuring liquid water content with hot wires and collecting drops on sticky tape. Optical probes were just coming into general use, but while they relieved graduate students of manual drop sizing, they offered no meaningful increase in sample volume (in fact their sample volume was kept purposely small so only one drop at a time could be in it). The sample area of the FSSP, the standard optical probe for the past 25 years, is about 0.004 cm^2, so on a typical flight leg of 100 km, 0.04 m^3 of cloud volume is sampled. At that rate, assuming an aircraft speed of 100 m/s, it would take 800,000 years, or about 8 ice ages, to sample 1 km^3 of cloud. Such sample volumes are so small that they raise serious issues of statistical significance. Only by assuming spatial homogeneity can such data be extrapolated to a whole cloud; but I challenge anyone to find a spatially homogeneous cloud.

A little-discussed issue is how to bootstrap 1D aircraft data into a 3D picture of cloud. This is not as simple as one might think. At the very least, the 1D view will be biased. A famous math problem asks, "given a random distribution of nonoverlapping circles on a plane, determine the probability distribution of circle diameters by laying a straight line on the plane and measuring the chords where it intersects

---

[24] another example of general tool creation making unexpected new science possible

the circles." It turns out that the answer inferred from the straight line is always biased. This is the simplest example I can find of "dimensional bootstrapping" – trying to extrapolate to higher dimensions from lower ones. The problem of unbiased dimensional bootstrapping may be impossible in principle, but with enough 1D views (lines), as in tomography, the problem seems to be tractable. It is faced in its fullest severity only when one has a single line and wants to extrapolate not just to 2D but to 3D, as in an aircraft probe of a cloud.

Abbott's famous book *Flatland* (1884) is a whimsical commentary on the difficulty of dimensional bootstrapping. The hero, A. Square, visits the 3D world and tries to comprehend it, but is judged insane when he reports his findings back in Flatland. The book reminds us that understanding things in 1D or 2D does not immediately provide correct generalizations to 3D and 4D. Qualitatively new phenomena occur when jumping to higher dimensions. A 4D sphere passing through our 3D world would appear like a ghost, growing from a point to full size and then shrinking and finally vanishing. No one would regard this as a trivial extension of a 3D sphere! We can paper over this difficulty by assuming that higher dimensions are just a mathematical cross-product of single dimensions, but such an assumption, tantamount to a miracle, should not be accepted at face value without extensive proof.

We were in better shape on the theoretical than on the observational front, due to the extensive work on 3D radiative transfer in the Manhattan Project and after World War II, motivated by nuclear weapon and nuclear reactor problems. Because the stakes were so high, there was a powerful incentive to get the models right, and with an almost limitless supply of funding, progress was rapid. Around that time, the field of radiative transfer, which had been entirely analytical since it was invented around 1905 by Schuster, bifurcated into an analytical branch and a Monte Carlo branch; the traditional analytical branch was strong, whereas the new Monte Carlo branch, while weak, grew rapidly in order to solve Cold War problems that did not yield easily to analytical methods. Both branches were woefully short of computer power and thus had to make many numerical approximations. I highly recommend Chap. 4 for a remarkably balanced introduction to both types of numerical radiative transfer methods and the strengths and weaknesses of each.

Computer power for Monte Carlo was extremely limited, however, and great emphasis was placed on either analytical brilliance (in the Russian school) or engineering fixes (in the Los Alamos school) to reduce variance.[25] It was a feat just to calculate 20,000 trajectories, far fewer than were needed to beat errors down below 10% except by extensive spatial and/or angular averaging. It was impossible to calculate enough realizations to do proper ensemble-averages. Random number generators, the heart of every Monte Carlo model, were suspect and sometimes wrong. There were pioneering Monte Carlo calculations for homogeneous cubic clouds (e.g., McKee and Cox, 1974), but the results were severely limited by computer power and ignorance of 3D cloud structure, and thus only a few general conclusions could be

---

[25] variance reduction is an attempt to milk the maximum information from every photon trajectory without biasing the outcome

reached – things we had more or less guessed already, such as that finite clouds reflected less than their 1D counterparts with the same liquid water.

There is a fascinating contrast between the Russian approach to Monte Carlo, which is highly mathematical, and the U.S. approach, which is highly practical. Russians object to variance reduction techniques that can't be justified rigorously. They claim these techniques are easy to think up, and mostly misbegotten. In the end, our ARM group implemented rigorous techniques[26] called "maximum-cross-section" and "local estimation" which enabled us to perform Monte Carlo calculations that we could only dream about a few years earlier.

The Russian approach to Monte Carlo states point-blank that Monte Carlo is just a way of solving the integral form of the radiative transfer equation. At root, the Russians say, Monte Carlo has nothing to do with "photons" propagating from one "point scatter" to the next; these are merely convenient (but potentially misleading) fictions for algorithmically calculating the Monte Carlo solution. Expressing the radiative transfer equation with scattering in integral form leads to a sum of integrals of increasing dimensionality, corresponding to higher and higher numbers of scatters (sometimes called the Neumann series). There are no good quadrature methods for such N-dimensional integrals except Monte Carlo and, for small N, a generalization of Trapezoidal Rule (Davis and Rabinowitz, 1984). All the elegant Gaussian quadrature methods for one-dimensional integrals are useless for N-dimensional integrals. So, in the Russian view, "photons" are merely drunken census takers, careening around the medium to get a decent-enough sample to do the N-dimensional integrals. Because the census takers are drunk and disorderly, their survey converges ever so slowly to the correct solution. But a good part is: bias can only creep in if the census takers are not sufficiently drunk – that is, if they actually try to purposefully control their wanderings. Getting them drunk enough corresponds to using a sufficiently good random number generator.

Americans may take umbrage at the idea that their loose language of photons and trajectories is misleading. But it is easy to get confused when one talks about specific typical wavelengths and specific types of scatterer. Suppose, for example, that one wants to solve the radiative transfer equation for pure air at solar wavelengths. Both the size of the scatterers (air molecules) and the distance between them are much smaller than typical wavelengths of sunlight. In that case, how can one speak of a point particle of light (a photon) caroming off a point scatterer? In fact, this situation is fully in the wave regime of light, and there are no definable trajectories in the Monte Carlo sense.[27] However, amazingly, Monte Carlo gives a perfectly valid and acceptable solution to this radiative transfer problem, even though the words used to describe the algorithm become meaningless and in fact false. Nothing could better indicate the true character of Monte Carlo: a way of solving the radiative transfer

---

[26] described in Marchuk et al. (1980)

[27] Of course the light waves excite each air molecule to emit dipole radiation in all directions, so the molecules can be viewed as point scattering centers. But a better mental model would be one of throwing a handful of pebbles into a calm pond rather than firing bullets into a vast empty space sparsely populated by point scatterers. It is also well to remember that Rayleigh originally treated this problem in the continuum view.

equation independent of wavelength, independent of type or size of scatterer, and independent of the model of matter being used (continuous, discrete, or raisin pudding). That is what allows Monte Carlo to be used by IR and microwave radiation scientists without guilt.

Monte Carlo methods really excel when the interest is mainly in the radiance escaping from the structure being sprayed with photons, or when only grand spatial and/or angular averages are needed. If one is interested in internal radiances, or in escaping radiances from every voxel at every angle, Monte Carlo will generally be too inaccurate for any reasonable number of initial photons. For example, you can launch 10M photons, but what if only 10 photons escape from a particular voxel in a particular direction? Then the error in radiance will be 30% or more.

Analytic methods, the alternative to Monte Carlo, solved the radiative transfer equation by a combination of special-function expansions and standard numerical methods for linear integral and/or differential equations.[28] This dichotomy continues to the present day: analytical methods on one side, deterministic but at the mercy of truncation errors; and Monte Carlo methods on the other side, simple to program (see the simple one-page Monte Carlo code in this book) but often with large stochastic errors.[29] Most analytic methods made some use of spherical harmonics and/or discrete ordinates ideas, although a few like the famous S-N method of the 1960s just discretized everything in sight and solved gigantic systems of linear equations. The actual 3D codes were usually classified because their DATA statements contained top secret material and optical parameters, but the general techniques, under various unmemorable names, were published mostly in the journal *JQSRT* in the 1960s and 1970s. The legendary Jerry Pomraning was a leader in this activity, and I remember that he was one of the few radiation scientists that my mentor, Burt Freeman, really looked up to. One characteristic of those early 3D methods, however, was a reliance on radical simplifying assumptions. Computer power was far too limited to solve the full 3D problem in all its glory – something we now take for granted with models like Spherical Harmonics Discrete Ordinate Method (SHDOM Evans, 1998), the ultimate fulfillment of many of the dreams of those days.

In the 1970s, we even lacked a proper geometry to describe clouds. It was nascent in the form of Mandelbrot's monofractals, popularized in his famous book *The Fractal Geometry of Nature*, but random monofractals turned out to be a poor representation of cloud liquid water structure although they could be tuned to create cloud images of great visual verisimilitude. The movie industry is probably still using monofractal clouds in animated films. But multifractals, adapted from new turbu-

---

[28] There was an odd detour after Chandrasekhar's book on radiative transfer was published in 1950. Following Chandrasekhar, scientists transformed the linear radiative transfer equation into a coupled set of nonlinear integral equations which were even less soluble than the original equations. For obvious reasons, this didn't last, but it did attract a lot of mathematicians who wrote several almost impenetrable books on radiative transfer and helped create a misimpression that radiative transfer was an avant garde branch of mathematics.

[29] Monte Carlo errors fall only as the inverse square root of the number of photons, except for the intriguing quasi-random number approach recommended by O'Brien (1992) but, for various reasons, little used.

lence cascade models developed in the 1980s, were necessary to describe cloud liquid water correctly.

We sent research aircraft into clouds regularly and tried not to think overly much about the biases involved. Among these biases were avoiding icing conditions, high winds, turbulence, rain, lightning, nighttime, high altitudes, and other conditions unpleasant or dangerous for humans. We mainly flew in benign clouds in the daytime. Pilots were contractually entitled to their days off, whether or not the clouds were good that day. Getting humans out of the plane will help remove some of these biases, in principle, but a concern with endangering the plane will not relieve other biases. Progress in UAV development has been slow, due as much to FAA restrictions as to technological limitations, but UAVs are certainly a far preferable solution, not least because time on station goes up from a few hours to, potentially, days (ARM has already done a 24-hr UAV research flight). Meanwhile, there has been some progress with crewed aircraft. NOAA armored aircraft now routinely gather data inside hurricanes. Daytime flights (typically clustered around local noon) have remained the norm, but the DYCOMS-II field program in 2001 (Stevens et al., 1999) figured out how to do night flights without incurring the usual punishment of long stand-downs afterwards. Cirrus experiments were nonexistent in the 1970s, but now cirrus clouds are routinely reached by instrumented jets, most recently in the phenomenal CRYSTAL-FACE field campaign, although instruments for measuring ice crystals are naturally less well developed than the time-tested instruments for liquid drops.

Another bias is the preference in cloud field programs for "ideal" clouds over the much more common "messy" clouds. Too many times I have overheard people say, "we aren't considering this dataset because the clouds were just too messy or complicated." Something about this always struck me wrong. Cloud modelers are perfectly content to calculate messy, complicated clouds on the computer. Why then prefer ideal ones in the field? Shouldn't we rather observe typical, messy clouds in the field, then distill what we observe into simple models on the computer? That's the normal process of science, whose main paradigm was established 400 years ago, when Newton started with the messy planetary motions and captured them inside simple dynamical laws. The whole process with clouds seems the opposite of that, namely: study only the simplest data, then fit it into the most complex theories. Just because it is easy to have complexity on the computer, doesn't mean it is the right thing to do. And just because it is hard to interpret observations of complicated clouds in the field, doesn't mean it should be eschewed.

It is possible to guess why cloud scientists avoid messy clouds in the field, even though current models could handle them. The main reason may be because so many measurement approaches to clouds make so many assumptions (e.g., constant drop number concentration for some radar retrieval algorithms), and naturally one would want to restrict observations to "ideal" clouds meeting the assumptions. A second reason may be to test models in limiting cases, an important principle of scientific software engineering. A third and little-discussed reason may be the dimensional bootstrapping problem mentioned above. Cloud scientists believe that it is less difficult to extrapolate measurements from 1D or 2D up to 3D for ideal clouds, because ideal clouds are more homogeneous, or at least vary in more predictable ways, than ordinary clouds. I suspect that ideal clouds are no more dimensionally bootstrap-pable than ordinary clouds, however, because turbulence, the great de-homogenizer, operates in all clouds.

Good as the reasons for targeting ideal clouds seem in principle, the practice raises disturbing issues. One can wait a week or more for ideal clouds, meanwhile passing up many ordinary cloud situations, and when ideal clouds do arrive, they often fall short of ideality anyway. Ideal clouds often seem like a vanishingly small subset of all clouds. In the end, our models have to work for a certain significant range of clouds, not just a vanishingly small subset. Even stratiform clouds are messy. Marine stratocumulus, the subject of many studies in this book because of its seeming simplicity, is full of small pockets of convection and drizzle, to say nothing of factors of four vacillations in optical depth over distances of 1 km or less.

Latterday field campaigns like SUCCESS, DYCOMS-II, and CRYSTAL-FACE also have begun to remedy an issue which quietly plagued earlier field outings: not getting enough data to make robust conclusions. Previously, a month in the field was lucky to net a handful of "golden days" where instruments and logistics worked well and the clouds were perfect or at least plentiful. One had to wonder whether issues of global climatic significance could be decided based on a few days of data – that is, whether an implicit assumption that "all clouds everywhere behaved like the ones sampled" was true. Nowadays, a combination of longer times in the field and better logistical organization nets a larger catch rate of data.

1.6 Current Cloud Observational Tools

Imagine my surprise, as I was working on this chapter, to open the *Boulder Daily Camera* of 26 Mar 2004 and find an article on a breakthrough in weather radar. Developed by a team of NCAR scientists, the new system piggybacks a 30-inch cloud radar dish on a 30-foot scanning rain radar dish. "Together, the two radars of the S-polka system can detect everything from baseball-sized hail to ice or water particles 10 microns in diameter," the article trumpets. This was a powerful reminder that we stand on the edge of finally being able to specify the 3D structure of a cloud, including not just the big particles and not just the small ones but all of them. A unified view of a cloud for the first time! The price: size, cost, and complexity. Tomography is also a possibility, one which deserves its own section, later. So the technology is there, if we think the cloud problem is worth the price.

If we ever want to "X-ray" a cloud, radar and microwave wavelengths are our only choice. Wavelengths longer than weather radar simply pass through a cloud as if it wasn't there. Infrared wavelengths are so strongly absorbed that they barely penetrate 50 meters into clouds. Solar wavelengths can penetrate even the thickest clouds, but multiple scattering scrambles the information they might otherwise provide about spatial structure.

Let us look at some examples of state-of-the-art cloud radars and microwave radiometers which go part way toward a complete X-ray of a cloud. Figure 1.9

Fig. 1.9. Modern microwave (*left column*) and radar (*right column*) instruments for the study of 3D clouds in the US DOE's ARM Program (*top row*) and in Europe (*bottom row*). The ARM two-channel microwave radiometer stands a little over 2 m tall and is modified from a Radiometrics Corporation commercial product. The ARM cloud radar was developed by NOAA Environmental Research Labs in Boulder. Both ARM instruments point vertically and measure 2D time-height profiles. The European 2-mm-wavelength radar and 22-channel microwave radiometer scan together in lockstep, rapidly enough to capture the 3D structure of clouds; both are described at http://www.meteo.uni-bonn.de/projekte/4d-clouds/tools/

shows the operational cloud radar and two-channel microwave radiometer deployed at ARM sites around the world. Both are the results of development that began in the 1970s at the NOAA Environmental Research Labs in Boulder. Also shown are a corresponding cloud radar and microwave radiometer used in a 2003 campaign in Holland as part of the European 4D-Clouds Project, an effort combining cloud measurements, radiative transfer modeling and dynamical modeling with field outings in 2001 and 2003.

The Europeans obviously learned a lot from the ARM experience and built next-generation instruments, which at this point the U.S. can only envy since there is little funding for major new surface cloud instrument development. Both their radar and microwave radiometer scan fast enough to potentially map out a full 3D cloud field, while the ARM instruments merely stare vertically and provide time-height slices through a 4D cloud field. And the Europeans have aggressively tackled the 4th dimension, time, by taking data every second. Their outhouse-sized two-ton microwave radiometer has 22 polarized channels and is capable of scanning in lockstep with the cloud radar! For years, we have been arguing in ARM about merely adding a third channel to our microwave radiometer to better capture thin clouds; imagine what we could do with 20 new channels! The Europeans have obviously showed us one future in surface cloud observations, albeit a very expensive one. It is of course well to remember that operational deployment of such large, complex instruments would present severe difficulties.

Not all instruments have to be so high-tech or expensive. There is still room for simple instruments and simple platforms. One example is the European tethered balloon with cloud physics instruments, shown in Fig. 1.10. This is something I lobbied for within ARM, but the FAA blocked us at every turn, carrying the bureaucratic mantra "just say no" to new heights. ARM did manage to fly tethered balloons in at least two field programs, although time on station was limited due to battery depletion.[30] For the Arctic sea-ice SHEBA experiment, I worked with Paul Lawson and Knut Stamnes to develop a new kind of tether which sends power up and brings data down using copper wire twisted together with the ordinary tether, eliminating the two heaviest parts of a typical tethered balloon payload (the battery and the transmitter). Much longer flights are possible with this unlimited supply of power. It would also be much easier to deploy multiple instruments spaced along such a tether, since each could draw power from the tether. In the end, only tethered balloons or cloud sondes can provide true in situ vertical profiles – at least until we develop a capability to hang a string of instruments more or less vertically below a slow-moving aircraft or dirigible.

Cloud aircraft instruments have come a long way from the 1970s, although hot wires and FSSPs are still lynchpins. But now there are many more instruments including Hermann Gerber's PVM which measures liquid water content and effective radius thousands of times a second, allowing unprecedented spatial resolution; the fast FSSP from the French allowing similar high spatial resolution in the drop size distribution; and the Cloud Particle Imager of Paul Lawson, particularly useful for looking at crystals in cirrus clouds. Several advanced spectrometers designed specifically for aircraft use have followed the path blazed by Alex Goetz's AVIRIS grating spectrometer and Bill Smith's HIS IR interferometer, which flew throughout most of the 1980s. Mike King's Cloud Absorption Radiometer brought the measurement of cloud absorption to a new level of sophistication, and lately has found unexpected use measuring surface BRDF! As a result of ARM's ARESE field campaigns of 1995 and 2000, we learned a lot more about how to use flux radiometers on aircraft, in particular about their thermal offsets, which also led to better surface flux mea-

---

[30] In a 1997 deployment, Jay Mace brought down spider webs and other biological material from the balloon that suggested to me a source of absorption in the atmosphere ignored in all radiation models.

Fig. 1.10. Tethered balloon with cloud physics instrument payload developed by the University of Utrecht for the ASTEX field campaign in 1992, shown here participating in a 2003 field outing as part of the European 4D Clouds Project

surements. By the time of CRYSTAL-FACE in 2003, cloud aircraft fairly bristled with new kinds of instruments, although some say that funding agencies still underappreciate the importance of continuous instrument development between field campaigns.

There were also big advances in coordination and observing strategies. Earlier cloud field programs had had a certain "sky-cowboying" aspect, chasing clouds hither and yon at the behest of the aircraft scientist. NASA's FIRE field program pioneered a more orderly strategy, following in the footsteps of the famous GATE program of the 1970s and various ocean programs where instruments were carefully deployed according to theoretical and statistical guidelines. FIRE coordinated multiple aircraft with satellites and surface instruments, presumably in order to validate aspects of the NASA satellite cloud climatology project called ISCCP but with real science goals as well. FIRE fell short in many areas, and I was too quick to point these out in the early 1990s, but with the perspective of time I now see that FIRE was pushing the envelope in so many areas that it was bound to stumble in some of them. Just coordinating multiple disparate aircraft alone was a big advance.

One issue in FIRE concerned me greatly in the late 1980s: I dubbed it "scale-babel." Many instruments used in field campaigns tended to take data at time, space, and angle scales which were convenient or natural for themselves, rather than at the natural scales of clouds. Apparently experimenters expected someone (not themselves) to coordinate these scales later. The alternative, which seemed to have escaped notice, was for everyone to coordinate scales in advance. Scale-babel was based on an untenable assumption of interpolability. It still goes on, and it is still a Bad Thing, but I shouldn't have expected FIRE to solve it – it is due to the obstinacy of highly individualistic experimenters, not to the program which hosts them. Scale-matching of (a) instrument with instrument, (b) instrument with natural cloud scales, and (c) overall experiment design with cloud modeling scales, improved a lot by the mid-1990s, however, and newer cloud field programs and EOS satellite instruments are more sensitive to the issue of scale-babel.[31]

The one area where little observational progress was made in the 1980s was the leap to 3D. Non-satellite measurements still remained determinedly 1D, whether it was 1D horizontal (aircraft flight tracks) or 1D vertical (microwave radiometers and radars, sondes). The aircraft flight tracks scrambled time and space because clouds evolved while the flight proceeded, while vertical microwave and radar (and lidar for thin clouds) at least capture the time dimension correctly, to build up a 2D time-height projection of a 4D cloud – time being the 4th dimension. Radars and microwave radiometers can be scanned to add more dimensions, as the Europeans have done in their 4D Clouds Project, but the cost involved is evident in Fig. 1.9.

Satellites gave us a 2D view, but it was hard to be sure what the satellite radiances were really telling us about clouds other than their fraction and their temperature. ISCCP (Rossow and Schiffer, 1999) continuously developed algorithms for giving cloud optical depth in addition to the obvious parameters of cloud fraction and cloud top temperature, but these optical depths had to rely on uncalibrated weather-satellite channels and it was well into the 1990s before a reliable dataset was in hand. In any case, satellite retrievals remain firmly grounded in 1D radiative transfer, with no prospect of change any time soon, so 3D effects remain scrambled into the retrievals in a poorly understood way. And satellites remain poor candidates to leverage us into the time dimension, since revisit times are at least 10 min for geostationary satellites[32] and 90 minutes for polar-orbiting satellites. Triana, a satellite built to observe Earth from the Earth-Sun Lagrange point L-1, would have given minute by minute time resolution of the whole Earth, but, sabotaged by politics, it sits in a warehouse like the Ark of the Covenant at the end of the first Indiana Jones movie.

---

[31] although there is still a tendency to tiptoe around scale-babel, and to throw whatever is ready to hand at the cloud problem whether or not it is scale-appropriate

[32] except in so-called "rapid scan mode" used for very limited regions experiencing severe storms

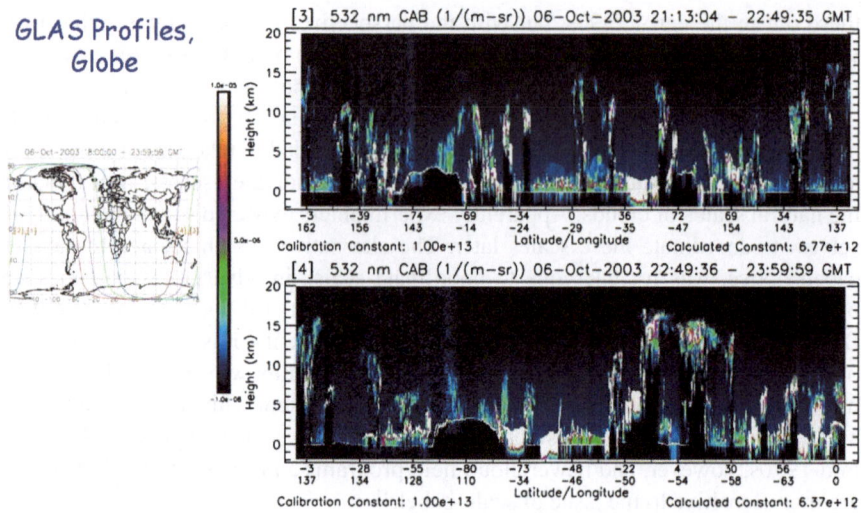

Fig. 1.11. Two complete passes of the Global Laser Altimetry System (GLAS) from the Arctic to the Antarctic and back on 6 Oct 2003. The inset map shows the satellite orbit segments. (*top*) Lidar backscatter at 532 nm (units: 1/(m-sr)) from 21:13 to 22:49 GMT. (*bottom*) Same as top but for 22:49 to 24:00 GMT. Note that the backscatter color scale is logarithmic. Aerosols as well as clouds can be seen, albeit with lower backscatter values. Apparent returns from below the Earth's surface may be due to multiple scattering in clouds, which delays the returning signal; this is not noise but actual information which can be used to learn more about clouds

The debate over "enhanced cloud absorption," which played out over the years since 1995, sparked two ARM aircraft field programs (ARESE I and II) to measure cloud shortwave absorption. ARESE I indicated how poorly prepared we were to deal with the real 4D complexities of the cloud problem. The aircraft couldn't remain stacked because of differing airspeeds. Aircraft roll, pitch, or yaw beyond a degree or two invalidated the flux radiometer data or required extrapolating it to perfect level. The aircraft were at different distances from the cloud so the flux radiometers saw vastly differing areas of cloud. This experience forced us to a new and higher level, both in quality of instruments, in theory, and in sampling strategy, and some gains were realized in ARESE II, but at the same time we realized how far we still have to go. Coordinating two aircraft (ARESE I), or one aircraft and a surface site (ARESE II), required a careful experiment design incorporating the very best 3D radiative transfer modeling. Marshak et al.'s (1997) work was seminal in this regard.

We round out our discussion of current satellite capabilities with two examples of new kinds of satellite cloud data. Both are full of mysteries that have barely begun to be studied. Figure 1.11 shows lidar profiles from the recently-launched IceSat,[33] designed solely to measure the elevation of the world's ice sheets, but, as a byproduct, measuring global vertical profiles of clouds – the first satellite lidar profiles since Pat

---

[33] http://glas.gsfc.nasa.gov/

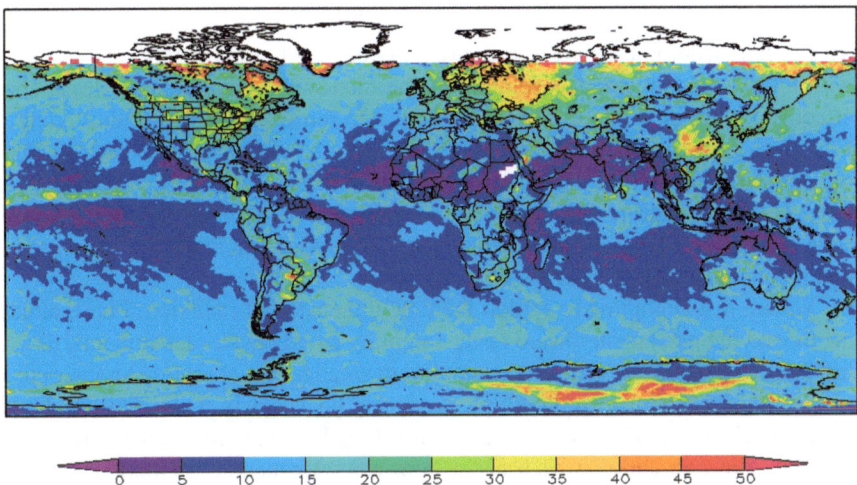

Fig. 1.12. Global map of monthly-average cloud optical depth retrieved using the Nakajima-King algorithm for MODIS data of Nov. 2003. The retrieval may be inaccurate for low optical depths and it saturates for large optical depths, but in the mid-range of values where most clouds live, it represents the state-of-the-art in 1D cloud retrievals

McCormick's pioneering Shuttle lidar mission of 1994.[34] IceSat is a good example of how tool creation can create benefits well beyond the originally intended application. We haven't had IceSat kind of cloud information in such abundance before, and so far it remains a relatively unmined resource. Much more such information will come from the CALIPSO[35] lidar to be launched in 2005 on the same rocket as CloudSat.

The second example of new kinds of satellite cloud data, Fig. 1.12, shows a recent retrieval of global cloud optical depth from MODIS using the most advanced 1D retrieval algorithm available, that of Nakajima and King (1990). This dataset is entering its 5th reprocessing and represents the combined effort of a huge team. As with IceSat, we simply haven't had such datasets before, and it will take a while for researchers to grow comfortable with them and exploit them fully. It is hard to learn much of a fundamental nature directly from such maps, which often seem to dissolve into Rohrschach blots – a quality they share with many GCM color contour plots. However, there certainly is much to be learned by compositing the data in clever ways and comparing it to surface and ISCCP retrievals.[36] Such work is ongoing. It is well to remember, though, that these data, like any cloud data retrieved using solar radiation, are only available in the daytime (and not even near sunrise or sunset, at which times the 1D retrieval breaks down); this diurnal bias may prevent accurate climatic conclusions from being drawn.

---

[34] Winker et al. (1996); http://www-lite.larc.nasa.gov/

[35] http://www-calipso.larc.nasa.gov/

[36] Jakob (2003) is exemplary in devising new ways to composite cloud information.

If you couple the optical depths from MODIS with the cloud liquid water path over oceans from microwave sensors (e.g., Greenwald et al., 1993), and are happy with the 10s-of-km spatial resolution of the microwave data, you can learn about effective radius as well. Of course, MODIS also retrieves effective radius from the Nakajima-King algorithm, so the two methods can be compared. This is an example of something that will be absolutely necessary to make progress on 3D clouds: coupling of several instruments. This was a cornerstone of ARM as well: a cooperative suite of instruments all co-measuring clouds, to produce an integrated data product that no one instrument could possibly provide.

We are just beginning to exploit the time dimension of radiation. I have already mentioned the movement toward faster-sampling cloud physics instruments, like the fast FSSP and Gerber PVM, and hence toward smaller spatial scales. Another movement is toward time-resolved multiple-scattering lidar (Davis et al., 1999). Lidar has of course always used pulses whose return is timed, so time-resolution here is nothing new. But conventional backscatter lidar cannot penetrate a cloud very far – a few optical depths at most. Multiple scattered photons sample a whole cloud and, when captured, carry information about their travels. Another advantage of multiple-scattering lidar is that it works at night.

Two efforts which sprang from ARM projects at Goddard are exploiting multiple-scattering lidar: the WAIL and THOR projects respectively at Los Alamos National Laboratory (Love et al., 2001) and at Goddard (Cahalan et al., 2005). They have learned to retrieve things like cloud optical and geometric depth from the returning photons' time history at a variety of angles away from the incident lidar beam. Both of these exciting efforts have led to new instruments – new tools for understanding 3D clouds. The necessity to understand time-dependent radiative transfer, something new for the vast majority of cloud radiation experts, is implicit in this new approach.

ARM sites, which everyone now takes for granted, were in 1990 a revolutionary development. No one had ever thought of creating a "permanent field program" to simultaneously furnish research and climate data. The novel idea underlying ARM was that continual data-gathering would provide long, well-calibrated datasets which could never be gathered in typically short field campaigns, and that discoveries of climatic importance were certain to emerge from this. At the same time, IOPs (Intensive Observational Periods) furnished an occasional field campaign milieu where extraordinary scientist attention was brought to bear on clouds for a briefer period of time. ARM had funding for both infrastructure and a science team, which were organized to interact closely with one another. It would be superfluous to list the many ways that ARM has promoted cloud science in the past 14 years. The concept has worked so well that other countries are now copying it, and ARM sites have been designated national facilities in the same category as astronomical observatories (with, naturally, an expanded purview).

One cornerstone of ARM, not possible on satellites, is to compare observations of the same quantity from multiple instruments. This was rare in field programs of old, where one had to trust a single instrument not to have errors or misfeatures. Even getting cloud base altitude correctly from three different instruments proved problematical in the early days of ARM, exposing our poor understanding of variables that we might have measured with but a single instrument in a field program of old.

Cloud models were much readier to go 3D than the observations. Some calculate full microphysics and some use parameterized microphysics where the drop size distribution is, for example, a sum of a few simple analytic functions (but never as simple as the one-humped Deirmendjian or Hansen-Travis distributions). Full microphysics slows down a model considerably and is thus not as popular. With the exponential rise in computer power during the 1980s, 3D cloud models finally came of age. Now there are a variety of models to pick from, with resolutions down to 10–20 m.[37] Radiation scientists are naturally taking advantage of this new capability to provide input to radiation models, and soon they will have to reciprocate and provide simple 3D radiation parameterizations to the cloud modelers.

1.7 Future Cloud Observational Tools

We tend to think of our field in isolation, and thus we miss common patterns of development across fields. These common patterns enable us to improve our crystal ball. In the cloud measurement field, progress generally follows the pattern in astrophysics: arrays of identical sensors; complementary sensors; and increased resolution in wavelength, angle, and time. Examples are: in arrays of identical sensors, tomography; in complementary sensors, any ARM site; in wavelength, O_2 A-band and sub-millimeter spectrometry; in angle, CAR and MISR; and in time, the fast FSSP and multiple-scattering lidar.

There are no observational magic bullets for clouds. We have sought them aggressively, but one after another has come up short. I remember when, in the ARM Program, we hoped that our cloud radar would be just such a magic bullet. That was before insects and other problems set in. Before long, we had to supplement the radar with other instruments. Our hope now rests in combinations of very different kinds of instruments, each filling a gap left by the others.

The ultimate combined set of instruments, although one that came together more or less accidentally, is the satellite A-train shown earlier in Fig. 1.5. All in all, the A-train is a formidable armada for cloud studies.

Armadas are one thing, but there are still new stand-alone instruments of great promise. My favorite is the aircraft-borne in-situ lidar (Fig. 1.13; cf. Evans et al. (2003)). This new lidar shoots sideways and, through intense scattering, nucleates an expanding, diffusing near-sphere of photons.[38] This near-sphere acts as a "short-range scan" extending the observational reach of an aircraft well beyond its fuselage. Detectors on the opposite wingtip measure the time-resolved returning photons to retrieve the average extinction coefficient in spheres of radii 25 m, 50 m, 75 m, etc., centered near the aircraft. When the expanding sphere reaches a cloud boundary, the

---

[37] At 10–20 m resolution, smaller than the photon mean-free-path in a typical cloud, 3D radiative transfer effects will have to be accounted for in future 3D cloud modeling.

[38] denser in the middle than at the edge, and with a fuzzy rather than a sharp edge

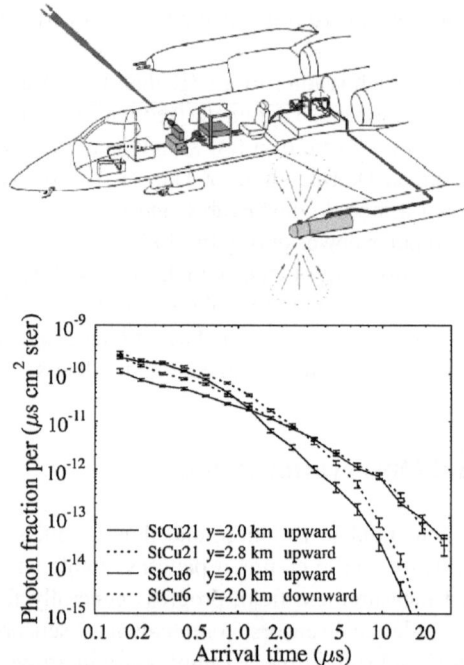

Fig. 1.13. (*Top*) Schematic design of an in-situ lidar system using multiple scattering of time-resolved returning photons to infer average extinction in expanding spheres around the aircraft, as well as the location of cloud boundaries. A laser shoots pulses to the right, while up- and down-looking detectors on the left wingtip record a time series of the multiply scattered light. (From Evans et al. (2003).) (*Bottom*) Monte Carlo-simulated in-situ lidar time series for up-pointing detectors at two locations within a stochastically-simulated 0.825-km-thick stratocumulus field (StCu21), and for up- and down-pointing detectors at a single location in a second 0.425-km-thick stratocumulus field (StCu6). In the simulation, the aircraft is near the top of StCu6 and near the middle of StCu21. The aircraft horizontal position is halfway into the 3.2 × 3.2 km periodically replicated cloud cell. The laser beamwidth is 2°. The two detectors are 8 m from the laser and have a full-width field of view of 0.5 rad. Time bins increase by a factor of $\sqrt{2}$ in order to accumulate more photons. Error bars indicate uncertainty due to Monte Carlo noise

curve of returning photons vs. time gradually steepens, since photons which would otherwise have scattered back now escape. At absorbing wavelengths, a retrieval of cloud absorption coefficient may also be possible.

Another new instrument is a multi-wavelength cloud extinctometer (Fig. 1.14). This is designed to measure extinction directly, in a multi-pass cell with an 8-m folded optical path. This is a much more direct measurement of extinction than the traditional method of convolving the drop size distribution with Mie theory, with at a much higher sampling rate. Hermann Gerber has developed a new Cloud Integrating Nephelometer that measures not only extinction but also scattering phase function asymmetry factor.

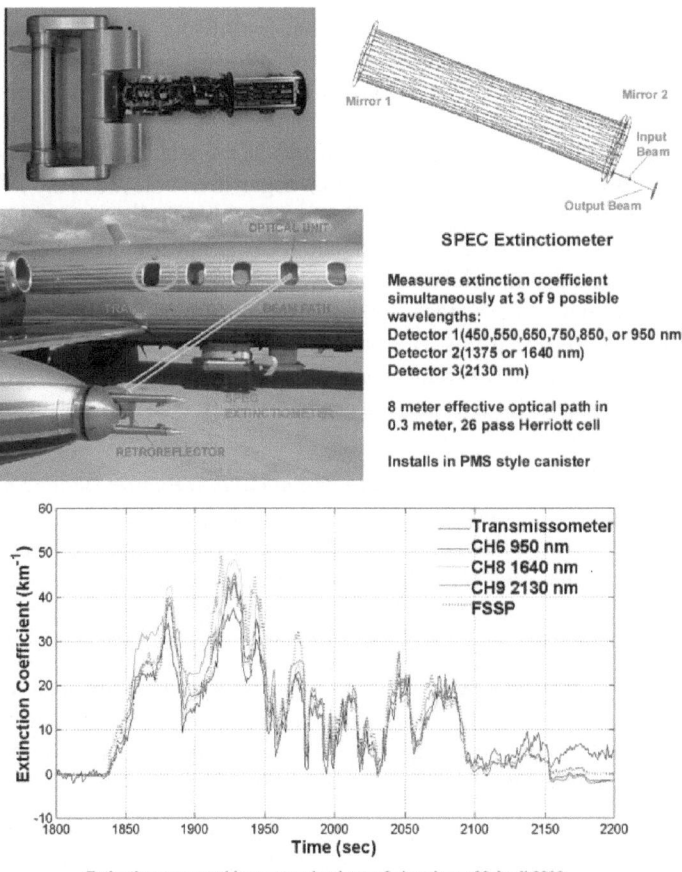

Fig. 1.14. (*Upper*) Multiple-pass three-channel cloud extinctometer (Zmarzly and Lawson, 2000) which can sample extinction at a far higher rate than a traditional FSSP since it does not need to wait for enough drops to be collected to form a stable, statistically significant drop distribution, if one even exists. The instrument is designed to fit in a standard cloud physics canister, shown slung under the aircraft in the middle photo. That aircraft also flew a Russian two-pass transmissometer which bounced a 532-nm laser beam from inside the plane off a retro-reflector on the side of the FSSP instrument. (*Lower*) This plot compares extinction coefficient vs. time measured in a wave cloud over Colorado on 26 April 2000 by the Russian transmissometer; also plotted are corresponding data from the new extinctometer at 3 wavelengths, and values inferred indirectly from the FSSP using its measured drop size distribution in Mie theory. (Original curves were in color but for present purposes it is not essential to distinguish them.)

Why the interest in extinction? Cloud optical depth, the most important optical property of a cloud for radiation, is still a poorly measured quantity in my opinion. We have nothing that would rigorously qualify as a direct *in situ* measurement since optical depth is a vertical integral and all cloud research aircraft travel horizontally.[39] The only rigorous remote measurement would be to measure the transmission of the Sun (or a man-made beam) through thin-enough clouds;[40] other remote measurements make the 1D-cloud and other assumptions leading to as-yet poorly quantified errors.

Uncertainty about cloud optical depth lies at the heart of what I once dubbed "the cloud-albedo paradox" – something that began to trouble me in the 1970s when I made my first calculations of adiabatic liquid water content,[41] integrated it vertically, and found that a cloud only 200 m thick could easily have an optical depth of 10. At the time, I thought that the global-annual-average value was about 10, so, since most clouds seemed much thicker than 200 m,[42] and must therefore have optical depths larger than 10, I regarded this as a paradox. [Using the latest ISCCP value of 3.7 (Rossow and Schiffer, 1999) rather than 10 makes the paradox much worse.] After sitting on my concern for a number of years, I finally made this statement near the end of Wiscombe et al. (1984):

> "[. . .] a paradox seems to be developing in cloud radiation studies: namely, that optical depths computed from seemingly reasonable liquid water content profiles based on actual field measurements reach values of several hundred for even moderately [1–2 km] thick clouds [. . .] To get the correct value of planetary albedo, the 50% cloud cover may only, on average, have a 50% albedo, a value which is consistent with optical depths on the order of 10."

In that paper, I suggested unaccounted-for large drops as a possible resolution (packing the same liquid water into fewer, larger drops lowers the optical depth), but the community has not embraced that idea. I wrote again of my concern in Wiscombe and Ramanathan (1985):

> "The second mystery concerns cloud optical depth. Cloud-physics models and observations give us liquid-water amounts from which we can compute optical depths. Observations of reflected solar radiation from space, on the other hand, allow us to infer optical depths. These two ways of inferring optical depth can differ by up to an order of magnitude!"

Furthermore, if 3.7 is the average cloud optical depth, then according to Bohren et al.'s (1995) criterion that a cloud has to reach an optical depth of 10 to obscure the Sun, one should be able to see the Sun through almost all clouds! Nothing could be further from common experience.

---

[39] Tethered balloons could be outfitted with miniature extinctometers spaced along the tether. Free balloons or dropsondes carrying a single extinctometer could travel more vertically but the cloud would evolve during their ascent or descent.

[40] Bohren et al. (1995) showed that "thin enough" for solar wavelengths means optical depth 10, roughly.

[41] This is the liquid water in a rising parcel of air starting at the surface and not entraining any surrounding air; it is an upper limit but no more than twice the actual value in many clouds. Cloud research aircraft have for decades searched for "pure adiabatic cores" in cumulus clouds, and some have actually been found.

[42] The sad truth is that we don't know the global probability distribution of cloud geometric thickness; IceSat, CloudSat and CALIPSO will provide the first such information.

This issue of optical depth well illustrates the estrangement between the cloud physics and cloud radiation communities. The radiation community seems unfazed by weak, wimpy clouds with mean optical depth of 3.7. Yet real clouds typically explode into existence and rapidly seek a fairly substantial optical depth above 5 or 10. Yes, a cloud must pass through a thinner stage while growing and dying, but it spends only a small fraction of its lifetime in those phases, and the rest of the time it is fairly thick – or as thick as the local water supply will allow. If the optical depth of a typical cloud were graphed as a function of time, it would look like a plateau with steep cliffs at the front and back end. Even water-starved popcorn cumulus clouds, so ubiquitous in the subtropics, explode up to optical depths in the 5–9 range (Coakley et al., 2005) and remain there for a long time. And it is no good blaming cirrus: cirrus typically overlie lower clouds of more substantial optical depth, and cases of pure cirrus don't occur often enough to drag the whole planet's cloud optical depth below 4.

An explanation that I favor is that there is no paradox because we are comparing apples with oranges: true optical depth with reflectional optical depth.[43] True optical depth could be measured by a fleet of drone aircraft stacked 10 m apart, and reporting extinction as they fly through a cloud. Integrate the reported extinctions vertically and presto, true optical depth! This would also be the optical depth predicted by a perfect cloud model. "Reflectional optical depth" is retrieved from cloud reflection and is the value that makes the planet's radiation balance come out right. It owes zero allegiance to non-radiational reality, including cloud physics. Cloud reflection reaches a near-asymptote when the true optical depth reaches 50–100. But true cloud optical depth can, invisibly to satellite sensors, grow into the hundreds (or even thousands in a big storm). So reflectional optical depth will be biased low compared to true optical depth, all else being equal. This bias may help explain the paradox – but my personal belief is that it doesn't go far enough, because clouds with optical depth over 100 only cover a few percent of the planet at any given moment. Something else is afoot as well, I think.

Living with two different optical depths is not really disturbing. The issue only becomes damaging if you deny the duality and expect the two optical depths to agree. But it will take some education for climate modelers to realize that the growing archives of satellite-retrieved cloud optical depth are not the same as the optical depths that their cloud physico-dynamical parameterizations will produce. (The related duality in cloud fraction would be just as important, except that climate models do not yet have a physically-based parameterization for cloud fraction.)

Alexander Marshak has developed a new method, briefly described in Chap. 14, that will help us to better understand reflectional cloud optical depth (Marshak et al., 2000, 2004). Current 1D retrieval methods assume no 3D effects. Marshak has invented the first surface-based[44] retrieval method that can recover reflectional optical

---

[43] Stephens (1988b) has already proved a similar duality in cloud fraction: there is a true cloud fraction, and there is a cloud fraction needed to make the radiation come out right, and they are never the same.

[44] Barker et al. (2002a) have extended Marshak's method to work from above clouds.

depths even in a 3D broken-cloud situation. His method uses surface reflection of sunlight from vegetation as a source of illumination of the cloud base. It is not only an important breakthrough, but it provides a reflectional optical depth from beneath a cloud instead of from above; if the two differ markedly, it will be further evidence that optical depth, in spite of its superficially trivial appearance, is a quantity with multiple personalities.

Right through the 1990s, cloud observing systems and strategies were not developed ab initio from cloud theory. This is changing, however. Pincus et al. (2005) illustrates the power of this new approach. They asked: to what degree can 3D radiative transfer effects be estimated from the 2D time-height profiles of an ARM or other surface radar? They worked entirely in a virtual world, sampling a Large-Eddy-simulated field of small cumulus clouds with a simulated vertically-pointing cloud radar. They made two sets of Monte Carlo radiative transfer calculations: one for the full cloud structure and one for the cloud structure inferred from the radar. The differences between these Monte Carlo calculations allowed them to disentangle the relative importance of dimensionality, sampling, and the frozen turbulence assumption.

Let me close this section with a plea for high measurement time resolution consonant with time scales of cloud dynamics. Many instruments for looking at clouds take data at a lazy pace appropriate for clear sky but not clouds. One-minute or longer averages are typical; yet in one minute a tremendous amount of cloud variability passes overhead with even a moderate wind. Our group lobbied hard at the inception of ARM just to get 20-sec sampling by the microwave radiometer, a lynchpin instrument for cloud studies and radar retrievals. Many instruments like lidars and radars can sample rapidly, but the high-frequency fluctuations caused by clouds are often viewed as just noise and not analyzed with powerful statistical tools. I would argue that there is much information in that noise.

1.8 Tomography: The Ultimate Solution?

Tomography is a way of imaging the 3D structure of an object by probing it with multiple beams and analyzing the multiple radiation beams emerging from it. You can shoot the incident beams yourself, let the Sun do it, or let infrared emission do it. The radiation can be anything from X-rays to microwaves, or even protons and muons in exotic forms of tomography. Transmission tomography is the simplest method, pioneered in the medical field. Emission tomography is harder, and multiple scattering tomography is the hardest.

In the 1980s, Jack Warner and his collaborators wrote a memorable series of papers on using microwave emission tomography to image 3D clouds (Warner et al., 1985, 1986; Twomey, 1987). It was the capstone of Warner's long career in cloud physics. While many past papers have somewhat faded from my memory, this series stood out in sharp relief because it offered a vision of the Holy Grail – complete knowledge of the liquid water structure of any cloud. Figure 1.15 shows cartoons of the three cloud tomography methods proposed by Warner and his collaborators. Also

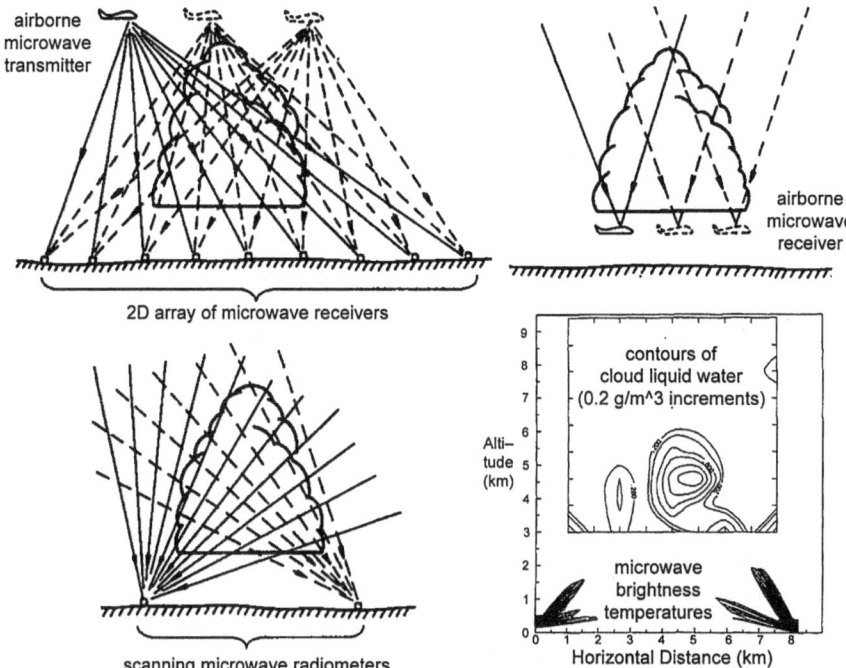

Fig. 1.15. Various strategies proposed by Warner et al. (1985, 1986) for "X-raying" a cloud using microwave tomography, and the results of one experiment. In the top row, an aircraft flies either over or under a cloud: if over, it emits microwave radiation in many directions toward the ground where receivers measure the fraction transmitted; if under, it receives natural microwave emission from the cloud. In the *bottom* row left, an array of scanning radiometers receive natural microwave emission from the cloud. The various movements (scanning and aircraft flight) are assumed to take place much faster than the time scale of cloud evolution. The bottom row right shows the results of the first-ever tomography of a real cloud in Aug 1983 over Boulder, Colorado; the lines radiating from the surface radiometers show the strength of the measured microwave emission, and the contours in the internal box (roughly enclosing the observed cloud) show the values of liquid water content retrieved from those emission measurements. Remember that microwaves are not appreciably absorbed or emitted by ice, so no useful values are retrieved in the upper reaches of the cloud

shown is the historic first tomographic retrieval of cloud liquid water, obtained from two surface microwave radiometers waiting for clouds to pass in between.[45]

In self-consistency tests using a 3D microwave radiative transfer model to produce simulated radiances, then applying tomographic retrieval methods to those

---

[45] It was yet another case of "clouds abhor a cloud field program" – a month of operation netted only three cases, two of which were marginal. Only the third case is shown.

radiances, Warner was able to recover a known distribution of liquid water even with a fairly primitive algorithm.[46]

The Warner and Twomey papers were of course just a glimmer of hope. Nevertheless, if we want to image 3D clouds, and especially if we want to add the time dimension to test cloud models, tomography is the only game in town, although perhaps we will need to supplement microwaves with other wavelength regimes in order to measure ice better (cf. Yodh and Chance (1995) and Klose et al. (2002), for highly scattering optical wavelengths). The cloud tomography idea has lain dormant since the Warner and Twomey papers, perhaps due to complexity[47] and technological immaturity, but microwave technology and spectroscopy have advanced considerably since 1986. When it is finally decided that we are ready to properly test cloud models and/or remote sensing retrievals of cloud properties, then it will be time to revisit tomography.

I have made a small hobby of collecting clippings about the use of tomography in other science fields. Astronomers are using gravitational tomography to look for invisible mass in the universe. Rock geophysicists use tomography by lowering Ground Profiling Radar (GPR) down multiple boreholes, or dragging GPR sleds over soil to study soil moisture vs. depth. The list is extensive. Soon the cloud subject may be one of the last holdouts ignoring tomography.

What are the alternatives to tomography? We could fill a cloudy sky with manned research aircraft. The recent CRYSTAL-FACE experiment in Florida, with six aircraft, probably represented the epitome of this approach. It is far too expensive and logistically complex a way to collect 3D cloud data regularly, and in the end six aircraft is far too few even if all of them had been put into a cloud at the same time (which never happened). But manned aircraft field campaigns are a vicious circle from which little thought is given to escaping. Inertia favors repetition of the past. By following that road, we will continue to fail to capture the 3D or 4D structure of clouds predicted by cloud-resolving models, thus failing to validate them, and one cannot help recalling the myth of Sisyphus.

One avenue for escaping this vicious circle is to develop fleets of small UAVs carrying miniaturized cloud physics instruments. Small-UAV technology is becoming available now due to pioneers like the Aerosonde Corporation (Holland et al., 2001) and NASA Wallops, which has become NASA's official UAV testing center. NASA is just starting a program to develop science instruments for small UAVs. I have helped start an SBIR project to miniaturize cloud physics instruments for small UAVs, and the Phase II designs look not just promising but really exciting. Formation flying a fleet of small UAVs inside a cloud may prove challenging, however it will undoubtedly prove easier than flying large crewed aircraft in formation, and eventually we could penetrate clouds which crewed aircraft avoid. Anyway, we are entering the steeply rising part of the technology curve for small UAVs. Already NASA has

---

[46] Caveats: errors were up to 10% of the maximum liquid water; ice could not be seen (it's transparent in the microwave); rain invalidated the retrieval; and the cloud could evolve significantly in the 2–3 min needed to complete a scan.

[47] it involves either multiple surface sites and/or multiple aircraft

shown it can power a model airplane with laser beams,[48] and this may prove only a foretaste of what is to come.

Fleets of small UAVs could help validate the complex tomographic remote sensing methods. Tomographic methods could in turn help validate simpler remote sensing methods like single cloud radars. This would establish an "audit trail" going back to trusted in situ cloud measurements. Meanwhile, I'm sure my colleagues can think of many other uses for fleets of small UAVs with such instrumentation. Each one is a tiny miracle of efficiency, and the cost savings in the long run are substantial.

1.9 Cloud Structure Modeling: Introduction

Let us now turn to scale, the other main theme of this chapter. My concern with scale, formed in the 1980s, more or less set my sails in the cloud area. A lot of my work, and the work I have urged upon bright young people in the field, has had a strong undercurrent of scale. I never worried about whether cloud radiation would progress from 1D to 3D – that was the natural course of evolution – but I did worry that the cloud structure required for input to 3D models would be unknown at the appropriate scales.

Cloud structure models underlie everything we do in 3D cloud radiation. For at least half of my career in cloud radiation, up through the mid-1980s, this subject was little discussed. Structure models were the simplest possible: plane-parallel slab clouds with no spatial variation and analytic one-humped exponential-tailed drop size distributions loosely based on a few aircraft flights. This made sense from a reductionist point of view, and also from an Occam's Razor point of view. Efforts to salvage this simplicity while moving into a 3D world, mainly by postulating cubic clouds with no spatial or drop distribution variation, got us another decade further before fading. By the end of the 1980s, it became clear that we required cloud structure models based more squarely on cloud observations, accounting properly for scaling including giving at least a cursory acknowledgment to cloud turbulence, and hopefully more acceptable to cloud physicists so that a constructive dialogue could be started where none had previously existed.

1.10 Cloud Structure Modeling: Luke Howard

I admit to almost complete ignorance of the 1802 origin of cloud structure modeling, and indeed I had not thought of the traditional classification scheme as a "model" because it was qualitative and I considered myself a quantitative modeler. In mid-2003, all that changed when I stumbled across an ad for a $6 remainder book in the Daedalus Books catalog titled "The Invention of Clouds" (Hamblyn, 2001). What cloud person could resist a title like that? So I ordered the book and wound up reading it cover to cover, a rare luxury for me these days. Hamblyn is a remarkably poetic

---

[48] not in clouds of course, although microwave power beaming may be an alternative there

science writer and the author of a quote at the beginning of this chapter. His book describes the whole history of the traditional cloud classification scheme – cumulus, cirrus, stratus, etc. – launched in an 1802 London lecture by Luke Howard. Howard's invention was much more revolutionary than I had realized. Thus, since the present book is appearing not long after the 200th anniversary of Howard's famous lecture, I intended to celebrate the Howard saga at greater length. But an article by Graeme Stephens (Stephens, 2003) anticipated me, and so I commend his article to the reader. However, I will still give a condensed account of the saga.

Before 1750 or so, people held a static view of the universe: nothing had changed since the dawn of time. Clouds were irrelevant in such a world. Only the growing recognition of a dynamic universe following the work of Newton and Galileo permitted clouds to become a legitimate subject of study. But even then, people despaired of any kind of cloud classification because of clouds' highly transitory and mutable nature. Several classifications were put forward based on visual appearance, but all failed because they merely described the static visual appearance of clouds – lumpy, bumpy, clumpy, and such like. Then Luke Howard electrified the European community, scientists and general public alike, with his 1802 London lecture. What captured the public imagination was that Howard tied clouds firmly to dynamics. He saw clouds as the same underneath, all subject to the same transformational processes, in spite of their infinitude of forms. He brought clouds into physics. And to this day, clouds remain the prime dynamic element in meteorology. Indeed, without clouds and the attendant hydrologic cycle, Earth's meteorology would be a much duller business.

Howard's work led eventually to the International Cloud Atlas, first published in 1896. It has gone through seven further English editions, the most recent appearing in two volumes in 1995. It is still solidly based on Howard's fundamental classification, with minor additions (Sc, Ac, As), and remains the definitive work on the nomenclature of clouds. The current tripartite structure by altitude, with 10 cloud types, is actually simpler than in 1896!

In the long run, no one is interested in the 3D structure of every cloud that forms on Earth. Clouds are the most changeable component of the Earth system, and indeed until Luke Howard's work, no one even imagined that they could be caged inside a simple classification scheme. In the future we will, I am sure, rely upon empirical statistical measures that summarize what is important about a cloud's "3D-ness." However, there is still no agreement about what these measures are, and what characteristics of 3D clouds can be safely disregarded. We will have to find this out. In the farther future, we can aspire to "Laws of Clouds" wherein these empirical results are put on a firmer foundation, but even then, I suspect, Howard's classification will still be with us as a kind of umbrella model.

1.11 Cloud Structure Modeling: After Luke Howard

In cloud research, as in climate research, everything is called a "model" no matter whether it aims for a crude or a nearly-exact description. Dyson (1999) points out the vital distinction:

> *"A theory is a construction, built out of logic and mathematics, that is sup-posed to describe the actual universe in which we live. A model is a con-struction that describes a much simpler universe, including some features of the actual universe, and neglecting others. Theories and models are both useful tools for understanding nature. But they are useful in different ways [. . .] A theory is useful because it can be tested by comparing its predictions with observations of the real world [. . .] A model is useful because its behav-ior is simple enough to be predicted and understood [. . .] On well-trodden ground we build theories. On the half-explored frontier we build models."*

Not knowing the difference between a theory and a model has led some people to view simple fractal-turbulent models of liquid water content and cloud-resolving "models" as somehow akin, or even competitive so that one has to vanquish the other. They are not. They serve different functions, so neither displaces the other. The goal in a cloud-resolving model is perfection and completeness, within the limits of pre-dictability. To achieve true perfection, cloud-resolving theories must eventually sim-ulate fractal behavior – since such behavior is observed – but even if they do, fractal models will remain as useful as they are today. A fractal model sums up thousands of observed details into a few simple numbers and equations and thus enables one to generate thousands of cloud realizations in a few seconds of computer time. Cloud-resolving "models" attempt to explain all the details which are merely summarized in a fractal model; details it can't explain are handled by a "parameterization", which is actually a true model! Even the most detailed cloud "models" today have parame-terizations and are thus not full theories, but they are as close as we can get, so I call them "near-theories."

Now let us look at the varieties of cloud models. The 2003 electronic version of the *Encyclopedia Brittanica* identifies four basic models of a cloud: dynamical, ther-modynamical, microphysical, and visual. I would add a fifth type, "fractal-turbulent", described below. Luke Howard's models were the first and simplest dynamical mod-els, albeit qualitative; dynamical models today are quantitative, of course, solving some version of the equations of fluid mechanics. The Clausius-Clapeyron relation wasn't known in Howard's day, but it plays a vital role in the thermodynamic model, which is based on cooling and mixing of air masses. Pure thermodynamic models hid inside GCM cloud parameterizations for many years; they simply created a strat-iform cloud when the relative humidity crossed a certain threshold   no dynamics or microphysics was involved. The microphysical model, in its simplest form, is based on a rising adiabatic air parcel with 100% relative humidity in which water drops are calculated explicitly as they condense, coalesce, and break up.

Compound models combine two basic model types; for example, cloud-resolving models combine dynamic and thermodynamic models but parameterize the microphysics; and 1D rising-parcel models combine microphysics and thermodynamics but parameterize the dynamics. The near-theories combine three of the basic model types: dynamic, thermodynamic, and microphysical. Such near-theories were designed mainly to study precipitation and unfortunately are too slow to run for the vast 1000-km cloud systems that affect cloud radiation so profoundly; they are mainly used to simulate small, intensely-convecting cloud systems.

Visual models are used mainly by weather observers and cloud radiation scientists. These include the Euclidean-shape models and even the early fractal models that just simulated cloud shape not internal structure. Visual models also survive in the bowels of GCM radiation parameterizations, in the form of "cloud fraction" and "cloud overlap."

Since this is a book on 3D cloud radiation, we should take special note of the fact that cloud radiation parameterizations in GCMs, an ultimate application for some of this work, exist in a unique netherworld of *visually-oriented* cloud structure modeling. Cloud fraction/overlap descriptors take a dimensionally decoupled view, with cloud fraction describing the horizontal dimensions and cloud overlap describing the vertical dimension. Often different groups of scientists provide algorithms for the two descriptors! In my experience, the physically-based cloud modelers don't take cloud fraction or cloud overlap seriously and don't validate their models against it, except when forced. Their models are not visually oriented except when producing conference movies. Thus, they probably feel that they wouldn't know how to fix their models if there were disagreements with the cloud fraction/overlap description, nor that such fixes would necessarily improve their models. Worse, they don't actively participate in improving these GCM cloud structure schemes because such schemes are so alien to their day-to-day research.

The cloud fraction/overlap structure model arose as part of the haphazard evolution of clouds in GCMs. Once GCMs abandoned fixed climatological clouds, they were adrift without a clear route forward.[49] Relative humidity, convective, and prognostic liquid water parameterizations replaced fixed clouds, but were not designed to provide cloud structure information for radiation parameterizations. Fractal geometry, necessary to correctly describe cloud structure, failed to penetrate the GCM world. If we were to start ab initio today, we would never follow this haphazard historical route. We would instead take an integral view of the three spatial dimensions, and at the same time develop cloud parameterizations that furnished enough information for radiation. In clouds, turbulence always couples the vertical and horizontal dimensions. The greater horizontal than vertical extent of non-cumuliform clouds suggests some kinds of approximations, but not ones which entirely divorce vertical from horizontal variability as the cloud fraction/overlap scheme does.

Other than ramifying Howard's scheme, no progress was made in the understanding of cloud spatial structure right up until the 1980s. No "laws of cloud structure" were discovered or even sought. The near-theories produced cloud structure in a brute

---

[49] super-parameterization (Randall et al., 2003) now offers such a route

force way, but with no advance in understanding. There was no assurance that they produced realistic 3D cloud structures anyway, since these couldn't be measured by any available instruments.[50]

Lacking any real theory of cloud spatial structure, Euclidean shapes, the assumption of maximum ignorance, dominated early 3D cloud radiative transfer. These Euclidean models only specified the exterior form of clouds; they had nothing to say about their interior structure. One pioneering example was McKee and Cox (1974), who did Monte Carlo radiation calculations for homogeneous cubic clouds. This spawned a cottage industry, leading by the mid-1980s to randomly-spaced arrays of randomly-sized cubes requiring 8 or more parameters to specify. Another early example was Appleby and van Blerkom (1975), who used a 2D railroad-tie model for the clouds of Jupiter that gave the first 3D radiative transfer explanation of observed absorption line formation (which 1D theory had failed to account for).

Enter fractals. Fractals were invented by Mandelbrot in the 1970s (following lines of mathematical thought going back to 1900). His most famous book (Mandelbrot, 1982), like his personal web site at Yale, continues to delight and madden readers. Mention fractals in any group of scientists, and you are sure to get a wide spectrum of reactions. Those who have heard messianic sermons on fractals will be apt to cite their limitations, which are many. Others will call attention to the somewhat empirical nature of fractals, springing as they do, and as Mandelbrot readily admits in his book, from empirical investigations and log-log plots of data in a wide variety of fields. Once one gets past such cavils, however, one is left with a sense of awe at an intellectual edifice of great descriptive power and simplicity. If one invokes Occam's Razor to decide between competing descriptions, the fractal description usually wins decisively. It gives us, for the first time, a tool which describes the spatial structure of a host of natural phenomena from galaxy clusters to clouds.

Fractals had no impact on cloud structure modeling until the keystone paper of Lovejoy (1982). This three-page paper in *Science* launched a revolution in thinking about cloud structure as deep as Howard's, and more useful to radiative transfer modelers. Using satellite images, Lovejoy showed (see Fig. 1.16) over a range of scales from 1 to 1000 km that the area of clouds did not go as their perimeter squared (true for all Euclidean objects) but as their perimeter to the power 3/2. Later Cahalan (1991) (see also Cahalan and Joseph, 1989) generalized this result down to Landsat scales of about 30 m, finding however that the exponent changed abruptly from 1.5 to another value below 1 km. This is called a "scale break."

The issue of scale breaks later became contentious, with Lovejoy and his followers maintaining that there were no scale breaks anywhere at any time, and the rest of the community taking a somewhat less dogmatic position. To many, it probably seemed like a tempest in a teapot, but underneath was a vital issue, namely, what are the limits of the fractal picture of cloud structure? A given fractal picture is probably good only over a scale range where the priority order of acting physical processes is not changing. A scale break is a sign that new physics has come to the fore, or

---

[50] weather radar could map out 3D rain structure, but was blind to non-precipitating cloud drops

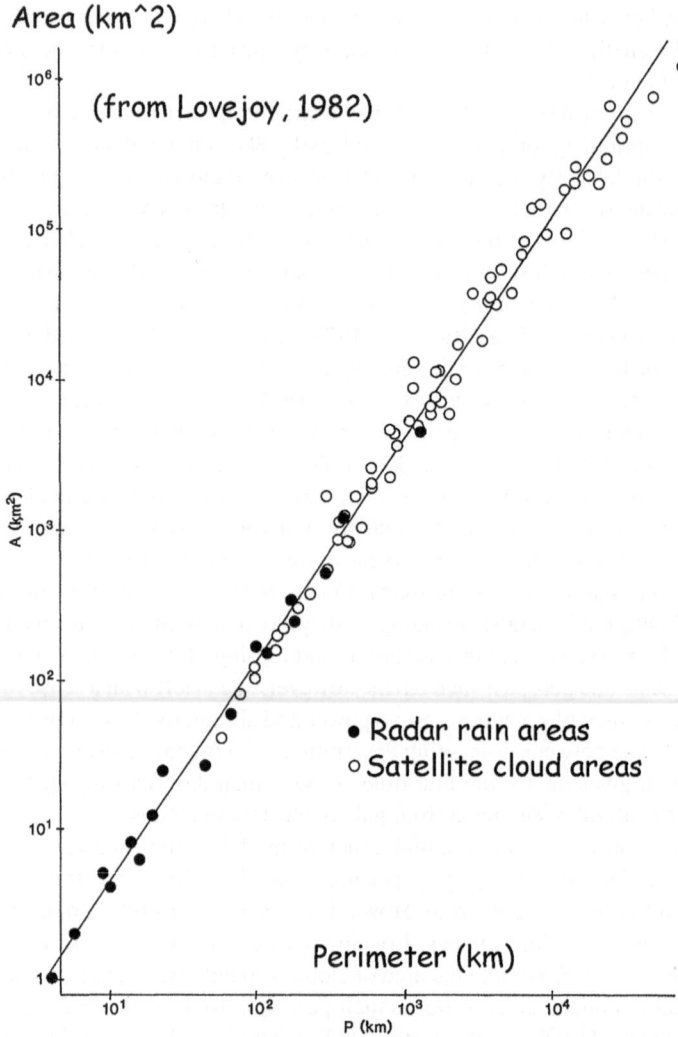

Fig. 1.16. The famous Lovejoy (1982) plot showing that clouds and rain areas on satellite images behave like fractals. If cloud shapes were Euclidean – circles, squares, and so forth – their area would increase as their perimeter squared, but instead their area increases roughly as the perimeter to the power 3/2

that the processes have re-ordered in the priority list. The fractal picture does not break down, but it has to be modified to account for scale breaks and is no longer so elegantly simple because more parameters must be specified. While scale breaks are a fact of life in clouds, the remarkable thing is the large range of scales between scale breaks – three orders of magnitude in the area-perimeter case and also in the measurements of cloud liquid water content (Davis et al., 1994).

Our group at Goddard became early adopters of multifractal models of cloud structure which are richer and more complex than the monofractal models in Mandelbrot's book *The Fractal Geometry of Nature*. We developed our own variant, the bounded cascade model, while Lovejoy preferred a model which jumped into Fourier space, applied a power-law scaling, then jumped back to real space again. We even have a fractal cloud web page.[51] It took years for the community to absorb the new models, and some are still not comfortable with them. But as the 1980s ended, the number of Euclidean cloud papers began to decline dramatically, virtually dying out by the mid-1990s.[52] The multifractal model was simply superior in three important respects: (1) the number of free parameters was considerably smaller, satisfying Occam's Razor; (2) the internal (liquid water) structure of a cloud could be modeled to agree with measurements, which all showed a fractal structure; and (3) it captured aspects of the turbulent cascade which we know occurs in clouds. Indeed, our multifractal cloud models were closely related to the new fractal cascade models developed by turbulence theorists in the 1980s.

Fractional cloudiness was harder to capture, since the multifractal models produced a positive optical depth everywhere; our group just applied a threshold below which optical depth was set to zero. Other groups used other methods. Cloud top bumpiness also required additions to the model – our group used fractional Brownian motion – although it rarely proved of importance for stratiform clouds. In the end, we used to joke that our fractal cloud models were a joint work of the three dwarfs Lumpy, Clumpy, and Bumpy, with Lumpy handling the internal variation of liquid water, Clumpy handling the gaps between clouds, and Bumpy handling the cloud tops.

One of the most advanced fractal-like cloud models is that of Evans and Wiscombe (2004), which generates 3D fields from cloud profiles obtained from vertically pointing radar. We call this a "data generalization model" since it matches statistics of observed input fields as closely as possible, and it can generate an arbitrary cloud field satisfying the observed fractal characteristic of cloud extinction. Venema (priv. comm.), working with the 4-D Clouds Project in Europe, has developed a general method which uses 1D or 2D input from a ground-based profiler and searches through a space of 3D cloud models to find the best match capturing the profiler statistics.

Fractal models are sufficiently mature now to qualify as the fifth and newest type of basic cloud model: I would dub them "fractal-turbulent" although they have also been called "fractal-statistical" because some of their parameters can be interpreted in terms of means, variances, correlations, and other more esoteric statistical properties of clouds. Fractal-turbulent models not only contain real physics, through mimicking turbulent cascades, but they embody scaling, one of the profoundest concepts

---

[51] http://climate.gsfc.nasa.gov/~cahalan/FractalClouds/FractalClouds.html/

[52] The driver then was the need for domain-average fluxes. Euclidean shapes may again be useful for helping make sense of point-wise radiances in 3D cloud remote-sensing, cf. Davis (2002) for an example using spherical clouds.

of contemporary physics. This naturally leads us to consider the issue of scale, which, after tools, is the other large theme of this chapter.

Recently, Petty (2002) came up with a new kind of cloud structure model which he calls the "Independently Scattering Cloudlet model." It doesn't fit neatly into either the Euclidean or fractal category, and in fact cannot be fractal without deep modification. It composes a cloud from small units which are however much bigger than the single-scattering "elementary volume" at the root of conventional radiative transfer. To prosper, Petty's model will have to capture the observed characteristics of clouds as well and as simply as fractal models. If it can do so, it certainly offers an attractive simplification of the radiative transfer.

1.12 Scale

Much of the discussion of scale has already taken place, mainly under the guise of "cloud structure." There are only a few somewhat disconnected points left to make.

"Scaling" is a term that is thrown around rather loosely, but in the multifractal world it has a real meaning: namely, a phenomenon with no preferred scale. Power-law statistics and scaling go together like love and marriage, because a power-law, unlike an exponential, has no scale parameter anywhere. In spite of occasional superficial similarities, there is a vast gulf between power-law spatial statistics and their exponential counterparts. If a phenomenon is scaling, as cloud liquid water is over several orders of magnitude of spatial scale, then the type of model you choose to simulate it must reflect this characteristic. Choosing an exponential model introduces a preferred scale where often none exists; it is not the choice of maximum ignorance, as often thought, but actually the wrong choice in such cases.

Scaling ideas are crucial; they brought elegance and simplicity to a field where distributions of Euclidean shapes were rapidly multiplying the number of needed parameters and creating situations where it mattered if your photons hit this edge or that corner. Arrays of Euclidean objects tended to have artificial preferred scales (like "the mean cube size"), and this just doesn't happen in natural clouds.

Stephens (1988a,b) was the first to introduce the idea of a scale hierarchy in 3D cloud radiative transfer. He Fourier-transformed the radiative transfer equation in space and grouped scales into various Fourier components. Then he looked at how the transfer equation for each scale group couples to that for larger and smaller scale groups. He concluded that "multiple scattering acts to filter out the smaller scale contributions." Stephens' theory is not the same as the theory of scaling as it arose in physics and particularly in turbulence, but it was a giant step away from implicit assumptions of homogeneity.

Stephens' conclusion that multiple scattering smoothes out small scales was confirmed by Cahalan's (1991) discovery of the Landsat scale break. The break occurred in a plot of the power spectrum of Landsat radiance versus spatial scale, shown in Fig. 12.2 of this book, and consisted in a sudden steepening of the curve below 250 m, indicating a strong smoothing effect. There were competing physical and instrumental theories for the scale break, but eventually the powerful analyses of Marshak et al.

(1995) and Davis et al. (1997) won the day. They greatly extended and rigorized the concept of "radiative smoothing" and, from 3D diffusion theory, derived a remarkably simple expression for the radiative smoothing scale (250 m in the Landsat case) as the harmonic mean of the cloud geometric thickness and the transport mean free path. Radiative smoothing theory was one of the major contributions to 3D cloud radiation science in the last decade and was directly responsible for the development of the new multiple-scattering lidars, discussed earlier.

Super-parameterization, also called MMFs or "multi-scale modeling framework" (Randall et al., 2003), embeds cloud resolving models within the several-hundred-km grid cells of a traditional GCM. MMFs resolve the cloud circulations explicitly down to a scale of a few km, and so represent some aspects of the spatial structure explicitly. Early MMFs have limitations, such as that clouds can't move from one GCM column to the next, and they embed 2D rather than 3D cloud-resolving models, but there are solutions to these problems if enough resources are made available. At least there is a foreseeable path forward, which is more than can be said of other approaches to clouds in GCMs. When fully implemented, MMFs will provide cloud output over three orders of magnitude in scale – a veritable gold mine for fractal-statistical analysis since scaling has not been built in ab initio yet should arise naturally if the model has verisimilitude. MMFs offer tremendous job opportunities to 3D cloud radiative transfer modelers, since on a 1-km scale no cloud can reasonably be approximated as 1D. They also make much less use of cloud fraction/overlap assumptions, which in turn can cause much less damage.

There is also renewed attention to scales below those resolved in MMFs. The review paper of Shaw (2003) exemplifies the struggle to understand what is going on, microphysically, at small scales in clouds. This area remains somewhat in its infancy. Fast-responding instruments like the Gerber PVM and the fast FSSP have only in the past decade given us a glimpse of clouds at the centimeter level. The first question is: is the small-scale structure Poissonian, i.e., perfectly random? This question was being bruited about in the early 1990s[53], but the answers were contradictory and the data just weren't good enough to decide. Now the answer seems to be that clouds are far from Poissonian at small scales. That being the case, the next question is, what effect does this have on larger-scale cloud optical properties like optical depth and absorption? This is a subject of active research.

The DYCOMS-II field campaign in stratocumulus (Stevens et al., 1999) emphasized that even the flattest, most homogeneous-looking Sc harbors immense variability at all scales. Thus the infinite plane-parallel cloud of radiative transfer fantasy finds no realization even in this most stratiform of all clouds – as was already discovered during the first FIRE field phase. This is a fitting note on which to end our discussion of scale.

---

[53] Who can forget "inch clouds" (Baker, 1992; Baumgardner et al., 1993)? Inch clouds, or more specifically centimeter-scale structure, fell out of favor for a while, but Pinsky and Khain (2003) indicate a potential revivial.

1.13 3D Radiative Transfer

A great part of this book is about 3D radiative transfer: fundamentals, algorithms, and applications. There is no need to paraphrase that material; paraphrasing great writing, like removing notes from a Mozart composition, can only deminish it.Yet I felt a need to introduce the subject, especially since I have acted as spur and mentor to several bright young people who made important contributions. I am also continually asked by young investigators what subjects they should be looking into and writing proposals about. Thus I will do as I have done above – talk about the subject's past including my experiences and impressions, then summarize where we are now. Radiation challenge problems for the future are given in Sect. 1.17.

First, the past. One impression from looking through my collection of historic papers on 3D cloud radiation, was how little really endured. There were gallant, even heroic, efforts. But as in scattering by nonspherical particles, one could do gargantuan calculations to achieve lilliputian results. Clean, simple, yet general conclusions were elusive, at best. What was lacking, in the end, was any feeling that one had made a breakthrough in understanding. Instead, the "progress" was like that in World War I trench warfare – a yard at a time, with losses as well as gains. Large subsets of literature became obsolete and are remembered only by old elephants like me. For example, many early 3D Monte Carlo papers simply didn't have the computer horsepower to use enough photons, or enough cloud realizations, to obtain a clean, statistically significant result, and are now little read. Euclidean cloud studies are also little read although much cited today because their model of cloud structure is so disconnected from today's multifractal, wavelet, and scaling models. And as Evans (1998) has noted, "there are many hopelessly inefficient ways to compute 3D radiative transfer." These inefficient methods now lie in the dustbin of history as Evans' SHDOM method has pretty much swept the field.

The past also saw a mini-rush to parameterization, an ancient curse on the radiation field. Much of it depended on Euclidean concepts like face, corner, aspect ratio, and mean cloud size – more useful for cities than for clouds – which have little utility in a fractal cloud universe where the variation of optical depth within a cloud is just as important as the shape of gaps between clouds.

As I look over my collection of old notes and papers on 3D cloud radiation, I am struck with the enormous struggle to simplify the 5D problem (two angles plus three spatial dimensions) enough to get some results. And this did not even consider the 6th dimension, time! The simplifications in cloud shape have been discussed above. Clouds were assumed internally homogeneous to further simplify the spatial aspect of the problem. The angle dimensions were typically simplified using delta-Eddington and diffusion approximations, requiring the cloud to be reasonably optically thick.[54]

Several pioneering papers on 3D cloud radiation modeling stood out in my admittedly eclectic sampling; they give further early references. First, Weinman and

---

[54] Diffusion approximations still have great utility, and have for example been used to understand multiple scattering lidar observations of real 3D clouds.

Swartztrauber (1968) used quasi-analytic radiative transfer methods to study a flat cloud with sinusoidally varying cloud optical depth and isotropically scattering particles. This typified the many prescient contributions of Jim Weinman, who always preferred elegance to brute force numerical solutions.

Second, Harshvardhan and Thomas (1984), one of the best of the Euclidean-era papers, focuses on a single quantity, the effective cloud fraction, and repeatedly plots it vs. the true cloud fraction (as viewed normally). Even in this paper, however, one is bedeviled by Euclidean artifacts like edges and faces that become low enough to start shadowing. Also, the proposed parameterization requires "the probability that the Sun will directly illuminate cloud sides", another Euclidean concept, like shadowing, that becomes hard to define in a world of fractal clouds.

Third, Davies (1984) showed that 3D cloud angular reflectance patterns tend to fall between those for a lambertian plane and a lambertian sphere; he also showed that viewing a patchy cloud scene at 60° is the most useful for estimating the flux, because the geometric details of the cloud field are the least important at 60°.

Fourth, Barker and Davies (1989) showed the surprisingly long-range effects on surface insolation of a surface albedo discontinuity under a cloud. This paper reminded us of the crucial interaction between 2D surface albedo variation and 3D cloud variation. Barker et al. (2002b) later returned to this subject, expanding upon Marshak et al.'s (2000) seminal work on using the "red-edge" near-discontinuity in vegetation albedo to retrieve optical depths of 3D clouds from the surface. Chiu et al. (2004), prompted by speculations that horizontally varying surface albedo increase cloud absorption, studied the problem in stark relief by postulating a black-white checkerboard surface.

I have already mentioned the work of Appleby and van Blerkom (1975) showing that the actual shape of absorption lines observed in the clouds of Jupiter depends on the 3D structure of the clouds. This was a novel idea at the time: that 3D spatial structure could have a marked effect on the shape of spectra.[55]

Finally, as the issue of "enhanced shortwave cloud absorption" (Cess et al., 1995) fades into history, I want to highlight its importance as a spur to the modern development of 3D radiative transfer. Much of the funding and support that led to the advanced situation portrayed in this book was a direct consequence of that 1995 claim, since, if there was enhanced absorption, most original bets (including mine) were that it was mainly due to ignoring 3D effects in the 1D models of the time. To our surprise, it turned out that 3D effects could not explain much enhanced absorption, but the advances made in discovering that fact led to permanent improvements in all the tools of the field.

What sorts of tools do we have for 3D radiative transfer nowadays? Monte Carlo methods of course date back to the 1940s. However, they only came of age for 3D cloud problems recently, when billions of photon trajectories became possible. It was interesting to look back at the growth in number of trajectories used. In 1947, Von Neumann used 100 neutrons colliding 100 times each, which took 5 hours on an ENIAC computer. Plass and Kattawar (1968), McKee and Cox (1974), and Davies

---

[55] See Chap. 13 of this volume for a contemporary take on this idea using the oxygen A-band.

(1984) all used between 30K and 100K photons. Bill Ridgway of Goddard set a new speed record in 1991: 36M photons/Cray hour (but using various other computers available to us at that time, the same code could be up to 50x slower). By 1994, our group was achieving 100M photons/Cray hour. These gains brought Monte Carlo power within the reach of laptop computers for many applications, since 100K photons is often enough for domain-average fluxes.

Thus, in Monte Carlo, advances in computer speed have made a qualitative difference; going from 10K to 1M photons buys a factor of 10 error reduction, which can make the difference between a 10% and a 1% radiance error. Increasing the number of photons from 10M to 1B would only buy a further factor of 10 reduction in the error, however, even though it is 990M more photons! And perhaps reducing radiance error below 1% is overkill anyway, since radiance instruments rarely achieve that accuracy. Thus we are now on the flat rather than the steep part of the Monte Carlo progress curve for 3D clouds.

Concomitant with computer speed increases have been algorithmic improvements in the random number generators at the heart of every Monte Carlo code. Press et al. (2000) tells the fascinating story of bad early generators with hidden correlations and biases that might not have ruined a 10K-photon run but could be disastrous for a 1B-photon run.[56] These problems have receded now, although each time the number of photons increases by a factor of ten, one must reconsider them. In particular, one should always worry about how well the Fortran intrinsic random number generator, not designed explicitly for unbiased Monte Carlo work, will function as we move into an era of routine 1B-photon runs.

The other large class of 3D radiative transfer methods (what I call analytical-numerical) take a more traditional approach: they approximate integrals by sums, derivatives by differences, take Fourier transforms or make spherical harmonic expansions, and generally employ the standard grab-bag of classical applied mathematics and numerical analysis methods. The end result is often the need to solve large sets of linear equations or eigen-problems. Errors are due to truncation (of infinite series, or of approximations to derivatives) rather than, as in Monte Carlo, statistical fluctuations in random number selection. The special functions used for representing radiances are chosen for their orthogonality properties and, in many cases, are not particularly apt for representing radiances from turbulent scaling structures like clouds. Thus, the expansions can take many terms to converge.

The best modern representative of analytical-numerical methods is SHDOM (Evans, 1998), a well-documented program enjoying increasingly wide use due to its great flexibility and generality. According to Evans, SHDOM "is the first explicit radiative transfer model efficient enough to perform broadband 3D atmospheric radiative transfer for significant sized domains." It was also the first model to incorporate adaptive gridding (putting more grids where needed) and even incorporated prescient features like horizontally-varying surface reflectance. SHDOM represents a point of

---

[56] No one has systematically explored the consequences of poor random number generators in cloud Monte Carlo, although anecdotal evidence indicates that the cloud application is rather forgiving.

perfection, the culmination of two decades of struggles epitomized by the work of Stephens in the 1980s (e.g., Stephens, 1988a,b), who emphasized the "many unsolved numerical and mathematical issues that hopefully will challenge researchers for some time to come."

The 3D cloud radiation field has reached that state of maturity which prompts not one but two model intercomparisons! The first, the Intercomparison of 3D Radiation Codes,[57] led by Bob Cahalan, uses 3D cloud structures taken from radar data, Landsat, and cloud-resolving models. The second, led by Howard Barker (Barker et al., 2003), compares cloud radiation results from many GCM radiation packages, with an emphasis on interpretation and handling of unresolved 3D cloud effects which GCMs attempt to account for by various cloud fraction/overlap assumptions. According to Barker, overlap assumptions don't work very well when compared to full 3D simulations and he concludes ". . . a paradigm shift is due for modeling 1D solar fluxes for cloudy atmospheres."

Cloud fraction is another nebulous concept which, like cloud overlap, has a powerful visual appeal combined with an extreme difficulty of application in radiative transfer practice. The papers of Stephens (1988a,b), still not fully appreciated, showed that the "cloud fraction" approach used to leverage 1D theory into 3D was doomed to ultimate failure because the "radiative cloud fraction" needed to fudge 1D theory would never be the same as the cloud fraction we could define operationally (e.g., from a sky-filling constellation of drones all reporting whether they were in cloud or not). In any case, taking a linear combination of clear and cloudy radiation with radiative cloud fraction as the weighting factor only has a prayer of working for fluxes; it is utterly ludicrous for radiances. But we plod onward with cloud fraction, by this time a sacred cow that cannot be killed no matter how much it fouls our yard, and hope only that the next generation will have the courage to give it the quietus it deserves.

The old joke about the weather, that "everyone complains about it but no one does anything about it", applies as well to cloud fraction. Lest I be lumped in with the nabobs of negativism, let me offer a positive alternative: abandon cloud fraction in favor of one or more fundamental measures of radiative turbulence. Cloud fraction is, after all, only used by the radiation community, and it is only used to get the radiation right – that is, to integrate either spatially or temporally, or both, across a regime where the radiation is turbulent. If the highly developed tools of statistical turbulence analysis, much improved in the 1980s, are applied to radiative turbulence, we may find universal parameters with a sound measurement basis that can be used to deal with patchy cloudiness. Such studies will require much higher sampling rates than the one-minute averages often used in surface radiation measurements. New techniques for measuring column-averaged atmospheric turbulence using cheap GPS instruments may help us leap across this gap from "cloud fraction" to "radiative turbulence."

Lidar scientists pioneered the time dimension of radiation. They had to time-resolve returning photon pulses at the microsecond level. But they remained an is-

---

[57] http://i3rc.gsfc.nasa.gov/

land in a sea of static, time-invariant radiation work. Now, however, a few cloud radiation scientists are solving the time-dependent version of the radiative transfer equation and applying it to remote sensing of clouds. Davis and Marshak (2002) provide an excellent example of this new trend. Cahalan et al. (2005) describe a new lidar concept in which a laser pulse is directed vertically down into a cloud from an aircraft high above and then the time-resolved multiple-scattered photons from increasingly wide angles away from the vertical are used to retrieve the cloud's geometric thickness. Love et al. (2001) had implemented the same concept with dramatically different – and already evolving (Polonsky et al., 2005) – technology and retrieval methodology in a ground-based configuration that yields both geometrical- and optical thicknesses of the cloud.[58] Finally, Evans et al. (2003) have proposed an "in-situ" cloud lidar for use by a microphysics-sampling aircraft (see Fig. 1.13). All three efforts share a common origin in ARM projects at Goddard. What is important is that all three lidars are backed up by extensive time-dependent 3D diffusion and Monte Carlo calculations using realistic models of cloud structure; this would have been unthinkable just a decade ago!

There is a duality between wavelength and time through the Laplace transform of the time-dependent radiative transfer equation (cf. Min and Harrison, 1999; Heidinger and Stephens, 2000; Portmann et al., 2001, and the chapter by Stephens et al. in this volume). The photon-path-length-distribution (PPLD) is the inverse Laplace transform of the ratio I/I_0 where I is radiance at an absorbing wavelength and I_0 is radiance at a non-absorbing wavelength. Extremely high spectral resolution measurements, such as in the oxygen A-band, can give us the PPLD which in turn can tell us much about cloud optical properties including absorption. Indeed, this would be the next step beyond the current way of specifying clouds (optical depth and effective drop radius). PPLD theory is in place, and surface and aircraft instruments have been built and deployed to measure it. Momentum is definitely building in the PPLD area in spite of the deletion of oxygen A-band instruments from the CloudSat and CALIPSO satellites.

Another time dimension is that of cloud dynamics, and that is on the order of seconds. For most of my career, cloud radiation scientists have not tried to understand or exploit this dynamical time-variation, but to freeze it – so-called "snapshot mode." In this, we became increasingly out of sync with cloud-resolving modelers who of course regarded the time evolution of clouds as a primary goal. The fact that clouds are tightly "wired together" in time was not even a part of the cloud radiation mindset, nor was any use whatsoever made of this fact. Most cloud radiation modelers don't even worry whether radiation instruments are adapted to cloud dynamics time scales – which may partially explain the ubiquitous "one-minute average." Sooner

---

[58] The general concept of "off-beam" cloud lidar was introduced by Davis et al. (1999) who also reported the first detection of highly-scattered lidar photons. They used a standard (on-beam) research lidar system peering through the roof of Building 22 at Goddard but purposely misaligned the transmitted beam with respect to the vertical axis of the receiver's field-of-view by up to 12°. At that point, they lost the pulse-integrated multiple scattering signal in the solar background. At night, they could have gone much further, and maybe even resolved the shape of the stretched pulse.

or later cloud radiation scientists must acknowledge that radiation is not just a series of independent snapshots, but snapshots which are tightly glued together in time as well as in space.

When Anthony Davis first joined our ARM group in 1992, he told us of his thesis work with Shaun Lovejoy at McGill University showing non-exponential transmission in wildly multifractal media. For a decade, we would badger him to publish that work, or argue with him that clouds were not intermittent enough to exhibit the phenomenon. His fascination with non-exponential transmission helped reawaken my slumbering interest in the real nature of the drop distribution and its spatial variation. Finally Alexander Marshak got interested, and now, a decade later, they have not only published an extensive paper on the subject (Davis and Marshak, 2004) but joined a group of young turks who are challenging some of the sacred foundations of cloud radiative transfer – the notions of an "elementary volume" and of a Poisson (homogeneous) spatial distribution of cloud drops. They have taken the old question "are clouds like Swiss cheese, with actual holes, or like lumpy yogurt?" to a new level: they show that the answer depends on drop size – creamy yogurt for small drops, more and more holey Swiss cheese as drops grow over about 14 microns in radius. The literature on this subject is growing exponentially, but a few recent examples include Kostinski and Jameson (2000); Kostinski (2002); Shaw et al. (2002); Mishchenko et al. (2004); Marshak et al. (2005); Knyazikhin et al. (2005).

This re-examination of fundamentals is long overdue. The fact that clouds have significant spatial structure on all scales down to millimeters is not a simple or trivial overlay on notions of homogeneity but a radically different way of looking at clouds, and every assumption containing a hidden sub-assumption of homogeneity must be challenged ... beginning with "elementary volume." Homogeneity is a perfectly natural assumption in many fields, but when we see a plume of smoke rising from a cigarette, we have to realize that all the complex folding and stretching we see exists also inside a cloud, concealed by the greater optical depth. We continue to hope that it can be captured by some simple assumption, but assuredly homogeneity will not be among the candidates.

As a closing note, 3D radiative transfer has been vital in showing us the situations in which 1D approaches actually work. Before, we just guessed and hoped about this. We have learned, for example, that the Independent Column Approximation (ICA) coupled with a proper distribution of cloud optical depths works in some situations – the better, the larger the domain average. The ICA performs well even in certain GATE simulations of scattered cumulus fields for domain averages (Barker et al., 1999). For 3D cloud effects in the UV, Meerkotter and Degunther (2001) report "the ICA causes maximum uncertainties up to 100% for a spatial resolution of 1 km, 10% for a resolution of 15 km and below 5% for a resolution greater than 30 km." This is a quiet, little heralded gain from our newfound power in 3D cloud radiation that allows us to leave behind the 1D vs. 3D battle and use each where appropriate.

1.14 Turbulent Radiative Transfer

I have always been fascinated by the turbulent character of radiation, reflecting the turbulence in the cloud medium. Stephens (1988b) gave the first formal development, decomposing the radiance into mean and fluctuating components and working out the consequences for the radiative transfer equations, which now, just as in fluid dynamics, require some form of closure assumption. But because many of our radiation instruments are so slow or average their data over such long periods, or because we can't follow the full 4D development of the cloud medium, this aspect of radiation remains underexplored. Why do we care? Aside from the simple fact that it is a new aspect of radiation whose understanding might lead to unexpected spinoffs, the analogy with fluid turbulence teaches us that being turbulent is a qualitative difference. It matters that radiances are intermittent and not like white or red noise, and that derivatives we confidently use in our differential equations only exist at scales we generally do not resolve either computationally or observationally. The very essence of the phenomenon is changed and, as a result, so should the mathematical representation we choose for it.

We have barely begun to understand the implications of the turbulence of surface radiation caused by clouds. The time resolution of our surface instruments is usually too coarse to capture it, or they are fast enough but the turbulence is hidden by the long-time-averages usually reported. The 3D radiation models are ready to study the problem, as this book demonstrates, but the cloud variables needed as input by those models are not available from current observations – and certainly not second by second as would be needed to simulate the turbulent spikiness in measured radiances and fluxes.

We are also becoming more aware of the turbulent character of the drop size distribution, which for decades was treated as a homogeneous variable by cloud radiation modelers and remote sensing experts. Knyazikhin et al. (2005) and Marshak et al. (2005) have shown that the conventional assumption, that number of drops of a given radius is a power law in volume with exponent unity, is false; their analysis of in situ FSSP data indicates the exponent depends on drop size, falling increasingly below unity for drops larger than 14 microns. This can affect the radiative transfer in significant ways.

1.15 Laws of Clouds

Save only for turbulence, there is a general impression that we know all the important physical laws – dynamic, thermodynamic, microphysical – governing clouds. So, from a reductionist point of view, the problem is solved. It should just be a matter of software engineering to work out the details. Yet we are continually surprised by clouds' behavior, and we can't seem to capture them inside a tight theoretical box.

I think the problem is that currently known laws only weakly constrain clouds. These laws act more like inequalities than equalities, forbidding clouds from certain regions of phase space but allowing too much free range within the rest of phase

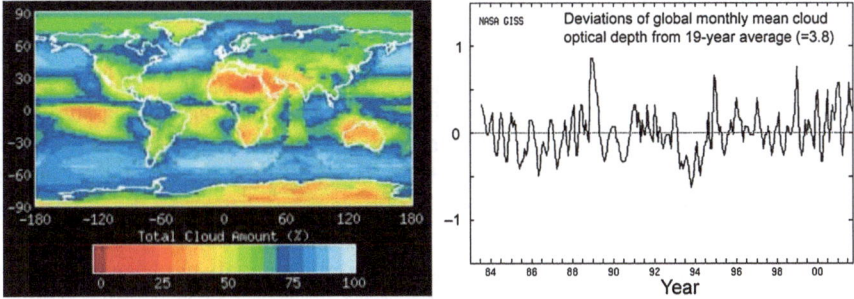

Fig. 1.17. The longest truly global cloud fraction and cloud optical depth data available, from the International Satellite Cloud Climatology Project (ISCCP). (*Left*) Average cloud fraction for 1983 to 2001; the persistent areas of cloud and no-cloud indicate that clouds are constrained by overall laws. (*Right*) Deviations of global-averaged monthly-mean cloud optical depth from the 19-year average of 3.8; the relative stability of cloud optical depth cannot be accidental but must be governed by unknown laws. A similar plot of cloud fraction would not show such stability but rather a long-term decrease of about 6% since the mid-1980s, cause unknown

space. This free range could only be narrowed by knowing more macroscopic Laws of Clouds than we do now. The search has barely begun.

There is a kind of urgency to the search, because until the Laws of Clouds are better known, our efforts to predict global change will remain stymied. The climate community has long known that clouds are the lever Archimedes sought. Clouds can move the climate in any direction – and without macroscopic Laws of Clouds in hand to say otherwise, clouds will continue to be used as tuning knobs for pet theories and explanations for climate changes. Recent examples include the brouhahas over global dimming (Liepert, 2002), Earthshine (Palle et al., 2003), and the Iris Hypothesis (Lin et al., 2002), all involving speculations on clouds as prime mover.

At the dawn of the satellite era, when we first saw that clouds appeared quasi-organized on vast scales, there was hope that there might be simple governing laws. Figure 1.17, showing cloud fraction averaged over 18 years (Rossow and Duenas, 2004, Fig. 2), does not show a uniform gray cloud cover, as one might naively expect from the Earthbound perspective that clouds come and go daily, weekly, and seasonally. There are strong gradients between regions of high and low cloudiness, and the gradients occur over relatively short distances compared to planetary scales. It is most significant that these gradients remain sharp even in an 18-year average, when one might have expected them to smear out as the climate vacillated – as El Niños came and went, and so forth. This cloudiness pattern must reflect rather tight control by underlying variables and processes which are stable in the face of mild climate fluctuations. We see here the smile of that elusive Cheshire Cat, the Laws of Clouds.

What do I mean by the Laws of Clouds? Just this: simple overall principles that constrain what clouds can and cannot do – a kind of thermodynamics for clouds.

An early example was Paltridge's (1975) climate model based on maximum entropy production.

What are some other concrete examples of such Laws? In the early days of climate modeling, in complete ignorance of clouds, people made arbitrary assumptions which, they hoped, somewhat bookended the problem: Fixed Cloud Temperature and Fixed Cloud Altitude (Schneider, 1972), meaning that temperature or altitude remained the same as climate changed. There was no evidence that these were correct, but it was at least a start.

Scaling laws are another example. The radiation community has been the primary force behind discovery of such laws. The cloud modeling community still seems rather uninvolved, perhaps because their models have such a small range of scales compared to Nature. Yet I think all would agree that knowing cloud scaling laws would be a great benefit, not least because it might put certain parameterizations on a more solid footing.

Another example springboards off the finding of Coakley et al. (2005) that "for pixel-scale cloud fractions between 0.2 and 0.8, optical depth, droplet effective radius, and column droplet number concentration decrease slowly with decreasing cloud cover fraction. The changes are only about 20–30% while cloud cover fraction changes by 80%." What we are seeing here, in my view, is the competition for water in a water-starved situation (signaled by actual gaps between clouds) leading to a quasi-equilibrium optical depth tightly constrained between roughly 5 and 8. This curious phenomenon is not yet a Law, but may point the way toward one.

A recent example is Hartmann's proposed Fixed Anvil Temperature law (Hartmann and Larson, 2002), which is that cirrus anvil temperature in the tropics is conserved during climate change. Hartmann seems particularly forthcoming in looking for underlying Laws: Hartmann et al. (2001) tackle the problem of why the radiation balance in the tropics is almost neutral.[59] This fact has been known at least since the ERBE era of the 1980s, but who bothered to look deeply or ask, why? It is the nature of science that discovery of Laws favors the prepared mind – and the mind which can see patterns as clues to underlying Laws, not mere inexplicable accidents. After all, everyone had observed the motion of the planets before Newton, but only Newton discerned the operation of a single law to bind them all.

Very few people, however, seem to be pursuing such simple Laws of Clouds. Perhaps this is partly due to discouragement. There were attempts beginning in the 1960s to put empirical relationships (such as between cloud fraction and relative humidity) into global models. These attempts, while plausible, were less than successful – albeit long-lived – and were premature considering that they had little observational underpinning. As a result, effort was redirected to putting prognostic liquid water into global models, which gained momentum in the 1990s even though the basic scheme was proposed much earlier by Sundqvist (1978; cf. also Sundqvist et al. (1989)). Prognostic cloud schemes represented an effort to insert a Law of Clouds that, among many benefits, made clouds continuous in time (rather than blinkers as formerly) and allowed them to move from one grid cell to the next. Yet, in another

---

[59] longwave and shortwave contributions cancel

intercomparison of GCMs in 2000, the climatic response to doubled CO_2 remained as stubbornly uncertain as in 1990, and the uncertainty was still due to differing cloud treatments. The crucial 2–5°C gap in predictions of global average surface temperature change due to doubled CO_2 barely closed at all.

Some people may simply feel that we won't be able to discern the Laws of Clouds until we have observations that can test current cloud models. While I would never gainsay such a fundamental principle of science as comparison to observation, the history of science shows that the great theoretical leaps forward were often based on a thin helping of observational data and a gigantic dollop of intuition.

The Laws of Clouds can also tell us what cloud data to gather. This is a chicken and egg situation, but science has always had to bootstrap itself in this way. We are already running up against practical limits in cloud data-gathering: number of aircraft we can field, number of surface sites with expensive active sensors, cost of tomography, and so on. Future data-gathering exercises for clouds should aim toward some proposed Law or empirical relation that needs testing. Random data-gathering is never a very effective method of moving forward.

If we knew the Laws of Clouds, would we vividly need to explicitly manifest clouds in climate models? I still remember a conversation with Suki Manabe, the legendary climate modeler, in the 1970s. He said that he would prefer not to manifest clouds explicitly in his models, as long as the three important functions of clouds were calculated correctly: precipitation, latent heat release (with or without precipitation), and radiation. He would be perfectly happy to have model rain coming down from a model clear sky, he said! At the time, he was using fixed climatological clouds in his models, a practice he defended vigorously for many years thereafter on the grounds that this was better than any of the schemes he had seen for calculating clouds.[60] I often found Manabe's idea preferable to a cloud paradigm dependent on the concepts of cloud fraction and cloud overlap. I doubt if anything will stop the momentum behind generating explicit clouds in climate models – but that should just be a waystation toward finding the Laws of Clouds that fulfill Manabe's three functions.

One thing that inhibits the search for Laws of Clouds is that the subject is still pursued as somewhat of a problem in geography. Discussions of clouds are rarely couched in terms of general principles but of specific cases in specific places at specific altitudes. Clouds certainly change with region, but underneath this variety, as Luke Howard knew, lies a cloud brotherhood which is much stronger than superficial differences. We need to understand clouds at a unified rather than a region by region geographic level.

Perhaps also the known equations of clouds need to be re-formulated to mirror the observed fractal geometry of clouds. Clouds, like strange attractors, exist "in between" the normal integer dimensions of our Euclidean world. We don't formulate the equations to reflect the scaling-fractal nature of clouds. Perhaps the equations

---

[60] Indeed, well into the 1990s GCM clouds blinked on and off like Christmas lights, an embarrassment so great that GCMers never showed any animations of their calculated cloud fields.

require fractional derivatives mirroring the fractal world in which clouds live. Yet the textbook equations are derived by drawing infinitesimal Euclidean boxes and reasoning about what is going in and out the faces of the boxes. No idealization could be further from the true situation. Lagrangian models portray the situation more truly, but when the geometry becomes too folded and stretched, models always retreat (re-initialize) to comfortable Euclidean boxes. Our Euclidean mindset is a bed of Procrustes preventing us from seeing the true geometry of clouds.

Why do I discuss Laws of Clouds in a book on 3D radiation? Isn't that the job of the cloud physicists and dynamicists? Yes and no. I see little evidence of a search for new Laws among the builders of cloud models; they tend to be fascinated by large complex software development rather than the search for new laws. Most of the examples of proposed new Laws have come from radiation scientists or from scientists outside the formal confines of cloud physics. One recent example is the discovery by cloud radiation scientists Knyazikhin et al. (2005) and Marshak et al. (2005) of an empirical scaling law governing clustering of drops in clouds. This remarkable underlying scaling law now promises to cast light on problems as wide-ranging as warm rain[61] and the aerosol indirect effect.[62] Kostinski and Shaw (2001) provide some excellent speculations on possible clustering mechanisms.

Suppose the cloud physics agenda is successful without finding any new laws or even reformulating their equations to be more geometrically a propos, and that cloud-resolving models are miraculously able to replicate any behavior we can measure. Would that be a satisfying conclusion to the cloud problem? From an engineering point of view, yes; but from a discovery point of view, a model as impenetrable as Nature herself offers little comfort. It is well to recall Feigenbaum's warning about numerical simulation (Horgan, 1996): "... people want to have fancier and fancier computers to simulate fluids. There is something to be learned, but unless you know what you're looking for, you're not going to see anything." You still need the "third eye" to tease out some general principle from a welter of data, whether those data come from Nature or from a large model. These will remain rare gifts even as cloud models become common.

I feel fortunate that the article by Jakob (2003) on evaluating cloud parameterizations came out during the writing of this chapter. I tend to fall prey to pessimism when contemplating what Jakob calls our "lack of strategy" and "lack of coherence" in previous approaches to the cloud problem, and his insightful and courageous contribution gives me hope. Jakob criticizes past comparisons of model to predicted clouds in an average sense and says "these studies cannot provide crucial insights into the reasons for the model failures" and "all one can learn is where, geographically, the general problems are." Jakob advocates a "compositing" way of looking at data or model output: examples include (a) sorting the data by dynamical regime, and (b) co-plotting multiple cyclones around the point of highest cloud optical depth.

---

[61] Warm rain has no ice phase. The speed of its evolution, a mere 20 minutes, has long eluded cloud physics models.

[62] in simplest terms, the effect of aerosol particles on cloud formation, persistence, precipitation, and death

Composites contain enough cases to be able to pick out a "typical" case. In using case studies from field programs, the biggest and often insurmountable problem is selecting a typical case on which to base a change in the parameterization. Jakob advocates using numerical weather prediction models rather than climate GCMs because "the large-scale flow is captured more realistically" so that errors can be more easily ascribed to the cloud parameterization. It is from deep thinking like Jakob's that the Laws of Clouds will emerge.

Another good reason to know the Laws of Clouds more completely is that, barring a miracle, we will never have paleo-cloud data. Clouds leave no trace in any of the layered records we use to infer past climate until they rain or snow, which tells us little about their radiative effect. Clouds have likely been very different in the past, since they can't occur without aerosol, and we know aerosol has been very different in the past. Thus, paleoclimate simulations will always have a cloud question mark hanging over them.

1.16 Why Cloud Radiation is So Hard

"The reality is that clouds are highly complex turbulent media in which physical, chemical, and probably biological, processes proceed at varying rates, on different scales, and interact with each other. We have yet to grapple with such complexity, although hopefully it is not beyond our wit to do so."

Hobbs (1991)

Clouds are fascinating, yet frustratingly difficult. The 3D cloud problem is a poster child for the kind of hard, interdisciplinary problem that Earth science increasingly faces. It is difficult to name an Earth phenomenon that changes faster on such a vast range of spatial scales. There is difficult cross-cutting physics, and there is difficult technology to somehow "X-ray" a cloud rapidly enough to see its internal structure change from moment to moment.

We remotely sense clouds with electromagnetic radiation that always, maddeningly, gives ambiguous and/or incomplete information. I remember being struck, in the first FIRE field campaign in marine stratocumulus off San Diego, how clouds that had appeared unbroken from our aircraft appeared broken on Landsat images – a consequence of "contrast stretching" which made the less white areas of the image turn black. This simple optical illusion taught me a good lesson about the dangers of remote sensing retrievals, which always contain buried and sometimes embarrassing assumptions.

We model clouds with sophisticated physics and chemistry theories and vast amounts of computer time, and yet know that, as of this writing, we can't hope to simulate a single individual cloud in all its details, or observe it sufficiently well to incisively test the model's predictions. We can still only hope for, at best, a statistically correct prediction. Thus, it seems, the study of 4D clouds (time being the 4th dimension) comes as close to an ultimate act of hubris as the study of the cosmos.

According to an apocryphal story, Heisenberg was asked what he would ask God, given the opportunity. His reply: "When I meet God, I am going to ask him two questions: Why relativity? And why turbulence? I really believe he will have an answer for the first." And he was speaking merely of ordinary homogeneous-gas turbulence. Clouds are turbulent near-colloids with phase change![63]

And clouds are by far the fastest component of the Earth system. If you want to learn about clouds, you have to sample fast. Many of our observing systems including those on satellites, whether by design or because of technological limitations, do not even come close to sampling clouds at the necessary speed.

Clouds are cited, sometimes with despair, sometimes with pride, as pre-eminent examples of a huge range of scales, from aerosol particles smaller than 0.1 micron that nucleate cloud drops, up to the 1000-km scales on which the biggest cloud systems are organized. This is of course an opportunity as well as a burden, and nowhere are the opportunities for discovering Laws of Clouds greater than in the scaling arena.

Cloud properties have proven devilishly hard to retrieve remotely – even "cloud fraction", which grew from 50% in my early career to 67% in my later career. This growth is mainly due to changes (hopefully improvements) in remote sensing strategies. Clouds' 3D character, and other factors, conspire to make us worry whether even the simplest characterizations like cloud fraction and cloud optical depth are "apparent" or "real," "reflectional" or "operational." One can only envy the relative simplicity of validating retrievals of land properties (e.g., Kustas et al., 2003, for soil moisture). Their properties don't change from minute to minute while they are measuring them. They must deal with the same kinds of upscaling issues as for clouds, but they don't have to worry about whether their variables are "apparent" or "real;" they can reach out and touch them.

Clouds are equally hard to characterize in situ. The instruments for so doing are, by this point, quite technologically advanced, but their sample volumes, save only for in situ lidar, are woefully small. Clouds are just too big! Characterizing the full 3D volume of even the smallest puffy cumulus, or its complete time dependence, remains beyond our reach until we step up at least to tomography if not to some even more advanced technology.

The whole "enhanced cloud absorption" debate was kicked off by several scientists who certainly overstated their case – a common enough occurrence in the heat of passion – but it was met by an equally passionate effort to stuff clouds back into the comfortable model bags they no longer, for a while, seemed to fit into. That has largely been accomplished, but just in case you think everything is just fine, consider Fig. 1.18. This shows what I have dubbed a "Radiation Hole." The two curves in the figure each represent averages of groups of radiometers of very different design, which happened to be participating in an ARM intercomparison. There were oral legends of such Holes in the surface radiation community, but they had been dismissed as instrument artifacts. This Hole, however, was unambiguous, and lasted

---

[63] I call them near-colloids because, as the DYCOMS-II (Stevens et al., 2003) and EPIC (Bretherton et al., 2004) field campaigns found, many cloud systems are on the edge of drizzling and thus often have falling droplets.

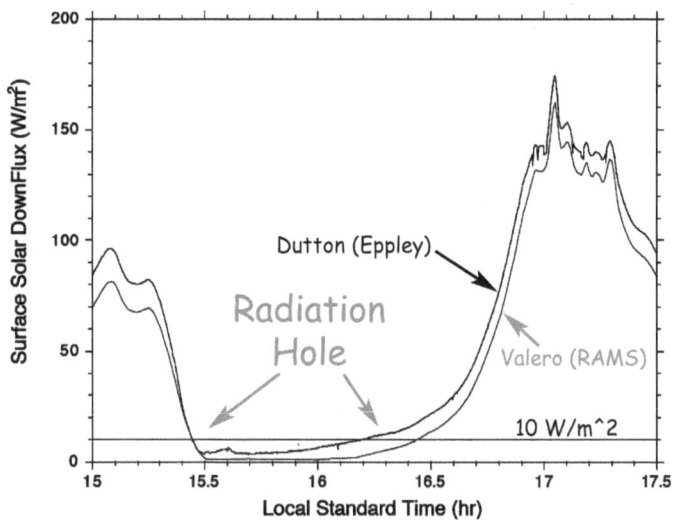

Fig. 1.18. Measured downfluxes of broadband shortwave radiation on Aug. 3, 1998, at Boulder, Colorado. Each curve is an average of several well-calibrated radiometers involved in an intercomparison between the Scripps radiometers of Francisco Valero and the NOAA radiometers of Ellsworth Dutton. The $10 \, \text{W/m}^2$ horizontal line is drawn simply to help delineate the temporal extent of the Radiation Hole; it represents the outer limit of measurability for standard field radiometers, given their inherent errors and thermal offsets, but not for these research-quality radiometers. The Hole, lasting about 1/2 hr between 3:30 and 4:00 pm, was seen during a mild thunderstorm. Radiation models are hard-pressed to explain this Hole with any reasonable cloud input parameters

over half an hour. There was a mild raincloud overhead, but nothing even close to the big thunderstorms of the Great Plains. 1D radiation models would require the cloud to have an optical depth of 500 or more – an unlikely occurrence over Boulder, Colorado, with its dry altiplano climate and mile-high altitude. 3D effects might explain some of the effect, but it cannot be dismissed with simple handwaving about "photons escaping out the cloud sides." It is likely that our cloud radiation models still don't work well for such extreme cases.

The full-time study of clouds involves only a tiny fraction of the people engaged in the atmospheric science enterprise. What kind of people are attracted to such a nearly impossible problem? They certainly require many unusual characteristics, among them being an ability to live with an overwhelming lack of information, a tolerance for falling discouragingly short of a complete solution, and a kind of heroism in continuing to attack when sometimes all hope seems lost. Indeed, I sometimes see these cloud warriors as Spartans defending the pass at Thermopylae – doomed, but fighting gallantly to the last – and keeping clouds safe from the barbarians who want to use them as arbitrary levers in the battle over global warming.

The authors of this book are among these heroes. They are pioneers in the study of 3D cloud radiation. Their chapters summarize the deep contributions they have made. I commend them to your attention.

1.17 Cloud Challenge Questions for the 21st Century

In 1900, the great mathematician Hilbert posed 23 mathematical questions to his colleagues as challenges for the 20th century. Some were very general, some very specific. Some were relatively easy and solved within a few years, some so difficult that they remain unsolved today – although the effort to solve them often led to important new mathematics anyway. In one stroke, Hilbert more or less set the agenda for large parts of 20th century mathematics. I have no pretensions to Hilbert's status, but I felt I could not leave the cloud subject without offering a few challenges of my own, most of which assume the field finally steps up to cloud tomography. As with Hilbert's challenges, the mere effort to attack these problems will certainly lead to unexpected advances:

- Aim to measure the 4D evolution of a few cloud cases perfectly enough, and at the right time and space scales, to incisively test both cloud-resolving models and 3D cloud radiation models. Do not turn cloud field campaigns into an example of the myth of Sisyphus by always going with insufficient resources to accomplish this goal.
- Make photon path distribution the central concept of cloud radiation. This lessens the artificial distinction between 1D and 3D and gets closer to the heart of the problem. See Chap. 13 for an introduction to the subject.
- Use Observing System Simulation Experiments to design statistically significant cloud field campaigns, then hew unswervingly to the calculated strategies.
- Coordinate the measurement scales (time, space, angle, wavelength) of instruments in field campaigns in advance. Doing it ex post facto is very expensive and often nearly impossible.
- Make better use of mini-UAVs and balloons in cloud field campaigns to help validate tomography and provide a 4D view of a cloud. The 4D view can never be achieved with large crewed aircraft alone.
- Increase sampling volumes of aircraft cloud instruments from cubic cm to cubic m and eventually to cubic km, beginning with liquid water content and extinction.
- Make use of multiple scattering off-beam and in-situ cloud lidars to probe larger cloud volumes rapidly, and find ways to beat down solar background.
- Formally intercompare the growing number of methods for reconstructing 3D or 4D clouds from dimensionally-challenged observations like 1D aircraft flights or vertical radar profiles. Use predictions of spectral surface fluxes as the criterion to judge models.
- Measure proper in situ vertical profiles in clouds for credible validation of radar and other purposes. Stop deluding ourselves that a descending aircraft gives a true vertical profile.

- Develop an instrument or a method to directly measure radiative heating rates in clouds, without the double-subtraction inherent in the present method.
- Make more use of Raman lidar remote sensing of cloud liquid water. This is an elegant measurement that could teach us much about initial cloud evolution and aerosol indirect effect.
- Explore other cloud phenomena, like fluorescence, which like Raman scattering involve change of wavelength. Fluorescence is a powerful new method for studying plankton from space.
- Explore cloud biology, both for its intrinsic interest and for its effect on cloud radiation.
- Make more use of polarization in surface and aircraft radiation measurements of clouds and cloud shadows. Cloud polarization is generally small, but ratios like degree of polarization can be much more accurately measured than radiances and may contain subtle new kinds of information. Use the partial polarization caused by sunlight to illuminate clouds from underneath and see what can be learned from this.
- Improve remote sensing retrieval of cloud properties to account for the fact that cloud pixels are strongly correlated both in time and space, often over vast distances and long times. GCMs recognize this fact, and thus remote sensing and GCMs lack comparability at a fundamental level. Comparisons between the two are difficult to ascribe meaning to, when remote sensing lacks the time and space coherence built into the equations in GCMs.
- Develop "cloud-shadow remote sensing."
- Speed up cloud radiance and flux instruments to capture the turbulent temporal character of cloud radiation and determine what can be learned from this sort of data. A good place to start would be solar direct-beam measurements at the surface.
- How much does reflectional cloud optical depth differ from true optical depth (the integral of the vertical profile of extinction)?
- Wavelets are a natural way to represent the scaling nature of cloud structure variables like liquid water. Develop wavelet expansion solutions to the 3D radiative transfer equation to mesh naturally with the true nature of the input variables.
- Develop equations for cloud evolution that reflect the underlying fractal geometry of clouds.
- Can cloud fraction, used only for radiation purposes and not by any other community, be made a prognostic variable in climate models? Or, alternately, can it be eliminated entirely in favor of a more robust measure of how clouds affect radiation?
- What set of parameters is necessary and sufficient to specify a cloud's 3D-ness for climate model parameterization purposes? Think beyond "cloud fraction and overlap"!
- What is the real nature of cloud gaps – their statistical structure, their pair correlations, etc. How can cloud gap structure be correctly modeled?
- What causes Radiation Holes?

- What causes the remaining biases in radiation model estimates of cloud shortwave absorption?
- How do the empirically-observed and size-dependent laws of drop clustering affect cloud radiation? As a corollary, find the consequences of Beer's Law violation in clouds.
- What are the consequences if the radiative-transfer assumption of "elementary volume" fails?
- Make greater use of stochastic radiative transfer theory to explain time-averaged cloud radiation observations. This theory has been almost completely ignored by the 3D cloud radiation field.
- Make greater use of time-dependent radiative transfer and fast-response photon detectors.
- Expose the unstated, often deeply buried assumptions or mental models about cloud scale and structure that underlie the design of cloud instruments and field programs – assumptions like stationarity, for example. How valid are these assumptions and mental models?

1.18 Epilogue

I wanted to end this chapter with a tribute to one of our long-suffering editors, Alexander Marshak, and give a final remonstrance. One of my favorite photos, Fig. 1.19, shows Dr. Marshak and myself puzzling over "a strange alien object." Dr. Marshak, like myself, came from an applied mathematics background, so surely for us this odyssey into field campaigns was even stranger than for other theoreticians. Yet we both persevered and wound up having an effect on the course of instrument development and even field program design.

What lessons can we draw from this photo? First, that radiation theoreticians have gotten out from behind their computers and into the field – something for which the ARM Program should take enormous credit. Second, that theoreticians have a hidden but often vital role to play in instrument development. In this case, we are gazing at an advanced type of flux radiometer built by Francisco Valero, one which ARM funded for a while but did not ultimately deploy; nevertheless it helped improve the state of the art in instruments we did deploy, and I for one was glad that I vigorously supported it in ARM's early days. Experimenters will sometimes growl and seem to rebuff any advice, but theoreticians must persevere because their input is so absolutely vital. Indeed, surface radiation measurements languished for decades because of lack of theoretician involvement. Photos like Fig. 1.19 must remain symbolic of the cloud radiation subject if it is to remain vibrant.

Fig. 1.19. Theoreticians Alexander Marshak (*left*) and the author puzzling over a strange alien object at the Oklahoma ARM site, which after considerable study they determined to be an "instrument", in this case a set of pyroelectric broadband flux radiometers designed by Francisco Valero of Scripps. ARM brought many sallow-complexioned theoreticians out from behind their computer screens and into "the field" with considerable salutary effects on the cloud radiation subject

References

Abbott, E. (1884). *Flatland: A Romance of Many Dimensions,* Revised 1952 edition. Dover, New York (NY).

Ackerman, T.P. and G.M. Stokes (2003). The Atmospheric Radiation Measurement program. *Phys. Today,* **56**, 38–44.

Appleby, J.F. and D. van Blerkom (1975). Absorption line studies of reflection from horizontally inhomogeneous layers. *Icarus,* **24**, 51–69.

Arking, A. (1991). The radiative effects of clouds and their impact on climate. *Bull. Amer. Meteor. Soc.,* **71**, 795–813.

Baker, B. (1992). Turbulent entrainment and mixing in clouds: A new observational approach. *J. Atmos. Sci.,* **49**, 387–404.

Barker, H.W. and J.A. Davies (1989). Multiple reflections across a linear discontinuity in surface albedo. *Internat. J. Climatology,* **9**, 203–214.

Barker, H.W., G.L. Stephens, and Q. Fu (1999). The sensitivity of domain-averaged solar fluxes to assumptions about cloud geometry. *Quart. J. Roy. Meteor. Soc.,* **125**, 2127–2152.

Barker, H.W., A. Marshak, W. Szyrmer, A. Trishchenko, J.-P. Blanchet, and Z. Li (2002a). Inference of cloud optical properties from aircraft-based solar radiometric measurements. *J. Atmos. Sci.,* **59**, 2093–2111.

Barker, H.W., R. Pincus, and J.-J. Morcrette (2002b). The Monte Carlo independent column approximation: Application within large-scale models. In *Proceedings from the GCSS Workshop.* Kananaskis, Alberta, Canada.

Barker, H.W., G.L. Stephens, P.T. Partain, J.W. Bergman, B. Bonnel, K. Campana, E.E. Clothiaux, S. Clough, S. Cusack, J. Delamere, J. Edwards, K.F. Evans, Y. Fouquart, S. Freidenreich, V. Galin, Y. Hou, S. Kato, J. Li, E. Mlawer, J.-J. Morcrette, W. O'Hirok, P. Räisänen, V. Ramaswamy, B. Ritter, E. Rozanov, M. Schlesinger, K. Shibata, P. Sporyshev, Z. Sun, M. Wendisch, N. Wood, and F. Yang (2003). Assessing 1D atmospheric solar radiative transfer models: Interpretation and handling of unresolved clouds. *J. Climate,* **16**, 2676–2699.

Baumgardner, D., B. Baker, and K. Weaver (1993). A technique for the measurement of cloud structure on centimeter scales. *J. Atmos. Oceanic Technol.,* **10**, 557–565.

Bohren, C.F., J.R. Linskens, and M.E. Churma (1995). At what optical thickness does a cloud completely obscure the sun? *J. Atmos. Sci.,* **52**, 1257–1259.

Bretherton, C., T. Uttal, C. Fairall, S. Yuter, R. Weller, D. Baumgardner, K. Comstock, R. Wood, and G. Raga (2004). The EPIC stratocumulus study. *Bull. Amer. Meteor. Soc.,* DOI: 10.1175/BAMS-85-7-967.

Cahalan, R.F. (1991). Landsat observations of fractal cloud structure. In *Nonlinear Variability in Geophysics.* Kluwer, Inc., D. Schertzer and S. Lovejoy (eds.). pp. 281–295.

Cahalan, R.F. and J.H. Joseph (1989). Fractal statistics of cloud fields. *Mon. Wea. Rev.,* **117**, 261–272.

Cahalan, R.F., M. McGill, J. Kolasinski, T. Várnai, and K. Yetzer (2005). THOR - cloud THickness from Offbeam lidar Returns. *J. Atmos. Ocean. Tech.,* **22**, 605–627.

Cess, R.D., G.L. Potter, J.-P. Blanchet, G.J. Boer, S.J. Ghan, J.T. Kiehl, S.B.A. Liang, J.F.B. Mitchell, D.A. Randall, M.R. Riches, E. Roeckner, U. Schlese, A. Slingo, K.E. Taylor, W.M. Washington, R.T. Wetherald, and I. Yagai (1989). Interpretation of cloud-climate feedback as produced by 14 atmospheric general circulation models. *Science,* **245**, 513–516.

Cess, R.D., M.-H. Zhang, P. Minnis, L. Corsetti, E.G. Dutton, B.W. Forgan, D.P. Garber, W.L. Gates, J.J. Hack, E.F. Harrison, X. Jing, J.T. Kiehl, C.N. Long, J.-J Morcrette, G. L. Potter, V. Ramanathan, B. Subasilar, C.H. Whitlock, D.F. Young, and Y. Zhou (1995). Absorption of solar radiation by clouds: Observations versus models. *Science,* **267**, 496–499.

Chiu, J.-Y., A. Marshak, and W.J. Wiscombe (2004). The effect of surface heterogeneity on cloud absorption estimate. *Geophys. Res. Lett.,* **31**, L15105.

Coakley, J.A., M. Friedman, and W. Tahnk (2005). Retrieval of cloud properties for partly cloudy imager pixels. *J. Atmos. Ocean. Tech.,* **22**, 3–17.

Davies, R. (1984). Reflected solar radiances from broken cloud scenes and the interpretation of scanner measurements. *J. Geophys. Res.,* **89**, 1259–1266.

Davis, A.B. (2002). Cloud remote sensing with sideways-looks: Theory and first results using Multispectral Thermal Imager (MTI) data. In *SPIE Proceedings: Algorithms and Technologies for Multispectral, Hyperspectral, and Ultraspectral*

Imagery VIII. S.S. Shen and P.E. Lewis (eds.). S.P.I.E. Publications, Bellingham, WA, pp. 397–405.

Davis, A.B. and A. Marshak (2002). Space-time characteristics of light transmitted through dense clouds: A Green function analysis. *J. Atmos. Sci.*, **59**, 2714–2728.

Davis, A.B. and A. Marshak (2004). Photon propagation in heterogeneous optical media with spatial correlations: Enhanced mean-free-paths and wider-than-exponential free-path distributions. *J. Quant. Spectrosc. Radiat. Transfer*, **84**, 3–34.

Davis, A., A. Marshak, W.J. Wiscombe, and R.F. Cahalan (1994). Multifractal characterizations of nonstationarity and intermittency in geophysical fields: Observed, retrieved, or simulated. *J. Geophys. Res.*, **99**, 8055–8072.

Davis, A., A. Marshak, R.F. Cahalan, and W.J. Wiscombe (1997). The LANDSAT scale-break in stratocumulus as a three-dimensional radiative transfer effect, Implications for cloud remote sensing. *J. Atmos. Sci.*, **54**, 241–260.

Davis, A.B., R.F. Cahalan, J.D. Spinhirne, M.J. McGill, and S.P. Love (1999). Off-beam lidar: An emerging technique in cloud remote sensing based on radiative Green-function theory in the diffusion domain. *Phys. Chem. Earth (B)*, **24**, 757–765.

Davis, P.J. and P. Rabinowitz (1984). *Methods of numerical integration*. Academic Press, New York (NY), 2nd edition.

Deirmendjian, D. (1969). *Electromagnetic Scattering on Spherical Polydispersions*. Elsevier, New York (NY).

Deirmendjian, D. (1975). Far-infrared and submillimeter wave attenuation by clouds and rain. *J. Appl. Meteor.*, **14**, 1584–1593.

Diner, D.J., J.C. Beckert, T.H. Reilly, C.J. Bruegge, J.E. Conel, R.A. Kahn, J.V. Martonchik, T.P. Ackerman, R. Davies, S.A.W. Gerstl, H.R. Gordon, J.-P. Muller, R.B. Myneni, P.J. Sellers, B. Pinty, and M.M. Verstraete (1998). Multiangle Imaging Spectroradiometer MISR: Description and experiment overview. *IEEE Trans. Geosci. and Remote Sens.*, **36**, 1072–1087.

Dutton, E.G. (1993). An extended comparison between LOWTRAN7 computed and observed broadband thermal fluxes. *J. Atmos. Oceanic Tech.*, **10**, 326–336.

Dyson, F. (1999). *The Sun, the Genome, & the Internet: Tools of Scientific Revolutions*. Oxford Press, New York (NY).

Ellingson, R.G. (1982). On the effects of cumulus dimensions on longwave irradiance and heating rate calculations. *J. Atmos. Sci.*, **39**, 886–896.

Ellingson, R.G. and Y. Fouquart (1991). The intercomparison of radiation codes in climate models (ICRCCM): An overview. *J. Geophys. Res.*, **96**, 8925–8927.

Ellingson, R.G. and W.J. Wiscombe (1996). The Spectral Radiance Experiment (SPECTRE): Project description and sample results. *Bull. Amer. Meteor. Soc.*, **77**, 1967–1985.

Evans, K.F. (1998). The spherical harmonics discrete ordinate method for three-dimensional atmospheric radiative transfer. *J. Atmos. Sci.*, **55**, 429–446.

Evans, K.F. and W.J. Wiscombe (2004). An algorithm for generating stochastic cloud fields from radar profile statistics. *Atmos. Res.*, **72**, 263–289.

Evans, K.F., R.P. Lawson, P. Zmarzly, D. O'Connor, and W.J. Wiscombe (2003). In situ cloud sensing with multiple scattering lidar: Simulations and demonstration. *J. Atmos. and Oceanic Tech.*, **20**, 1505–1522.

Gilgen, H. and A. Ohmura (1999). The Global Energy Balance Archive (GEBA). *Bull. Amer. Meteor. Soc.*, **80**, 831–850.

Goodstein, R. (2004). *Out of Gas: The End of the Age of Oil.* Norton & Co., New York (NY).

Greenwald, T., G. Stephens, T. Vonder Haar, and D. Jackson (1993). A physical retrieval of cloud liquid water over the global oceans using Special Sensor Microwave/Imager (SSM/I) observations. *J. Geophys. Res.*, **98**, 18,471–18,488.

Gu, Y. and K.-N. Liou (2001). Radiation parameterization for three-dimensional inhomogeneous cirrus clouds: Application to climate models. *J. Climate*, **14**, 2443–2457.

Hamblyn, R. (2001). *The Invention of Clouds: How an Amateur Meteorologist Forged the Language of the Skies.* Farrar, Straus and Giroux, New York (NY).

Hansen, J.E. and L.D. Travis (1974). Light scattering in planetary atmospheres. *Space Sci. Rev.*, **16**, 527–610.

Harshvardhan and R. Thomas (1984). Solar reflection from interacting and shadowing cloud elements. *J. Geophys. Res.*, **89**, 7179–7185.

Hartmann, D. and K. Larson (2002). An important constraint on tropical cloud-climate feedback. *Geophys. Res. Lett.*, **29**, 1951, 12–1–4.

Hartmann, D., L. Moy, and Q. Fu (2001). Tropical convective clouds and the radiation balance at the top of the atmosphere. *J. Climate*, **14**, 4495–4511.

Harwit, M. (2003). The growth of astrophysical understanding. *Physics Today*, **56**, 38–43.

Hasselmann, K. (1976). Stochastic climate models, Part 1: Theory. *Tellus*, **28**, 473–485.

Heidinger, A. and G.L. Stephens (2000). Molecular line absorption in a scattering atmosphere. II: Application to remote sensing in the O_2 A-band. *J. Atmos. Sci.*, **57**, 1615–1634.

Hobbs, P. (1991). Research on clouds and precipitation: Past, present and future, Part II. *Bull. Amer. Meteor. Soc.*, **72**, 184–191.

Holland, G., P. Webster, J. Curry, G. Tyrell, D. Gauntlett, G. Brett, J. Becker, R. Hoag, and W. Vaglienti (2001). The aerosonde robotic aircraft: A new paradigm for environmental observations. *Bull. Amer. Meteor. Soc.*, **82**, 889–902.

Horgan, J. (1996). *The End of Science.* Addison-Wesley, Reading (MA).

Hunt, G.E. and I.P Grant (1969). Discrete space theory of radiative transfer and its application to problems in planetary atmospheres. *J. Atmos. Sci.*, **26**, 963–972.

Jakob, C. (2003). An improved strategy for the evaluation of cloud parameterizations in GCMs. *Bull. Amer. Meteor. Soc.*, **84**, 1387–1401.

Klose, A., U. Netz, J. Beuthan, and A. Hielscher (2002). Optical tomography using the time-independent equation of radiative transfer. Part I: Forward model. *J. Quant. Spectrosc. Radiat. Transfer*, **72**, 691–713.

Knyazikhin, Yu., A. Marshak, W.J. Wiscombe, J. Martonchik, and R.B. Myneni (2002). A missing solution to the transport equation and its effect on estimation of cloud absorptive properties. *J. Atmos. Sci.*, **59**, 3572–3585.

Knyazikhin, Yu., A. Marshak, M.I. Larsen, W.J. Wiscombe, J. Martonchik, and R.B. Myneni (2005). Small-scale drop size variability: Impact on estimation of cloud optical properties. *J. Atmos. Sci.*, in press.

Kostinski, A. (2002). On the extinction of radiation by a homogeneous but spatially correlated random medium: Review and response to comments. *J. Opt. Soc. Amer. A*, **19**, 2521–2525.

Kostinski, A. and A. Jameson (2000). On the spatial distribution of cloud particles. *J. Atmos. Sci.*, **57**, 901–915.

Kostinski, A. and R.A. Shaw (2001). Scale-dependent droplet clustering in turbulent clouds. *J. Fluid Mech.*, **434**, 389–398.

Kustas, W., T. Jackson, J. Prueger, J. Hatfield, and M. Anderson (2003). Remote sensing field experiments evaluate retrieval algorithms and land-atmosphere modeling. *EOS Trans. AGU*, **84**, 485 and 492–493.

Liepert, B. (2002). Observed reductions of surface solar radiation at sites in the United States and worldwide from 1961 to 1990. *Geophys. Res. Lett.*, **29**, 1421, doi:10.1029/2002GL014910.

Liepert, B., A. Anderson, and N. Ewart (2003). Spatial variability of atmospheric transparency in the New York metropolitan area in summer. *AGU Fall Meeting.*

Lin, B., B. Wielicki, L. Chambers, Y.-X. Hu, and K. Xu (2002). The Iris Hypothesis: A negative or positive cloud feedback? *J. Climate*, **15**, 3–7.

Love, S.P., A.B. Davis, C. Ho, and C.A. Rohde (2001). Remote sensing of cloud thickness and liquid water content with Wide-Angle Imaging Lidar (WAIL). *Atmos. Res.*, **59-60**, 295–312.

Lovejoy, S. (1982). The area-parameter relation for rain and clouds. *Science*, **216**, 185–187.

Manabe, S. and R.T. Wetherald (1967). Thermal equilibrium of the atmosphere with a given distribution of relative humidity. *J. Atmos. Sci.*, **24**, 241–259.

Mandelbrot, B.B. (1982). *The Fractal Geometry of Nature*. W. H. Freeman, New York (NY).

Marchuk, G., G. Mikhailov, M. Nazaraliev, R. Darbinjan, B. Kargin, and B. Elepov (1980). *The Monte Carlo Methods in Atmospheric Optics*. Springer-Verlag, New York (NY).

Marshak, A., A. Davis, W.J. Wiscombe, and R.F. Cahalan (1995). Radiative smoothing in fractal clouds. *J. Geophys. Res.*, **100**, 26,247–26,261.

Marshak, A., A. Davis, W.J. Wiscombe, and R.F. Cahalan (1997). Inhomogeneity effects on cloud shortwave absorption measurements: Two-aircraft simulations. *J. Geophys. Res.*, **102**, 16,619–16,637.

Marshak, A., Yu. Knyazikhin, A.B. Davis, W.J. Wiscombe, and P. Pilewskie (2000). Cloud - vegetation interaction: Use of normalized difference cloud index for estimation of cloud optical thickness. *Geophys. Res. Lett.*, **27**, 1695–1698.

Marshak, A., Yu. Knyazikhin, K.D. Evans, and W.J. Wiscombe (2004). The "RED versus NIR" plane to retrieve broken-cloud optical depth from ground-based measurements. *J. Atmos. Sci.*, **61**, 1911–1925.

Marshak, A., Yu. Knyazikhin, M.L. Larsen, and W.J. Wiscombe (2005). Small-scale drop size variability: Empirical models for drop-size-dependent clustering in clouds. *J. Atmos. Sci.*, **62**, 551–558.

McKee, T.B. and S.K. Cox (1974). Scattering of visible radiation by finite clouds. *J. Atmos. Sci.*, **31**, 1885–1892.

Meerkotter, R. and M. Degunther (2001). A radiative transfer case study for 3-D cloud effects in the UV. *Geophys. Res. Lett.*, **28**, doi: 10.1029/2000GL011932.

Min, Q.-L. and L.C. Harrison (1999). Joint statistics of photon pathlength and cloud optical depth. *Geophys. Res. Lett.*, **26**, 1425–1428.

Mishchenko, M.I., J.W. Hovenier, and D. Mackowski (2004). Single scattering by a small volume element. *J. Opt. Soc. Am. A*, **21**, 71–87.

Nakajima, T. and M.D. King (1990). Determination of optical thickness and effective radius of clouds from reflected solar radiation measurements: Part I: Theory. *J. Atmos. Sci.*, **47**, 1878–1893.

Nunez, M., K. Fienberg, and C. Kuchinke (2005). Temporal structure of the solar radaition field in cloudy conditions: Are retrievals of hourly averages from space possible? *J. Appl. Meteor.*, **44**, 167–178.

O'Brien, D.M. (1992). Accelerated quasi Monte Carlo integration of the radiative transfer equation. *J. Quant. Spect. Radiat. Transfer*, **48**, 41–59.

Ockert-Bell, M.E. and D.L. Hartmann (1992). The effect of cloud type on Earth's energy balance: Results for selected regions. *J. Climate*, **5**, 1157–1171.

Ohmura, A., E. Dutton, B. Forgan, C. Froelich, H. Gilgen, H. Hegner, A. Heimo, G. Konig-Langlo, B. McArthur, G. Muller, R. Philipona, R. Pinker, C. Whitlock, K. Dehne, and M. Wild (1998). Baseline Surface Radiation Network (BSRN/WCRP): New precision radiometry for climate research. *Bull. Amer. Meteor. Sci.*, **79**, 2115–2136.

Palle, E., P.R. Goode, V. Yurchyshyn, J. Qiu, J. Hickey, P. Rodriguez, M.-C. Chu, E. Kolbe, C.T. Brown, and S.E. Koonin (2003). Earthshine and the Earth's albedo: 2. Observations and simulations over three years. *J. Geophys. Res.*, **108**, 4710, doi:10.1029/2003JD003611.

Paltridge, G.W. (1975). Global dynamics and climate – a system of minimum entropy exchange. *Quart. J. Roy. Meteor. Soc.*, **101**, 475–484.

Petty, G.W. (2002). Area-average solar radiative transfer in three-dimensionally inhomogeneous clouds: The independently scattering cloudlets model. *J. Atmos. Sci.*, **59**, 2910–2929.

Pierluissi, J., K. Tomimaya, and R.B. Gomez (1987). Analysis of the LOWTRAN transmission functions. *Appl. Opt.*, **18**, 1607–1612.

Pincus, R., C. Hannay, and K.F. Evans (2005). The accuracy of determining three-dimensional radiative transfer effects in cumulus clouds using ground-based profiling instruments. *J. Atmos. Sci.*, in press.

Pinsky, M. and A. Khain (2003). Fine structure of cloud droplet concentration as seen from the Fast-FSSP measurements. Part II: Results of in situ observations. *J. Appl. Meteor.*, **42**, 65–73.

Plass, G. and G. Kattawar (1968). Monte Carlo calculations of light scattering from clouds. *Appl. Opt.*, **7**, 415–419.

Polonsky, I.N., S.P. Love, and A.B. Davis (2005). Wide-Angle Imaging Lidar (WAIL) deployment at the ARM Southern Great Plains site: Intercomparison of cloud property retrievals. *J. Atmos. Ocean. Tech.*, **22**, 628–648.

Portmann, R.W., S. Solomon, R.W. Sanders, J.S. Daniel, and E. Dutton (2001). Cloud modulation of zenith sky oxygen path lengths over Boulder, Colorado: Measurement versus model. *J. Geophys. Res.*, **106**, 1139–1155.

Potter, G. and R.D. Cess (2004). Testing the impact of clouds on the radiation budgets of 19 atmospheric general circulation models. *J. Geophys. Res.*, **109**, 1139–1155, doi:10.1029/2003JD004018.

Prata, A.J. and P.J. Turner (1997). Cloud-top height determination using ATSR data. *Remote Sens. Envir.*, **59**, 1–13.

Press, W., S. Teukolsky, W. Vettering, and B. Flannery (2000). *Numerical Recipes in Fortran: The Art of Scientific Computing*. Cambridge University Press, Cambridge (UK), 2nd edition.

Ramanathan, V., R.D. Cess, E.F. Harrison, P. Minnis, B.R. Barkstrom, E. Ahmad, and D. Hartmann (1989). Cloud-radiative forcing and climate: Results from the Earth-radiation Budget Experiment. *Science*, **243**, 57–63.

Randall, D., M. Khairoutdinov, A. Arakawa, and W. Grabowski (2003). Breaking the cloud parameterization deadlock. *Bull. Amer. Meteor. Soc.*, **84**, 1547–1564.

Randall, D., S. Krueger, C. Bretherton, J. Curry, P. Duynkerke, M. Moncrieff, B. Ryan, D. Starr, M. Miller, W. Rossow, G. Tselioudis, and B. Wielicki (2004). Confronting models with data: The GEWEX cloud systems study. *Bull. Amer. Meteor. Soc.*, **84**, 455–469.

Rossow, W.B. and E. Duenas (2004). The International Satellite Cloud Climatology Project (ISCCP) web site. *Bull. Amer. Meteor. Soc.*, **85**, 167–172.

Rossow, W.B. and R.A. Schiffer (1999). Advances in understanding clouds from ISCCP. *Bull. Amer. Meteor. Soc.*, **11**, 2261–2287.

Schneider, S. (1972). Cloudiness as a global climatic feedback mechanism. *J. Atmos. Sci.*, **29**, 1413–1422.

Senior, C.A. and J.F.B. Mitchell (1993). CO_2 and climate: The impact of cloud parametrizations. *J. Climate*, **6**, 393–418.

Shaw, R.A. (2003). Particle-turbulence interactions in atmospheric clouds. *Annu. Rev. Fluid Mech.*, **35**, 183–227.

Shaw, R.A., A. Kostinski, and D. Lanterman (2002). Super-exponential extinction of radiation in a negatively-correlated random medium. *J. Quant. Spect. Radiat. Transfer*, **75**, 13–20.

Sigl, G. (2001). Ultrahigh-energy cosmic rays: Physics and astrophysics at extreme energies. *Science*, **291**, 73–79, doi:10.1126/science.291.5501.73.

Stamnes, K., S.-C. Tsay, W.J. Wiscombe, and K. Jayaweera (1988). Numerically stable algorithm for discrete-ordinate-method radiative transfer in multiple scattering and emitting layered media. *Appl. Opt.*, **27**, 2502–2509.

Stephens, G.L. (1988a). Radiative transfer through arbitrary shaped optical media, I: A general method of solution. *J. Atmos. Sci.*, **45**, 1818–1836.

Stephens, G.L. (1988b). Radiative transfer through arbitrary shaped optical media, II: Group theory and simple closures. *J. Atmos. Sci.*, **45**, 1837–1848.

Stephens, G.L. (2003). The useful pursuit of shadows. *American Scientist*, **91**, 442–449.

Stephens, G.L., D. Vane, R. Boain, G. Mace, K. Sassen, Z. Wang, A. Illingworth, E. O'Connor, W. Rossow, S. Durden, S. Miller, R. Austin, A. Benedetti, C. Mitrescu, and CloudSat Science Team (2002). The CloudSat mission and the A-Train: A new dimension of space-based observations of clouds and precipitation. *Bull. Amer. Metereol. Soc.*, **83**, 1771–1790.

Stevens, B., C.-H. Moeng, and P.P. Sullivan (1999). Large-Eddy simulations of radiatively driven convection: Sensitivities to the representation of small scales. *J. Atmos. Sci.*, **56**, 3963–3984.

Stevens, B., D.H. Lenschow, G. Vali, H. Gerber, A. Bandy, B. Blomquist, J.-L. Brenguier, C.S. Bretherton, F. Burnet, T. Campos, S. Chai, I. Faloona, D. Friesen, S. Haimov, K. Laursen, D.K. Lilly, S.M. Loehrer, S.P. Malinowski, B. Morley, M.D. Petters, D.C. Rogers, L. Russell, V. Savic-Jovcic, J.R. Snider, D. Straub, M.J. Szumowski, H. Takagi, D.C. Thornton, M. Tschudi, C. Twohy, M. Wetzel, and M.C. van Zanten (2003). Dynamics and chemistry of marine stratocumulus: DYCOMS-II. *Bull. Amer. Meteor. Soc.*, **84**, 579–593.

Sundqvist, H. (1978). A parameterization scheme for non-convective condensation including prediction of cloud water content. *Q. J. Roy. Meteor. Soc.*, **104**, 677–690.

Sundqvist, H., E. Berge, and J. E. Kristjansson (1989). Condensation and cloud parameterization studies with a mesoscale Numerical Weather Prediction model. *Mon. Wea. Rev.*, **117**, 1641–1657.

Takara, E.E. and R.G. Ellingson (2000). Broken cloud field longwave scattering effects. *J. Atmos. Sci.*, **57**, 1298–1310.

Twomey, S. (1987). Iterative nonlinear inversion methods for tomographic problems. *J. Atmos. Sci.*, **44**, 3544–3551.

Vonder Haar, T. and V. Suomi (1971). Measurements of Earth's radiation budget from satellites for a five-year period. *J. Atmos. Sci.*, **28**, 305–314.

Warner, J., J. Drake, and P. Krehbiel (1985). Determination of cloud liquid water distribution by inversion of radiometric data. *J. Atmos. Oceanic Technol.*, **2**, 293–303.

Warner, J., J. Drake, and J. Snider (1986). Liquid water distribution obtained from coplanar scanning radiometers. *J. Atmos. Oceanic Technol.*, **3**, 542–546.

Weinman, J.A. and P.N. Swartztrauber (1968). Albedo of a striated medium of isotropically scattering particles. *J. Atmos. Sci.*, **34**, 642–650.

Wielicki, B.A., R.D. Cess, M.D. King, D.A. Randall, and E.F. Harrison (1995). Mission to Planet Earth - Role of clouds and radiation in climate. *Bull. Amer. Meteor. Soc.*, **76**, 2125–2153.

Winker, D., R. Couch, and M.P. McCormick (1996). An overview of LITE: NASA's Lidar In-space Technology Experiment. *Proc. IEEE*, **84**, 164–180.

Wiscombe, W.J. (1975). Solar radiation calculations for Arctic summer stratus conditions. In *Climate of the Arctic*. G. Weller and S. Bowling (eds.). University of Alaska Press, Fairbanks (AK).

Wiscombe, W.J. (1983). Atmospheric radiation: 1975-1983. *Rev. Geophys. Space Phys.*, **21**, 997–1021.

Wiscombe, W.J. and V. Ramanathan (1985). The role of radiation and other renascent subfields in atmospheric science. *Bull. Amer. Meteor. Soc.*, **66**, 1278–1287.

Wiscombe, W.J. and R. Welch (1986). Reply. *J. Atmos. Sci.*, **43**, 401–407.

Wiscombe, W.J., R. Welch, and W. Hall (1984). The effect of very large drops on cloud absorption I. Parcel models. *J. Atmos. Sci.*, **41**, 1336–1355.

Yodh, A. and B. Chance (1995). Spectroscopy and imaging with diffusing light. *Phys. Today*, **48**, 34–40.

Zmarzly, P.M. and R.P. Lawson (2000). An optical extinctometer for cloud radiation measurements and planetary exploration. Technical Report Fulfillment of Contract NAS5-98032, NASA GSFC.

Suggested Further Reading

Davies, R. (1978). The effect of finite cloud geometry on the 3D transfer of solar irradiance in clouds. *J. Atmos. Sci.*, **35**, 1712–1725.

Hasler, A. (1981). Stereoscopic observations from geosynchronous satellites. *Bull. Amer. Meteor. Soc.*, **62**, 194–212.

Kummerow, C. and J.A. Weinman (1988). Determining microwave brightness temperatures from precipitating horizontally finite and vertically structured clouds. *J. Geophys. Res.*, **93**, 3720–3728.

Li, Z., M. Cribb and A. Trishchenko (2002). Impact of surface inhomogeneity on solar radiative transfer under overcast conditions. *J. Geophys. Res.*, **107** (D16), 10.1029/2001JD00976.

Oreskes, N., K. Schrader-Frechette and K. Belitz (1994). Verification, validation and confirmation of numerical models in the Earth sciences. *Science*, **263**, 641–646 (and Letters to *Science*, **264**, 329–331).

Pinnick, R., S. Jennings, P. Chylek, C. Ham and W. Grandy (1983). Backscatter and extinction in water clouds. *J. Geophys. Res.*, **88**, 6787–6796.

Plank, V. (1969). The size distribution of cumulus clouds in representative Florida populations. *J. Appl. Meteor.*, **8**, 46–67.

Twomey, S. (1976): The effects of fluctuations in liquid water content on the evolution of large drops by coalescence. *J. Atmos. Sci.*, **33**, 720–723.

Venema, V., S. Crewell and C. Simmer (2003): Surrogate cloud fields with measured cloud properties. In *Proceedings of Inter. Symp. on Tropos. Profiling*, 14–20 September 2003, Leipzig (Germany), 303–305.

Weckwerth, T., D.B. Parsons, S.E. Koch, J.A. Moore, M.A. Le Mone, B.B. Demoz, C. Flamant, B. Geerts, J. Wang and W.F. Feltz (2004): An overview of the International H2O Project (IHOP_2002) and some preliminary highlights. *Bull. Amer. Meteor. Soc.*, **85**, 253–277.

Welch, R. and W. Zdunkowski, 1981: The radiative characteristics of noninteracting cumulus cloud fields, Part I: Parameterization for finite clouds. *Contrib. Atmos. Phys.*, **54**, 258–272.

Recommended Web Sites

- A short history of Luke Howard:
  http://www.islandnet.com/~see/weather/history/howard.htm/
- A variety of cloud information:
  http://cloudsat.atmos.colostate.edu/outreach/clouds.php/
- European 4D-Clouds Project:
  http://www.meteo.uni-bonn.de/projekte/4d-clouds/
- Orbiting Wide-angle Light-collectors (OWL) cosmic ray satellites:
  http://owl.gsfc.nasa.gov/
- Flatland, a romance of many dimensions:
  http://www.alcyone.com/max/lit/flatland/
- A Course on Fractal Geometry (Frame, Mandelbrot, Neger):
  http://classes.yale.edu/fractals/
- Daisyworld workshop honoring Garth Paltridge:
  http://www.cogs.susx.ac.uk/daisyworld/ws2002_overview.html

List of Acronyms and Abbreviations

AIDJEX	Arctic Ice Dynamics Joint EXperiment
AMS	American Meteorological Society
ARESE	ARM Enhanced Shortwave Experiment
ARM	Atmospheric Radiation Measurement
ARPA	Advanced Research Projects Agency
ATRAD	ATmospheric RADiation (code)
AVHRR	Advanced Very High Resolution Radiometer
AVIRIS	Airborne Visible Infra-Red Imaging Spectrometer
BRDF	Bidirectional Reflectance Distribution Function
BOMEX	Barbados Oceanographic and Meteorological EXperiment
CALIPSO	Cloud-Aerosol Lidar and Infrared Pathfinder Satellite Observations
CAR	Cloud Absorption Radiometer
CCC	Canadian Climate Centre
CERES	Clouds and the Earth's Radiant Energy System
CRYSTAL	Cirrus Regional StudY of Tropical Anvils and cirrus Layers
CRYSTAL-FACE	CRYSTAL-Florida Area Cirrus Experiment
DOE	U.S. Department of Energy
DISORT	DIScrete ORdinate radiative Transfer (code)
DYCOMS-II	DYnamics and Chemistry Of Marine Stratocumulus – Phase II
ECHAM	EC(MWF) forecast models, modified and extended in HAMburg
ECMWF	European Center for Medium-Range Weather Forecasting
ERBE	Earth Radiation Budget Experiment
FAA	Federal Aviation Administration
FIRE	First ISCCP Regional Experiment
FSSP	Forward Scattering Spectrometer Probe
GATE	Global atmospheric research program Atlantic Tropical Experiment
GCM	Global Climate Model
GFDL	Geophysical Fluid Dynamics Laboratory
GISS	Goddard Institute for Space Studies
GLAS	Global Laser Altimetry System
GOES	Geostationary Operational Environmental Satellites
GPR	Ground Profiling Radar
HIS	High-resolution Interferometeric Sounder
ICRCCM	Inter-Comparison of Radiation Codes in Climate Models
IPCC	Intergovernmental Panel on Climate Change
ISCCP	International Satellite Cloud Climatology Program
I3RC	Intercomparison of 3-dimensional Radiation Codes
JQSRT	Journal of Quantitative Spectroscopy and Radiative Transfer
LMD	Laboratoire de Météorologie Dynamique
LOWTRAN	LOW-resolution TRANsmittance and background radiance (code)

MISR	Multiangle Imaging Spectro-Radiometer
MIT	Massachusetts Institute of Technology
MODIS	MODerate resolution Imaging Spectroradiometer
NASA	National Aeronautics and Space Administration
NCAR	National Center for Atmospheric Research
NOAA	National Oceanic and Atmospheric Administration
NSF	National Science Foundation
OWL	Orbiting Wide-angle Light collectors (cosmic ray satellites)
PVM	Particle Volume Monitor
SBIR	Small-Business Innovation Research program
SHDOM	Spherical Harmonics Discrete Ordinate Method (code)
SHEBA	Surface HEat Balance of the Arctic
SPECTRE	SPECTral Radiance Experiment
SUCCESS	SUbsonic aircraft: Contrail and Cloud Effects Special Study
THOR	THickness from Off-beam Returns (lidar system)
TRMM	Tropical Rainfall Measuring Mission
UAV	Untenanted Aerial Vehicles
UKMO	United Kingdom Meteorological Office
WAIL	Wide-Angle Imaging Lidar

2

Observing Clouds and Their Optical Properties

E.E. Clothiaux, H.W. Barker and A.V. Korolev

2.1   Introduction: The Nature of the Problem 93

2.2   In Situ Measurements of Cloud Particles 99

2.3   From Cloud Particles to the Atmospheric Radiation Field 106

2.4   Radiative Properties of Atmospheric Particles 115

2.5   Wavelength Dependence of Radiation Sources and Atmospheric
      Transmission 123

2.6   A Remote Sensing View of Cloud Structure 124

2.7   Concluding Remarks 146

References 146

2.1 Introduction: The Nature of the Problem

As we go about our business each day of our lives, we observe clouds of varying shapes, sizes, brightness levels and altitudes. Indeed, clouds are so much a part of our lives that we intuitively take them to be known, easily-characterized components of our daily experience. However, when it comes to describing quantitatively the microphysical and radiative properties of clouds on a global scale, so that we might observationally determine their impact upon, or changes in, the atmosphere, we immediately discover that clouds, our constant companions in life, are not so easily characterized.

One might ask the question why observational characterization of clouds is important. One answer is that a quantitative, global characterization of the three-dimensional (3D) spatial distribution of clouds over an extended period of time would allow us to quantify more accurately the impact of clouds on the Earth's radiation budget, as well as changes in the radiation budget of Earth that result from changes in cloud properties. A second answer is that a high quality observational database of 3D cloud properties for an extended period of time, at least at a few locations, is necessary for improving the treatment of clouds in numerical cloud re-solving, weather prediction and global climate models. For example, improving or assessing numerical model heating rate computations requires characterization of the fidelity of the model fields to observations and many issues arise regarding both the model fields and the observations in these kinds of studies.

Surprisingly, and in spite of the fact that we deal with clouds on a daily basis, to date there is no universal definition of a cloud. Usually, a cloud is referred to as an ensemble of liquid and/or ice particles suspended in the atmosphere (e.g., AMS, 2000). However, such a definition does not characterize a cloud in any way that can be quantified and hence it must be considered as incomplete. For example, is an ensemble of particles having a concentration of one per cubic centimeter, or one per cubic meter, or even one per cubic kilometer worthy of the label "cloud"? And what moment of a particle size distribution should be used in the definition of a cloud, the zeroth, second, third, or sixth moment?

Currently, each scientific community studying clouds employs its own definition based on the instruments and methods it is using. The operational definition of cloud that we use in casual life may be quite different from the definition used by investi-gators who use radar-, lidar-, or satellite-based definitions of a cloud. For example, clouds invisible to the human eye are easily detected by lidars, and clouds invisible to radars can produce strong signals in microwave radiometers. Ultimately, the defin-ition of a cloud depends on the threshold sensitivity of the instruments used in cloud studies. Hereafter, we refer to a cloud as an ensemble of liquid and/or ice particles that change the properties of the electromagnetic radiation field of interest.

Tropospheric clouds are dynamic, continuously changing objects. The forma-tion, evolution and spatial distribution of cloud particles depend on many processes, such as turbulence, vertical motions, entrainment and mixing with out-of-cloud air, intensity of upwelling and downwelling radiation, and chemical and physical prop-erties of liquid and ice nuclei. Interactions between cloud particles and the above processes are quite complex, affecting the spatial and temporal variations of cloud microphysical parameters, such as particle concentration, extinction coefficient, and water content. The spatial scale of inhomogeneity of cloud microphysics ranges from thousands of kilometers, for the case of cyclones, down to meters and centimeters (e.g., Brenguier, 1993; Korolev and Mazin, 1993; Gerber et al., 2001). The temporal variations of local cloud parameters are defined by a so called time of phase relax-ation, typically changing from seconds for liquid clouds to tens of minutes for ice clouds (Korolev and Mazin, 2003). The time of phase relaxation and the lifetime of an individual cloud particle are usually much less than the lifetime of the whole cloud. The characteristic lifetime of clouds varies from several minutes (e.g., cumu-lus humilis) to several days (e.g., clouds associated with fronts and cyclones). As these properties of clouds illustrate, clouds are characterized by a cascade of spatial and temporal scales.

With existing ground- and satellite-based cloud remote sensing systems we are currently unable to quantify the 3D distribution of clouds and their properties, even over a relatively small region. Present-day satellite sensors do allow us to identify those atmospheric columns that contain clouds and the top heights of the highest clouds in the cloudy columns. Using ground-based sensors, we are able to quantify the vertical locations of cloud particles in a narrow column of the atmosphere and estimate the vertically integrated amount of liquid water associated with the cloud particles in the column. In addition, many scientists are working on retrieving column ice-water particle amounts using either ground- or satellite-based observations and algorithms for vertically distributing the retrieved column-integrated water amounts are under development.

Overall, however, a synergy of the appropriate ground- and satellite-based observations that can produce, for example, reliable estimates of the amount of liquid and ice water that advects into a domain with the dimensions of a global climate model (GCM) grid cell, e.g., 200 km by 200 km, has yet to take place. To demonstrate just how variable, both spatially and temporally, cloud properties are and the difficulties these variations introduce into the retrieval of cloud properties using remotely sensed data, we present and discuss ground-based observations from the United States (US) Department of Energy (DOE) Atmospheric Radiation Measurement (ARM) program (Stokes and Schwartz, 1994; Ackerman and Stokes, 2003), satellite-based observations from the US National Aeronautics and Space Administration (NASA) Earth Observing System (EOS) Terra satellite program (Kaufman et al., 1998), and aircraft-based in situ observations that are accessible to one of the authors.

To illustrate some of the problems in the retrieval of cloud properties consider a satellite image of a typical boundary-layer cloud deck located just west of southern California over the eastern Pacific Ocean. The size of the domain in Fig. 2.1a is 122.88 km by 122.88 km and each individual pixel in the image represents an area at the surface of the Earth of 480 m by 480 m. Considering that global climate models have resolutions ranging from approximately 50 km to 500 km, these models cannot explicitly characterize the distribution of cloud in Fig. 2.1a. At best, they are capable of producing parameterized estimates of the cloud fraction within the grid column, so that the distribution of clouds in Fig. 2.1b is represented in global climate models as Fig. 2.1c. At this point in the example we have not encountered any insurmountable problems: global climate models produce cloud fraction estimates within each of their grid cells and satellite data with spatial resolutions of 1 km or better provide, for the most part, an observational means for quantifying the vertically-integrated cloud fractions on a global scale. Difficulties appear, however, both in the observations and models when the cloud fraction and variability of cloud condensate for each atmospheric layer and the spatial correlations of clouds between vertical layers must be taken into account, as is the case for atmospheric radiation heating rate computations.

For example, looking directly down upon the cloud in Fig. 2.1a, we find that 86% of the pixels contain cloud according to a certain threshold, whereas the remaining 14% of the pixels are cloud free (Fig. 2.1b). We represent this state-of-affairs in the global climate model as Fig. 2.1c. One possible configuration of clouds that can lead to a cloud fraction of 86% is illustrated in the left column of Fig. 2.1d, where two partially overlapping cloud layers, each with a cloud fraction of 60%, collectively

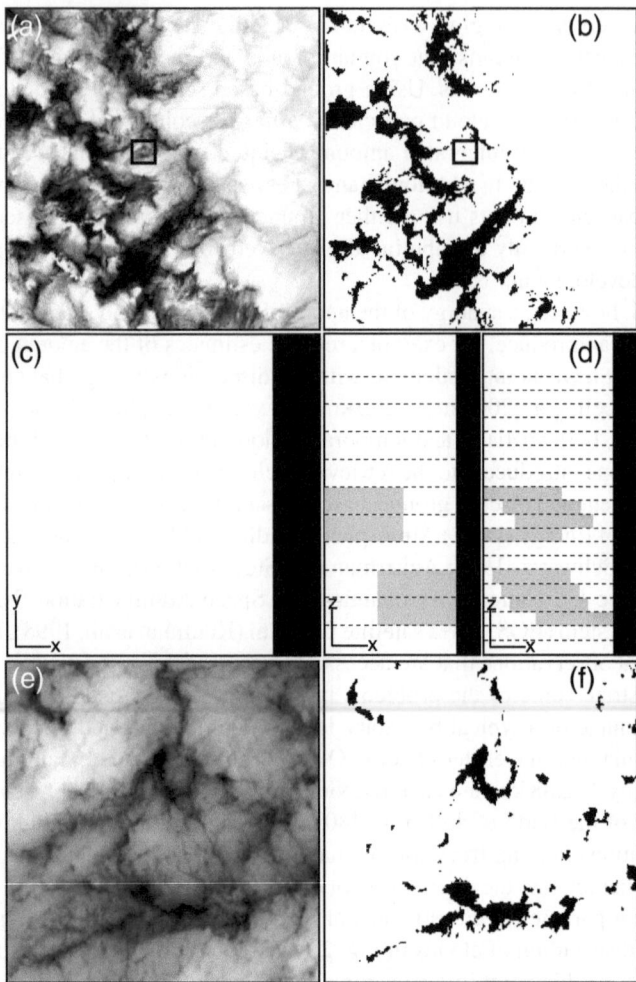

Fig. 2.1. (a) A mid-visible image of boundary-layer clouds off the coast of southern California obtained by the LandSat satellite sensor. **(b)** A partitioning of the image in (a) into cloudy and cloud-free atmospheric columns using a simple threshold – pixel radiances above 65 radiometric units (on a scale from 0 to 255 units) are taken to be cloudy, while pixel radiances less than or equal to 65 radiometric units are taken to be cloud free; the cloud fraction for this scene is 86%. **(c)** Representation of a global climate model grid-column cloud fraction of 86% (*white*) and cloud-free column percentage of 14% (*black region on the right*). **(d)** Given the same volumes of cloudy air in two global climate model grid columns, the overlap of the cloudy layers depends on many factors and assumptions, including the number of vertical layers. Again, the black strips on the right sides of the two panels represent a clear-sky fraction of 14%. **(e)** A blow-up of the pixel radiances within the black delineated square in (a); the domain size is 7.68 km by 7.68 km with a pixel size of 30 m by 30 m. **(f)** A partitioning of the image in (e) into cloudy and cloud-free atmospheric columns using the same threshold as in (b)

yield an overall cloud fraction of 86%. There is nothing unique about the choice of layers and cloud fractions in the left column of Fig. 2.1d and there are many combinations of cloud layers with different cloud fractions that can lead to an overall cloud fraction of 86%.

Suppose that we double the number of global climate model vertical layers while keeping the overall cloud fraction and the cloud fraction per vertical layer through the atmosphere fixed. We now have the situation in the right hand column of Fig. 2.1d. The total volume of cloudy air in the right hand column of Fig. 2.1d is the same as in the left hand column of Fig. 2.1d and the cloud fraction for each model vertical layer is 60%. We could arrange the cloudy layers in the right hand column to physically match the distribution in the left hand column, or we could, as we have illustrated, arrange the cloudy layers in a different configuration while keeping the total cloud fraction and the cloud fraction per layer fixed. Again, there are many ways to arrange the cloud layers and the cloud fraction per layer to obtain an overall cloud fraction of 86%. Just how cloud layers should be vertically aligned within a global climate model grid cell must ultimately be based upon schemes that take observations into account (Chap. 9). But, as we will demonstrate, developing an accurate observational database of the 3D distribution of clouds is difficult at best, even with today's ground- and satellite-based observational capabilities.

The black box illustrated within Fig. 2.1a contains 16 by 16 pixels covering a domain of 7.68 km by 7.68 km, each pixel having a spatial resolution of 480 m by 480 m, which is typical of the spatial resolutions of present-day satellite instruments whose data are used in analyses of cloud properties. At this moderate resolution the brightness levels of the pixels fall into three groups: relatively dark clear-sky pixels, low- and moderately-reflecting cloud pixels and brightly reflecting cloud pixels. If we now expand these 256 pixels into their 65536 full-resolution 30 m by 30 m pixels, a much richer cloud structure emerges (Fig. 2.1e) with more clear-sky gaps now apparent in the cloud field (Fig. 2.1f). At a pixel resolution of 30 m by 30 m we are now at the spatial scale of numerical cloud resolving models and within the range of spatial scales for which radiation smoothing is dominant (Chap. 12). As Fig. 2.1e illustrates, at these scales the distribution of cloud water is non-trivial. In fact, as a result of radiative smoothing, the radiance field illustrated in Fig. 2.1e is actually smoother than the underlying two-dimensional field of vertically integrated cloud-liquid water amounts (mass per unit area, e.g., g m^{-2}) that produces it.

There is a fundamental difference in the radiative properties of the domains illustrated in Fig. 2.1a and Fig. 2.1e. As the domain horizontal size decreases and the horizontal dimensions of the domain approach the vertical dimension of the domain, the horizontal transport of radiation across the lateral boundaries of the domain becomes more important. As one-dimensional (1D) radiative transfer theory, with the dimension of importance being oriented vertically, does not account for the horizontal transport of radiation, we find larger errors in domain-averaged heating rates using 1D theory as the domain size decreases. Therefore, while 1D radiative transfer may be adequate for estimating the domain-averaged heating rates for the domain illustrated in Fig. 2.1a, it will produce significant errors for the domain illustrated in Fig. 2.1e for column sizes approaching 30 m by 30 m (see Chaps. 12 and 6).

Moreover, 1D radiative transfer theory is completely inadequate for estimating radiances emanating from the 30 m by 30 m columns towards space (Chap. 11). That is, for small domains with high spatial resolutions, such as domains characteristic of high spatial resolution cloud resolving model simulations, 3D radiative transfer theory is a more appropriate theory for both heating rate and radiance calculations. Evidence supporting this statement will be presented over and again in the following chapters.

The internal variability in the distribution of cloud water within a cloud layer, as illustrated in Fig. 2.1e, is another challenge for global climate models. Not only must global climate models contain realistic assumptions about vertically-displaced, overlapping cloud layers, they must also contain realistic assumptions about the horizontal distribution of cloud water within each layer. If all clouds were composed of uniform densities of identical liquid water drops, the difficulties in accounting properly for the radiative effects of clouds in models would be limited primarily to estimating cloud fractions for each layer and the overlap of clouds from one layer to the next. But clouds consist of a myriad array of liquid- and ice-particle sizes, with the ice particles coming in a seemingly endless variety of shapes. Moreover, the shapes, sizes and number densities of cloud particles, in addition to their 3D spatial distributions, affect the transfer of radiation through them.

The intent of this chapter is to provide an observational overview of the properties of clouds. We hope to drive home the points that clouds are highly variable in space, both in their total amounts of water and the form that water takes, and change rapidly in time. Given the strong wavelength dependencies of the radiative properties of the gases in which clouds are embedded, as well as the wavelength dependencies of cloud-particle radiative properties, the atmospheric radiation field throughout a cloudy atmospheric region can be quite complicated with dramatic changes as a function of wavelength, which we attempt to illustrate. As a result of the spatial and temporal complexity of clouds and their associated wavelength-dependent radiation fields, there are difficulties involved in trying to map observationally the 3D distribution of clouds, even over a small region of the atmosphere, and we discuss what some of these difficulties are. All of these properties of clouds make atmospheric radiance, irradiance and heating rate studies a challenge – the subject of this book.

At present the only observational approach for characterizing detailed cloud properties is to use in situ aircraft probes to sample cloud particles. To illustrate the variety of cloud-particle types, their associated water contents and their rapid variations in space, we make use of aircraft observations that were collected during a number of field campaigns that took place in high northern hemisphere latitudes (Sect. 2.2). Moving from in situ aircraft observations of cloud particles to ground- and satellite-based observations of cloud properties, we must first introduce a few concepts from radiative transfer theory, as the remotely sensed cloud properties are based on radiometric measurements. To this end we discuss radiance, particle cross-sections, transmissivity, particle scattering phase functions and particle emission (Sect. 2.3). In Sect. 2.4 we illustrate the wavelength-dependencies of atmospheric gas and cloud particle absorption and scattering cross-sections, which, together with gas and particle species concentrations, determine the wavelength-dependence of the atmospheric transmissivity (Sect. 2.5). Sections 2.3–2.5 serve as a preamble to our discussion of collocated ground- and satellite-based radiance measurements, allowing us to illustrate meaningfully aspects of the spatial and temporal changes in the radiation field associated with clouds (Sect. 2.6). The examples that we use also clearly illustrate that the clear- and cloudy-sky downwelling radiation fields at the surface and upwelling radiation fields at the top of atmosphere are strong functions of wavelength. Using information presented in Sects. 2.4–2.6, we touch upon issues that make ground- and satellite-based remote sensing of cloud properties so difficult (Sect. 2.7), thereby justifying this book and bringing this chapter to a close.

2.2 In Situ Measurements of Cloud Particles

Satellite images of clouds, like Fig. 2.1, show inhomogeneities in the radiation field as a result of scattering by cloud tops. The satellite data suggest that the characteristic scale of cloud-top inhomogeneity varies from tens of meters to hundreds of kilometers (Loeb et al., 1998). We would be wise to expect that the characteristic scale of inhomogeneity of cloud microphysical parameters inside clouds would be no less than that obtained from the satellite observations of cloud tops. The best way to characterize in-cloud variability of the microphysical parameters is to use airborne in situ measurements. Modern aircraft-based instruments are capable of cloud particle measurements with high accuracy and spatial resolution, capturing the shapes and sizes of cloud particles as well as different moments of their size distribution, such as concentration, extinction coefficient, and liquid- and ice-water contents. While aircraft-based microphysical instrumentation provides the most detailed measurements of cloud properties, the in situ measurements have some significant limitations.

One of the fundamental limitations of in situ measurements is related to their small sample volume along a thin line following the aircraft flight track. Typically, the cloud volume that is sampled by airborne microphysical instruments along a line 100 m long varies from about 10 cm^3 to 1 m^3. As a result, the representativeness of the cloud particles sampled to those comprising the entire cloud element from which the sample is drawn is always an important source of uncertainty in the interpretation of the measurements. A second limitation is related to the relatively long time span between measurements during characterization of the same cloud. For example, a vertical sounding of a convective cloud from 1 km to 8 km of altitude may take about one hour. This period of time is comparable to the characteristic lifetime of the whole cloud. Consequently, towards the end of the measurements the cloud will be sampled at a different age and with different characteristics compared to the beginning measurements in the cloud.

Ground-based active remote sensing systems sample volumes of air with typical vertical dimensions of tens to hundreds of meters and horizontal dimensions of less than a meter to several tens of meters. Satellite-based remote sensing systems for cloud applications have spatial resolutions ranging from a few tens of meters to a kilometer or more. Relating aircraft in situ probe measurements to ground- and satellite-based observations would be much easier if the aircraft measurements characterized all of the particles in the volumes of air sampled by the ground- and satellite-based instruments. However, this is not the case, with the aircraft only able to sample a small fraction of the cloudy air sensed by the ground- and satellite-based systems. In turn, the ground-based systems sample only a fraction of the cloudy air that influences the satellite measurements. Presently, there are no readily available solutions to the sampling differences between the various aircraft-, ground- and satellite-based sensors and one should keep these differences in mind when comparing measurements from them.

The aircraft measurements presented below were collected by a variety of devices that have evolved over the years. Since the mid 1970s, Particle Measuring Systems (PMS Inc., Colorado) has developed a set of airborne instruments for measurement of the sizes and concentrations of cloud particles while simultaneously recording their images (Knollenberg, 1976). These instruments, owing to their accuracy and high reliability, have become part of most scientific projects and campaigns on measurements of cloud microphysical parameters. Among the most popular PMS probes are the Forward Scattering Spectrometer Probe (FSSP) for measurement of cloud droplet size distributions in the size range from 0.5 μm to 95 μm and the Optical Array Probes (OAP) for recording particle shadow binary images in different size ranges. The OAP-2DC, OAP-2DG and OAP-2DP record particle sizes from 25–800 μm, 10–1600 μm, and 200–6400 μm, respectively. Other instruments are employed for measurements of integral cloud microphysical parameters, such as extinction coefficient and ice- and liquid-water content. The King probe (PMS Inc.; King et al., 1978) and Particulate Volume Monitor (PVM; Gerber Scientific Inc., Virginia; Gerber et al., 1994) have become conventional instruments for measurements of cloud-liquid water. The Nevzorov probe is usually used for airborne measurements of liquid- and ice-water contents (Korolev et al., 1998). The description of other relevant instruments for cloud measurements can be found in Baumgardner et al. (2002).

Liquid-water clouds have particles with the simplest shapes, i.e., spheres, and in situ measurements are focused on quantifying the cloud drop-size distribution (number of particles per unit volume of air per drop radius interval, usually in units of $cm^{-3} \mu m^{-1}$) together with its moments. For radiative transfer studies the important moments of the size distribution are the Liquid Water Content (L_{wc}), or mass (in units of g) of liquid water per unit volume (in units of m^{-3}), total number of drops per unit volume (in units of cm^{-3}), mean radius of the drops (in units of μm) and a measure of the width of the drop distribution. Examples of in situ measurements in liquid-water stratus are presented in Fig. 2.2 for flights during the Atlantic Stratocumulus Transition Experiment (ASTEX – June 1992; Davis et al., 1994), the First ISCCP Regional Experiment (FIRE) Intensive Field Observations (FIRE87 – 16 July 1987; Davis et al., 1996) and regional experiments in the environs of Moscow, Russia (6 February 1984). Our motivation for showing the first two sets of measurements is that they were made during two important field campaigns whose data have been widely used, while the third set of measurements was made by one of the authors, thereby allowing us to link the drop-size distributions to their moments.

The cloud liquid water content data obtained from these flights (Fig. 2.2a,b,e) are highly variable with significant changes over a range of spatial scales. Applying

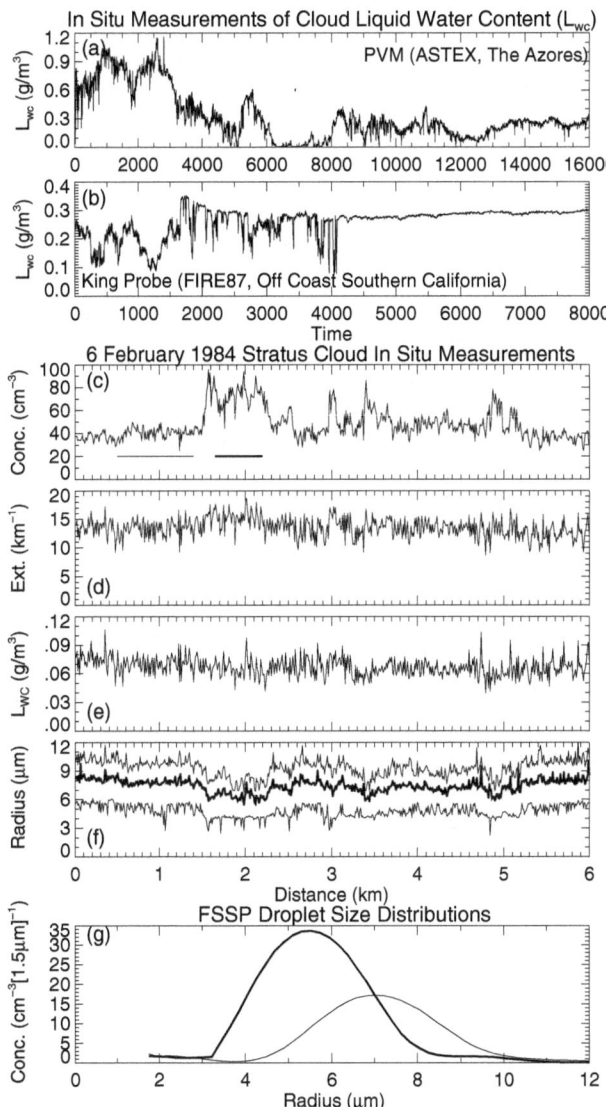

Fig. 2.2. Airborne cloud-liquid water content measurements in stratus clouds during (**a**) AS-TEX (June 1992) using the PVM-100 probe and (**b**) FIRE87 (16 July 1987) using the King probe at an altitude of 625 m. Spatial variations of (**c**) droplet number concentrations, (**d**) extinction coefficient, (**e**) liquid water content, and (**f**) droplet mean radius (*thick line*) with the 10 and 95 percentiles (*thin lines*) deduced from FSSP measurements in stratus-stratocumulus over the Moscow region on 6 February 1984 at an altitude of 1600 m and temperature of $-3°C$. (**g**) cloud drop-size distributions averaged over the two periods highlighted in panel (**c**)

Fourier analysis to them, we would most likely find that their power density versus spatial frequency follows a power-law (see Appendix). Moreover, as a result of dynamical processes (Davis et al., 1999; Jeffery, 2001), we would also most likely find that the variations in the liquid water content fields generally have higher amplitudes at the smaller scales, i.e., 8 m in Fig. 2.2a, 5 m in Fig. 2.2b and approximately 10 m in Fig. 2.2e.

For the flight in the environs of Moscow on 6 February 1984 Fig. 2.2c–f shows the spatial variations of droplet number concentration (Fig. 2.2c), extinction coefficient (Fig. 2.2d), liquid water content (Fig. 2.2e), and the mean radius of the cloud drop-size distribution (Fig. 2.2f) along with its 10 and 95 percentile radii, indicating the boundaries of the droplet size distribution inside which 85% of all droplets are contained. The two averaged droplet size distributions in Fig. 2.2g were obtained from the two periods indicated by horizontal lines in Fig. 2.2c. All of these cloud microphysical parameters show significant variations throughout the flight. While the highest spatial resolution variations are, in part, the result of the small sampling volume of the FSSP and warrant care in their interpretation, the variations at the larger spatial scales are indicative of changes in cloud properties. For example, compare the observations during the two time periods illustrated in Fig. 2.2c. During these two time periods, the liquid water content (Fig. 2.2e) does not significantly change. However, from the first to the second period the cloud drops show a marked decrease in size (Fig. 2.2f). That is, consistent with a constant liquid water content and a decrease in the cloud drop sizes, we find a dramatic increase in the cloud droplet number density (Fig. 2.2c). For fixed liquid water content, as the number of drops increases the combined cross-sectional area of all of the drops also increases. Hence, the extinction coefficient of the drops increases from the first to the second time period (Fig. 2.2d). Hence, these two regions of the cloud will have different radiative properties, with different transmissivities and reflectivities.

The liquid-water stratus clouds discussed in the context of Fig. 2.2 have some of the simplest properties of any cloud type, notwithstanding the variations in cloud-drop properties on a range of spatial scales. To demonstrate this point consider the Meteorological Service of Canada aircraft flight on 23 January 1998 over Lake Ontario in a deep frontal system consisting of liquid, ice and mixed phase clouds (Figs. 2.3 and 2.4). While the clouds below 2 km were mainly liquid, the clouds above 2 km were either glaciated or mixed-phase (Fig. 2.3a–c). Ice and liquid water content along the flight track exhibited high spatial variation, with the spatial correlation between ice and liquid water content in mixed phase clouds usually close to zero (Korolev et al., 2003). For most of the flight the airplane stayed in clouds with irregular ice particles (Fig. 2.3e). Analysis of a large data set of OAP-2D imagery has shown that irregular shape (Fig. 2.4b) is a dominating habit of cloud ice particles (Korolev et al., 2000). Dendrites (Fig. 2.4c) and needles (Fig. 2.4d) occurred in cells with a characteristic scale of a few kilometers (Fig. 2.3f–g). Several times during the flight the aircraft encountered freezing drizzle (Fig. 2.3d and Fig. 2.4a).

In spite of the fact that the OAP-2DC provides binary (i.e., black-and-white) low pixel resolution (i.e., 25 μm) imagery, Fig. 2.4 nonetheless provides a glimpse of the variety and complexity of ice particle structures. The Cloud Particle Imaging (CPI)

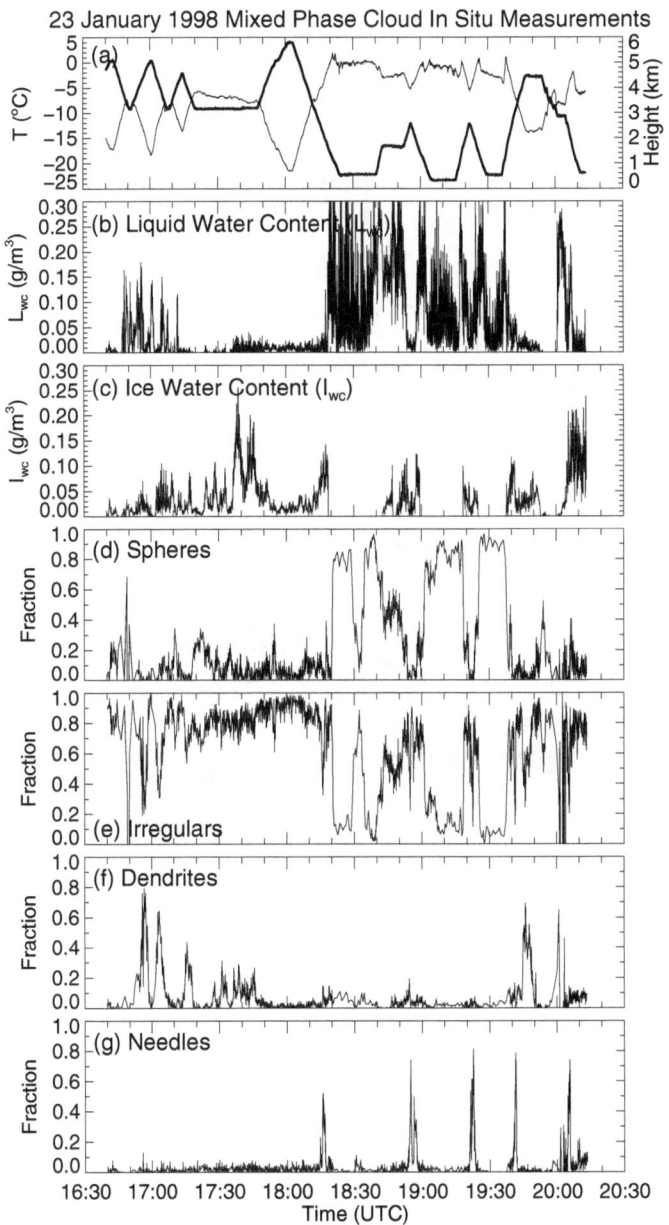

Fig. 2.3. (**a**) Flight-track altitudes (*thick line*) and temperatures (*thin line*) for the flight on 23 January 1998 in the environs of Toronto. Time series of the (**b**) liquid and (**c**) ice water contents were measured by the Nevzorov probe along the flight track. The fraction of the particles along the flight track that were (**d**) spheres, (**e**) irregularly shaped ice crystals, (**f**) dendrites and (**g**) needles were obtained from analysis of OAP-2DC measurements of the type illustrated in Fig. 2.4

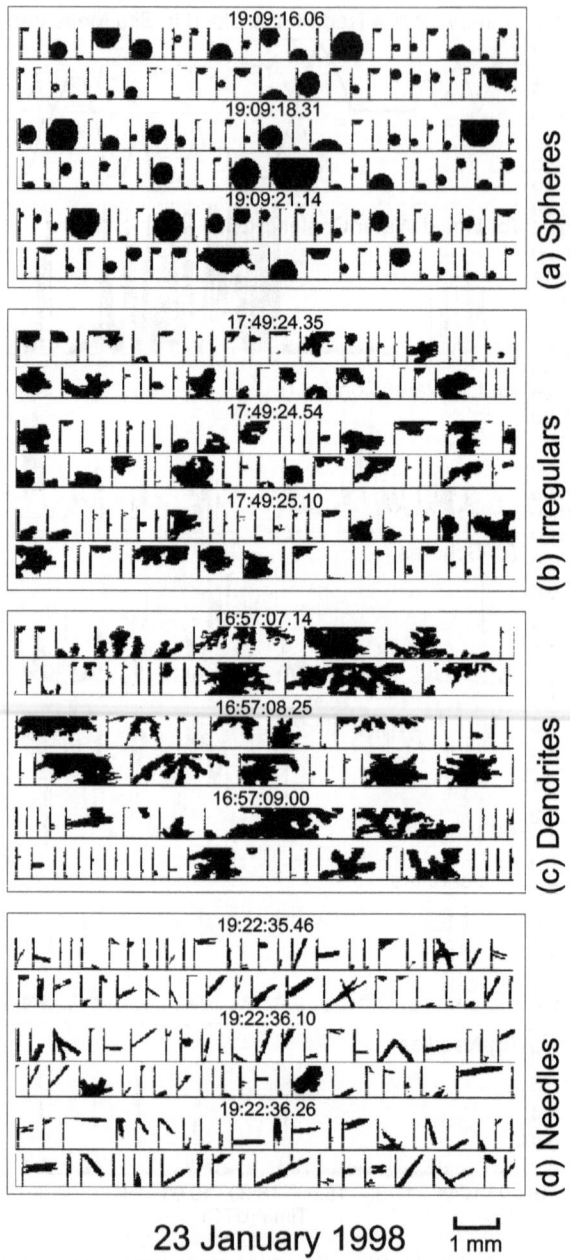

Fig. 2.4. Binary shadow images of cloud particles measured by the OAP-2DC during the flight on 23 January 1998 over Lake Ontario. The particle shapes are classified as (**a**) spheres, (**b**) irregular particles, (**c**) dendrites, i.e., planar crystals with complex branching structures and their aggregates, and (**d**) needles. The flight-track legs along which they were collected is illustrated in Fig. 2.3a

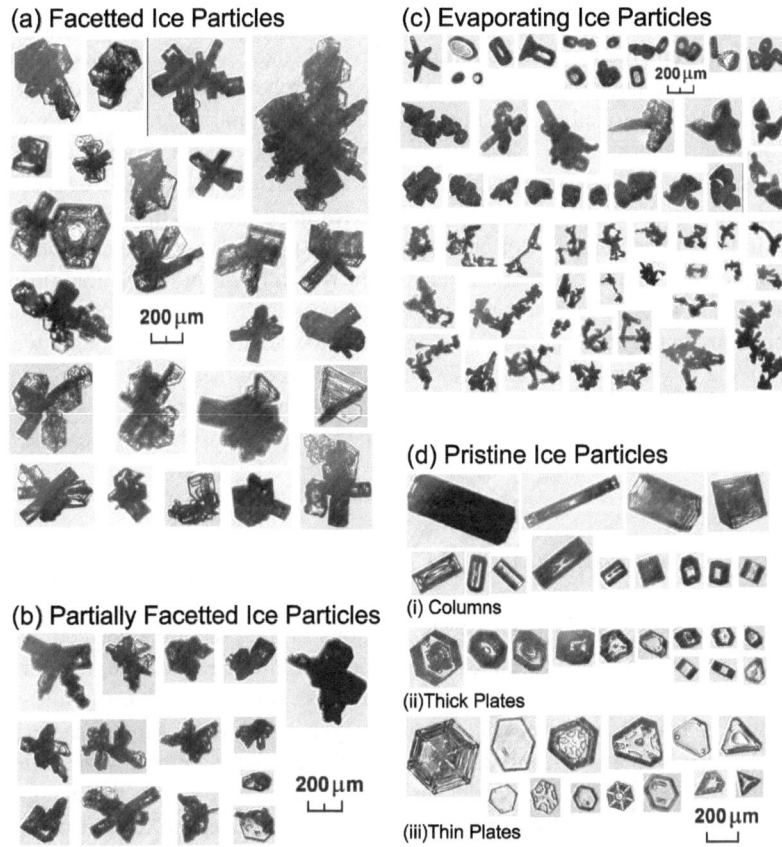

(a) Facetted Ice Particles

200 µm

(b) Partially Facetted Ice Particles

200 µm

(c) Evaporating Ice Particles

200 µm

(d) Pristine Ice Particles

(i) Columns

(ii)Thick Plates

(iii)Thin Plates

200 µm

Fig. 2.5. During the FIRE Arctic Cloud Experiment (Uttal et al., 2002), a Cloud Particle Imaging (CPI) probe was carried onboard research aircraft sampling clouds of interest. Different types of cloud ice particles imaged by the CPI, including (**a**) facetted ice particles, (**b**) partially facetted ice particles, (**c**) evaporating ice crystals, and (**d**) pristine ice particles, provide a glimpse at the complexity of ice particle shapes and sizes in arctic ice clouds

probe is capable of imaging cloud particles with a much greater number of gray scales (i.e., 256 levels) and with a much higher pixel resolution of 2.3 µm. The photographic quality CPI imagery enables us to capture internal ice-particle features and lends depth to the images, allowing one to glean information on the 3D structures of ice particles. For example, during April 1998 of the FIRE Arctic Cloud Experiment (ACE) the CPI was flown through a variety of cloud conditions, capturing with great clarity the geometric properties of a myriad ice crystals (Fig. 2.5).

The data and images illustrated in Figs. 2.2–2.5 provide a glimpse at just how variable cloud properties can be on a range of spatial scales. A detailed analysis of relations and correlation between cloud particle sizes, concentration, extinction coefficient and water content can be found in Korolev et al. (2001). In particular, the water contents associated with liquid- and ice-water particles, as well as the cloud-particle types and their fractional mixing ratios, all change in dramatic and significant ways throughout the cloud fields sampled by the probes. Since the geometric shapes of cloud particles, together with their size-dependent number densities, in a volume of cloudy air determine the radiative properties of that volume of air, knowledge of the 3D spatial distribution of liquid- and ice-water particles and their sizes is essential to understanding the impact the cloud particles have on the atmospheric radiation field. In turn, the complexity of cloud-particle mixtures in mixed-phase clouds leads to complex atmospheric radiation fields from which retrieval of detailed properties of the cloud particles themselves is nearly impossible. To begin to understand these connections between cloud particles and their associated radiation fields we must introduce a few concepts from radiative transfer theory (see Chap. 3 for a detailed introduction to radiative transfer with an emphasis on its 3D aspects), as well as some of the wavelength-dependent radiative properties of atmospheric gases and cloud particles.

2.3 From Cloud Particles to the Atmospheric Radiation Field

As the satellite- and aircraft-based observations in the previous two sections illustrate, clouds are spatially variable; as we know from our experience, they are temporally variable as well. What is not so obvious, however, is that the mid-visible radiation emanating from a cloud field to our eyes is not necessarily characteristic of the radiation at other wavelengths. For example, consider a cloud with sufficiently many particles that multiple scattering of photons at mid-visible wavelengths becomes important. Since absorption by liquid- and ice-water particles at mid-visible wavelengths is negligible, these photons can diffuse significant distances through the cloud before exiting it. Therefore, mid-visible radiances from clouds recorded by ground- and satellite-based sensors actually contain photons that have migrated into the fields of view of the sensors from regions not directly imaged by the sensors. At wavelengths for which particles are strongly absorbing and solar radiation is dominant, most of the radiance reflected to a satellite-based sensor from an optically thick cloud has only been scattered once by the cloud particles and originates primarily from those cloud particles towards the boundary of the cloud that is closest to the sensor. At these wavelengths the radiance leaving the base of an optically thick cloud towards a ground-based instrument would be quite small. At longer wavelengths, in the microwave region, most clouds are not opaque and the radiance emanating from a cloud is composed of photons generated by the cloud together with photons that are incident on the cloud and have been transmitted through it. If our eyes were sensitive to microwave radiation or radiation at wavelengths for which cloud absorption is important, the appearance of our world would be markedly different. This is not necessarily "bad" in any sense, as these differences provide us information. For example, Chap. 13 demonstrates how radiances at two or more closely spaced wavelengths, with absorption significant at some of the wavelengths and negligible at others, can be used to infer 3D cloud structure information.

Mapping time-varying, 3D cloud structures to their spectrally-dependent radiation fields requires quantification of the physical properties of the particles that compose the cloud. Moreover, atmospheric heating rates and upwelling top-of-atmosphere, as well as downwelling surface, irradiances and radiances depend upon the location of clouds within the molecular atmosphere in which they are embedded, as many of the molecules in the atmosphere have their own wavelength-dependent radiative properties that interact with cloud-leaving radiation. We have the situation of 3D cloud structures embedded in an essentially 1D vertically-varying molecular atmosphere, with the radiative properties of the clouds correlated with the molecular radiative properties over some wavelength intervals but not over others. The result is an intricate atmospheric radiation field, whereby information on 3D cloud structures can be extracted from it at some wavelengths but not at other wavelengths.

To illustrate the coupling of the atmospheric radiation field to 3D cloud structures embedded in a vertically-varying molecular atmosphere, we will make use of ground- and satellite-based measurements from the DOE ARM program and NASA EOS Terra satellite, respectively. But to understand measurements from these systems, as well as the reasons that the atmospheric radiation field depends so strongly on wavelength, we must have an understanding of some of the most basic concepts of radiative transfer. So, in the next few sections we present, in a relatively straightforward framework, some key concepts of atmospheric radiation and we subsequently use these concepts to illustrate the spectrally- and spatially-dependent nature of the atmospheric radiation field and its link to the underlying clouds. A much more complete and rigorous discussion of atmospheric radiative transfer theory follows in Chap. 3 while detailed descriptions of computational models for the transport of radiation through 3D cloud fields can be found in Chaps. 4 and 5. Chapters 6 and 10 present models that target specifically domain-average radiative fluxes.

2.3.1 Basic Elements of Atmospheric Radiative Transfer

The amount of electromagnetic radiation penetrating all regions of the atmosphere, impinging on the surface and leaving the Earth system to space is a function of the 3D spatial distribution of matter together with the radiative properties of that matter. As the radiative properties of the matter across the surface and throughout the atmosphere change in time, so too does the distribution of electromagnetic radiation throughout the Earth system, leading to changing surface and atmospheric heating rates and to a rich set of downwelling radiances at the surface and upwelling radiances to space. Given knowledge of the general radiative properties of matter, the radiances leaving the atmosphere to the surface and space can, in turn, be used to retrieve information about the matter in the atmosphere and surface that leads to these radiances in the first place.

The spatial distribution of liquid and ice water in cloudy atmospheres is complicated, even for what one might consider to be the simplest of cloud types (e.g., Sect. 2.1). Considering the radiative properties of gas molecules in conjunction with the properties of water particles, the situation is even more complicated as the wavelength dependence of the radiative properties of these two types of matter are correlated at some wavelengths and not others, leading to distributions of electro-magnetic radiation throughout the atmosphere that are a function of wavelength. Consequently, estimating heating rates through any, even the simplest, cloudy atmosphere is not straightforward, as readers will learn as they delve into chapters subsequent to this one.

To understand the observational data that we will present one must have a rudimentary understanding of radiance and the interaction of radiation with matter. To this end we start with a definition of radiance and then move on to a discussion of particle cross-sections and atmospheric transmission. We then briefly consider particle scattering phase functions and particle emission of radiation. While the concepts of radiance, cross-sections, transmission, scattering phase functions and emission will be described in full detail in Chap. 3, we included discussions of them here to facilitate understanding of the observational data that we present in Sect. 2.6.

After introducing the key elements of radiative transfer needed for our purposes, we consider the radiative properties of molecules, isolated water spheres and distributions of liquid-water spheres in order to provide insight into the basic radiative properties of the atmosphere. Considering the two primary sources (i.e., sun and Earth) of electromagnetic radiation in the Earth system, together with the wavelength dependence of atmospheric transmission, we are led naturally to separation of the electromagnetic spectrum into the shortwave, longwave and microwave regions. With knowledge of the importance of these spectral regions, we present and describe some of the state-of-the-art satellite- and ground-based measurements currently available of cloudy atmospheres over a variety of locations. The inherently 3D nature of clouds and their associated radiation fields will emerge as the most salient feature of these observations.

2.3.2 Radiance

A straightforward approach to illustrating radiance is contained in Fig. 2.6, where we partition the flow of electromagnetic radiation, i.e., photons, through the Earth atmosphere into those wavelengths for which the radiation originates primarily from the sun (shortwave radiation; Fig. 2.6a) and those wavelengths for which the radiation originates primarily from the Earth (longwave radiation; Fig. 2.6b). The concept of radiance is illustrated by the dashed-line, slightly expanding tubes in the figures that originate from matter and denote a directed beam of electromagentic radiation leaving the matter. To a good enough approximation for our present purposes, the magnitude of the radiance at area A_{in} for long, narrow tubes originating from A_{in} is

$$I_\lambda = \frac{E_{\Delta\lambda}}{(\Delta t)(A_{in})(A_{out}/r^2)(\Delta\lambda)} \,, \tag{2.1}$$

where r is the tube length, A_{in} and A_{out} are the areas through which the radiation enters and exits the tube, respectively, and $E_{\Delta\lambda}$ is the amount of energy between wavelengths $\Delta\lambda = \lambda_{max} - \lambda_{min}$ that flows into the tube in the time interval Δt. The quantity A_{out}/r^2 in (2.1) is the solid angle $\Delta\Omega$ subtended by the exit of the

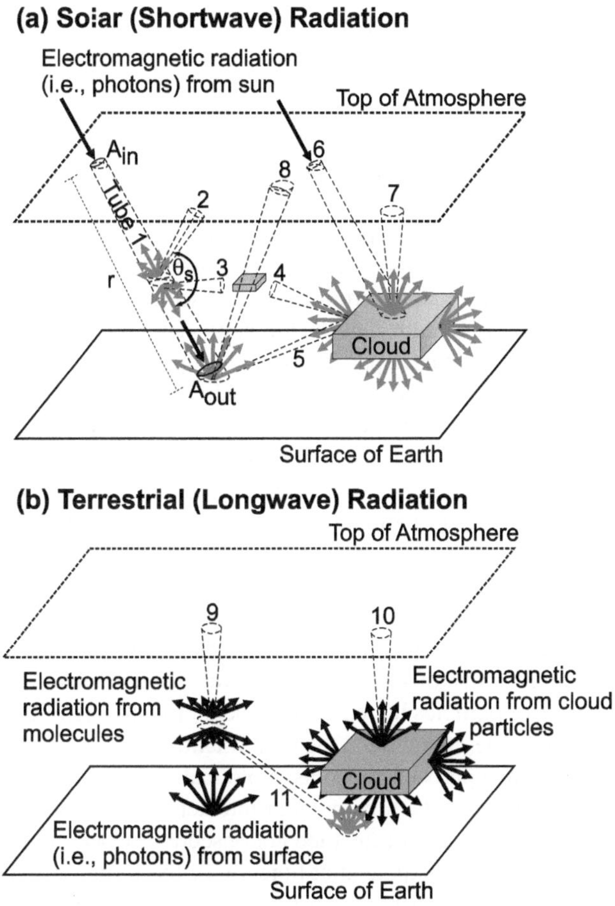

(a) Solar (Shortwave) Radiation

(b) Terrestrial (Longwave) Radiation

Fig. 2.6. Cartoon illustration of radiances passing through the atmosphere of the Earth for (**a**) radiation originating from the sun (i.e., shortwave radiation) and (**b**) radiation originating from the Earth (i.e., longwave radiation). The *dashed-line* tubes represent radiances originating from a variety of sources and propagating through atmospheric molecules and cloud particles. The black arrows represent radiances from the source, while the *gray* arrows represent scattered radiation. The areas associated with tube 1 on the left side of (a) are used to provide an approximate definition for radiance. The angle θ_s in (a) is the scattering angle, while its associated azimuth angle ϕ_s, which represents the angle about the central axis of tube 1 at which the photon is scattered, is not drawn

tube from its entrance and is generally given in units of steradians (sr). Therefore, the dimensions for radiance are energy per unit time per unit area per unit solid angle per unit wavelength interval ($J\,s^{-1}\,m^{-2}\,sr^{-1}\,m^{-1}$ in the SI system of units). For tube 1 in Fig. 2.6a the entrance and exit areas are illustrated by the solid circles, while the magnitude and direction of the solar radiance associated with this tube are depicted by the black solid arrow entering the tube from the top of the atmosphere.

2.3.3 Particle Cross-Sections and Transmission

The radiance does not change along the beam represented by a tube unless the energy associated with it interacts with matter. In this case the radiance is decremented by the fraction of $E_{\Delta\lambda}$ that is either absorbed or scattered by the matter along the extent of the tube. To quantify the interaction of radiation with matter the concept of cross-section is attached to the constituents of the matter (i.e., atmospheric gas molecules, aerosol particles, liquid-water drops and ice particles). Since matter may either absorb or scatter radiation differently as the wavelength of the radiation incident upon it changes, the matter has both absorption $s_{a,\lambda}$ and scattering $s_{s,\lambda}$ cross-sections that are a function of the wavelength λ of the radiation. The total, i.e., extinction, cross-section $s_{e,\lambda}$ associated with each particle type is simply the sum $s_{a,\lambda} + s_{s,\lambda}$.

In Fig. 2.7a–f we have taken a cross-sectional view of the solar radiance tube in Fig. 2.6a, where we are located at the exit area A_{out} and we are looking up into the tube to the entrance area A_{in}. In these figures the extinction cross-section for each particle is represented by a single black dot and all of the black dots in any one figure represent the total number of particles in the tube through which the radiation must pass. Note that for less than about 5600 particles in the tube (Fig. 2.7a) the fraction of the tube cross-sectional area A_{in} obstructed with particle extinction cross-sections is linear in the number of particles (Fig. 2.7g). That is, each time a particle is added to the tube the probability of its cross-section partially overlapping another particle cross-section along the line of sight through the tube is negligible.

As the number of particles increases, they begin to have significant probabilities of overlap along the line of sight and the total fraction of the tube obstructed by extinction cross-sections approaches one asymptotically (Fig. 2.7b–f,g). The amount of radiation that penetrates the tube without interacting with matter, i.e., the transmission, is simply the fraction of the tube not obscured by particle extinction cross-sections. As it turns out, an exponential function of the form $\exp(-\sigma_{e,\lambda}r)$ is a reasonable model for the transmission, where the extinction coefficient $\sigma_{e,\lambda} = ns_{e,\lambda}$ and n is the average number of particles per unit volume along the path. Note that in Fig. 2.7g the transmission computed numerically and the exponential function model of it diverge slightly as the particle number increases as a result of approximations in the numerical simulations. For particles with non-zero cross-sections the transmission goes to zero before the number of particles goes to infinity, but for the ideal exponential law the transmission goes to zero only as the number of particles goes to infinity.

The model above for the transmission is sufficient for treating scattering by air molecules, as well as scattering and absorption by homogeneous collections of cloud particles. However, gas molecule absorption cross-sections $s_{a,\lambda}$ are a function of pressure, temperature and gas amount, implying a path dependence in their transmission:

$$T_{\mathrm{dir},a,\lambda} = \exp\left(-\int_0^r \sigma_{a,\lambda}(r')\,dr'\right). \qquad (2.2)$$

Moreover, if gas molecules are in an excited state E_e with the possibility of transition to a lower state E_g with the release of a photon of energy $E_e - E_g = h\,c/\lambda$,

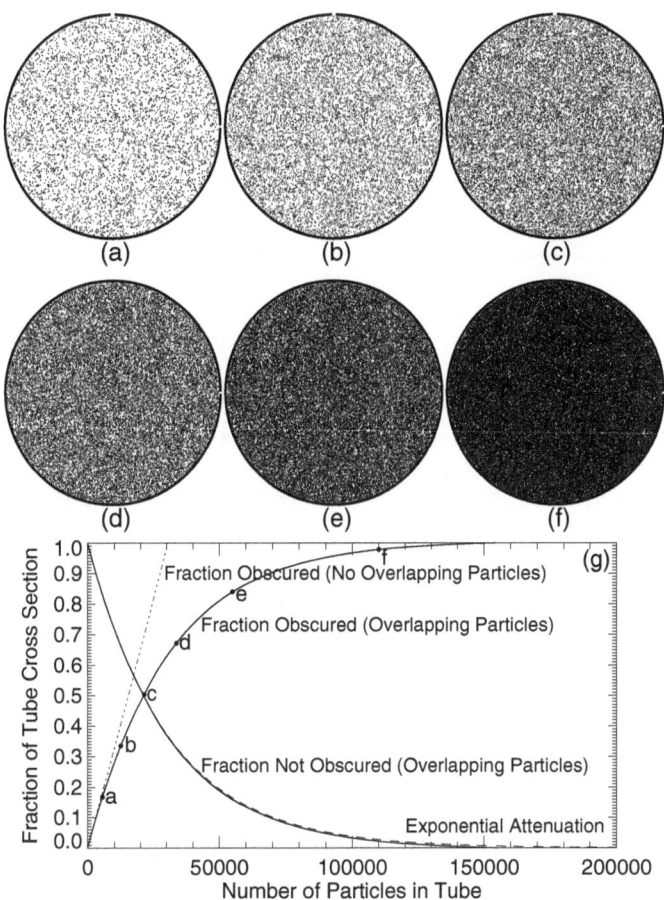

Fig. 2.7. The big circles in (**a**)–(**f**) represent cross-sectional areas across a radiance tube, while each black dot (not drawn to scale) within a big circle represents the extinction cross-section of a particle independently and randomly located within the radiance tube between the entrance area A_{in} and the exit area A_{out}. For the results in this figure the ratio of the diameter of a circle based on the particle cross-section to the diameter of the tube was 0.01. The number of particles for each of the circles in (**a**)–(**f**), along with the fraction of the radiance tube, i.e., the fraction of the tube cross-sectional area, obscured by particle radiometric cross-sections is indicated in (**g**). The fraction of the cross-sectional area blocked without particle overlap is indicated by the *dotted line*, while the fraction of the cross-sectional area blocked with overlapping particles is indicated by the *upper solid line*. An exponential function (*dashed line*) is an accurate model for the fraction of the cross-sectional area of the tube not blocked by particle extinction cross-sections when particles are allowed to overlap. As a result, the exponential function (*dashed line*) lies almost on top of the simulated cross-sectional area not blocked by overlapping particles (*lower solid line*)

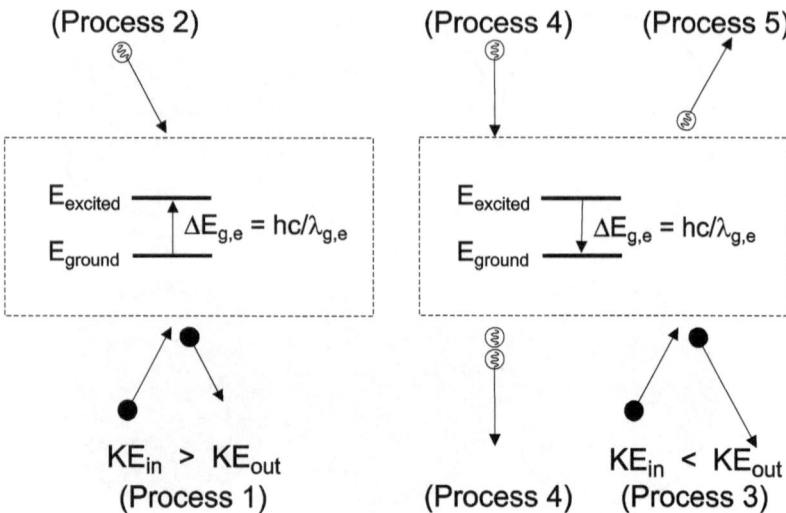

Fig. 2.8. Cartoon illustration of the processes by which radiation and matter interact. Particles (*black circles*) can transfer energy to (Process 1) and from (Process 3) internal energy states of other particles (*dashed boxes*) through collisions, while particles can absorb photons (*white circles*), or radiation (Process 2), emit radiation on their own accord (Process 5) and be stimulated by radiation to emit radiation (Process 4)

where c is the speed of light (in units of m s^{-1}) and h is Planck's constant (in units of J s), the passage of photons of wavelength λ through the gas can actually stimulate excited gas molecules to de-excite with the emission of radiation of wavelength λ (cf. Process 4; Fig. 2.8). Stimulated emission prevents the radiance from decreasing at the rate predicted by (2.2). To account for this nuance of attenuation by gas molecules we define an "effective" absorption cross-section $s'_{a,\lambda}$ ($< s_{a,\lambda}$) that accounts for stimulated emission. In terms of $s'_{a,\lambda}$ the absorption coefficient in (2.2) becomes $\sigma_{a,\lambda} = n s'_{a,\lambda}$.

2.3.4 Particle Scattering Phase Functions

Given that a photon has a scattering interaction with matter, the scattering phase function $p_\lambda(\theta_s, \phi_s)$ describes the probability of the photon scattering into a particular direction (θ_s, ϕ_s), where (θ_s, ϕ_s) represent the angle of photon travel after the scattering event *relative* to the direction of photon travel before the scattering event (cf. Fig. 2.6, Tube 1). The scattering phase function provides the mechanism for quantifying how the loss of radiance in one tube provides an increment in the radiance associated with other tubes.

The scattering phase function is defined as the ratio

$$p_\lambda(\theta_s, \phi_s) = \frac{\Delta P_{s,\Delta\lambda}(\theta_s, \phi_s)/\Delta\Omega}{P_{s,\Delta\lambda}(\text{Total})/4\pi} , \qquad (2.3)$$

where $P_{s,\Delta\lambda}(\text{Total})$ is the total power across the wavelength interval $\Delta\lambda$ centered on λ that is scattered into all possible directions (hence 4π sr), and $\Delta P_{s,\Delta\lambda}(\theta_s, \phi_s)$ is that part of the total scattered power that is scattered into the small solid angle $\Delta\Omega$ centered on the direction (θ_s, ϕ_s). Note that a value of $p_\lambda(\theta_s, \phi_s) \equiv 1$ implies that the scattering is isotropic-like in the direction (θ_s, ϕ_s), while a value of $p_\lambda(\theta_s, \phi_s) = 0$ implies no scattered power in the direction (θ_s, ϕ_s). There is no theoretical upper bound on $p_\lambda(\theta_s, \phi_s)$ since the phase function can go to infinity if all of the scattered power is directed into some infinitesimally small solid angle in some direction.

Consider tube 8 in the scattering atmosphere of Fig. 2.6a. The radiance leaving its exit area to space is now dependent on the attenuated solar direct beam scattering from the cloud particles within tube 8 (similar to the solar radiance in tube 6 scattering within the bigger cloud to the right with some of the total power scattered into tube 7), the attenuated solar direct beam (tube 1) scattering from the surface into tube 8, scattering from air molecules contained in tube 8 that are both below and above the cloud particles in tube 8 (similar to the contributions to tube 2 from tube 1) and scattering into tube 8 from clear-air (tube 3) and cloud (tube 4) scattering in the vicinity of tube 8. The situation is further complicated by such processes as scattering from the cloud in tube 8 to the cloud to the right of it and back. Horizontal transport of radiation in a cloudy scene depends on the arrangement of cloud particles in the scene, their number densities, the ratio of the particle absorption to extinction coefficients and the particle phase functions.

As we illustrated for shortwave radiation, with the source of photons outside of the atmosphere (Fig. 2.6a), scattering processes lead to diverse exchanges of radiation between objects in the atmosphere and on the surface, especially when the objects are inhomogeneous across the scene. For longwave radiation (Fig. 2.6b) the situation can be more complicated because each of the objects in the scene becomes a source of radiation. We must now account for the emission from the inhomogeneous objects across the scene as well as the subsequent scattering of this radiation from these same objects (see Chap. 10).

2.3.5 Particle Emission

Stimulated emission of electromagnetic radiation is but one means whereby matter can exchange energy with its surroundings through radiation. The other processes for such exchange are illustrated in Fig. 2.8. A particle, represented by the black circle in the figure, can collide with matter either transfering some of its kinetic energy to internal energy of the matter (Process 1) or removing internal energy from the matter (Process 3) in the form of particle kinetic energy. Photons, represented by the open circles, can be absorbed by the matter (Process 2), spontaneously emitted by the matter (Process 5) or produced by stimulated emission (Process 4) as we have just described.

Taking into account both the probabilities of excited states in matter being occupied and the number of states in matter with equivalent energies, we find that emission of radiation from matter goes to zero asymptotically at both long and small wavelengths with a peak in the emission that is characteristic of the temperature of the matter. Moreover, as Fig. 2.8 illustrates, absorption (Process 2) and emission (Processes 4 and 5) of photons with wavelength $\lambda_{g,e}$ from matter are related processes insofar as the matter must have an energy level transition $\Delta E_{g,e}$ corresponding to the wavelength $\lambda_{g,e}$. Even though matter might contain energy level transitions corresponding to photon energies of $hc/\lambda_{g,e}$, the matter will neither absorb nor emit photons of this wavelength if it has no way of interacting with the photons in the radiation field. Interactions of matter with electromagnetic radiation are mediated through electric and magnetic fields associated with the matter and a set of rules that guarantee conservation of energy, momentum, etc., during the interaction process.

The results of these interactions and rules is that if the transmission $T_{dir,a,\lambda}$ through an absorbing medium with temperature T is given by (2.2) in some particular direction, the absorptivity A_λ of the medium along this direction is

$$A_\lambda = 1 - \exp\left(-\int_0^r \sigma_{a,\lambda}(r')dr'\right) \tag{2.4}$$

and emission of radiation from the medium along this direction is given by $A_\lambda B_\lambda(T)$, assuming constant temperature T. The quantity

$$B_\lambda(T) = \left(\frac{2hc^2}{\lambda^5}\right)\left(\frac{1}{\exp[hc/(\lambda k_b T)] - 1}\right) \tag{2.5}$$

is the Planck function and it has the same dimensions as radiance. Since the absorptivity A_λ is dimensionless, the emitted radiation $A_\lambda B_\lambda(T)$ is a radiance. Note that for any object for which the absorptivity $A_\lambda = 1$ the radiance emitted by the object is simply $B_\lambda(T)$. Examples of the Planck function for sun- and Earth-like temperatures of 5917 K and 255 K, respectively, are illustrated in Fig. 2.9.

With only absorption and emission processes in an isothermal medium, solving for the radiance emanating from any tube is straightforward as all one must know

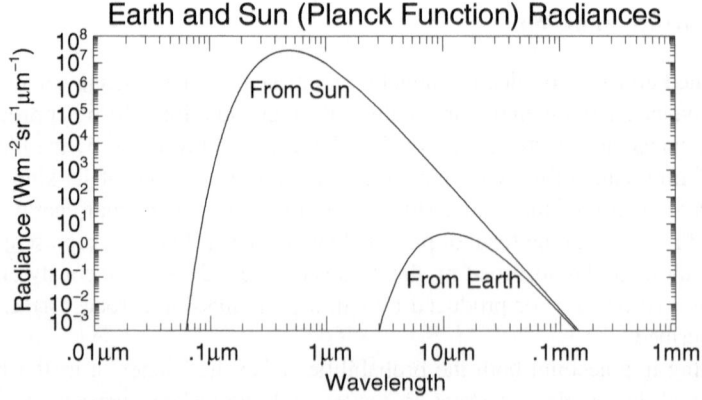

Fig. 2.9. Radiances that result from treating the sun and Earth as blackbody (i.e., $A_\lambda = 1$) radiators respectively with temperatures of 5917 K and 255 K

is the radiance at the entrance area A_{in} of the tube, the variation of the absorption coefficient $\sigma_{a,\lambda}$ in (2.2) and (2.4) along the tube and the value of $B_\lambda(T)$. If the temperature along the tube varies, solving for the radiance is only slightly more difficult as one must now allow $B_\lambda(T)$ to vary along the path (see Chaps. 3 and 10). Only when scattering processes become important does computing the radiance leaving a tube become truly difficult, as now the radiance can have contributions from radiance beams in other directions that pass through the tube and interact with the particles in it.

2.4 Radiative Properties of Atmospheric Particles

In terms of the radiometric concepts introduced in Sect. 2.3, we find that atmospheric particles have wavelength-dependencies in their scattering and absorption cross-sections, their scattering phase functions and their emission. The three types of particles important to atmospheric radiative transfer are gases, aerosols and cloud particles. Since the impacts of aerosols on atmospheric radiative transfer fall somewhere between those from molecules and cloud particles, we focus our attention on the radiative properties of gas molecules and cloud particles. In particular, we illustrate some of the properties of their scattering and absorption cross-sections, as well as their scattering phase functions, as a preamble to interpreting the observational data that follows.

2.4.1 Atmospheric Molecules

In Fig. 2.10a we present the scattering cross-sections of air molecules, with the properties of molecular nitrogen and oxygen contributing the most to the behavior of the scattering cross-section versus wavelength. The effective absorption cross-sections of the seven most important absorbing atmospheric air molecules, computed at standard pressure and temperature of 1013 mb and 288 K, respectively, follow in Fig. 2.10b–h. To generate these effective absorption cross-sections we used the Line-By-Line Radiative Transfer Model (LBLRTM) developed by Tony Clough and his colleagues (e.g., Clough et al., 1992; Clough and Iacono, 1995). We present the results in Fig. 2.10b–h as effective absorption cross-sections, i.e., absorption coefficient divided by molecule number density n, in order to emphasize the strength of the interaction per molecule.

As Fig. 2.10a illustrates, scattering cross-sections of air molecules are negligible for wavelengths greater than 1 μm and become increasingly important as the wavelength of the radiation drops from 1.0 μm to 0.3 μm. Note that the variation of the air molecule scattering cross-section with wavelength is smooth. Ozone has large absorption cross-sections at wavelengths less than 0.3 μm (Fig. 2.10d), while the absorption cross-sections of the seven major absorbing molecules are relatively small in the visible region of the spectrum from 0.3–0.7 μm, apart from weak absorption by ozone and water vapor. All seven of the major absorbing gases, except for carbon dioxide, have significant absorption cross-sections across the wavelength range

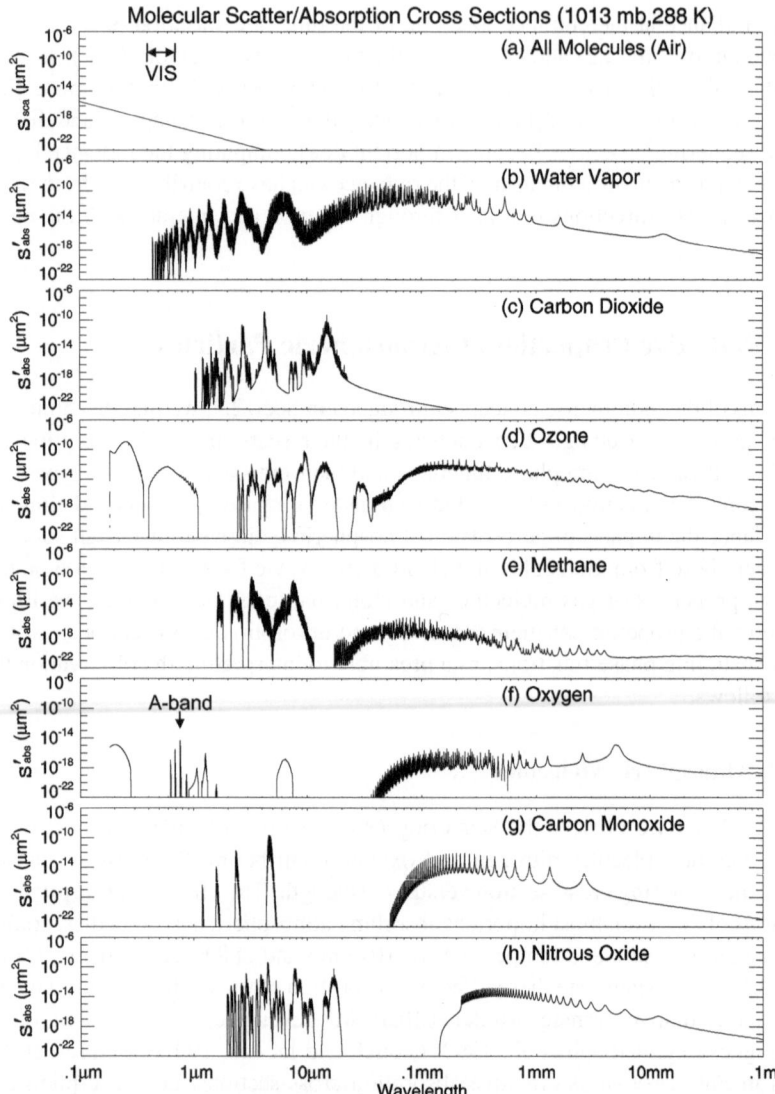

Fig. 2.10. (a) The average scattering cross-section for all atmospheric molecules and (b) – (h) the effective absorption cross-sections for the seven major absorbing constituents in the atmosphere of Earth at a standard temperature of 288 K and a standard pressure of 1013 mb. The effective absorption cross-sections are obtained by dividing the absorption coefficient by the number density of the relevant molecule. Note that these panels illustrate radiometric cross-sections and have nothing to do with the physical cross-sections of the molecules. Moreover, the cross-sections alone do not indicate their importance to radiative processes; one must multiply them by the relevant molecule concentration and photon path length through the atmosphere to obtain their optical thicknesses and transmissivities, which are the quantities of interest. The O_2 A-band, used extensively in Chap. 13, is highlighted

from approximately 50 μm to 1 mm. The absorption cross-sections of all seven gases generally decrease with increasing wavelength beyond wavelengths of about 1 mm. In the near and thermal infrared regions of the spectrum, from 0.7 μm to 20 μm, the absorption cross-sections of the gases are variable with strong wavelength-dependencies that change from one molecule to the next. Note that the absorption cross-sections of a molecule do not indicate their importance to radiative processes. One must multiply them by the relevant molecule concentration and photon path-length through the atmosphere to obtain their corresponding optical thicknesses and transmissivities, which are the quantities of interest.

To this end we used LBLRTM to compute the total atmospheric transmission through a mid-latitude summer atmosphere (Fig. 2.11a). These calculcations included computation of the absorption $\sigma_{a,\lambda} = n s'_{a,\lambda}$ and scattering $\sigma_{s,\lambda} = n s_{s,\lambda}$ coefficients for each molecular species in each atmospheric layer, their associated absorption and scattering optical thicknesses $\Delta\tau_{a,\lambda} = \sigma_{a,\lambda} \Delta z$ and $\Delta\tau_{s,\lambda} = \sigma_{s,\lambda} \Delta z$, where Δz is layer thickness, the total optical thickness per layer, which is the sum of all absorption and scattering optical thicknesses in each layer (i.e., $\Delta\tau_{e,\lambda} = \sum_{\text{species}}[\Delta\tau_{a,\lambda} + \Delta\tau_{s,\lambda}]$), and finally the total optical thickness of the entire atmosphere (i.e., $\tau_{e,\lambda} = \sum_{\text{layers}} \Delta\tau_{e,\lambda}$). Inserting the atmospheric optical thickness into

$$T_{\text{dir},a,\lambda} = \exp(-\tau_{e,\lambda}) , \qquad (2.6)$$

we arrive at the atmospheric transmissions depicted in Fig. 2.11a. As the wavelength of the radiation decreases from 0.1 m to 1 mm, absorption by water vapor and oxygen increases. At wavelengths shorter than approximately 1 mm, water vapor absorption is so strong that the atmospheric transmission is negligible until wavelengths of about 20 μm are reached. From 20 μm down to approximately 0.9 μm the atmospheric transmission increases in distinct regions, starting from the "window region" from 8–12 μm to 3.5–4.0 μm to 2.0–2.5 μm to a peak transmission close to one between strong water vapor absorption between 0.9 μm and 1.35 μm. Shortwave of 0.9 μm, there is absorption by oxygen in well-defined spectral regions (cf. Chap. 13) and weak, but continuous, absorption by ozone in the range 0.4–0.7 μm. Below 0.3 μm, absorption by ozone, oxygen and nitrogen molecules is sufficiently strong to make the atmospheric transmission negligible in this part of the electromagnetic spectrum. The smooth decrease in the transmissivity from 0.5 μm to 0.3 μm is a result of scattering by air molecules.

For remote sensing applications there are several important features about the transmission results in Fig. 2.11a. First, the molecular atmosphere becomes transparent as the wavelength of the radiation increases beyond approximately 1 mm. Second, for wavelengths from approximately 20 μm to 1 mm, the molecular atmosphere is opaque. Finally, for wavelengths less than 20 μm atmospheric transmissivity approaches one in distinct bands. To obtain information about a molecular constituent in the atmosphere using electromagnetic radiation, the radiation must interact with the constituent in some manner. Moreover, the most information in the radiation field about the constituent is contained at wavelengths where constituent absorption increases from weak to strong amounts. Alternatively, if one is trying to assess

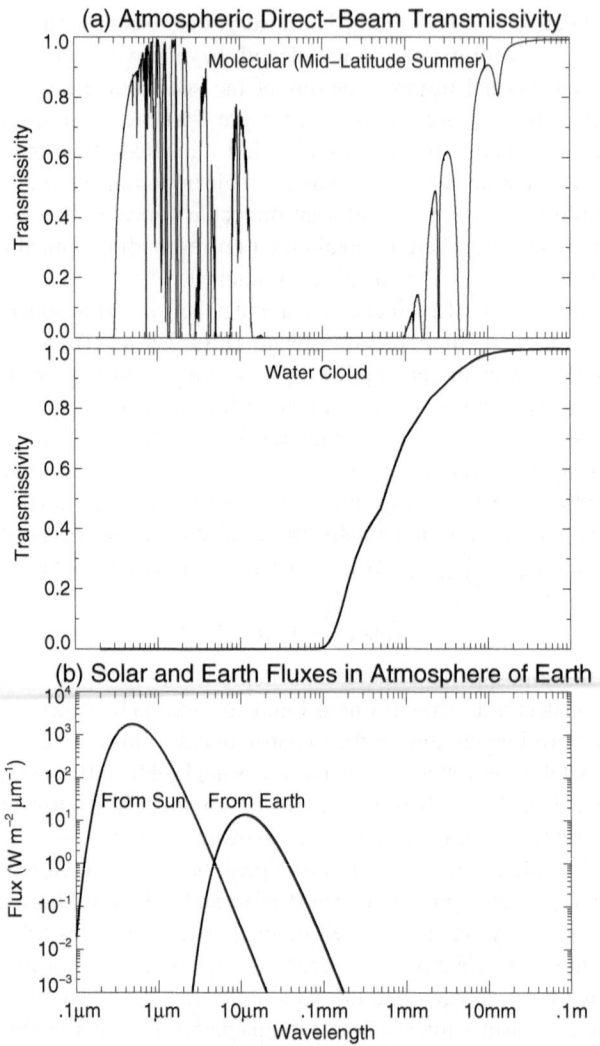

Fig. 2.11. (a) Transmissivity as a function of the wavelength of radiation through a cloud-free, or molecular, atmosphere and a 500 m thick stratus cloud with 175 spherical liquid-water drops per cubic centimeter that are lognormally distributed in size with the parameters given in Fig. 2.12. Molecular transmissivities for wavelengths shorter than 0.1 mm have been smoothed for presentation purposes. (b) The fluxes per unit wavelength interval incident on the Earth atmosphere that result by treating the sun and Earth as blackbody radiators with temperatures of 5917 K and 255 K, respectively. Note that clouds are opaque in the shortwave and longwave spectral regions, where the fluxes have their largest magnitudes, while molecules are relatively transparent in several wavelength intervals in the shortwave, longwave and microwave regions of the electromagnetic spectrum

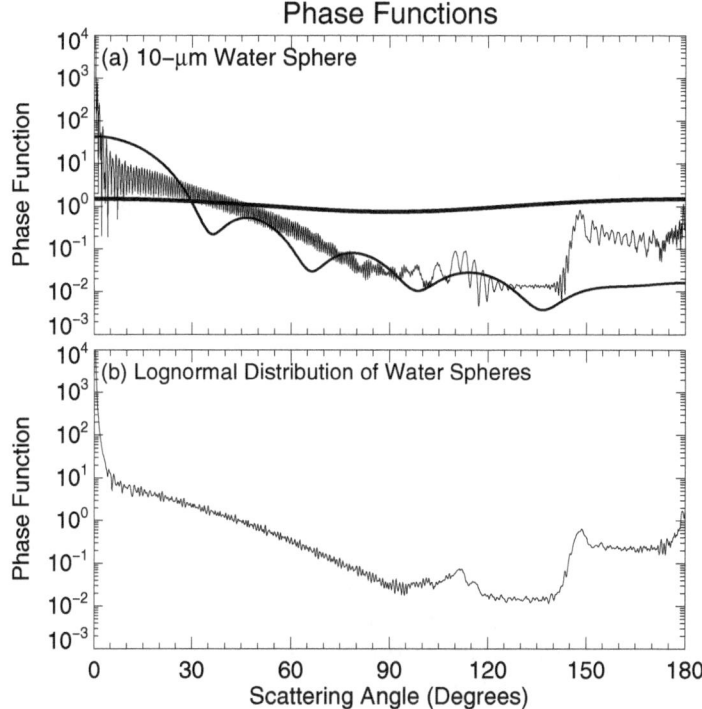

Fig. 2.12. (a) Scattering phase function for a $10\,\mu m$ radius liquid-water sphere for radiation with wavelengths of $0.2\,\mu m$ (*thin line*), $10\,\mu m$ (*medium line*) and 1 mm (*thick line*). (b) Scattering phase function for a lognormal distribution of drops with an effective radius $r_e = 7.5\,\mu m$ and a logarithmic width $\sigma_{log} = 0.35$; the wavelength of the radiation is $0.2\,\mu m$

the presence or absence of some constituent, such as cloud particles, embedded in a molecular atmosphere using electromagnetic radiation, one would use radiation at "window" wavelengths for which the transmission through the molecular atmosphere approaches one while interactions of the radiation with the constituent are yet strong.

Scattering by molecules is approximately the same in all directions (thick line, Fig. 2.12a). That is, molecular scattering is quasi-isotropic with only double the amount of scattering in the forward and backward directions as compared to scattering at $90°$.

2.4.2 Water Particles

The situation for cloud scattering is much different. The scattering phase functions of a sphere with radius $10\,\mu m$ at three different wavelengths are represented by the curves in Fig. 2.12. While all of the phase functions in Fig. 2.12a are strictly for scattering of radiation of different wavelengths from a $10\,\mu m$ radius sphere and were generated by the spherical particle scattering code of Toon and Ackerman (1981), the thick curve for scattering of 1 mm wavelength radiation from the 10 μm radius liquid-water sphere, i.e., scattering of radiation with a wavelength that is large compared to the particle size, well-represents the scattering phase function of air molecules for all wavelengths of radiation of importance to atmospheric radiative transfer. In terms of the directionality of scattering by spherical water drops, for particles much larger than the wavelength of the incident radiation the scattering is strongly forward peaked. As the particle size decreases relative to the wavelength of the radiation, the magnitude of the forward peak decreases until the scattering is equally likely in the forward and backward directions.

Using electromagnetic calculations of scattering and absorption by water spheres (i.e., "Mie" calculations; Bohren and Huffman, 1983; Toon and Ackerman, 1981) with radii of 0.1 μm, 10 μm and 1.0 mm, we obtained the results illustrated in Figs. 2.13 and 2.14. One salient feature in the figures is the differences in the absorption cross-sections of liquid- and ice-water spheres. For example, the absorption cross-sections of 10 μm radius ice particles dip between wavelengths of 20–30 μm, while the absorption cross-sections of 1 mm radius ice particles decrease significantly from a wavelength of 0.1 mm to a wavelength of 10 mm. These decreases in the absorption cross-sections of ice particles are not present in the absorption cross-sections of liquid-water spheres. As a result, information is present in the radiances at these wavelengths to distinguish liquid-water drops from ice-water particles.

The spheres of radii 0.1 μm, 10 μm and 1.0 mm are on the order of aerosol, liquid-cloud and precipitation-particle sizes, respectively. While the scattering cross-sections of 0.1 μm water particles exhibit the same tendencies across the visible region of the spectrum as for molecular scattering, the 10 μm and 1 mm radius spheres have scattering cross-sections across both the visible and infrared regions of the spectrum that are approximately twice the geometric cross-sectional area of the spheres. As a result, cloud drop densities, even for the smallest of cloud elements, are typically large enough to produce large optical depths across the visible and infrared regions of the electromagnetic spectrum. For example, a continental-type stratus cloud with a density of $175\,\mathrm{cm}^{-3}$ and a vertical extent of 500 m has transmissivities approaching 0 for all wavelengths less than 0.1 mm (Fig. 2.11a). While precipitation particles have large cross-sections, under most environmental conditions there are not that many of them in a vertical column and their resulting optical thicknesses are generally low across the electromagnetic spectrum. However, for long horizontal paths through the atmosphere the total path optical thickness of precipitating drops can become quite large for wavelengths less than approximately 10–30 mm.

For example, consider the raindrop size distribution for a 17 mm h^{-1} rain shower near Hilo, Hawaii, that Pruppacher and Klett (1997) illustrate in their Fig. 2.28. Inspecting this figure, we find there are approximately 8 raindrops per cubic meter with radii between 0.8 mm and 1.2 mm. The extinction coefficient for wavelengths less than 1 mm that result from these drops is approximately $(8\,\mathrm{m}^{-3})(2)\,[\pi(0.001\,\mathrm{m})^2]$, or $5 \times 10^{-5}\,\mathrm{m}^{-1}$. So, 90% of the energy in a laser pulse with a wavelength of 0.5 μm will propagate 2 km through these drops without any interaction, while to reduce the laser beam to 10% of its original power the pathlength must exceed 46 km. Scanning radars in support of precipitation studies typically have wavelengths around 50 mm

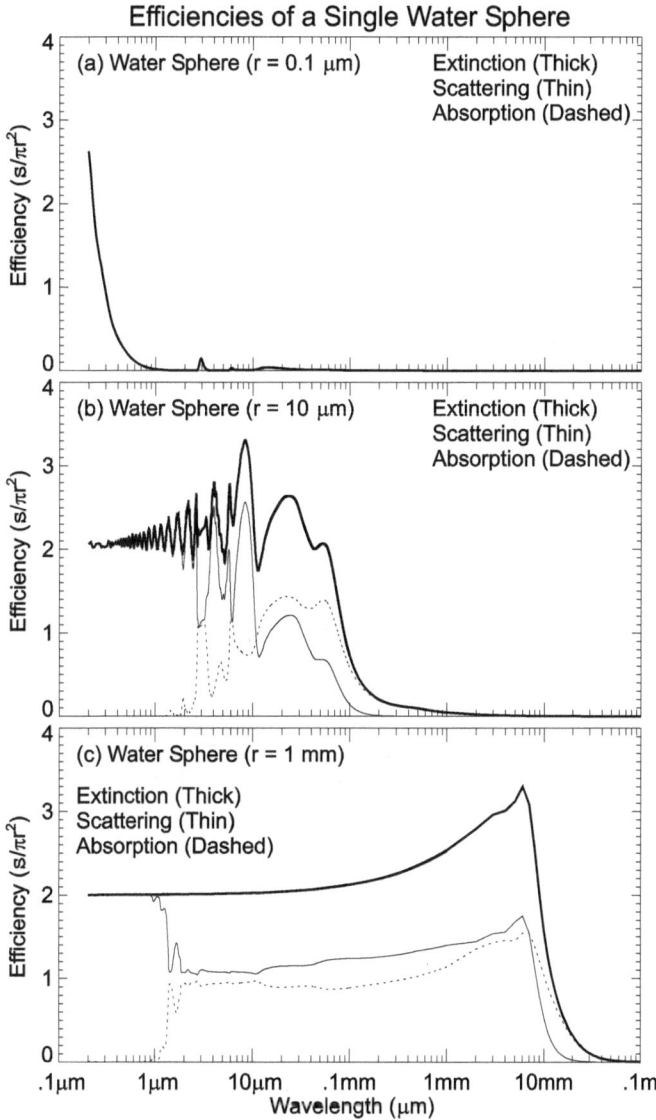

Fig. 2.13. Extinction (*thick line*), scattering (*thin line*) and absorption (*dashed line*) efficiencies, i.e., the ratio of the relevant radiometric cross-section to the cross-sectional area of the particle, for a liquid-water sphere of radius (**a**) 0.1 μm, (**b**) 10 μm and (**c**) 1 mm

(C-band) and 100 mm (S-band), where attenuation by the raindrops is significantly less than for wavelengths less than 1 mm. At these large wavelengths attenuation by drizzle and moderate rain is insignificant and yet the power that is backscattered by raindrops 100–200 km from the radar is still sufficient to be detected by radars operating at these wavelengths.

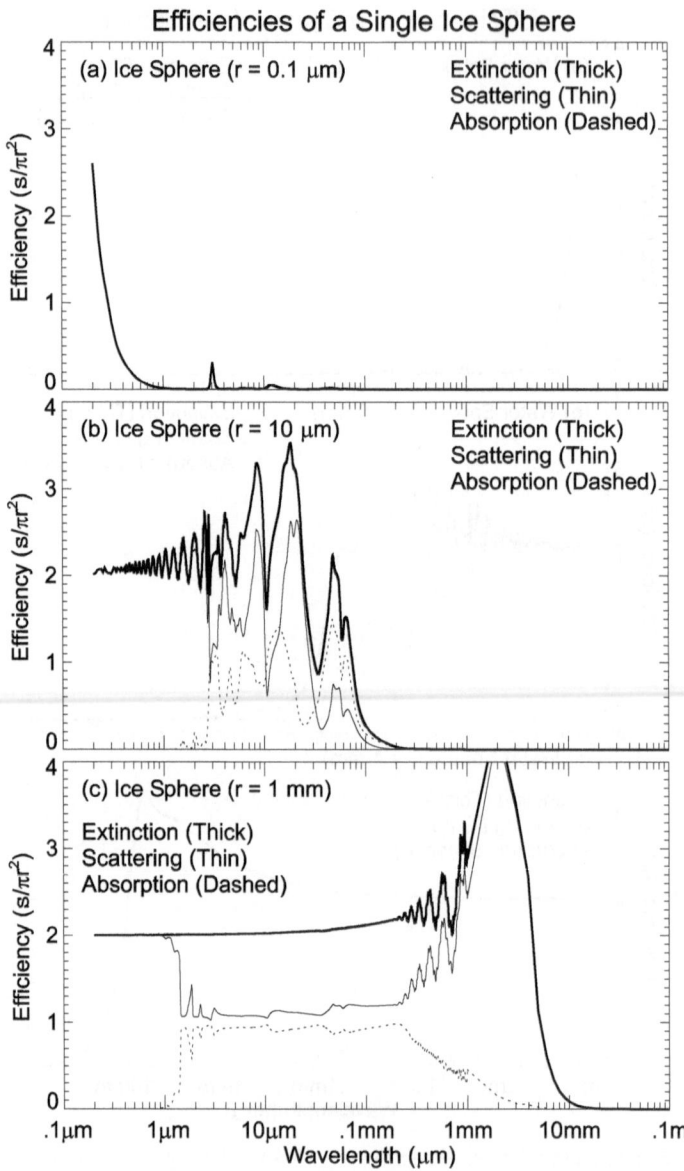

Fig. 2.14. Same as Fig. 2.13, but for ice-water spheres

2.4.3 Distributions of Water Particles

The phase functions and cross-sections illustrated in Figs. 2.12a, 2.13 and 2.14 are for single sized spheres. In reality, clouds consist of drops with varying sizes. To illustrate the scattering and absorption cross-sections that result from a collection of differently sized drops consider a lognormal distribution of drops with effective

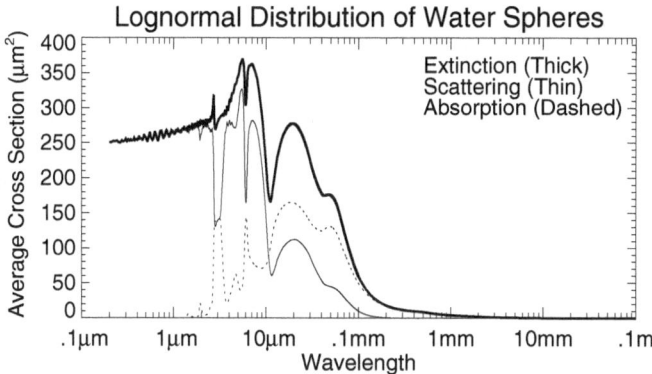

Fig. 2.15. Extinction (*thick line*), scattering (*thin line*) and absorption (*dashed line*) efficiencies, i.e., the ratio of the relevant radiometric cross-section to the cross-sectional area of the particle, for liquid-water spheres with a lognormal distribution of radii with the same parameters as for Fig. 2.12b

radius r_e of 7.5 μm and logarithmic width $\sigma_{\log}$ of 0.35, where the effective radius is defined as

$$r_e = \frac{\int_0^\infty n(r)r^3 dr}{\int_0^\infty n(r)r^2 dr} \ . \tag{2.7}$$

In the above equation r is the cloud-drop radius (in units of μm) and $n(r)$ represents the lognormally distributed drops (in units of $m^{-3}\,\mu m^{-1}$) given by

$$n(r) = \frac{N_t}{\sqrt{2\pi}\sigma_{\log}r}\exp\left(\frac{-[\ln(r/r_{n,\log})]^2}{2\sigma_{\log}^2}\right), \tag{2.8}$$

where N_t is the total number of drops per unit volume and $r_{n,\log}$ is the median diameter of the distribution of drops. As Figs. 2.12b and 2.15 illustrate, the oscillations in the scattering phase function and absorption and scattering cross-sections for the individual drops are significantly reduced for the drop distribution. For the lognormally distributed drops the oscillations from the differently sized drops occur at slightly different wavelengths and their average effect is to wash out the oscillations. Paradoxically, in this case having differently sized drops in a cloud potentially makes treatment of the radiative properties simpler.

2.5 Wavelength Dependence of Radiation Sources and Atmospheric Transmission

Inspection of Figs. 2.10–2.15, which illustrate wavelength-dependent changes in the important radiation sources and all of the important radiometric quantities of gases and cloud particles, leads us to formulate a number of important points regarding atmospheric radiative transfer. During daylight hours, most of the radiation in the

Earth/atmosphere system with wavelengths shorter than approximately 5 μm originates from the sun, while for wavelengths greater than 5 μm most radiation is created by some component of the Earth system (Fig. 2.11b). We call the former type of radiation shortwave radiation and the latter type longwave radiation. This designation, however, is not exact in specifying the source of the radiation, as some photons in the shortwave originate from the Earth and some in the longwave originate from the sun (Fig. 2.11b).

For radiation calculations we generally treat the flow of solar and terrestrial radiation separately. For solar calculations the major contributions to atmospheric heating rates are for wavelengths between approximately 0.1–5.0 μm, while the terrestrial calculations comprise wavelengths from approximately 4–60 μm. While energy budget calculations consist of performing calculations for radiation with wavelengths shorter than 60 μm, a cloudy atmosphere is most transparent at wavelengths that exceed 1 mm. Consequently, while longwave radiation with wavelengths that exceed 1 mm, which we call microwave radiation, is not important to the energy balance of Earth, this region is important for remote sensing. Ground- and satellite-based radiance measurements in the microwave region will have contributions from each part of the total atmospheric column within the field of view of the sensor. Hence, microwave sensor measurements can provide information about the constituents throughout the atmospheric column together with some integral radiative properties of these constituents.

While molecular scattering is confined to wavelengths shorter than approximately 1 μm (Fig. 2.10a), molecular absorption must be considered for all shortwave and longwave radiation processes (Fig. 2.10b–h). While liquid- and ice-water particle scattering processes dominate absorption processes for most shortwave radiation, both processes are important in the longwave region of the spectrum (Figs. 2.13–2.15). While "complete-column" remote sensing can be accomplished in the microwave region of the spectrum (Fig. 2.11a), the greatest sensitivity to the presence of cloud particles is obtained at shorter wavelengths where particle extinction efficiencies asymptote to a value of two (Figs. 2.13–2.15).

The consequences of these properties of atmospheric constituents, together with their diverse and inhomogeneous distributions in space and time, make the discipline of atmospheric radiative transfer challenging. The chapters that follow describe in detail the methods and approaches that have been, and are being, developed both to compute atmospheric heating rates and to retrieve the properties of atmospheric constituents. But before moving on to these topics, we first provide an observation-based illustration of the 3D nature of atmospheric radiative processes using state-of-the-art observations from NASA EOS satellite instruments and DOE ARM ground-based instruments.

2.6 A Remote Sensing View of Cloud Structure

With the onset of the United States Global Change Research Program (USGCRP) in the late 1980s and early 1990s NASA developed the Mission to Planet Earth, now called the Earth Science Enterprise, whose goal was to monitor globally the Earth from space with unprecendented spectral and angular radiance information at a relatively high spatial resolution (Wielicki et al., 1995). The goal was to make these measurements from multi-instrument satellite platforms so that the resulting observations could be easily combined in synergistic studies of the Earth environment. The first satellite platform launched by the NASA Earth Science Enterprise was called Terra. Terra was launched into a sun-synchronous orbit with a morning equatorial crossing. Terra now has an afternoon equatorial crossing counterpart called Aqua with some, but not all, of the same instruments. While NASA developed the Earth Observing System (EOS) within the Earth Science Enterprise, the DOE developed the Atmospheric Radiation Measurement program (Stokes and Schwartz, 1994; Ackerman and Stokes, 2003), the ground-based equivalent of the NASA satellite-based EOS program. The Atmospheric Radiation Measurement (ARM) program now maintains multi-instrument sites in the Tropical Western Pacific (TWP), the Southern Great Plains (SGP) of the United States and the North Slope of Alaska (NSA). In the discussion that follows we use observations from two of the Terra sensors in conjunction with ground-based measurements from two of the DOE ARM sites.

The NASA EOS Terra satellite platform contains three imagers for cloud studies, the Multi-angle Imaging SpectroRadiometer (MISR; Diner et al., 2002), the MODerate resolution Imaging Spectroradiometer (MODIS; King et al., 2003; Platnick et al., 2003) and the Advanced Spacebourne Thermal Emission and Reflection (ASTER; Logar et al., 1998) radiometer. Of these three imagers, MISR and MODIS are the primary instruments for observing and quantifying the properties of the atmosphere on a global scale. The primary characteristics of MISR are that it consists of nine bore-sighted cameras that view Earth objects at nine different angles, with one nadir-directed camera and four cameras in both the forward and aft directions with view zenith angles of $26.1°$, $45.6°$, $60.0°$, and $70.5°$. Each of the nine cameras has four channels centered on wavelengths of $0.446\,\mu m$ (blue), $0.558\,\mu m$ (green), $0.672\,\mu m$ (red), and $0.867\,\mu m$ (near-infrared) with an average band-pass of approximately $0.033\,\mu m$ for each of the four channels. The cross-track spatial resolution of all camera data is 275 m, while the along-track spatial resolution of the MISR radiances varies from 214 m at nadir to 707 m for the most oblique viewing cameras. The sampling of the MISR radiances is 275 m both cross- and along-track. The MISR does have a local mode operational paradigm in which all camera radiances are reported without averaging on its 275 m by 275 m collection grid. However, to reduce data rates in its standard operational mode all of the MISR off-nadir radiances, except for the red-band radiance data, are averaged over groups of 4 by 4 pixels to produce a standard product on a 1100 m by 1100 m sampling grid.

While the MODIS has only a nadir-directed field of view, its across-track swath width is 2330 km, as compared to the 360 km swath of MISR (Fig. 2.16). The spatial resolutions of the MODIS radiances at nadir are 250 m, 500 m or 1000 m depending upon the wavelength of the radiation. Towards the edge of the MODIS scans the across-track spatial resolution of the MODIS radiances decreases by approximately a factor of two. The MODIS contains 36 spectral channels, four of which

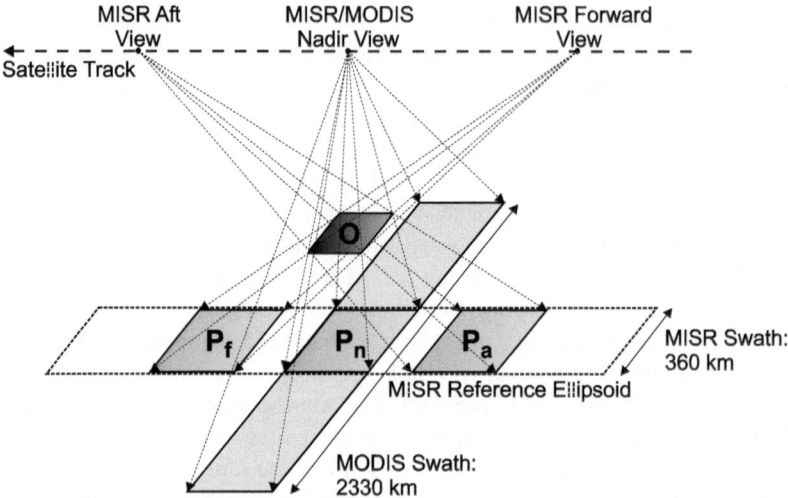

Fig. 2.16. Viewing geometries of the MISR forward, nadir and aft cameras with projection of the image radiances to the World Geodetic System 1984 (WGS84) ellipsoid, a hypothetical smooth ellipsoid surface that represents mean sea-level coordinates everywhere, including for regions with terrain. Note that as the object (**O**) is lowered towards the surface the projected forward-camera (P_f) and aft-camera (P_a) images approach the location of the nadir-camera projection (P_n). The location of the MODIS image radiances relative to the MISR ones are only approximate, as the MODIS radiances are not referenced to the MISR reference ellipsoid and the scanning, or "wisk-broom," geometry of MODIS is different from the "push-broom" configuration of the MISR cameras

are quite close in wavelength to the four MISR channels: $0.469 \pm 0.010\,\mu m$ (band 3), $0.555 \pm 0.010\,\mu m$ (band 4), $0.645 \pm 0.025\,\mu m$ (band 1) and $0.8585 \pm 0.0175\,\mu m$ (band 2). (The "uncertainties" in the wavelengths actually represent the bandwidths of the MODIS spectral channels.) The spatial resolution of bands 1 and 2 is 250 m, while the spatial resolution of bands 3 and 4 is 500 m.

To illustrate the 3D nature of clouds we use the red and near-infrared bands from MISR, together with the following spectral channels from MODIS: $1.375 \pm 0.015\,\mu m$ (band 26), $1.640 \pm 0.012\,\mu m$ (band 6), $2.130 \pm 0.025\,\mu m$ (band 7), $3.750 \pm 0.090\,\mu m$ (band 20), $6.715 \pm 0.180\,\mu m$ (band 27) and $12.020 \pm 0.250\,\mu m$ (band 32). There are several motives for choosing these MODIS channels. First, molecular scattering is insignificant at these wavelengths (Figs. 2.10a and 2.11a). As Fig. 2.11a also illustrates, all of them, except for the $1.375\,\mu m$ and $6.715\,\mu m$ channels, occur at wavelengths for which atmospheric gaseous transmissivity is close to unity. Therefore, radiances to space at these wavelengths will originate predominately from clouds and the surface. At $1.375\,\mu m$ and $6.715\,\mu m$ water-vapor absorption is important, so that photons with wavelengths in the $1.375\,\mu m$ channel, which mostly originate from the sun on the daylight side of Earth, will be absorbed by atmospheric water vapor and photons with wavelengths in the $6.715\,\mu m$ channel, which are primarily from the Earth, both originate from and are absorbed by water vapor. While solar photons in the 0.672 μm (red) and 0.867 μm (near-infrared) channels are scattered by liquid- and ice-water clouds (Figs. 2.13–2.15), they are not absorbed by these clouds. However, vegetated surfaces are much more reflective at a wavelength of 0.867 μm as compared to 0.672 μm (see Chap. 14).

The shortwave channels at 1.640 μm, 2.130 μm and 3.750 μm complement the red and near-infrared channels in that liquid- and ice-water particles are absorbing at these wavelengths. Of these three channels the 3.750 μm channel has the largest ratio of absorption to extinction cross-section for cloud-sized particles (e.g., Figs. 2.13b and 2.14b). For all three channels cloud-sized ice particles are more absorbing than liquid-water particles of the same radius. Interpretation of the radiances at 3.750 μm is complicated by the mixture of Earth- and solar-emitted photons that contribute to these radiances during daylight hours. Finally, the 12.020 μm channel radiances provide the best direct measure of surface and optically thick cloud temperatures, as this channel occurs in a relativey transparent spectral region of the molecular atmosphere (Fig. 2.11a) and is near the peak of the Planck function at terrestrial temperatures (Fig. 2.11b).

2.6.1 Clouds Over the Arctic

A view of the Arctic in the environs of Greenland as captured by the MODIS 0.645 μm channel on 12 June 2001 is illustrated in the histogram-equalized radiances of Fig. 2.17a, where histogram-equalized radiances are generated by mapping the original radiance values to pixel intensities that are proportional to the rank of the original radiance values in the histogram of radiances generated from the entire scene. Properly identifying the rough topography, snow, sea ice and clouds in the scene, which is a pre-requisite to computing accurately the atmospheric radiation field, is not easily accomplished with this channel alone. For example, consider the contents of the white box towards the top of Fig. 2.17a, just to the north and east of Greenland. The MISR zenith-directed 0.672 μm channel radiances, which are similar to the radiances illustrated in Fig. 2.17a, for this box are illustrated in Fig. 2.18a. Complementing these radiances with the MISR 70.5° forward-scattering radiances (Fig. 2.18b) and the MODIS channels with high atmospheric transmissivity (Figs. 2.18c–f), a much more detailed and informationally rich picture of the scene emerges.

Inspecting the bottom third of the images in Fig. 2.18, we find 1) low-level clouds that are both opaque (region 1) and transparent (region 2) in the zenith-directed 0.672 μm channel with the transparent clouds tending toward opaqueness in all other channels, 2) some surface temperatures that are warmer than all cloud temperatures (region 3), 3) other surface temperatures that are as cold as the opaque low-level clouds (region 4) and 4) a tenuous mid-level cloud (region 5), partially transparent in the 0.672 μm channel but not in the other channels, that casts a shadow on the lower-level clouds (e.g., Fig. 2.18b). Comparing Fig. 2.18a–b, we find that clouds throughout this scene scatter much more radiation in the forward direction than the surface ice features. The highest altitude cloud features at the top of the images (region 6)

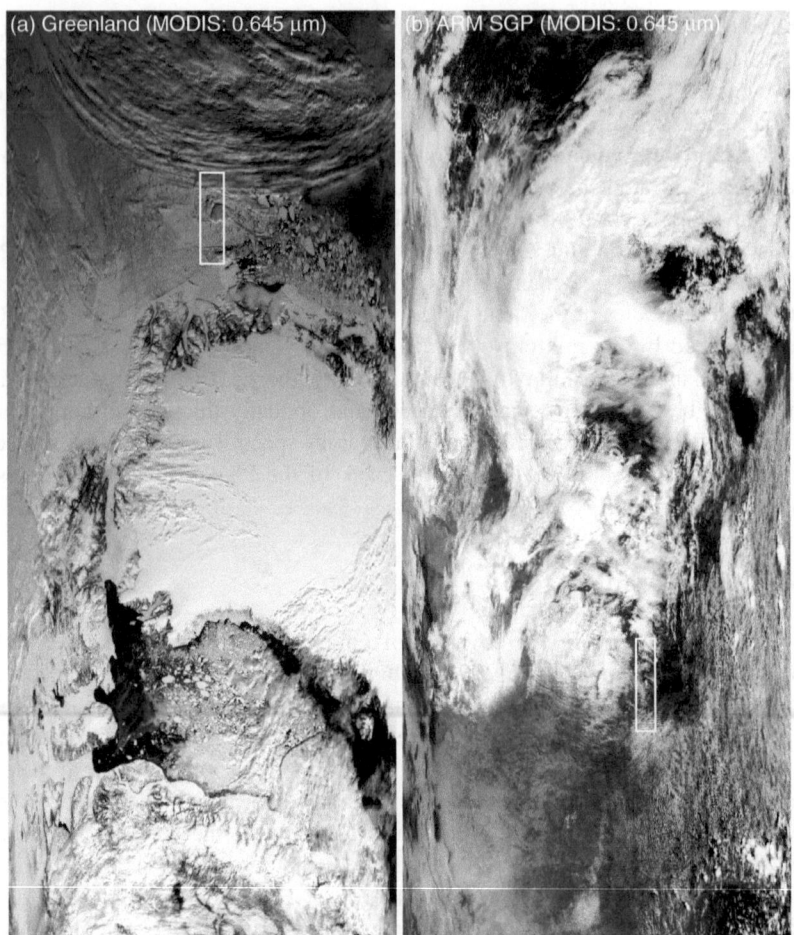

Fig. 2.17. MODIS 0.645 μm radiance imagery for **(a)** Greenland and Baffin Bay on 12 June 2001 around 17:00 UTC (Orbit 7898) and **(b)** the ARM SGP site on 7 September 2001 at approximately 17:15 UTC (Orbit 9165). The images in both (a) and (b) have been histogram-equalized, i.e., radiance values are mapped to pixel intensities that are proportional to the rank of the radiance value in the histogram of radiances generated from the entire scene, in order to enhance all of the features across the image. Black regions represent low values of radiance and white regions represent high values of radiance. For these two images the pixel resolution is approximately 4 km by 4 km and the domain size is approximately 2330 km by 6060 km

can be identified as the whitest regions in Fig. 2.18c, where solar 1.375 μm radiation is scattered back to space before entering the lower reaches of the atmosphere where water-vapor concentrations are higher, and the darkest regions in Fig. 2.18f, corresponding to the coldest objects in the image.

Of the many images from MISR and MODIS that we have viewed to date this particular scene highlights many of the interwoven connections between atmospheric

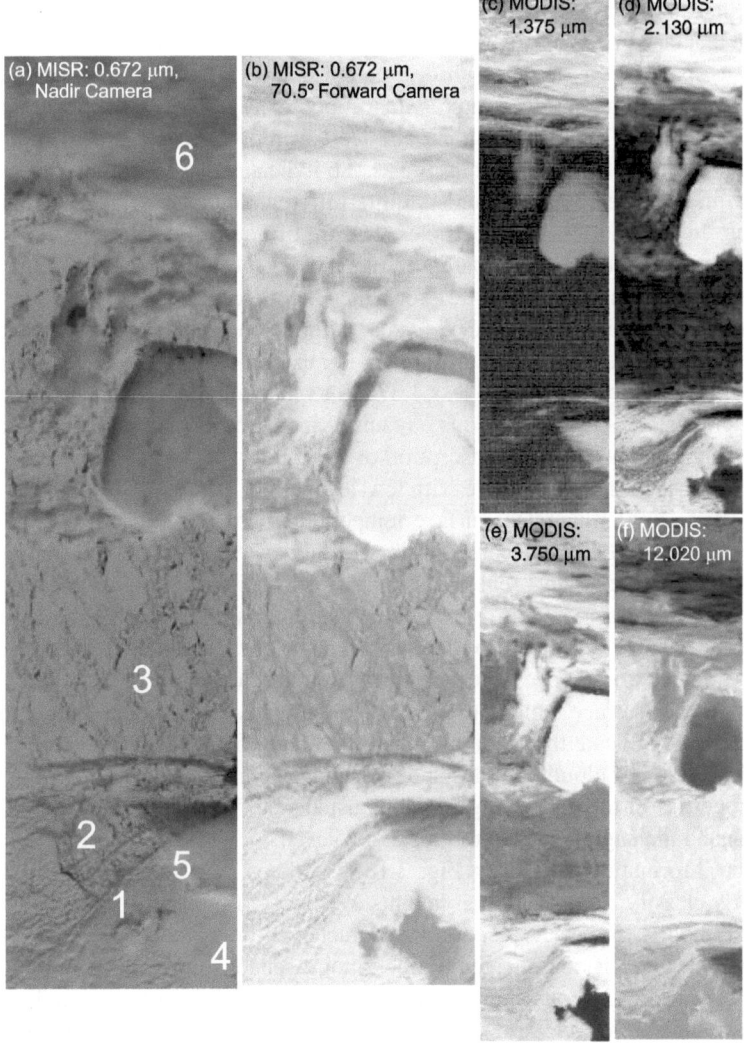

Fig. 2.18. MISR and MODIS radiance imagery for the region demarcated by the white box in Fig. 2.17a. The MISR radiances include the 0.672 μm radiance imagery for the (**a**) nadir and (**b**) 70.5° forward viewing cameras, while MODIS imagery is presented for the (**c**) 1.375 μm, (**d**) 2.130 μm, (**e**) 3.750 μm and (**f**) 12.020 μm radiances. The MISR radiances in (a)–(b) have not been scaled with radiance increasing from black to white, while the MODIS radiances in (c)–(f) have been histogram-equalized for contrast enhancement, again with radiance increasing from black to white. Note that the image in (a) is quite a bit darker than the image in (b) because forward scattering from the clouds is much larger than zenith-directed scattering from clouds and surface ice. The pixel resolution of the MISR images is 550 m by 550 m, while the pixel resolution of the MODIS images is 1000 m by 1000 m. The domain size for all of these images is approximately 80 km by 320 km

molecules and clouds and the resulting radiation fields. As Figs. 2.18a–b illustrate, the molecular atmosphere is transparent at a wavelength of $0.672\,\mu m$ so that radiances to space at this wavelength originate primarily from scattering of solar radiation by clouds and surface features. Inspection of Fig. 2.18a illustrates that distinguishing between clouds and snow- and ice-covered surfaces can be difficult using mid-visible nadir radiances, while Fig. 2.18b indicates that the angular dependence of the mid-visible radiation field is quite different for clouds and the underlying surface. In fact, atmospheric particles that are difficult to detect in nadir radiance imagery can be much more apparent in radiance imagery from more oblique views.

At wavelengths, e.g., $1.375\,\mu m$, for which the molecular atmosphere is not transparent, the solar radiation reflected to space from clouds now depends upon the altitudes of the clouds within the atmosphere. For example, Fig. 2.18c clearly illustrates that low altitude clouds reflect less $1.375\,\mu m$ solar radiation to space than high altitude clouds. This phenomenon is a result of increased photon pathlengths through water vapor and hence more absorption of $1.375\,\mu m$ solar radiation for photons entering regions of the atmosphere with low-level clouds.

As Fig. 2.18f illustrates, surface temperatures are not necessarily higher than those of cloud particles and can vary significantly over relatively small spatial scales. This state-of-affairs is apparent in a comparison of regions 1 (low-level cloud), 3 (relatively warm surface) and 4 (relatively cold surface) in the figure. If we assume that the surfaces in regions 3 and 4 reflect $3.750\,\mu m$ solar radiation in the same way, then the low values of radiance in region 4 relative to region 3 in Fig. 2.18e are an indicator of lower surface temperatures in region 4 relative to region 3. Taken together, Figs. 2.18a–f indicate that the atmospheric radiation field is a strong function of the 3D distribution of cloud particles and their placement in the molecular atmosphere, the properties of the underlying two-dimensional surface, the wavelength of the radiation and the angle at which the scene is viewed.

The information content in Fig. 2.18 is perhaps now sufficient to allow us to properly classify clouds and their heights over most of the Arctic during most of the daylight hours. However, quantitative characterization of cloud optical thicknesses and particle sizes is complicated by the inhomogeneous distribution of cloud properties over heterogeneous snow- and ice-covered land and water. Given that we can quantify the surface and atmospheric properties for this scene, computational estimates of the resulting radiation field and its associated heating rates are going to be extremely time consuming to produce as the scene is quite variable. Reducing scene complexity, while preserving important scene statistics for computing unbiased domain-averaged radiative properties, is possible and is the subject of much ongoing research.

2.6.2 Clouds Over the ARM SGP Site

On 7 September 2001 EOS Terra passed over the ARM SGP site at about the time a front was clearing the region (Fig. 2.17b). The clouds at this time were in their "typical" comma-shaped pattern around the low pressure center, stretching from the top right (northeast) to bottom left (southwest) corner of the image. The clouds at the bottom right corner of the image are located along the Texas/Gulf of Mexico coastline and the white box is located just to the east of the ARM SGP site. Along the entire extent of the main cloud structure in the image the cloud elements are multi-level and nonhomogeneous, even at the smaller scales. The post-frontal cumulus clouds trailing higher level clouds within the region indicated by the white box are a case in point (Fig. 2.19).

Our choice of spectral channels in Fig. 2.19 is slightly different from Fig. 2.18. As the MISR red and near-infrared linearly-scaled radiances in Fig. 2.19a–b indicate, vegetated land surfaces (region 1) are much more reflective at near-infrared wavelengths than red wavelengths and water surfaces (region 2) are absorbing at both wavelengths. Comparing the MISR red image (Fig. 2.19a) with the MODIS histogram-equalized blue radiances in Fig. 2.19c, we find much the same thing, although surface water (region 2) is not as absorbing relative to surface land at blue wavelengths as for the other wavelengths. Although there are important differences in the radiances illustrated in Fig. 2.19a–c as a result of scattering by molecules and aerosols, the appearances of the cloud elements at these different wavelengths are quite similar, as we would expect from the radiative properties of cloud particles illustrated in Figs. 2.13–2.15 and despite histogram-equalization of the MODIS radiances.

Comparing the MODIS histogram-equalized $0.469\,\mu m$, $2.130\,\mu m$, $6.715\,\mu m$ and $12.020\,\mu m$ radiances (Fig. 2.19c–f), we find that the appearance of the scene changes significantly from one wavelength to the next and in markedly different ways. For the $12.020\,\mu m$ channel image (Fig. 2.19f) the three salient features of the image are the high-level cold clouds in the top half of the image (e.g., dark areas in region 3), the low-level warm clouds in the bottom half of the image (e.g., gray areas in region 4) and the transparent regions of the atmosphere throughout the image where surface-leaving radiances make it to space (e.g., whitish areas, region 1). Emission and attenuation features evident in the $6.715\,\mu m$ channel radiances resulting from water vapor and cloud particles (Fig. 2.19e) are consistent with high clouds to the north (region 3) and low clouds to the south (region 4). Note, however, that attenuation of surface-leaving radiance with emission of radiation at colder temperatures evident in the $6.715\,\mu m$ radiance image is more uniform across the top half of Fig. 2.19e than in Fig. 2.19f, indicating that upper-level water vapor is more uniform than the upper-level cloud coverage. The distribution of bright, cold cloud (Fig. 2.19c,f) takes on a more mottled appearance in the $2.130\,\mu m$ radiance image (Fig. 2.19d, region 5), as now differing absorption between cloud-liquid drops and cloud-ice particles becomes important.

Our overall assessment of the imagery in Fig. 2.19 is that there is little that is homogeneous in this scene. The distributions of water vapor and liquid- and ice-cloud particles are highly variable and they are occurring over a surface with different reflectances from one wavelength to the next. This variability in the cloud and surface properties leads to a spatial radiation field that changes dramatically from one wavelength to the next, as illustrated in Fig. 2.19. Spectrally-dependent radiances are also observed in time series of ground-based measurements obtained from ARM SGP site sensors for this same event (Figs. 2.20 and 2.21).

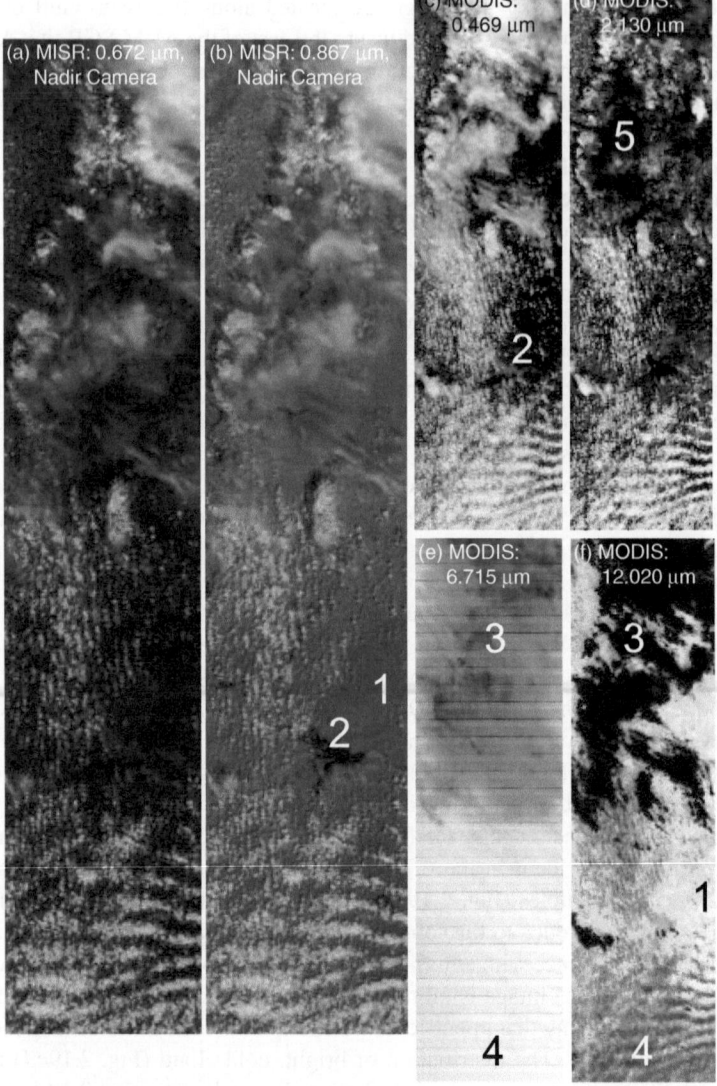

Fig. 2.19. MISR and MODIS radiance imagery for the region demarcated by the white box in Fig. 2.17b. The MISR radiances include the (**a**) 0.672 μm and (**b**) 0.867 μm radiance imagery for the nadir camera, while MODIS imagery is presented for the (**c**) 0.469 μm, (**d**) 2.130 μm, (**e**) 6.715 μm, where the stripping is a known artifact of the multi-sensor calibration, and (**f**) 12.020 μm radiances. The MISR radiances in (a)–(b) have been scaled linearly so that black represents zero radiance and white represents the maximum value of radiance in the image. The MODIS radiances in (c)–(f) have been histogram-equalized for contrast enhancement, again with radiance increasing from black to white. The pixel resolution of the MISR images is 550 m by 550 m, while the pixel resolution of the MODIS images is 1000 m by 1000 m. The domain size for all of these images is approximately 80 km by 320 km

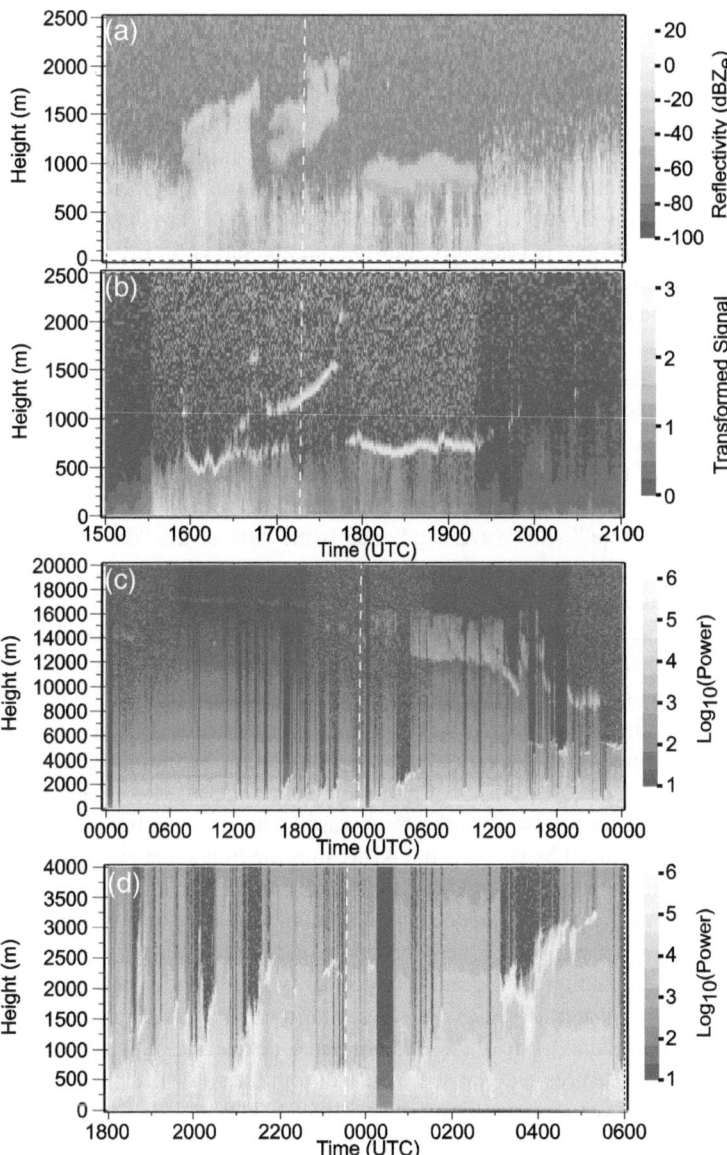

Fig. 2.20. Ground-based observations of clouds with profiling instruments over both the ARM SGP site on 7 September 2001 collected with (**a**) the millimeter-wavelength cloud radar (Moran et al., 1998; Clothiaux et al., 2000) and (**b**) a Vaisala laser ceilometer and the ARM TWP Nauru site on 22–23 November 2000 collected with (**c**) the micropulse lidar (Spinhirne, 1993). (**d**) is an enlarged version of (c) from 18:00 UTC 22 November to 06:00 UTC 23 November with clouds over 4000 m deleted. The dashed vertical lines indicate the passage of the Terra satellite (cf. Figs. 2.17b, 2.19, 2.22a and 2.23)

Locations of the clouds in the vertical column above the ARM SGP site are illustrated by the millimeter-wave cloud radar (MMCR) and Vaisala ceilometer (VCEIL) returns presented in Fig. 2.20a–b, respectively, and the Vaisala ceilometer cloud-base height retrievals illustrated in Fig. 2.21a. Downwelling radiances at the surface within narrow spectral intervals located in the shortwave near-infrared (i.e., 1.00 μm), longwave-infrared (i.e., 10.65 μm) and longwave-microwave (i.e., 9.55 mm) regions of the electromagnetic spectrum, all wavelengths at which the molecular atmosphere is relatively transparent, are shown in Fig. 2.21b–d, whereas the downwelling total, direct and diffuse shortwave irradiances (i.e., fluxes) and longwave irradiances measured at the surface are presented in Fig. 2.21e–f, respectively. For the measurements in Fig. 2.21e "shortwave" corresponds to a spectral interval from approximately 0.3–3.0 μm, while the "longwave" measurements in Fig. 2.21f are over the interval from 4.0–50.0 μm. By "direct" shortwave irradiance we are referring to solar photons that impinge upon the sensor without interacting with any constituents of the atmosphere, while "diffuse" shortwave irradiance refers to solar photons that reach the sensor after one or more interactions with atmospheric constituents. The "total" shortwave irradiance is the sum of the "direct" and "diffuse" irradiances. Longwave surface irradiance is composed of terrestrially-emitted photons within the spectral interval from 4.0–50.0 μm that reach the sensor.

To facilitate understanding of the data presented in Fig. 2.20, which were generated by active remote sensing radar and lidar, we now briefly discuss how these instruments operate. Both radar and lidar generate a pulse of electromagnetic radiation of the appropriate wavelength and then direct this pulse of radiation into a narrow, vertically oriented cone directly above the instrument. While not generally considered as such, the directed pulses of radiation from radar and lidar can be interpreted as a radiance. These pulses propagate upwards with a transmission given by (2.2), but with $\sigma_{a,\lambda}$ replaced by $\sigma_{e,\lambda}$. At the ARM sites the lidars operate at wavelengths of 0.524 μm (micropulse lidar, or MPL) and 0.905 μm (VCEIL), wavelengths for which absorption is not important. Hence, attenuation of the lidar pulse is generally a result of scattering by air molecules, aerosols and cloud particles. At the radar wavelength of 8.66 mm attenuation of the pulse is primarily the result of absorption by water vapor and liquid-water drops, as well as scattering by ice-water particles.

When a pulse leaves one of these instruments, an accurate electronic "stop watch" is started that measures the elapsed time from initiation of the pulse to the time of return of the power that is scattered back to the instrument. (Note that the overall two-way trip transmissivity is simply the one-way transmissivity squared.) The elapsed time of a power return and the speed of light are subsequently used to infer the distance to the atmospheric particles scattering energy back to the instrument; hence, the data in Fig. 2.20 are height resolved. Each of the data points in Fig. 2.20 is a measure of the amount of power scattered back to the instrument. For calibrated radars, like the ARM radar, the backscattered power is converted to an estimate of the radar reflectivity η, or the total backscattering cross-section of the atmospheric particles per unit volume (in units of $mm^2\ m^{-3}$). In the radar literature the radar equivalent reflectivity factor Z_e is generally used in place of η, where Z_e is defined by $Z_e = (\lambda^4 \eta)/(\pi^5 K^2)$ (in units of $mm^6\ m^{-3}$). In this relationship λ is the

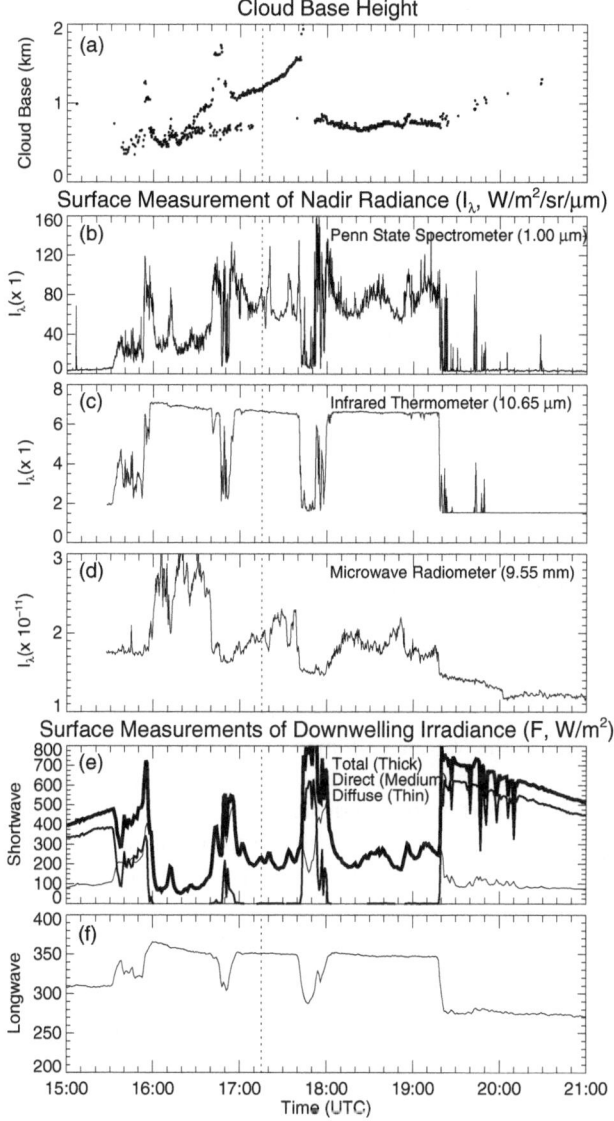

Fig. 2.21. Ground-based radiometric observations at the ARM SGP site on 7 September 2001 during the same period as shown in Fig. 2.20. Observations include (**a**) cloud-base height (Vaisala ceilometer), (**b**) downwelling 1.00 μm radiance (Pennsylvania State University narrow field of view spectrometer), (**c**) downwelling 10.65 μm radiance (Heimann infrared thermometer), (**d**) downwelling 9.55 mm radiance (microwave radiometer; Liljegren, 1994; Liljegren et al., 2001; Westwater et al., 2001), (**e**) downwelling total, direct and diffuse short-wave irradiance (unshaded pyranometer, pyrheliometer and shaded pyranometer, respectively; Michalsky et al., 1999; Long and Ackerman, 2000) and (**f**) downwelling longwave irradiance (pyrgeometer; Philipona et al., 2001). The dashed vertical lines indicate the passage of the Terra satellite

wavelength of the radiation in the radar pulse and K is a physical constant related to the properties of water. Since Z_e can span many orders of magnitude, it is generally presented in imagery on the logarithmic scale $10 \log_{10} Z_e$ and labeled with the units of dBZ_e, as in Fig. 2.20a.

Unlike the ARM radar, the ARM Vaisala ceilometer and micropulse lidar are not calibrated. In the case of the Vaisala ceilometer proprietary software that processes the lidar backscattering power returns transforms the signal so as to maximize the performance of algorithms that Vaisala developed to identify cloud base. Since we do not have access to this proprietary software, we simply label the Vaisala ceilometer returns as a "Transformed Signal" with no associated units. For the micropulse lidar the backscattering power is recorded as the number of photon counts per microsecond. In Figs. 2.20c–d we present the data on the logarithmic scale of $10 \log_{10} (\text{Power})$ with no associated units.

For the power returns illustrated in Fig. 2.20a–b, which are from clouds approximately 500 m thick, assume the cloud drops have a radius of 6 μm and a number density of $300 \, \text{cm}^{-3}$, which are numbers appropriate for some continental boundary-layer clouds. The extinction cross-sections for particles of this size at wavelengths of 8.66 mm (MMCR) and 0.910 μm (VCEIL) are approximately $0.93 \, \mu\text{m}^2$ and $640 \, \mu\text{m}^2$, respectively (e.g., Fig. 2.13). Computing the transmission through the clouds at these wavelengths, we find that the Vaisala ceilometer transmitted pulse is completely attenuated, i.e., scattered, by the cloud while 97% of the millimeter-wave cloud radar pulse is transmitted through the cloud with a loss of 3% resulting from absorption of radiation. The white-colored returns in Fig. 2.20b are the entry point of the Vaisala ceilometer pulse into the cloud, where most of the photons in the pulse are first scattered, some of them backscattered to the Vaisala ceilometer receiver as part of the return signal.

The Vaisala ceilometer returns actually drop in magnitude from the ground to cloud base because the source of these returns is scattering by air molecules and the number of air molecules decreases exponentially with height. However, this feature of the Vaisala ceilometer returns is not readily apparent in Fig. 2.20b. The speckle throughout the image arises primarily from solar photons that are scattered into the Vaisala ceilometer receiver by atmospheric molecules, though the magnitude of this effect is small at a wavelength of 0.910 μm. The exponential drop with height in laser-return power arising from molecular scattering is much better demonstrated in the micropulse lidar 0.524 μm wavelength returns presented in Fig. 2.20c–d. Moreover, notice that during the nighttime hours (i.e., 0600–1800 UTC), and especially at altitudes between 18–20 km, the background speckle in Fig. 2.20c is significantly reduced, as we would expect.

For these nonprecipitating clouds the millimeter-wave cloud radar returns from cloud particles start at cloud base and extend upwards towards cloud top with reflectivity values that range from $-40 \, dBZ_e$ to $-20 \, dBZ_e$. The strong white-colored radar returns from the ground to cloud base arise not from molecular scattering, as this is completely inconsequential at the millimeter-wave cloud radar wavelength (e.g., Fig. 2.10a), but rather from large particles and insects. To understand the consequences of this fact, again assume that the cloud above the clutter is composed of

$6\,\mu m$ radius drops and suppose that the $6\,\mu m$ radius drops are coalesced into larger 1 mm radius drops. The drop density for these large 1 mm radius drops would be approximately $65\,m^{-3}$ and the transmissivity through this collection of drops would be about 82% at wavelengths of both 8.66 mm and $0.910\,\mu m$. That is, the "cloud" of large particles would now be fairly transparent at both wavelengths. However, the overall backscattering cross-section per unit volume decreases by a factor of about 2.5×10^{-3} at $0.910\,\mu m$ while it increases by a factor of 6.5×10^6 at 8.66 mm.

In Fig. 2.20a the increase in reflectivity from cloud to sub-cloud layer is about $20\,dBZ_e$ and not $70\,dBZ_e$ as would be the case if the "cloud" of 1 mm radius water drops given above were located in the sub-cloud layer. However, if we assume that the sub-cloud layer is composed of one 1 mm radius insect, or water particle, per millimeter-wave cloud radar sample volume of approximately $1000\,m^3$, the transmissivity of this layer approaches 100% at both wavelengths. The backscattering cross-section of this layer at a wavelength of $0.910\,\mu m$ has now decreased by a factor of 3.95×10^{-8} relative to the cloud layer of $6\,\mu m$ radius drops. However, for radiation with wavelengths of 8.66 mm the backscattering cross-section of this sub-cloud layer is now a factor of 103, or approximately $20\,dBZ_e$, greater than the reflectivity of the cloud layer above. These transmissivities and backscattering cross-sections are in keeping with the observations illustrated in Fig. 2.20a–b. In summary, a few large drops have a significant effect on radar reflectivities but they are completely inconsequential for lidars operating at visible to near-infrared wavelengths.

The observations in Fig. 2.20a–b exhibit one of the fundamental problems with active remote sensing. The micropulse lidar and Vaisala ceilometer systems operate at wavelengths in the mid-visible and near-infrared. As such, the radiation from them interacts with cloud particles in much the same way as does solar radiation. What we learn from these laser systems regarding the radiative properties of clouds will be directly applicable to much of the shortwave spectrum as well. However, clouds often exceed optical depths of 5 or so, leading to complete attenuation of the laser beams and making retrieval of cloud properties throughout the depth of the atmosphere problematic.

While most of the radiation from radars operating at microwave wavelengths penetrate most clouds, these radars have the problem that their backscattering power returns are dominated by the largest cloud particles. Since the largest cloud particles are relatively few in number, at least within most cloud elements, they are not the cloud particles of greatest importance to the shortwave and longwave radiation budgets. As a result, knowing the exact 3D distribution of the larger particles in a cloud field is now critically important to interpreting correctly the radar returns from the cloud field and making an accurate inference of the shortwave and longwave radiative properties of the cloud field.

Downwelling radiances at the surface within narrow spectral intervals centered on wavelengths of $1.00\,\mu m$, $10.65\,\mu m$ and 9.55 mm for this cloudy period are illustrated in Fig. 2.21b–d. At microwave wavelengths of 9.55 mm the clouds for this case are not opaque (e.g., Fig. 2.11a) and their emissivities are not close to one. As a consequence, increases in cloud-liquid water in the column over the ARM SGP site lead to increases in emissivity and hence larger downwelling radiances at the surface.

As Fig. 2.21d illustrates, the three main cloud elements during this period exhibit a general decrease in liquid water.

At a wavelength of 10.65 μm the situation is much different. The transmissivity of the cloud is now close to zero (e.g., Fig. 2.11a) with a resulting emissivity that is close to one for all three cloudy periods. Hence, the downwelling radiance at the surface is approximately the value of the Planck function evaluated at the temperature of cloud base. As the cloud-base height increases during the first two periods (Fig. 2.21a), the temperature decreases with a corresponding drop both in the value of the Planck function at cloud base and the downwelling 10.65 μm radiance at the surface (Fig. 2.21c). As cooler air is moving into the environs of the ARM SGP site with the passage of the front, the temperature at cloud base during the third cloudy period is slightly less than for the first period, with an associated drop in downwelling radiance at the surface. Compared to the radiances at 9.55 mm, those at 10.65 μm are quite smooth, showing significant variations only during periods when there are gaps in the clouds.

The largest changes in radiance over short time scales occur for the downwelling radiances at 1.00 μm (Fig. 2.21b). The large ripples in the 1.00 μm radiances result from inhomogeneities in the cloud field that produce variations in the solar radiation scattered by the clouds to the surface. In fact, both the detailed 3D cloud geometry and the nature of its illumination by solar radiation impact the temporal variations in the 1.00 μm radiances. Downwelling radiance at the surface at a wavelength of 9.55 mm depends on the column-integrated amount of cloud-liquid water. As illumination effects are no longer relevant at 9.55 mm, temporal variations in the 9.55 mm radiances are a bit smoother than for the radiances at 1.00 μm. At a wavelength of 10.65 μm the temperature of the drops at cloud base is the important quantity. As long as the cloud-base height is relatively stable, the 10.65 μm radiances emitted by the cloud to the surface will not change that much and will also exhibit smaller temporal fluctuations than the 1.00 μm radiances.

For small vertically integrated cloud-liquid water amounts, the radiances at 1.00 μm and 9.55 mm are correlated, while at large values they are anti-correlated. This relationship between the two sets of radiances is best illustrated from 17:50–18:20 UTC. Closely inspecting Fig. 2.21b,d from 17:50–18:00 UTC, we find that a small 0.1×10^{-11} W m^{-2} sr^{-1} μm^{-1} increase in the 9.55 mm radiance corresponds approximately to a 140 W m^{-2} sr^{-1} μm^{-1} increase in the 1.00 μm radiance. During this period, a cloud element with small vertically integrated cloud-liquid water amounts is passing over the site. Because the cloud element is relatively transmissive, it actually increases, relative to clear-sky molecular scattering, the amount of 1.00 μm radiance scattered downwards into the vertically pointing sensor. From 18:00–18:20 UTC the vertically integrated cloud-liquid water increases, driving up the 9.55 mm radiance. During this same period, however, the 1.00 μm radiance decreases. With the increasing cloud-liquid water path cloud transmission at a wavelength of 1.00 μm is decreasing. So, while more 1.00 μm radiance from the sun is scattered by these clouds, less of the scattered radiance makes it through the cloud to the sensor.

The longwave irradiances illustrated in Fig. 2.21f show much the same trends as the 10.65 μm radiances in Fig. 2.21c. Unlike the 1.00 μm radiances, the downwelling shortwave total irradiance is generally anticorrelated with the 9.55 mm radiance for all vertically integrated cloud-liquid water amounts. As the cloud-liquid water path increases, the shortwave transmission decreases and the amount of shortwave irradiance reaching the surface decreases. As irradiances are integrated quantities over the hemisphere of downwelling radiances, they are smoother than the downwelling zenith radiances. Inspecting all of the data in Fig. 2.21, we are led to conclude that the radiation field is highly variable in time as well as space and the variations are once again spectrally-dependent.

2.6.3 Clouds Over the ARM TWP Nauru Site

Tropical cloud fields are much different from those in the mid-latitudes and the Arctic. The cloud field over the ARM TWP Nauru site imaged by the MISR nadir camera on 22 November 2000 at 23:34 UTC (Fig. 2.22a) is typical of what we generally find for this site. As the micropulse lidar images in Fig. 2.20c–d illustrate, trade cumulus under high altitude cirrus occurs throughout the two-day period centered on the time of the MISR overflight. Moreover, during this two-day period, clouds are also found to occur at altitudes of 2000 m, 5000 m and 9000 m (Fig. 2.20c). The 3D character of the cloud field is further illustrated by comparing the MISR 70.5° aft-viewing camera image (Fig. 2.22b) with the MISR nadir image (Fig. 2.22a). As Fig. 2.16 illustrates, surface features on the reference ellipsoid are mapped to the same location by all nine cameras, while clouds close to the surface are mapped to nearly the same location on the reference ellipsoid if the clouds do not move significantly or grow vertically between camera views. As cloud height increases, the differences in pixel locations of the cloud in the nine MISR camera image projections to the reference ellipsoid become greater.

In Fig. 2.22a–b the black dot within one of the white boxes indicates a single location on the island of Nauru. Manual measurements of the coordinates of the black dots in the nadir and aft images indicated that they were located in identical positions on the reference ellipsoid, demonstrating that camera alignments and image projections to the surface ellipsoid were well-characterized. The dashed lines through the figures represent the direction of the MISR satellite track, while the four white dots that are also enclosed by white boxes, one just to the northeast of Nauru, one to the northwest of Nauru, one to the southeast of Nauru in the vicinity of trade cumulus and one to the far southwest of Nauru at the southwest corner of a high-altitude cirrus cloud, represent well-defined cloud features. Based on the displacement of the white dots between the two images in the along-the-satellite-track direction, we find that the trade cumulus represented by the white dots to the northwest and southeast of Nauru are the lowest in altitude while the cirrus cloud to the southwest of Nauru is the highest in altitude. The cloud just to the northeast of Nauru is intermediate in altitude.

Motion of the clouds associated with the white dots is most evident for the intermediate level cloud, where the cloud has clearly shifted cross-track to the northwest between the two camera views. The trade cumulus clouds have also moved to the northwest during this period, but by smaller amounts. Low-altitude winds from the

Fig. 2.22a. The MISR nadir camera 0.672 μm radiance image for the ARM TWP Nauru site at 23:34 UTC on 22 November 2000. The image has been histogram-equalized for contrast enhancement. The pixel resolution of the image is 550 m by 550 m, while the domain size of the image is approximately 260 km by 360 km

southeast in the trade regions of the tropical southern Pacific Ocean are not uncommon, consistent with what we find in the MISR images. An important distinction between the trade cumulus to the northwest and southeast of Nauru is their vertical development. The trade cumulus to the northwest of Nauru has approximately the same extent along the direction of the projection, or satellite-flight track, in both the MISR nadir and aft images, whereas the trade cumulus to the southeast of Nauru

(b) MISR: ARM TWP-Nauru (0.672 μm, 70.5° Aft Camera)

Fig. 2.22b. Same as Fig. 2.22a, but for the MISR 70.5° aft viewing camera 0.672 μm radiance image

has a much larger projected area in the MISR aft image. This result indicates that the trade cumulus to the southeast has much more vertical development than the trade cumulus to the northwest. In fact, the projected area of the trade cumulus to the southeast is much larger for the MISR aft image as compared to the MISR forward image, indicating that this cumulus cloud has developed vertically during the time (7 min) between MISR forward and aft camera views. Comparing the low- and mid-level cloud structures throughout the MISR nadir and aft images demonstrates a range of cloud heights, many of which exhibited vertical development during the MISR overflight.

While the MISR images provide a glimpse into the 3D structure of the clouds over the TWP Nauru site, they also show that the oblique view of the MISR aft camera is much more sensitive to thin cirrus compared to the MISR nadir view. This is particularly evident in the regions to the northwest of Nauru. The distribution of thin cirrus, as well as the estimates of cloud-top heights, inferred from MISR radiances are in keeping with what is seen in the four MODIS images presented in Fig. 2.23. High thin cirrus clouds are evident in the $1.375\,\mu m$ radiance image (Fig. 2.23a, region 1) and they correlate well with the slight decreases in the $12.020\,\mu m$ radiances that are expected in regions of thin cirrus over open water (Fig. 2.23d). For optically thick cirrus clouds the correspondence between the $1.375\,\mu m$ and $12.020\,\mu m$ radiances are striking (regions labelled 2 in Fig. 2.23a).

As expected, the $6.715\,\mu m$ radiances (Fig. 2.23c) are lowest in regions of high-altitude, optically thick clouds (regions labelled 2 in Fig. 2.23a), increasing in value in those regions of low clouds and clear sky (regions labelled 3 in Fig. 2.23a). Note, however, that water vapor does lead to signficant differences in the spatial distribution of the $6.715\,\mu m$ radiances compared to the $12.020\,\mu m$ radiances. The changes in brightness of the $1.640\,\mu m$ radiances (Fig. 2.23b) across the top of the cloud field in region 4 (Fig. 2.23b) indicate the presence of either a combination of liquid and ice particles or ice particles of different sizes in this region.

The ARM TWP Nauru ground-based measurements for 22–23 November 2000 are illustrated in Fig. 2.20c–d and Fig. 2.24. As Fig. 2.24a illustrates, the measurements are dominated by the effects of low-level trade cumulus passing over the site under an umbrella of high altitude cirrus. The fluctuations in the $8.86\,\mu m$ (Fig. 2.24b) and $9.55\,mm$ (Fig. 2.24c) radiances, as well as both the shortwave (Fig. 2.24d–e) and longwave (Fig. 2.24e) irradiances, are dominated by these trade cumulus. The sensor most capable of detecting the high thin cirrus above 16 km is the micropulse lidar (Fig. 2.20c–d) as the ARM millimeter-wave cloud radar does not have the sensitivity to detect these small particle clouds. When trade cumulus attenuates the micropulse lidar pulse, the thin cirrus goes undetected by the radar and lidar instrument suite. During these brief periods, as well as for extended periods of thin cirrus above low-level stratus, one potential way of inferring the presence of the cirrus from the ground may be the oxygen A-band approach discussed in Chap. 13.

To obtain cloud-property information through the depth of a cloudy column remote sensing at microwave wavelengths is an attractive approach because clouds are not optically thick at these wavelengths. Active remote sensing at microwave wavelengths is a viable method for retrieving height-resolved cloud properties. However, there are subtleties in such an approach and we have discussed some of them in the context of the ARM SGP site data. Passive remote sensing at microwave wavelengths is useful for retrieving column-integrated quantities, such as cloud liquid-water paths, but there are complications in these methods as well, which we now discuss.

Consider the middle of the day at the ARM TWP Nauru site from 22:00 UTC 22 November through 02:00 UTC 23 November (Fig. 2.20d and Fig. 2.24). Throughout this period, trade cumulus constantly cross the line of sight from the pyrheliometer to the sun. At these times the solar direct beam is attenuated significantly (Fig. 2.24d),

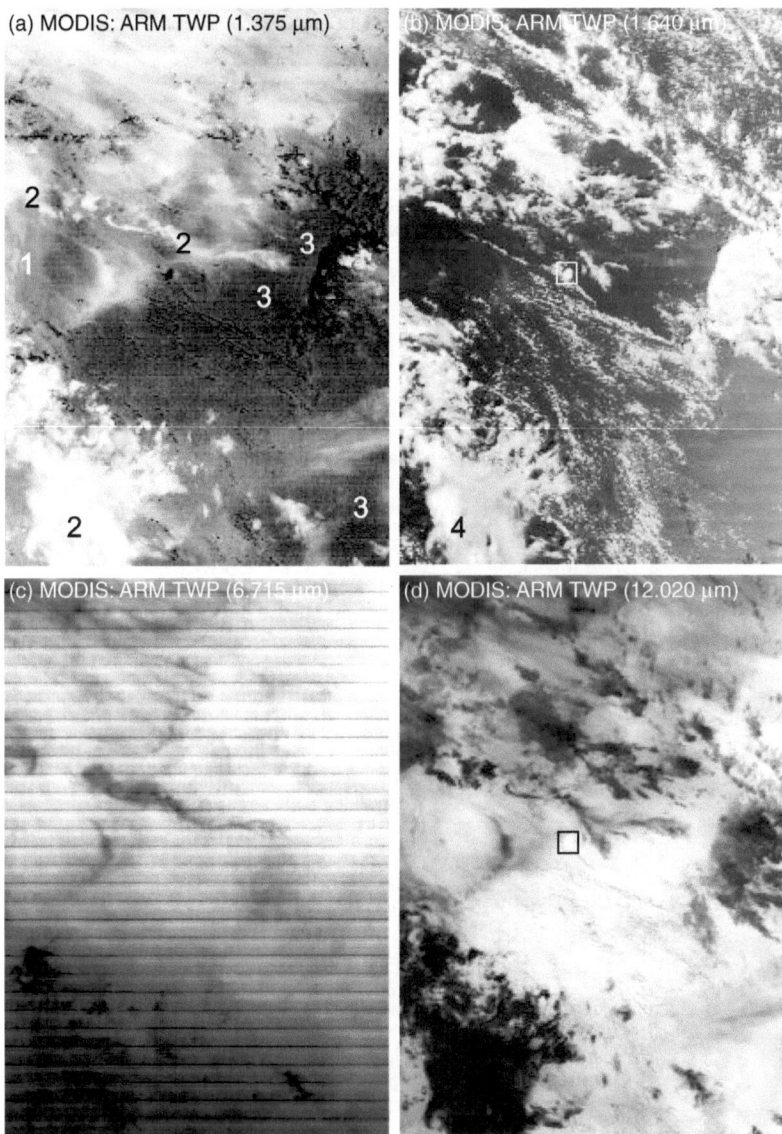

Fig. 2.23. MODIS (**a**) 1.375 μm, (**b**) 1.640 μm, (**c**) 6.715 μm, where the stripping is a known artifact of the multi-sensor calibration, and (**d**) 12.020 μm radiance imagery for the ARM TWP Nauru site at 23:34 UTC on 22 November 2000 corresponding to the MISR radiance images in Fig. 2.22a. The images have been histogram-equalized for contrast enhancement with radiance increasing from black to white. The pixel resolution of these images is 1 km by 1 km, while the domain size of theses images is approximately 220 km by 310 km. The island of Nauru is indicated by a white box in (b) and a black box in (d)

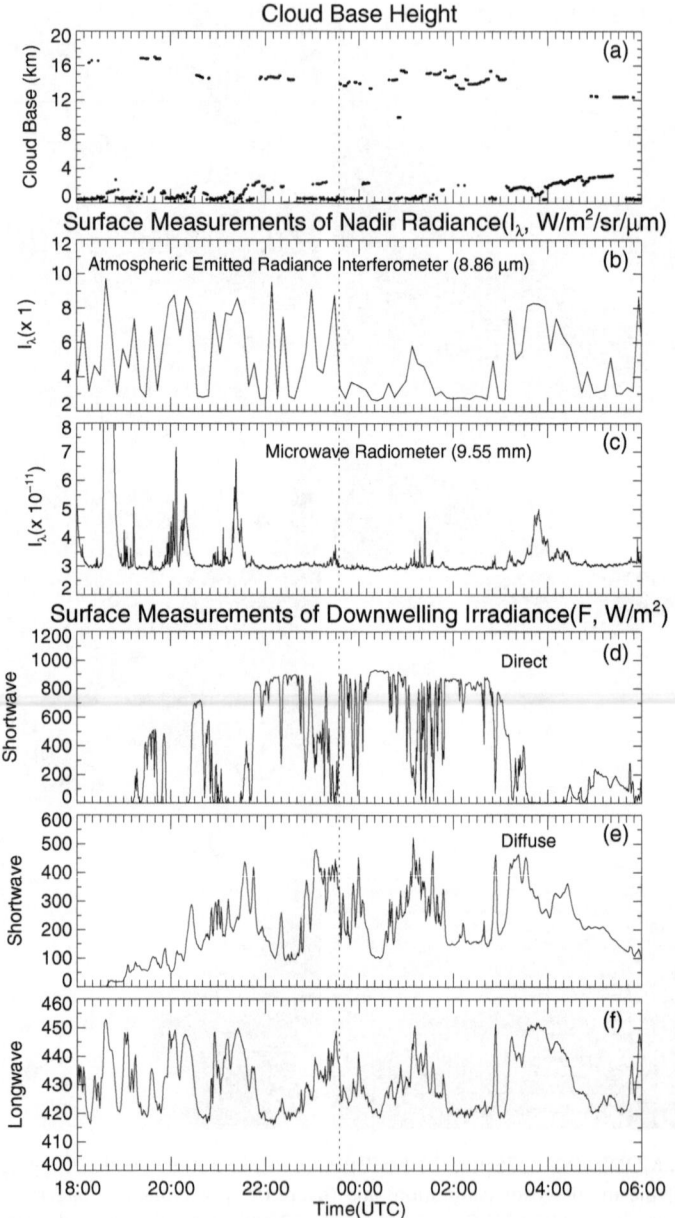

Fig. 2.24. Ground-based observations at the ARM TWP Nauru site on 22–23 November 2000 of (**a**) cloud-base height (micropulse lidar), (**b**) downwelling 8.86 µm radiance (Atmospheric Emitted Radiance Interferometer; Feltz et al., 2003; Turner et al., 2003), (**c**) downwelling 9.55 mm radiance (microwave radiometer), (**d**) downwelling direct shortwave irradiance (pyrheliometer), (**e**) downwelling diffuse shortwave irradiance (shaded pyranometer) and (**f**) downwelling longwave irradiance (pyrgeometer). The *dashed vertical lines* indicate the passage of the Terra satellite

with some of the radiation removed from the direct beam enhancing the shortwave diffuse irradiance (Fig. 2.24e). Cloudy periods also increase downwelling longwave radiances (Fig. 2.24b) and irradiances (Fig. 2.24f) at the surface as a result of emission of radiation from the clouds.

The correlation between the downwelling shortwave diffuse (Fig. 2.24e) and longwave (Fig. 2.24f) irradiances at the surface is an indication that the trade cumulus optical depths and liquid water contents are relatively small. If they were not small, we would have results similar to those at the ARM SGP site (Fig. 2.21e–f). The boundary-layer clouds at the ARM SGP site are sufficiently thick that they are opaque in the longwave spectral region (Fig. 2.21f) and most solar radiation incident upon them does not penetrate through them (Fig. 2.21e). During the cloudy periods at the ARM SGP site, the solar direct beam at the surface is negligible and the downwelling shortwave diffuse irradiance is approximately 100–300 W m^{-2}, i.e., only a small fraction of the incident solar direct beam irradiance. Consequently, for these thick clouds there is not much of a correlation between the downwelling shortwave diffuse and longwave irradiances at the surface.

The clear sky 9.55 mm radiances are smaller at the ARM SGP site (e.g., 20:00–21:00 UTC in Fig. 2.21d) as compared to the ARM TWP site (e.g., 00:20–00:30 UTC in Fig. 2.24c). This is a result of there being less water vapor at cooler temperatures over the ARM SGP site. The boundary-layer cloud contributions to the 9.55 mm radiances at the ARM SGP site are clearly above the clear sky background and exhibit a marked decrease over the three cloud events. Since the 9.55 mm radiance is proportional to the cloud liquid-water path, we conclude that the liquid water contents of these clouds decrease over the three cloud events. The decreases in the cloud-liquid water are sufficient to increase the shortwave diffuse irradiance (Fig. 2.21e) but insufficient to change the longwave opacity of the clouds (Fig. 2.21f).

For the period from 22:00 UTC 22 November through 02:00 UTC 23 November at the ARM TWP Nauru site there are significant fluctuations in the downwelling longwave radiances (Fig. 2.24b), longwave irradiances (Fig. 2.24f) and shortwave direct (Fig. 2.24d) and diffuse (Fig. 2.24e) irradiances. However, the clouds producing these fluctuations often do not produce corresponding changes in the microwave radiances (Fig. 2.24c). That is, during this period many of the trade cumulus clouds contain too little liquid water to produce a significant increase in the microwave radiance above its clear sky level. For these thin boundary-layer clouds, which occur with high frequency at all of the ARM sites (e.g., Sengupta et al., 2003), there is not sufficient information in the microwave radiances to retrieve the liquid-water paths of these clouds. We again have the problem that the information content in the microwave spectral region during cloudy periods is not sufficient to infer the shortwave and longwave radiative properties of the clouds at these times. Recognition of these difficulties is driving the development of cloud property retrievals that use shortwave and longwave spectral measurements, but the accuracies to be obtained by these retrievals, as well as the ones that use microwave radiances, are still being investigated at this time (e.g., Min and Harrison, 1999; Marshak et al., 2000; Barker and Marshak, 2001; Daniel et al., 2002; Savigny et al., 2002; Crewell and Löhnert, 2003; Löhnert and Crewell, 2003; Marchand et al., 2003; Turner et al., 2003).

2.7 Concluding Remarks

Of the Arctic, ARM SGP and ARM TWP Nauru cloud fields illustrated in this chapter, the ones over Nauru contain the greatest number of overlapping cloud layers. Yet, the individual cloud layers over the ARM SGP site and in the environs of Greenland exhibit a variety of different structures. Building global observational databases of both radiatively important cloud-property statistics for individual cloud layers and the statistics of overlapping cloud layers is the focus of much current research. While cloud-property retrievals that use microwave radiances are the most straightforward, they are not without limitations and difficulties. As a result, retrievals that use short-wave and longwave-infrared radiances and irradiances are in development. However, the 3D structures of clouds will affect these retrievals at predominately scattering wavelengths because of horizontal transport of radiation, which complicates considerably the retrievals. At absorbing wavelengths in these spectral regions clouds are often opaque and there is limited information about all of the clouds in a vertical column of the atmosphere. In summary, there is no "free lunch" in the retrieval of cloud microphysical properties.

Developing a global observational database of the 3D properties of clouds is only one step towards the proper treatment of clouds and radiation in numerical models and understanding the impact of clouds on the radiation budget of Earth. The second step is to incorporate properly the observed cloud properties into the radiation calculations that are used in numerical models and in studies of the radiation budget of Earth. Underlying the successful completion of both of these steps is a proper treatment of spectrally-dependent radiative transfer through 3D cloud fields.

References

Ackerman, T.P. and G.M. Stokes (2003). The Atmospheric Radiation Measurement program. *Phys. Today*, **56**, 38–44.

AMS (2000). *American Meteorological Society, Glossary of Meteorology*, T.S. Glickman (ed.). Allen Press, Boston (MA), 2nd edition.

Barker, H.W. and A. Marshak (2001). Inferring optical depth of broken clouds above green vegetation using surface solar radiometric measurements. *J. Atmos. Sci.*, **58**, 2989–3006.

Baumgardner, D., Gayet J.-F., H. Gerber, A. Korolev, and C. Twohy (2002). Clouds: Measurement techniques in situ. In *Encyclopedia of Atmospheric Sciences*. J.R. Holton, J.A. Curry and J. Pyle (eds.). Academic Press, London, UK, pp. 489–498.

Bohren, C.F. and D.R. Huffman (1983). *Absorption and Scattering of Light by Small Particles*. John Wiley and Sons, New York (NY).

Brenguier, J.-L. (1993). Observation of cloud microstructure at centimeter scale. *J. Appl. Meteor.*, **32**, 783–793.

Clothiaux, E.E., T.P. Ackerman, G.G. Mace, K.P. Moran, R.T. Marchand, M.A. Miller, and B.E. Martner (2000). Objective determination of cloud heights and radar reflectivities using a combination of active remote sensors at ARM CART sites. *J. Appl. Meteor.*, **39**, 645–665.

Clough, S.A. and M.J. Iacono (1995). Line-by-line calculations of atmospheric fluxes and cooling rates II: Application to carbon dioxide, ozone, methane, nitrous oxide and the halocarbons. *J. Geophys. Res.*, **100**, 16,519–16,535.

Clough, S.A., M.J. Iacono, and J.-L. Moncet (1992). Line-by-line calculation of atmospheric fluxes and cooling rates: Application to water vapor. *J. Geophys. Res.*, **97**, 15,761–15,785.

Crewell, S. and U. Löhnert (2003). Accuracy of cloud liquid water path from ground-based microwave radiometry – 2. Sensor accuracy and synergy. *Radio Science*, **38**, 10.1029/2002RS002634.

Daniel, J.S., S. Solomon, R.W. Portmann, A.O. Langford, C.S. Eubank, and E.G. Dutton (2002). Cloud liquid water and ice measurements from spectrally resolved near-infrared observations: A new technique. *J. Geophys. Res.*, **107**, Art. No. 4599.

Davis, A., A. Marshak, W.J. Wiscombe, and R.F. Cahalan (1994). Multifractal characterizations of nonstationarity and intermittency in geophysical fields: Observed, retrieved, or simulated. *J. Geophys. Res.*, **99**, 8055–8072.

Davis, A., A. Marshak, W.J. Wiscombe, and R.F. Cahalan (1996). Scale-invariance in liquid water distributions in marine stratocumulus, Part I, Spectral properties and stationarity issues. *J. Atmos. Sci.*, **53**, 1538–1558.

Davis, A.B., A. Marshak, H. Gerber, and W.J. Wiscombe (1999). Horizontal structure of marine boundary-layer clouds from cm- to km-scales. *J. Geophys. Res.*, **104**, 6123–6144.

Diner, D.J., J.C. Beckert, G.W. Bothwell, and J.I. Rodriguez (2002). Performance of the MISR instrument during its first 20 months in earth orbit. *IEEE Trans. Geosci. and Remote Sens.*, **40**, 1449–1466.

Feltz, W.F., W.L. Smith, H.B. Howell, R.O. Knuteson, H. Woolf, and H.E. Revercomb (2003). Near-continuous profiling of temperature, moisture and atmospheric stability using the atmospheric emitted radiance interferometer (AERI). *J. Appl. Meteor.*, **42**, 584–597.

Gerber, H., B.G. Arends, and A.S. Ackerman (1994). New microphysical sensor for aircraft use. *Appl. Opt.*, **31**, 235–252.

Gerber, H., J.B. Jensen, A.B. Davis, A. Marshak, and W.J. Wiscombe (2001). Spectral density of cloud liquid water content at high frequencies. *J. Atmos. Sci.*, **58**, 497–503.

Jeffery, C.A. (2001). Investigating the small-scale structure of clouds using the delta-correlated closure: Effect of particle inertia, condensation/evaporation and intermittency. *Atmos. Res.*, **59**, 199–215.

Kaufman, Y.J., D.D. Herring, K.J. Ranson, and G.J. Collatz (1998). Earth Observing System AM1 mission to earth. *IEEE Trans. Geosci. and Remote Sens.*, **36**, 1045–1055.

King, M.D., W.P. Menzel, Y.J. Kaufman, D. Tanre, B.C. Gao, S. Platnick, S.A. Ackerman, L.A. Remer, R. Pincus, and P.A. Hubanks (2003). Cloud and aerosol properties, precipitable water and profiles of temperature and water vapor from MODIS. *IEEE Trans. Geosci. and Remote Sens.*, **41**, 442–458.

King, W.D., D.A. Parkin, and R.J. Handsworth (1978). A hot-wire water device having fully calculable response characteristics. *J. Appl. Meteor.*, **17**, 1809–1813.

Knollenberg, R.G. (1976). Three new instruments for clouds physics measurements: The 2D spectrometer, the forward scattering spectrometer probe and the active scattering aerosol spectrometer. In *Proceedings of Int. Cloud Phys. Conf.* held in Boulder, Colorado, Amer. Meteor. Soc., Boston (MA), 444–461.

Korolev, A.V. and I.P. Mazin (1993). Zones of increased and decreased droplet concentrations in stratiform clouds. *J. Appl. Meteor.*, **32**, 760–773.

Korolev, A.V. and I.P. Mazin (2003). Supersaturation of water vapor in clouds. *J. Atmos. Sci.*, **60**, 2957–2974.

Korolev, A.V., J.W. Strapp, G.A. Isaac, and A.N. Nevzorov (1998). The Nevzorov airborne hot-wire LWC-TWC probe: Principle of operation and performance characteristics. *J. Atmos. Oceanic Technol.*, **15**, 1495–1510.

Korolev, A.V., G.A. Isaac, and J. Hallett (2000). Ice particle habits in stratiform clouds. *Q. J. Roy. Meteor. Soc.*, **126**, 2873–2902.

Korolev, A.V., G. Isaac, I. Mazin, and H. Barker (2001). Microphysical properties of continental stratiform clouds. *Quart. J. Roy. Meteor. Soc.*, **127**, 2117–2151.

Korolev, A.V., G.A. Isaac, S. Cober, and J.W. Strapp (2003). Microphysical characterization of mixed-phase clouds. *Quart. J. Roy. Meteor. Soc.*, **129**, 39–66.

Liljegren, J.C. (1994). Two-channel microwave radiometer for observations of total column precipitable water vapor and cloud liquid water path. In *Proceedings of Fifth Symposium on Global Change Studies*, Nashville, Tennessee, 262–269.

Liljegren, J.C., E.E. Clothiaux, G.G. Mace, S. Kato, and X.Q. Dong (2001). A new retrieval for cloud liquid water path using a ground-based microwave radiometer and measurements of cloud temperature. *J. Geophys. Res.*, **106**, 14,485–14,500.

Loeb, N.G., T. Várnai, and D.M. Winker (1998). Influence of subpixel-scale cloud-top structure on reflectances from overcast stratiform cloud layers. *J. Atmos. Sci.*, **55**, 2960–2973.

Logar, A.M., D.E. Lloyd, E.M. Corwin, M.L. Penaloza, R.E. Feind, T.A. Berendes, K.-S. Kuo, and R.M. Welch (1998). The ASTER polar cloud mask. *IEEE Trans. Geosci. and Remote Sens.*, **36**, 1302–1312.

Löhnert, U. and S. Crewell (2003). Accuracy of cloud liquid water path from ground-based microwave radiometry – 1. Dependency on cloud model statistics. *Radio Science*, **38**, 10.1029/2002RS002654.

Long, C.N. and T.P. Ackerman (2000). Identification of clear skies from broadband pyranometer measurements and calculation of downwelling shortwave cloud effects. *J. Geophys. Res.*, **105**, 15,609–15,626.

Marchand, R., T.P. Ackerman, E.R. Westwater, S.A. Clough, K. Cady-Pereira, and J.C. Liljegren (2003). An assessment of microwave absorption models and retrievals of cloud liquid water using clear-sky data. *J. Geophys. Res.*, **108**, 4773, doi:10.1029/2003JD003843.

Marshak, A., Yu. Knyazikhin, A.B. Davis, W.J. Wiscombe, and P. Pilewskie (2000). Cloud – vegetation interaction: Use of normalized difference cloud index for estimation of cloud optical thickness. *Geophys. Res. Lett.*, **27**, 1695–1698.

Michalsky, J., E. Dutton, M. Rubes, D. Nelson, T. Stoffel, M. Wesley, M. Splitt, and J. DeLuisi (1999). Optimal measurement of surface shortwave irradiance using current instrumentation. *J. Atmos. Oceanic Technol.*, **16**, 55–69.

Min, Q.-L. and L.C. Harrison (1999). Joint statistics of photon pathlength and cloud optical depth. *Geophys. Res. Lett.*, **26**, 1425–1428.

Moran, K.P., B.E. Martner, M.J. Post, R.A. Kropfli, D.C. Welch, and K.B. Widener (1998). An unattended cloud-profiling radar for use in climate research. *Bull. Amer. Meteor. Soc.*, **79**, 443–455.

Philipona, R., E.G. Dutton, T. Stoffel, J. Michalsky, I. Reda, A. Stifter, P. Wendling, N. Wood, S.A. Clough, E.J. Mlawer, G. Anderson, H.E. Revercomb, and T.R. Shippert (2001). Atmospheric longwave irradiance uncertainty: Pyrgeometers compared to an absolute sky-scanning radiometer, atmospheric emitted radiance interferometer and radiative transfer model calculations. *J. Geophys. Res.*, **106**, 28,129–28,141.

Platnick, S., M.D. King, S.A. Ackerman, W.P. Menzel, B.A. Baum, J.C. Riedi, and R.A. Frey (2003). The MODIS cloud products: Algorithms and examples from Terra. *IEEE Trans. Geosci. and Remote Sens.*, **41**, 459–473.

Pruppacher, H.R. and J.D. Klett (1997). *Microphysics of Clouds and Precipitation: Second Revised and Enlarged Edition with an Introduction to Cloud Chemistry and Cloud Electricity*. Kluwer Academic Publishers, Dordrecht, The Netherlands.

Savigny, C. von, A.B. Davis, O. Funk, and K. Pfeilsticker (2002). Time-series of zenith radiance and surface flux under cloudy skies: Radiative smoothing, optical thickness retrievals and large-scale stationarity. *Geophys. Res. Lett.*, **29**, 1825, doi:10.1029/2001GL014153.

Sengupta, M., E.E. Clothiaux, T.P. Ackerman, S. Kato, and Q.L. Min (2003). Importance of accurate liquid water path for estimation of solar radiation in warm boundary layer clouds: An observational study. *J. Climate*, **16**, 2997–3009.

Spinhirne, J.D. (1993). Micropulse lidar. *IEEE Trans. Geosci. and Remote Sens.*, **31**, 48–55.

Stokes, G.M. and S.E. Schwartz (1994). The Atmospheric Radiation Measurement (ARM) program: Programmatic background and design of the cloud and radiation test bed. *Bull. Amer. Meteor. Soc.*, **75**, 1201–1221.

Toon, O.B. and T.P. Ackerman (1981). Algorithms for the calculation of scattering by stratified spheres. *Appl. Optics*, **20**, 3657–3660.

Turner, D.D., S.A. Ackerman, B.A. Baum, H.E. Revercomb, and P. Yang (2003). Cloud phase determination using ground-based AERI observations at SHEBA. *J. Appl. Meteor.*, **42**, 701–715.

Uttal, T., J.A. Curry, M.G. McPhee, D.K. Perovich, R.E. Moritz, J.A. Maslanik P.S. Guest, H.L. Stern, J.A. Moore, R. Turenne, A. Heiberg, M.C. Serreze, D.P. Wylie O.G. Persson, C.A. Paulson, C. Halle, J.H. Morison, P.A. Wheeler, A. Makshtas, H. Welch M.D. Shupe, J.M. Intrieri, K. Stamnes, R.W. Lindsey, R. Pinkel, W.S. Pegau, T.P. Stanton, and T.C. Grenfeld (2002). Surface heat budget of the Arctic Ocean. *Bull. Amer. Meteor. Soc.*, **83**, 255–275.

Westwater, E.R., Y. Han, M.D. Shupe, and S.Y. Matrosov (2001). Analysis of integrated cloud liquid and precipitable water vapor retrievals from microwave radiometers during the Surface Heat Budget of the Arctic Ocean project (2001). *J. Geophys. Res.*, **106**, 32,019–32,030.

Wielicki, B.A., R.D. Cess, M.D. King, D.A. Randall, and E.F. Harrison (1995). Mission to Planet Earth – Role of clouds and radiation in climate. *Bull. Amer. Meteor. Soc.*, **76**, 2125–2153.

Part II

Fundamentals

3

A Primer in 3D Radiative Transfer

A.B. Davis and Y. Knyazikhin

3.1   **Introduction** 153

3.2   **Radiometric Quantities** 157

3.3   **Sinks** 164

3.4   **Scattering** 170

3.5   **Propagation** 178

3.6   **Sources** 181

3.7   **Local Balance** 188

3.8   **Global Balance** 194

3.9   **Outgoing Radiation** 203

3.10 **Green Functions to Reciprocity via Adjoint Transport** 213

3.11 **Summary and Outlook** 224

References 228

3.1 Introduction

3.1.1 Three-Dimensional?

Technically speaking, only the two-stream model in homogeneous (or layered) plane-parallel, cylindrical, or spherical geometries can be truly one-dimensional (1D) because there is no angular dependence to worry about, only the axial flow of radiant energy in a highly symmetric medium with equally symmetric source distributions. By strict mathematical standards, azimuthally-averaged or -symmetric radiative transfer in a plane-parallel medium is already 2D (one spatial and one angular coordinate). By the same token, it is patently 3D if there is also azimuthal variation (one extra angular coordinate) as, e.g., when solar illumination is off-zenith. However, it is generally understood that, independently of the how angles are treated, all plane-parallel radiative transfer (RT) theory is called "1D"; at most stratification in the vertical z direction is allowed. So only spatial variability counts here. Then what about patently 2D cases, so often used in sensitivity studies, where optical properties and/or sources vary at most in the horizontal x and vertical z directions? Well, this is still called "3D" RT, for the legendary simplicity. In short, when we say we are treating 3D radiative transfer it only means that we are making no assumptions about the translational or rotational symmetry of the optical medium's macro-structure nor about the sources of radiation. To make things worse, we will see that the most general 3D problem in RT is exactly solvable as long as there is no scattering: only emission and absorption are present and no coupling exists between the radiation beams. Mathematically speaking, this solution is a simple 1D integration beam-by-beam, where opposite directions count separately (since they are not coupled). And then there is the possibility of time-dependence.[1]

Having somewhat clarified and somewhat obfuscated what is meant by "dimension" in the RT literature, we can ask about the history of RT theory that acknowledges that we live in a 3D world. This question of chronology breaks into two more specific ones covered in the next few paragraphs. First, how did we get to modern radiometry and formulate the radiative transfer equation (RTE)? Then, skipping much on the solution of the RTE in slab geometry with angular details (for planetary or stellar atmospheres) or spherical geometry in a 2-stream mode (for stellar interiors), how did 3D radiative transfer per se develop from the dawn of scientific computing to circa 1980 in application to the *natural* sciences (atmospheric and, to some extent, astrophysical questions)? We cover the first topic simply by tracing a thread through the contributions of many celebrated scientists, primarily to build historical context. The second topic is covered with detailed references to the seminal papers by the pioneers of 3D radiative transfer because we have far more than occasionally found it refreshing to go back to the early publications in our field.

We have decided, somewhat arbitrarily, that post-1980 literature is best covered in the specialized chapters of this volume. We have also decided that applications to *engineered* systems is another story altogether, an interesting one in its own right that we could not do justice to. We will simply acknowledge that the engineering community has had to struggle with 3D radiation transport, primarily from thermal sources, in increasingly intricate geometries. One is bound to find significant overlap between our concerns and theirs. Indeed, both atmospheric scientists and engineers will start with simple geometries either because they are tractable or because they are viable designs. However, in the end, both will have to consider the complexity of how turbulent reacting flows interact with radiation. It might prove very rewarding for both communities to draw more on each other's experience with 3D RT.

---

[1] Time-dependent, equivalently pathlength-based, 3D RT has always been around but it is now becoming important in the applications (cf. Chaps. 12–13). Since time is not just another dimension (causality oblige), the physicist's "3+1 D" shorthand is better than talking about "4D" RT.

3.1.2 From Radiometry to Radiative Transfer

As far as we know, the earliest physically correct analyses of radiometric data (i.e., based on the intuitive notion of radiant energy conservation) were by Galileo Galilei (1564–1642) and Johannes Kepler (1571–1630), discussing their respective observations of the Moon and of Mars. This is of course only about the propagation of radiant energy – whatever they understood that to be – across empty space; so the problem at hand is fully 3D but in the simple case where there is no scattering, nor absorption for that matter. Let us acknowledge the forefathers of general-purpose (hence 3D) radiometry: Lambert, Bouguer, de Beer, Helmholtz, and others. We must mention in passing the founders of particle transport theory on which modern RT is based: Maxwell and Boltzmann, who worked in the earliest years of the (modern) atomic theory of matter when it was still highly controversial. Then come the pioneers of RT per se, i.e., with the complication of scattering: Schuster, Schwarzschild, Eddington, and Peierls. They were soon followed by the giants of 1D RT theory: Milne, Sobolev, Ambartsumian, and Chandrasekhar. The onset of the nuclear age brought us phenomenal advances in computational transport theory driven by the 3D geometry of weapons and reactors. We commemorate from this period the brilliant contributions by von Neumann, Ulam, Metropolis, Teller, Marshak, Davison, Vladimirov, Germogenova, and others.

On a parallel track, we can trace scientific progress in "elementary" radiation-matter interaction, defined operationally as what provides RT with its emission, absorption, and scattering coefficients and terms. This is in fact the bridge between RT and mainstream optics, drawing on both sides of its celebrated duality between waves and particles. Here the modern era opens arguably with Leonardo da Vinci's (1452–1519) notes on smoke plumes and unfolds with Newton, Descartes and Huygens. The fundamental link between spectroscopy and thermal physics was established by Fraunhofer, Kirchhoff, Planck, and Einstein. There are too many important contributions of early quantum theorists and experimentalists to attempt even a partial list that is meaningful. Because scattering is what makes RT so interesting and challenging, especially in a 3D setting, we will recall the classic work, still in use, by Rayleigh, Lorenz, Mie, and Raman.

Computing absorption and/or scattering coefficients and emission terms is one thing, and deriving the full RTE from first principles in optics is another. The difficulty hinges on the connection between the radiance field that plays a central photon transport theory and the fundamental quantities of scalar or, better still, electromagnetic (EM) wave theory. The crux of the matter is the loss of wave theoretical (i.e., amplitude and phase) information in the spatial coarse-graining to scales of at least a few wavelengths where a statistical description of the wave field applies. For remarkable – and still on-going – efforts to bridge this gap between radiometry and optics, we refer to Ishimaru (1975) who works from scalar waves, Wolf (1976) who works from vector waves in the frame of classic or quantum EM theory, and Mishchenko (2003) who starts with Maxwell's EM equations and carries polarization throughout. To this day, the theory of radiative transfer we are concerned with in this chapter and volume remains a phenomenology, a powerful one even though it is not yet rigorously connected to optics per se.

3.1.3 Atmospheric 3D Radiative Transfer: The Early Years ($\lesssim$1980)

By the second half of the 20th century, the stage is set for 3D RT as we presently understand it, that is, in application to astrophysical or geophysical rather than man-made systems; we are also interested in theoretical studies of abstract media that are based on at least some analytical work on the 3D RT equation or an approximation thereof. With these selection rules, we have traced the beginning of 3D RT to Richards' investigation (Richards, 1956) of a point-source in a homogeneous scattering slab medium that is finitely thick, not in boundary-free 3D space, while Giovanelli and Jefferies (1956) looked at variable sources in more generality. Around the same time, Chandrasekhar (1958) considered a collimated "pencil-beam" source impinging on a uniform semi-infinite medium. However, Giovanelli's (1959) paper stands out as the earliest study of 3D variability effects as we still think of them most often: the slab medium is internally variable and results are compared to the prediction of a standard 1D (internally uniform) model. During the 1960s, the first 3D RT papers appeared in the atmospheric literature per se: Romanova (1968a,b) on the pencil-beam problem in a uniform medium, Weinman and Swartztrauber (1968) on uniformly illuminated media with a horizontal sine-wave structure. In the 1970s, we continue to see the same two classes of problem addressed with increasing sophistication. On the one hand, we have pencil-beams (now readily materialized with laser technology) illuminating a uniform scattering plane-parallel medium (Romanova, 1971a,b), or the closely related (essentially adjoint) problem of surface albedo blurring by the intervening atmosphere (Odell and Weinman, 1975; Otterman and Fraser, 1979; Kaufman, 1979). On the other hand, we have uniformly illuminated but internally variable slabs (van Blerkom, 1971; Avaste and Vainikko, 1974; McKee and Cox, 1974; Appleby and van Blerkom, 1975; Romanova, 1975; McKee, 1976; Aida, 1977a,b; Wendling, 1977), or simply non-plane-parallel media such as upright cylinders with circular sections (B&A, 1977) or perpendicular parallelepipeds (Davies and Weinman, 1977; Davies, 1978). These are the two extreme situations for predominantly scattering media, most often with solar illumination at the upper boundary: optically thin and thick cases, respectively for aerosol and cloud problems. Another transport regime of considerable interest is large optical thickness in predominantly absorbing media, most often with internal sources. This scenario applies to the thermal spectrum (Weinman and Swartztrauber, 1968) and to microwaves where the strongly 3D structure of rain matters. The first line of attack here is to neglect scattering altogether; after that, just a few successive orders-of-scattering makes for almost exact models.

The methodologies used in the early studies of strongly scattering optically thick media were almost invariably Monte Carlo simulation for numerical results (if any) and either the diffusion or small-angle approximations for the analytical work (if any). The noteworthy exceptions were (1) Chandrasekhar's (1958) pencil-beam study in purely scattering media which used neither approximations nor numerics but established the formal connection between horizontal transport away from the beam and the problem of an absorbing/scattering medium under uniform illumination problem, and (2) Avaste and Vainikko's (1973) "mean-field" theory for a stochastic binary (cloudy/clear) medium with a random (Poissonian) distribution of transitions. Two other notable publications were Cannon's (1970) article, a penetrating analysis of numerical results on line transfer in a 2D medium using a finite-difference technique to solve the RT equation (not an approximation), and the compilation by Mullamaa et al. (1972), a poorly distributed report (even in translation), where the linear mixture of 1D results that eventually became known as the Independent Pixel/Column Approximation (IPA or ICA) was first introduced, at least in the former Soviet Union.

This brings us up to the English edition of Marchuk et al.'s landmark volume on the Monte Carlo technique (Marchuk et al., 1980). Developments beyond 1980 are better covered in the specialized chapters that follow. At this cusp, we will also mention the paper by Ronnholm et al. (1980) who reinvented the important IPA/ICA technique for the benefit of the Western literature. The IPA/ICA is used extensively in Chaps. 6, 8, 9, and 12. The main purpose of this volume however is to go beyond the IPA/ICA, either analytically or computationally.

3.1.4 Overview

This introductory chapter is organized as follows. In the next section, we review the basic concepts of radiometry and radiative transfer (RT) that are prerequisite for the following sections and chapters. Before formulating the radiative transfer equation (RTE) in Sect. 3.7, we follow a logical but physically backwards flow from detectors (Sect. 3.2) and sinks (Sect. 3.3) to sources (Sect. 3.6), via scattering (Sect. 3.4) and propagation (Sect. 3.5). Once we have the RTE in hand (Sect. 3.7), we examine boundary conditions and integral formulations (Sect. 3.8). At that point, numerical solutions of a couple of 3D RT problems are presented, primarily to illustrate less familiar boundary shapes (non-flat lower boundary and horizontally-finite clouds). Green functions, adjoint RT theory and reciprocity are covered in Sect. 3.10. We summarize in Sect. 3.11 and offer our perspective on the future of research into the fundamental aspects of RT theory. A compendium of Suggested Reading, with running commentary, supplements the usual list of References, and there is some inevitable overlap between the two resources. At the end of the volume, we have compiled in tabular form the most common Notations as well as some useful constants and definitions.

3.2 Radiometric Quantities

We recall and apply the definitions of all the important quantities used in radiometry and, from there, RT theory. Radiometry is essentially a theory of light detection in the sense of photon gathering, just before conversion into electrical current or charge, heat, or whatever else that can become an instrument reading.

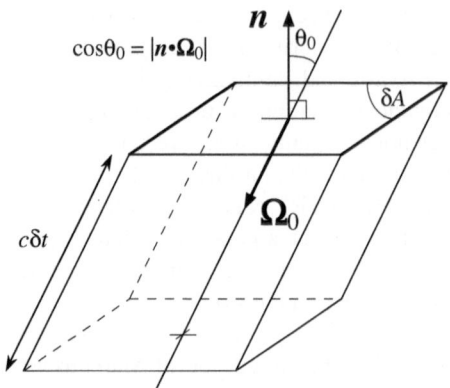

$$\cos\theta_0 = |n \cdot \Omega_0|$$

Fig. 3.1. Flux transfer by an oblique collimated beam

3.2.1 Flux/Irradiance in a Collimated Beam

The most basic quantity in radiometry is "flux," a.k.a. "irradiance" (or photon "current density", coming from general particle transport theory). It is at once an observable that can be sampled at any point with the proper equipment and a field that exists everywhere, like gravity. Figure 3.1 shows a simple experiment where a collimated beam impinges on a detection area δA for a certain time interval δt. Our goal is to count the number of light quanta that are detected by crossing the surface, each carrying energy in the amount of $h\nu$ (where h is the Planck constant and ν the frequency). If δA and δt are small enough, this number δN is certainly proportional to the kinetic volume in the figure; specifically,

$$\delta N = \frac{\delta E}{h\nu} \propto \delta V = \cos\theta_0 \delta A \times c\delta t \tag{3.1}$$

where c is the speed of light in the optical medium[2] and θ_0 is the incidence angle of the beam away from the normal to the small/flat detection surface. The dependencies on δA and δt are fully expected while the "$\cos\theta_0$" factor takes a little more thought (δA has to be projected perpendicularly to the beam to get δN right). This is known as Lambert's cosine law of radiometry and it is in fact a requirement for radiometers to follow this law which, in practice, is not so easy to achieve at large incidence angles.

Some radiometric devices count photons, others respond to radiant energy, so we allow for both possibilities in (3.1). The proportionality factor in (3.1),

$$f_{\text{col}} = \lim_{\delta A, \delta t \to 0} \frac{\delta N \ (\text{or } \delta E)}{\cos\theta_0 \delta A \times c\delta t}, \quad \text{in m}^{-3} \ (\text{or J/m}^3), \tag{3.2}$$

---

[2] If there are significant variations of the index of refraction across the transport medium of interest, not counting microscopic scattering centers, then several aspects of RT need to be modified.

is thus the density of photons (or radiant energy) in space at the point where δN was obtained propagating in the given beam. It is a characteristic property of the beam – its strength – as is its direction of propagation, Ω_0 in Fig. 3.1. The other quantities relate the specifics of its measurement, either the outcome δN (or δE) or the controlled parameters δA (aperture) and δt (exposure).

A more conventional characterization of beam strength is by the flux it transports, which is given by

$$F_0 = c f_{\text{col}}, \quad \text{in m}^{-2}\text{s}^{-1} \text{ (or W/m}^2\text{)} . \tag{3.3}$$

The result of the above measurement is thus

$$\frac{\delta E}{\delta t} = F_n^{(\pm)}(\Omega_0)\delta A = \cos\theta_0 F_0 \delta A = |n \cdot \Omega_0| F_0 \delta A \tag{3.4}$$

where the subscript n identifies the orientation of the detector and the superscript $(\pm)$ the direction from which the beam is coming, specifically $\pm = \text{sign}(n \cdot \Omega)$. In the case of Fig. 3.1, the outcome is $(-)$.

To illustrate, we imagine an isotropic point-source of power P (in W or photons/s) and a detector at some distance d subtending a solid angle $\delta\Omega = \cos\theta\delta A/d^2$; see Fig. 3.2. The reading of the device is

$$\frac{\delta E}{\delta t} = P\frac{\delta\Omega}{4\pi} = \frac{P}{4\pi d^2}\cos\theta\delta A . \tag{3.5}$$

By comparison with (3.4), we have

$$F_0 = \frac{P}{4\pi d^2} . \tag{3.6}$$

So flux diminishes with distance, as required by the overall conservation of energy flowing through spheres of any radius d. Strictly speaking, this well-known "$1/d^2$" decay applies only in absence of absorbing/scattering material; otherwise, it is only one of several terms (as we will see in Sects. 3.3 and 3.8).

3.2.2 Radiance/Intensity in a Diffuse Light Field

The experiment in Fig. 3.3 is a generalization of that in Fig. 3.1 where exposure time is now represented by a stop-watch icon rather than by a kinetic volume. Light is

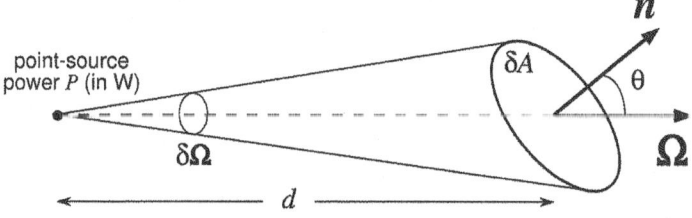

Fig. 3.2. Flux from a distant point-source transferred through an optical vacuum

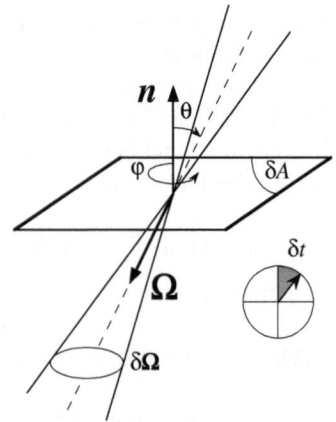

Fig. 3.3. Radiance in a diffuse light field

now admitted into $\delta V = \delta A \cos \theta \times c\delta t$, but only from a small but finite solid angle $\delta\Omega$ around Ω. The outcome is now

$$\delta N(\Omega) = \frac{\delta E}{h\nu} \propto \delta V \times \delta\Omega = (\delta A \cos \theta \times c\delta t) \times \delta\Omega , \qquad (3.7)$$

and the relevant *diffuse* beam property is

$$f_{\text{dif}} = \lim_{\delta V, \delta\Omega \to 0} \frac{\delta N \text{ (or } \delta E)}{\delta V \times \delta\Omega}, \quad \text{in m}^{-3}\text{sr}^{-1} \text{ (or J/m}^3\text{/sr)} \qquad (3.8)$$

in comparison with the *collimated* beam property in (3.2).

Here again, a more conventional characterization of beam strength uses radiance or (specific) intensity[3]

$$I(\Omega) = cf_{\text{dif}}, \quad \text{in m}^{-2}\text{s}^{-1}\text{sr}^{-1} \text{ (or W/m}^2\text{/sr)} \qquad (3.9)$$

and the associated measurement outcome is

$$\delta E = |\boldsymbol{n} \cdot \boldsymbol{\Omega}| \, I(\Omega)\delta\Omega\delta A\delta t . \qquad (3.10)$$

From this point on, it is important to bear in mind that polarization selection and wavenumber filters may be used in conjunction with radiometers. So the most general description of the light field anywhere in space-time calls for an intensity I dependent on all of the quantum mechanical parameters of the photon population:

---

[3] In this volume, we have adopted standard notations for radiance/intensity $I_\lambda(\boldsymbol{x}, \boldsymbol{\Omega})$ and irradiance/flux $F_\lambda(\boldsymbol{x})$ from the astrophysical and transport-theoretical literatures because they are also well used in the geophysics community. However, readers more familiar with some remote-sensing textbooks will recognize respectively L_λ and E_λ.

- wavenumber ν (or energy $E = h\nu$);
- direction of travel $\mathbf{\Omega}$ (or momentum $\mathbf{p} = (E/c)\mathbf{\Omega}$);
- statistical state of polarization (or spin).

In this volume we will be concerned exclusively with the first two and, in this chapter, mostly with the second. The most popular representation of polarization uses Stokes' radiance "vector" where $I(\mathbf{\Omega})$ is complemented by three other quantities. For more details, we refer the interested reader to Chandrasekhar (1950).

So far, we have always been at some position, presumably in 3D space, making radiometric measurements. Now imagine a diffuse source at a certain distance d from the detector, as illustrated in Fig. 3.4. The throughput in radiant energy can be evaluated in two different ways:

$$\delta E = I_{\det}(\mathbf{\Omega})\delta A_{\det} \cos \theta_{\det} \delta \mathbf{\Omega}_{\det} \delta t \ ,$$
$$\delta E = I_{\mathrm{src}}(\mathbf{\Omega})\delta A_{\mathrm{src}} \cos \theta_{\mathrm{src}} \delta \mathbf{\Omega}_{\mathrm{src}} \delta t \ ,$$

respectively from the detector's and source's viewpoints, where

$$\delta \mathbf{\Omega}_{\det} = \delta A_{\mathrm{src}} \cos \theta_{\mathrm{src}} / d^2 \ ,$$
$$\delta \mathbf{\Omega}_{\mathrm{src}} = \delta A_{\det} \cos \theta_{\det} / d^2 \ .$$

This shows that, by definition, radiance is conserved across optical vacuum,

$$I_{\det}(\mathbf{\Omega}) = I_{\mathrm{src}}(\mathbf{\Omega}) \ . \tag{3.11}$$

Apart from showing that the quantity "radiance" was basically designed to be conserved along a beam in optical vacuum, we see that it is productive to think of radiance as a 5-dimensional field $I(\mathbf{x}, \mathbf{\Omega})$ for a given wavelength and (optionally) state of polarization. Above, we considered the photon flow between $\mathbf{x}_{\mathrm{src}}$ and $\mathbf{x}_{\det}$.

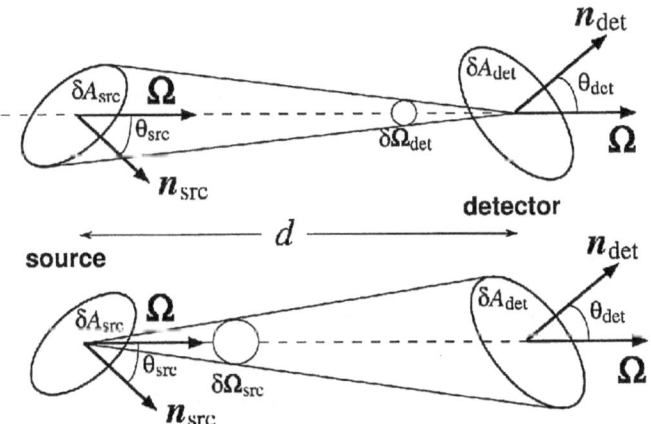

Fig. 3.4. Conservation of radiance in a beam across an optical vacuum

This brings us to the question of "pixels" in Earth/planet observation as well as astronomy. They are in practice identified and populated (say) with photon counts by varying Ω *from* the observer's position x_{det} in space. That is in principle how an image is collected remotely. Often however, Ω is in a small subregion of direction space Ξ and an individual pixel's solid angle is of course much smaller. In this case, it is convenient to think of Ω as the constant direction *towards* the distant observer while x_{pix} (a.k.a. x_{src}) scans the image's pixels in some convenient reference plane in position space (such as cloud top).[4]

We can now revisit the concept of flux from the previous subsection without the assumption of a collimated beam. Indeed, by comparing (3.10) and (3.4) we can define the *element* of flux

$$\delta F_n^{(-)}(x, \Omega) = |n \cdot \Omega| \, I(x, \Omega) \delta\Omega \; . \tag{3.12}$$

At this point, we need an analytical representation of the beam direction Ω on the unit sphere Ξ. We will use both Cartesian and spherical (pole at $\hat{z}$) coordinates:

$$\Omega(\theta, \varphi) = \begin{pmatrix} \Omega_x \\ \Omega_y \\ \Omega_z \end{pmatrix} = \begin{pmatrix} \sin\theta \cos\varphi \\ \sin\theta \sin\varphi \\ \cos\theta \end{pmatrix} = \begin{pmatrix} \sqrt{1-\mu^2} \cos\varphi \\ \sqrt{1-\mu^2} \sin\varphi \\ \mu \end{pmatrix} \; , \tag{3.13}$$

for $\mu = \cos\theta = \Omega_z \in [-1, +1]$, and $\varphi = \tan^{-1}(\Omega_y/|\Omega_x|) + \pi \, \text{sign}[\Omega_y](\text{sign}[\Omega_x] - 1)/2 \in (-\pi, +\pi]$. From there, the element of solid angle is given simply by

$$d\Omega = d\mu d\varphi = \sin\theta d\theta d\varphi \; . \tag{3.14}$$

This enables us to define the two *hemispherical* fluxes with respect to an arbitrary plane at any point in space:

$$F_n^{(\pm)}(x) = \int_{\pm n \cdot \Omega > 0} |n \cdot \Omega| \, I(x, \Omega) d\Omega \; . \tag{3.15}$$

These can in turn be combined algebraically to define the *net* flux in any direction:

$$F_n(x) = F_n^{(+)}(x) - F_n^{(-)}(x) = \int_{4\pi} (n \cdot \Omega) \, I(x, \Omega) d\Omega \; . \tag{3.16}$$

In classic plane-parallel – often called one-dimensional (1D) – RT, there is only an interest in vertical fluxes (assuming the slab is horizontal), obtained for $n = \hat{z}$. In 3D RT, there is also an interest in horizontal fluxes, $n = \hat{x}$, or $n = \hat{y}$.

---

[4] We have ignored here complications due to finite stand-off distance and detector motion during the imaging that arise for moderate- to low-resolution systems with large swaths. In this case, each pixel has its $\{x, \Omega\}$-pair and both vectors must be somehow "georegistered."

Consider two extreme situations that we will encounter frequently in the following chapters and where we need to know how to relate radiance/intensity and irradiance/flux:

- Collimated beam: $f_{\mathrm{dif}} = f_{\mathrm{col}}\delta(\mathbf{\Omega} - \mathbf{\Omega}_0)$ in (3.8) where f_{col} was defined in (3.2). Using (3.3) and (3.9), we have

$$I(\mathbf{\Omega}) = F_0\delta(\mathbf{\Omega} - \mathbf{\Omega}_0) \ . \tag{3.17}$$

- Isotropic (Lambertian) emittance into a hemisphere by a surface element: $I(\mathbf{\Omega}) \equiv I_{\mathrm{L}}$, $\forall \mu > 0$, $\forall \varphi \in [0, 2\pi)$. The associated hemispherical flux is therefore

$$F_{\mathrm{L}} = \pi I_{\mathrm{L}} \ . \tag{3.18}$$

There is a popular non-dimensional representation of radiance in solar problems, especially for satellite imaging analysis, that makes use of both these examples. If the mono-directional radiance field in (3.17) is incident on a scattering medium, then a field of diffusely reflected radiance is generated that we will denote $I_{\mathrm{TOA}}(\mu, \varphi)$, with $\mu > 0$. In atmospheric applications, the uppermost level is colloquially called the Top-Of-Atmosphere (or "TOA"). As we will see further on in our discussion of "secondary" sources, the albedo of a surface (or of a plane-parallel medium) is defined as the ratio of outgoing-to-incoming fluxes, measured perpendicular to the surface (or upper boundary). We now assume that the surface (boundary) is horizontal. Then the incoming flux is $\mu_0 F_0$, a quantity we will frequently encounter where $\mu_0 = \cos\theta_0$. We do not necessarily know the out-going flux, a hemispherical integral. In fact, often we have only one directional sample of the out-going radiance distribution, say, the nadir radiance (propagating vertically upward) in every pixel of a satellite image $I_{\mathrm{TOA}}(\mathbf{\Omega} = \hat{\mathbf{z}})$. However, with a Lambertian hypothesis, we can use (3.18) to predict the flux and, from there, we can define the *apparent* albedo of the medium (generally a surface/atmosphere composite). This is known as the "Bidirectional Reflectance Factor" or

$$\mathrm{BRF} = \frac{\pi I_{\mathrm{TOA}}(\hat{\mathbf{z}})}{\mu_0 F_0} \ . \tag{3.19}$$

Note that the BRF, unlike the original out/in flux-ratio concept, is not bounded between 0 and 1; notwithstanding, this is often called "TOA reflectance" in satellite remote sensing. Sections 3.6.2 and 3.9 cover reflection properties of surfaces and atmosphere-surface systems in more detail, including angular integrals that are flux ratios and are between 0 and 1.

3.2.3 Scalar/Actinic and Vector Fluxes

So far, we have illustrated the operational principles of radiometric measurement using radiance $I(x, \mathbf{\Omega})$ which will generally depend on both position x and direction $\mathbf{\Omega}$. Other quantities can be defined by integration over direction-space. There are both theoretical and practical reasons for doing this.

We start with the *actinic* (a.k.a. scalar) flux

$$J(x) = \int_{4\pi} I(x, \Omega) d\Omega \qquad (3.20)$$

which can be related to photon (or radiant energy) density. We already encountered a photon density in (3.2) but here it is understood, as usual, to be irrespective of direction of travel:

$$U(x) = J(x)/c, \quad \text{in m}^{-3} \text{ (or J/m}^3\text{)} . \qquad (3.21)$$

Next in the hierarchy, we have the *vector* flux

$$F(x) = \int_{4\pi} \Omega I(x, \Omega) d\Omega = \begin{pmatrix} F_x \\ F_y \\ F_z \end{pmatrix} \qquad (3.22)$$

where $F_x = F_{\hat{x}}$ in (3.16), etc. This vector field tells us about the mean flow of radiation in space. It can be used to compute the outcome of the generic radiometric measurement of net flux described in (3.16). Specifically, we have

$$F_n(x) = n \cdot F(x) . \qquad (3.23)$$

In essence, $J(x)$ and $F(x)$ represent respectively the monopolar/isotropic (0th-order) and dipolar (1st-order) components of the radiance field $I(x, \Omega)$ in a spherical-harmonic expansion. So there are obviously higher-order terms that add more and more angular details; they will be used extensively in the following chapter. Only the 2nd-order term has a special name through its connection with the radiation pressure tensor, cf. Mihalas (1979).

3.3 Sinks

We consider all the important mechanisms for removal of photons from a population of interest. In an inward zoom, we go from boundaries to bulk, to a point. We then consider detailed processes unfolding along a beam. At that point, we will have a closer look at what is going on inside the elementary kinetic volume.

3.3.1 Boundary Losses

Consider some region M (cf. Fig. 3.5). We can compute the energy budget in steady state from the radiance field at its boundary denoted (as in mathematical topology) by ∂M. To that effect, we use integrals over the resulting elements of flux:

$$\left. \frac{\delta E}{\delta t} \right|_{\text{out}(+)/\text{in}(-)} = \int_{x \in \partial M} dS(x) \int_{\pm n(x) \cdot \Omega > 0} |n(x) \cdot \Omega| \, I(x, \Omega) d\Omega \geq 0 \qquad (3.24)$$

given: $I(x, \Omega), x \in \partial M, \Omega \cdot n(x) < 0$

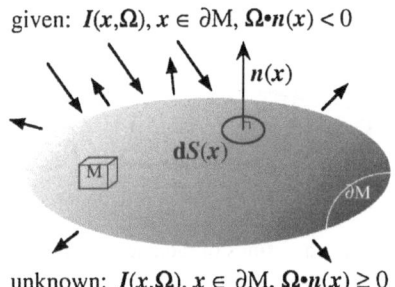

unknown: $I(x, \Omega), x \in \partial M, \Omega \cdot n(x) \geq 0$

Fig. 3.5. Steady-state radiative energy budget of a 3D macroscopic region M with a convex boundary ∂M

where $dS(x)$ is an element of the boundary of the region. As far as M is concerned incoming $(-)$ flux is a gain and outgoing $(+)$ flux is a loss.

From (3.24) and various definitions, the net radiative budget for region M is

$$
\left.\frac{\delta E}{\delta t}\right|_{\text{in}} - \left.\frac{\delta E}{\delta t}\right|_{\text{out}} = - \int_{x \in \partial M} dS(x) \int_{4\pi} n(x) \cdot \Omega I(x, \Omega) d\Omega
$$

$$
= - \int_{x \in \partial M} F(x) \cdot n(x) dS(x) = \int_{M} (-\nabla \cdot F) dx , \quad (3.25)
$$

where the last step used the divergence theorem for the vector field $F(x)$. If there are neither sources nor sinks *inside* M, the result of (3.25) will clearly be null. Since M is an arbitrary volume, this establishes that radiation flows are irrotational (divergence-free) in conservative optical media. In other words, flux lines start and end at the boundaries where all the sources and sinks are to be found.

We now assume we are in the case with internal sources only, i.e., $\delta E_{\text{in}}/\delta t = 0$ and $\delta E_{\text{out}}/\delta t > 0$. For instance, think of the Sun or a planet in the thermal part of the EM spectrum. Then, for all practical purposes, the boundary ∂M is absorbing the energy produced in the bulk of M, none is entering from the boundaries, hence the notion of "absorbing" boundary conditions introduced in Sect. 3.8 below.

3.3.2 Bulk Losses

We return again to Fig. 3.5, this time in the absence of sources in the bulk of M (so they must all be accounted for with $\delta E_{\text{in}}/\delta t$). We can estimate the total absorptance in the region, namely,

$$
A = 1 - \frac{\delta E_{\text{out}}/\delta t}{\delta E_{\text{in}}/\delta t} = \frac{\int_{M}(-\nabla \cdot F) dx}{\delta E_{\text{in}}/\delta t} \geq 0 . \quad (3.26)
$$

The inequality is certainly true in the shortwave (solar) spectrum where the source is at the upper boundary of the medium. So the net effect of the Sun is always a *heating*

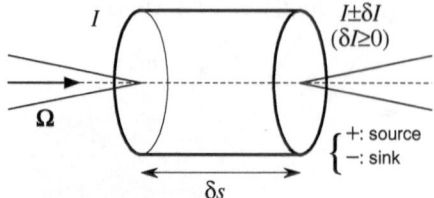

Fig. 3.6. Accounting for sources and sinks of radiance in a beam interacting with matter over a short distance

of the atmosphere/surface system, a loss for the incoming solar beam (and a gain for other sorts of energy in M). How much and where this heating occurs is discussed in more detail in Chap. 9 but it is fair to say that the effect of clouds is far from well-understood, and this is at least partially due to 3D RT effects in the observations as well as in the radiation physics.

By contrast, in the long-wave (thermal infrared) spectrum, the sources are internal so the sign of (3.26) can go either way depending on the wavelength, the region of interest, and overall (vertical and horizontal) atmospheric structure. Chapter 10 will provide some insight into this important 3D RT problem. The net effect, which has to balance solar heating in the climate system, is of course a *cooling*. This is a net loss for M which is radiating at its boundary ∂M.

3.3.3 Local Loss

The simplest description of matter-radiation interaction is photon depletion when a narrow beam crosses an optical medium, cf. Fig. 3.6 with the "−" sign representing a net loss across a distance δs (we assume $\delta I \geq 0$). Noting that the surface used in Sect. 3.3.1 is in fact quite general, we have basically expressed here the flux-divergence theorem in (3.25) for an "elementary" volume inside the medium. Along the horizontal cylinder the net transport is 0; to the left, there is an in-flux; to the right, an out-flux. So the divergence integral is simply the difference from left to right.

Operationally, we have

$$\delta I \propto I \times \delta s \tag{3.27}$$

and the proportionality constant, defined as

$$\sigma = \lim_{\delta s \to 0} \frac{\delta I / I}{\delta s}, \quad \text{in } m^{-1}, \tag{3.28}$$

is the *extinction coefficient* or simply "extinction." This inherent optical property of matter is non-negative (except in laser cavities, and other situations where stimulated emission dominates the underlying quantum physics).

Much of 3D RT is predicated on σ's propensity to vary with position x in the atmosphere. Vertical variability of σ is a given because of its strongly stratified structure and of course solar and thermal sources as well as sinks are unevenly distributed vertically. So *atmospheric* RT is generally considered to become 3D when σ varies in one (or both) horizontal direction(s). In this case, σ is often left uniform in the vertical, but sources and/or boundaries will still drive vertical gradients in radiance. There are notable exceptions to this rule since horizontal variability in the radiance field can be excited in a uniform atmosphere

- by non-uniform boundary illumination as used, e.g., in "off-beam" lidar techniques (Davis et al., 1999), for active cloud remote sensing at optical wavelengths, or
- by non-uniform surface albedo as used, e.g., in modeling "pixel adjacency" effects mediated by aerosol particulates, in passive remote sensing in the solar spectrum (Lyapustin and Knyazikhin, 2002).

Non-flat terrain, even without an overlaying atmosphere, is also 3D RT problem attracting considerable attention, as demonstrated further on.

Time-dependence of σ is never a concern here because the time for photons to propagate through the system (tens of μs at most) is short by comparison to the turn-over time in any atmospheric dynamics. More importantly, σ can depend on photon state variables: direction Ω, frequency ν, and polarization. In this volume, we will account fully for the former, touch on the second (mostly in Chaps. 9–10), and neglect the latter completely.

3.3.4 Loss Along a Beam

The calculus problem in (3.28), namely,

$$\mathrm{d}I/I = \mathrm{d}\ln I = -\sigma(x)\mathrm{d}s , \tag{3.29}$$

is easily solved.

First define *optical* distance as the running integral of σ along the given beam direction Ω_0 from some given starting point x_0:

$$\tau(d;x_0,\Omega_0) = \int_0^d \sigma(x_0 + \Omega_0 s)\mathrm{d}s . \tag{3.30}$$

To address the problem of *cumulative* extinction, we will consider $\{x_0, \Omega_0\}$ to be fixed parameters. When it is not convenient to put them in sub-indices, we will separate parameters from the independent variables, in this case d, by a semi-colon. An alternative notation for optical distance emphasizes only the starting and ending points is

$$\tau(x_0,x) = \|x - x_0\| \int_0^1 \sigma(\xi x_0 + (1 - \xi)x)\mathrm{d}\xi . \tag{3.31}$$

One can easily switch from one representation to the other using $x = x_0 + \Omega_0 d$, or else $\tau(x_0,x) = \tau(d;x_0,\Omega_0)$ where $d = \|x - x_0\|$ and $\Omega_0 = (x - x_0)/d$.

The solution of the ordinary differential equation (ODE) in (3.29) is therefore

$$I(d; x_0, \Omega_0) = I(0; x_0, \Omega_0) \exp[-\tau(d; x_0, \Omega_0)] . \tag{3.32}$$

This is the exponential law of direct transmission with respect to *optical* distance. Consider a uniform medium where optical distance is simply

$$\tau(d; x_0, \Omega_0) = \sigma d, \ \forall x_0, \forall \Omega_0 ; \tag{3.33}$$

thus

$$I(d) = I_0 \exp[-\sigma d] . \tag{3.34}$$

This is Beer's law of exponential transmission with respect to *physical* distance, sometimes called the Lambert-Bouguer-Beer law to be more historically correct. It is obviously of more limited applicability than (3.32).

For future reference, we will define a general notation for *direct* transmission between two arbitrary points x_0 and x:

$$T_{\text{dir}}(x_0 \to x) = \exp[-\tau(x_0, x)] . \tag{3.35}$$

The arrow is used in the notation for the argument of T_{dir} to emphasize causality: the photons were at x_0 before going to x. This is not to be interpreted as a dependence on the direction of propagation which would violate reciprocity in a fundamental way. Even in vegetation canopies (cf. Chap. 14) where extinction can depend on direction, we have $\sigma(x, \Omega) = \sigma(x, -\Omega)$. So it is understood that $T_{\text{dir}}(x_0 \to x) = T_{\text{dir}}(x \to x_0)$ since $\tau(x_0, x) = \tau(x, x_0)$.

Optical distance *across* a medium is called optical "thickness" and sometimes (less correctly) optical "depth" (which should vary with z, normally away for a source and/or boundary). Opaque objects such as clouds and fog layers have, by definition, considerable optical thickness. Equivalently, the amount of directly transmitted or "uncollided" light predicted in (3.35) with positions on either side of the medium will be somewhere between small and negligible. For an empirical investigation of how optically thick this means, from a human observer's perspective, we refer to Bohren et al. (1995).

3.3.5 A Look Inside the Elementary Kinetic Volume

Extinction Mechanism

We now study the detailed mechanism of extinction illustrated schematically in Fig. 3.7. This is about a population of streaming photons colliding with a static population of massive particles. Here, "static" is with respect to the speed of light of course, while "massive" is in comparison with photon mass-equivalent energy $h\nu/c^2$ where $h\nu$ is at the most an eV or so in energy units for solar problems. This is important because, otherwise, efficient momentum transfer between radiation and matter would make the collision cross-sections dependent on the light field and the

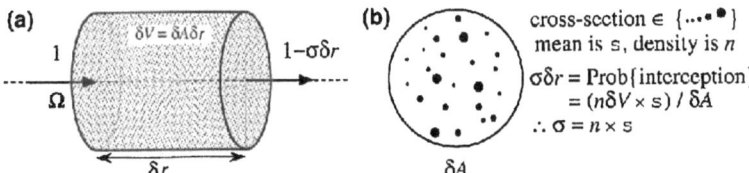

Fig. 3.7. Mechanism of optical extinction by a dilute medium of scattering/absorbing particles: (a) Geometrical parameters of the kinetic volume, and (b) What the volume looks like to the incoming photon beam

whole RT problem becomes patently nonlinear.[5] In all atmospheric applications, the smallest particles are diatomic molecules with already many MeV of mass in energy-equivalent units. So all we have to do is estimate the number of particles in the sample volume $\delta A \times \delta r$ in Fig. 3.7: $\delta N = n\delta A\delta r$ where n is the ambient particle density. Multiplying this by the (mean) cross-section s and dividing by δA yields the element of probability for an interaction which, by definition (3.28), is $\sigma\delta r$ and should be $\ll 1$. We thus find

$$\sigma = s \times n .\tag{3.36}$$

In this sense, extinction is the interaction cross-section per unit of volume, equivalently, the probability of collision per unit of length.

For cloud droplets, density n as well as the mean cross-section s are highly variable in space – 3D RT oblige! – and in time. This variability notwithstanding, it is good to have some typical numbers in mind. The density of (activated) cloud condensation nuclei or "CCN" is often quoted as hundreds to thousands per cm^3 in marine and continental air-masses respectively, so we can use that as an estimate of droplet concentration. At visible (VIS) to near-IR (NIR) wavelengths, we have

$$s \approx 2\pi\langle r^2\rangle\tag{3.37}$$

where r is the droplet radius and $\langle \cdot \rangle$ denotes an average carried over the distribution of droplet radii. The factor of 2 is the asymptotic value of the "efficiency factor" in Lorenz-Mie theory for scattering dielectric spheres that are much larger than the wavelength (cf. Sect. 3.4.4 and Chap. 2).

If we are to make an equivalent monodisperse assumption for the droplets, given the amount of condensed water $(4\pi\rho_w\langle r^3\rangle n/3$ where ρ_w is the density of water), it is best to use the "effective" droplet radius

$$r_e = \frac{\langle r^3\rangle}{\langle r^2\rangle} .\tag{3.38}$$

---

[5] The RT equation can become nonlinear in other ways than by momentum transfer. The quantized energy levels of absorbing atoms or molecules can depend on the photon population in non-LTE situations. This happens frequently in tenuous astrophysical media and in photochemically active regions of the atmosphere.

In terrestrial liquid water clouds, r_e is $\approx 10\,\mu m$, give or take a factor of 2 or so. Using (3.37), this puts the extinction coefficient σ in (3.36) for clouds in a range from almost nil (aerosol levels) to 0.1 or even $1\,m^{-1}$.

An independent way of estimating this range is to use the observed optical depths of cloud layers to obtain a vertically-averaged σ. Optical depth is simply optical distance τ measured vertically from cloud bottom to cloud top and it ranges from somewhat less than 10 to several 100 in the bulk of the cloud. This is for physical thickness d in (3.33). Again excluding cloud edges, we can take d in the range from a few hundred meters to a couple of km. The lower end for d gives us back our upper limit for σ and we anticipate less for an average, say $\tau/d = 25/0.5 = 50\,km^{-1} = 0.05\,m^{-1}$.

Absorption vs. Scattering

Upon collision with an atmospheric particle, a photon can be either absorbed or scattered. In both cases, it is a loss for the beam; in the latter case, it becomes a source for another beam (cf. Sect. 3.4). So the extinction cross-section (per particle) has to be broken down into its scattering and absorption components, $s = s_s + s_a$, and similarly for the extinction coefficient in (3.36):

$$\sigma = \sigma_s + \sigma_a . \tag{3.39}$$

The conventional representation of this breakdown uses the *single-scattering albedo*:

$$\varpi_0 = \sigma_s/\sigma \leq 1 , \tag{3.40}$$

and single-scattering *co*-albedo,

$$1 - \varpi_0 = \sigma_a/\sigma . \tag{3.41}$$

It is noteworthy that in nuclear reactor theory, the counterpart of ϖ_0 describes the mean number of neutrons produced after collision with a nucleus and is typically larger than unity, and that is precisely what makes sustained chain reactions possible. So in this context σ_a can be formally taken as negative (anti-absorption).

In atmospheric RT, scattering and absorption can be traced to both gaseous constituents (i.e., molecules) and particulates (i.e., aerosol and cloud droplets). All coefficients depend on wavelength λ. The spectral features of gases tend to vary faster with λ, especially for absorption. This is discussed, as needed, in various parts of this volume.

3.4 Scattering

Scattering is the process that makes 3D RT such a challenge because photon transport through a scattering medium is a fundamentally nonlocal process, as will be shown in Sect. 3.8. We describe here the basic concepts and popular models for photon scattering. When we get to our brief survey of physical theories of light-particle interaction, it will become clear that we can not treat absorption and scattering separately. So, although the new quantity introduced here is the phase function, we will revisit the partition of extinction σ into σ_s and σ_a.

3.4.1 The (Poorly-Named) Scattering Phase Function

Figure 3.8 illustrates the redistribution of radiant energy between different beams through scattering. Our goal is to estimate the element of scattered flux δF_s. It is surely proportional to the small solid angle into which the scattering occurs $\delta\Omega$ and to the small loss of flux δF_0 incurred when the incoming photons cross the sample volume (conditional to scattering rather than absorption); the latter term is equal to the scattering coefficient times the small length δs. In summary, we have

$$\delta F_s \propto \delta F_0 \times \delta\Omega = F_0 \sigma_s \delta s \times \delta\Omega . \tag{3.42}$$

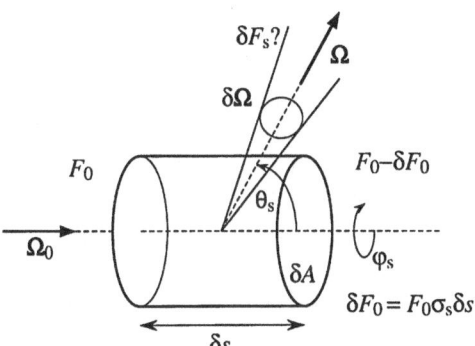

Fig. 3.8. Schematic of scattered flux and radiance

We define the scattering *phase function* as

$$p(\boldsymbol{x}, \Omega_0 \to \Omega) = \lim_{\delta F_0, \delta\Omega \to 0} \frac{\delta F_s}{\delta F_0 \times \delta\Omega}, \quad \text{in sr}^{-1} . \tag{3.43}$$

The explicit notation tells us that this property will generally depend on position $\boldsymbol{x}$. Using the above definitions, the integral of $p(\boldsymbol{x}, \Omega_0 \to \Omega)$ over all final directions Ω will be unity (since the sum of all the δF_s in Fig. 3.8 has to equal δF_0).[6] As a first example, we take everywhere isotropic scattering:

---

[6] It is important to note that there is another popular normalization convention for the phase function, often denoted $P(\cdot)$. Even in this volume both conventions and notations are used. The phase function's integral is then equated to 4π; in this case, it is a non-dimensional quantity and $d\Omega$ is always divided by 4π wherever $P(\cdot)$ is used.

$$p(x, \Omega_0 \to \Omega) \equiv 1/4\pi . \tag{3.44}$$

More general formulations include changes in polarization and wavenumber mediated by scattering. In the former case, one needs a phase matrix; in the latter case, one talks about inelastic scattering since photon energy is changed (and consequently the energy of the scattering entity too, by an equal amount in the opposite direction).

We note in passing that these so-called "phase functions" and "phase matrices" have very little to do with "phases" in the wave (or *coherent*) optics sense of the word since here energies are added and subtracted, not the complex amplitudes used in EM as well as scalar-wave theory. In this respect, we recall that all of RT theory is entirely about *incoherent* optics while (coherent) wave theory contributes at most scattering and absorption cross-sections, one particle at a time. The origin of the "phase function" terminology in fact goes back to early lunar and planetary astronomy were the "phase angle" is defined, following the deflection of the light rays, as the angle between the axis going from the Sun to the celestial body of interest and the line between the said celestial body and the Earth. It is therefore the equivalent of the scattering angle $\theta_s = \cos^{-1}(\Omega_0 \cdot \Omega)$. In the course of the Moon's monthly "phases," it varies from 0 at new Moon (in a solar eclipse configuration if *exactly* 0) to π at full Moon (in a lunar eclipse configuration if *exactly* π). The astronomical phase function's purpose is simply to capture the dependence of total planetary brightness (hence photometry) not explained by celestial mechanics, i.e., relative distances. For a given body (hence radius), phase angle is the dominant term but albedo, and the regional variability thereof, also matter.

As for extinction, we can have a closer look at the mechanics of scattering at the individual collision level. To isolate the inherent property of the scattering medium, we compute

$$\lim_{\delta s, \delta\Omega \to 0} \frac{\delta F_s/F_0}{\delta s \times \delta\Omega} = \sigma_s(x)p(x, \Omega_0 \to \Omega) = n(x) \times \frac{ds_s}{d\Omega}(x, \Omega_0 \to \Omega) \tag{3.45}$$

where the last expression is obtained by straightforward generalization of (3.36) to *differential* cross-sections, again averaged over the population of particles in the sample volume sorted by size and/or type.

By energy (flux) conservation, we have

$$\int_{4\pi} p(x, \Omega_0 \to \Omega)d\Omega \equiv 1, \ \forall\Omega , \tag{3.46}$$

and for any x where scattering occurs. By reciprocity (cf. Sect. 3.10.3), we have $p(x, -\Omega \to -\Omega_0) = p(x, \Omega_0 \to \Omega)$, hence

$$\int_{4\pi} p(x, \Omega_0 \to \Omega)d\Omega_0 \equiv 1, \ \forall\Omega_0 , \tag{3.47}$$

and for any x. In the remainder of this section, we will assume the spatial variability the phase function is implicit, and drop x from its arguments.

3.4.2 Phase Functions with Axial Symmetry

In most atmospheric applications, ice clouds being a notable exception, it is reasonable to assume that scattering is axi-symmetric around the incoming beam. Mathematically,

$$p(\mathbf{\Omega}_0 \rightarrow \mathbf{\Omega}) \equiv p(\mathbf{\Omega}_0 \cdot \mathbf{\Omega}) = p(\mu_s) , \tag{3.48}$$

where the scattering angle θ_s is given by $\mu_s = \cos\theta_s = \mathbf{\Omega}_0 \cdot \mathbf{\Omega}$.

This enables an expansion of the phase function in spherical harmonics without the complication of azimuthal terms:

$$p(\mu_s) = \left(\frac{1}{4\pi}\right) \sum_{l \geq 0} \omega_l P_l(\mu_s) , \tag{3.49}$$

where the coefficient is often factored as $\omega_l = (2l + 1)\eta_l$. These coefficients can be computed from

$$\eta_l = \frac{\omega_l}{2l + 1} = 2\pi \int_{-1}^{+1} P_l(\mu_s) p(\mu_s) \mathrm{d}\mu_s . \tag{3.50}$$

The orthogonality relation of the Legendre polynomials is used here, that is,

$$\int_{-1}^{+1} P_n(x) P_{n'}(x) \mathrm{d}x = \frac{\delta_{nn'}}{n + 1/2} \tag{3.51}$$

where $\delta_{nn'}$ is the Kronecker symbol ($= 1$ if $n = n'$, $= 0$ otherwise). Specific values of the polynomials can be obtained efficiently by recursion, but their analytical expressions are best derived from the generating function

$$\Phi(x, z) = \sum_{n \geq 0} P_n(x) z^n = (1 - 2xz + z^2)^{-1/2} \tag{3.52}$$

for any z inside the unit circle of the complex plane. Using

$$P_n(x) = \frac{1}{n!} \left(\frac{\partial}{\partial z}\right)^n \Phi(x, z)\Big|_{z=0} , \tag{3.53}$$

we find

$$\begin{aligned} P_0(x) &= 1, \\ P_1(x) &= x , \\ P_2(x) &= (3x^2 - 1)/2 , \end{aligned} \tag{3.54}$$

and so on.

We have $\eta_0 = \omega_0 = 1$ by conservation for any phase function,[7] and the only non-vanishing coefficient for isotropic scattering in (3.44) and (3.49). Also of considerable interest is

$$g = \eta_1 = \frac{\omega_1}{3} = 2\pi \int_{-1}^{+1} \mu_s p(\mu_s) d\mu_s \,, \tag{3.55}$$

the *asymmetry factor*, or mean cosine of the scattering angle that is obviously between -1 and $+1$. This correctly presents the phase function as a probability density function (PDF) in angle space. Any deviation of the phase function from isotropy corresponds to a directional correlation between incident and scattered photons.

3.4.3 Henyey–Greenstein Models

The most popular 1-parameter model for single-scattering in atmospheric radiation and elsewhere is by far the Henyey–Greenstein or "HG" phase function

$$p_{HG}(g; \mu_s) = \left(\frac{1}{4\pi}\right) \frac{1 - g^2}{(1 + g^2 - 2g\mu_s)^{3/2}} \tag{3.56}$$

which, like the expression "phase function" itself, comes to us from astronomy. It was indeed proposed first by Henyey and Greenstein (1941) to model scattering by interstellar dust, i.e., the stellar astronomer's counterpart of aerosol as a nuisance in surface remote sensing in the solar spectrum. Interstellar dust grains also have in common with aerosol huge spatial variability in quantity and in quality. As for the aerosol, they cause trouble for one kind of observation but have inherent interest in other studies: aerosol matters in climate, cloud physics and pollution; interstellar dust matters in life-cycles of stars and planets.

In spherical harmonics, (3.56) yields

$$\eta_l = g^l \,. \tag{3.57}$$

Indeed, $4\pi p_{HG}(z; x)$ is identical to $\sum_{n \geq 0}(2n + 1)P_n(x)z^n = 2\partial \Phi(x, z)/\partial z + \Phi(x, z)$ from (3.52); the above coefficients then follow by comparison with (3.49).

A related 3-parameter model is the *double* Henyey–Greenstein or "DHG" phase function

$$p_{DHG}(g_f, g_b, f; \mu_s) = f \times p_{HG}(g_f; \mu_s) + (1 - f) \times p_{HG}(-g_b; \mu_s) \,. \tag{3.58}$$

We have $g = fg_f - (1 - f)g_b$, and so on (for higher-order spherical harmonics). Two other constraints beyond this expression for g can be invoked to uniquely determine all three parameters.

---

[7] If we think of $\varpi_0 p(\theta_s)$ as a non-normalized phase function, then its integral over 4π is the single-scattering albedo $\varpi_0 \leq 1$ and its first of possibly many Legendre coefficients is $\omega_0 \leq 1 \ldots$ hence the frequent use of ω_0 to denote the single-scattering albedo.

3.4.4 Physical Theories for Scattering and Absorption

The above HG phase functions are convenient models but they have no physical basis. More accurate computations of scattering properties from first (EM or other) principles yield Rayleigh and, for spherical particles, Lorenz-Mie phase functions. However, not all optically important particles in the atmosphere are tiny nor spherical, far from it. Scattering and absorption of course come together in a physically correct theory at the single particle level; basically they come as direct consequences of the existence of interfaces with a discontinuity in the complex index of refraction m, which generally has real ($\neq 1$) and imaginary (≥ 0) parts.

Rayleigh Scattering by Molecules

Rayleigh scattering can be computed using the classic theory of equilibrium thermodynamical fluctuations in molecular density around the mean n, or semi-classical or pure quantum mechanics. This invariably leads to the cross-section (per molecule)

$$s_{\text{Ray}}(\lambda) = \frac{24\pi^3}{n^2\lambda^4}\left(\frac{m^2-1}{m^2+2}\right)^2\left(\frac{6+3\delta}{6-7\delta}\right) \tag{3.59}$$

where λ is the wavelength, m is the index of refraction of dry air at STP, and δ is its depolarization ratio, a weakly λ-dependent term accounting for the anisotropy of certain (tri-atomic) air molecules. At solar wavelengths, δ can be set to ≈ 0.031. We also have $m - 1 \approx 2.781 \times 10^{-4} + 5.67 \times 10^{-3}/\lambda^2$, where λ is expressed in μm.

To a first approximation, scattering by clear air is isotropic. However, an accurate calculation of Rayleigh differential cross-section leads to

$$p_{\text{Ray}}(\mu_s) = \frac{3}{16\pi}(1 + \mu_s^2) . \tag{3.60}$$

Equivalently, we have $\eta_0 = 1$ and $\eta_2 = 1/10$ with all other Legendre coefficients in (3.49) vanishing.[8]

Lorenz-Mie Scattering by Cloud Droplets

Being too small (by definition) for their shapes to be affected by gravity and/or hydrodynamic flow around them, cloud water droplets are almost perfectly spherical. This means that Lorenz-Mie theory can accurately describe their absorption and scattering properties as long as they do not contain insoluble (or undissolved) particles. The conventional representation of Lorenz-Mie extinction (total) and scattering cross-sections in the monodisperse case are

$$s_{e,s}(\lambda, r) = Q_{e,s}(m_\lambda, 2\pi r/\lambda) \times \pi r^2 \tag{3.61}$$

---

[8] We refer to Lilienfeld (2004) for a thoughtful account of the historical origins of the Rayleigh differential cross-section $s_{\text{Ray}}(\lambda) \times p_{\text{Ray}}(\mu_s)$ and the explanation of the blue-sky phenomenon and its polarization.

where r is the droplet radius, and $2\pi r/\lambda$ is known as the "size parameter." The non-dimensional functions $Q_{e,s}$ are efficiency factors that also depend on wavelength through changes in the index of the real and imaginary parts of the refraction index m_λ. Absorption cross-section is obtained from $s_a = s_e - s_s$. A representation similar to (3.61) exists for the differential cross-section for scattering $ds_s/d\Omega$ used to compute the phase function in (3.45).

For large $2\pi r/\lambda$ and no absorption, $Q_e \approx Q_s$ approaches 2 (see Chap. 2). Recalling that droplet radii range from a few µm to a few tens of µm, this is not a bad approximation at non-absorbing wavelengths in the VIS/NIR spectrum. Cross-sections of scattering/absorbing spheres are complemented by empirical representations of polydisperse droplet populations $dN(r)/dr$, given typically in $cm^{-3}\mu m^{-1}$, to yield usable extinction, scattering and absorption coefficients:

$$\sigma_{s,a}(\lambda) = \pi n \int_{r_{min}}^{r_{max}} r^2 Q_{s,a}(m_\lambda, 2\pi r/\lambda) d\Pr(r) , \qquad (3.62)$$

where (total) droplet density n is the integral of $dN(r)/dr$ over all possible r values and $d\Pr(r) = (dN(r)/dr) \times dr/n$. In the approximation where $Q_e = Q_s \approx 2$, we have

$$\sigma = \sigma_s \approx 2\pi \langle r^2 \rangle n , \qquad (3.63)$$

as was already used in (3.36)–(3.37). Similar averaging over $ds_s/d\Omega$ yields the Lorenz-Mie scattering phase function $p_{Mie}(\mu_s)$ which the underlying EM theory naturally produces in terms of spherical harmonics.

Figure 3.9 shows, on the one hand, the natural outcome of Lorenz-Mie theory (values of the Legendre coefficients) in panel (a) and, on the other hand, the reconstruction of the phase function in angle space in panel (b). The droplet population is the "C1" standard (Deirmendjian, 1969) and the wavelength is 1.064 µm. We note the relatively slow decay in Legendre coefficients. We also note the strong forward peak caused by diffraction; its width (in radians) is inversely proportional to the size parameter. In contrast with this inherently scalar or EM wave phenomenon, we also see a peak at the "rainbow" deflection angle that, for the most part, is explained by geometrical optics with one total internal reflection inside the droplet.

We have also plotted in Fig. 3.9 two approximations using the simple- and double-HG models from (3.56) and (3.58) respectively. In the former case, we just set $g = 0.848$. In the later case, we can match the 2nd- and 3rd-order Legendre coefficients too; this leads to $g_f = 0.879$, $g_b = 0.9835$, and $f = 0.983$, with the result in Fig. 3.9b that the backscatter peak at $\theta_s = \pi$ is captured on a relative scale. Alternatively, we can fit the height and position of the maximum in ω_l; this leads to $g_f = 0.977$, $g_b = -0.625$, and $f = 0.633$, with the result in Fig. 3.9b that the diffraction peak at $\theta_s = 0$ is better reproduced by adding two forward HG phase functions. There are of course other possibilities.

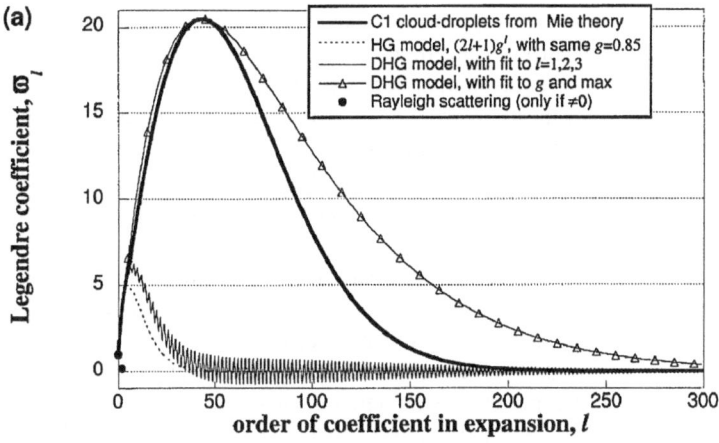

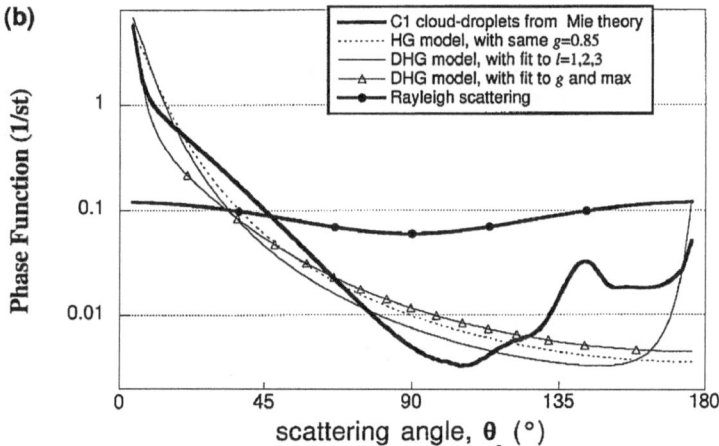

Fig. 3.9. Rayleigh (molecular) phase function and Lorenz-Mie (cloud droplet) phase function with Henyey-Greenstein approximations. (**a**) Legendre coefficients in the $\omega_l = (2l + 1)\eta_l$ representation that multiply the $P_l(\mu_s) \in [-1, +1]$. (**b**) Angular values. Notice the variation over 3+ orders-of-magnitude for the C1 phase function. As the scattering angle increases, we see: (1) the strong forward-scattering peak caused by diffraction at $\theta_s \lesssim 1/10$ rad ≈ 6 dog and readily observed in the "silver lining" phenomenon; (2) the maximum causing the rainbow phenomenon at $\theta_s \approx 140°$; and (3) the backscattering peak responsible for the "glory" effect at θ_s close to $180°$, the anti-solar direction

Scattering and Absorption by Non-Spherical Particles

Not all clouds are made of liquid droplets. Cirrus and mixed-phase clouds contain ice-particles with a myriad shapes (cf. Chap. 2). Some crystals inherit very regular geometry from the 6-fold symmetry induced by the hydrogen bond in ice; others are extremely random, and everything in between has been observed. It suffices to state here that scattering properties, especially phase functions, of distributions of large non-spherical particles are qualitatively different from Lorenz-Mie theoretical predictions using "equivalent sphere" assumptions. In the range of (very large) size-parameters relevant to solar and even to large extent thermal atmospheric RT, geometric optics has been used quite successfully to predict scattering properties of non-spherical particles (Liou, 2002). The volume by Mishchenko et al. (2000) is a recent and comprehensive source of information on single-scattering theory for non-spherical particles, ice crystals or other.

In the lower troposphere, aerosol particles play an important role in its optics and radiation budget, and so do particulates injected by large volcanic eruptions into the swift circulations in the stratosphere. Because of its role in the microphysics and life-cycle of clouds, the climate community has developed a strong interest in the anthropogenic component of the aerosol. In some regions/seasons, it is by far the dominant one with dramatic consequences on air quality as well as global and local climate (Ramanathan et al., 2002). Among man-made aerosol, black carbon is highly absorbing, hence very important for the solar radiation budget and how it is partitioned between the atmosphere and the surface. Black-carbon particles have notoriously convoluted shapes, best modeled as randomly aggregated fractal objects over a wide range of scales that includes the wavelength (at least in the early phases of the particle's life). Because these particles would allegedly dominate the nuclear winter scenarios investigated in the 1980s, their scattering and absorption properties were computed quite a while ago by Berry and Percival (1986).

3.5 Propagation

We presented scattering as a random choice of new direction of propagation for the photon. After emission and between collisions (resulting in either a scattering or a final absorption) or escape, there is also an inherent randomness in photon propagation. We define here a few statistical quantities needed to characterize photon transport per se.

3.5.1 Photon Free Path Distributions

We will be using several kinds of averages in this chapter. We have already used $\langle \cdot \rangle$ to denote an average over the "disorder" of the cloud droplets which can have a variety of sizes. Those averages that concern photon scattering and propagation events deserve a special notation, which we borrow from the probability literature: $\mathcal{E}(\cdot)$ which stands for (mathematical) *expectation* of the random variable in the argument. Thus, we can recast the asymmetry factor in (3.55) as

$$g = \mathcal{E}(\mathbf{\Omega}_0 \cdot \mathbf{\Omega}) = \int_{4\pi} (\mathbf{\Omega}_0 \cdot \mathbf{\Omega}) \mathrm{d}\Pr(\mathbf{\Omega}|\mathbf{\Omega}_0) \tag{3.64}$$

where $\mathrm{d}\Pr(\mathbf{\Omega}|\mathbf{\Omega}_0) = p(\mathbf{\Omega}_0 \cdot \mathbf{\Omega})\mathrm{d}\mathbf{\Omega}$ is an element of probability. We use the "|" in a PDF to separate the random variable from the given (fixed) quantities.

From (3.32), but dropping the "0" subscripts for simplicity, we can derive *direct transmission*

$$T_{dir}(s;\boldsymbol{x},\boldsymbol{\Omega}) = \exp[-\tau(s;\boldsymbol{x},\boldsymbol{\Omega})] = \Pr\{\text{step} \geq s|\boldsymbol{x},\boldsymbol{\Omega}\} \qquad (3.65)$$

by taking the ratio $I_{out}/I_{in} = I(\cdot;s)/I(\cdot;0)$. We have also expressed that this is the probability that a photon does *not* suffer any kind of collision in an experiment over the *fixed* distance s, starting at $\boldsymbol{x}$ in direction $\boldsymbol{\Omega}$. Now think of the photon's free path or "step" to its next collision as a random variable. Since $T_{dir}(s;\boldsymbol{x},\boldsymbol{\Omega})$ is the probability that this random variable exceeds s, the PDF of s is defined by

$$p(s|\boldsymbol{x},\boldsymbol{\Omega})ds = dP(s|\boldsymbol{x},\boldsymbol{\Omega}) = \Pr\{s \leq \text{step} < s + ds|\boldsymbol{x},\boldsymbol{\Omega}\} . \qquad (3.66)$$

Using (3.65) and (3.30), this leads to

$$p(s|\boldsymbol{x},\boldsymbol{\Omega}) = -\left(\frac{d}{ds}\right)P(s|\boldsymbol{x},\boldsymbol{\Omega}) = \sigma(\boldsymbol{x}+\boldsymbol{\Omega}s)\exp[-\tau(s;\boldsymbol{x},\boldsymbol{\Omega})] . \qquad (3.67)$$

The above notations $p(\cdot)$ and $P(\cdot)$ are not to be confused with the variously normalized phase functions introduced in Sect. 3.4.1 above for volume scattering and in Sect. 3.6.2 below for surface scattering (i.e., bidirectional reflection). We note however that both free path distributions and phase functions are PDFs that play closely interlaced roles in the photon transport process: here we move (propagate) photons to a new position while phase functions move them into a new direction (of propagation). So the shared notations can serve as a reminder of this shared probabilistic meaning. We are confident that context will resolve any ambiguity.

Consider the case of *uniform* extinction σ, the only quantity required in the problem at hand. The resulting free path distribution is given by

$$p(s|\sigma) = \sigma e^{-\sigma s} , \qquad (3.68)$$

as follows directly from (3.67), or using Beer's exponential transmission law in (3.34).

3.5.2 Mean-Free-Path

A fundamental quantity in transport theory (for light quanta or any other type of particle) is the *mean-free-path* or "MFP"

$$\ell(\boldsymbol{x},\boldsymbol{\Omega}) = \mathcal{E}(s|\boldsymbol{x},\boldsymbol{\Omega}) = \int_0^\infty s\, dP(s|\boldsymbol{x},\boldsymbol{\Omega}) \qquad (3.69)$$

which, as indicated, will generally depend on the pair $\{\boldsymbol{x},\boldsymbol{\Omega}\}$ in the 3D case. Reconsidering the uniform-σ case in (3.68), we find

$$\ell = \mathcal{E}(s) = 1/\sigma . \qquad (3.70)$$

So there is such a thing as *the* mean-free-path in homogeneous media, but not in 3D media. One can talk about $1/\sigma(x)$ as a *local* MFP in 3D media. However, at a given x it will only occasionally coincide with $\ell(x, \Omega)$ in (3.69) for certain choices of Ω. We prefer to call this a 3D field of *pseudo*-MFP values. By averaging (3.69) over $\{x, \Omega\}$, on can define the *mean* mean-free-path, which is necessarily larger than the inverse of the mean extinction (e.g., Davis and Marshak, 2004).

Equation (3.70) provides us with a more descriptive interpretation of optical distance, at least for homogeneous media, as given in (3.33):

$$\tau = \sigma d = d/\ell, \tag{3.71}$$

is just physical distance d in units of MFPs. If d is the thickness of (i.e., distance across) the medium, we are looking at the ratio of the two fundamental scales in the RT problem. The solution of the problem will clearly reflect a different flavor of transport physics depending on whether τ is smaller or τ is larger than unity:

- if $\tau \ll 1$, photons will tend to "stream" (move ballistically along straight lines);
- if $\tau \gg 1$, photons will tend to "diffuse" (move along convoluted paths akin to random walks).

In typical 3D RT problems, there are regions where optical thickness is large and others where it is small, at least on a relative scale. Davis and Marshak (2001) show that this sets up horizontal fluxes in predictable patterns they recognize as "channeling" events, using language introduced by Cannon (1970).

3.5.3 Other Moments of the Free Path Distribution

Higher-order moments of the free path distribution are also of interest:

$$\mathcal{E}(s^q|x, \Omega) = \int_0^\infty s^q dP(s|x, \Omega). \tag{3.72}$$

Free path moments of arbitrary order $q > -1$ can be computed from the exponential distribution in (3.68) for homogeneous media, and we find

$$\mathcal{E}(s^q) = \Gamma(q+1)/\sigma^q = \Gamma(q+1)\ell^q \tag{3.73}$$

where $\Gamma(\cdot)$ is Euler's Gamma function:

$$\Gamma(x) = \int_0^\infty t^{x-1}e^{-t}dt. \tag{3.74}$$

Recall that, for integer values, $\Gamma(n+1) = n!$, $n \geq 0$. So, in particular, the root-mean-square (RMS) free path is

$$\sqrt{\mathcal{E}(s^2)} = \sqrt{2}/\sigma = \sqrt{2}\,\mathcal{E}(s). \tag{3.75}$$

It is larger than the MFP in (3.70), as required by Schwartz's inequality. Free path variance $\mathcal{D}(s) = \mathcal{E}(s^2) - [\mathcal{E}(s)]^2$ is equal to $\mathcal{E}(s)^2$, a characteristic of the exponential distribution. Davis and Marshak (2004) show that, since $\mathcal{E}(s^q) > [\mathcal{E}(s)]^q$ in general 3D media for any $q > 1$, free path distributions are always *wider* than the exponential ones based on the MFP.

3.6 Sources

In this section, we introduce explicitly the dependence of all radiative quantities and most optical properties on wavelength λ or wavenumber $\nu = 1/\lambda$ (adopting spectroscopic usage) that has been implicit so far. Even if nothing else does, source terms will drive this dependence in atmospheric applications. A wide variety of sources are found in the bulk of optical media as well as on their boundaries. We call these *primary* sources. Furthermore, volume scattering and surface reflection are at once sinks and sources, depending on which beam one is talking about. We will call these *secondary* sources.

3.6.1 Volume Sources

General Definition

We return to Fig. 3.6 used already to define the extinction of I with no strict need for an incoming beam this time (i.e., $I = 0$ is a possibility); we focus however on the "+" sign in the exiting radiance. This describes a situation where photons are generated inside the sample volume, thus adding

$$\delta I_\nu \propto \delta s \tag{3.76}$$

to the existing population, if any. As usual, the proportionality constant has a name and an important role in RT theory. Define

$$Q_\nu(x, \Omega) = \lim_{\delta s \to 0} \frac{\delta I_\nu}{\delta s}, \quad \text{in } \mathrm{m}^{-3}\mathrm{s}^{-1}\mathrm{sr}^{-1}(\mathrm{cm}^{-1})^{-1} \text{ (or } \mathrm{W/m^3/sr/cm^{-1}) \tag{3.77}$$

as the (volume) source term.[9] Two contrasting and important examples follow.

Solar Photon Injection

Rather than "incoming" at the upper boundary, we can use what we have learned about propagation and scattering in previous sections to model the "injection" of sunlight into the bulk of the medium after a first scattering or surface reflection; see

---

[9] The reader will know from context how to distinguish the source term introduced here and the Lorenz-Mie efficiency factor $Q_{e,s,a}$ introduced in Sect. 3.4.1.

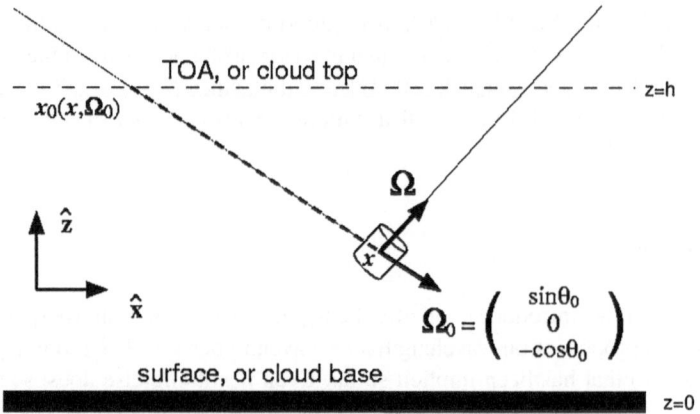

Fig. 3.10. Volume injection of solar flux in a plane-parallel medium

Fig. 3.10. Note that in this case, the radiance field is split between the direct and diffuse components, and this source term feeds only the latter. We have

$$Q_{\odot v}(\mathbf{\Omega}_0; \mathbf{x}, \mathbf{\Omega}) = F_{0v} \exp[-\tau_v(\mathbf{x}_0(\mathbf{x}, \mathbf{\Omega}_0), \mathbf{x})]\, \sigma_v(\mathbf{x})\, \varpi_{0v}(\mathbf{x})\, p_v(\mathbf{x}, \mathbf{\Omega}_0 \to \mathbf{\Omega})$$
(3.78)

where F_{0v} is the spectral value of the solar constant and $\mathbf{x}_0(\mathbf{x}, \mathbf{\Omega}_0)$ is the point where the solar beam of interest starts at the TOA or cloud top. For a plane-parallel cloud $\{z \in \mathbb{R}^3 : 0 < z < h\}$ and solar rays coming in, as is often assumed, along the x-axis (negative-to-positive direction) we have $\mathbf{x}_0 = (x - (h - z)/\mu_0, y, h)^{\mathrm{T}}$ where $\mu_0 \in (0, 1]$ is the cosine of the sun angle. As similar expression as (3.78) can be written for a direct transmission through the atmosphere and a reflection at the lower boundary.

The relatively long expression in (3.78) is really just a sequence of probabilities. Given a solar $\{\mathbf{\Omega}_0, v\}$-photon impinging on the top of the cloudy layer, we have the following events in causal order:

- transmission from impact point $\mathbf{x}_0$ to $\mathbf{x}$;
- interception at point of interest $\mathbf{x}$;
- scattering (rather than absorption);
- scattering from solar beam direction $\mathbf{\Omega}_0$ into the beam of interest $\mathbf{\Omega}$.

Thermal Emission

In local thermal equilibrium (LTE), the rate of emission equals the rate of absorption (Kirchhoff's law). From there, we can write the source term for thermal emission:

$$Q_{\mathrm{T}v}(\mathbf{x}, \mathbf{\Omega}) = \sigma_{av}(\mathbf{x})\, B_v[T(\mathbf{x})], \quad \forall \mathbf{\Omega} ,$$
(3.79)

where $T(\mathbf{x})$ is the local absolute temperature and $B_v(T)$ is Planck's function.

We used these two examples of source term partially because of their contrasting mathematical expressions but also because of their paramount importance in both remote sensing and climate applications. The Earth's climate system is essentially an engine that converts radiative "fuel" $Q_{\odot v}(\mathbf{\Omega}_0; \mathbf{x}, \mathbf{\Omega})$ into atmospheric, oceanic and all kinds of other motions, leaving the radiative "exhaust" $Q_{Tv}(\mathbf{x})$ to dissolve in the cold universe. Solar photons intercepted by the Earth are high-energy and collimated, hence low-entropy, while their thermal counterparts emitted by the Earth are low-energy, hence more numerous, and isotropic. So they are carrying away the excess of entropy required to maintain the climate.

Multiple Scattering

Scattering, like absorption, depletes a beam in terms of direct transmission. However, unlike absorption, the same scattering replenishes other beams. So it is productive to see scattering as a source of radiance. From (3.42)–(3.43), but in terms of scattered radiance, we have

$$\delta I_{vs} \approx \delta F_{vs}/\delta\Omega \approx F_{0v}\sigma_{sv}(\mathbf{x})p_v(\mathbf{x}, \mathbf{\Omega}_0 \to \mathbf{\Omega})\delta s \ . \tag{3.80}$$

Replacing $F_0 v$ by $I_v(\mathbf{x}, \mathbf{\Omega}_0)d\mathbf{\Omega}_0$ and integrating over all incidence directions (denoted more traditionally as $\mathbf{\Omega}'$ rather than $\mathbf{\Omega}_0$), we obtain

$$S_v(\mathbf{x}, \mathbf{\Omega}) = \lim_{\delta s \to 0} \frac{\delta I_{vs}}{\delta s} = \sigma_{sv}(\mathbf{x}) \int_{4\pi} p_v(\mathbf{x}, \mathbf{\Omega}' \to \mathbf{\Omega})I_v(\mathbf{x}, \mathbf{\Omega}')d\mathbf{\Omega}' \ . \tag{3.81}$$

This is known as the source *function* in multiple scattering theory. It is not to be confused with the (spectral) source *term* in (3.77), Q_v, especially since they have the same physical units.

3.6.2 Boundary Sources

General Definition

What if photons are emitted in direction $\mathbf{\Omega}$ from a boundary point $\mathbf{x}_S$ with normal $\mathbf{n}(\mathbf{x}_S)$? We need a modified mathematical description of the photon creation at the surface of a medium, or at its interface with another medium. By reconsidering (3.76) and (3.77), we now have an addition to the existing photon population, if any, given by

$$\delta E_v = h v \times \delta N_v \propto |\mathbf{n}(\mathbf{x}_S) \bullet \mathbf{\Omega}|\delta A\delta t\delta\Omega \tag{3.82}$$

where we have reverted to the elementary quantities used in Sect. 3.2 since there is no δs here to define a volume.

The proportionality constant again has a name and, furthermore, it has the same physical units as radiance. Define

$$f_v(\boldsymbol{x_S}, \boldsymbol{\Omega}) = \lim_{\delta A, \delta t, \delta\Omega \to 0} \frac{\delta N_v \ (\text{or } \delta E_v)}{|\boldsymbol{n}(\boldsymbol{x_S}) \bullet \boldsymbol{\Omega}|\delta A \delta t \delta\Omega}, \quad \text{in } \mathrm{m^{-2}s^{-1}sr^{-1}} \ (\text{or } \mathrm{W/m^2sr^{-1}})$$

(3.83)

as the surface source term. This field plays a critical role further on in the formulation of boundary conditions for the general 3D RT problem.

Example of Thermal Emission

By its definition, the spectral radiance coming from the surface of a black body at temperature T is (1) isotropic and (2) given by the Planck function $B_v(T)$. So $f(\boldsymbol{x_S}, \boldsymbol{\Omega}) \equiv B_v[T(\boldsymbol{x_S})]$ in (3.83). Most natural surfaces are however not purely black: they are at least partially reflective in amounts that generally depend on wavenumber. In other words, they have specific spectral emissivities $\varepsilon_v(\boldsymbol{x_S})$, generally position-dependent, defined by

$$f_v(\boldsymbol{x_S}, \boldsymbol{\Omega}) = \varepsilon_v(\boldsymbol{x_S})B_v[T(\boldsymbol{x_S})], \ \forall\boldsymbol{\Omega} .$$

(3.84)

By comparison of (3.84) above with (3.79) for bulk thermal emission, we see that (non-dimensional) emissivity is for surfaces what the absorption coefficient (in units of inverse length) is for volumes. This captures the fact that surface sources have the same units as radiance while volume sources are radiance "gained" per unit of length.

Surface emission is of course a powerful resource in thermal sensing of surface properties from aircraft or satellite. This exercise is however predicated on the detector- and/or algorithm-based ability for "$\varepsilon - T$" separation, and the correction for atmospheric effects. Part of the "$\varepsilon - T$" separation problem is that the "$\forall\boldsymbol{\Omega}$" in (3.84) is in fact an idealization and for even quite fine observation scales ε_v is actually function of $\boldsymbol{\Omega}$ as well as of $\boldsymbol{x_S}$. This non-thermodynamical dependence captures unresolved surface heterogeneity and roughness effects that can for a large part be modeled with 3D radiative transfer, as shown further on.

Bidirectional Reflectance Distribution Function, and Related Quantities

We now need to formulate mathematically what happens at the surface of a medium in the frequently encountered situation where it has a reflecting property. This is not a source of photons per se but, like the scattering process, it behaves as a sink for outgoing beams ($\boldsymbol{\Omega} \bullet \boldsymbol{n}(\boldsymbol{x_S}) \geq 0$) and a source for in-coming ones ($\boldsymbol{\Omega} \bullet \boldsymbol{n}(\boldsymbol{x_S}) < 0$). The classic paper on textured surface radiometry is by Minnaert (1941) while the standard reference for definitions and nomenclature for reflecting surfaces is by Nicodemus et al. (1977).

The local bidirectional reflectance distribution function (or "BRDF") is defined as the ratio of reflected radiance per unit of incoming irradiance at a surface point $\boldsymbol{x_S} \in \partial\mathrm{M}$. Consider a small area δA around $\boldsymbol{x_S}$ and an element of solid angle $\delta\Omega$ around the direction $\boldsymbol{\Omega}$ into which the photons are reflected. An amount δE_{ref} of radiant energy is detected, and we define the BRDF as:

$$\rho_v(x_S, \Omega_0 \to \Omega) = \lim_{\delta A, \delta t, \delta\Omega \to 0} \frac{\delta E_{\text{ref}}}{\mu_0 F_{0v} \delta A \delta t \delta\Omega} = \frac{I_v(x_S, \Omega)}{\mu_0 F_{0v}}, \quad \text{in sr}^{-1} \quad (3.85)$$

where F_{0v} is the incoming collimated flux and $\mu_0 = |n(x_S) \cdot \Omega_0|$ the associated cosine of the zenith angle. Assuming there are no sub-surface radiative fluxes, the BRDF obeys Helmholtz's reciprocity relation: $\rho_v(x_S, \Omega_0 \to \Omega) = \rho_v(x_S, -\Omega \to -\Omega_0)$, cf. Sect. 3.10.3.

In plane-parallel geometry $n(x_S) = \hat{z}$ and the BRF (bidirectional reflectance factor) in (3.19) is just a non-dimensionalized BRDF for a specific reflection event, $\pi\rho_v(x_S, \Omega_0 \to \hat{z})$. The BRF can of course be defined for any reflection angle, not just towards the zenith:

$$\text{BRF} = \frac{\pi I_v(x_S, \Omega)}{\mu_0 F_{0v}} = \pi\rho_v(x_S, \Omega_0 \to \Omega) . \quad (3.86)$$

This quantity is becoming a standard product for a new generation of global imaging spectro-radiometers, such as the Polarization and Directionality of the Earth's Reflectance Instrument (POLDER), the Along-Track Scanning Radiometer-2 (ATSR-2), and the Multiangle Imaging Radiometer Spectro-Radiometer (MISR). These instruments have acquired and continue to acquire this angular signature of reflected radiation from individual scenes, with spatial resolutions ranging from kilometers to hundreds of meters (Diner et al., 1999).

Spectral *planar* albedo α_v, as the ratio of outgoing- to incoming-fluxes, is a non-dimensional quantity:

$$\alpha_v(x_S, \Omega_0) = \int_{n(x_S) \cdot \Omega > 0} (n(x_S) \cdot \Omega) \rho_v(x_S, \Omega_0 \to \Omega) \, d\Omega , \quad (3.87)$$

where $n(x_S) \cdot \Omega = \mu$ if the surface is horizontal ($n(x_S) \equiv \hat{z}$). For locally Lambertian surfaces, the BRF and BRDF are independent of both angles: $\rho_v(x_S, \Omega_0 \to \Omega) \equiv \alpha_v(x_S)/\pi$. This makes the quantity $\pi\rho_v(\cdots)$ easy to interpret in the applications as the (non-dimensional) albedo a Lambertian reflector would have to possess in order to yield the same radiance under the same illumination conditions. For actual Lambertian surfaces, α_v is of course independent of Ω as well.

Spectral *spherical* albedo a_v is obtained by averaging the planar albedo over the hemisphere of possible irradiance angles weighted by $|\mu_0|$, as required by incoming photon flux conservation:

$$a_v(x_S) = \frac{1}{\pi} \int_{n(x_S) \cdot \Omega_0 < 0} |n(x_S) \cdot \Omega_0| \alpha_v(x_S, \Omega_0) d\Omega_0 , \quad (3.88)$$

where $n(x_S) \cdot \Omega_0 = \mu_0$ if the surface is horizontal. This is the ratio of reflected to incoming fluxes for an isotropic sky; equivalently, this is the overall albedo of a planet uniformly covered with the given planar albedo. Lambertian surfaces yield $a_v = \alpha_v$ which, in this case, is independent of the in-coming direction.

Kirchhoff's law of detailed balance (conservation) of radiation during surface-environment exchanges under LTE tells us that $\varepsilon_v(\pmb{x}_S)$ defined in (3.84) is given by

$$\varepsilon_v(\pmb{x}_S) = 1 - a_v(\pmb{x}_S) \tag{3.89}$$

for all v in the thermal spectrum, and we will show in the next section that strict thermal equilibrium precludes directional effects. "Black" bodies indeed get their name from the requirement that $a_v \equiv 0$ (absolutely no reflection) to obtain $\varepsilon_v = 1$ for all v. But, even for non-black materials, this applies only in the case of ideal micro-uniform surfaces. Natural surfaces have texture (roughness and heterogeneity) and its effect on emissivity is captured at scales of interest in remote sensing, even at the finest resolution, by assuming a local/directional surface emissivity model with $\varepsilon_v(\pmb{x}_S, \pmb{\Omega})$ for $\pmb{n}(\pmb{x}_S) \cdot \pmb{\Omega} > 0$. We can equate this with $1 - \alpha_v(\pmb{x}_S, -\pmb{\Omega})$ in (3.87) by invoking reciprocity (exchanging the places of $\pmb{\Omega}_0$ and $\pmb{\Omega}$ while changing their signs). The surface will reflect – and therefore not emit – a fraction $\alpha_v(\pmb{x}_S, -\pmb{\Omega})$ of the incoming flux into direction $-\pmb{\Omega}$ when subjected to an isotropic diffuse illumination, which is precisely what a thermally-balanced environment would look like to the surface.

Reflection is sometimes called "surface scattering" and we can indeed draw a fruitful analogy here with the scattering phase function presented in Sect. 3.4.1, and then used in Sect. 3.6.1, for an elementary *volume*. We can similarly define a phase function for *surface* reflection or scattering using

$$I_v(\pmb{x}_S, \pmb{\Omega}) = \alpha_v(\pmb{x}_S, \pmb{\Omega}) \int\limits_{\pmb{n}(\pmb{x}_S) \cdot \pmb{\Omega}' < 0} p_{Sv}(\pmb{x}_S, \pmb{\Omega}' \rightarrow \pmb{\Omega}) I(\pmb{x}_S, \pmb{\Omega}') \mathrm{d}\pmb{\Omega}' , \tag{3.90}$$

for any $\pmb{\Omega}$ such that $\pmb{n}(\pmb{x}_S) \cdot \pmb{\Omega} > 0$. Notice how $\alpha_v(\pmb{x}_S, \pmb{\Omega})$ plays the role of the scattering probability $\sigma_{sv}(\pmb{x})$ in (3.81) or, better still, the non-dimensional single-scattering albedo ϖ_{0v} since $\sigma_{sv} = \varpi_{0v}\sigma_{ev}$. Like scattering phase functions and BRDFs, $p_{Sv}(\pmb{x}_S, \pmb{\Omega}' \rightarrow \pmb{\Omega})$ is expressed in sr^{-1}. Comparing this definition with (3.85)–(3.87), we see that

$$p_{Sv}(\pmb{x}_S, \pmb{\Omega}' \rightarrow \pmb{\Omega}) = |\pmb{n}(\pmb{x}_S) \cdot \pmb{\Omega}'| \rho_v(\pmb{x}_S, \pmb{\Omega}' \rightarrow \pmb{\Omega}) / \alpha_v(\pmb{x}_S, \pmb{\Omega}) . \tag{3.91}$$

To conserve fluxes, the integral of $p_{Sv}(\pmb{x}_S, \pmb{\Omega}' \rightarrow \pmb{\Omega})$ over the lower hemisphere ('ed angles) is required to be unity.

For illustration purposes, consider two extreme types of reflecting surface that we will assume uniform and horizontal for simplicity:

• Lambertian (diffuse, isotropic) reflection illustrated on the right-hand side of Fig. 3.11b; this leads to

$$p_{Sv}(\pmb{\Omega}' \rightarrow \pmb{\Omega}) = |\mu'|/\pi . \tag{3.92}$$

• Specular (metallic, mirror) reflection as on l.-h. side of Fig. 3.11b; this yields

$$p_{Sv}(\pmb{\Omega}' \rightarrow \pmb{\Omega}) = \delta(\mu' + \mu)\delta(\varphi' - \varphi) . \tag{3.93}$$

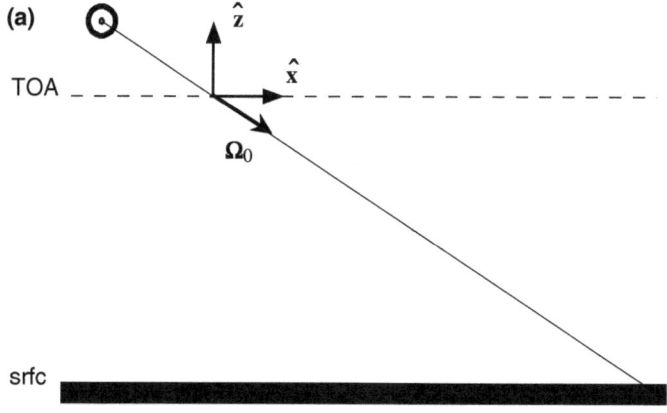

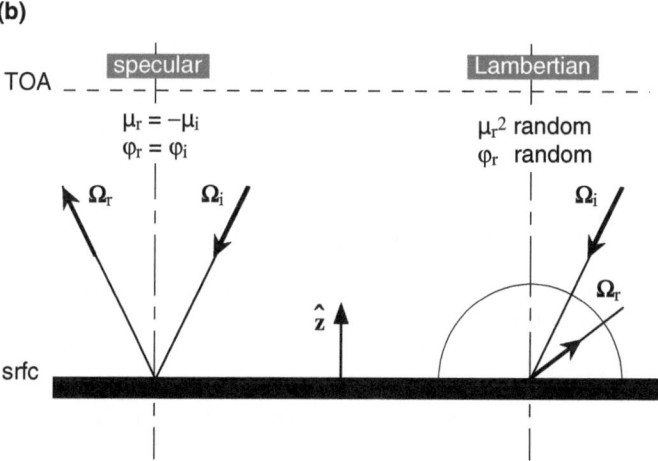

Fig. 3.11. Boundary sources for an infinite slab. (**a**) Irradiance by a (solar) collimated beam from above. (**b**) Reflective surface below, either specular (l.-h. side) or Lambertian (r.-h. side). Subscript "i" designates the incident beams, and "r" is for their reflected counterparts. Note that to generate random vectors uniformly distributed (isotropic) in the upper hemisphere, one exploits (3.92): the probability density of φ is uniform over $[0, 2\pi)$ while for μ the uniform measure over $[0, 1]$ is $2\mu d\mu = d\mu^2$

Steady Irradiance in Plane-Parallel Geometry: Collimated or Diffuse, Uniform or Localized

We introduce here the short-hand $\vec{x} = (x, y)^{\mathrm{T}}$ for Cartesian coordinates, hence

$$\boldsymbol{x} = \begin{pmatrix} \vec{x} \\ z \end{pmatrix}. \qquad (3.94)$$

The boundaries of the plane-parallel (or "slab") medium are set at $z = 0$ and $z = h$ and they can act as radiation sources.

We need to describe how the Sun excites incoming radiance at a cloud top, a collimated but spatially uniform irradiance, cf. Fig. 3.11a. Mathematically, we have

$$\begin{cases} I_v(\vec{x}, h, \mathbf{\Omega}) = F_{0v}\delta(\mathbf{\Omega} - \mathbf{\Omega}_0), & \mu < 0 \\ I_v(\vec{x}, 0, \mathbf{\Omega}) = 0, & \mu > 0 \end{cases}, \forall \vec{x}. \tag{3.95}$$

This assumes that $I_v(x, \mathbf{\Omega})$ is the *total* radiance field (i.e., not separated into diffuse and direct components). Note that the boundaries are still radiation sinks for outgoing beams, as described in Sect. 3.3.

Another useful example is steady isotropic illumination from a localized source below at $\vec{x}_0$:

$$\begin{cases} I_v(\vec{x}, h, \mathbf{\Omega}) \equiv 0, & \mu < 0 \\ I_v(\vec{x}, 0, \mathbf{\Omega}) = F_{0v}\delta(\vec{x} - \vec{x}_0)\mu/\pi, & \mu > 0 \end{cases} \tag{3.96}$$

where, again, the boundaries are sinks for out-going radiation. We will see such sources in the theory of RT Green functions covered Sect. 3.10.1.

The interested reader can also write descriptions for other combinations of boundary source properties: uniform and diffuse, localized and collimated, possibly moved to the opposite side.

Reflection in Plane-Parallel Geometry: Lambertian, Specular, or Otherwise

What happens at the *lower* boundary of a plane-parallel medium? Using the surface phase function in (3.90)–(3.91), we define

$$\begin{cases} I_v(\vec{x}, h, \mathbf{\Omega}) \equiv 0, & \mu < 0 \\ I_v(\vec{x}, 0, \mathbf{\Omega}) = \alpha_v(\vec{x}, \mathbf{\Omega}) \int_{\mu' < 0} p_{Sv}(\vec{x}, \mathbf{\Omega}' \to \mathbf{\Omega})I(\vec{x}, 0, \mathbf{\Omega}')d\mathbf{\Omega}', & \mu > 0 \end{cases}. \tag{3.97}$$

Real surfaces are of course not pure cases of Lambertian or specular behavior used until now as examples. Combinations are possible and other types of BRDF can be introduced. A popular 3-parameter representation of the BRDF for many natural surfaces is given by Rahman et al. (1993).

Finally, the linearity of RT with respect to sources can be invoked to break down complex problems with boundary and/or volume sources and one or more reflecting surfaces into a non-trivial combination of problems with purely absorbing boundaries and others with properly chosen boundary sources. More details are provided in Sect. 3.10.1 and in Chap. 14.

3.7 Local Balance

Looking back, we have studied how photons are created (Q), transported (σ and $p(\mathbf{\Omega}' \to \mathbf{\Omega})$), destroyed ($\sigma_a$) or lost ($\partial M$), and finally detected (I, J, and $\boldsymbol{F} \cdot \boldsymbol{n}$). We collect here the positive and negative contributions to the photon population in an elementary volume and thus obtain at last the RT equation or "RTE", in its basic integro-differential form. We then derive the continuity equation for radiant energy and pause for a few thoughts on radiative transfer in the greater scheme of things.

3.7.1 Integro-Differential Radiative Transfer Equation

Returning once more to Fig. 3.6, we see that position along the beam $\{x, \mathbf{\Omega}\}$ can be represented in general as $x + \mathbf{\Omega}s$ and positions infinitesimally close to x by $x + \mathbf{\Omega}\delta s$ where $\delta s \to 0$. Therefore,

$$\lim_{\delta s \to 0} \frac{\delta I}{\delta s} = \mathbf{\Omega} \cdot \nabla I \tag{3.98}$$

in notations independent of any particular coordinate-system. This operator is known as a directional (or advective) derivative and quantifies change in $I(x, \mathbf{\Omega})$ near x in direction $\mathbf{\Omega}$. In Cartesian coordinates, we have

$$\nabla = \left(\frac{\partial}{\partial x}, \frac{\partial}{\partial y}, \frac{\partial}{\partial z} \right)^{\mathrm{T}} = \begin{pmatrix} \partial_x \\ \partial_y \\ \partial_z \end{pmatrix}, \tag{3.99}$$

hence

$$\mathbf{\Omega} \cdot \nabla = \Omega_x \frac{\partial}{\partial x} + \Omega_y \frac{\partial}{\partial y} + \Omega_z \frac{\partial}{\partial z}. \tag{3.100}$$

The steady-state radiative transfer equation (RTE) is

$$\mathbf{\Omega} \cdot \nabla I = -\sigma(x) I(x, \mathbf{\Omega}) + S(x, \mathbf{\Omega}) + Q(x, \mathbf{\Omega}) \tag{3.101}$$

where we have collected the r.-h. terms from Sections 3.3.3, 3.6.1 and 3.6.1 respectively, and given them the appropriate sign ($+$ for a gain, $-$ for a loss). Dependence on frequency ν is again made implicit since it is omnipresent. Note that we retrieve $I = $ constant along the beam if $\sigma \equiv 0$ which, in turn, implies $S \equiv 0$ as well as $Q \equiv 0$, at least for the common sources in the atmosphere described in the previous section.

Grouping all terms dependent on radiance I, we can write the RTE formally as

$$L I = Q \tag{3.102}$$

where

$$L = \mathbf{\Omega} \cdot \nabla + \sigma(x) - \sigma_{\mathrm{s}}(x) \int_{4\pi} p(x, \mathbf{\Omega}' \to \mathbf{\Omega})[\cdot] \mathrm{d}\mathbf{\Omega}' \tag{3.103}$$

is the integro-differential *linear* transport operator. The mathematical structure of the RTE is that of an infinite system of *coupled* 1st-order partial differential equations (PDEs) parameterized by $\mathbf{\Omega} \in \Xi$. The next chapter is entirely devoted to methods of numerical solution of the RTE complemented with boundary conditions to be described in the next section.

When the phase function is azimuthally symmetric, it is often helpful to combine the two last terms of L into a single integral operator

$$L = \mathbf{\Omega} \cdot \nabla + \sigma(\mathbf{x}) \int_{4\pi} \{\delta(\mathbf{\Omega}' - \mathbf{\Omega}) - \varpi_0(\mathbf{x}) p(\mathbf{x}, \mathbf{\Omega}' \cdot \mathbf{\Omega})\} [\cdot] d\mathbf{\Omega}' \qquad (3.104)$$

Both the "delta-M rescaling" used in spherical harmonics (Wiscombe, 1977) and the "maximum cross-section" method used in Monte Carlo (Marchuk et al., 1980) exploit this operator identity. These tricks are both invoked in the next chapter to improve numerical accuracy and/or computational efficiency. The angular kernel in the above equation has Legendre coefficients $1 - \varpi_0(\mathbf{x}) \eta_l(\mathbf{x})$, $l \geq 0$, since those of a Dirac δ centered on $\theta_s = 0$ are all unity in (3.50). A simple way to see this is to set $g = 1$ for the HG phase function in (3.57).

Consider the case where volume sources vanish ($Q \equiv 0$), and volume sinks also vanish ($\sigma = \sigma_s$, scattering is conservative). It is interesting to notice that it is the *non-isotropic* part of the radiance field that drives the spatial gradients. Indeed, if the radiance field I is independent of $\mathbf{\Omega}$, then the two last terms in (3.103) cancel, as does the r.-h. side of (3.102). So the directional gradients vanish identically. Conversely, if the directional gradients vanish, then I is a fixed point of the angular transform $I(\mathbf{\Omega}) \mapsto \int_{4\pi} p(\mathbf{\Omega}' \rightarrow \mathbf{\Omega}) I(\mathbf{\Omega}') d\mathbf{\Omega}'$ for *any* $\mathbf{\Omega}$; equivalently, it is in the null space of the angular integral transformation in (3.104). If $p(\mathbf{\Omega}' \rightarrow \mathbf{\Omega})$ is not $\delta(\mathbf{\Omega}' - \mathbf{\Omega})$, this implies that I is isotropic (independent of $\mathbf{\Omega}$).

As another example of this two-way connection between spatial gradients and non-isotropic radiance fields (hence net fluxes), consider *exact* thermodynamical equilibrium (TE), i.e., uniform temperature T. In this case, $I_v \equiv B_v(T)$ where we have restored the dependence on v explicitly. Moreover, $Q_v = \sigma_{av} B_v(T)$ and the isotropic radiance yields $S = \sigma_{sv} J_v / 4\pi = \sigma_{sv} B_v$. So, as expected, gradients vanish, and the RTE reduces to the identity $0 = \sigma_v(I_v - B_v)$ for any single-scattering albedo $\varpi_0 = \sigma_{sv}/\sigma_v$ and phase function under the important condition that $\sigma_v \neq 0$ (i.e., non-transparent matter is present). In *local* thermal equilibrium (LTE), we only require that $Q_v(\mathbf{x}) = \sigma_{av} B_v[T(\mathbf{x})]$; so the gradients in T will generate an anisotropy in $I(\mathbf{x}, \mathbf{\Omega})$, and the fluctuations of $I(\mathbf{x}, \mathbf{\Omega})$ will not follow those of $B_v[T(\mathbf{x})]$ exactly.

In summary, radiation transport per se results from a intricate balance of spatial and angular variability in $I(\mathbf{x}, \mathbf{\Omega})$ as controlled by the RTE.

3.7.2 Radiant Energy Conservation and Local Heating/Cooling Rates

By integrating (3.101) over all possible directions, we obtain an expression for the conservation of (as well as conversion to/from) radiant energy, irrespective of the direction it is traveling in. Explicitly, using definitions from Sect. 3.2.3, we have

$$\nabla \cdot \mathbf{F} = -\sigma_a(\mathbf{x}) J(\mathbf{x}) + q(\mathbf{x}) \qquad (3.105)$$

where

$$q(\boldsymbol{x}) = \int_{4\pi} Q(\boldsymbol{x}, \boldsymbol{\Omega}) \mathrm{d}\boldsymbol{\Omega}$$

$$= \begin{cases} 4\pi\sigma_\mathrm{a}(\boldsymbol{x}) B_\mathrm{v}[T(\boldsymbol{x})], & \text{for thermal emission} \\ \sigma_\mathrm{s}(\boldsymbol{x}) \exp[-\tau(\boldsymbol{x}_0(\boldsymbol{x}, \boldsymbol{\Omega}_0), \boldsymbol{x})] F_0, & \text{for solar-beam injection} \end{cases}. \tag{3.106}$$

Also, if $\{J, \boldsymbol{F}\}$ are only modeling the diffuse field, i.e., solar flux injection is modeled with $Q(\boldsymbol{x}, \boldsymbol{\Omega})$, then another term is needed to capture the energy absorbed from the directly transmitted beam. By direct evaluation, we have

$$J_\mathrm{dir}(\boldsymbol{x}) = \int_{4\pi} I_\mathrm{dir}(\boldsymbol{x}, \boldsymbol{\Omega}) \mathrm{d}\boldsymbol{\Omega} = F_0 e^{-\tau(\boldsymbol{x}_0(\boldsymbol{x}, \boldsymbol{\Omega}_0), \boldsymbol{x})} .$$

Thus

$$\nabla \bullet \boldsymbol{F}_\mathrm{dir} = -\sigma_\mathrm{a}(\boldsymbol{x}) \exp[-\tau(\boldsymbol{x}_0(\boldsymbol{x}, \boldsymbol{\Omega}_0), \boldsymbol{x})] F_0 . \tag{3.107}$$

There is a practical meaning for total radiative flux divergence in the computation of absorptance A in (3.26) for the source-free case and in the local energy conservation law in (3.105): conversion to and from thermal energy. In other words, we get:

- cooling if $\nabla \bullet \boldsymbol{F} > 0$ as, e.g., in the LTE problem when $J(\boldsymbol{x}) < 4\pi B_\mathrm{v}(T(\boldsymbol{x}))$ in (3.105)–(3.106); and
- heating if $\nabla \bullet \boldsymbol{F} < 0$ as, e.g., for $\boldsymbol{F}_\mathrm{dir}$ in (3.107).

The algebraically-valued *heating* rate is given by

$$\frac{\mathrm{d}T}{\mathrm{d}t} = \frac{1}{\rho C_p}(-\nabla \bullet \boldsymbol{F}), \quad \text{in K/s (or } \times 3600 \text{ K/hr, or } \times 86400 \text{ K/day)} \tag{3.108}$$

where ρ is the ambient mass density and C_p is the specific heat at constant pressure.

The heating/cooling rate in (3.108) is usually computed after full spectral integration, and only makes real physical sense as a time change in kinetic temperature if all non-radiative contributions to the local energy budget are included. Notwithstanding, it is conventional in climate science at least to further divide $\mathrm{d}T/\mathrm{d}t$ into "shortwave" (solar) and "longwave" (thermal, terrestrial) components. In principle, one can preserve all the spectral information by leaving the "specific" /cm^{-1} units in F_v and in B_v (or the /μm units in F_λ and B_λ); these units will carry over to $(-\nabla \bullet \boldsymbol{F})$ and to $\mathrm{d}T/\mathrm{d}t$. In practice, the simpler r.-h. side(s) of (3.105) (and of (3.107), as required) is (are) of course used to compute the flux divergence field(s) in (3.108).

The local rate of deposition of radiant energy, $-\nabla \bullet \boldsymbol{F} = \sigma_\mathrm{a}(\boldsymbol{x}) J(\boldsymbol{x})$ in the absence of bulk sources, is used in (3.108) for a concern in climate or cloud-system dynamics. There are other important applications, especially in photochemistry where some judicious spectral sampling and integration is implied: ozone production, chlorophyll activity, etc. In vegetation remote sensing, it is commonly known as "FPAR," fraction of photosynthetically active radiation.

3.7.3 A Few Thoughts on Climate, Remote Sensing, and Beyond

At this point, we are about midway through the chapter and we have finally juxta-
posed the two most fundamental elements of climate physics: solar heating and IR
cooling in the Earth's thin but vital atmosphere. This is essentially all climate mod-
elers want from RT, $-\nabla \cdot \boldsymbol{F}_{\mathrm{v}}$ in (3.108) integrated across the solar-through-thermal
spectrum. This radiative quantity – along with a few other energy exchange terms
that the first law of thermodynamics tells us to look at – will tell the model(er)s what
happens next in the evolution of climate system or some portion thereof, maybe a
single cloud or a plant stand. This energy budget is assessed at the smallest spatial
and temporal scales the models can or modelers want to resolve. In turn, the climate
system dynamics will modify the scenario given to the embedded RT solver: the
changing temperature T appears in the thermal source term, the solar source $\mu_0 F_0$
is modulated by the diurnal cycle (shutting off completely at night), and the various
density fields that determine the absorption and scattering properties in the RTE will
also evolve. We are therefore in an endless feed-back loop. There are many tools used
in the difficult task of creating new knowledge about the climate system in which we
live. Modeling is one way, a way that computer technology has enhanced considerably over the past decades. Remote observations (radiance fields sampled in space,
time, direction, and across the EM spectrum) are another way, a way that has been
considerably enhanced – at least in sheer volume – by satellite technology.

For all practical purposes, the "fuel" running the complex climate machine is
short-wave radiation, flowing towards the Earth in neatly collimated (high-energy/
low-entropy) photon beam. At the same time, the "exhaust" from the climate machine is the (low-energy/high-entropy) radial flow of long-wave photons. So radiation is essential to the balance of the climate system. It is therefore incumbent on
RT experts to deliver their very best estimation of the Earth system's 3D radiation
budget at all the spatial and temporal scales that matter for all operational modeling
frameworks – and, going from GCMs to Large Eddy Simulations (LESs), this range
of scales is huge. In remote sensing also the geophysical retrievals are only as good
as the RT used to process the measured radiances. Here again, the radiances are captured over a wide range of scales by present and future sensor systems. So, to deliver
accurate Earth system diagnostics from remote observations, RT experts are required
to work with both resolved and unresolved variability. From both the energetic and
the diagnostic perspectives, this is a tall order!

Maybe this is a good time to take a short pause from the science of RT and
engage in some more lofty thoughts? It is interesting to note that when we finally
touch the essence of a physical science like RT, we find principles that been articulated very clearly in a very different era and in an altogether different culture.
Looking at (3.105) as would Capra (1991), we see the interactions of Brahma-the-
Creator (q), of Vishnu-the-Preserver ($\nabla \cdot \boldsymbol{F}$), and of Shiva-the-Destroyer ($-\sigma_{\mathrm{a}} J$).[10]
This metaphor based on the core trinity from the Hindu pantheon applies even better

---

[10] Alternatively, one can picture RT as a glorified version of book-keeping where, instead
of bean-counting, we photon-count: income (q), cash-flow and -transfers ($\nabla \cdot \boldsymbol{F}$), and ex-
penses ($-\sigma_{\mathrm{a}} J$ and boundaries).

to our deeper formulation of RT using the linear transport equation, augmented for the circumstances with potentially inelastic collisions: $L_v I_v = Q_v$, where I_v is the full spectrum of an ever-moving pool of radiant energy, Q_v (Brahma, Creation) is its source, and its fate is controlled by $L_v = \mathbf{\Omega} \cdot \nabla[\cdot]$ (Vishnu, Preservation) $+ \sigma_v[\cdot]$ (Shiva, Destruction) $- \sigma_{sv} \int \int p(v', \mathbf{\Omega}' \rightarrow v, \mathbf{\Omega})[\cdot]dv'd\mathbf{\Omega}'$ (again Shiva, who is also worshiped as the God of Transformation).[11] As noted earlier, Lord Shiva's Dance (that is, the intertwined processes of extinction/propagation, scattering/reflection and absorption/escape) is what makes radiative transfer so interesting, and such a challenge in the real 3D world.

It is fascinating to see that we have a continuity equation in (3.105) that can be evaluated using only $J_v(\mathbf{x})$, the simplest radiation transport quantity. In the end, that is all that counts for the dynamics of the *material universe*. Now $J_v(\mathbf{x})$ derives from the full 3D radiative transfer equation for $I_v(\mathbf{x}, \mathbf{\Omega})$ in (3.102)–(3.103). Radiance $I_v(\mathbf{x}, \mathbf{\Omega})$ is a more subtle quantity than $J_v(\mathbf{x})$ not only mathematically: it is what feeds our insatiable need to explore the universe via remote observation. This exploration by remote sensing calls for all sorts of instruments that basically extend our senses. The data these instruments produce are ultimately distilled into new *information* (i.e., geophysical properties), often with the help of sophisticated inverse RT theory. By any standard, this is a more elevated plane than the material one. As a general rule, we are not content with gathering information and distilling it into *knowledge*; we eventually take some *action*. That is just human nature and, in fact, this end is invariably what justifies the often costly means of the scientific and technological enterprise in observation and computation. Now this action can play out in domestic affairs or in foreign policy, with any combination of economical, legislative, regulatory, diplomatic or military ramifications. This action can be good or bad for our environment at large, including our fellow human beings. Is it *wise* or *unwise*?

This, dear reader, is the threshold at which we must stop. We can only cross the threshold of judging an action, taken or planned, as informed citizens of a nation or of the world, and not as scientists. This seems obvious in the abstract, but is not that easy since we all have issues we deeply care about. Science and politics should not be mixed. Nothing less than the credibility of the scientific community in the eyes of the public is at stake. We can only encourage our fellow citizen-scientists to look at the state of the world and the actions of those in power with the same mixture of open-mindedness and critique that spawns good science. Closer to home, we must resist external (overt or covert) or internal (even unconscious) pressures to arrive at predetermined conclusions that are politically correct.[12] At least that is our credo. As RT experts, it is our modest hope that judgment error can minimized by better physics-based interpretations of $I_v(\mathbf{x}, \mathbf{\Omega})$ samples captured by radiometers

---

[11] Use $-\sigma_v\delta(v' - v)\delta(\mathbf{\Omega}' - \mathbf{\Omega}) + \sigma_{sv} \int \int p(v', \mathbf{\Omega}' \rightarrow v, \mathbf{\Omega})[\cdot]dv'd\mathbf{\Omega}'$ to unify Shiva's destructive (extinction) and transformational (scattering) actions into a single operation.

[12] Climate and environmental science are unfortunately prone to this process. The Earth system is so complex and the data so sparse – in spite of heroic observation efforts – that opposing views can be substantiated under present levels of uncertainty in modeling and in analysis.

and imagers, and by better estimations of $J(x)$ in complex multi-physics models of the climate system or key parts of it. Both endeavors indeed support potentially far-reaching decisions. Both endeavors will occupy us for the remainder of this volume.

3.8 Global Balance

We cover the general boundary conditions (BCs) that are needed, beyond the RTE, to specify completely the radiance field $I(x, \Omega)$. Our commitment to 3D RT requires us to consider non-plane-parallel media, that may or may not be internally homogeneous, in some detail. A natural and interesting counterpoint to BCs are "escaping" boundary fields because, on the one hand, they are all that can be observed remotely (a task for radiances) and, on the other hand, they control the radiation budget of the medium (a task for fluxes). Finally, we derive the *formal* solution of the RTE and the two widely-used integral formulations of the RTE (with BCs necessarily included).

3.8.1 Boundary Conditions

A complication arises in prescribing BCs if the medium M, defined as the domain where extinction $\sigma(x)$ is strictly positive, is not convex. That is because of re-entering rays, and we want to be able to specify exactly radiation is going *into* the medium a priori but generally do not know what is coming *out* of it. This issue is basically geometrical and is best dealt with simply by allowing for vanishing extinction $\sigma(x)$ and extending the definition of M to its "convex hull." That is what becomes of M if it is covered by an imaginary sheet of rubber. For instance, take a doughnut-shaped optical medium. When wrapped (but not shrink-wrapped) in imaginary cellophane, the resulting convex medium will have region of zero extinction where the hole used to be.

We can therefore always assume that M is an open convex subset of $\mathbb{R}^3$ and we denote the closed set of all its boundary points as in mathematical analysis by "∂M." We can now express the most general BC for the RTE as

$$I(x, \Omega) = f(x, \Omega), \ x \in \partial M, \ \Omega \cdot n(x) < 0 . \tag{3.109}$$

Along with the RTE, including its own source term, this determines the radiance field uniquely. In some applications, we must also consider (internal) reflection properties at the boundaries. This gives rise to constraints that couple various out-going and in-coming beams at the inside surface of ∂M, as described in Sect. 3.6.2. As already pointed out, surface reflection processes act formally like a special kind of scattering.

In Fig. 3.12, to which we will return for further discussion momentarily, we have illustrated the case of a smooth ∂M where $n(x)$ exists everywhere. We address boundary points where $n(x)$ does not exist further on. We only require that the "measure" of that set be zero, which basically means that they intercept vanishingly few incoming or outgoing beams. What can we say about fractal cloud boundaries where $n(x)$ exists almost nowhere? This is a very relevant question for real clouds

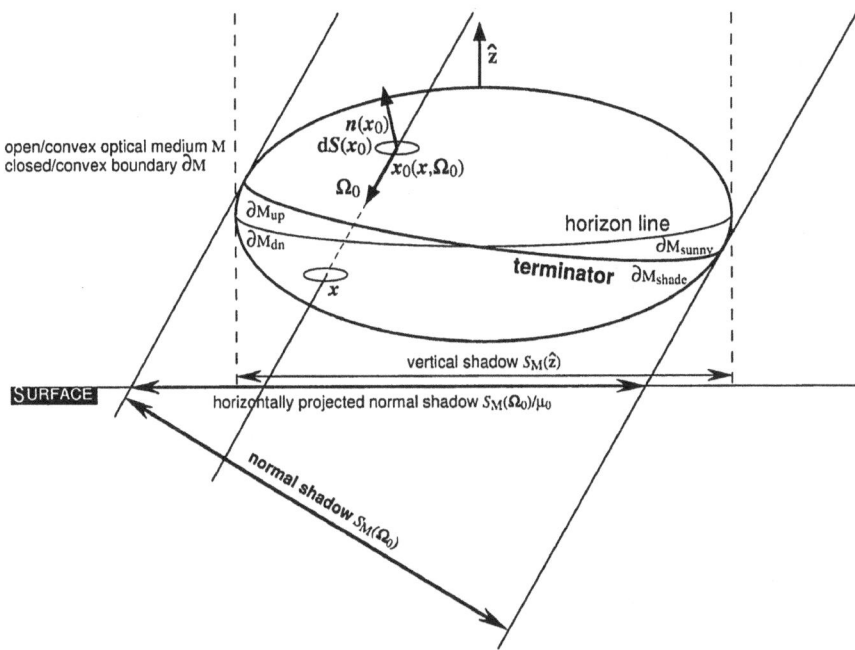

Fig. 3.12. Illumination of a horizontally finite medium with a collimated beam that may be oblique with respect to a surface (terminator and horizon lines defined further on)

(Lovejoy, 1982). Such shapes are necessarily non-convex, so it is a perfect example of where the convex hull is used; it will be made of triangular facets, so we are back to a situation where $n(x)$ exists almost everywhere.

Plane-Parallel Media: The Many Ways They can be Hosts to 3D Radiative Transfer

Section 3.6.2 on steady boundary sources describes typical BCs for slab geometry. They can be combined to have sources at both top and bottom, or none at all (so-called "absorbing" BCs). It suffices that the boundary or volume sources be spatially variable to necessitate the 3D RTE; internal variability of optical properties is therefore not always a requirement. Sometimes it is necessary to consider surface reflection, as described in Sect. 3.6.2. Here again the properties of the bulk of the medium and the sources can be uniform and just variability of the surface reflectivity is enough to excite the horizontal gradients in the 3D RTE. These scenarios are germane to cloud lidar studies and aerosol adjacency effects respectively.

So, the RTE in (3.101) and one of these BC scenarios entirely determine $I(x, \Omega)$ in the plane-parallel medium

$$M = \{x \in \mathbb{R}^3 : 0 < z < h\} \tag{3.110}$$

described by given optical properties $\sigma(x)$, $\varpi_0(x)$, $p(x, \Omega' \cdot \Omega)$, and volume sources $Q(x, \Omega)$. These quantities appear in the various terms on the r.-h. side of the RTE. Now, in practice, the variability of the optical properties is often specified only over a finite domain $[0, L_x) \times [0, L_y) \times (0, h)$. The vertical limits $z = 0$ and $z = h$ receive the usual treatment, while "periodic" BCs are applied horizontally. Therefore, to determine radiance in the basic cell,

$$M = \{ x \in \mathbb{R}^3 : 0 \leq x < L_x, \ 0 \leq y < L_y, \ 0 < z < h \}, \qquad (3.111)$$

we require

$$\left. \begin{cases} I(x, 0, z, \Omega) = I(x, L_y, z, \Omega), \ 0 \leq x < L_x \\ I(0, y, z, \Omega) = I(L_x, y, z, \Omega), \ 0 \leq y < L_y \end{cases} \right\}, \ 0 < z < h, \ \forall \Omega, \qquad (3.112)$$

inside the medium.

We often need to consider, at least formally, semi-infinite media, i.e., the limit $h \to \infty$ in (3.110). This is of course an idealization, albeit a useful one. In this case, we only need to specify BCs on the boundary at $z = 0$, which may be viewed as a top (e.g., of an ocean) or a bottom (e.g., of an extended atmosphere).

We can even think of the atmosphere-surface system as a stratified semi-infinite medium with its upper boundary being the TOA. This TOA can be set for convenience at a fixed altitude ($z = 0$ or h or whatever), with or without incoming radiation, and all the rest is about *internal* sources and scattering/reflection processes. From then on, surface emission is assimilated to an internal source confined to a manifold $z = z_S(x, y)$ and directed toward the upward side of the said manifold while surface reflection is assimilated to a special kind of oriented scattering that occurs on the same manifold. Below the surface, extinction is formally viewed as infinite; so there is no need to go there, radiatively speaking. See Fig. 3.13. All we have excluded here is topologically complicated terrain that can not be modeled with an analytical or digital elevation model of the form $z_S(x, y)$. We thus exclude over-hangs, caves and tunnels since these would call for multi-valued functions $z_S(x, y)$.

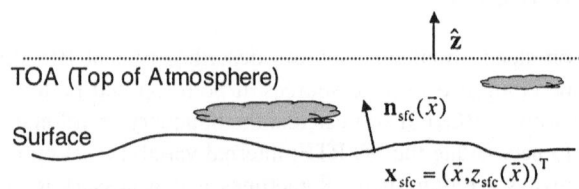

Fig. 3.13. A 3D radiative transfer problem in variable-altitude terrain with or without an atmosphere over overlying it, with or without horizontal variability (such as clouds) in it. This scenario is considered general enough for most present needs, including small-scale modeling of a rough surface's angular properties

The beauty of this formulation is that we are no longer limited to flat or even convex terrain. We have already accommodated in Sect. 3.6.2 the possibility of non-uniform reflectivity and emissivity properties for the special case $z_S(\vec{x}) \equiv 0$ (under the assumption that TOA is at $z = h > 0$). For this more general situation, we only assume that the *inward* normal vector $\boldsymbol{n}(\overrightarrow{x})$ exists (almost everywhere) and that it is "open to the sky" (i.e., $\boldsymbol{n}(\overrightarrow{x}) \cdot \hat{\boldsymbol{z}} > 0$ which follows directly from the single-valuedness of $z_S(\overrightarrow{x})$). In summary, we have to solve the 3D RTE at all points in

$$\mathrm{M} = \{\boldsymbol{x} = (\overrightarrow{x}, z_S(\overrightarrow{x}))^{\mathrm{T}} \in \mathbb{R}^3 : z_S(\overrightarrow{x}) < z < h\} \tag{3.113}$$

where $h \geq \max_{\overrightarrow{x}}\{z_S(\overrightarrow{x})\}$, subject to the constraints

$$\left\{ \begin{array}{l} f(\boldsymbol{x}, \boldsymbol{\Omega}) = \varepsilon(\boldsymbol{x}, \boldsymbol{\Omega})B[T(\boldsymbol{x})] \text{ in (3.109))} \\ I(\boldsymbol{x}, \boldsymbol{\Omega}) = \alpha(\boldsymbol{x}, \boldsymbol{\Omega}) \int\limits_{\boldsymbol{n}(\overrightarrow{x})\cdot\boldsymbol{\Omega}'<0} p_S(\boldsymbol{x}, \boldsymbol{\Omega}' \to \boldsymbol{\Omega})I(\boldsymbol{x}, \boldsymbol{\Omega}')\mathrm{d}\boldsymbol{\Omega}' \end{array} \right\}, \ \forall \boldsymbol{x} = (\overrightarrow{x}, z_S(\overrightarrow{x}))^{\mathrm{T}},$$

$$\forall \boldsymbol{\Omega} \text{ such that } \boldsymbol{\Omega} \cdot \boldsymbol{n}(\overrightarrow{x}) > 0 , \tag{3.114}$$

where the dependence on ν is made implicit. There is also a standard BC at $z = h$. Even though we require inward normal vectors to exist *almost everywhere*, there can be very many facets in the terrain model that quickly change their orientation $\boldsymbol{n}(\overrightarrow{x})$ as well as their optical properties, ε, α, p_S in (3.114). So we now we have the possibility of modeling rough terrain that is fractal-like over a large range of scales.

This is a necessary complication in many important applications, some of them in planetary science where there is in fact no atmosphere at all. One application of 3D radiative transfer driven only by rough terrain effects is to compute, starting with a deterministic or stochastic description of a uniformly emissive but rough surface, the macroscopic angular dependence of the "effective" emissivity in (3.114). The resulting model for $\varepsilon(\boldsymbol{\Omega})$ could be used as a parameterization of unresolved small-scale variability in a subsequent flat-surface plane-parallel computation. The same remark applies to the macroscopic models for reflective properties of surfaces with complex internal structure (cf. Chap. 14 on vegetation canopies).

An example is given in Fig. 3.14 where we show the angular dependence of the apparent emissivity enhancement caused by surface roughness. Although radiosity methods (e.g., Siegel and Howell, 1981) are also popular for 3D RT problems where only surfaces interact, we used a straightforward Monte Carlo scheme (cf. Chap. 4) to compute these results. To illustrate this well-known systematic effect of small-scale terrain variability, we used a surface made of an unresolved array of closely-packed circular "craters" with a (power-law) size distribution such that they fill 2D space completely.[13] This way, the response of the unresolved ensemble of craters is the same as that of a single one as long as they all have the same radius-to-depth (or "aspect") ratio. The surface was maintained at a constant temperature T and its *uniform* emissivity ε takes the three indicated values while we changed the aspect ratio of the craters. As illustrated, we had hemispherical craters (aspect ratio is unity),

---

[13] The rims of all these space-filling craters form an "Apollonian" fractal investigated by Mandelbrot (1982) and others. Its fractal dimension is ≈ 1.3058.

(a)

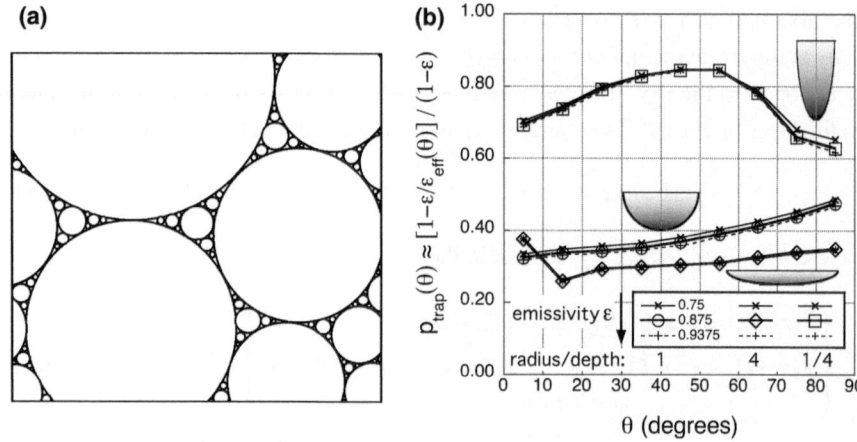

(b)

Fig. 3.14. Angular dependence of effective emissivity for a uniform but variable-height surface, an array of closely-packed ellipsoidal craters. (**a**) A schematic of the unresolved variability. (**b**) Computations performed using a straightforward Monte Carlo ray-tracing technique with the 3 indicated values of ε and the 3 indicated values of the aspect ratio (i.e., radius-over-depth at the center). The effect is systematic and present at all viewing angles. Plotted here are estimates of $p_{\text{trap}}(\theta)$ from (3.116). The good collapse of the curves for a relatively wide range of $\alpha = 1 - \varepsilon$ shows that the semi-analytical model is quite accurate

shallow craters (aspect ratio is 4), and deep wells (aspect ratio is 1/4). Emissivity $\varepsilon = 0.75$, 0.875, and 0.9375; hence albedo $\alpha = 1 - \varepsilon = 0.25$, 0.125, and 0.0675. We computed $\varepsilon_{\text{eff}}(\theta)$ as a function of θ and the optical (ε) and structural (aspect ratio) parameters of the problem.

As a first approximation, we can reduce this problem to the estimation of mean probability $p_{\text{trap}}(\theta)$ for a photon propagating at zenith angle θ to remain trapped in the cavity. Escape probability $1 - p_{\text{trap}}(\theta)$ could thus be defined as $1/2\pi$ times the solid-angle of open sky viewed from a point on the surface and averaged over that part of the crater that is seen from viewing angle θ. We can also derive $p_{\text{trap}}(\theta)$ from our Monte Carlo results for $\varepsilon_{\text{eff}}(\theta)$ using a nonlinear model. Indeed, if $I_n(\theta)$ is the radiance contributed to the observation after one surface emission followed by $n \geq 0$ internal reflections, then

$$I_{n+1}(\theta) \approx I_n(\theta) \times \alpha \times p_{\text{trap}}(\theta)$$

and total radiance is

$$I(\theta) = \sum_{n \geq 0} I_n(\theta) \approx \frac{I_0(\theta)}{1 - \alpha p_{\text{trap}}(\theta)} \qquad (3.115)$$

where $I_0(\theta) = \varepsilon B_v(T)$. Hence

$$\varepsilon_{\text{eff}}(\theta) = \frac{I(\theta)}{B_v(T)} = \frac{\varepsilon}{1 - (1 - \varepsilon)p_{\text{trap}}(\theta)} > \varepsilon . \qquad (3.116)$$

To assess the accuracy of this simple estimation of emissivity enhancement we can solve the above equation for $p_{\text{trap}}(\theta)$. In our model for rough terrain, it should not depend much on ε for a given θ and crater aspect ratio. This is done in Fig. 3.14b and we see that the data collapse is better than a few %.

This heuristic physical argument based on successive orders of reflection/scattering and leakage can be made mathematically rigorous using the eigen-analysis of the global transport operator; see Chap. 14 for the details and an application to RT in plant canopies.

Generalization to Horizontally Finite Media

Another interesting class of 3D RT problems involve horizontally finite media that may or may not be internally variable. For instance, the popular case of rectangular parallelepipeds or "cuboids" would have M defined by (3.111) but without the lateral recycling of radiance described in (3.112). Geometrical – actually, topological – considerations are in order to specify BCs as well as partition the boundary fields further on. All of what is said here is general enough to contain the plane-parallel media treated above in the limit of infinite aspect ratio $\min\{L_x, L_y\}/h$, where M becomes the slab in (3.110).

If only volume sources are considered, then $f(x, \Omega) = 0$ (absorbing BCs) while for a solar beam (not "injected" through the volume source term) with flux F_0 and incidence direction Ω_0, we have

$$f(x, \Omega) = F_0 \delta(\Omega - \Omega_0) 1_{\partial M_{\text{sunny}}(\Omega_0)}(x) \tag{3.117}$$

where $1_{\partial M_{\text{sunny}}(\Omega_0)}(x)$ is the indicator function[14] for the subset of ∂M where the solar rays enter M. We will call it the illuminated or "sunny" side of M.

Even if the solar beam source is modeled by "injection" in to the bulk of M, it is useful to define $\partial M_{\text{sunny}}(\Omega_0)$. Specifically, we have

$$\partial M_{\text{sunny}}(\Omega_0) = \underline{\{x \in \partial M : n(x) \cdot \Omega_0 \leq 0\}} . \tag{3.118}$$

Notice the inclusion of $n(x) \cdot \Omega_0 = 0$ here, which makes $\partial M_{\text{sunny}}$ a *closed* set, at least if $n(x)$ exits everywhere (as, for instance, in Fig. 3.12). If $n(x)$ does not exist everywhere (as on the conspicuous edges that appear in Fig. 3.15), then we use the "closure" of the set, i.e., the set itself plus the limit points of all possible infinite sequences belonging to the set. We have used an underscore to designate the closure operation. Closure adds no new points in the everywhere smooth boundary case in Fig. 3.12). However, in cases like in Fig. 3.15, all points on edges from which there is a unobstructed view of the sun, even if at grazing angles, are added to $\{x \in \partial M : n(x) \cdot \Omega_0 \leq 0\}$.

By extension, we have

$$\partial M_{\text{shady}}(\Omega_0) = \{x \in \partial M : n(x) \cdot \Omega_0 > 0\} = \partial M \backslash \partial M_{\text{sunny}}(\Omega_0) \tag{3.119}$$

---

[14] The indicator function $1_S(x)$ of a set S is $= 1$ if $x \in S$, and $= 0$ otherwise.

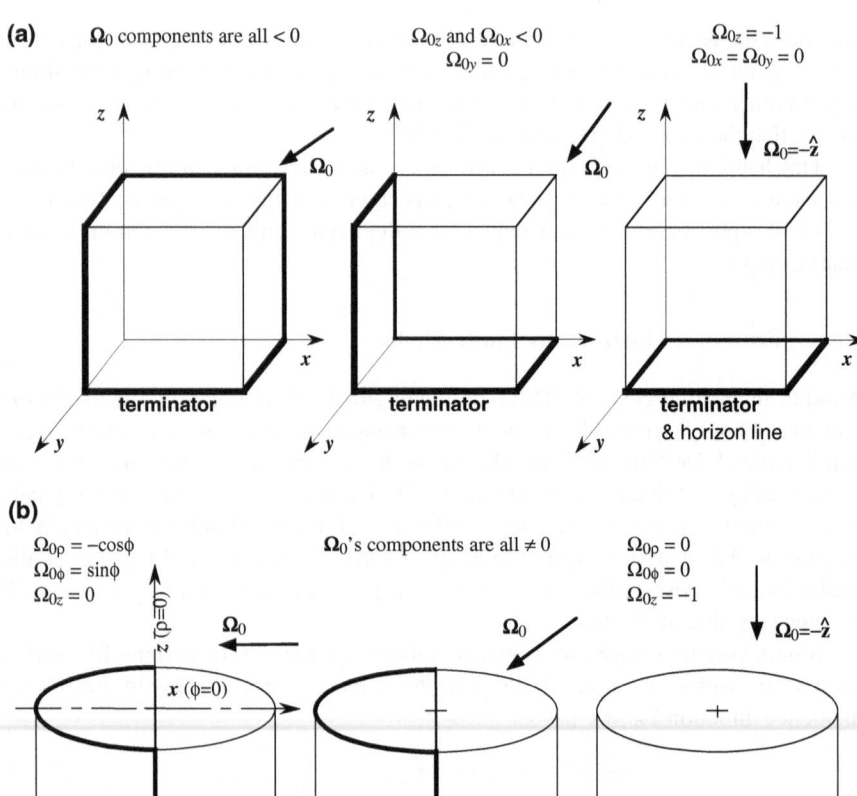

Fig. 3.15. Reflection versus transmission in isolated cloud models (terminator curves defined in (3.139): (**a**) cuboidal and (**b**) cylindrical shapes

which is an *open* set that is the non-illuminated or "shady" side of the boundary of M in Fig. 3.12. The direction of the asymmetry with respect to the points where $n(x) \cdot \Omega_0 = 0$ is not arbitrary, and is justified physically further on. Convexity of M guarantees that ∂M_{sunny} and ∂M_{shady} are both singly connected.

3.8.2 Exact and Formal Solutions of the 3D RTE

Suppose $S(x, \Omega) \equiv 0$ in (3.101), equivalently $\sigma_s(x) \equiv 0$ in (3.81), i.e., no scattering. The RTE is then a doubly infinite system of *independent* ODEs, one class for each Ω in Ξ and then one for each two-dimensional (2D) vector $\vec{x} \in S(\Omega)$, the associated 2D projection of M along $\boldsymbol{\Omega}$ (as used previously, e.g., in Fig. 3.12 to compute projected areas of finite clouds). Each of these ODEs can be written as

$$\frac{\mathrm{d}I}{\mathrm{d}s} = -\sigma(\boldsymbol{x})I + Q(\boldsymbol{x}, \boldsymbol{\Omega}) \tag{3.120}$$

where Q is the known volume source term and $\boldsymbol{x}$ is at a variable distance s away from an arbitrary point in M along on the fixed direction $-\boldsymbol{\Omega}$. This ODE is immediately integrable, yielding

$$I(\boldsymbol{x}, \boldsymbol{\Omega}) = f(\boldsymbol{x}_{\partial M}(\boldsymbol{x}, \boldsymbol{\Omega}), \boldsymbol{\Omega}) \exp[-\tau(s_{\partial M}(\boldsymbol{x}, \boldsymbol{\Omega}); \boldsymbol{x}, -\boldsymbol{\Omega})]$$
$$+ \int_0^{s_{\partial M}(\boldsymbol{x}, \boldsymbol{\Omega})} Q(\boldsymbol{x} - \boldsymbol{\Omega}s, \boldsymbol{\Omega}) \exp[-\tau(s; \boldsymbol{x}, -\boldsymbol{\Omega})]\mathrm{d}s \tag{3.121}$$

where

- $\boldsymbol{x}_{\partial M}(\boldsymbol{x}, \boldsymbol{\Omega})$ is the (unique) piercing point of the beam $\{\boldsymbol{x}, -\boldsymbol{\Omega}\}$ with the (convex) boundary ∂M, and
- $s_{\partial M}(\boldsymbol{x}, \boldsymbol{\Omega}) = \|\boldsymbol{x} - \boldsymbol{x}_{\partial M}(\boldsymbol{x}, \boldsymbol{\Omega})\|$ is the distance from $\boldsymbol{x}$ to ∂M along $-\boldsymbol{\Omega}$.

The first term is thus a given BC in (3.109) followed by direct transmission to $\boldsymbol{x}$ (cf. Sect. 3.3.3) and, in the second term, $\tau(s; \boldsymbol{x}, -\boldsymbol{\Omega})$ is the optical distance through the 3D medium from $\boldsymbol{x}$ to a *backwards* running point $\boldsymbol{x} - \boldsymbol{\Omega}s$ again from (3.30).

The steady-state 3D RT problem in the absence of volume scattering and reflective surfaces is thus reduced to the numerical implementation of (3.121). Our expression of this limited but exact solution is a bit heavy notation-wise, but its physical interpretation is simple. For instance, in remote sensing applications we are primarily interested in radiance escaping the medium, i.e., when $\boldsymbol{x}$ is a near-side boundary point visible from the detector and $\boldsymbol{\Omega}$ is the direction of said detector. In that configuration, the interpretation of (3.121) is that the radiometrically available information is dominated by the sources at one mean-free-path, plus or minus another MFP or so, into the medium.[15] We recall from (3.79) that, in LTE thermal RT, the source goes as the local absorption (and, here, also extinction) coefficient. Consequently, going backwards along the beam, as suggested in (3.121), the contribution to the measured value of $I(\boldsymbol{x}, \boldsymbol{\Omega})$ is relatively small as long as we are in the optically thin region $\{s \in \mathbb{R}^+ : \tau(s; \boldsymbol{x}, -\boldsymbol{\Omega}) \ll 1\}$ where $\exp(-\tau) \approx 1$. Sources embedded in the optically thick region $\{s \subset \mathbb{R}^+ : \tau(s; \boldsymbol{x}, -\boldsymbol{\Omega}) \gg 1\}$ also contribute little to the medium due to the exponentially decaying weight in the integral. Exceptionally bright and/or non-LTE sources can of course compensate the effects of tenuousness (poor emissivity) and opacity (poor transmissivity) at any distance into the medium.

In the presence of volume scattering and reflective surfaces, we can formally equate the source *term* Q in (3.120) to the source *function* S which in fact depends on the unknown $I(\boldsymbol{x}, \boldsymbol{\Omega})$. Then (3.121) is called the "formal" solution of the RTE.

---

[15] To make this rule-of-thumb universal, consider that the surface at the far-side of the medium with respect a space-based detector in Fig. 3.13 is the upper boundary of a region with infinite extinction. Its contribution here is determined by (3.84) and the numerical value of the total optical path through the overlying atmosphere.

3.8.3 Integral Radiative Transfer Equations

By substituting the expression in (3.81) for S into the formal solution (3.121) where we also make the substitution $Q \mapsto S + Q$, we find

$$I(\boldsymbol{x}, \boldsymbol{\Omega}) = I_{fQ}(\boldsymbol{x}, \boldsymbol{\Omega})$$

$$+ \int_0^{s_{\partial M}(\boldsymbol{x}, \boldsymbol{\Omega})} \sigma_s(\boldsymbol{x} - \boldsymbol{\Omega}s) \exp[-\tau(s; \boldsymbol{x}, -\boldsymbol{\Omega})]$$

$$\times \int_{4\pi} p(\boldsymbol{x} - \boldsymbol{\Omega}s, \boldsymbol{\Omega}' \to \boldsymbol{\Omega}) I(\boldsymbol{x} - \boldsymbol{\Omega}s, \boldsymbol{\Omega}') d\boldsymbol{\Omega}' ds \qquad (3.122)$$

where $I_{fQ}(\boldsymbol{x}, \boldsymbol{\Omega})$ is the boundary- and/or volume- "forcing" term given by (3.121) as it stands. This defines the integral form of the RTE where we recognize the cumulative contributions of "up-stream" elements: the positional argument is $-\boldsymbol{\Omega}s$ and scattering is into the beam of interest.

We have thus obtained a self-contained integral equation for the general radiative transfer problem which can however be written more simply at the cost of making the 3-dimensional integral in (3.122) look as if it was 5-dimensional. Specifically, one makes use of

$$d\boldsymbol{x}' = s^2 ds d\boldsymbol{\Omega}' \qquad (3.123)$$

to convert the resulting double (line and angle) integral(s) over $ds d\boldsymbol{\Omega}'$ into a volume integral over M. Then, noting that $s = \|\boldsymbol{x}' - \boldsymbol{x}\|$ and using the identity

$$\int_M [\cdot] d\boldsymbol{x}' = \int_{4\pi} \int_M [\cdot] \delta\left(\boldsymbol{\Omega}' - \frac{\boldsymbol{x}' - \boldsymbol{x}}{\|\boldsymbol{x}' - \boldsymbol{x}\|}\right) d\boldsymbol{x}' d\boldsymbol{\Omega}' , \qquad (3.124)$$

we obtain

$$I(\boldsymbol{x}, \boldsymbol{\Omega}) = \int_{4\pi} \int_M \mathcal{K}_I(\boldsymbol{x}', \boldsymbol{\Omega}' \to \boldsymbol{x}, \boldsymbol{\Omega}) I(\boldsymbol{x}', \boldsymbol{\Omega}') d\boldsymbol{x}' d\boldsymbol{\Omega}' + I_{fQ}(\boldsymbol{x}, \boldsymbol{\Omega}) \qquad (3.125)$$

where the 5-dimensional transport kernel is

$$\mathcal{K}_I(\boldsymbol{x}', \boldsymbol{\Omega}' \to \boldsymbol{x}, \boldsymbol{\Omega}) = \sigma_s(\boldsymbol{x}') p(\boldsymbol{x}', \boldsymbol{\Omega}' \to \boldsymbol{\Omega}) \delta\left(\boldsymbol{\Omega}' - \frac{\boldsymbol{x}' - \boldsymbol{x}}{\|\boldsymbol{x}' - \boldsymbol{x}\|}\right) \frac{\exp[-\tau(\boldsymbol{x}', \boldsymbol{x})]}{\|\boldsymbol{x}' - \boldsymbol{x}\|^2} .$$

$$(3.126)$$

For numerical implementation, (3.122) is the only route.

Of course, if S is actually known, then (3.121), again with the substitution $Q \mapsto S + Q$, can be used to infer I. Hence the idea of formulating another integral equation, this time for the source function $S(\boldsymbol{x}, \boldsymbol{\Omega})$, by performing the reverse substitution of (3.121) into (3.81). Through similar manipulations as above, this leads to the so-called "ancillary" equation which reads as

$$S(\boldsymbol{x}, \boldsymbol{\Omega}) = S_{fQ}(\boldsymbol{x}, \boldsymbol{\Omega}) + \sigma_s(\boldsymbol{x}) \int_{4\pi} p(\boldsymbol{x}, \boldsymbol{\Omega}' \to \boldsymbol{\Omega})$$

$$\times \int_0^{s_{\partial M}(\boldsymbol{x}, \boldsymbol{\Omega})} \exp[-\tau(s; \boldsymbol{x}, -\boldsymbol{\Omega})] S(\boldsymbol{x} - \boldsymbol{\Omega}s, \boldsymbol{\Omega}') \mathrm{d}s \mathrm{d}\boldsymbol{\Omega}' \qquad (3.127)$$

where

$$S_{fQ}(\boldsymbol{x}, \boldsymbol{\Omega}) = \sigma_s(\boldsymbol{x}) \int_{4\pi} p(\boldsymbol{x}, \boldsymbol{\Omega}' \to \boldsymbol{\Omega}) I_{fQ}(\boldsymbol{x}, \boldsymbol{\Omega}') \mathrm{d}\boldsymbol{\Omega}' . \qquad (3.128)$$

is the known forcing term. Again we see in (3.127) the up-stream integration but with fewer terms in the spatial integral. This turns out to be significant advantage in numerical implementations and the relatively minor price to pay is that, after $S(\boldsymbol{x}, \boldsymbol{\Omega})$ is obtained, there is one last application of the formal solution (3.121) to derive $I(\boldsymbol{x}, \boldsymbol{\Omega})$.

Here again, a more compact rewriting of (3.127) is possible:

$$S(\boldsymbol{x}, \boldsymbol{\Omega}) = \int_{4\pi} \int_M \mathcal{K}_S(\boldsymbol{x}', \boldsymbol{\Omega}' \to \boldsymbol{x}, \boldsymbol{\Omega}) S(\boldsymbol{x}', \boldsymbol{\Omega}') \mathrm{d}\boldsymbol{x}' \mathrm{d}\boldsymbol{\Omega}' + S_{fQ}(\boldsymbol{x}, \boldsymbol{\Omega}) \qquad (3.129)$$

where the transport kernel is now

$$\mathcal{K}_S(\boldsymbol{x}', \boldsymbol{\Omega}' \to \boldsymbol{x}, \boldsymbol{\Omega}) = \sigma_s(\boldsymbol{x}) \, p(\boldsymbol{x}, \boldsymbol{\Omega}' \to \boldsymbol{\Omega}) \, \delta\left(\boldsymbol{\Omega}' - \frac{\boldsymbol{x}' - \boldsymbol{x}}{\|\boldsymbol{x}' - \boldsymbol{x}\|} \right) \frac{\exp[-\tau(\boldsymbol{x}', \boldsymbol{x})]}{\|\boldsymbol{x}' - \boldsymbol{x}\|^2} .$$

$$(3.130)$$

The only difference between the kernels $\mathcal{K}_S$ and $\mathcal{K}_I$ is the dependence of the scattering properties on the ending point rather than the starting point of the displacement modeled by the kernels, cf. $\tau(\boldsymbol{x}', \boldsymbol{x})$. Both the Spherical-Harmonics Discrete-Ordinate Method (SHDOM) and the backward Monte Carlo method described in Chap. 4 capitalize on this remark.

Finally, we note that the full 5-dimensional formalism used in the above integral equations is primarily useful in theoretical considerations (existence and analytical properties of solutions, etc.). Only 3-dimensional formulations are used in numerical implementations. Indeed, the identity in (3.124) can be used to get rid of the angular integral it was used to create. Pre-scattering direction $\boldsymbol{\Omega}'$ then becomes a simple function of $\boldsymbol{x}' - \boldsymbol{x}$, namely, its direction in Ξ, and $\boldsymbol{\Omega} \in \Xi$ becomes essentially a control parameter for the fields and the kernels while the 3D integration over $\boldsymbol{x}'$ is done (and probably iterated).

3.9 Outgoing Radiation

The photon population that leaves an optical medium through its outer boundary is of particular interest. This is in part because it offers a means of assessing the radiant energy budget of the medium; recall the arguments given in Sect. 3.3.1 where we were considering an arbitrary region, not necessarily the whole medium, in Fig. 3.12.

Maybe even more importantly, this is because escaping radiation can be detected remotely and it tells us volumes about the structure and properties of the medium at some significant stand-off distance. All of observational astrophysics and remote sensing is indeed predicated on this simple fact.

3.9.1 Plane-Parallel Media with Horizontally Variable Structure and/or Sources

Varying radiances and fluxes at the boundaries are of special interest in 3D RT. For the plane-parallel medium in (3.110), we are looking at the up- and down-welling radiances

$$\left\{ \begin{array}{l} \pi I(\overrightarrow{x}, h, \mathbf{\Omega}(\mu, \varphi))/\mu_0 F_0, \ \mu \geq 0 \\ \pi I(\overrightarrow{x}; 0, \mathbf{\Omega}(\mu, \varphi))/\mu_0 F_0, \ \mu \leq 0 \end{array} \right\}, \ \forall \overrightarrow{x}, \tag{3.131}$$

given here in the natural non-dimensional representation introduced in (3.19). In this standard normalization, radiance is represented as the non-dimensional BRF in (3.19) and, more generally, (3.86) in Sect. 3.6.2. This interpretation works similarly for transmittance if one thinks of a *bi*-Lambertian surface, i.e., that reflects *and* transmits isotropically. For fluxes, we are interested in

$$\left\{ \begin{array}{l} R(\overrightarrow{x}) = F_z^{(+)}(\overrightarrow{x}, h)/\mu_0 F_0, \quad \text{reflectance field} \\ T(\overrightarrow{x}) = F_z^{(+)}(\overrightarrow{x}, 0)/\mu_0 F_0, \quad \text{transmittance field} \end{array} \right\}, \forall \overrightarrow{x}, \tag{3.132}$$

recalling from (3.15) that the normal vectors are always oriented outward from M, hence the "(+)" subscripts assigned to *both* hemispherical fluxes. They are relative to the local normal and not the absolute upward z-axis.

Analogous definitions apply to the periodically replicated media in (3.111). In that case at least, it is easy to define the domain-average quantities:

$$\left\{ \begin{array}{l} R = \overline{R(\overrightarrow{x})} = \displaystyle\int_{[0,L_x] \times [0,L_y]} R(\overrightarrow{x}) d\overrightarrow{x} / L_x L_y, \quad \text{(mean) reflectance} \\[2ex] T = \overline{T(\overrightarrow{x})} = \displaystyle\int_{[0,L_x] \times [0,L_y]} T(\overrightarrow{x}) d\overrightarrow{x} / L_x L_y, \quad \text{(mean) transmittance} \end{array} \right. , \tag{3.133}$$

where we use the overscore to designate a spatial average of a field. The above definitions naturally extend to the case of an infinite domain by taking the limit of arbitrarily large L_x and L_y.

3.9.2 Generalization to Horizontally Finite Media

"R/T" Partition by Illumination and Position at Escape

The simplest plane-parallel definition to emulate here is in (3.132). We thus define the *normalized* flux fields

$$\begin{cases} R(\boldsymbol{x}) = F_{\boldsymbol{n}(\boldsymbol{x})}^{(+)}(\boldsymbol{x})/(|\boldsymbol{n}(\boldsymbol{x}) \bullet \boldsymbol{\Omega}_0|F_0), \ \boldsymbol{x} \in \partial \mathrm{M}_{\mathrm{sunny}}(\boldsymbol{\Omega}_0), \qquad \text{reflectance} \\[2mm] T(\boldsymbol{x}) = F_{\boldsymbol{n}(\boldsymbol{x})}^{(+)}(\boldsymbol{x})/(|\boldsymbol{n}(\boldsymbol{x}_0(\boldsymbol{x}, \boldsymbol{\Omega}_0)) \bullet \boldsymbol{\Omega}_0|F_0), \ \boldsymbol{x} \in \partial \mathrm{M}_{\mathrm{shady}}(\boldsymbol{\Omega}_0), \quad \text{transmittance} \end{cases}$$
$$(3.134)$$

where $\boldsymbol{x}_0(\boldsymbol{x}, \boldsymbol{\Omega}_0)$ is defined, in analogy with Sect. 3.6.1 for the plane-parallel medium, as the unique point on the sunny side of the cloud where the beam $\{\boldsymbol{x}, -\boldsymbol{\Omega}_0\}$ pierces $\partial \mathrm{M}_{\mathrm{sunny}}(\boldsymbol{\Omega}_0)$, cf. Fig. 3.12.

The overall responses to boundary (or otherwise modeled mono-directional) illumination are

$$\begin{cases} R = \int\limits_{\partial \mathrm{M}_{\mathrm{sunny}}(\boldsymbol{\Omega}_0)} F_{\boldsymbol{n}(\boldsymbol{x})}^{(+)}(\boldsymbol{x}) \mathrm{d}S(\boldsymbol{x}) \, / \, [F_0 S_{\mathrm{M}}(\boldsymbol{\Omega}_0)], \quad \text{for reflectance} \\[4mm] T = \int\limits_{\partial \mathrm{M}_{\mathrm{shady}}(\boldsymbol{\Omega}_0)} F_{\boldsymbol{n}(\boldsymbol{x})}^{(+)}(\boldsymbol{x}) \mathrm{d}S(\boldsymbol{x}) \, / \, [F_0 S_{\mathrm{M}}(\boldsymbol{\Omega}_0)], \quad \text{for transmittance} \end{cases}$$
$$(3.135)$$

where

$$S_{\mathrm{M}}(\boldsymbol{\Omega}_0) = \boldsymbol{\Omega}_0 \bullet \int\limits_{\partial \mathrm{M}_{\mathrm{shady}}(\boldsymbol{\Omega}_0)} \boldsymbol{n}(\boldsymbol{x}) \mathrm{d}S(\boldsymbol{x}) = -\boldsymbol{\Omega}_0 \bullet \int\limits_{\partial \mathrm{M}_{\mathrm{sunny}}(\boldsymbol{\Omega}_0)} \boldsymbol{n}(\boldsymbol{x}) \mathrm{d}S(\boldsymbol{x}) \quad (3.136)$$

is the area of the *normal* geometrical shadow of the medium M under collimated illumination from $\boldsymbol{\Omega}_0$, as illustrated in Fig. 3.12. Notice that, in contrast with the $\{R, T\}$ pair in (3.133), the one in (3.135) is not made of straightforward averages of the fields in (3.134). They are weighted averages using local solar irradiance (i.e., the denominators in (3.134)) so that the spatial integral is, as indicated, in non-normalized fluxes; then the totals are properly normalized by the integral of the weights in (3.136).

In Fig. 3.16a, we illustrate the quantities in (3.135) using spherical media

$$\mathrm{M} = \{\boldsymbol{x} \in \mathbb{R}^3 : x^2 + y^2 + z^2 < r^2\} . \qquad (3.137)$$

No absorption was assumed so we have $R + T = 1$ by conservation and this suggests that we only need to study the ratio R/T to determine both quantities. Both isotropic and forward-peaked (Henyey–Greenstein, $g = 0.85$) scattering phase functions were used in the computations. Solar illumination strength on the (say) upper hemisphere of the boundary

$$\partial \mathrm{M}_{\mathrm{sunny}} = \{\boldsymbol{x} \in \partial \mathrm{M} : \boldsymbol{\Omega}_0 \bullet \boldsymbol{n}(\boldsymbol{x}) < 0\}$$

was modulated by the appropriate $|\boldsymbol{n}(\boldsymbol{x}) \bullet \boldsymbol{\Omega}_0|$ term. Also the boundary sources were either collimated along $\boldsymbol{\Omega}_0$ or diffuse, i.e., isotropic on the local inward hemisphere of directions $\{\boldsymbol{\Omega} \in \Xi : \boldsymbol{\Omega} \bullet \boldsymbol{n}(\boldsymbol{x}) < 0, \boldsymbol{x} \in \partial \mathrm{M}_{\mathrm{sunny}}\}$. Optical thickness is measured across the diameter of the sphere

$$\tau_{\mathrm{diam}} = 2\sigma r$$

and varied by factors of 2 from 0.125 to 64 (for $g = 0$) or to 512 (for $g = 0.85$). A Monte Carlo scheme (cf. Chap. 4) with 10^5 histories was used for each case.

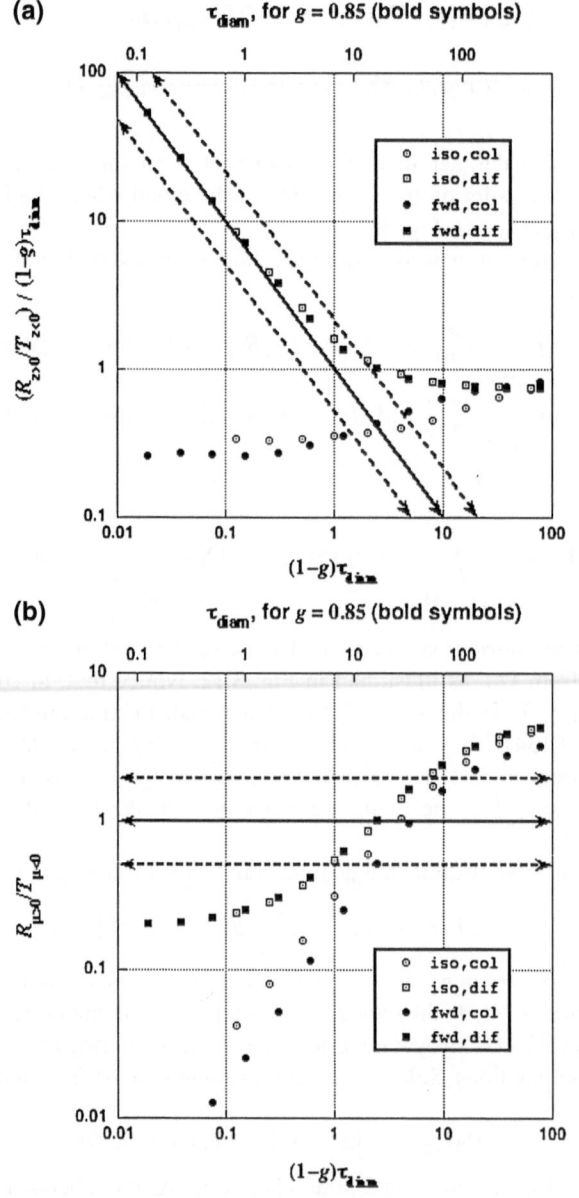

Fig. 3.16. Reflection and transmission of spherical cloud models. (**a**) Partition of R and T according to position at escape (hence the subscripts describing the sign of z). (**b**) Partition of R and T according to direction at escape (hence the subscripts describing the sign of μ). In both partitions, the thick/solid line indicates where $R = T = 1/2$ (since $R + T = 1$ here); the *dashed lines* on either side indicate where $R/T = 2^{\pm 1}$, hence $R = 1/3$ or $R = 2/3$ and conversely for T. Empty symbols are for isotropic scattering (iso) and full symbols are for forward-peaked scattering (fwd)

Davis (2002) derives an exact diffusion theoretical expression for R/T for such purely scattering spherical media:

$$\frac{R}{T} = \frac{(1-g)\tau_{\text{diam}}}{2\chi} \quad\quad (3.138)$$

where the numerical prefactor $1/2\chi$ is between 0.70 to 0.86. To be precise, χ is the "extrapolation length" expressed in transport MFPs, $[(1-g)\sigma]^{-1}$. Photon diffusion theory (cf. Chap. 6) is an approximation to RT that is expected to work well in the bulk of opaque and highly scattering media, i.e., at more than a transport MFP or so from boundaries and/or sources. This prediction is clearly confirmed by the numerical solutions of the RTE plotted in Fig. 3.16a for large optical thicknesses. Indeed, the vertical axis is R/T divided by the "rescaled" optical diameter $(1-g)\tau_{\text{diam}}$ and it goes to the anticipated value as τ_{diam} increases without bound. This is irrespective of the angular pattern of the illumination or the phase function. The collapse of the curves for $g = 0$ and $g = 0.85$ is particularly good for the case of isotropic sources, as is expected in this more preemptively diffusive situation, and thus more fully consistent with the way Davis (2002) set up his BCs and sources. Since $R \to 1$ in this limit, we have $T \propto 1/(1-g)\tau_{\text{diam}}$. As noted by Davis (2002), this law does not seem to be sensitive to cloud geometry: a similar law is indeed obtained in slab geometry. This insensitivity to outer cloud geometry is exploited further on.

We have also highlighted in Fig. 3.16a the locus of points where $R = T = 1/2$. In the optically thin limit ($\tau_{\text{diam}} \ll 1/(1-g)$), we see that the two angular models for the sources give very different results, both easily explained. Diffuse sources appropriately distributed on the upper boundary yield $R \approx T \approx 1/2$ in spherical geometry because in this limit scattering is no longer a concern, only ballistic trajectories matter. In contrast, the collimated beam scenario crosses the $R = T$ line at $(1-g)\tau_{\text{diam}} \approx 2.5$ and R/T goes to another limit determined by the opposite approximation of diffusion, single-scattering and quasi-linear transmission laws. In this latter approximation, we indeed expect $R = 1 - T \propto \tau_{\text{diam}}$ with a prefactor that will be sensitive to the phase function as well as to cloud geometry.

In the above natural enough definitions, we have used only the solar direction to partition the boundary fields according to the position on the boundary where the light emerges. Thus, introducing a ground surface and the associated vertical direction, "reflected" light can reach the ground and "transmitted" light can return to space, even if the Sun is at zenith. This is an essential feature of isolated (horizontally finite) clouds, and a topological impossibility in plane-parallel geometry. The important demarcation line is not so much the "horizon" line in Fig. 3.12 as the "terminator," a terminology we borrow from planetary astronomy. It is defined as the following intersection of the two closed sets

$$\partial M_{\text{term}}(\boldsymbol{\Omega}_0) = \partial M_{\text{shady}}(\boldsymbol{\Omega}_0) \cap \partial M_{\text{sunny}}(\boldsymbol{\Omega}_0) . \quad\quad (3.139)$$

This definition is preferable to

$$\partial M_{\perp}(\boldsymbol{\Omega}_0) = \{\boldsymbol{x} \in \partial M : \boldsymbol{n}(\boldsymbol{x}) \cdot \boldsymbol{\Omega}_0 = 0\} \quad\quad (3.140)$$

because of possible degeneracy, that is, situations where whole facets with finite areas are part of the proposed demarcation set.[16] If this is the case, then some of the out-going flux is then neither reflected nor transmitted – it is quite literally "side-leaked." This is a popular notion but unfortunately it can be quantified precisely only in very special cloud/Sun geometries where $\partial M_\perp$ has a finite area. These special illumination conditions that "resonate" with outer cloud geometry may have received more attention because they happen to be attractive simplifications for the RT modeling.

An often-used but pathological geometry for cloud modeling is the popular cuboidal medium. The cuboid's pathology results from the possibility of solar rays grazing finite areas on its surface. The resolution of its terminator according to (3.139) under all possible illuminations is illustrated in Fig. 3.15a, while finite cylinders are treated in Fig. 3.15b. Note that the sunny/shady asymmetry in (3.118)–(3.119) is chosen so that, when present, side-leakage is grouped with reflectance. This is physically justified in order to minimize the number of escaping photons with a low order-of-scattering binned as transmission. Indeed, physical intuition tells us that significant contributions from low orders-of-scattering is the hallmark of reflectance. Another physical reason for including $\partial M_\perp(\mathbf{\Omega}_0)$ (when it is finite) with $\partial M_{\text{sunny}}(\mathbf{\Omega}_0)$, hence side-leakage (when it exists) with reflection, is obtained by slightly perturbing the illumination direction from the special (resonant) case that endows $\partial M_\perp(\mathbf{\Omega}_0)$ with a finite area. That will generally collapse $\partial M_\perp(\mathbf{\Omega}_0)$ onto $\partial M_{\text{term}}(\mathbf{\Omega}_0)$ as defined in (3.139), and put its area into that of $\partial M_{\text{sunny}}(\mathbf{\Omega}_0)$.

In summary, our manipulation of set-topological concepts support the physics of RT because topology is about point "proximity" and "connectedness." Spatial connection is also what transport theory is very much about.

The motivation behind discussions of side-leakage is to find a simple mechanistic model for an obvious effect of 3D geometry. Partially because the notion of side-leakage cannot be transmuted into a mathematically robust construct, Davis and Marshak (2001) have advocated the more physical concept of photon "channeling" as a better way to describe the elementary interaction between a steady, initially uniform photon flow and a spatial disturbance. Channeling is based on flow (i.e., vector-flux field) geometry rather than boundary geometry and therefore applies equally to internally variable media and to homogeneous media that are not plane-parallel because their horizontal extension is finite.

"R/T" Partition by Horizontal Orientation and Direction at Escape

As if the above complications were not enough, definitions of reflection and transmission direction based on direction $\mathbf{\Omega}$ rather than position x are also possible and natural for horizontally finite (as well as infinite) media. Furthermore, these definitions are more likely to use the vertical rather than solar direction because of surface radiation budget considerations, as well as remote sensing. Unlike in plane-parallel

---

[16] The important distinction between the sets defined in (3.139) and (3.140) is that the former *always* has measure zero and the latter *can* have a non-vanishing measure.

geometry, the results are not the same here. Referring again to Fig. 3.12 and looking down along $-\hat{\mathbf{z}}$ at the scene from far above, we can define

$$\partial M_{up} = \partial M_{sunny}(\mathbf{\Omega}_0 = -\hat{\mathbf{z}}) \tag{3.141}$$

and corresponding ∂M_{shady} is ∂M_{dn}, while ∂M_{term} in (3.139) becomes $\partial M_{horizon}$. From this specific vantage point, we have

$$\begin{cases} \pi I(\mathbf{x}, +\hat{\mathbf{z}})/\mu_0 F_0, & \mathbf{x} \in \partial M_{up} \cap \partial M_{sunny}, & \text{nadir radiance from} \\ & & \text{illuminated part} \\ \pi I(\mathbf{x}, +\hat{\mathbf{z}})/\mu_0 F_0, & \mathbf{x} \in \partial M_{up} \cap \partial M_{shady}, & \text{nadir radiance from} \\ & & \text{shadowed part} \end{cases}, \tag{3.142}$$

the second case being impossible in slab geometry. A similar two-way partition will exist for zenith radiance

$$\pi I(\mathbf{x}, -\hat{\mathbf{z}})/\mu_0 F_0, \mathbf{x} \in \partial M_{dn} ; \tag{3.143}$$

a sub-sample of this field along a line could be interpreted as the sequence of readings of vertically-pointing ground-based radiometer as the mean wind advects the cloud by. Analogous definitions can be spelled out for all other directions.

Encouraged by the apparent robustness of the thick-cloud limiting behavior of R/T with respect to cloud geometry, Davis (2002) applies his result for spherical non-absorbing clouds to the remote determination of the optical thickness of real-world finite clouds observed in high (5 m) resolution satellite imagery. At a purely scattering solar wavelength, all that is required is an estimate of the mean-flux ratio R/T to infer at least a rough – or "effective" – value of $(1 - g)\tau_{diam}$.[17] This estimation of R/T is easily achieved for opaque isolated or broken cumulus clouds, especially viewed obliquely with respect to the Sun. As shown in Fig. 3.17, it is not hard to find the terminator in high-resolution images, being the relatively sharp transition between bright (reflective) and dark (transmissive) sides of the cloud. Then, since we are interested in highly scattered photons for both R and T, a Lambertian assumption that links radiance in any particular direction and flux from the corresponding surface (in this case, a cloud boundary) is not unreasonable. Therefore a radiance ratio can be equated with R/T in (3.138). Taking $g \approx 0.85$ as usual for liquid clouds, the ensuing values of τ_{diam} are in the range 20 to 50; accounting for in-cloud and cloud-to-cloud variability, this range is not unrealistic for the type of cloud present in the scene.

Boundary radiances given for all positions and directions can be used collectively to define overall responses to collimated illumination with respect to the zenith direction: How much radiation reaches the ground (even if it is coming from the illuminated side of the cloud)? How much goes back to space (even if coming from the shaded side of the cloud)? The answers are

---

[17] One should say this is for the *equivalent* homogeneous sphere but that caveat is almost always omitted in descriptions of operational cloud remote sensing when applying standard plane-parallel theory to clouds that are more-or-less stratiform.

Fig. 3.17. Los Alamos (NM) scene with broken clouds captured with the Multispectral Thermal Imager (MTI) from a viewing angle of about 60°. This grey-scale image was produced from a true-color rendering of the scene based on 3 narrow channels at 484, 558, 650 nm. General characteristics of the MTI instrument and orbit are given by Weber et al. (1999). For a determination of mean optical thicknesses for three selected clouds with different outer sizes, see Davis (2002)

$$\begin{cases} R_\uparrow = \int\limits_0^{+1} \mu d\mu \int\limits_0^{2\pi} d\varphi \int\limits_{n(x)\bullet\Omega(\mu,\varphi)\geq 0} I(x,\Omega(\mu,\varphi))dS(x) \,/\, [F_0 S_M(\Omega_0)] \\[2em] T_\downarrow = \int\limits_{-1}^{0} |\mu| d\mu \int\limits_0^{2\pi} d\varphi \int\limits_{n(x)\bullet\Omega(\mu,\varphi)>0} I(x,\Omega(\mu,\varphi))dS(x) \,/\, [F_0 S_M(\Omega_0)] \end{cases},$$

$$(3.144)$$

respectively for reflection back to space and for transmission to the surface. The denominators are simplified expressions for the incoming flux $\mu_0 F_0$ measured along the vertical times the projection along the horizontal of the illuminated boundary $S_M(\Omega_0)/\mu_0$.

Figure 3.16b illustrates the definitions in (3.144) for the same four sequences of spherical clouds described above in connection with the companion figure in panel (a) under the simple assumption that $\mathbf{\Omega}_0 = -\hat{\mathbf{z}}$. Note that in this partition of R versus T, we have no a priori reason to divide the ratio $R_\uparrow / T_\downarrow$ by the optical thickness, but we of course still have $R_\uparrow + T_\downarrow = 1$ (by photon conservation). We observe the same excellent collapse of the curves for $g = 0$ and $g = 0.85$ when plotted against $(1-g)\tau_{\text{diam}}$ as in Fig. 3.16a. However, this only occurs when the spatially-modulated boundary sources are generated diffusely. Logically, the *isotropic* scattering media under collimated illumination merge with their isotropically illuminated counterparts in the large τ_{diam} limit. However, this is clearly not the case for forward scattering media; we attribute this to systematic reduction (enhancement) of $R_\uparrow$ ($T_\downarrow$) by low-orders of scattering that are always present in reflected light but dominate in the periphery (near-terminator) of the cloud. The small τ_{diam} behavior here is exactly the same as for R/T according to position at escape in Fig. 3.16a, only without dividing by the independent variable τ_{diam}.

Discussion

It is physically obvious (and rigorously proven by reversing the angular and surface integrals) that absorptions computed from $1 - R - T$ in (3.135) and $1 - R_\uparrow - T_\downarrow$ in (3.144) are equal. The main point here is that solar sources in finite clouds yield only two kinds of light: reflected and transmitted, just as in plane-parallel geometry. Transmitted light comes from the boundary points *far* from the sources and, at least for optically thick media, is characterized by high orders-of-scattering and correspondingly low radiance levels. Reflected light comes from the boundary points *close* to the sources and, for optically thick media, is characterized by a broad distribution of orders-of-scattering (from a single scattering to at least as many as it typically takes to be transmitted); this circumstance will naturally correlate with relatively high levels of radiance. If the definitions of transmission and reflection are predicated on direction with respect to the vertical direction rather than boundary topology and source direction, then they will be mixtures of the above more physical definition based on position. This is true even if the axis of symmetry of the finite cloud (if it exists) is aligned with the vertical. Only in plane-parallel geometry do the definitions coincide.

A far-reaching consequence of the physics-based partition of escaping radiation by connection to the source is that, apart from a set of photon beams of measure zero (propagating exactly horizontally), there is no such thing as "side-leakage" from a finite cloud. There is only reflected and transmitted light. However, transmitted photons may be traveling downward or upward, thus possibly misleading satellite remote-sensing algorithms that are hard-coded to think that clouds have to be bright. Similarly, reflected photons may be traveling upward or downward, thus contributing possibly very strongly to surface fluxes. To illustrate this last effect, Fig. 3.18 shows time-series of the direct broad-band (BB) solar flux, measured normal to the beam, and the total BB down-welling surface flux collected over two days in Boulder (CO). The diffuse down-welling flux was also sampled. The first day (July 10, 2003) was

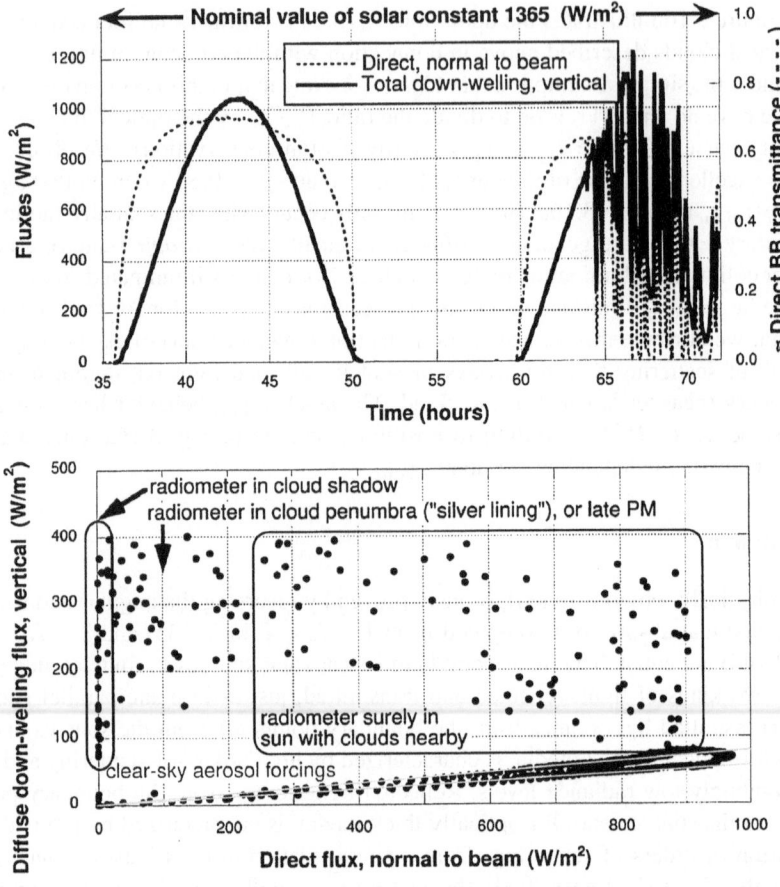

Fig. 3.18. Broken clouds enhance total down-welling surface flux far beyond the clear-sky direct contribution. (*Upper*) The total vertical down-welling flux can vastly exceed the direct flux normal to the beam. (*Lower*) A rough partition of the scatter plot of diffuse/vertical vs. direct/normal fluxes from upper panel. The data, courtesy of John Augustine and Gary Hodges from NOAA, and was collected in Boulder (CO) on July 10-11, 2003, and previously used by Chýlek et al., (2004) in a study of broken cloud impact on clear-air property retrievals

clear, the next day ended with an episode of broken cloudiness. We first note in the upper panel that in the presence of broken clouds the total vertical flux can vastly exceed its clear-sky counterpart and even the direct flux measured perpendicular to the beam (i.e., without the usual μ_0 factor). The lower panel shows, on the one hand, how the diffuse down-welling flux is driven almost linearly by the direct solar flux (single scattering dominates and optical distances are small enough to set $\exp(-\tau) \approx 1 - \tau$) and, on the other hand, how much the diffuse BB component is non-linearly enhanced by broken clouds. At non-absorbed VIS-NIR wavelengths, scattering by clouds can cause total transmittance to exceed unity. This is occurs in fact often enough, even for BB transmittance, to survive in day-long averages at least at sub-tropical locations (Dutton et al., 2004).

Finite clouds in spherical refractive atmospheres open up even more interesting paradoxes with direct illumination of cloud base. This is a frequent and often spectacular display of radiance for ground-based observers located near the terminator of planet Earth, locally identified as sunset and sunrise.

3.10 Green Functions to Reciprocity via Adjoint Transport

We cover formal Green function theory for the RTE and relate it to the adjoint RT problem; both are essential to a number of numerical techniques (Chaps. 4, 5) and remote-sensing applications (Chaps. 12, 14) in 3D RT. Green functions also provide the natural framework for introducing the *reciprocity principle* which may or may not apply in a given 3D atmosphere-surface system depending on its inherent properties and the conditions surrounding the observations.

3.10.1 Green Functions in Radiative Transfer Theory

Definitions

Consider a 3D absorbing and scattering medium M bounded by a non-reflecting and non-emitting boundary ∂M. The *volume* Green function $G_V(x, \Omega; x', \Omega')$ is the radiative response of M at a point x, in direction Ω, to a monodirectional point-source located at a *given* point x' in M, continuously emitting photons in a *given* direction Ω'. The volume Green function satisfies the RTE (3.102)–(3.103) with a delta function source term Q, i.e.,

$$\Omega \cdot \nabla G_V(x, \Omega; x', \Omega') + \sigma(x) G_V(x, \Omega; x', \Omega')$$
$$= \sigma_s(x) \int_{4\pi} p(x, \Omega'' \to \Omega) G_V(x, \Omega''; x', \Omega') d\Omega''$$
$$+ \delta_V(x - x') \delta(\Omega - \Omega') \tag{3.145}$$

with homogeneous (no entering radiance) BCs. Here $\delta(\Omega - \Omega')$, in sr^{-1}, and $\delta_V(x - x')$, in m^{-3}, are Dirac delta-functions. Note that $\delta_V(x - x') \delta(\Omega - \Omega')$ is a volume source normalized by its power. The volume Green function, therefore, is expressed in $\text{m}^{-2}\text{sr}^{-1}$. It should be also noted that the point x' and the direction of the monodirectional source Ω' are *parameters* in the RTE; that is, the determination of the complete Green function requires the solution of (3.146) for each and every point $x' \in M$ and direction $\Omega' \in \Xi$.

In the "operator" language introduced in (3.102)–(3.103), (3.146) can be written simply as $L G_V = \delta$ where the delta-function source term takes care of all the photon state-variables of immediate interest (position, direction). After performing a spherical harmonic decomposition in angle space and a 3D Fourier transform in position space (i.e., a continuous decomposition on harmonic functions), this concise formulation becomes $\tilde{L}\,\tilde{G}_V = 1$ where the tilde designates a transformed entity. So, it is not surprising that in some literatures the Green function is denoted $G_V = L^{-1}$, i.e., as the inverse of a linear operator. This expresses the fact that knowing G_V or knowing L (including knowledge of the spatial distribution of the optical properties in M) are formally equivalent. The Green function is therefore called the "fundamental" solution or "resolvant" of the problem at hand, in this case, the RT problem. In this operator formalism, the solution to the general linear transport problem, $L\,I = Q$ (subject to homogeneous BCs), is $I = L^{-1}Q = G_V Q$ which is short-hand for a 5-dimensional integral over (i.e., a superposition of) the source's positions and directions.

In the case of purely absorbing media ($\sigma_s(x) \equiv 0$), the solution to (3.146) already derived in Sect. 3.8.2 can be given in explicit form using the volume Green function (Case and Zweifel, 1967). Bearing in mind that it is designed to be the kernel of 5-dimensional integral, the Green function is best written as

$$G_V(x, \Omega; x', \Omega') = \frac{\exp[-\tau(x', x)]}{\|x' - x\|^2}\delta(\Omega' - \Omega)\delta\left(\frac{x' - x}{\|x' - x\|} - \Omega\right) . \qquad (3.146)$$

This follows from the exact "no-scattering" solution in (3.121) with no boundary sources ($f(\cdot) \equiv 0$ on ∂M) and using the identities in (3.123)–(3.124). The resemblance of the Green function in (3.146) with the kernels (3.126) and (3.130) of the integral forms of the RTE covered in the previous section is not accidental: the scattering quantities $\sigma_s p(\Omega' \to \Omega)$ are replaced by $\delta(\Omega' - \Omega)$. Also, and contrary to the ones used previously in this section, the last δ-function is non-dimensional. Indeed, it is paired with dx' (in m^{-3}) and, in its argument, x' is divided by the distance $\|x' - x\|$. Finally, the $\|x' - x\|^{-2}$ term does not express an algebraic decay (in addition to $\exp[-\tau(x', x)]$). The source is indeed monodirectional, in which case we know from previous sections that it is only affected by extinction term $\exp[-\tau(x', x)]$. Rather, this algebraic term comes from the Jacobian required to go from the one-dimensional integral in $s = \|x' - x\|$, as mandated by the directional derivative in (3.146), to a three-dimensional integral in dx' by bringing in a solid angle integral in $d\Omega$.

The *surface* Green function, $G_S(x, \Omega; x_S, \Omega')$, is the solution to the transport equation in (3.146) but without the volume source term. However, radiation is flowing into the medium through the surface ∂M, as described by the inhomogeneous BC

$$G_S(x, \Omega; x_S, \Omega') = \delta_S(x - x_S)\delta(\Omega - \Omega'), \ x \in \partial M, \ \Omega \cdot n(x) < 0 , \qquad (3.147)$$

i.e., a point source at $x_S \in \partial$M emitting with unitary power in the direction Ω'. Here $\delta_S(\cdot)$ is a two-dimensional δ function (in m^{-2}). Thus, $G_S(x, \Omega; x_S, \Omega')$ is the radiative response of the medium M at a point x, in direction Ω, to a collimated boundary source. Because sources can be located on the boundary, the volume and surface Green functions are related:

$$G_S(x, \Omega; x_S, \Omega') = |n(x_S) \cdot \Omega'|\, G_V(x, \Omega; x_S, \Omega') , \qquad (3.148)$$

as is shown further on.

In terms of these two Green functions, we may write the general solution to the RTE (3.101) with arbitrary volume source $Q(x, \Omega)$ and BCs (3.109) with sources given by $f(x_S, \Omega)$. We obtain

$$
I(x, \Omega) = \int\limits_M \int\limits_{4\pi} G_V(x, \Omega; x', \Omega') Q(x', \Omega') dx' d\Omega'
$$

$$
+ \int\limits_{x_S \in \partial M} dS(x_S) \int\limits_{n(x_S) \cdot \Omega < 0} G_S(x, \Omega; x_S, \Omega') f(x_S, \Omega') d\Omega' . \quad (3.149)
$$

The first term in (3.149) is the solution of the 3D transport equation with the volume sources $Q(x, \Omega)$ and no incoming radiance. The second term describes the 3D radiation field in M generated by sources $f(x, \Omega)$ distributed over the boundary ∂M.

The Green function concept was originally developed in neutron transport theory several decades ago (Bell and Glasstone, 1970). It has enabled the reformulation of the radiative transfer problems in terms of some "basic" subproblems and to express the solution to the transport equation with arbitrary sources and boundary conditions as a superposition of the solutions of the basic subproblems. We now demonstrate with a relevant example for RT in the atmosphere-surface system.

Illustration with Cloud-Surface Radiative Interaction

Consider a cloudy or aerosol layer bounded from below by a non-uniform Lambertian surface. This is a problem of considerable interest in satellite remote sensing of surface properties (Lyapustin and Knyazikhin, 2002) as well as ground-based remote sensing of clouds (Marshak et al., 2000). Radiation-cloud-surface interaction can be described by the RTE with zero volume sources and BCs given in Sect. 3.6.2. The intensity $I(x, \Omega)$ can be represented as a sum of two components: the radiation calculated for a "black" surface, $I_{blk}(x, \Omega)$, and the remaining radiation, $I_{rem}(x, \Omega)$; that is,

$$
I(x, \Omega) = I_{blk}(x, \Omega) + I_{rem}(x, \Omega) . \quad (3.150)
$$

In (3.150), the second component accounts for the radiation field excited by multiple surface-cloud interactions. It satisfies $LI_{rem} = 0$, with a homogeneous (zero-incoming radiance) BC on the upper boundary, and

$$
I_{rem}(x_S, \Omega) = \pi^{-1} \alpha(x_S) F(x_S) \quad (3.151)
$$

at the lower boundary $z_S = 0$ where $\alpha(x_S)$ is the variable surface albedo, assumed Lambertian, and $F(x_S)$ is a variable down-welling hemispherical flux assumed, for the moment, to be a given quantity. Note that since the geometry of this medium is plane-parallel, as in Sect. 3.6.2, we could use the "split" notation $x_S = (\vec{x}_S, z_S)^T$ here, but we will continue to use notation that applies to the most general medium geometry.

The remaining radiation can be expressed through the surface Green function as

$$
I_{\text{rem}}(\boldsymbol{x}, \boldsymbol{\Omega}) = \frac{1}{\pi} \int\limits_{z_{\text{S}}=0} \alpha(\boldsymbol{x}_{\text{S}}) F(\boldsymbol{x}_{\text{S}}) \left[\int\limits_{\mu'<0} G_{\text{S}}(\boldsymbol{x}, \boldsymbol{\Omega}; \boldsymbol{x}_{\text{S}}, \boldsymbol{\Omega}') d\boldsymbol{\Omega}' \right] dS(\boldsymbol{x}_{\text{S}}) . \quad (3.152)
$$

In (3.152), the integral over upward directions describes the radiation field in M generated by an isotropic point-source $\pi^{-1}\delta(\boldsymbol{x} - \boldsymbol{x}_{\text{S}})$ located at the point $\boldsymbol{x}_{\text{S}} \in \partial\text{M}$. Given the downward flux field $F(\boldsymbol{x}_{\text{S}})$ at the lower boundary, the remaining radiance I_{rem} can be evaluated from (3.152). The field $F(\boldsymbol{x}_{\text{S}})$ itself depends on I_{rem} and thus (3.152) alone provides a full description of surface-cloud interactions. Combining (3.150) and (3.152), one obtains a *two*-dimensional integral equation for the unknown total flux $F_z^{(-)}(\boldsymbol{x}_{\text{S}})$ for $\boldsymbol{x}_{\text{S}} \in \partial\text{M}$ (meaning here the plane $z_{\text{S}} = 0$):

$$
F_z^{(-)}(\boldsymbol{x}) = \int\limits_{z_{\text{S}}=0} \alpha(\boldsymbol{x}_{\text{S}}') R(\boldsymbol{x}_{\text{S}}, \boldsymbol{x}_{\text{S}}') F_z^{(-)}(\boldsymbol{x}_{\text{S}}') dS(\boldsymbol{x}_{\text{S}}') + F_{z,\text{blk}}^{(-)}(\boldsymbol{x}) . \quad (3.153)
$$

This unknown flux accounts for what the cloud transmits as well as all the multiple surface-cloud interactions. Here $F_{z,\text{blk}}^{(-)}$ is the downward flux at the bottom of the medium calculated for the "black" surface problem and acts as a source term in the integral equation. $R(\boldsymbol{x}_{\text{S}}, \boldsymbol{x}_{\text{S}}')$ is the downward flux at $\boldsymbol{x}_{\text{S}} \in \partial\text{M}$ generated by the point-wise and isotropic source $\pi^{-1}\delta(\boldsymbol{x}_{\text{S}} - \boldsymbol{x}_{\text{S}}')$ located at $\boldsymbol{x}_{\text{S}} \in \partial\text{M}$; it acts as a kernel for the integral equation in (3.153). Multiplying (3.152) by $d\boldsymbol{\Omega}$ and integrating over the $\mu > 0$ hemisphere, and then by identification with the 1st term in (3.153), we see that this kernel is given by a double angular integral of the surface Green function:

$$
R(\boldsymbol{x}_{\text{S}}, \boldsymbol{x}_{\text{S}}') = \int\limits_{\mu>0} \mu d\boldsymbol{\Omega} \int\limits_{\mu'<0} \pi^{-1} G_{\text{S}}(\boldsymbol{x}_{\text{S}}, \boldsymbol{\Omega}; \boldsymbol{x}_{\text{S}}', \boldsymbol{\Omega}') d\boldsymbol{\Omega}' , \quad (3.154)
$$

for any pair of surface points $(\boldsymbol{x}_{\text{S}}, \boldsymbol{x}_{\text{S}}')$. Notice that we are preserving angular symmetry between the isotropic source at $\boldsymbol{x}_{\text{S}}'$ and the resulting field at $\boldsymbol{x}_{\text{S}}$. If moreover the atmosphere is horizontally uniform, then we are sure that the Green function in (3.154) will depend only on $||\boldsymbol{x}_{\text{S}} - \boldsymbol{x}_{\text{S}}'||$. This makes the integral in (3.153) a straightforward convolution product.

We now return to the aerosol "adjacency" and cloud remote sensing problems that motivated this exercise. Because the above horizontal uniformity assumption is not unreasonable for the aerosol atmosphere, we have reduced the full 3D RT problem of assessing the mixture of albedo values $\alpha(\boldsymbol{x}_{\text{S}})$ in the observations $I(\boldsymbol{x}, \boldsymbol{\Omega})$, e.g., from space, to an integral equation that is easily solved in Fourier space. For more details about this adjacency mitigation strategy in surface property remote sensing, including generalization to non-Lambertian ground, we refer to the paper by Lyapustin and Knyazikhin (2002). In the case of clouds, the same horizontal uniformity assumption is of course highly questionable. Nonetheless, by working with two wavelengths where the clouds have similar scattering properties – but the ground's reflection properties not – one can minimize the impact of 3D RT effects in the observations and apply 1D RT theory locally to infer cloud properties. More details about this 3D mitigation technique are in the papers by Marshak et al. (2000, 2004) as well as in Chaps. 12 and 14.

Inverse Problems

Green functions play an important role in developing algorithms for retrieving coefficients in the RTE from radiation leaving a medium, in other words, performing an optical tomography. Choulli and Stefanov (1996) and Antyufeev and Bondarenko (1996) reported that, under quite general conditions on the sources, the 3D fields of *total* cross-section (per unit of length), $\sigma(x)$, and the *differential* scattering cross-section, $\sigma_s(x)p(x, \Omega \to \Omega')$, can be uniquely retrieved from boundary-field measurements. This result indicates that there is a one-to-one correspondence between the complex 3D structure of a given domain M of space bounded by a non-reflecting surface ∂M and the outgoing boundary radiation field $I(x_S, \Omega)$, $x_S \in \partial M$, $n(x_S) \cdot \Omega > 0$. The following interpretation of the Green function underlies the derivation of this property.

The volume and surface Green functions describe the radiative response of the medium M to a source concentrated at an isolated spatial point and emitting photons in one direction. A Dirac δ-function is naturally used to describe such a source. The theory of distributions developed by Schwartz (1950) justifies the use of such functions in describing and solving physical problems. Since the BC is expressed in terms of a Schwartz distribution, the solution to the transport equation is a distribution too. Schwartz's theory distinguishes two types of functions, "regular" and "singular" distributions.

There is a one-to-one correspondence between usual functions (with a well-defined value for each value of its argument) and regular distributions; thus, an ordinary function can be regarded as a special case of a distribution. The Dirac δ-function is the simplest example of a singular distribution. No ordinary function can be identified with it and it is only defined under integral operations.

Generally speaking, the solution of the RTE can be expressed as a sum of regular and singular distributions. The singular component must be treated separately because numerical techniques cannot deal with bone fide distributions. A technique to separate the singular components from (3.146) is based on the following result (Germogenova, 1986; Choulli and Stefanov, 1996; Antyufeev and Bondarenko, 1996). For a 3D medium, the radiances due to uncollided and single-scattered photons from a point-wise monodirectional source, denoted respectively G_0 and G_1, are singular distributions while the remaining multiply-scattered field is described by a regular distribution G_{ms}. The Green function is therefore the sum of three components, two singular and one regular:

$$G = G_0 + G_1 + G_{ms} . \tag{3.155}$$

The singular components make the above mentioned one-to-one correspondence between observable radiance fields and optical properties possible. This generalizes the classic idea of tomographic reconstruction based only on the uncollided (a.k.a. directly-transmitted) radiance in G_0 and opens the possibility of using reflected radiance to perform 3D reconstruction. Application of this technique to describe radiation regimes in clouds and vegetation canopies is discussed respectively by Knyazikhin et al. (2002) and Zhang et al. (2002). More on this will be found in Chap. 14.

Observational access to G_0 and G_1 assumes an optically thin medium and quite sensitive detectors for the latter component. In contrast, we anticipate that in optically dense weakly-absorbing media G_{ms} will be the dominant term and its removal from the measurements of boundary radiances can leave estimates of the singular component $G_0 + G_1$ at par with the instrumental noise level. In this situation, radically different techniques that capitalize on G_{ms} and photon "diffusion-wave" theory can be invoked (Yodh and Chance, 1995, and references therein). Here exact 3D reconstruction is of course not an option, but large-enough and strong-enough inhomogeneities can be detected.

As an illustration in the time-domain, standard (or "on-beam") lidar is based on the illumination of a cloud boundary with a pulsed laser and the opposite side is detected and positioned at the range where the single-scattering signal G_1 all but vanishes. However, this assumes that there is enough signal to measure after the two-way transmission and that it is not significantly contaminated by multiply scattered light, i.e., this technique applies only to optically thin clouds. By contrast, "off-beam" lidar techniques apply to optically thick clouds since they exploit the multiple-scattering returns in G_{ms}; their spatio-temporal distributions are used to detect the presence and position of the opposite cloud boundary (Davis et al., 1999).

3.10.2 Adjoint Radiative Transfer

Definitions

Adjoint equations and their solutions play an important role in radiative transfer theory. In particular, they are widely used in perturbation theory and variational calculations relating to the behavior of 3D optical media. The properties of the solutions of the adjoint RTE are also used in the development of efficient Monte Carlo codes, as explained in Chap. 4.

The adjoint RTE can be written formally as

$$L^+ I^+ = Q^+ , \tag{3.156}$$

where L^+ is the adjoint integro-differential linear transport operator,

$$L^+ = -\mathbf{\Omega} \cdot \nabla + \sigma(\mathbf{x}) - \sigma_s(\mathbf{x}) \int_{4\pi} p(\mathbf{x}, \mathbf{\Omega} \to \mathbf{\Omega}')[\cdot] d\mathbf{\Omega}' , \tag{3.157}$$

and Q^+ is the adjoint source term. The following differences should be noted between L^+ in (3.157) and L in (3.103):

1. the Lagrangian derivative has the opposite sign, and
2. the incident and scattering directions have been interchanged, i.e., $\boldsymbol{\Omega}' \to \boldsymbol{\Omega}$ in (3.103) becomes $\boldsymbol{\Omega} \to \boldsymbol{\Omega}'$ in (3.157).

Physically, we are considering here the time-reversed photon flow. This gives us the hint that adjoint sources Q^+ describe the position of detectors while the adjoint transport operator L^+ takes the photons backwards in time to their actual sources. This makes the space-angle distribution I^+ of adjoint "photons" an estimate of how much a given position-direction matters for a given radiometric observation – often in a remote region – modeled by Q^+. We are thus looking for the solution of (3.157) satisfying the adjoint BCs, namely,

$$I^+(\boldsymbol{x}, \boldsymbol{\Omega}) = f^+(\boldsymbol{x}, \boldsymbol{\Omega}), \; \boldsymbol{x} \in \partial \mathrm{M}, \; \boldsymbol{\Omega} \boldsymbol{\cdot} \boldsymbol{n}(\boldsymbol{x}) > 0 \, . \tag{3.158}$$

Note that this boundary condition is formulated for *outgoing* directions. This makes physical sense if there are detectors at the boundaries. If there are not ($f^+ \equiv 0$), then escaping photons will no longer influence detectors inside the medium (where $Q^+ \neq 0$).

To better capture the notions of "weight" and "influence" used here to give physical meaning to the adjoint radiance field, some authors following Marchuk (1964) have came to call I^+ "importance." Adjoint equations are used many fields of dynamical modeling to analyze nonlinear tele-connections. In meteorology, this can be done by looking at the clusters of backwards trajectories which, in turn, has influenced data assimilation methodology. In 3D RT, one can think of the 3D "component" of the radiance field as response to a perturbation of uniformity in extinction. It is therefore not surprising that adjoint RT theory – and indeed adjoint Green functions introduced below – play a key role in the perturbative approach to 3D RT, cf. Box et al. (1988, 2003), Polonsky et al. (2003), and Chap. 5.

Some Useful Identities

To describe the relationship between solutions of the standard (or "forward") and adjoint radiative transfer equations, the inner product of two RT functions $f(\boldsymbol{x}, \boldsymbol{\Omega})$ and $g(\boldsymbol{x}, \boldsymbol{\Omega})$ is introduced:

$$< f, g > = \int_{\mathrm{M}} \int_{4\pi} f(\boldsymbol{x}, \boldsymbol{\Omega}) g(\boldsymbol{x}, \boldsymbol{\Omega}) \mathrm{d}\boldsymbol{\Omega}\mathrm{d}\boldsymbol{x} \, . \tag{3.159}$$

We note that the notation (f, g) is also used for this functional scalar product.

Now let $I(\boldsymbol{x}, \boldsymbol{\Omega})$ be the solution of the forward problem, i.e., I satisfies the RTE $LI = Q$ and the generic BCs in (3.109). This equation is now multiplied by I^+ and (3.156) by I; the resulting expressions are subtracted and the difference is integrated over M and Ξ. Taking into account the identity

$$< \boldsymbol{\Omega} \boldsymbol{\cdot} \nabla I, I^+ > = - < I, \boldsymbol{\Omega} \boldsymbol{\cdot} \nabla I^+ >$$
$$+ \int_{\partial \mathrm{M}} \mathrm{d}S(\boldsymbol{x}) \int_{4\pi} \boldsymbol{n}(\boldsymbol{x}) \boldsymbol{\cdot} \boldsymbol{\Omega} I(\boldsymbol{x}, \boldsymbol{\Omega}) I^+(\boldsymbol{x}, \boldsymbol{\Omega}) \mathrm{d}\boldsymbol{\Omega} \, , \tag{3.160}$$

one obtains the basic relationship between solutions of the forward and adjoint RTE, namely,

$$
<Q, I^+> - <I, Q^+> = \int_{\partial M} dS(x) \int_{n(x)\cdot\Omega>0} n(x)\cdot\Omega I(x,\Omega)f^+(x,\Omega)d\Omega
$$

$$
- \int_{\partial M} dS(x) \int_{n(x)\cdot\Omega<0} |n(x)\cdot\Omega| f(x,\Omega)I^+(x,\Omega)d\Omega .
$$

(3.161)

In the case of the homogeneous BCs, no incoming photons ($f \equiv 0$) and no outgoing adjoint radiance ($f^+ \equiv 0$), (3.161) can be simplified to $< Q, I^+ > = < I, Q^+ >$, or explicitly

$$
\int_M \int_{4\pi} Q(x,\Omega)I^+(x,\Omega)dxd\Omega = \int_M \int_{4\pi} Q^+(x,\Omega)I(x,\Omega)dxd\Omega .
$$

(3.162)

Connection with Green Functions

By substituting $Q(x,\Omega) = \delta_V(x-x_1)\delta(\Omega-\Omega_1)$ in (3.162), hence $I(x,\Omega) \equiv G_V(x,\Omega;x_1,\Omega_1)$, one obtains

$$
I^+(x_1,\Omega_1) = \int_M \int_{4\pi} Q^+(x,\Omega)G_V(x,\Omega;x_1,\Omega_1)dxd\Omega .
$$

(3.163)

Thus, $I^+(x_1,\Omega_1)$ is a Q^+-weighted integral response of the medium to a monodi-rectional point-source. So the adjoint solution $I^+(x_1,\Omega_1)$ is indeed a measure of the importance for the medium's response of a photon leaving from $\{x,\Omega\}$. By further substituting $Q^+(x,\Omega) = \delta_V(x-x_2)\delta(\Omega-\Omega_2)$ into (3.163), the relation between the forward and adjoint volume Green functions is obtained:

$$
G_V(x_2,\Omega_2;x_1,\Omega_1) = G_V^+(x_1,\Omega_1;x_2,\Omega_2) .
$$

(3.164)

This symmetry makes physical sense since one goes from the linear transport op-erator in (3.103) to its adjoint counterpart in (3.154) by reversing the Lagrangian flow.

Equation (3.161) yields a useful result when the solution of the forward problem is a volume Green function, and that of the adjoint problem is a surface Green func-tion. We set $Q(x,\Omega) = \delta_V(x-x_1)\delta(\Omega-\Omega_1)$ and $f \equiv 0$ for the forward problem, and $Q^+ \equiv 0$, $f^+(x,\Omega) = \delta_S(x-x_S)\delta(\Omega-\Omega_2)$, $x_S \in \partial M$, $n(x_S)\cdot\Omega_2 > 0$, for the adjoint problem. Substituting these into (3.161) results in

$$
G_S^+(x_1,\Omega_1;x_S,\Omega_2) = (n(x_S)\cdot\Omega_2)\, G_V(x_S,\Omega_2;x_1,\Omega_1)
$$

(3.165)

where x_S is on the boundary. Using (3.163) leads to (3.148) since Ω' in the for-ward problem is equated with $-\Omega_2$ in the adjoint problem. A relationship between the surface Green function and its adjoint can be derived from (3.161) in a similar manner.

3.10.3 Reciprocity Principle

Formulation with Green Functions

Intensity $G(x_2, \Omega_2; x_1, \Omega_1)$ at x_2 in direction Ω_2 due to a point source at x_1 emitting in direction Ω_1 can be related to the intensity at x_1 in direction $-\Omega_1$ due to a source at x_2 emitting in direction $-\Omega_2$ by means of the RTE. Such reciprocity relations often prove useful in relating the solution of a particular problem to that of a simpler problem or to one for which the solution is known. The adjoint RTE can be used to derive reciprocity relations. In the following, we will assume that

$$p(x, -\Omega' \to -\Omega) = p(x, \Omega \to \Omega') , \tag{3.166}$$

which is certainly the case if the scattering phase function depends only on the scattering angle $\cos^{-1}(\Omega \cdot \Omega')$. Even without the axi-symmetric scattering, this is a reasonable assumption in most atmospheric applications. It should be noted, however, that this property does not generally hold true in the case of radiative transfer in vegetation canopies; see Chap. 14.

Consider a 3D absorbing and scattering medium M bounded by a non-reflecting surface ∂M. Let Q and f be the volume and boundary sources, respectively. Intensity $I(x, \Omega)$ of the 3D radiation field satisfies the RTE (3.101) and general BCs (3.109). If a function I^+ is defined such that $I^+(x, \Omega) = I(x, -\Omega)$, then $I^+(x, \Omega)$ satisfies the adjoint RTE (3.156) with volume and boundary sources defined as (Bell and Glasstone, 1970)

$$Q^+(x, \Omega) = Q(x, -\Omega) \text{ and } f^+(x, \Omega) = f(x, -\Omega) . \tag{3.167}$$

In the case of the "free-surface" boundary condition of no incoming photons ($f \equiv 0$) and no outgoing adjoint flux ($f^+ \equiv 0$), the right-hand side of (3.164) can be replaced by $G_V(x_1, -\Omega_1; x_2, -\Omega_2)$, i.e.,

$$G_V(x_2, \Omega_2; x_1, \Omega_1) = G_V(x_1, -\Omega_1; x_2, -\Omega_2) . \tag{3.168}$$

This states that the intensity $I(x_2, \Omega_2)$ at x_2 in the direction Ω_2 due to a point source at x_1 emitting in direction Ω_1 is the same as the intensity $I(x_1, \Omega_1)$ at x_1 in the direction $-\Omega_1$ due to a point source at x_2 emitting in direction $-\Omega_2$. Thus, according to (3.168), the intensity is the same in two situations depicted in Fig. 3.19. The relation in the form of (3.168) is referred to as the *optical reciprocity theorem* (Bell and Glasstone, 1970).

By virtue of (3.148), we also have a reciprocity in the surface Green functions:

$$G_S(x_{S2}, \Omega_2; x_{S1}, \Omega_1) = G_S(x_{S1}, -\Omega_1; x_{S2}, \quad \Omega_2) \tag{3.169}$$

for any two points on ∂M. Note that the source directions in the second argument pair are oriented inward ($n(x_{S1}) \cdot \Omega_1 < 0$) while the detection directions in the first argument pair are oriented outward ($n(x_{S2}) \cdot \Omega_2 > 0$).

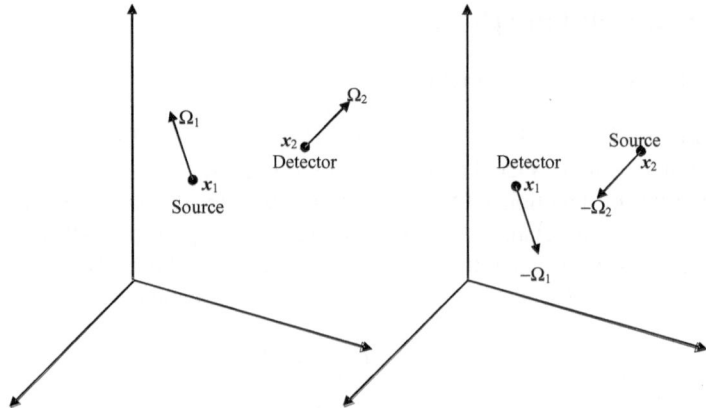

Fig. 3.19. The reciprocity principle: as expressed in (3.168), switching detector and source and inverting the directions of propagation yield the same result for the Green function

Illustration with an Atmosphere-Surface System

Consider again the problem of a cloudy or aerosol layer bounded from below by a non-uniform and non-Lambertian reflecting surface at $z = 0$. In this situation, a fraction of the radiation can be reflected back into the layer by the ground according to some spatially varying BRDF, $\rho(x_S, \Omega' \rightarrow \Omega)$, where Ω' is in the downward hemisphere of directions and Ω is in the upward one. Going back to (3.90)–(3.91), the BRDF is normalized in such a way that

$$I(x_S, \Omega) = \int_{\mu' < 0} \rho(x_S, \Omega' \rightarrow \Omega) |\mu'| I(x_S, \Omega') d\Omega' , \qquad (3.170)$$

for all x_S in the plane $z = 0$ and all $\mu > 0$. Under what conditions on ρ does the optical reciprocity theorem apply to such a composite atmosphere-surface medium? First, the right-hand side of (3.161) should vanish in order to obtain (3.162) and, as a consequence, the relation (3.164), and hence (3.168). Second, the conditions (3.167) should be imposed to obtain the relationship $I(x, -\Omega) = I^+(x, \Omega)$ between solutions of the forward and adjoint RTE. The former condition is satisfied if the solution of the adjoint RTE satisfies the following BC (Germogenova, 1986)

$$I^+(x_S, \Omega) = \int_{\mu' > 0} \rho(x_S, \Omega \rightarrow \Omega') I^+(x_S, \Omega') |\mu'| d\Omega' , \qquad (3.171)$$

for all x_S and $\mu > 0$. In our example, the functions f and f^+ are given by the right-hand sides of (3.170) and (3.171), respectively. Therefore, the equality $f^+(x, -\Omega) = f(x, \Omega)$ takes place if and only if

$$\int_{\mu'<0} \rho(x_S, -\Omega \to -\Omega') I^+(x_S, -\Omega') |\mu'| d\Omega' = \int_{\mu'<0} \rho(x_S, \Omega' \to \Omega) I(x_S, \Omega') |\mu'| d\Omega' \,.$$

(3.172)

Under this condition, $I(x_S, \Omega) = I^+(x_S, -\Omega)$ and, therefore, the identity (3.172) holds true if

$$\rho(x_S, -\Omega \to -\Omega') = \rho(x_S, \Omega' \to \Omega) \,, \tag{3.173}$$

which is analogous to our a priori assumption in (3.166) about the scattering phase function.

Thus, the condition (3.173) should be imposed on the BRDF to extend the validity of the reciprocity principle to media with reflecting boundaries. Note that the scattering phase function that appears in the RTE for vegetation canopies is not, as a rule, rotationally invariant. Besides, this function is generally asymmetric, i.e., $\rho(x, -\Omega' \to -\Omega) \neq \rho(x, \Omega \to \Omega')$. The BRDF of vegetation canopies, therefore, does not necessarily follow (3.173). This means that the reciprocity principle may not be applicable in the case of an atmosphere/vegetation-canopy system.

Violation of Directional Reciprocity over Finite Domains

Ultimately, the general reciprocity relations we have uncovered in Green functions are traceable to the microscopic reversibility of each and every reflection and scattering event. If, at a given locale, a certain change of propagation direction $\Omega_1 \to \Omega_2$ can occur then so can $\Omega_2 \to \Omega_1$, with equal probability. That is the meaning of the conditions (3.166) and (3.173) for reciprocity. In this respect, the 3D RT reciprocity relations we uncovered are special cases of Onsager's general reciprocity relations for kinetic systems, not necessarily in equilibrium.

This also opens up the question of purely directional reciprocity as more symmetry is imposed on the system, or as the wealth of information in the Green function is selectively degraded. For instance, if the (cloud-surface) medium is plane-parallel and horizontally homogeneous, then the surface Green functions in (3.169) depend only on $x_1 - x_2$. If, furthermore, the source direction is normal to the surface (i.e., $\Omega_2 = (0, 0, -1)^T$) or, generally speaking, an axi-symmetric illumination pattern, then only the modulus $\|x_1 - x_2\|$ matters. This of course includes isotropic illumination, equivalently, an average Green function for all possible incoming directions. One can also integrate the surface Green function in (3.169) over one spatial variable and then average the result over the other; more precisely, one carries this average over a finite domain which is gradually extended to infinity (unless periodic BCs are encountered first). Assuming x_1 and x_2 were on the same surface or TOA boundary, this yields for the domain-average albedo

$$R(\Omega_1; \Omega_2) = R(-\Omega_2; -\Omega_1) \tag{3.174}$$

in natural notations, where Ω_2 is inward-oriented at the source and Ω_1 is outward-oriented at the detector. We thus recover Chandrasekhar's (1950) angular reciprocity relations for homogeneous plane-parallel media as a special case.

It was once believed that the reciprocity relation in (3.174) could be used in the real 3D atmosphere-surface system. The motivation was to faster build "angular models" which are used to infer hemispheric TOA fluxes from observed radiances in ERBE/CERES-like missions[18] that monitor the Earth's radiation budget. Di Girolamo et al. (1998) showed beyond any doubt that this is not a good idea since in a 3D system reciprocity applies either for infinitesimal areas around the two points, or for the domain averages, but not for the intermediate pixel-scale which is just an attribute of the observation system. There is further discussion of 1D RT reciprocity relation abuse in Chap. 11.

3.11 Summary and Outlook

We have surveyed the definitions of the fundamental quantities used in general – that is, 3D – radiative transfer from the standpoint of classical particle transport theory. We have looked at how these quantities relate to one another, including reciprocity relations and how these relations break down in 3D media over finite domains. We have shown how the key quantities are constrained by radiant energy conservation in the steady-state radiative transfer equation under various guises, along with its associated boundary conditions. Green functions are introduced and illustrated by showing how the general (and highly relevant) atmosphere/surface problem can be reduced to the two simpler problems of an atmosphere over an absorbing surface and a convolution using a Green function kernel.

We have thus introduced the basic tools used in the remainder of this volume. In mathematical short-hand, they compactly contain all the necessary information about the systematic biases that the 3D world inflicts upon forward radiative transfer and inversions based on 1D modeling. Recalling the historical background painted in broad strokes in Sect. 3.1, these biases have been documented extensively over the past three or more decades. Many of these biases will be discussed in the following chapters. Several exciting developments in the fundamental aspects of 3D radiative transfer, some with tutorial value, have occurred over the same period. For lack of space, they were not covered in this chapter, but some at least are covered elsewhere in this volume. A non-exhaustive list based largely, but not entirely, on our own research is:

- scale-separation conditions for the applicability of the radiative transfer equation and its connection to scalar- and vector-wave optics;
- photon transport theory in spherical atmospheres with or without significant radial gradients in the index of refraction;
- critical examination of the applicability of an "ensemble" distribution of droplet sizes motivated by the real-world observation that the largest droplets are so rare one may not be able to define a density for them;

---

[18] ERBE: Earth Radiation Budget Experiment; CERES: Clouds and the Earth's Radiant Energy System.

- introduction of a new term into the radiative transfer equation to account for the rare encounters of photons with the very large cloud droplets that do not have a well-defined density;
- eigen-analysis of the integral transport operator over a 3D optical medium and its application to escape probability calculation;
- various derivations of Fick's law for photon diffusion from the 3D radiative transfer equation, leading to the powerful 3D diffusion approximation;
- multiple forward-peaked scattering recast as a random walk on the sphere (of propagation directions) and derivation of the Eddington/van de Hulst rescaling of extinction without diffusion or asymptotic theory;
- proof that effective (or "mean-field") transport kernels in random 3D media are *never* exactly exponential and, in the relevant case of spatially correlated media, that they have longer-than-exponential tails;
- criteria that predict the onset of strong 3D effects on the scale of the *actual* mean-free-path which can be considerably larger than the inverse of the mean extinction.
- definition of the elementary interaction of a steady-state photon flow with a spatial disturbance in scattering media as a Cannon (1970) "channeling" event, and its application to domain-average biases;
- order-of-scattering decomposition (the Neumann series) and its connection with Markov chain theory in Kolmogorov's formulation of transition probabilities;
- time-dependent radiative transfer theory and the "Equivalence Theorem" approach to absorption processes and diffusion-based formulation;
- derivation of a general expression used in Chap. 12 for mean photon path length as volume-angle integral of the temporal Green function;
- asymptotic and exact diffusion theoretical formulas for the spatial and temporal characteristics of diffusely reflected and transmitted light on cloud parameters;
- extension of classic/Gaussian diffusion theory to highly variable media using Lévy-stable distributions and transport by "anomalous" diffusion.

Some of this missing material could be forged into a more advanced appraisal of the fundamentals of 3D radiative transfer that could in turn become the backbone of a phenomenology of 3D effects in photon transport. Some of this material could also be used to build a theoretical framework to articulate better-informed strategies in computational or observational projects. The reader is therefore encouraged to use the suggested reading list below to further his/her awareness of the fundamental issues in 3D radiative transfer. The present authors have always found added-value for their institutional research assignments in remote sensing and/or radiation budget estimation by revisiting the fundamentals.

As an example, in Sect. 3.10.1 we remarked that the decomposition of the Green function into its singular components (0th- and 1st-order scattering fields) and its regular component (higher-order scattered photons) has proved useful in tomography. That breakdown of the Green function has recently been applied to the characterization of radiative transfer regimes in 3D clouds (Knyazikhin et al., 2002) as well as in 3D vegetation canopies (Zhang et al., 2002).

To further illustrate this process of knowledge percolating from the fundamentals to applied radiative transfer, we gave special attention here to the radical departures from plane-parallel geometry embodied by horizontally finite clouds in isolation. We were thus forced to revisit the conventional partition of escaping solar radiation. In particular, we do away with the notion of "side-leakage" in favor of reflection since cloud "sides" can only be identified unambiguously in very special shape/illumination configurations. Even when they can be identified, the topology (proximity metrics) of cloud sides and radiation sources will cause them to be crossed by a significant population of escaping photons that have undergone relatively few orders-of-scattering. This indeed mimics the behavior of cloud top, the undisputed origin of reflected light. The conceptual cost here is to accept that *reflected* photons can reach the surface (i.e., for all climatic purposes, be "transmitted"). This is in fact an everyday observation: visualize the bright side of a cumulus under fair weather conditions or a distant cumulonimbus generated by deep convection. Similarly, we need to recognize that some of the highly scattered photons *transmitted* through the dimmer side of a finite cloud will eventually reach space (i.e., be "reflected"). This again is a frequent observation by Earth-monitoring satellites that can resolve small broken or isolated clouds, often from sun-synchronous orbits that exclude exactly overhead sun. Pixel-by-pixel processing of such imagery would probably misidentify a pixel from the shady side of a cloud for lack of brightness. If (e.g., by using thermal emittance) the pixel was properly classified as cloudy, then the optical depth would be vastly underestimated. The simpler "reflection-or-transmission" partition of solar photon fate proposed here was successfully put into application by one of us (Davis, 2002) to infer optical depths of cumulus clouds from high-resolution satellite images. In short, this somewhat rude reminder of common-sense observation of our 3D world populated with finite-sized clouds – and support from elementary considerations in set topology – has clarified the role of these clouds in the Earth's radiation energetics and taught us how to better interpret satellite data.

This is just one example of why the fundamentals of radiative transfer are still a vibrant area of research. An area that is indeed pressured to advance by the climate community as well as the remote sensing community. Anticipating on the topics of Chaps. 6–10 (Part III) and 11–14 (Part IV), we conclude by assessing the needs of these two major stake-holders of radiative transfer theory:

- *Climate Science.* It is commonplace to state that clouds are a major source of uncertainty in current climate system models. About all we know for sure is that low/warm/opaque clouds cool the climate (solar effects dominate) and therefore mitigate the global greenhouse effect, especially if the clouds are as extensive as typical marine stratocumulus systems; in contrast, their high/cold/semi-transparent counterparts trap heat (thermal effects dominate) and therefore contribute to the global greenhouse effect. Cirrus layers fulfill all the conditions for the latter effect and are also very pervasive at all latitudes. So, interestingly, the net effect of clouds on the global climate is small but strong regional effects can be expected. This underscores the importance of accurate representations of clouds and of their radiative properties in Global Climate Models (GCMs). It is fair to say that, along with the most common tri-atomic molecules (and some more complex ones), clouds regulate the climate system's vast heat reservoirs (oceans, land masses, and cryosphere). As part of this mechanism, clouds are active participants in the complex dynamics of the hydrological cycle that may be stressed anthropogenically in ways we understand very poorly. Paradoxically, clouds are *never* plane-parallel and homogeneous in Nature but *always* plane-parallel and homogeneous in climate models. This is the case even for the most current ones, simply because it makes them amenable to the simple two-stream approximation in radiative transfer (or one of its numerous variants). One wonders why clouds are not mentioned as often as aerosols, let alone greenhouse gases, as a source of concern in "big" climate science, at the IPCC level where research priorities are formulated. It is true that clouds are an inherent part of the climate system rather than a "forcing" that one (at least in principle) can control. But another part of the explanation surely comes from the necessity to "tune" GCMs to the climatologies of outgoing long- and short-wave fluxes obtained from satellites. These datasets are constantly being improved by NASA's Earth Radiation Budget Experiment (ERBE) and Clouds and the Earth's Radiant Energy System (CERES) programs and by initiatives from other agencies worldwide. Since cloud-radiation interactions are essentially the first and last lines of defense in the Earth's climate system as it interacts radiatively with the rest of the Universe, the corresponding parameterizations are the obvious candidates for dialing the "right" CERES/ERBE-based fluxes. Indeed cloud optical properties can be used to obtain essentially any answer: unlike aerosols for instance, they give climate modelers a full dynamical range. Now the tuning of cloud optical depth is justified primarily by uncertainty in the parameterization of cloud physics rather than that of the radiative transfer. This state of affairs is nonetheless rather discouraging for the cloud-radiation modeling community because even one of its poorest representations of clouds can be used to yield the desired answer. What would happen if an independently validated, hence fundamentally non-tunable, cloud-radiation scheme were delivered to the GCM community?

- *Remote Sensing Science.* The 2-stream particle transport model was first formulated and solved analytically by Schuster (1905); as this volume goes to press, we celebrate its first 100 years of loyal service to the atmospheric radiation community. The first computationally viable multi-stream solution for homogeneous slab geometry was obtained well over 50 years ago by Chandrasekhar and co-workers using discrete ordinates. These solutions for plane-parallel optical media are still the workhorses in GCM-based climate modeling and in operational remote sensing of cloud properties respectively. New photon properties such as polarization and total path (from off-beam lidar or O_2 spectroscopy) are being explored at the same time as usage of the more familiar ones, wavelength and direction/pixel-space, is being pushed to new heights. Indeed, hyperspectral is superceding multispectral sampling of wavelength and sub-meter resolutions are now available, at least commercially. The increasing cost of space assets – by sheer numbers if not by the unit – demands ever more *realistic* end-to-end modeling of existing and future observation systems. This modeling activity will undoubtedly usher in an entirely new class of physics-based algorithms for remote sensing that exploit rather than neglect spatial variability of the atmosphere/surface system. If nothing else, the harsh economics of programmatic investment in space-based Earth science will force the horizontally homogeneous plane-parallel atmosphere/cloud/surface model into retirement because new theory is very inexpensive compared to new hardware.

We again encourage the reader to delve into the suggested reading listed below with a running commentary. Some of the entries offer damage mitigation strategies for the widespread use of 1D standard models in applied radiative transfer (including "effective" properties and other corrections). Others offer outright alternatives in the form of new transport equations. Yet others describe new instrumental designs using both passive and active modalities that exploit rather than neglect the effects 3D radiative transfer.

References

Aida, M. (1977a). Reflection of solar radiation from an array of cumuli. *J. Met. Soc. Japan*, **55**, 174–181.

Aida, M. (1977b). Scattering of solar radiation as a function of cloud dimensions and orientation. *J. Quant. Spectrosc. Radiat. Transfer*, **17**, 303–310.

Antyufeev, V.S. and A.N. Bondarenko (1996). X-ray tomography in scattering media. *SIAM J. Appl. Math.*, **56**, 573–587.

Appleby, J.F. and D. van Blerkom (1975). Absorption line studies of reflection from horizontally inhomogeneous layers. *Icarus*, **24**, 51–69.

Avaste, O.A. and G.M. Vainikko (1974). Solar radiative transfer in broken clouds. *Izv. Acad. Sci. USSR Atmos. Oceanic Phys.*, **10**, 1054–1061.

Barkstrom, B.R. and R.F. Arduini (1977). The effect of finite size of clouds upon the visual albedo of the earth. In *Radiation in the Atmosphere*. H.-J. Bolle (ed.). Science Press, Princeton (NJ), pp. 188–190.

Bell, G.I. and S. Glasstone (1970). *Nuclear Reactor Theory*. Van Nostrand Reinholt, New York (NY).

Berry, M.V. and I.C. Percival (1986). Optics of fractal clusters such as smoke. *Optica Acta*, **33**, 577–591.

Bohren, C.F., J.R. Linskens, and M.E. Churma (1995). At what optical thickness does a cloud completely obscure the sun? *J. Atmos. Sci.*, **52**, 1257–1259.

Box, M.A., Keevers, M., and B.H.J. McKellar, (1988). On the perturbation series for radiative effects. *J. Quant. Spectrosc. Radiat. Transfer*, **39**, 219–223.

Box, M.A., S.A.W. Gerstl, and C. Simmer (1989). Computation of atmospheric radiative effects via perturbation theory. *Beitr. Phys. Atmosph.*, **62**, 193–199.

Box, M.A., Polonsky, I.N., and A.B. Davis (2003). Radiative transfer through inhomogeneous turbid media: Implementation of the adjoint perturbation approach at the first-order. *J. Quant. Spectrosc. Radiat. Transfer*, **78**, 85–98.

Cannon, C.J. (1970). Line transfer in two dimensions. *Astrophys. J.*, **161**, 255–264.

Capra, F. (1991). *The Tao of Physics*. Shambhala Publ., Boston (MA), 3rd edition.

Case, K.M. and P.F. Zweifel (1967). *Linear Transport Theory*. Addison-Wesley, Reading (MA).

Chandrasekhar, S. (1950). *Radiative Transfer*. Oxford University Press, reprinted by Dover Publications (1960), New York (NY).

Chandrasekhar, S. (1958). On the diffuse reflection of a pencil of radiation by a plane-parallel atmosphere. *Proc. Natl. Acad. Sci. U.S.A.*, **44**, 933–940.

Choulli, M. and P. Stefanov (1996). Reconstruction of the coefficient of the stationary transport equation from boundary measurements. *Inverse Problems*, **12**, L19–L23.

Davies, R. (1978). The effect of finite geometry on the three-dimensional transfer of solar irradiance in clouds. *J. Atmos. Sci.*, **35**, 1712–1725.

Davies, R. and J.A. Weinman (1977). Results from two models of the three dimensional transfer of solar radiation in finite clouds. In *Radiation in the Atmosphere*. H.-J. Bolle (ed.). Science Press, Princeton (NJ), pp. 225–227.

Davis, A.B. (2002). Cloud remote sensing with sideways-looks: Theory and first results using Multispectral Thermal Imager (MTI) data. In *SPIE Proceedings: Algorithms and Technologies for Multispectral, Hyperspectral, and Ultraspectral Imagery VIII*. S.S. Shen and P.E. Lewis (eds.). S.P.I.E. Publications, Bellingham, WA, pp. 397–405.

Davis, A.B. and A. Marshak (2001). Multiple scattering in clouds: Insights from three-dimensional diffusion/P_1 theory. *Nuclear Sci. and Engin.*, **137**, 251–280.

Davis, A.B. and A. Marshak (2004). Photon propagation in heterogeneous optical media with spatial correlations: Enhanced mean-free-paths and wider-than-exponential free-path distributions. *J. Quant. Spectrosc. Radiat. Transfer*, **84**, 3–34.

Davis, A.B., R.F. Cahalan, J.D. Spinhirne, M.J. McGill, and S.P. Love (1999). Off-beam lidar: An emerging technique in cloud remote sensing based on radiative Green-function theory in the diffusion domain. *Phys. Chem. Earth (B)*, **24**, 757–765.

Deirmendjian, D. (1969). *Electromagnetic Scattering on Spherical Polydispersions*. Elsevier, New York (NY).

Di Girolamo, L., T. Várnai, and R. Davies (1998). Apparent breakdown of reciprocity in reflected solar radiances. *J. Geophys. Res.*, **103**, 8795–8803.

Diner, D.J., G.P. Asner, R. Davies, Yu. Knyazikhin, J.-P. Muller, A.W. Nolin, B. Pinty, C.B. Schaaf, and J. Stroeve (1999). New directions in Earth observing: Scientific application of multi-angle remote sensing. *Bull. Amer. Meteor. Soc.*, **80**, 2209–2228.

Dutton, E.G., A. Farhadi, R.S. Stone, C.N. Long, and D.W. Nelson (2004). Long-term variations in the occurrence and effective solar transmission of clouds as determined from surface-based total irradiance observations. *J. Geophys. Res.*, **109**, D03204, doi:10.1029/2003JD003568.

Germogenova, T.A. (1986). *The Local Properties of the Solution of the Transport Equation* (in Russian). Nauka, Moscow (Russia).

Giovanelli, R.G. (1959). Radiative transfer in non-uniform media. *Aust. J. Phys.*, **12**, 164–170.

Giovanelli, R.G. and J.T. Jefferies (1956). Radiative transfer with distributed sources. *Lond. Phys. Soc. Proc.*, **69**, 1077–1084.

Henyey, L.C. and J.L. Greenstein (1941). Diffuse radiation in the galaxy. *Astrophys. J.*, **93**, 70–83.

Ishimaru, A. (1975). Correlations functions of a wave in a random distribution of stationary and moving scatterers. *Radio Science*, **10**, 45–52.

Kaufman, Y.J. (1979). Effect of the Earth's atmosphere on contrast for zenith observation. *J. Geophys. Res.*, **84**, 3165–3172.

Knyazikhin, Yu., A. Marshak, W.J. Wiscombe, J. Martonchik, and R.B. Myneni (2002). A missing solution to the transport equation and its effect on estimation of cloud absorptive properties. *J. Atmos. Sci.*, **59**, 3572–3585.

Lilienfeld, P. (2004). A blue sky history. *Optics and Photonics News (OPN)*, **15**, 32–39.

Liou, K.-N. (2002). *An Introduction to Atmospheric Radiation*. Academic Press, San Diego (CA), 2nd edition.

Lovejoy, S. (1982). The area-parameter relation for rain and clouds. *Science*, **216**, 185–187.

Lyapustin, A.I. and Yu. Knyazikhin (2002). Green's function method for the radiative transfer problem. 2. Spatially heterogeneous anisotropic surface. *Applied Optics*, **41**, 5600–5606.

Mandelbrot, B.B. (1982). *The Fractal Geometry of Nature*. W. H. Freeman, New York (NY).

Marchuk, G. (1964). Equation for the value of information from weather satellites and formulation of inverse problems. *Kosm. Issled.*, **2**, 462–477.

Marchuk, G., G. Mikhailov, M. Nazaraliev, R. Darbinjan, B. Kargin, and B. Elepov (1980). *The Monte Carlo Methods in Atmospheric Optics*. Springer-Verlag, New York (NY).

Marshak, A., Yu. Knyazikhin, A.B. Davis, W.J. Wiscombe, and P. Pilewskie (2000). Cloud – vegetation interaction: Use of normalized difference cloud index for estimation of cloud optical thickness. *Geophys. Res. Lett.*, **27**, 1695–1698.

Marshak, A., Yu. Knyazikhin, K.D. Evans, and W.J. Wiscombe (2004). The "RED versus NIR" plane to retrieve broken-cloud optical depth from ground-based measurements. *J. Atmos. Sci.*, **61**, 1911–1925.

McKee, T.B. (1976). Simulated radiance patterns for finite cubic clouds. *J. Atmos. Sci.*, **33**, 2014–2020.

McKee, T.B. and S.K. Cox (1974). Scattering of visible radiation by finite clouds. *J. Atmos. Sci.*, **31**, 1885–1892.

Mihalas, D. (1979). *Stellar Atmospheres*. Freeman, San Francisco (CA), 2nd edition.

Minnaert, M. (1941). The reciprocity principle in lunar photometry. *Astrophys. J.*, **93**, 403–410.

Mishchenko, M.I. (2003). Radiative transfer theory: From Maxwell's equations to practical applications. In *Wave Scattering in Complex Media: From Theory to Applications*. B.A. van Tiggelen and S.E. Skipetrov (eds.). Kluwer Academic, Dordrecht (the Netherlands), pp. 367–414.

Mishchenko, M.I., J.W. Hovenier, and L.D. Travis (2000). *Light Scattering by Non-Sperical Particles*. Academic Press, San Diego (CA).

Mullamaa, Ü.-A.R., M.A. Sulev, V.K. P oldmaa, H.A. Ohvril, H.J. Niilisk, M.I. Allenov, L.G. Chubakov, and A.E. Kuusk (1972). Stochastic Structure of Cloud and Radiation Fields, Ü.-A. R. Mullamaa (ed.). IPA, Acad. Sci. Est. SSR, Tartu, 282 pp (in Russian, English translation: 1975. Technical Report TT F-822, NASA Technical Translation, Washington (DC).

Nicodemus, F.E., J.C. Richmond, J.J. Hsia, I.W. Ginsberg, and T. Limperis (1977). *Geometrical Considerations and Nomenclature for Reflectance*. National Bureau of Standards, NBS Monograph No. 160.

Odell, A.P. and J.A. Weinman (1975). The effect of atmospheric haze on images of the Earth's surface. *J. Geophys. Res.*, **80**, 5035–5040.

Otterman, J. and R.S. Fraser (1979). Adjacency effects on imaging by surface reflection and atmospheric scattering: Cross radiance zenith. *Appl. Opt.*, **18**, 2852–2860.

Polonsky, I.N., M.A. Box, and A.B. Davis (2003). Radiative transfer through inhomogeneous turbid media: Implementation of the adjoint perturbation approach at the first-order. *J. Quant. Spectrosc. Radiat. Transfer*, **78**, 85–98.

Rahman, H., B. Pinty, and M.M. Verstraete (1993). Coupled surface-atmosphere reflectance (CSAR) model. 2. Semiempirical surface model usable with NOAA Advanced Very High Resolution Radiometer data. *J. Geophys. Res.*, **98**, 20,791–20,801.

Ramanathan, V., P.J. Crutzen, A.P. Mitra, and D. Sikka (2002). The INDian Ocean EXperiment and the Asian brown cloud. *Curr. Sci.*, **83**, 947–955.

Richards, P.I. (1956). Scattering from a point-source in plane clouds. *J. Opt. Soc. Am.*, **46**, 927–934.

Romanova, L.M. (1968a). Light field in the boundary layer of a turbid medium with strongly anisotropic scattering illuminated by a narrow beam. *Izv. Acad. Sci. USSR Atmos. Oceanic Phys.*, **4**, 1185–1196 (in Russian), 679–685 (English translation).

Romanova, L.M. (1968b). The light field in deep layers of a turbid medium illuminated by a narrow beam. *Izv. Acad. Sci. USSR Atmos. Oceanic Phys.*, **4**, 311–320 (in Russian), 175–179 (English translation).

Romanova, L.M. (1971a). Effective size of the light spot on the boundaries of a thick turbid medium illuminated by a narrow beam. *Izv. Acad. Sci. USSR Atmos. Oceanic Phys.*, **7**, 410–420 (in Russian), 270–277 (English translation).

Romanova, L.M. (1971b). Some characteristics of the light field generated by a point-collimated stationary light source in clouds and fog. *Izv. Acad. Sci. USSR Atmos. Oceanic Phys.*, **7**, 1153–1164 (in Russian), 758–764 (English translation).

Romanova, L.M. (1975). Radiative transfer in a horizontally inhomogeneous scattering medium. *Izv. Acad. Sci. USSR Atmos. Oceanic Phys.*, **11**, 509–513.

Ronnholm, K., M.B. Baker, and H. Harrison (1980). Radiation transfer through media with uncertain or random parameters. *J. Atmos. Sci.*, **37**, 1279–1290.

Schuster, A. (1905). Radiation through a foggy atmosphere. *Astrophys. J.*, **21**, 1–22.

Schwartz, L. (1950). *Théorie des Distributions*, 2 vols. Hermann, Paris (France).

Siegel, R. and J.R. Howell (1981). *Thermal Radiation Heat Transfer*. McGraw-Hill, New York (NY), 2nd edition.

van Blerkom, D.J. (1971). Diffuse reflection from clouds with horizontal inhomo-geneities. *Astrophys. J.*, **166**, 235–242.

Weber, P.G., B.C. Brock, A.J. Garrett, B.W. Smith, C.C. Borel, W.B. Clodius, S.C. Bender, R. Rex Kay, and M.L. Decker (1999). Multispectral Thermal Imager mission overview. *SPIE Proceedings*, **3753**, 340–346.

Weinman, J.A. and P.N. Swartztrauber (1968). Albedo of a striated medium of isotropically scattering particles. *J. Atmos. Sci.*, **34**, 642–650.

Wendling, P. (1977). Albedo and reflected radiance of horizontally inhomogeneous clouds. *J. Atmos. Sci.*, **34**, 642–650.

Wiscombe, W.J. (1977). The delta-M method: Rapid yet accurate radiative flux cal-culations for strongly asymmetric phase functions. *J. Atmos. Sci.*, **34**, 1408–1422.

Wolf, E. (1976). New theory of radiative energy transfer in free electromagnetic fields. *Phys. Rev. D*, **13**, 869–886.

Yodh, A. and B. Chance (1995). Spectroscopy and imaging with diffusing light. *Phys. Today*, **48**, 34–40.

Zhang, Y., N. Shabanov, Yu. Knyazikhin, and R.B. Myneni (2002). Assessing the information content of multiangle satellite data for mapping biomes. II. Theory. *Remote Sens. Environ.*, **80**, 435–446.

Suggested Reading

This selected bibliography has been sorted into a few broad categories at the cost of some repetitions between them and with the above reference list (quoted explicitly in the main text).

(1) *Books:* There now many excellent volumes on atmospheric radiation but they tend to treat only plane-parallel media when multiple scattering is included. For our 3D RT purposes, they remain a good resource for all matters of radiation interaction with the gases and particulates that compose the Earth's atmosphere. We also list here a few classic monographs that cover the fundamentals of RT, or particle trans-port generally speaking, in more detail than most atmospheric radiation volumes and, in some cases, treat at least one geometry other than plane-parallel, even if it remains highly symmetric. The books by Ishimaru and Wolf contain elements of the delicate derivation of the RT equation from scalar or EM wave theory while Mihalas takes an interesting "photon as material particle" route through Boltzmann's transport equa-tion. By chronological order:

Chandrasekhar, S. (1950). *Radiative Transfer*. 393 pp., Oxford University Press, London (United Kingdom): reprinted by Dover (1960), New York (NY).

Davison, B. (1958). *Neutron Transport Theory*. 450 pp., Oxford University Press, London (United Kingdom).

Vladimirov, V.S. (1963). *Mathematical Problems in the One-Velocity Theory of Par-ticle Transport*, Tech. Rep. AECL-1661, Atomic Energy of Canada Ltd., Chalk River, Ontario.

Case, K.M. and P.F. Zweifel (1967). *Linear Transport Theory*. Addison-Wesley Publ. Co., Reading (MA).

Bell, G.I. and S. Glasstone (1970). *Nuclear Reactor Theory*. 619 pp., Van Nostrand Reinholt, New York (NY).

Pomraning, G.C. (1973). *The Equations of Radiation Hydrodynamics*. 288 pp., Oxford-Pergamon Press, New York (NY).

Preisendorfer, R.W. (1978). *Hydrological Optics*, NOAA-PMEL (Hawaii).

Ishimaru, A. (1978). *Wave Propagation and Scattering in Random Media*. 2 vols., Academic Press, New York (NY).

Mihalas, D. (1979). *Stellar Atmospheres*. 2nd ed., xvii+632 pp., Freeman, San Francisco (CA).

van de Hulst, H.C. (1980). *Multiple Light Scattering: Tables, Formulas, and Applications*. 2 vols., Academic Press, San Diego (CA).

Welch, R.M., S.K. Cox and J. M. Davis (1980). *Solar Radiation and Clouds*. Meteorological Monograph Series, Vol. 17 (No. 39), American Meteorological Society, Boston (MA).

Siegel, R. and J.R. Howell (1981). *Thermal Radiation Heat Transfer*. 2nd ed., xvi+862 pp., McGraw-Hill, New York (NY).

Bohren, C.F. and D.R. Huffman (1983). *Absorption and Scattering of Light by Small Particles*. xiv+530 pp., Wiley, New York (NY).

Lenoble, J. (ed.) (1985). *Radiative Transfer in Scattering and Absorbing Atmospheres: Standard Computational Procedures*. A. Deepak Publ., Hampton (VA).

Germogenova, T.A. (1986). *The Local Properties of the Solution of the Transport Equation* (in Russian). 272 pp., Nauka, Moscow (Russia).

Goody, R.M. and Y.L. Yung (1989). *Atmospheric Radiation: Theoretical Basis*. xiii+519 pp., Oxford University Press, New York (NY).

Lewis, E.E. and W.F. Miller, Jr. (1993). *Computational Methods of Neutron Transport*. xvi+401 pp., American Nuclear Society, La Grange Park (IL).

Stephens, G.L. (1994). *Remote Sensing of the Lower Atmosphere: An Introduction*. xvi+523 pp., Oxford University Press, New York (NY).

Lenoble, J. (1993). *Atmospheric Radiative Transfer*. 532 pp., A. Deepak Publ., Hampton (VA).

Thomas, G.E. and K. Stamnes (1999). *Radiative Transfer in the Atmosphere and Ocean*. 546 pp., Cambridge University Press, New York (NY).

Wolf, E. (2001). *Selected Works of Emil Wolf, with Commentary*. x+661 pp., World Scientific Co., Singapore.

Liou, K.N. (2002). *An Introduction to Atmospheric Radiation*. 2nd Ed., xiv+583 pp., Academic Press, San Diego (CA).

Kokhanovsky, A.A. (2004). *Light Scattering Media Optics, Problems and Solutions*. 3rd ed., Springer, Heidelberg (Germany).

(2) *RT in the time-domain:* Particle transport, even with steady sources and sinks, unfolds by its very definition in time. Although time-dependence is out of the scope of the present chapter it is used at least implicitly in Chaps. 12 (on net horizontal transport) and 13 (on solar photon path statistics). Time/pathlength-based radiometry,

whether time-resolved or via fine-resolution spectroscopy, has such tremendous potential for remote sensing and climate diagnostics that we confidently predict it will become routine in the relatively near future. The following publications in this broad category are grouped as theoretical, O_2-oriented, lidar-oriented, lightning-oriented, and then listed chronologically. The avid reader is also encouraged to search the bio-medical imaging literature for pathlength-based methods of probing soft tissue with diffusing NIR photons; he/she will find interesting parallels with the multiple-scattering cloud lidar techniques.

Theory:

Irvine, W.M. (1964). The formation of absorption bands and the distribution of photon optical paths in a scattering atmosphere. *Bull. Astron. Inst. Neth.*, **17**, 266–279.

Ivanov, V.V. and S.D. Gutshabash (1974). Propagation of brightness wave in an optically thick atmosphere. *Physika Atmosphery i Okeana*, **10**, 851–863.

Davis, A. and A. Marshak (1997). Lévy kinetics in slab geometry: Scaling of transmission probability. In *Fractal Frontiers*, M.M. Novak and T.G. Dewey (eds.), World Scientific, Singapore, pp. 63–72.

Davis, A.B. and A. Marshak (2001). Multiple scattering in clouds: Insights from three-dimensional diffusion/P_1 theory. *Nucl. Sci. Eng.*, **137**, 251–288.

Platnick, S. (2001). A superposition technique for deriving photon scattering statistics in plane-parallel cloudy atmospheres. *J. Quant. Spectrosc. Radiat. Transfer*, **68**, 57–73.

Davis, A.B. and A. Marshak (2002). Space-time characteristics of light transmitted by dense clouds: A Green function analysis. *J. Atmos. Sci.*, **59**, 2713–2727.

Oxygen-band/line spectroscopy:

Pfeilsticker, K., F. Erle, O. Funk, H. Veitel and U. Platt (1998). First geometrical pathlength distribution measurements of skylight using the oxygen A-band absorption technique – I, Measurement technique, atmospheric observations, and model calculations. *J. Geophys. Res.*, **103**, 11,483–11,504.

Pfeilsticker, K. (1999). First geometrical pathlength distribution measurements of skylight using the oxygen A-band absorption technique - II, Derivation of the Lévy-index for skylight transmitted by mid-latitude clouds. *J. Geophys. Res.*, **104**, 4101–4116.

Min, Q.-L. and L.C. Harrison (1999). Joint statistics of photon pathlength and cloud optical depth. *Geophys. Res. Lett.*, **26**, 1425–1428.

Stephens, G.L. and A. Heidinger (2000). Molecular line absorption in a scattering atmosphere – Part I: Theory. *J. Atmos. Sci.*, **57**, 1599–1614.

Heidinger, A. and G.L. Stephens (2000). Molecular line absorption in a scattering atmosphere – Part II: Application to remote-sensing in the O_2 A-Band. *J. Atmos. Sci.*, **57**, 1615–1634.

Min, Q.-L., L.C. Harrison and E.E. Clothiaux (2001). Joint statistics of photon pathlength and cloud optical depth: Case studies. *J. Geophys. Res.*, **106**, 7375–7385.

Portman, R.W., S. Solomon, R.W. Sanders, J.S. Daniels and E.G. Dutton (2001). Cloud modulation of zenith sky oxygen photon path lengths over Boulder: Measurement versus model. *J. Geophys. Res.*, **106**, 1139–1155.

Heidinger, A. and G.L. Stephens (2002). Molecular line absorption in a scattering atmosphere – Part III: Path length characteristics and the effects of spatially heterogeneous clouds. *J. Atmos. Sci.*, **59**, 1641–1654.

Min, Q.-L., L.C. Harrison, P. Kiedron, J. Berndt and E. Joseph (2004). A high-resolution oxygen A-band and water vapor band spectrometer. *J. Geophys. Res.*, **109**, D02202, doi:10.1029/2003JD003540.

Multiple-scattering in lidar signals from clouds (beyond the small-angle approximation):

Winker, D.M. and L.R. Poole (1995). Monte-Carlo calculations of cloud returns for ground-based and space-based LIDARs. *Applied Physics B – Lasers and Optics*, **B60**, 341–344.

Winker, D.M., R.H. Couch and M.P. McCormick (1996). An overview of LITE: NASA's Lidar In-space Technology Experiment. *Proc. IEEE*, **84**, 164–180.

Miller, S.D. and G.L. Stephens (1999). Multiple scattering effects in the lidar pulse stretching problem. *J. Geophys. Res.*, **104**, 22,205–22,219.

Davis, A.B., R.F. Cahalan, J.D. Spinhirne, M.J. McGill and S.P. Love (1999). Off-beam lidar: An emerging technique in cloud remote sensing based on radiative Green-function theory in the diffusion domain. *Phys. Chem. Earth (B)*, **24**, 757–765.

Davis, A.B., D.M. Winker and M.A. Vaughan (2001). First retrievals of dense cloud properties from off-beam/multiple-scattering lidar data collected in space. In *Laser Remote sensing of the atmosphere: Selected Papers from the 20th International Conference on Laser Radar*, A. Dabas and J. Pelon (eds.), École Polytechnique, Palaiseau (France), pp. 35–38.

Kotchenova, S.Y., N.V. Shabanov, Y. Knyazikhin, A.B. Davis, R. Dubayah and R.B. Myneni (2003). Modeling lidar waveforms with time-dependent stochastic radiative transfer theory for remote estimations of forest biomass. *J. Geophys. Res.*, **108**(D15), 4484, doi:1029/2002JD003288.

Evans, K.F., R.P. Lawson, P. Zmarzly, D. O'Connor and W.J. Wiscombe (2003). In situ cloud sensing with multiple scattering lidar: Simulations and demonstration. *J. Atmos. Ocean Tech.*, **20**, 1505–1522.

Optical lightning studies:

Thomason, L.W. and E.P. Krider (1982). The effects of clouds on the light produced by lightning. *J. Atmos. Sci.*, **39**, 2051–2065.

Koshak, W.J., R.J. Solakiewicz, D.D. Phanord and R.J. Blakeslee (1994). Diffusion model for lightning radiative transfer. *J. Geophys. Res.*, **99**, 14,361–14,371.

Light, T.E., D.M. Suszcynsky, M.W. Kirkland and A.R. Jacobson (2001). Simulations of lightning optical waveforms as seen through clouds by satellites. *J. Geophys. Res.*, **106**, 17,103–17,114.

(3) *Lateral photon transport:* For more ease, we have made separate lists for optically thin and thick media, with applications respectively to dense clouds and aerosol layers.

Transmission and reflection from optically thin medium over a reflecting surface: We list here a few investigations of the so-called aerosol "adjacency" effect that appeared in the 1980s or thereafter. Earlier references are listed in the main text on the closely related problem of transmission through optically thin scattering media.

Mekler, Y. and Y.J. Kaufman (1980). The effect of Earth's atmosphere on contrast reduction for a nonuniform surface albedo and "two-halves" field. *J. Geophys. Res.*, **85**, 4067–4083.

Otterman, J., S. Ungar, Y. Kaufman and M. Podolak (1980). Atmospheric effects on radiometric imaging from satellites under low optical thickness conditions. *Remote Sens. Environ.*, **9**, 115–129.

Tanré, D., M. Herman and P.-Y. Deschamps (1981). Influence of the background contribution upon space measurements of ground reflectance. *Appl. Optics.*, **20**, 3676–3684.

Kaufman, Y.J. (1982). Solution of the equation of radiative-transfer for remote-sensing over nonuniform surface reflectivity. *J. Geophys. Res.*, **87**, 4137–4147.

Diner, D.J. and J.V. Martonchik (1984). Atmospheric transfer of radiation above an inhomogeneous non-Lambertian ground: 1 – Theory. *J. Quant. Spectrosc. Radiat. Transfer*, **31**, 97–125.

Diner, D.J. and J.V. Martonchik (1984). Atmospheric transfer of radiation above an inhomogeneous non-Lambertian ground: 2 – Computational considerations and results. *J. Quant. Spectrosc. Radiat. Transfer*, **31**, 279–304.

Takashima, T. and K. Masuda (1992). Simulation of atmospheric effects on the emergent radiation over a checkerboard type of terrain. *Astrophys. Space Sci.*, **198**, 253–263

Reinersman, P.N. and K.L. Carder (1995). Monte Carlo simulation of the atmospheric point-spread function with an application to correction for the adjacency effect. *Appl. Optics.*, **34**, 4453–4471.

Lyapustin, A.I. (2001). Three-dimensional effects in the remote sensing of surface albedo. *IEEE Trans. Geosc. and Remote Sens.*, **39**, 254–263.

Lyapustin, A.I. and Y. Kaufman (2001). Role of adjacency effect in the remote sensing of aerosol. *J. Geophys. Res.*, **106**, 11,909–11,916.

Lyapustin, A.I. and Y. Knyazikhin (2002). Green's function method for the radiative transfer problem. II. Spatially heterogeneous anisotropic surface. *Appl. Optics.*, **41**, 5600–5606.

Transmission and reflection fields for optically thick media illuminated with localized sources: Chapter 12 is largely devoted to this case of lateral transport in clouds and, here again, there is fruitful searching to be done in the medical optics literature that will reveal interesting parallels with independent atmospheric endeavors.

Weinman, J.A. and M. Masutani (1987). Radiative transfer models of the appearance of city lights obscured by clouds observed in nocturnal satellite images. *J. Geophys. Res.*, **92**, 5565–5572.

Stephens, G.L. (1986). Radiative transfer in spatially heterogeneous, two-dimensional anisotropically scattering media, *J. Quant. Spectrosc. Radiat. Transfer*, **36**, 51–67.

Stephens, G.L. (1988). Radiative transfer through arbitrary shaped optical media, Part 1 – A general method of solution. *J. Atmos. Sci.*, **45**, 1818–1835.

Ganapol, B.D., D.E. Kornreich, J.A. Dahl, D.W. Nigg, S.N. Jahshan and C.A. Temple (1994). The searchlight problem for neutrons in a semi-infinite medium. *Nucl. Sci. Eng.*, **118**, 38–53.

Marshak, A., A. Davis, W.J. Wiscombe and R.F. Cahalan (1995). Radiative smoothing in fractal clouds. *J. Geophys. Res.*, **100**, 26,247–26,261.

Kornreich, D.E. and B.D. Ganapol (1997). Numerical evaluation of the three-dimensional searchlight problem in a half-space. *Nucl. Sci. Eng.*, **127**, 317–337.

Davis, A., A. Marshak, R.F. Cahalan and W.J. Wiscombe (1997). The Landsat scale-break in stratocumulus as a three-dimensional radiative transfer effect: Implications for cloud remote sensing. *J. Atmos. Sci.*, **54**, 241–260.

Davis, A.B., R.F. Cahalan, J.D. Spinhirne, M.J. McGill and S.P. Love (1999). Off-beam lidar: An emerging technique in cloud remote sensing based on radiative Green-function theory in the diffusion domain. *Phys. Chem. Earth (B)*, **24**, 757–765.

Romanova, L.M. (2001). Narrow light beam propagation in a stratified cloud: Higher transverse moments. *Izv. Atmos. Oceanic Phys.*, **37**, 748–756.

Platnick, S. (2001). Approximations for horizontal photon transport in cloud remote sensing problems. *J. Quant. Spectrosc. Radiat. Transfer*, **68**, 75–99.

Davis, A.B. and A. Marshak (2001). Multiple scattering in clouds: Insights from three-dimensional diffusion/P_1 theory. *Nucl. Sci. Eng.*, **137**, 251–288.

Davis, A.B. and A. Marshak (2002). Space-time characteristics of light transmitted by dense clouds: A Green function analysis. *J. Atmos. Sci.*, **59**, 2713–2727.

Polonsky, I.N. and A.B. Davis (2004). Lateral photon transport in dense scatering and weakly absorbing media of finite thickness: Asymptotic analysis of the space-time Green functions. *J. Opt. Soc. Amer. A*, **21**, 1018–1025.

(4) *New transport models and RT equations:* Chapter 7 is devoted to "stochastic" RT, which usually means in binary Markovian binary media. It covers in detail mean-field theory for a special kind of random medium that has a vast literature in its own right in atmospheric optics, astrophysics and neutronics. Here we list some other kinds of mean-field investigations that produce new transport equations for multiple scattering, or new kernels for propagation between the scatterings.

Borovoi, A.G. (1984). Radiative transfer in inhomogeneous media. *Dok. Akad. Nauk SSSR*, **276**, 1374–1378. (English version in *Sov. Phys. Dokl.*, **29**(6).)

Stephens, G.L. (1988). Radiative transfer through arbitrarily shaped media, Part 2 – Group theory and closures. *J. Atmos. Sci.*, **45**, 1836–1848.

Evans, K.F. (1993). A general solution for stochastic radiative transfer. *Geophys. Res. Lett.*, **20**, 2075–2078.

Davis, A. and A. Marshak (1997). Lévy kinetics in slab geometry: Scaling of transmission probability. In *Fractal Frontiers*, M.M. Novak and T.G. Dewey (eds.), World Scientific, Singapore, pp. 63–72.

Cairns, B., A.W. Lacis and B.E. Carlson (2000). Absorption within inhomogeneous clouds and its parameterization in general circulation models. *J. Atmos. Sci.*, **57**, 700–714.

Kostinski, A.B. (2001). On the extinction of radiation by a homogeneous but spatially correlated random medium. *J. Opt. Soc. Amer. A*, **18**, 1929–1933.

Shaw, R.A., A.B. Kostinski and D.D. Lanterman (2002). Super-exponential extinction of radiation in a negatively-correlated random medium. *J. Quant. Spectrosc. Radiat. Transfer*, **75**, 13–20.

Knyazikhin, Y., A. Marshak, W.J. Wiscombe, J. Martonchik and R.B. Myneni (2002). A missing solution to the transport equation and its effect on estimation of cloud absorptive properties. *J. Atmos. Sci.*, **59**, 3572–3585.

Davis, A.B. and A. Marshak (2004). Photon propagation in heterogeneous optical media with spatial correlations: Enhanced mean-free-paths and wider-than-exponential free-path distributions. *J. Quant. Spectrosc. Radiat. Transfer.*, **84**, 3–34.

(5) *New instrument and/or algorithm designs that exploit 3D RT:* Here we exclude high-resolution O_2 band/line spectroscopy because it has an entry of its own in the above item (2), and a whole chapter further on in this volume by Stephens, Heidinger, and Gabriel (Chap. 13). It is to be considered as a new and promising technique even though it has been under consideration and even testing for a long while. The references by von Savigny et al., Marshak et al. and Barker et al. use a simple zenith- or nadir-looking radiometer that records radiance time-series at one or more non-absorbed solar wavelengths as the clouds are advected above the ground station or scanned from an aircraft. Davis et al., Love et al. and Evans et al. use a pulsed laser as a source and examine the information about clouds contained in the "off-beam" (i.e., multiple-scattering) returns.

Savigny, C. von, O. Funk, U. Platt and K. Pfeilsticker (1999). Radiative smoothing in zenith-scattered skylight transmitted through clouds to the ground. *Geophys. Res. Lett.*, **26**, 2949–2952.

Davis, A.B., R.F. Cahalan, J.D. Spinhirne, M.J. McGill and S.P. Love (1999). Off-beam lidar: An emerging technique in cloud remote sensing based on radiative Green-function theory in the diffusion domain. *Phys. Chem. Earth (B)*, **24**, 757–765.

Marshak, A., Y. Knyazikhin, A.B. Davis, W.J. Wiscombe and P. Pilewskie (2000). Cloud-vegetation interaction: Use of normalized difference cloud index for estimation of cloud optical thickness. *Geophys. Res. Lett.*, **27**, 1695–1698.

Love, S.P., A.B. Davis, C. Ho and C.A. Rohde (2001). Remote sensing of cloud thickness and liquid water content with Wide-Angle Imaging Lidar (WAIL). *Atm. Res.*, **59–60**, 295–312.

Barker, H.W. and A. Marshak (2001). Inferring optical depth of broken clouds above green vegetation using surface solar radiometric measurements. *J. Atmos. Sci.*, **58**, 2989-3006.

Barker, H.W., A. Marshak, W. Szyrmer, A. Trishchenko, J.-P. Blanchet, and Z. Li (2002). Inference of cloud optical depth from aircraft-based solar radiometric measurements. *J. Atmos. Sci.*, **59**, 2093-2111.

Evans, K.F., R.P. Lawson, P. Zmarzly, D. O'Connor and W.J. Wiscombe (2003). In situ cloud sensing with multiple scattering lidar: Simulations and demonstration. *J. Atmos. Ocean Tech.*, **20**, 1505-1522.

Marshak, A., Yu. Knyazikhin, K.D. Evans and W.J. Wiscombe (2004). The "RED versus NIR" plane to retrieve broken-cloud optical depth from ground-based measurements. *J. Atmos. Sci.*, **61**, 1911-1925.

(6) *Further considerations on surface reflectance properties and/or reciprocity:* A few recent studies worth consulting are listed below, to go beyond the references given in Sect. 3.10.3.

Loeb, N.G. and R. Davies (1996). Observational evidence of plane parallel model biases: Apparent dependence of cloud optical depth on solar zenith angle. *J. Geophys. Res.*, **101**, 1621-1634.

Di Girolamo, L. (1999). Reciprocity principle applicable to reflected radiance measurements and the searchlight problem. *Appl. Optics*, **38**, 3196-3198.

Diner, D.J., G.P. Asner, R. Davies, Y. Knyazikhin, J.P. Muller, A.W. Nolin, B. Pinty, C.B. Schaaf and J. Stroeve (1999). New directions in Earth observing: Scientific application of multi-angle remote sensing. *Bull. Amer. Meteor. Soc.*, **80**, 2209-2228.

Knyazikhin, Y. and A. Marshak (2000). Mathematical aspects of BRDF modeling: Adjoint problem and Green's function. **Remote Sens. Rev. 18**, 263-280.

Martonchik, J.V., C.J. Bruegge and A. Strahler (2000). A review of reflectance nomenclature used in remote sensing. *Remote Sens. Rev.*, **19**, 9-20.

Snyder, W.C. (2002). Definition and invariance properties of structured surface BRDF. *IEEE Trans. Geoscience Rem. Sensing*, **40**, 1032-1037.

(7) *More studies on cloud models with non-plane-parallel geometry:* In the introduction to the main text, we covered the historical (pre-1980) period where horizontally finite (fundamentally non-plane-parallel) clouds where a popular topic, especially using the diffusion approximation. In the 1990s, the trend was to return to plane-parallel geometry for the outer geometry but the models were rife with internal 2D or 3D variability. We predict a renewed interest in horizontally finite clouds to cope with cumulus-type clouds. In Sect. 3.9.2, we discussed a recent application by one of the authors (Davis, 2002) to remote-sensing that capitalized on a closed-form diffusion theoretical result for spherical clouds. Here are a few references from the 1980s where horizontally finite clouds are investigated in isolation or in (random or regular) arrays, always going beyond diffusion theory.

Welch, R. and W. Zdunkowski (1981). The radiative characteristics of noninteracting cumulus cloud fields, Part I – Parameterization for finite clouds. *Contrib. Atmos. Phys.*, **54**, 258–272.

Welch, R. and W. Zdunkowski (1981). The radiative characteristics of noninteracting cumulus cloud fields, Part II – Calculations for cloud fields. *Contrib. Atmos. Phys.*, **54**, 273–285.

Harshvardhan and J. Weinman (1982). Infrared radiative transfer through a regular array of cuboidal clouds. *J. Atmos. Sci.*, **39**, 431–439.

Harshvardhan and R. Thomas (1984). Solar reflection from interacting and shadowing cloud elements. *J. Geophys. Res.*, **89**, 7179–7185.

Welch, R.M. and B.A. Wielicki (1984). Stratocumulus cloud field reflected fluxes: The effect of cloud shape. *J. Atmos. Sci.*, **41**, 3085–3103

Preisendorfer, R.W. and G.L. Stephens (1984). Multimode radiative transfer in finite optical media, I: Fundamentals. *J. Atmos. Sci.*, **41**, 709–724.

Stephens, G.L. and R.W. Preisendorfer (1984). Multimode radiative transfer in finite optical media, II: Solutions. *J. Atmos. Sci.*, **41**, 725–735.

Joseph, J. and V. Kagan (1988). The reflection of solar radiation from bar cloud arrays. *J. Geophys. Res.*, **93**, 2405–2416.

(8) *Selected readings on the Independent Pixel/Column Approximation (IPA/ICA):* This is the simple idea of applying 1D RT to the column under every (computational) grid-point or (satellite) pixel in the horizontal projection of a 3D cloud field. We see this as a prediction for the 2D horizontal field of reflected or transmitted fluxes or of the heating rate at a given level. Some authors consider the IPA/ICA to include the next step which consists in spatially or statistically averaging this predicted field. At any rate, the now popular IPA/ICA terminology was introduced during the 1990s to describe an already common practice. In a sense, this is the default approach to radiative budget estimation in climate models as well as in remote sensing operations when unresolved variability is ignored. If reasonable assumptions are made about the unresolved variability the averaging can, at least under some circumstances, predict the domain-average quite accurately. So the IPA/ICA has become a real workhorse in contemporary 3D RT. It was mentioned in the main text only in connection with two early publications of primarily historical interest: one that appeared in the former Soviet Union (Mullamaa et al., 1972), and another in the West (Ronnholm et al., 1980), later and of course independently. In this volume alone, the IPA/ICA is used extensively in Chaps. 6, 8, 9 and 12 either as a benchmark (from which to measure "true" 3D RT effects mediated by horizontal flux divergences and convergences) or as a framework (for producing domain-average properties by accounting for the variability but ignoring the horizontal fluxes). Below is a sampler of studies where the IPA/ICA is applied (some even before the abbreviations were adopted), assessed (by comparison with more accurate 3D RT methods), and improved upon (without sacrificing efficiency).

Applications of the IPA/ICA:

Stephens, G.L. (1985). Reply (to Harshvardan and Randall). *Mon. Wea. Rev.*, **113**, 1834–1835.

Stephens, G.L., P.M. Gabriel and S.-C. Tsay (1991). Statistical radiative transport in one-dimensional media and its application to the terrestrial atmosphere. *Transp. Theory and Statis. Phys.*, **20**, 139–175.

Cahalan, R.F., W. Ridgway, W.J. Wiscombe, T.L. Bell and J.B. Snider (1994). The albedo of fractal stratocumulus clouds. *J. Atmos. Sci.*, **51**, 2434–2455.

Barker, H.W. (1996). A parameterization for computing grid-averaged solar fluxes for inhomogeneous marine boundary layer clouds – Part 1, Methodology and homogeneous biases. *J. Atmos. Sci.*, **53**, 2289–2303.

Barker, H.W., B.A. Wielicki and L. Parker (1996). A parameterization for computing grid-averaged solar fluxes for inhomogeneous marine boundary layer clouds – Part 2, Validation using satellite data. *J. Atmos. Sci.*, **53**, 2304–2316.

Oreopoulos, L. and H.W. Barker (1999). Accounting for subgrid-scale cloud variability in a multi-layer 1D solar radiative transfer algorithm. *Quart. J. Roy. Meteor. Soc.*, **125**, 301–330.

Deviations from the IPA/ICA:

Cahalan, R.F., W. Ridgway, W.J. Wiscombe, S. Gollmer and Harshvardhan (1994). Independent pixel and Monte Carlo estimates of stratocumulus albedo. *J. Atmos. Sci.*, **51**, 3776–3790.

Marshak, A., A. Davis, W.J. Wiscombe and G. Titov (1995). The verisimilitude of the independent pixel approximation used in cloud remote sensing. *Remote Sens. Environ.*, **52**, 72–78.

Marshak, A., A. Davis, W.J. Wiscombe and R.F. Cahalan (1995). Radiative smoothing in fractal clouds. *J. Geophys. Res.*, **100**, 26,247–26,261.

Chambers, L., B. Wielicki and K.F. Evans (1997). On the accuracy of the independent pixel approximation for satellite estimates of oceanic boundary layer cloud optical depth. *J. Geophys. Res.*, **102**, 1779–1794.

Davis, A., A. Marshak, R.F. Cahalan and W.J. Wiscombe (1997). The Landsat scale-break in stratocumulus as a three-dimensional radiative transfer effect, Implications for cloud remote sensing. *J. Atmos. Sci.*, **54**, 241–260.

Titov, G.A. (1998). Radiative horizontal transport and absorption in stratocumulus clouds. *J. Atmos. Sci.*, **55**, 2549–2560.

Davis, A.B. and A. Marshak (2001). Multiple scattering in clouds, Insights from three-dimensional diffusion/P_1 theory. *Nucl. Sci. Eng.*, **137**, 251–288.

Savigny, C. von, A.B. Davis, O. Funk and K. Pfeilsticker (2002). Time-series of zenith radiance and surface flux under cloudy skies: Radiative smoothing, optical thickness retrievals and large-scale stationarity. *Geophys. Res. Lett.*, **29**, 1825–1828.

Corrections to the IPA/ICA:

Gabriel, P.M. and K.F. Evans (1996). Simple radiative-transfer methods for calculating domain-averaged solar fluxes in inhomogeneous clouds. *J. Atmos. Sci.*, **53**, 858–877.

Marshak, A., A. Davis, R.F. Cahalan and W.J. Wiscombe (1998). Nonlocal independent pixel approximation: Direct and inverse problems. *IEEE Trans. Geosc. and Remote Sens.*, **36**, 192–205.

Faure, T., H. Isaka and B. Guillemet (2001). Neural network analysis of the radiative interaction between neighboring pixels in inhomogeneous clouds. *J. Geophys. Res.*, **106**, 14465–14484.

Polonsky, I.N., M.A. Box and A.B. Davis (2003). Radiative transfer through inhomogeneous turbid media: Implementation of the adjoint perturbation approach at the first-order. *J. Quant. Spectrosc. Radiat. Transfer*, **78**, 85–98.

Cornet, C., H. Isaka, B. Guillemet and F. Szczap (2004). Neural network retrieval of cloud parameters of inhomogeneous clouds from multispectral and multiscale radiance data: Feasibility study. *J. Geophys. Res.*, **109**, D12203, doi:10.1029/2003JD004186.

Finally, we note that the "adjacency" effect covered under item (3) is a 3D RT process that nonlinearly mixes surface reflectances in satellite imagery. This radiometric mixing mediated by the ambient aerosol is what defeats the clear-sky equivalent of the IPA: satellite pixels can no longer be analyzed separately to infer surface properties. Because the aerosol atmosphere is optically thin, methods used in that context are interestingly different from those favored by the cloud radiation community.

4

Numerical Methods

K.F. Evans and A. Marshak

4.1 Explicit Numerical Methods 244

4.2 Statistical Modeling or Monte Carlo Methods 261

4.3 Examples of Radiances for the 3D Cloud Fields and Comparison with SHDOM 274

References 278

Suggested Reading 280

The general form of the radiative transfer equation (RTE) cannot be solved analytically, and thus numerical methods must be applied. The solution is more difficult than for many other linear equations in physics because 1) the RTE is a boundary value problem with only partial information on each boundary (i.e., the incident radiance is known, but the outgoing radiance is not), 2) the RTE is a mixed integro-differential equation, and 3) the monochromatic RTE lives in a five dimensional space (three independent spatial variables and two angular variables). It is this high dimensionality of the underlying space that makes three-dimensional radiative transfer solutions very computer intensive. There are many numerical solution methods of the 3D RTE, but they fall into two general classes: deterministic (or explicit) and statistical (or Monte Carlo) methods. Explicit methods solve for the whole radiance field (or the source function from which radiances can be derived). The radiation field is explicitly represented in some discrete fashion, and the elements of the field are iteratively adjusted until agreement with the RTE is achieved. After the iterations the desired radiative quantities (e.g., particular radiances or fluxes) are calculated from the radiation field elements. By contrast, Monte Carlo methods estimate the desired radiative quantities statistically with some level of confidence that depends on the number of photon trajectories simulated and the variance of the estimate. In addition, Monte Carlo methods provide us naturally with the implicit "dynamics" of the radiative transfer process captured in order-of-scattering and photon pathlength decompositions.

4.1 Explicit Numerical Methods

This section discusses the general features of explicit 3D radiative transfer solution methods. The methods considered are those with arbitrary accuracy, depending on the chosen resolution, not those that are inherently approximate (covered in Chap. 6). The key choice of how to discretize the radiation field is discussed at length. The basic characteristics of two leading numerical methods are described. First, we introduce important concepts by describing the most common plane-parallel radiative transfer method.

4.1.1 Discrete Ordinates in Plane-Parallel Radiative Transfer

The unpolarized monochromatic plane-parallel solar radiative transfer equation is

$$\mu \frac{\mathrm{d}I(\tau, \mu, \phi)}{\mathrm{d}\tau} = I(\tau, \mu, \phi) - \frac{\varpi_0}{4\pi} \int_0^{2\pi} \int_{-1}^1 P(\mu, \phi; \mu', \phi') I(\tau, \mu', \phi') \mathrm{d}\mu' \mathrm{d}\phi'$$

$$- \frac{\varpi_0}{4\pi} P(\mu, \phi; -\mu_0, \phi_0 + \pi) \, F_0 e^{-\tau/\mu_0} , \qquad (4.1)$$

where $I(\tau, \mu, \phi)$ is the radiance in direction (μ, ϕ) at optical depth τ. (Here $\mu = 1$ is up and τ increases downward.) The first term on the right hand side is the sink due to attenuation, the second term is the source due to light scattered into the direction of travel, and the third term is the source of diffuse radiation. The second and third terms together are called the source function. The term on the left hand side is sometimes called the streaming term because it describes streaming (or advection) of radiation in the absence of sources or sinks.

The usual way to discretize the angular variables in plane-parallel radiative transfer is to expand in a Fourier series in azimuth angle and use discrete angles in zenith angle. Chandrasekhar (1950) introduced the idea of discrete ordinates in radiative transfer using Gaussian quadrature. The Fourier series in azimuth converts the radiance from $I(\tau, \mu, \phi)$ to $I_m(\tau, \mu)$:

$$I(\tau, \mu, \phi) = \sum_{m=0}^{M} I_m(\tau, \mu) \cos m(\phi_0 - \phi) . \qquad (4.2)$$

If the scattering phase function is only a function of the scattering angle θ_s, then the phase function can be expressed in terms of a Legendre series

$$P(\cos \theta_s) = \sum_{l=0}^{N} \chi_l \mathcal{P}_l(\cos \theta_s) , \qquad (4.3)$$

where χ_l are the $N + 1$ Legendre coefficients and $\mathcal{P}_l$ are the Legendre polynomials (see Chap. 3). Using the Legendre series representation for the phase function allows us to calculate the Fourier transform of the phase function with the addition theorem of spherical harmonics:

$$P(\mu, \phi; \mu', \phi') = P(\cos \theta_s)$$

$$= \sum_{m=0}^{N} \sum_{l=m}^{N} \chi_l \, a_{lm} \, \mathcal{P}_l^m(\mu) \mathcal{P}_l^m(\mu') \cos m(\phi' - \phi) , \tag{4.4}$$

where $\mathcal{P}_l^m(\mu')$ are associated Legendre functions and

$$a_{lm} = (2 - \delta_{0m}) \frac{(l-m)!}{(l+m)!}$$

where δ_{ij} is a Kronecker-delta symbol. Substituting the Fourier series for the radiance $I(\tau, \mu, \phi)$ and the addition theorem for the phase function into the RTE gives

$$\mu \frac{\mathrm{d}I_m(\tau, \mu)}{\mathrm{d}\tau} = I_m(\tau, \mu) - \frac{\varpi_0}{2} \sum_{l=m}^{N} a_{lm} \chi_l \mathcal{P}_l^m(\mu) \int_{-1}^{+1} \mathcal{P}_l^m(\mu') I_m(\tau, \mu') \mathrm{d}\mu'$$

$$- F_0 e^{-\tau/\mu_0} \frac{\varpi_0}{4\pi} \sum_{l=m}^{N} a_{lm} \chi_l \mathcal{P}_l^m(\mu) \mathcal{P}_l^m(-\mu_0) . \tag{4.5}$$

The sum over azimuth mode m was eliminated in the scattering integral over ϕ by the orthogonality of $\cos m(\phi - \phi')$. Thus, due to the addition theorem of spherical harmonics the Fourier azimuth modes of the RTE separate, leaving $N + 1$ separate equations ($m = 0, \ldots, N$). This decoupling of the azimuthal modes is the reason a Fourier series is used to discretize the azimuthal angle in the plane-parallel RTE.

The next task is to discretize the cosine of the zenith angle (μ) so the remaining scattering integral becomes a summation (unfortunately there is no way to make the RTE decouple in the zenith angle). Gaussian quadrature approximates an integral by a sum

$$\int_{-1}^{1} f(\mu) \mathrm{d}\mu \approx \sum_{j=1}^{N_\mu} w_j f(\mu_j) , \tag{4.6}$$

where μ_j are the *discrete ordinates* and w_j are quadrature weights. The RTE with the discrete ordinates is

$$\pm \mu_j \frac{\mathrm{d}I_{mj}^\pm(\tau)}{\mathrm{d}\tau} = I_{mj}^\pm(\tau) - \frac{\varpi_0}{2} \sum_{j'=1}^{N_\mu} w_{j'} \left[P_{mjj'}^{\pm+} I_{mj'}^+ + P_{mjj'}^{\pm-} I_{mj'}^- \right] + \mathcal{S}(\pm \mu_j) \tag{4.7}$$

where the radiance is a vector $I_{mj}^\pm(\tau) = I_m(\tau, \pm \mu_j)$, the phase function is a matrix

$$P_{mjj'}^{\pm+} = \sum_{l=m}^{N_\mu} (2 - \delta_{0m}) \frac{(l-m)!}{(l+m)!} \chi_l \mathcal{P}_l^m(\pm \mu_j) \mathcal{P}_l^m(\mu_{j'}) , \tag{4.8}$$

and $\mathcal{S}$ is the discrete source term. The pluses refer to upwelling, and the minuses to downwelling.

The plane-parallel RTE is now discretized in angle so that the radiance is a vector at each optical depth, $I^{\pm}(\tau)$, with the $+$ referring to upwelling discrete ordinates and the $-$ referring to downwelling ordinates. The RTE is then an ordinary differential matrix equation

$$M \frac{\mathrm{d}}{\mathrm{d}\tau} \begin{pmatrix} I^+ \\ I^- \end{pmatrix} = \begin{pmatrix} I^+ \\ I^- \end{pmatrix} - \begin{pmatrix} P^{++} & P^{+-} \\ P^{-+} & P^{--} \end{pmatrix} \begin{pmatrix} I^+ \\ I^- \end{pmatrix} - \begin{pmatrix} S^+ \\ S^- \end{pmatrix} , \qquad (4.9)$$

where M is a diagonal matrix with $\pm\mu_j$ entries, P is the discrete ordinate phase function matrix (which includes the $\varpi_0/2$ and weights w_j), and S is the source vector. The matrices have $N_\mu \times N_\mu$ entries.

Now we have a discrete system in the angular variables, though not in the optical depth variable. The optical depth variable could be discretized with a uniform grid and the derivative in the matrix RTE expressed with a finite difference. Then we would have an algebraic system to solve with radiance vectors $I_m(\tau_k, \mu_j)$ of length $N_\tau \times N_\mu$ for each Fourier azimuthal mode m. This system could be solved, for example, by Gauss-Seidel iteration.

The more commonly used methods are the doubling-adding method (Grant and Hunt, 1969) and the eigenmatrix method (Stamnes et al., 1988; Thomas and Stamnes, 1999). Both start with the matrix differential equation and hence give identical solutions even though they have very different solution methods. The doubling-adding method uses the linearity of the RTE to express the radiance vectors outgoing from a layer (I_0^+ and I_1^-) in terms of the incident radiance vectors (I_0^- and I_1^+) with the so-called interaction principle:

$$I_0^+ = T^+ I_1^+ + R^+ I_0^- + S^+ \qquad I_1^- = T^- I_0^- + R^- I_1^+ + S^- \qquad (4.10)$$

where T is the transmission matrix, R is the reflection matrix, and S is the source vector (e.g., solar source). The transmission and reflection matrices and source vectors are related simply to the terms in the matrix RTE for an infinitesimally optically thin sublayer. The "doubling" formula, derived from the interaction principle, are then used to compute the T, R, and S properties for a thick homogeneous layer with a procedure that doubles the optical depth in each step. Similiar "adding" formulas are used to combine layers with different properties. The interaction principle can then be used to find the desired outgoing radiance.

The eigenmatrix method (e.g., DISORT – a DIScrete Ordinates Radiative Transfer code) solves the differential matrix RTE using standard linear algebra techniques. The homogeneous matrix RTE (i.e., with $S^{\pm} = 0$) is turned into an eigenvalue problem by seeking solutions of the form $I^{\pm} = G^{\pm} e^{-k\tau}$. There are N_μ pairs of eigenvalues $\pm k$ and corresponding eigenvectors $G^{\pm}$. The general solution for a single homogeneous layer is

$$I^{\pm} = \sum_{j=1}^{N_\mu} C_{-j} G_{-j}^{\pm} e^{k_j \tau} + \sum_{j=1}^{N_\mu} C_j G_j^{\pm} e^{-k_j \tau} + Z_0^{\pm} e^{-\tau/\mu_0} . \qquad (4.11)$$

The particular solution is found by substituting $Z_0^{\pm} e^{-\tau/\mu_0}$ in the matrix RTE and solving the resulting matrix equation for $Z_0^{\pm}$. The $2N_\mu$ constants $C_{\pm j}$ are found from the boundary conditions (the two radiance vectors incident on the layer) after they have been rescaled to avoid numerical ill-conditioning. The number of floating point operations for a single layer increases as $N_\mu^3 \times M$ for both the doubling-adding and eigenmatrix methods, though the eigenmatrix method has fewer operations.

This section has illustrated for the plane-parallel system that there are two fundamental aspects of an explicit method: first, the discretization of the radiation field, and second, the numerical solution method that solves the RTE in that discrete representation. Even essentially exact numerical solution methods such as the doubling-adding and eigenmatrix methods only solve the original RTE approximately because the numerical system they solve is a discrete representation of the radiance field. Their accuracy relative to true solution depends on the angular resolution as determined by the the number of quadrature zenith angles and Fourier azimuthal modes.

4.1.2 Discretization of the Radiation Field

The last section showed the two main methods by which the radiance field can be discretized: spectrally and with discrete values. For the plane-parallel system the azimuthal angle was represented spectrally with a truncated Fourier series, while the zenith angle was represented with discrete ordinates. In three-dimensional radiative transfer the spatial dimensions cannot be solved analytically, as the optical depth is in plane-parallel transfer, and so they must be discretized as well. Thus there are two angular dimensions (μ and ϕ) and three spatial dimensions (x, y, and z) to represent in the computer in one of two basic ways. We will now show the implications of the choice of discretization method for the angular and spatial variables.

The monochromatic 3D RTE with no internal sources is

$$\sin\theta\cos\phi\frac{\partial I(\boldsymbol{x},\boldsymbol{\Omega})}{\partial x} + \sin\theta\sin\phi\frac{\partial I(\boldsymbol{x},\boldsymbol{\Omega})}{\partial y} + \cos\theta\frac{\partial I(\boldsymbol{x},\boldsymbol{\Omega})}{\partial z}$$
$$= \sigma(\boldsymbol{x})\left[-I(\boldsymbol{x},\boldsymbol{\Omega}) + \frac{\varpi_0(\boldsymbol{x})}{4\pi}\int P(\boldsymbol{x},\boldsymbol{\Omega}\cdot\boldsymbol{\Omega}')\,I(\boldsymbol{x},\boldsymbol{\Omega}')\,d\boldsymbol{\Omega}'\right] , \quad (4.12)$$

where $\boldsymbol{\Omega}$ is the direction cosine vector with components $(\Omega_x, \Omega_y, \Omega_z) = (\sin\theta\cos\phi, \sin\theta\sin\phi, \cos\theta)$. The difference from the plane-parallel equation is the addition of the first two streaming terms on the left hand side. These terms depend on the azimuthal angle (with $\cos\phi$ or $\sin\phi$), and so a Fourier series in ϕ no longer results in separate equations for each azimuthal mode m. Instead the equation for the mth mode couples to the $m-1$ and $m+1$ modes.

Discrete Ordinates

The simplest way to discretize the angular variables is to use discrete ordinates for both zenith and azimuth angles. To keep the notation general we will express the discrete ordinates in terms of direction cosines, $\boldsymbol{\Omega}_n$ and leave for later how to choose the particular directions. Associated with each discrete direction is an integration weight, w_n, for the scattering integral. The discrete ordinate RTE for the radiance I_n in the nth ordinate direction is then

$$\boldsymbol{\Omega}_n \cdot \nabla I_n(\boldsymbol{x}) = \sigma(\boldsymbol{x}) \left[-I_n(\boldsymbol{x}) + \frac{\varpi_0(\boldsymbol{x})}{4\pi} \sum_{n'=1}^{N_\Omega} w_{n'} P(\boldsymbol{x}, \boldsymbol{\Omega}_n \cdot \boldsymbol{\Omega}_{n'}) I_{n'}(\boldsymbol{x}) \right] . \quad (4.13)$$

Note that the computation of the scattering integral takes a few times N_Ω^2 floating point operations, since there are N_Ω incident and outgoing discrete ordinates. The streaming term is very simple in discrete ordinates, being simply multiplication of the radiance gradient by the direction cosine vector. The boundary conditions for discrete ordinates are quite straightforward; for example, at a nonreflecting boundary the radiance is simply specified for the incident discrete ordinates, $I(\boldsymbol{x}_{\text{bnd}}, \boldsymbol{\Omega}_n)$, where $\boldsymbol{\Omega}_n \cdot \boldsymbol{n} < 0$, $\boldsymbol{n}$ being the outward normal.

Spherical Harmonics

The appropriate spectral representation for the angular variables is an expansion in spherical harmonics. The spherical harmonic radiance coefficients I_{lm}^+ and I_{lm}^- are defined by

$$I(\boldsymbol{x}, \mu, \phi) = \sum_{m=0}^{M} \sum_{l=m}^{L, L+m} \left[I_{lm}^+(\boldsymbol{x}) Y_{lm}^+(\mu, \phi) + I_{lm}^-(\boldsymbol{x}) Y_{lm}^-(\mu, \phi) \right] , \quad (4.14)$$

where m is the azimuthal mode, l is the meridional mode, and $Y_{lm}^\pm(\mu, \phi)$ are orthonormal spherical harmonic functions. The classical spherical harmonic functions are defined by

$$Y_{lm}^+(\mu, \phi) = \gamma_{lm} \mathcal{P}_l^m(\mu) \cos(m\phi) \qquad Y_{lm}^-(\mu, \phi) = \gamma_{lm} \mathcal{P}_l^m(\mu) \sin(m\phi) \quad (4.15)$$

where $\mathcal{P}_l^m(\mu)$ are associated Legendre functions and

$$\gamma_{lm} = \left(\frac{2l+1}{2\pi(1+\delta_{0m})} \frac{(l-m)!}{(l+m)!} \right)^{1/2} .$$

The series truncation is called "triangular" if the l sum goes to L and "parallelogram" if it goes to $L+m$ ("rhomboidal" for $M = L$). For two-dimensional media with variations only in x and z, if the solar azimuthal angle is chosen to be $\phi = 0$ or $\phi = \pi$, then only the cosine ϕ terms are needed (i.e., $I_{lm}^- = 0$), so the number of terms is reduced by about half.

The angular functions in the streaming terms of the RTE (4.12) result in coupling between spherical harmonic terms. For example, multiplying I_{lm}^+ by $\cos\theta$ results in $I_{l-1,m}^+$ and $I_{l+1,m}^+$ terms. This makes the streaming terms much more complicated than in the discrete ordinate approach. The scattering integral, however, is greatly simplified as long as the phase function depends only on the scattering angle. Using the addition theorem of spherical harmonics and the orthogonality of spherical harmonics, the scattering integral transforms to

$$\int_0^{2\pi}\int_{-1}^1 \left[\frac{1}{4\pi}\int_0^{2\pi}\int_{-1}^1 P(\mu,\phi,\mu',\phi')\,I(\mu',\phi')\,\mathrm{d}\mu'\mathrm{d}\phi'\right] Y_{lm}^{\pm}(\mu,\phi)\mathrm{d}\mu\mathrm{d}\phi$$

$$=\frac{\chi_l}{2l+1}I_{lm}^{\pm}\,. \tag{4.16}$$

In spherical harmonics space the scattering integral is simply multiplication by the Legendre phase function coefficient, rather than a summation! The 3D RTE in spherical harmonics is then

$$a_{lm}^{--}\frac{\partial I_{l-1,m-1}^{\pm}}{\partial x}+a_{lm}^{+-}\frac{\partial I_{l+1,m-1}^{\pm}}{\partial x}+a_{lm}^{-+}\frac{\partial I_{l-1,m+1}^{\pm}}{\partial x}+a_{lm}^{++}\frac{\partial I_{l+1,m+1}^{\pm}}{\partial x}$$

$$+\,b_{lm}^{--}\frac{\partial I_{l-1,m-1}^{\mp}}{\partial y}+b_{lm}^{+-}\frac{\partial I_{l+1,m-1}^{\mp}}{\partial y}+b_{lm}^{-+}\frac{\partial I_{l-1,m+1}^{\mp}}{\partial y}+b_{lm}^{++}\frac{\partial I_{l+1,m+1}^{\mp}}{\partial y}$$

$$+\,c_{lm}^{-}\frac{\partial I_{l-1,m}^{\pm}}{\partial z}+c_{lm}^{+}\frac{\partial I_{l+1,m}^{\pm}}{\partial z}=-\sigma\left[1-\frac{\varpi_0\chi_l}{2l+1}\right]I_{lm}^{\pm}\,, \tag{4.17}$$

where the x dependence of the radiance and optical properties is understood. The values of the a, b, and c coefficients can be obtained from recurrence relations for spherical harmonics (the ones that refer to radiance terms $I_{lm}^{\pm}$ beyond the l and m truncation limits are set to zero). There are only two c coefficients because the $\partial I/\partial z$ streaming terms are independent of the azimuth angle ϕ and hence there is no coupling to adjacent m terms.

The boundary conditions for spherical harmonics are more complicated than for discrete ordinates because each spherical harmonic term involves both upwelling and downwelling radiation. Furthermore, the truncated spherical harmonic series can take on specified values only at a relatively small number of directions, not the whole incident hemisphere. There are two approaches to applying boundary conditions, both involving linear constraint equations on the spherical harmonic terms ($I_{lm}^{\pm}$) at the boundaries. The first is to apply the boundary conditions at a set of specified angles, Ω_j, using (4.14). The second approach, due to Marshak (1947), constrains the odd hemispheric moments of the radiance field at the top and bottom boundaries. In 3D radiative transfer this generalizes to

$$\int_{\Omega\bullet n<0} \left[I(x_{\mathrm{bnd}},\Omega)-I_{\mathrm{bnd}}(x_{\mathrm{bnd}},\Omega)\right]\,Y_{lm}^{+}(\Omega)\,\mathrm{d}\Omega=0\,, \tag{4.18}$$

for $l-m$ odd, where I_{bnd} is the specified incident radiance at the boundaries. Substituting in (4.14) for $I(x_{\mathrm{bnd}},\Omega)$ and using Gaussian quadrature for the integration over μ results in a set of linear equations constraining the $I_{lm}^{\pm}(x_{\mathrm{bnd}})$.

Comparison of Discrete Ordinates and Spherical Harmonics

We can now compare, from a computational point of view, the discrete ordinate and spherical harmonic approaches to discretizing the angular aspects of the RTE. In both cases the radiance field at a point is represented by a finite length vector, I_n for discrete ordinates and $I_{lm}^{\pm}$ for spherical harmonics. For the same angular resolution, the lengths of these vectors are comparable. The RTE can be written as a matrix partial differential equation

$$M_x \frac{\partial I(x)}{\partial x} + M_y \frac{\partial I(x)}{\partial y} + M_z \frac{\partial I(x)}{\partial z} = C(x)I(x) + S(x) , \qquad (4.19)$$

with the streaming term on the left hand side, the extinction and scattering integral expressed with the C matrix, and an internal source expressed with the S vector. In the spherical harmonic basis the scattering matrix C is diagonal (4.17), while in the discrete ordinate basis it is a full matrix (4.13). This means that the amount of computation for the scattering integral is much higher in discrete ordinates than in spherical harmonics, especially when using high angular resolution (a long radiance vector I). In discrete ordinates the streaming matrices $M_{x,y,z}$ are diagonal, while in spherical harmonics they are not diagonal, but still sparse, with at most four off-diagonal lines. One could say that when scattering is the dominant process the spherical harmonic basis is most natural, and if streaming is the dominant process then discrete ordinates is the natural basis.

Fourier Spatial Discretization

Now consider the discretization of the spatial variables, which we restrict to the Cartesian coordinates, x, y, and z. There are three basic approaches to spatial discretization: spectral, discrete grid, and finite element. First let us investigate the spectral approach. Periodic horizontal boundary conditions are often chosen, so one might consider Fourier transforming x and y. The horizontal part of the radiance vector field can be represented in a Fourier series by

$$I(x, y, z) = \sum_{k_x=0}^{N_x-1} \sum_{k_y=0}^{N_y-1} \hat{I}_{k_x,k_y}(z) e^{2\pi i (x k_x/L_x + y k_y/L_y)} , \qquad (4.20)$$

where $L_x \times L_y$ is domain size. Expressing (4.19) in Fourier series results in the horizontal spectral RTE

$$\frac{2\pi i k_x}{L_x} M_x \hat{I}_{k_x,k_y} + \frac{2\pi i k_y}{L_y} M_y \hat{I}_{k_x,k_y} + M_z \frac{d\hat{I}_{k_x,k_y}}{dz}$$

$$= \sum_{k_x'=0}^{N_x-1} \sum_{k_y'=0}^{N_y-1} \hat{C}_{k_x-k_x',k_y-k_y'} \hat{I}_{k_x',k_y'} + \hat{S}_{k_x,k_y} , \qquad (4.21)$$

where $\hat{C}_{k_x,k_y}(z)$ is the Fourier transform of the optical properties matrix $C(x)$ and $\hat{S}_{k_x,k_y}(z)$ is the Fourier transform of the source vector. The streaming terms have been simplified by transforming the horizontal derivatives into simple multiplication.

The horizontal spectral discretization has converted the 3D RTE into an ordinary differential equation that can be solved with methods similiar to the plane-parallel ones (e.g., the eigenmatrix method). This apparent simplification comes at a tremendous cost. The multiplication by optical properties $(C(x)I(x))$ transforms into a convolution. If the convolution is performed explicitly then the computational burden is enormous. For example, for a domain with $N_x = N_y = 100$ the convolution requires 10^8 matrix multiplications (remember that C is a matrix and I a vector whose length depends on the angular discretization). The convolution could be performed with fast Fourier transforms (i.e., transforming back to x and y to perform the multiplications), but this is still much more computation than non-spectral methods.

Spatial Grid

The most common and straightforward way to discretize the RTE is with a spatial grid. The spatial derivatives in the streaming terms are then computed with some form of finite differencing. First order and second order finite differencing schemes are common. A simple second order scheme for a uniform grid is

$$M_x \frac{I_{i+1,jk} - I_{i-1,jk}}{2\Delta x} + M_y \frac{I_{i,j+1,k} - I_{i,j-1,k}}{2\Delta y} + M_z \frac{I_{ij,k+1} - I_{ij,k-1}}{2\Delta z}$$
$$= C_{ijk} I_{ijk} + S_{ijk} , \tag{4.22}$$

where I_{ijk} is the radiance vector at grid point $(x = i\Delta x, y = j\Delta y, z = k\Delta z)$. This particular finite difference scheme has difficulties at the top and bottom boundaries where z_{k+1} and z_{k-1} don't exist, so other schemes are used in practice (e.g., see Sect. 4.1.3). Higher order finite difference schemes are probably not warranted given the errors of the angular discretization. Discrete ordinate approaches can use the integral formulation of the RTE, and then there are no derivatives to finite difference. However, the source function and grid cell boundary radiances still must be interpolated across the grid cells, and these approximations can be thought of as a type of finite differencing scheme. The grid approach to spatial discretization is computationally efficient because the right hand side of (4.22) is simply multiplication and the streaming terms couples only adjacent grid points.

Finite element methods use local basis functions, with nonzero values only inside small non-overlapping volumes with simple shapes (e.g., tetrahedral). The "trial" functions vary smoothly within the volumes (e.g., piecewise linearly). The spatial discretization of the radiance field is represented by the coefficients of the local basis functions. For example, these coefficients could be the values of the radiance at the four vertices (nodes) of the tetrahedron while the basis functions vary linearly from one to zero along the tetradedron edges. The derivatives in the RTE streaming term are readily expressed in terms of the finite element coefficients. Since the basis functions are local, the finite element representation of the RTE is sparse, similiar to the finite difference grid approach. The finite element method allows considerable flexibility for complex boundary geometries encountered in engineering computations.

Examples

We now give some examples in the literature of the different discretization approaches that have been implemented in multi-dimensional atmospheric radiative transfer by describing some methods. Several methods have been developed that use Fourier transforms in x and y and discrete ordinates in μ to reduce the RTE to an ordinary differential equation in z. The adding-doubling technique was the solution method in Martonchik and Diner (1985) and Stephens (1988), while a Ricatti transformation of the RTE to an initial value problem was performed in Gabriel et al. (1993). These methods have only been computationally feasible for small 2D (x and z) problems for the reasons discussed above. Horizontal Fourier transforms are practical for the related radiative transfer problem of a horizontally homogeneous atmosphere above an inhomogeneous surface (e.g., Lyapustin and Muldashev, 2001).

Many methods use discrete ordinates in μ and ϕ and a spatial grid for x, y, and z. The solution methods almost invariably involve iterating the radiance field by calculating the new discrete ordinate radiances from the source function calculated with the previous iteration's radiances. The various solution methods are called Λ iteration (Stenholm et al., 1991), successive order-of-scattering (Liou and Rao, 1996), or Picard iteration (Kuo et al., 1996). The widely used Discrete Ordinate Method (DOM) or S_N method is discussed in Sect. 4.1.3.

The spherical harmonic spatial grid (SHSG) method (Evans, 1993) solves solar and thermal atmospheric radiative transfer problems for two-dimensional media. The angular part of the radiance field is discretized in spherical harmonics and the spatial part with a grid. Centered finite differencing is used in x and two-point trapezoidal differencing is used in z; both are second order accurate. Marshak boundary conditions are implemented. A conjugate gradient method is used to iteratively solve the coupled linear equations from the discretized RTE. The number of computational operations is proportional to the product of the number of grid points, the number of spherical harmonic terms, and the number of iterations. The number of iterations is typically a few hundred, but increases with the area of low extinction values (clear sky) as this expands the eigenvalue range of the matrix operator. Sharp extinction transitions (cloud edges) also cause spurious oscillations in the radiation field. These problems with a purely spherical harmonic approach led to the development of the spherical harmonic discrete ordinate method (SHDOM). SHDOM, which is described in Sect. 4.1.4, uses a spatial grid and both types of angular discretizations during the solution process.

More complex spherical harmonic approaches have been developed in the neutron transport community. For example, the EVENT model (de Oliveira, 1986), transforms the RTE to a second-order transport equation similiar to the diffusion equation (cf. Chap. 5) but for higher order angular structure expressed with an even parity spherical harmonic series. The EVENT model uses finite elements based on tetrahedral volumes. Standard sparse matrix solution methods are used to solve the finite element equations.

How the partial-differential/integral RTE is discretized to represent the radiation field in a computer is the key to understanding the computational efficiency of various methods. A typical 3D problem with modest angular resolution can easily take 40 million elements to represent the radiance field, e.g., $100 \times 100 \times 50$ grid points (see numerical example at the end of this chapter) with 80 discrete ordinates at each point. The linear systems resulting from discretizing the RTE must therefore be very sparse (in matrix language) so that the number of operations per iteration of the solver is manageable. Computationally viable discretization approaches all use a grid, or its close cousin finite elements, to represent the spatial part of the radiance field. Either discrete ordinates or a spherical harmonic representation may be used efficiently for the angular part of the radiance field. Spherical harmonics is better at high angular resolution, however, because the scattering integral takes of order N operations instead of N^2 for discrete ordinates (where N is the number of angular elements).

4.1.3 Discrete Ordinate Method (DOM)

The three-dimensional discrete ordinate or S_N method was first developed in the neutron transport community (Lathrop, 1966; Carlson and Lathrop, 1968). The S_N method has also been adopted and developed by the heat transfer community (Fiveland, 1988; Truelove, 1988; Modest, 1993). Its use in atmospheric radiation has been much more limited, mainly to practitioners that have crossed over from the neutron transport (Gerstl and Zardecki, 1985) and heat transport (Sanchez et al., 1994) communities. The standard DOM technique has been modified to handle optically thick cloudy cells for visualization purposes (Tofsted and O'Brien, 1998) and has been extended to solve the 3D polarized RTE and applied to microwave remote sensing of precipitation (Haferman et al., 1997).

The quadrature schemes for the discrete ordinate RTE (4.13) are usually chosen to satisfy certain properties. The directions Ω_n are chosen 1) to be symmetric so they are invariant under rotation by $90°$, 2) to exactly integrate the zeroth, first, and second moments,

$$\sum_{n=1}^{N_\Omega} w_n = 4\pi \quad \sum_{n=1}^{N_\Omega} w_n \Omega_n = 0 \quad \sum_{n=1}^{N_\Omega} w_n \Omega_n \cdot \Omega_n = \frac{4\pi}{3} \,, \qquad (4.23)$$

and 3) to exactly integrate the first moment over the x, y, and z half ranges,

$$\sum_{\Omega_x > 0} w_n \Omega_{nx} = \pi \quad \sum_{\Omega_y > 0} w_n \Omega_{ny} = \pi \quad \sum_{\Omega_z > 0} w_n \Omega_{nz} = \pi \,. \qquad (4.24)$$

A quadrature set for the S_N approximation has $N(N + 2)$ discrete ordinates over all eight octants. Table 4.1 gives the discrete ordinates in one octant for one possible quadrature set satifying the conditions above. The discrete ordinate directions for all octants are obtained from those in the table by choosing all eight possible positive or negative combinations, i.e., $(\pm\Omega_x, \pm\Omega_y, \pm\Omega_z)$.

The finite differencing of the RTE (4.13) is usually done with the concept of a rectangular volume element $\Delta x \times \Delta y \times \Delta z$ (Modest, 1993). The radiance along the

Table 4.1. Quadrature Sets, after Modest (1993)

Order	Ordinates			Weights
	Ω_x	Ω_y	Ω_z	w
S_4	0.2958759	0.2958759	0.9082483	0.5235987
	0.2958759	0.9082483	0.2958759	0.5235987
	0.9082483	0.2958759	0.2958759	0.5235987
S_6	0.1838670	0.1838670	0.9656013	0.1609517
	0.1838670	0.6950514	0.6950514	0.3626469
	0.1838670	0.9656013	0.1838670	0.1609517
	0.6950514	0.1838670	0.6950514	0.3626469
	0.6950514	0.6950514	0.1838670	0.3626469
	0.9656013	0.1838670	0.1838670	0.1609517
S_8	0.1422555	0.1422555	0.9795543	0.1712359
	0.1422555	0.5773503	0.8040087	0.0992284
	0.1422555	0.8040087	0.5773503	0.0992284
	0.1422555	0.9795543	0.1422555	0.1712359
	0.5773503	0.1422555	0.8040087	0.0992284
	0.5773503	0.5773503	0.5773503	0.4617179
	0.5773503	0.8040087	0.1422555	0.0992284
	0.8040087	0.1422555	0.5773503	0.0992284
	0.8040087	0.5773503	0.1422555	0.0992284
	0.9795543	0.1422555	0.1422555	0.1712359

nth discrete ordinate at the center P of the volume is denoted by I_{Pn}. The radiances at the faces of the volume are $I_{Tn}, I_{Bn}, I_{En}, I_{Wn}, I_{Nn}, I_{Sn}$, for the top $(+z)$, bottom $(-z)$, east $(+x)$, west $(-x)$, north $(+y)$, and south $(-y)$ faces. With this notation the finite difference discrete ordinate RTE is

$$\Omega_x \frac{I_{En} - I_{Wn}}{\Delta x} + \Omega_y \frac{I_{Nn} - I_{Sn}}{\Delta y} + \Omega_z \frac{I_{Tn} - I_{Bn}}{\Delta z} = \sigma(x_P)[-I_{Pn} + S_{Pn}] , \quad (4.25)$$

with

$$S_{Pn} = \frac{\varpi_0(x_P)}{4\pi} \sum_{n'=1}^{N_\Omega} w_{n'} P(x_P, \Omega_n \cdot \Omega_{n'}) I_{Pn'} + S_{Pn} , \quad (4.26)$$

where S_{Pn} is the source function at the center point x_P in direction n, and S_{Pn} is the value of any internal sources (e.g., single scattered solar source or thermal emission). Obviously, the radiances at the common faces of adjacent volumes are equal (e.g., I_{Tn} of the current cell is the same as I_{Bn} for the cell above). The volume element center radiance and the face radiances are related by

$$I_{Pn} = \gamma I_{Tn} + (1-\gamma) I_{Bn} = \gamma I_{En} + (1-\gamma) I_{Wn} = \gamma I_{Nn} + (1-\gamma) I_{Sn} , \quad (4.27)$$

which is known as "weighted diamond differencing" (Carlson and Lathrop, 1968). Often $\gamma = 1/2$ is chosen. The updating scheme may go unstable for $\gamma = 1/2$, while the less accurate choice of $\gamma = 1$ is apparently always stable.

It is standard practice in neutron transport (e.g., Gerstl and Zardecki, 1985) to compute the scattering integral summation in (4.26) by transforming the discrete ordinate radiances to spherical harmonics to take advantage of (4.16), and then transforming back to discrete ordinates. This is the method used in SHDOM, described in section 4.1.4, and results in the scattering integral summation taking of order $N_\Omega^{3/2}$ rather than N_Ω^2 operations.

The solution method for the finite difference equations is iteration, first calculating the volume center radiances I_{Pn} from the source function and face radiances using (4.25) and (4.27), and then calculating the cell center source function from the cell center radiances using (4.26). In order to update the volume center radiance, the radiances at the upstream faces must be known. Thus a specific "sweeping" pattern is employed for each octant of discrete ordinates. The cell updating starts at the appropriate boundary corner, where the inward pointing discrete ordinate radiances are known from the boundary conditions, and progresses to adjacent cells that are downstream according to the discrete ordinate direction. Let $I_{i,n}^{(x)}$, $I_{i,n}^{(y)}$, and $I_{i,n}^{(z)}$ be the incident (subscript 'i') radiances for direction n on the upstream faces in the x, y, and z directions, respectively. Then combining (4.25) and (4.27) and solving for the volume center radiance I_{Pn} gives the expression for updating the volume center radiance:

$$I_{Pn} = \frac{\gamma\varpi_P S_{Pn} + \dfrac{I_{i,n}^{(x)}|\Omega_x|}{\Delta x} + \dfrac{I_{i,n}^{(y)}|\Omega_y|}{\Delta y} + \dfrac{I_{i,n}^{(z)}|\Omega_z|}{\Delta z}}{\gamma\varpi_P + \dfrac{|\Omega_x|}{\Delta x} + \dfrac{|\Omega_y|}{\Delta y} + \dfrac{|\Omega_z|}{\Delta z}} . \tag{4.28}$$

The outgoing (subscript 'o') radiances on the downstream faces are then obtained by extrapolating the upstream and center radiances using (4.27), e.g., for the downstream x face

$$I_{o,n}^{(x)} = \frac{I_{Pn} - (1 - \gamma)I_{i,n}^{(x)}}{\gamma} . \tag{4.29}$$

A schematic diagram (Fig. 4.1) shows the sweeping order for updating the cells. Discrete ordinates with both Ω_x and Ω_z positive start each iteration in the lower left corner where the radiances on the west and bottom faces are presumed to be known from the boundary conditions. After updating the volume center radiance I_{Pn} for cell 1 and extrapolating to get the east and top face radiances, the algorithm moves on to cell 2. After all the cells touching the lower boundary are updated then the sweeping continues with the row of cells above, and so on. The updating proceeds similarly for discrete ordinates in other quadrants, starting with one of the other three corners. The generalization of the sweeping order to three dimensions should be clear. If the boundary conditions are periodic in the horizontal then the radiances on the x and y faces from previous iterations may be used for the starting cell. After the radiance updating is finished using (4.28) and (4.29) for all the discrete ordinates, the source function in each cell is updated from the discrete ordinate radiances with (4.26).

The DOM iterations given in (4.28) and (4.29) can lead to negative face radiances for large cells. Most codes will correct these unphysical negative radiances, but this effect can be a source of instability. The number negative radiances is minimized,

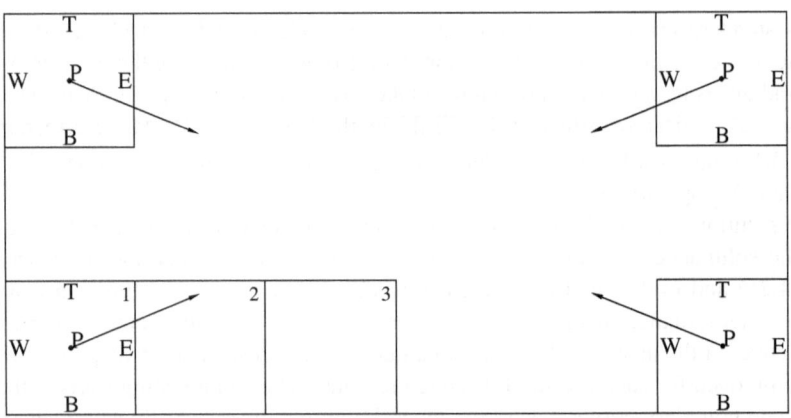

Fig. 4.1. Depiction of discrete ordinate volume elements showing the top (T), bottom (B), east (E), and west (W) faces surrounding the volume center point P. The four starting cells for sweeping with discrete ordinates in the four quadrants are shown

and the accuracy improved, if the grid cell size is kept within (Fiveland, 1988)

$$\Delta x < \frac{|\Omega_x|_{\min}}{\sigma(1-\gamma)}, \quad \Delta y < \frac{|\Omega_y|_{\min}}{\sigma(1-\gamma)}, \quad \Delta z < \frac{|\Omega_z|_{\min}}{\sigma(1-\gamma)}, \quad (4.30)$$

and therefore higher S_N approximations require finer meshes.

The source iteration method, i.e., iterating between the discrete ordinate radiances and the source function, is akin to a successive order of scattering method. Thus, the convergence is slow for large optical depths and single-scattering albedos near unity. Various iteration acceleration methods have been developed in the neutron transport literature such as diffusion synthetic acceleration (Larsen, 1982) and more recently transport synthetic acceleration (Ramone et al., 1997). There are also more advanced solution methods which can improve convergence dramatically (Balsara, 2001). The iterations proceed until a convergence criterion is reached, such as the largest fractional change in discrete ordinate radiances less than a specified value.

After the iterations are done, the desired radiative quantities may be computed from the discrete ordinate radiances. Hemispheric or actinic fluxes are calculated using quadrature integration of the discrete radiances. Radiances in a specified direction may be computed accurately by using the discrete ordinate radiances to calculate the source function in that direction and then using the integral form of the RTE to compute the outgoing radiance.

One serious problem with the discrete ordinate method is the "ray effect." The streaming of photons can occur only along the discrete ordinates. Thus radiation streaming away from localized sources through clear sky does so along discrete rays, and between the rays there is less light. The ray effect tends to produce spatial oscillations with some locations receiving more light from strong sources along the discrete ordinates, while other locations are between the ordinates and received less light. Obviously, the ray effect is reduced by having more ordinates (higher S_N

approximation). Scattering reduces ray effects as the radiation is redirected to new directions. The finite grid resolution acts to diffuse radiation because the radiation is considered constant on each cell face, so a coarse grid (perhaps paradoxically) reduces ray effects. Due to its oscillating nature the ray effect is much less apparent in integrals of the radiance field, so computing radiances by integrating source function diminishes the ray effect substantially.

4.1.4 Spherical Harmonics Discrete Ordinate Method (SHDOM)

SHDOM (Evans, 1998) is the most widely used explicit multi-dimensional radiative transfer model in the atmospheric sciences. This is probably because it is highly efficient, very flexible for atmospheric radiation problems, and publicly available.

SHDOM uses both the spherical harmonic and discrete ordinate representations of the angular aspects of the radiation field during different phases of the solution procedure. The spatial part of the fields is represented with a grid. Discrete ordinates are chosen because they more physically model the streaming of radiation through space. Spherical harmonics are chosen because they are more efficient for computing the scattering integral in the RTE and can more compactly represent the radiation field. Rather than storing radiances, SHDOM stores the source function with a spherical harmonic series at each grid point. The radiance field can be obtained easily from the source function with the integral form of the RTE. The length of the spherical harmonic series is adaptive, meaning that the truncation level varies from grid point to grid point. Only a few terms of the series are needed if the grid point has no scattering, the scattering is smooth (e.g., isotropic or Rayleigh), or the radiance field is smooth (e.g., inside optically thick scattering media). This reduction of memory usage is critical for 3D problems that take tens to hundreds of millions of numbers to represent the radiance field.

The SHDOM solution method is source iteration, which is the usual approach for the discrete ordinate (S_N) method. Each SHDOM iteration consists of four steps in which

1. the source function is transformed from spherical harmonics to discrete ordinates,
2. the source function is integrated in the RTE to obtain the radiances along discrete ordinates,
3. the radiance field is transformed to spherical harmonics, and
4. the source function in spherical harmonics is computed from the radiance field.

A flowchart illustrating the major components of the algorithm is shown in Fig. 4.2. To speed convergence for optically thick, conservative scattering media, a sequence acceleration based on geometrical convergence of the source function pattern is performed every other iteration.

Section 4.1.2 showed that the source function in spherical harmonic space could be computed in order N operations (where N is the number of spherical harmonic terms). One would think, however, that the transforms between spherical harmonic and discrete ordinate representations would negate this advantage over DOM. The explicit form of the transforms given below illustrates how the azimuthal and zenith

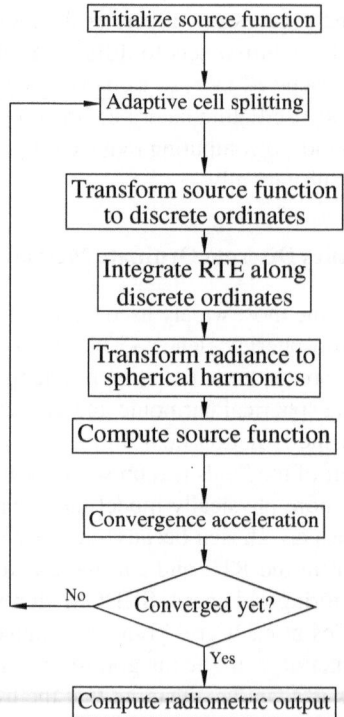

Fig. 4.2. Flowchart of the SHDOM algorithm: the 4-step iteration at the core of the algorithm is in large letters

angle parts partially separate, leading to a substantial reduction in the number of operations. The source function in spherical harmonics, $S_{lm}^{\pm}$, is transformed to discrete ordinates, $S(\mu_j, \phi_k)$, by

$$S_{jk} = \sum_{m=0}^{M} \cos(m\phi_k) \sum_{l=m}^{L} \gamma_{lm} P_l^m(\mu_j) S_{lm}^{+} + \sum_{m=1}^{M} \sin(m\phi_k) \sum_{l=m}^{L} \gamma_{lm} P_l^m(\mu_j) S_{lm}^{-},$$

(4.31)

using (4.14) and (4.15). The μ_j are obtained from Gaussian quadrature, while the ϕ_k are equally spaced, but the number at each μ_j ($N_{\phi,j}$) is reduced for $|\mu_j|$ near 1. This discrete ordinate set assures orthogonality of the spherical harmonic functions for the truncation $L = N_{\mu} - 1$ and $M = N_{\phi}/2 - 1$. The discrete ordinate radiance at each grid point is transformed to spherical harmonic space according to

$$I_{lm}^{\pm} = \sum_{j=1}^{N_{\mu}} w_j \gamma_{lm} P_l^m(\mu_j) \sum_{k=1}^{N_{\phi,j}} \hat{w}_{jk} \begin{bmatrix} \cos(m\phi_k) \\ \sin(m\phi_k) \end{bmatrix} I_{jk}, $$

(4.32)

where w_j are the Gauss-Legendre quadrature weights and $\hat{w}_{jk}$ are the azimuthal integration weights normalized appropriately. The azimuthal functions (e.g.,

$\cos(m\phi_k)$) do not depend on l, and therefore the Fourier transforms in ϕ can be performed separately. The associated Legendre functions do couple l and m, so the major computational cost is in the sum involving $\mathcal{P}_l^m(\mu_j)$. For more than about 12 azimuthal angles SHDOM uses an FFT for the azimuthal Fourier transform. If there are N discrete ordinates then the number of floating point operations for both transforms together is approximately $9N^{3/2}$, and less when using the azimuthal FFT (asymptotically $\sim 3N^{3/2}$). This compares with at least $2N^2$ operations for calculating the source function in discrete ordinates.

The second step in each iteration is to integrate the source function, S, in the formal solution of the RTE,

$$I(s) = \exp\left[-\int_0^s \sigma(s')\mathrm{d}s'\right] I(0) + \int_0^s \exp\left[-\int_{s'}^s \sigma(t)\mathrm{d}t\right] S(s')\sigma(s')\mathrm{d}s' , \quad (4.33)$$

to obtain the radiance I at each grid point. The extinction and the product $S\sigma$ are assumed to vary linearly across a grid cell. The source function is integrated backwards from a grid point along the discrete ordinate to an opposite cell face. The initial radiance, $I(0)$, and the $S\sigma$ product are bilinearly interpolated from the four face grid points to the ray piercing point. The integration in (4.33) is approximated by a formula that is accurate for small optical depth across the cell. A similiar sweeping pattern as described for the DOM is used to assure that the radiances are known at the entering cell faces.

The atmospheric extinction can vary tremendously from clear to thick clouds, and so an adaptive grid is implemented to give additional spatial resolution where it is needed. Grid cells in the original "base grid" are split in half according to a criterion based on the change in source function across a grid cell. Figure 4.3 shows the adaptive grid for a box cloud illuminated by solar radiation. The grid is refined along the edges of the cloud, especially on the sunlit side where the source function changes most rapidly. The fine scale features away from the cloud and shadow edges in the downwelling flux are due to ray effects (as discussed in Sect. 4.1.3).

After the source function iterations have converged, the desired radiometric quantities are computed. The hemispheric fluxes at each grid point are calculated by quadrature integration during the discrete ordinate phase of the iterations. The mean radiance and the net fluxes in x, y, and z are simply proportional to the first four terms of the spherical harmonic expansion of the radiance. The radiance in specified directions is calculated by integrating the source function with (4.33) through the medium. For solar problems with the delta-M method (Wiscombe, 1977), the TMS method (Nakajima and Tanaka, 1988) is used to compute the source function. This method replaces the scaled, truncated Legendre phase function expansion for the singly scattered solar radiation by the full, unscaled phase function expansion. The multiply scattered contribution still comes from the truncated phase function. The TMS method provides accurate radiances for directions away from the solar aureole region.

SHDOM can perform unpolarized multi-dimensional radiative transfer with either or both solar and thermal sources of radiation. Optical properties (extinction,

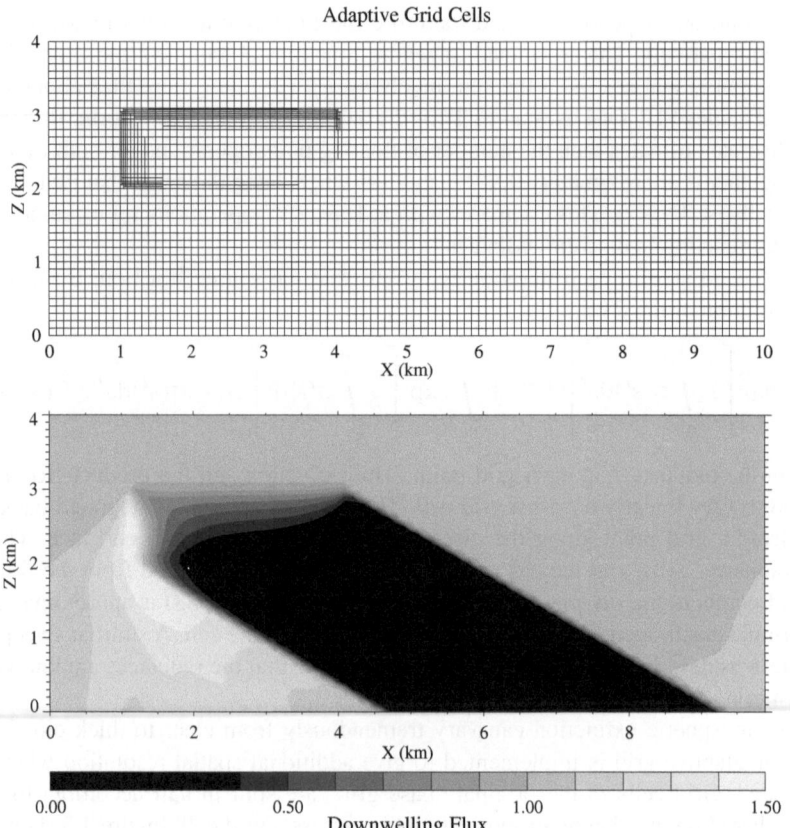

Fig. 4.3. An example of SHDOM output for a 3 km × 1 km uniform box cloud with optical depth of 20 and Mie phase function for $\lambda = 1.65$ μm and $r_e = 10$ μm. The solar zenith angle is $60°$, and the horizontal solar flux is unity. There is no surface reflection. The SHDOM parameters are $N_\mu = 8$, $N_\phi = 16$, and splitacc = 0.03. The top panel shows the base grid with the adaptive cells split around the edges of the box cloud. The bottom panel shows the downwelling flux

single-scattering albedo, and phase function) can be specified arbitrarily on the input grid. Radiative transfer across a spectral band with molecular absorption lines can be calculated with a k-distribution. The lower boundary condition may be Lambertian reflection or one of a few types of bidirectional reflectance distribution functions. The horizontal boundaries may be either periodic or open. SHDOM can also compute the solutions to the independent pixel (1D) and independent slice (2D) approximations. SHDOM is efficient enough to be able to solve 3D problems with modest angular resolution and a million grids points (with typical grid cell optical depths around unity) on modern computer workstations. As for any explicit multidimensional radiative transfer model, one has to carefully balance the spatial and angular resolution in SHDOM in order to achieve the desired accuracy most efficiently. Advice on how to achieve that balance is available in Evans (1998) and in the documentation available at the SHDOM Web site (http://nit.colorado.edu/shdom.html).

4.2 Statistical Modeling or Monte Carlo Methods

4.2.1 Introduction

The outline of this section is as follows. First we describe the main components of any Monte Carlo (MC) method for radiative transfer; they are simulations of photon free path and scattering angle. A simple example of a one-dimensional MC method where photons can go either forward or backward illustrates the above blocks; for clarity, a few lines long FORTRAN code that estimates reflectance, transmittance, and absorptance is provided. Then we discuss the general concept of the MC approach and its accuracy.

We will distinguish between "straightforward" MC that directly simulates photon trajectories from entering a cloud to exiting (or absorbing) and "local estimation" MC that calculates contributions from each order-of-scattering. While the first one is more applicable for estimating fluxes, the second one is appropriate for radiances. We will mostly focus on radiances. Based on direct and adjoint radiative transfer, we will discuss how to construct photon trajectories to estimate radiance in a small solid angle around a given direction. An example demonstrates this technique for calculations of nadir and zenith radiances. Finally, we describe the "maximum cross-section" technique that substantially simplifies the simulation of photon paths in any complex 3D medium.

Except for a few examples, none of the material is original. Most of the techniques, including forward and backward MC and the maximum cross-section method, can be found in Marchuk's (1980) book. A good review of the advantages and disadvantages of MC methods in comparison with other numerical methods is given by Lenoble (1985). A list of recommended literature on 3D Monte Carlo in cloudy atmospheres is provided at the end of the chapter.

The purpose of this section is to give a general picture of MC with an emphasis on the MC method as a solution to the monochromatic radiative transfer equation (RTE) for a given 3D extinction field. Some aspects of longwave and broadband MC methods and calculations can be found in Chaps. 9 and 10, respectively.

4.2.2 What is a Monte Carlo Method? Application to Radiative Transfer

The Monte Carlo method – or rather methods – is a technique for constructing probabilistic models of real processes to estimate certain average properties, e.g., mathematical expectations, variances and covariances. The main points of the MC method are: (*i*) generation of random numbers α uniformly distributed between 0 and 1; (*ii*) simulation of random values with more complicated distribution functions with the help of α; (*iii*) calculation of the quantities of interest for the simulated process via realizations of random values obtained in step (*ii*). In the present section we will deal with only the last two steps, as there are now many reliable and well-documented pseudo-random number generators available (L'Ecuyer, 1998).

4.2.3 Simulation of Random Numbers

Continuous Random Number

Let us assume that a continuous random variable ξ is defined by its probability density $p(x), x_{min} \leq x \leq x_{max}$. To simulate a realization of ξ, we have to solve the equation

$$F(\xi) = \alpha \tag{4.34}$$

where $F(x)$ is the cumulative distribution function defined as,

$$F(x) = \int_{x_{min}}^{x} p(x')dx' . \tag{4.35}$$

That is,

$$\xi = F^{-1}(\alpha) , \tag{4.36}$$

since,

$$\Pr\{\xi < x\} = \Pr\{F^{-1}(\alpha) < x\} = \Pr\{\alpha < F(x)\} = F(x) . \tag{4.37}$$

Example

Let us simulate the distance l a photon travels before interaction in a homogeneous medium; l is distributed with respect to the probability density $p(x) = \sigma \exp(-\sigma x)$, $\int_0^\infty p(x)dx = 1$, where the extinction coefficient σ is related to the mean-free-path or "mfp" $\ell = 1/\sigma$. Solving the equation

$$\int_0^l p(x')dx' = \alpha , \tag{4.38}$$

we find $l = -\ln(1 - \alpha)/\sigma$ or, equivalently,

$$l = -\ln(\alpha)/\sigma \tag{4.39}$$

since both α and $1 - \alpha$ are identically and uniformly distributed random numbers on the interval [0,1].

Discrete Random Number

Let us now assume that a discrete random variable, ξ, is defined by a table

$$\begin{pmatrix} x_1 \; x_2 \; \dots \; x_N \\ p_1 \; p_2 \; \dots \; p_N \end{pmatrix} \qquad (4.40)$$

where $\Pr(\xi = x_i) = p_i$; $\sum_{i=1}^{N} p_i = 1$. To simulate ξ we have to find the interval $\left(\sum_{i=1}^{j-1} p_i, \sum_{i=1}^{j} p_i \right)$ that α belongs to. In other words,

$$\Pr \left(\sum_{i=1}^{j-1} p_i \le \alpha \le \sum_{i=1}^{j} p_i \right) = p_j . \qquad (4.41)$$

Example

After each act of interaction (collision) between a photon and a particle in the optical medium, a photon can be either absorbed or scattered. Similar to (4.40), this can be written as

$$\begin{pmatrix} \text{scattering} \quad \text{absorption} \\ \varpi_0 \qquad\quad 1 - \varpi_0 \end{pmatrix} \qquad (4.42)$$

where ϖ_0 is the single-scattering albedo. According to (4.41), if $0 < \alpha < \varpi_0$, then $\xi = $ scattering; otherwise, $\xi = $ absorption.

4.2.4 An Example of 1D Monte Carlo for Radiative Transfer

In this section we apply the above examples to build a simple MC code to simulate radiative transfer along a line where a photon can fly only forward or backward (see Prog. 4.1). The medium is interval $[0, h]$ with a photon mean-free-path equal to ℓ, thus, optical depth $\tau = h/\ell$. When a photon encounters a particle, the probability of absorption is $1 - \varpi_0$. Upon each scattering, photons either continue forward or turn backward. The asymmetry parameter g gives the probability $p = (1 + g)/2$ that a photon goes forward and the probability $1 - p = (1 - g)/2$ that it jumps backward. Notice that g is still the mean cosine of the scattering angle which in 1D is either 0 or π.

We use a total of N_p photons that cannot scatter more than N_s times; for simplicity, after N_s scatterings a photon is called "lost." The code has two loops: an outer loop (from line 8 to line 31) over number of photons and an inner loop (from line 11 to line 29) over number of scatterings. A photon from its original position at $z = 0$ moves forward ($\mu = 1$); the distance it travels before an interaction (Δz) is simulated using (4.39) where mfp $\ell = 1/\sigma$ and "random(iseed)" is a realization of α. The new coordinate $z + \mu\Delta z$ is checked for escaping the medium as a transmitted (lines 14-16) or a reflected (lines 17-19) photons. If a photon is not absorbed at the collision (for simulations, see (4.42) and lines 21-23), its direction [forward with ($\mu_s = +1$)

```
1       subroutine mc1D(h,mfp,g,w0,N_p,N_s,iseed,R,T,A)
2       real h, mfp, g, w0, p, z, del_z, R, T, A, random
3       integer N_p,N_s,iseed,n_r,n_t,n_a,mu,mu_s,ip,is
c   INPUT
c h - geometrical thickness
c mfp - geometrical mean free path
c g - asymmetry parameter
c w0 - single-scattering albedo
c N_p - total number of photons
c N_s - maximum allowable number of scatterings per photon
c iseed - first seed for random number generator
c   OUTPUT
c R - reflectance; T - transmittance; A - absorptance
4       n_r = 0
5       n_t = 0
6       n_a = 0
7       p = (1.+g)/2.
8       do ip = 1, N_p
9          z = 0.
10         mu = 1
11         do is = 1, N_s
12            del_z = -mfp*log(random(iseed))
13            z = z+mu*del_z
14            if (z.gt.h) then
15               n_t = n_t+1
16               goto 1
17            else if(z.lt.0.) then
18               n_r = n_r+1
19               goto 1
20            endif
21            if(random(iseed).gt.w0) then
22               n_a = n_a+1
23               goto 1
24            else
25               mu_s = 1
26               if (random(iseed).gt.p) mu_s = -1
27               mu = mu*mu_s
28            endif
29         enddo
30 1     continue
31      enddo
32      R = n_r/float(N_p)
33      T = n_t/float(N_p)
34      A = n_a/float(N_p)
35      return
36      end
```

Programme 4.1: A FORTRAN Monte Carlo code on a line

or backward ($\mu_s = -1$)] is simulated using probability p defined by asymmetry factor g [see lines 25-27 based on (4.41)]. Finally, reflectance R, transmittance T, and absorptance A are estimated by averaging the outcome (lines 32-34).

If $h = 10$, mfp $\ell = 1$, $g = 0.85$, $\varpi_0 = 0.99$ then the code gives $R = 0.391$, $T = 0.515$, $A = 0.094$.[1] In this case, it is safe to take number of scatterings $N_s = 100$. However, at large τ, the average number of scatterings is asymptotically proportional to τ for reflected photons and to $(1-g)\tau^2$ for transmitted photons (see Chap. 12); so higher N_s will be needed for thicker clouds. If photons are lost, the MC method is biased. Rather than repeated increasingly long executions, one can use an unbiased method such as "Russian Roulette" or a random cut off of the trajectory (e.g., Sobol, 1974), at the cost of slightly increased variance, to deal with unusually long trajectories.

To estimate the accuracy of this method with respect to the number of photons, in the next section we discuss the general aspects of MC methods that follow directly from the main results of probabilistic theory (see, e.g., Papoulis, 1965).

4.2.5 General Concept of Monte Carlo Error Estimates

Let us assume that we obtained N independent realizations $x_i (i = 1, \ldots, N)$ of a random value ξ with finite mathematical expectation $E\xi$ and variance $D\xi$. If N is large enough, then the average of x_i has a normal distribution and the inequality

$$\left| E\xi - \frac{1}{N} \sum_{i=1}^{N} x_i \right| \le c_\beta \sqrt{\frac{D\xi}{N}} \tag{4.43}$$

is valid at a given confidence level β which defines the constant c_β. For example, $c_\beta = 0.67$ if $\beta = 0.5$, $c_\beta = 1.0$ if $\beta = 0.68$ (the "1-sigma" value), $c_\beta = 1.96$ if $\beta = 0.95$, and $c_\beta = 3$ if $\beta = 0.997$ (the famous rule of "3 sigmas"). More precisely (e.g., Papoulis, 1965),

$$\Pr\left\{ \left| E\xi - \frac{1}{N} \sum_{i=1}^{N} x_i \right| < c\sqrt{\frac{D\xi}{N}} \right\} \approx \Phi(c) = \frac{2}{\sqrt{2\pi}} \int_0^c \exp(-t^2/2)\,dt \tag{4.44}$$

and c_β is the solution of the equation,

$$\Phi(c) = \beta . \tag{4.45}$$

The variance $D\xi$ can be estimated as

---

[1] The exact answers are $R = 0.3919$, $T = 0.5141$, $A = 0.0941$, obtained from the analytical 2-stream/diffusion solution for "literal" 1D RT, i.e., particles moving on a segment. This solution is given in Chap. 5; see expressions (5.31) for T and (5.33) for R with $\tau_t = (1 - \varpi_0 g)h/\ell$, $S = \sqrt{(1 - \varpi_0)/(1 - \varpi_0 g)}$ (notice that there is no factor of 3 in literal 1D RT), and $\chi = 1$ (the exact value for mixed boundary conditions in 1D RT).

$$D\xi \approx \frac{N}{N-1} \left[\frac{1}{N} \sum_{i=1}^{N} x_i^2 - \left(\frac{1}{N} \sum_{i=1}^{N} x_i \right)^2 \right].$$ (4.46)

For example, based on (4.43), $\sqrt{D\xi/N}$ is an estimate of MC uncertainties with a confidence level $\beta = 0.68$. This means that after N trials, with 68% probability, the MC estimate, $\frac{1}{N} \sum_{i=1}^{N} x_i$, will have an error smaller than $\sqrt{D\xi/N}$ where variance $D\xi$ is estimated using (4.46). Similarly, if φ is a function of a random argument ξ defined by the probability density $p(x)$, the expected value $E\varphi$ is

$$E[\varphi(\xi)] = \int p(x)\varphi(x)\mathrm{d}x \approx \frac{1}{N} \sum_{i=1}^{N} \varphi(x_i)$$ (4.47)

where points x_i are distributed with respect to $p(x)$. The variance $D\varphi$ can be estimated as in (4.46) with $\varphi(x)$ instead of x. The above approximate equality is the key MC estimate. We will use it repeatedly through the rest of the chapter.

Example

In order to estimate some average characteristics of solar radiative transfer in clouds (exiting a cloud in a specific region of the boundary and of direction space), one can simulate the fate of a photon from its "birth" (entering the cloud) to its "death" (absorption or exiting the cloud). The random function $\varphi(x) \equiv x$ will characterize two possible outcomes

$$\varphi(x) \equiv x = \begin{cases} 1, & \text{success (exiting a cloud at the given region and direction)} \\ 0, & \text{failure (absorbing/exiting a cloud in other regions/directions).} \end{cases}$$ (4.48)

In this case, (4.47) yields

$$E[\varphi] \approx \frac{1}{N} \sum_{i=1}^{N} \varphi(x_i) = \frac{1}{N} \sum_{i=1}^{N} x_i = q$$ (4.49)

where q is the probability of success and can be interpreted as reflectance or transmittance. For such a Bernoulli trial, we know a priori that $D[\varphi] = q(1-q)$. If Δq is the absolute uncertainty of (4.49), the relative accuracy (in %) will be estimated from (4.43) as

$$100 \times \frac{\Delta q}{q} \approx 100 \times \sqrt{\frac{1-q}{Nq}} \ (\%).$$ (4.50)

The above method, called "straightforward" MC, works well if the estimated probability $0 < q < 1$ is large enough. For the straightforward MC code in Sect. 4.2.4 with the stated parameter values, $N = 100,000$ photons yields better than 1% accuracy for reflectance, transmittance and even absorptance. However, if $q \ll 1$, and an accuracy of 1% is required, it follows from (4.50) that we need $N \approx 10^4/q$ photon histories. Since at present, even modern computers cannot trace more than $N \approx 10^9$ histories in a reasonable computer time, q in (4.49) should not be smaller than 10^{-5}. Suppose we are trying to estimate radiance emerging from a field of $10^3 - 10^4$ pixels. The probability to exit in a small solid angle, say $1°$ around zenith, for each of these pixels is about $3 \times 10^{-6} - 10^{-7}$; so this task is computationally *impossible* without sacrificing spatial or angular resolution, or both. To summarize, straightforward MC is good enough for fluxes (solid angle 2π) across an arbitrary plane element while in most cases it fails to estimate radiances in small solid angles.

The only alternative to straightforward MC is a "weighted" method based on the solution of the RTE. Below we consider both the direct and the adjoint radiative transfer processes and explain how to obtain much better and faster MC estimates of radiance, once expressed as an integral like in (4.47). One of them which uses the direct RTE is called "forward" MC or "local estimation;" the other uses the adjoint radiative transfer process and is called "backward" MC. In contrast to forward MC where photons travel as they do in the real atmosphere, in backward MC they start at the detector and follow in the time-reversed direction along the photon path.

4.2.6 Forward Monte Carlo for Radiative Transfer

Based on monochromatic 3D RTE (4.12) and the Neumann series, we describe below the forward Monte Carlo technique that estimates radiance in a small solid angle around a given direction. We start from the integral RTE in optical medium M (see Chap. 3, (3.125))

$$U(x, \Omega) = \int_M \int_{4\pi} k[(x', \Omega') \to (x, \Omega)] \, U(x', \Omega') d\Omega' dx' + f(x, \Omega) \qquad (4.51a)$$

for the "collision density" – the product of extinction $\sigma(x)$ and radiance $I(x, \Omega)$,

$$U(x, \Omega) = \sigma(x) I(x, \Omega) . \qquad (4.51b)$$

Here $x = (x, y, z)$ and $\Omega = (\Omega_x, \Omega_y, \Omega_z) = (\sin\theta\cos\phi, \sin\theta\sin\phi, \cos\theta)$ where θ and ϕ are zenith and azimuthal angles, respectively. The kernel

$$k[(x', \Omega') \to (x, \Omega)]$$
$$= \varpi_0(x')\sigma(x)\frac{P(x', \Omega \cdot \Omega')}{4\pi} \frac{\exp[-\tau(x', x)]}{\|x - x'\|^2} \delta\left(\Omega - \frac{x - x'}{\|x - x'\|}\right) \qquad (4.52)$$

is the probability density of the transition from point x' and direction Ω' into point x and direction Ω, and the source function

$$f(x, \Omega) = \int_M k[(x', \Omega_0) \to (x, \Omega)] \, \sigma(x') \exp[-\tau(x', x_0)] dx' . \qquad (4.53)$$

$P(x, \Omega \cdot \Omega')/4\pi$ is the normalized scattering phase function at point x. For simplicity, for the rest of this chapter we assume that a phase function and the single-scattering albedo are the same for all of M, i.e., $P(x, \Omega \cdot \Omega') \equiv P(\Omega \cdot \Omega')$ and

$\varpi_0(\boldsymbol{x}) \equiv \varpi_0$, respectively. The generalization for space-dependent scattering phase functions and single-scattering albedos is straightforward. Note that (4.51a)–(4.52) can be obtained from (3.125)–(3.126) multiplying both parts of (3.125) by $\sigma(\boldsymbol{x})$ and substituting $\sigma_s(\boldsymbol{x}')/\sigma(\boldsymbol{x}')$ in (3.126) by $\varpi_0(\boldsymbol{x}')$. In other words, $\mathcal{K}_I$ in (3.126) and k in (4.52) are related as $k[(\boldsymbol{x}',\boldsymbol{\Omega}') \to (\boldsymbol{x},\boldsymbol{\Omega})] = \mathcal{K}_I\sigma(\boldsymbol{x})/\sigma(\boldsymbol{x}')$.

We assume that the boundary point $\boldsymbol{x}_0$ is illuminated by the incident beam in the direction $\boldsymbol{\Omega}_0$; $\tau(\boldsymbol{x}',\boldsymbol{x})$ is the optical distance between points $\boldsymbol{x}'$ and $\boldsymbol{x}$ along $\boldsymbol{\Omega} = (\boldsymbol{x}' - \boldsymbol{x})/\|\boldsymbol{x}' - \boldsymbol{x}\|$:

$$\tau(\boldsymbol{x}',\boldsymbol{x}) = \int\limits_0^{\|\boldsymbol{x}'-\boldsymbol{x}\|} \sigma(\boldsymbol{x} - \boldsymbol{\Omega}s)\mathrm{d}s \ . \tag{4.54}$$

Note that the factor $\|\boldsymbol{x}' - \boldsymbol{x}\|^{-2}$ in the definition of the radiative transfer kernel (4.52) characterizes an element of solid angle $\mathrm{d}\boldsymbol{\Omega}$. The δ-function term in (4.52) couples direction $\boldsymbol{\Omega}$ with points $\boldsymbol{x}$ and $\boldsymbol{x}'$ and thus $\boldsymbol{\Omega}' \cdot \boldsymbol{\Omega} = \boldsymbol{\Omega}' \cdot (\boldsymbol{x}-\boldsymbol{x}')/\|\boldsymbol{x}-\boldsymbol{x}'\|$ in the phase function P. Both kernel k and source function f in (4.53) are probability densities.

For simplicity, we rewrite the integral equation (4.51a) in operator form as

$$U = KU + f \tag{4.55}$$

where the integral operator K is defined by

$$[KU](\boldsymbol{x},\boldsymbol{\Omega}) = \int\limits_M \int\limits_{4\pi} k[(\boldsymbol{x}',\boldsymbol{\Omega}') \to (\boldsymbol{x},\boldsymbol{\Omega})] \, U(\boldsymbol{x}',\boldsymbol{\Omega}') \, \mathrm{d}\boldsymbol{\Omega}'\mathrm{d}\boldsymbol{x}' \ . \tag{4.56}$$

The solutions of (4.55) can be represented as a Neumann series (i.e., by orders of scattering)

$$U = f + Kf + K^2f + K^3f + \dots \ , \tag{4.57}$$

which converges (e.g., Marchuk et al., 1980) for all bounded media with $0 \le \varpi_0 \le 1$.

To simplify notations, we define the inner product of two functions a and b as,

$$(a,b) = \int\limits_M \int\limits_{4\pi} a(\boldsymbol{x},\boldsymbol{\Omega}) \, b(\boldsymbol{x},\boldsymbol{\Omega}) \, \mathrm{d}\boldsymbol{\Omega}\mathrm{d}\boldsymbol{x} \ . \tag{4.58}$$

Also, in addition to K, we introduce the adjoint to K operator defined as

$$[K^+g](\boldsymbol{x},\boldsymbol{\Omega}) = \int\limits_M \int\limits_{4\pi} k[(\boldsymbol{x},\boldsymbol{\Omega}) \to (\boldsymbol{x}',\boldsymbol{\Omega}')] \, g(\boldsymbol{x}',\boldsymbol{\Omega}') \, \mathrm{d}\boldsymbol{\Omega}'\mathrm{d}\boldsymbol{x}' \ , \tag{4.59}$$

hence $(Kf,g) = (f,K^+g)$. Now, substituting U by its Neumann series (4.57), we have

$$\begin{aligned}
U(\boldsymbol{x}^*,\boldsymbol{\Omega}^*) &= (U,\delta_{\boldsymbol{x}^*\boldsymbol{\Omega}^*}) \\
&= (f,\delta_{\boldsymbol{x}^*\boldsymbol{\Omega}^*}) + (Kf,\delta_{\boldsymbol{x}^*\boldsymbol{\Omega}^*}) + (K^2f,\delta_{\boldsymbol{x}^*\boldsymbol{\Omega}^*}) + \dots \\
&= f(\boldsymbol{x}^*,\boldsymbol{\Omega}^*) + (f,K^+\delta_{\boldsymbol{x}^*\boldsymbol{\Omega}^*}) + (Kf,K^+\delta_{\boldsymbol{x}^*\boldsymbol{\Omega}^*}) + \dots
\end{aligned} \tag{4.60}$$

where

$$\delta_{x^*\Omega^*}(x, \Omega) = \delta(x - x^*)\,\delta(\Omega - \Omega^*) \tag{4.61}$$

is the Dirac delta-function and

$$[K^+\delta_{x^*\Omega^*}](x, \Omega) = k[(x, \Omega) \to (x^*, \Omega^*)]$$
$$= \varpi_0\sigma(x^*)\,\frac{P(\Omega \cdot \Omega^*)\,\exp[-\tau(x^*, x)]}{4\pi\,\|x^* - x\|^2}\,\delta\left(\Omega^* - \frac{x^* - x}{\|x^* - x\|}\right) \tag{4.62}$$

is the probability density of transition from point x and direction Ω into point x^* and direction Ω^*.

Let us assume that we are interested in calculating radiance I at a given point x^* in an arbitrarily small solid angle $\Delta\Omega^*$ around direction Ω^*,

$$I_{\Delta\Omega^*}(x^*) = \int_{\Delta\Omega^*} I(x^*, \Omega)\,d\Omega = \frac{1}{\sigma(x^*)} \int_{\Delta\Omega^*} U(x^*, \Omega)\,d\Omega. \tag{4.63}$$

Integrating (4.60) over $\Delta\Omega^*$, and dividing both parts by $\sigma(x^*)$, we get

$$I_{\Delta\Omega^*}(x^*) = \int_{\Delta\Omega^*} \frac{f(x^*, \Omega)}{\sigma(x^*)}\,d\Omega + (\varpi_0 f, \Psi_f)$$
$$+ (\varpi_0 K f, \Psi_f) + (\varpi_0 K^2 f, \Psi_f) + \ldots \tag{4.64}$$

with "contribution" function Ψ_f defined as (subindex "f" stands for "forward"),

$$\Psi_f(x, \Omega) = \frac{P(\Omega \cdot \Omega^*)}{4\pi} \times \frac{\exp[-\tau(x^*, x)]}{\|x^* - x\|^2}\chi_{\Delta\Omega^*}(\Omega) \tag{4.65}$$

where

$$\chi_{\Delta\Omega^*}(\Omega) = \begin{cases} 1, & \Omega \in \Delta\Omega^* \\ 0, & \text{otherwise} \end{cases} \tag{4.66}$$

is the indicator function of the solid angle $\Delta\Omega^*$. Note that, up to this point we have used only identities, nothing to do with MC simulation yet. Now we use a MC technique to estimate the terms in (4.64) which are all integrals of the type (4.47).

Recalling the definition of f in (4.53) and applying MC estimate (4.47), we can get the first term in (4.64) as

$$\int_{\Delta\Omega^*} \frac{f(x^*, \Omega)}{\sigma(x^*)}\,d\Omega = \int_M \{\varpi_0\sigma(x)\exp[-\tau(x, x_0)]\}\,\Psi_f(x, \Omega_0)\,dx$$
$$\approx \frac{1}{N}\sum_{i=1}^N \Psi_f(x_{1i}, \Omega_0) \tag{4.67}$$

where points (x_{1i}, Ω_0) are simulated with respect to the probability density $p(x) = \varpi_0\sigma(x)\exp[-\tau(x, x_0)]$. We use x_0 and Ω_0 to denote the starting position and direction of the photon trajectory; these may be either deterministic or random (according to the pdf defined by the forcing term). The second and third terms in (4.64) can be estimated as follows,

$$(\varpi_0 f, \Psi_f) = \int_M \int_{4\pi} [\varpi_0 f(\boldsymbol{x}, \boldsymbol{\Omega})] \, \Psi_f(\boldsymbol{x}, \boldsymbol{\Omega}) \, \mathrm{d}\boldsymbol{\Omega}\mathrm{d}\boldsymbol{x} \approx \frac{1}{N} \sum_{i=1}^{N} \Psi_f(\boldsymbol{x}_{2i}, \boldsymbol{\Omega}_{2i}) \quad (4.68)$$

and

$$(\varpi_0 K f, \Psi) = \int_M \int_{4\pi} [\varpi_0 K f(\boldsymbol{x}, \boldsymbol{\Omega})] \, \Psi_f(\boldsymbol{x}, \boldsymbol{\Omega}) \, \mathrm{d}\boldsymbol{\Omega}\mathrm{d}\boldsymbol{x}$$

$$\approx \frac{1}{N} \sum_{i=1}^{N} \Psi_f(\boldsymbol{x}_{3i}, \boldsymbol{\Omega}_{3i}) \tag{4.69}$$

where points $(\boldsymbol{x}_{2i}, \boldsymbol{\Omega}_{2i})$ and $(\boldsymbol{x}_{3i}, \boldsymbol{\Omega}_{3i})$ are simulated with respect to the density $p(\boldsymbol{x}) = \varpi_0 f(\boldsymbol{x}, \boldsymbol{\Omega})$ and $\varpi_0 K f(\boldsymbol{x}, \boldsymbol{\Omega})$, respectively. The approximation of other integrals in (4.64) is obvious (increasing powers of K appear).

Physically, (4.67)-(4.69) estimate the contribution of the first, second and third scattering-order photons, respectively. Thus to estimate $I_{\Delta\Omega^*}(\boldsymbol{x}^*)$ we need to simulate the photon's trajectory $(\boldsymbol{x}_1, \boldsymbol{\Omega}_1) \rightarrow (\boldsymbol{x}_2, \boldsymbol{\Omega}_2) \rightarrow (\boldsymbol{x}_3, \boldsymbol{\Omega}_3) \rightarrow \ldots \rightarrow (\boldsymbol{x}_m, \boldsymbol{\Omega}_m)$ where m is the random order of the last scattering, always finite in a finite medium. Each point $(\boldsymbol{x}_k, \boldsymbol{\Omega}_k)$ is simulated with respect to the density $\varpi_0 K^{k-2} f (k = 2, 3 \ldots)$. For instance, in the solar problem, the first point $(\boldsymbol{x}_1, \boldsymbol{\Omega}_1)$ has $\boldsymbol{\Omega}_1 = \boldsymbol{\Omega}_0$ and $\boldsymbol{x}_1$ is simulated according to $\varpi_0 \sigma(\boldsymbol{x}) \exp[-\tau(\boldsymbol{x}, \boldsymbol{x}_0)]$ starting at a random $\boldsymbol{x}_0$ at cloud top. After each scattering at the point $\boldsymbol{x}_k$ the contribution $\Psi_f(\boldsymbol{x}_k, \boldsymbol{\Omega}_k)$ is included in the statistical estimation of I:

$$I_{\Delta\Omega^*}(\boldsymbol{x}^*) \approx \frac{1}{N} \sum_{i=1}^{N} \sum_{k=1}^{m(i)} \Psi_f(\boldsymbol{x}_{ki}, \boldsymbol{\Omega}_{ki})$$

$$= \frac{1}{N} \sum_{i=1}^{N} \sum_{k=1}^{m(i)} \frac{P(\boldsymbol{\Omega}_{ki} \cdot \boldsymbol{\Omega}^*)}{4\pi} \frac{\exp[-\tau(\boldsymbol{x}^*, \boldsymbol{x}_{ki})]}{\|\boldsymbol{x}^* - \boldsymbol{x}_{ki}\|^2} \chi_{\Delta\Omega^*} \left(\frac{\boldsymbol{x}^* - \boldsymbol{x}_{ki}}{\|\boldsymbol{x}^* - \boldsymbol{x}_{ki}\|} \right) . \tag{4.70}$$

Note that if Ψ in (4.70) were defined as in (4.48), namely, if Ψ were equal to either 1 or 0, depending on the success or failure of a photon to exit a cloud at a given point $\boldsymbol{x}^*$ in a small solid angle $\Delta\Omega^*$ around direction $\boldsymbol{\Omega}^*$, we would get the "straightforward" MC method which is much less efficient than the above forward MC with the "contribution" function Ψ defined by (4.65).

The statistical estimate (4.70) can be modified by including weights W_{ki}; e.g., if we simulate without absorption ($K^k f$ instead of $\varpi_0 K^k f$), then $W_{ki} = \varpi_0 W_{k-1i}$, $W_{1i} = 1$. Thus, the general form of the estimate of $I_{\Delta\Omega^*}(\boldsymbol{x}^*)$ can be written

$$I_{\Delta\Omega^*}(\boldsymbol{x}^*) \approx \frac{1}{N} \sum_{i=1}^{N} \sum_{k=1}^{m(i)} W_{ki} \Psi_f(\boldsymbol{x}_{ki}, \boldsymbol{\Omega}_{ki})$$

$$= \frac{1}{N} \sum_{i=1}^{N} \sum_{k=1}^{m(i)} \varpi_0^k \frac{P(\boldsymbol{\Omega}_{ki} \cdot \boldsymbol{\Omega}^*)}{4\pi} \frac{\exp[-\tau(\boldsymbol{x}^*, \boldsymbol{x}_{ki})]}{\|\boldsymbol{x}^* - \boldsymbol{x}_{ki}\|^2} \chi_{\Delta\Omega^*} \left(\frac{\boldsymbol{x}^* - \boldsymbol{x}_{ki}}{\|\boldsymbol{x}^* - \boldsymbol{x}_{ki}\|} \right) . \tag{4.71}$$

These algorithms are called "local estimates" in the MC literature (e.g., Marchuk et al., 1980). Note that if a detector is located inside the cloud, the contribution from the scattering points x close to x^* will lead to a large increase of the variance in (4.46) which in fact diverges as $N \to \infty$. In this case, local estimate (4.71) becomes inappropriate and a backward MC method (see next section) is needed. If, however, x^* is far away from the photons' trajectories (e.g., a satellite), and/or the distance between x^* and x_k, for which $\chi_{\Delta\Omega^*} = 1$, does not change much from scattering to scattering, the factor $\|x^* - x\|^{-2}$ can be omitted (see example below).

Example of Calculations of Nadir and Zenith Radiances as a Local Estimate

The upward or downward radiances I for each cell S on a horizontal grid, can be estimated by the flux of radiant energy across

- the upper boundary of S (at $z = h$) in the zenith direction (Ω_+), or
- the lower boundary of S (at $z = 0$) in the nadir direction (Ω_-),

$$
I_{\pm}(S) = \frac{\int_S I(x, \Omega_{\pm})dx}{\int_S dx} = E[\Psi(S, \Omega_{\pm})] \approx \frac{1}{N} \sum_{i=1}^{N} \Psi_i(S, \Omega_{\pm}) . \tag{4.72}
$$

Here $\Psi_i (i = 1, \ldots, N)$ are N independent realizations (photon trajectories) of a random function Ψ. For each realization (a trajectory), the random value $\Psi(S, \Omega_{\pm})$ is the contribution to the grid-point S into the direction $\Omega_{\pm}$ from all orders of scattering:

$$
\Psi(S, \Omega_{\pm}) \approx \sum_{k=1}^{m} \varpi_0^k \frac{P(\Omega_k \bullet \Omega_{\pm})}{4\pi} \chi_S(x_k) \begin{cases} \exp[-\sigma_S(h - z_k)], \text{zenith}(+) \\ \exp[-\sigma_S z_k], \text{nadir}(-) \end{cases} .
$$
$$\tag{4.73}$$

Here m is the (random) last scattering order of the photon trajectory, σ_S is the extinction (assumed vertically uniform) of the grid-point S, $x_k = (x_k, y_k, z_k)$ are the coordinates of the point of photon's kth scattering, Ω_k is its direction of propagation before this scattering event, and finally, $\chi_S(x_k)$ indicates whether the photon was in cell S or not at its kth scattering:

$$
\chi_S(x_k) = \begin{cases} 1, & x_k, y_k \in S \\ 0, & \text{otherwise} \end{cases} . \tag{4.74}
$$

Finally, we add two more comments. First, to omit the factor $\|x^* - x\|^{-2}$ in (4.73), we assumed that the (linear) dimensions of the cloud, both for S and h, are much smaller than the distance between cloud and detectors. However, in general, neglecting the factor $\|x^* - x\|^{-2}$ in (4.71) may lead to inaccurate results. Second, we note that a strongly forward peaked phase function can at some point Ω_k close to Ω^* (or $\Omega_{\pm}$ as in this example) yield a very large value that makes estimate (4.71) (or (4.72)) inaccurate. For this case, Antyufeev (1996) proposed a modification of the above calculation scheme that can substantially increase the efficiency of MC calculations. The proposed modification substitutes the original forward-peaked phase function by an approximation called a "pseudo-transport" where the phase function is replaced by a linear mixture of a peak-shaped function with a "regular" (smooth) one.

4.2.7 Backward Monte Carlo

As forward MC for radiative transfer is based on the integral RTE, backward MC is based on its adjoint counterpart. Let I^+ be the solution of the adjoint integro-differtial equation (see Chap. 3, (3.156)) with delta-function source and homogeneous boundary conditions; thus I^+ is also the solution of the adjoint integral equation

$$I^+ = K^+ I^+ + g \tag{4.75}$$

with the adjoint operator K^+ defined in (4.59) and

$$g(\boldsymbol{x}, \boldsymbol{\Omega}) = \exp[-\tau(\boldsymbol{x}, \boldsymbol{x}^*)]\delta(\boldsymbol{\Omega} - \boldsymbol{\Omega}^*) . \tag{4.76}$$

Using the reciprocity theorem (Chap. 3, Sect. 3.10) one can show that:

$$I(\boldsymbol{x}^*, \boldsymbol{\Omega}^*) = \int_M \int_{4\pi} I^+(\boldsymbol{x}, -\boldsymbol{\Omega}) \, F(\boldsymbol{x}, -\boldsymbol{\Omega}) \, d\boldsymbol{\Omega} \, d\boldsymbol{x} , \tag{4.77}$$

where

$$F(\boldsymbol{x}, \boldsymbol{\Omega}) = \varpi_0 \sigma(\boldsymbol{x}) \frac{P(\boldsymbol{\Omega}_0 \cdot \boldsymbol{\Omega})}{4\pi} \exp[-\tau(\boldsymbol{x}, \boldsymbol{x}_0)] \tag{4.78}$$

describes external illumination of the medium at the boundary point $\boldsymbol{x}_0$ and the direction $\boldsymbol{\Omega}_0$.

Representing I^+ by its Neumann series,

$$I^+ = g + K^+ g + K^{+2} g + K^{+3} g + \dots \tag{4.79}$$

and substituting (4.79) and (4.78) into (4.77), we get

$$
\begin{aligned}
I(\boldsymbol{x}^*, \boldsymbol{\Omega}^*) = &\int_M \int_{4\pi} g(\boldsymbol{x}, -\boldsymbol{\Omega}) \, F(\boldsymbol{x}, -\boldsymbol{\Omega}) \, d\boldsymbol{\Omega} \, d\boldsymbol{x} \\
&+ \int_M \int_{4\pi} [K^+ g](\boldsymbol{x}, -\boldsymbol{\Omega}) \, F(\boldsymbol{x}, -\boldsymbol{\Omega}) \, d\boldsymbol{\Omega} \, d\boldsymbol{x} \\
&+ \int_M \int_{4\pi} [K^{+2} g](\boldsymbol{x}, -\boldsymbol{\Omega}) \, F(\boldsymbol{x}, -\boldsymbol{\Omega}) \, d\boldsymbol{\Omega} \, d\boldsymbol{x} + \dots .
\end{aligned}
\tag{4.80}
$$

Unlike in the forward MC scheme, we are going to simulate here the adjoint trajectory according to the adjoint operator K^+, starting from the point $(\boldsymbol{x}^*, -\boldsymbol{\Omega}^*)$.

Similar to (4.67)-(4.69), (4.80) can be rewritten in form of a sum of terms that are "convenient" for MC estimates, i.e., looking like (4.47), namely

$$
\begin{aligned}
I(\boldsymbol{x}^*, \boldsymbol{\Omega}^*) = &\int_M \{\varpi_0 \sigma(\boldsymbol{x}) \exp[-\tau(\boldsymbol{x}, \boldsymbol{x}^*)]\} \Psi_b(\boldsymbol{x}, -\boldsymbol{\Omega}^*) \, d\boldsymbol{x} \\
&+ \int_M \int_{4\pi} \{\varpi_0 \sigma(\boldsymbol{x})[K^+ g](\boldsymbol{x}, -\boldsymbol{\Omega})\} \Psi_b(\boldsymbol{x}, -\boldsymbol{\Omega}) \, d\boldsymbol{\Omega} \, d\boldsymbol{x} \\
&+ \int_M \int_{4\pi} \{\varpi_0 \sigma(\boldsymbol{x})[K^{+2} g](\boldsymbol{x}, -\boldsymbol{\Omega})\} \Psi_b(\boldsymbol{x}, -\boldsymbol{\Omega}) \, d\boldsymbol{\Omega} \, d\boldsymbol{x} + \dots
\end{aligned}
\tag{4.81}
$$

where

$$\Psi_b(x, \Omega) = \frac{P(-\Omega_0 \cdot \Omega)}{4\pi} \exp[-\tau(x, x_0)] \tag{4.82}$$

is the contribution function for the backward MC. Now we are ready to use MC for estimating the integrals in (4.81).

We have

$$I(x^*, \Omega^*) \approx \frac{1}{N} \sum_{i=1}^{N} \Psi_b(x_{1i}, -\Omega^*) + \frac{1}{N} \sum_{i=1}^{N} \Psi_b(x_{2i}, -\Omega_{2i})$$

$$+ \frac{1}{N} \sum_{i=1}^{N} \Psi_b(x_{3i}, -\Omega_{3i}) + \ldots = \frac{1}{N} \sum_{i=1}^{N} \sum_{k=1}^{m(i)} \Psi_b(x_{ki}, -\Omega_{ki}) . \tag{4.83}$$

Here the first set of random points, $(x_{1i}, -\Omega_{1i}) = (x_{1i}, -\Omega^*)$, is simulated with respect to the density $\varpi_0 \sigma(x) \exp[-\tau(x, x^*)]$, the second set, $(x_{2i}, -\Omega_{2i})$, with respect to $\varpi_0 \sigma(x)[K^+g](x, -\Omega)$, and so on. Finally, as for forward MC, if we simulate "without absorption," its effect can be incorporated by using weights:

$$I(x^*, \Omega^*) \approx \frac{1}{N} \sum_{i=1}^{N} \sum_{k=1}^{m(i)} W_{ki} \Psi_b(x_{ki}, -\Omega_{ki})$$

$$= \frac{1}{N} \sum_{i=1}^{N} \sum_{k=1}^{m(i)} \varpi_0^k \frac{P(\Omega_{ki} \cdot \Omega_0)}{4\pi} \exp[-\tau(x_{ki}, x_0)] . \tag{4.84}$$

To summarize, for backward MC we simulate the adjoint trajectories starting from the point $(x^*, -\Omega^*)$ and at each collision, we calculate Ψ_b defined in (4.82). In contrast, for forward MC we start trajectories from the point (x_0, Ω_0) and at each collision the function Ψ_f defined in (4.65) is calculated. The main difference between Ψ_b and Ψ_f is a factor $\|x^* - x\|^{-2}$; hence, when the detector is inside the scattering medium, $D[\Psi_b] \ll D[\Psi_f]$ [see (4.46)] thus much faster convergence of the estimation (4.43). Another advantage of the backward MC scheme is that photons leave the detector in a given direction. Besides, in forward MC, not all scattering points x_k contribute to the estimate of $I_{\Delta\Omega^*}(x^*)$ but only those with $(x^* - x_k)/\|x^* - x_k\| \in \Delta\Omega^*$ as seen in (4.65)–(4.66). From the other side, since the backward Monte Carlo uses an adjoint trajectory, it normally yields radiance at only a single point and a single direction, whereas one trajectory in the forward MC can yield contributions to many detectors x_j^* in many directions Ω_j^*. Receiver position x^* and orientation Ω^* can be chosen at random and/or assigned weights to simulate "less local" estimates but experience proves that there is eventually a trade-off and forward (and even straight-forward) methods can become equally efficient.

4.2.8 The Maximum Cross-Section Method

Straightforward and forward (backward) MC schemes all call for the simulation of photon trajectories, from a source to a sink. This is achieved efficiently with the "Maximum Cross-Section Method" which involves transforming the integro-differential RTE (4.12) to (Marchuk et al., 1980, p.9)

$$
\mathbf{\Omega} \cdot \nabla I(\mathbf{x}, \mathbf{\Omega}) + \sigma_{\max} I(\mathbf{x}, \mathbf{\Omega})
$$
$$
= \sigma_{\max} \int_{4\pi} \left[\frac{\sigma(\mathbf{x})}{\sigma_{\max}} \varpi_0 \frac{P(\mathbf{\Omega} \cdot \mathbf{\Omega}')}{4\pi} + \left(1 - \frac{\sigma(\mathbf{x})}{\sigma_{\max}} \right) \delta(\mathbf{\Omega} - \mathbf{\Omega}') \right] I(\mathbf{x}, \mathbf{\Omega}') \, d\mathbf{\Omega}'
$$

(4.85)

where $\sigma_{\max} = \max_x \{\sigma(\mathbf{x})\}$ is the maximal extinction.

Equation (4.85) can be interpreted as the transport equation with constant extinction and a modified phase function equal to

$$
\begin{cases}
\varpi_0 P(\mathbf{\Omega} \cdot \mathbf{\Omega}')/4\pi, \text{ with probability } \sigma(\mathbf{x})/\sigma_{\max} & \text{(a "physical" scattering)} \\
\delta(\mathbf{\Omega} - \mathbf{\Omega}'), \text{ otherwise} & \text{(a "mathematical" scattering)}
\end{cases}
$$

(4.86)

In this method, the photon jumps immediately to its next scattering point rather than accumulating optical depth cell-by-cell and interpolating within the last one. This makes the computer time almost insensitive to (*i*) whether we use 1D, 2D or 3D geometry; (*ii*) the variability of $\sigma(\mathbf{x})$; and (*iii*) the number of cells. All three of these factors substantially slow the execution of standard MC for inhomogeneous media. However, if $\sigma_{\max}$ is very large, photon steps are very small and the method becomes impractical.

4.3 Examples of Radiances for the 3D Cloud Fields and Comparison with SHDOM

Our first example is a cloud field retrieved pixel-by-pixel from a LandSat scene with 128×128 pixels (Oreopoulos and Davies, 1998) used in the Intercomparison of 3D Radiation Codes project (I3RC; Cahalan et al., 2005). Figure 4.4 shows zenith (downward) and nadir (upward) radiances calculated with 10^8 and 10^9 photons, respectively, for solar zenith angle of 60°. It is clearly seen that the lower images (with 10^9 photons) have much smaller noise than the middle ones (with 10 times less photons). The majority of pixels in the upper images have an error less than 1%. The errors at the boundary are much larger and can exceed 5 and 10%. The absolute errors are about 0.005. As explained in Sect. 4.2.5, this means that with about 70% probability, the second decimal of the pixel-by-pixel radiances results is either correct or differs from the true value by unity.

The above MC results are in excellent agreement with SHDOM from Sect. 4.1.2; a scatter plot in Fig. 4.5 shows the same row # 101 as in Fig. 4.4 for both nadir and zenith radiances and both methods. The average difference between the two methods is less than 1% while the pixel-by-pixel differences are at the level of 2-3% with the absolute difference of about 0.01. This is very close to the level of MC noise. Panels (b) and (c) illustrate a 1.5 km fragment (from the same row) of zenith and

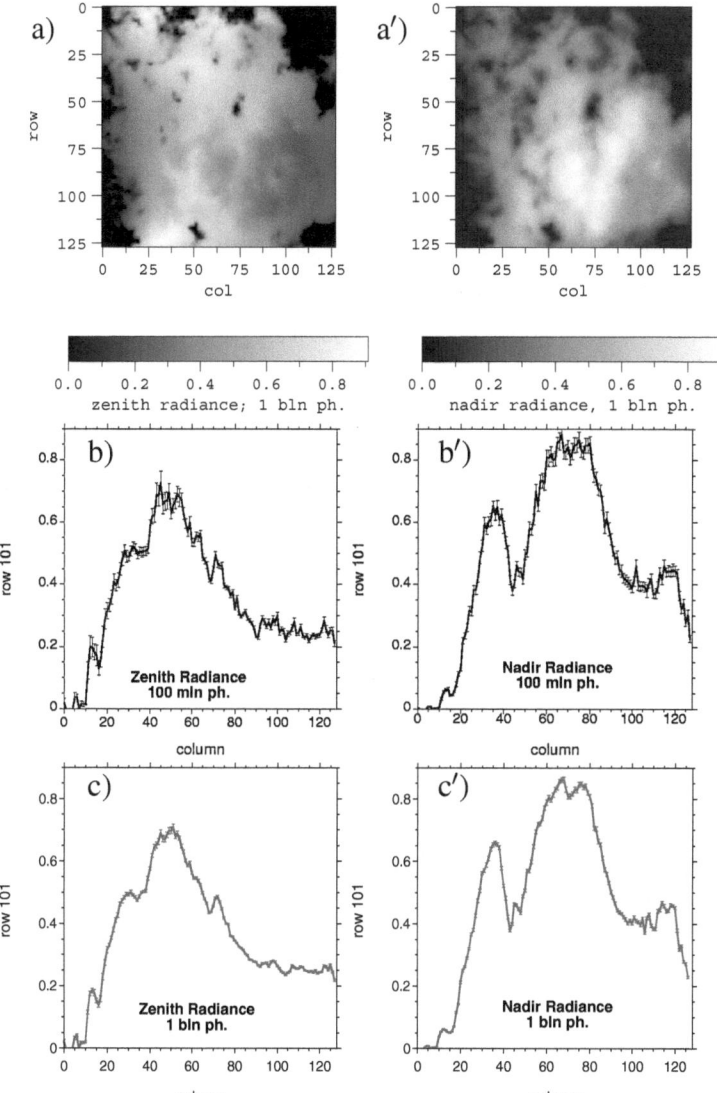

Fig. 4.4. Zenith and nadir radiances calculated with 10^8 and 10^9 photons, respectively. Solar zenith angle is $60°$ from the left. Henyey-Greenstein phase function with $g = 0.85$ and single-scattering albedo $\varpi_0 = 1$ is used. Surface is black. The horizontal grid size is 128×128 with 30 m pixels. (**a**) Zenith radiances calculated with 10^9 photons. The grayscale goes from 0 (*black*) to 0.91 (*white*). (a′) Nadir radiances calculated with 10^9 photons. The grayscale is the same as in panel (a). (**b**) Horizontal cut along the row # 101 shown in panel (a) from the field of zenith radiances calculated with 10^8 photons. With the "one-sigma," the error bars are also added. (b′) Same as in panel (b) but for nadir radiances. (**c**) Same as in panel (b) but for 10^9 photons. (c′) Same as in panel (b′) but for 10^9 photons

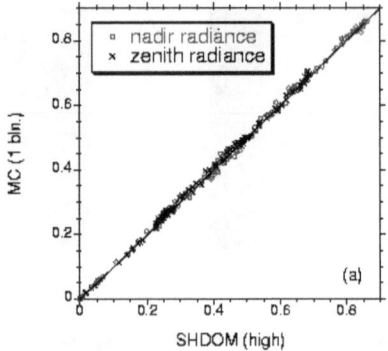

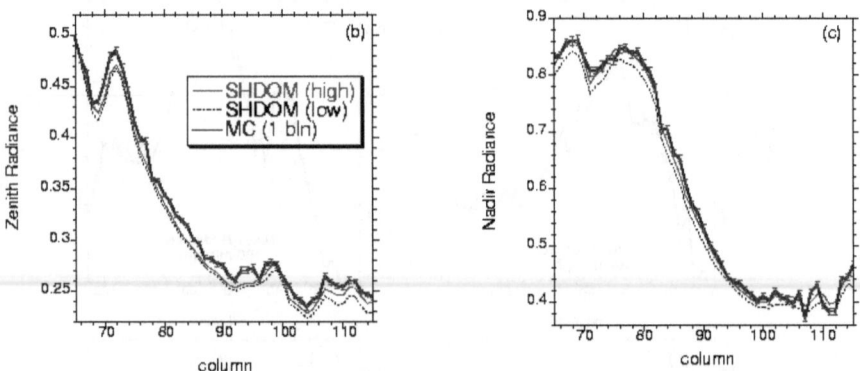

Fig. 4.5. Comparison between MC and SHDOM for row # 101 in Fig. 4.4: **(a)** scatter plot of higher ($N_\mu = 12$, $N_\phi = 24$) and lower ($N_\mu = 6$, $N_\phi = 12$) accuracy SHDOM results versus 10^9-photon MC, and 50 pixel (1500 m) fragment with MC and two SHDOM results for zenith **(b)** and nadir **(c)** radiance

nadir radiances, respectively. In addition to higher accuracy SHDOM ($N_\mu = 12$, $N_\phi = 24$), lower accuracy SHDOM calculations ($N_\mu = 6$, $N_\phi = 12$) are also shown. The convergence of SHDOM is well pronounced. We can see that SHDOM radiances are slightly smoother than its MC counterparts. This is because SHDOM (tri)linearly interpolates the extinction between grid points while the MC assumes uniform cells and has inherent noise. Note that the computer time used to run the lower accuracy SHDOM is 5 times shorter than that for higher accuracy; in case of the MC, computer time is obviously proportional to the number of photons. Finally, the 10^9-photon MC run was only 2–3 times slower than the high-accuracy SHDOM run. However, the MC time-performance versus SHDOM worsens for a more general situation of more asymmetric Mie phase functions and other (oblique) output directions.

The next example, also used in the I3RC, illustrates a cumulus cloud field from a Large Eddy Simulation (LES) model by Stevens et al. (1999). This cloud represents a continental shallow boundary layer and has cloud fraction 0.23 and grid structure

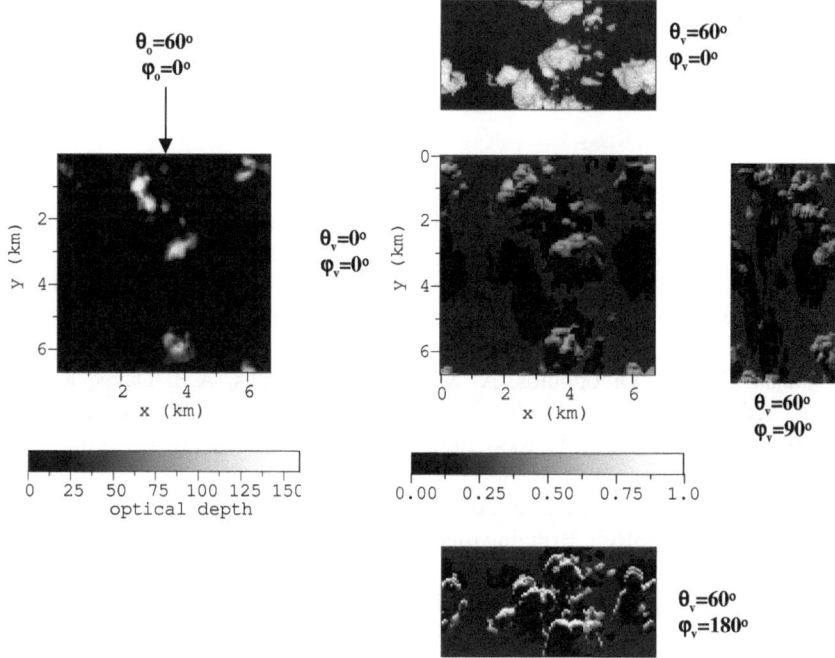

Fig. 4.6. Optical depth field and four bidirectional reflectances (at nadir and 60° zenith angle and 0°, 90°, 180° viewing azimuths) calculated with 5×10^8 photons. The average error is less than 2%. The leftmost panel shows optical depth at the near infrared wavelength of 2.13 μm. The cloud is illuminated from the north with solar zenith angle of 60°. Droplet scattering is described by the Mie phase function with effective radius $r_e = 10$ μm. Surface is assumed to be Lambertian with uniform surface albedo 0.2. Rayleigh scattering, molecular absorption and aerosols with vertically varying extinction are used to model the clear atmosphere. The grayscale goes from 0 (*black*) to 1 (*white*)

$100 \times 100 \times 36$ with $66.7 \times 66.7 \times 40$ m³ cells. The optical depth field for the near infrared wavelength of 2.13 μm is plotted on the leftmost panel of Fig. 4.6. Right panels in Fig. 4.6 illustrate four bidirectional reflectances at nadir and 60° zenith angle and 0°, 90°, 180° viewing azimuths.

Note that plotted on the same grayscale, out of four bidirectional reflectances, the back-scattering image ($\varphi_v = 0°$) is the brightest on average (mean $I = 0.32$) because it does not have any shadows. However, the brightest pixels (max $I = 1.1$) are in the forward-scattering direction ($\varphi_v = 180°$); this is not a 3D effect per se; rather it is due to the scattering phase function's forward peak. What is a 3D feature is a larger contrast between illuminated and shadowed pixels seen most dramatically in the forward-scattering direction. The nadir reflectance ($\theta_v = 0°$) and the "sideway" viewing angle ($\varphi_v = 90°$) are the darkest, having mean $I = 0.17$. The minimum values of all images are about the same (min $I = 0.015$) and are reached at shadowed pixels on the ground.

References

Antyufeev, V.S. (1996). Solution of the generalized transport equation with a peak-shaped indicatrix by the Monte Carlo method. *Russ. J. Numer. Anal. and Modeling*, **11**, 113–137.

Balsara, D. (2001). Fast and accurate discrete ordinate methods for multidimensional radiative transfer. Part I, basic methods. *J. Quant. Spectrosc. Radiat. Transfer*, **69**, 671–707.

Cahalan, R.F., L. Oreopoulos, A. Marshak, K.F. Evans, A.B. Davis, R. Pincus, K. Yetzer, B. Mayer, R. Davies, T.P. Ackerman, H.W. Barker, E.E. Clothiaux, R.G. Ellingson, M.J. Garay, E. Kassianov, S. Kinne, A. Macke, W. O'Hirok, P.T. Partain, S.M. Prigarin, A.N. Rublev, G.L. Stephens, F. Szczap, E.E. Takara, T. Várnai, G. Wen, and T.B. Zhuravleva (2005). The international Intercomparison of 3D Radiation Codes (I3RC): Bringing together the most advanced radiative transfer tools for cloudy atmospheres. *Bull. Amer. Meteor. Soc.*, to appear in Sept 2005 issue.

Carlson, B.G. and K.D. Lathrop (1968). Transport theory – the method of discrete ordinates. In *Computing Methods in Reactor Physics*. Gordon & Breach, New York (NY).

Chandrasekhar, S. (1950). *Radiative Transfer*. Oxford University Press, reprinted by Dover Publications (1960), New York (NY).

de Oliveira, C.R.E. (1986). An arbitrary geometry finite element method for multigroup neutron transport with anisotropic scattering. *Prog. in Nuclear Engin.*, **18**, 227–236.

Evans, K.F. (1993). Two-dimensional radiative transfer in cloudy atmospheres: The spherical harmonic spatial grid method. *J. Atmos. Sci.*, **50**, 3111–3124.

Evans, K.F. (1998). The spherical harmonics discrete ordinate method for three-dimensional atmospheric radiative transfer. *J. Atmos. Sci.*, **55**, 429–446.

Fiveland, W.A. (1988). Three-dimensional radiative heat-transfer solutions by the discrete-ordinates method. *J. Thermophysics and Heat Transfer*, **2**, 309–316.

Gabriel, P.M., S.-C. Tsay, and G.L. Stephens (1993). A Fourier-Riccati approach to radiative transfer. Part I: Foundations. *J. Atmos. Sci.*, **50**, 3125–3147.

Gerstl, S.A. and A. Zardecki (1985). Discrete-ordinate finite-element method for atmospheric radiative transfer and remote sensing. *Appl. Optics*, **24**, 81–93.

Grant, I.P. and G.E. Hunt (1969). Discrete space theory of radiative transfer i: Fundamentals. *Proc. Roy. Soc. London*, **A313**, 183–197.

Haferman, J.L., T.F. Smith, and W.F. Krajewski (1997). A multi-dimensional discrete-ordinates method for polarized radiative transfer. I. Validation for randomly oriented axisymmetric particles. *J. Quant. Spectrosc. Radiat. Transfer*, **58**, 379–398.

Kuo, K.-S., R.C. Weger, R.M. Welch, and S.K. Cox (1996). The Picard iterative approximation to the solution of the integral equation of radiative transfer. Part II: Three-dimensional geometry. *J. Quant. Spectrosc. Radiat. Transfer*, **55**, 195–213.

Larsen, E.W. (1982). Unconditionally stable diffusion synthetic acceleration methods for the slab geometry discrete ordinates equations. Part I: Theory. *Nuclear Sci. Engin.*, **82**, 47.

Lathrop, K.D. (1966). Use of discrete-ordinate methods for solution of photon transport problems. *Nuclear Sci. Engin.*, **24**, 381–388.

L'Ecuyer, P. (1998). Random number generation. In *The Handbook of Simulation*. J. Banks (ed.). Wiley and Sons, New York (NY), pp. 93–137.

Lenoble, J. (ed.) (1985). *Radiative Transfer in Scattering and Absorbing Atmospheres: Standard Computational Procedures*. Deepak Publishing, Hampton (VA).

Liou, K.-N. and N. Rao (1996). Radiative transfer in cirrus clouds. Part IV: On the cloud geometry, inhomogeneity, and absorption. *J. Atmos. Sci.*, **53**, 3046–3065.

Lyapustin, A.I. and T.Z. Muldashev (2001). Solution for atmospheric optical transfer function using spherical harmonic method. *J. Quant. Spectrosc. Radiat. Transfer*, **68**, 43–56.

Marchuk, G., G. Mikhailov, M. Nazaraliev, R. Darbinjan, B. Kargin, and B. Elepov (1980). *The Monte Carlo Methods in Atmospheric Optics*. Springer-Verlag, New York (NY).

Marshak, R.E. (1947). Note on the spherical harmonic methods as applied to the Milne problem for a sphere. *Phys. Rev.*, **71**, 443–446.

Martonchik, J.V. and D.J. Diner (1985). Three-dimensional radiative transfer using a Fourier-transform matrix-operator method. *J. Quant. Spectrosc. Radiat. Transfer*, **34**, 133–148.

Modest, M.F. (1993). *Radiative Heat Transfer*. McGraw-Hill, Inc., New York (NY).

Nakajima, T. and M. Tanaka (1988). Algorithms for radiative intensity calculations in moderately thick atmospheres using a truncation approximation. *J. Quant. Spectrosc. Radiat. Transfer*, **40**, 51–69.

Oreopoulos, L. and R. Davies (1998). Plane parallel albedo biases from satellite observations. Part I: Dependence on resolution and other factors. *J. Climate*, **11**, 919–932.

Papoulis, A. (1965). *Probability, Random Variables, and Stochastic Processes*. McGraw-Hill, New York (NY).

Ramone, G.L., M.L. Adams, and P.F. Nowak (1997). A transport synthetic acceleration method for transport iterations. *Nuclear Sci. Engin.*, **125**, 257.

Sanchez, A., T.F. Smith, and W.F. Krajewski (1994). A three-dimensional atmospheric radiative transfer model based on the discrete-ordinates method. *Atmos. Research*, **33**, 283–308.

Sobol, I.M. (1974). *The Monte Carlo Method*. The University of Chicago Press, Chicago (IL).

Stamnes, K., S.-C. Tsay, W.J. Wiscombe, and K. Jayaweera (1988). Numerically stable algorithm for discrete-ordinate-method radiative transfer in multiple scattering and emitting layered media. *Appl. Opt.*, **27**, 2502–2509.

Stenholm, L.G., H. Storzer, and R. Wehrse (1991). An efficient method for the solution of 3D radiative transfer problems. *J. Quant. Spectrosc. Radiat. Transfer*, **45**, 47–56.

Stephens, G.L. (1988). Radiative transfer through arbitrary shaped optical media, I: A general method of solution. *J. Atmos. Sci.*, **45**, 1818–1836.

Stevens, B., C.-H. Moeng, and P.P. Sullivan (1999). Large-Eddy simulations of radiatively driven convection: Sensitivities to the representation of small scales. *J. Atmos. Sci.*, **56**, 3963–3984.

Thomas, G. and K. Stamnes (1999). *Radiative Transfer in the Atmosphere and Ocean*. Cambridge University Press, New York (NY).

Tofsted, D.H. and S.G. O'Brien (1998). Physics-based visualization of dense natural clouds. I. Three-dimensional discrete-ordinates radiative transfer. *Appl. Optics*, **37**, 7718–7728.

Truelove, J.S. (1988). Three-Dimensional radiation in absorbing-emitting-scattering media using the discrete-ordinates approximation. *J. Quant. Spectrosc. Radiat. Transfer*, **39**, 27–31.

Wiscombe, W.J. (1977). The delta-M method: Rapid yet accurate radiative flux calculations for strongly asymmetric phase functions. *J. Atmos. Sci.*, **34**, 1408–1422.

Suggested Reading

● **for Explicit Methods:**

Evans, K.F. (1993). Two-dimensional radiative transfer in cloudy atmospheres: The spherical harmonic spatial grid method. *J. Atmos. Sci.*, **50**, 3111–3124.

Evans, K.F. (1998). The spherical harmonics discrete ordinate method for three-dimensional atmospheric radiative transfer. *J. Atmos. Sci.*, **55**, 429–446.

Gerstl, S.A. and A. Zardecki (1985). Discrete-ordinate finite-element method for atmospheric radiative transfer and remote sensing. *Appl. Optics*, **24**, 81–93.

Lewis, E.E. and W.F. Miller, Jr. (1993). *Computational Methods of Neutron Transport*. xvi+401 pp., American Nuclear Society, La Grange Park (IL).

Modest, M.F. (1993). *Radiative Heat Transfer*. McGraw-Hill, Inc., New York (NY).

Thomas G. and K. Stamnes (1999). *Radiative Transfer in the Atmosphere and Ocean*. Cambridge University Press, New York (NY).

● **for Monte Carlo Methods:**

Barker, H.W., J.-J. Morcrette and G.D. Alexander (1998). Broadband solar fluxes and heating rates for atmospheres with 3D broken clouds. *Quart. J. Roy. Meteor. Soc.*, **124**, 1245–1271.

Cahalan, R.F., W. Ridgway, W.J. Wiscombe, S. Gollmer and Harshvardhan (1994). Independent pixel and Monte Carlo estimates of stratocumulus albedo. *J. Atmos. Sci.*, **51**, 3776–3790.

Davies, R. (1978). The effect of finite geometry on the three-dimensional transfer of solar irradiance in clouds. *J. Atmos. Sci.*, **35**, 1712–1725.

Marchuk, G., G. Mikhailov, M. Nazaraliev, R. Darbinjan, B. Kargin and B. Elepov (1980). *The Monte Carlo Methods in Atmospheric Optics*. 208 pp., Springer-Verlag, New-York (NY).

McKee, T.B. and S.K. Cox (1974). Scattering of visible radiation by finite clouds. *J. Atmos. Sci.*, **31**, 1885–1892.

O'Brien, D.M. (1992). Accelerated quasi-Monte Carlo integration of the radiative transfer equation. *J. Quant. Spectrosc. Radiat. Transfer*, **48**, 41–59.

O'Hirok, W. and C. Gautier (1998). A three-dimensional radiative transfer model to investigate the solar radiation within a cloudy atmosphere. Part I: Spatial effects. *J. Atmos. Sci.*, **55**, 2162–2179.

Takara, E.E. and R.G. Ellingson (1996). Scattering effects on longwave fluxes in broken cloud fields. *J. Atmos. Sci.*, **53**, 1464–1476.

Titov, G.A., T.B. Zhuravleva, and V.E. Zuev (1997). Mean radiation fluxes in the near-IR spectral range: Algorithms for calculation. *J. Geophys. Res.*, **102 (D2)**, 1819–1832.

5

Approximation Methods in
Atmospheric 3D Radiative Transfer
Part 1:
Resolved Variability and Phenomenology

A.B. Davis and I.N. Polonsky

5.1   Introduction and Overview 283

5.2   Efficient/Approximate Computation
of Detailed 3D Radiation Fields 286

5.3   Efficient Computation of Detailed 3D Radiation Fields:
Discussion and Outlook 316

5.4   Three-Dimensional Radiation Transport Phenomenology:
A Case Study 322

5.5   Concluding Remarks 334

References 335

5.1 Introduction and Overview

In this chapter and the following one, we survey most of the solutions that have been proposed so far in approximation theory for three-dimensional (3D) radiative transfer (RT) for cloudy atmospheres. No single solution is a panacea because each application has its own needs and tolerance to error. Two broad categories of RT problems are, however, addressed in these pages.

1. The first type of problem follows directly from the program covered in the previous chapter, targeting the estimation of detailed 3D radiance fields when all the information about spatial variability is provided explicitly. The problem is to develop computational 3D RT models that deliver 3D distributions of radiative heating/cooling rates, trading the accuracy of Chap. 4's methods for far more efficiency.[1] These solutions apply directly to cloud system-resolving models (CSRMs) and large eddy simulations (LESs) where grid spacing ranges from

---

[1] In this Chapter, we equate "efficiency" with execution speed, basically CPU cycles.

tens of meters to a few kilometers. Such approximate but efficient 3D RT models will eventually be used in remote-sensing applications, but they will first have to make the computational leap from flux to radiance estimation. Either way, the variability is *resolved* at least down to some fine grid-scale.

2. The second, and in a sense complementary, type of challenge arises when the desired outcomes are vertical profiles of fluxes and heating-rate (flux divergences) averaged over large areas, hence both angular and spatial integrals of spectral radiance (and then spectral integration as well). The essential difference with the problems defined above, and addressed in Chaps. 3 and 4, is that now only a limited amount of information is available to guide necessarily statistical descriptions of optical variability at the *unresolved* scales. The main motivation and application here is the need for accurate radiative budgets in global climate models (GCMs) that have horizontal grid-spacings typically on the order of hundreds of kilometers and vertical (generally thermodynamically-weighted) grid-spacings on the order of tenths to a few kilometers.

Figure 5.1 presents a geometrically correct schematic (i.e., without any vertical exaggeration) of the situations for these two broad classes of 3D RT problems. Figure 5.1a shows a 2D cross-section through the tropospheric portion of a single GCM column, and for which horizontally averaged radiative fluxes are sought. Notice that the natural "zone of radiative influence" of density fluctuations in either the optical properties or radiation fields is on the order of the thickness of the troposphere. The large (horizontal/vertical) aspect ratio of the optical medium (essentially the clouded troposphere) justifies the averaging of many independent, subgrid-scale column computations to approximate profiles of domain-averaged fluxes. Moreover, radiant power flowing through the sides of GCM-size columns is small compared to what flows vertically (except maybe at very low sun). This justifies treating entire GCM grid-cell as an ensemble of neighboring yet radiatively independent columns, a powerful approach known as the Independent Column Approximation (ICA).

Figure 5.1b illustrates a few columns of a typical CSRM. Like GCM columns, radiative flux divergences are sought for each resolved grid-cell, but now their dimensions are on the order of or smaller than the above-mentioned zone of radiative influence. Since individual CSRM cells influence fluxes for neighboring cells, independent column-*by*-column computation of RT is inaccurate because it is so obviously unphysical.[2] The same goes for LES models except that their cells are even smaller and inaccuracies potentially more extreme. The issue of fluxes through the walls of CSRM and LES domains is usually side-stepped with cyclic horizontal boundary conditions for both optical properties and RT.

CSRMs and LESs are used to investigate cloud-scale processes. At their resolutions, some processes, such as convection, are resolved explicitly so they need not be parameterized. The RT, however, is still computed with legacy code, GCM schemes now applied independently to each resolved column. Physically this means that photon trajectories are constrained to individual columns (imagine the walls dividing

---

[2] The important quantity here is not (net) horizontal fluxes per se – they exist in uniform media under slant illumination – but horizontal flux-divergence.

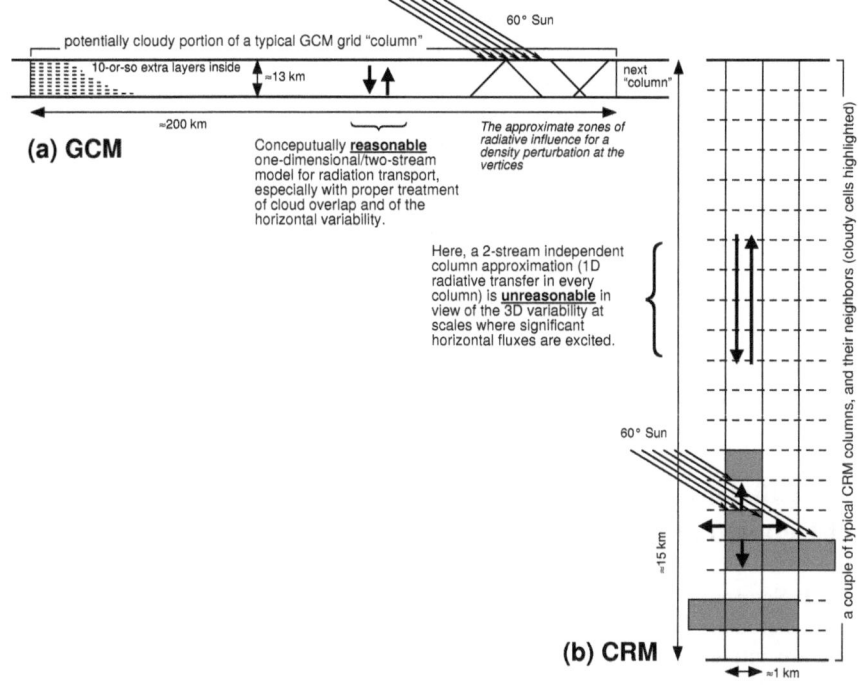

Fig. 5.1. Geometrically correct representations of the 3D cloudy atmosphere problems of interest in this chapter. (**a**) A transect through a GCM (Global Climate Model) grid-cell above the Earth's surface, showing a typical number of atmospheric layers. (**b**) A few columns extracted from a CRM (Cloud Resolving Model), now called a CSRM (Cloud *System* Resolving Model). More discussion in the text

neighboring column as perfect mirrors), even though they are now much taller than they are wide. Clearly, these dynamical models are in need of 3D RT schemes that, ideally, will consume only a few more CPU cycles than their two-stream counterparts, yet capture at least the first-order effects of 3D RT.

Cloud and cloud-system modeling efforts are not the only outlets for efficient approximations leading to improved radiation energetics (i.e., boundary fluxes and local heating rates). For instance, atmospheric photochemistry modelers are also interested in the spatial fluctuations of actinic flux driven by 3D cloud variability. The next level of challenge is to predict 3D radiance fields efficiently, largely to serve the remote sensing communities interested in clouds or in properties of surface (pixels) in view but near or between dense clouds. The ultimate goal here is to perform the *inverse* problem of inferring physical information about cloud/surface structure in the presence of 3D RT effects caused by the clouds. Ideas have been advanced recently on how to address at least parts of this problem.

In this chapter, we cover the first kind of approximation challenge in atmospheric 3D RT, the *deterministic* problem where we are given all the structural and optical detail there is to know about the medium. Section 5.2 is devoted to the forward 3D

RT problem starting with fluxes and working towards radiances (diffusion, discrete angles, perturbation, and hybrid/semi-empirical methods are covered). In Sect. 5.3, we discuss informally the role we anticipate for approximation theory in the near- and long-term future for forward (cloud-modeling) and inverse (remote-sensing) 3D RT. Section 5.4 exploits a specific cloud variability model (flat boundaries but a periodic sine-wave in extinction) for a tutorial in the phenomenology of 3D RT based on classic approximations (homogeneity, ICA and 3D diffusion). Finally, Sect. 5.5 offers some parting remarks on the broad topic of this chapter.

The following Chap. 6 by Barker and Davis covers the second class of approximation question in atmospheric 3D RT, the *probabilistic* problem where we are only asked to compute only domain-average means but only given a few statistical parameters of the medium (typically, cloud fraction, distribution of optical depths in cloudy portion, and correlation coefficients, possibly even correlation scales).

5.2 Efficient/Approximate Computation of Detailed 3D Radiation Fields

5.2.1 Scope

We are interested in determining the 3D spatial details of the radiation fields and maybe even its angular distribution to some extent … but we are not willing to pay the computational price associated with a numerical solution of the full-blown 3D RT equation.

This program calls for a class of approximations in computational 3D RT where we trade accuracy versus efficiency. This trade-off can be of course explored with the "exact" methods described in Chap. 4: spatial and/or angular grids can be coarsened and expansions can be severely truncated in the explicit methods, and relatively few photon histories can be traced in the random quadrature methods (a.k.a. Monte Carlo simulation techniques). We are not interested here in such approaches here, which is not to say that those possibilities should not be considered in some applications. It is just that there is nothing new to say about them. It is clear however that such straightforward techniques will be sub-optimal in many other applications because stripped-down explicit methods are not designed to work on coarse grids and, in this context, we generally want more than a large-scale average (which is all a limited Monte Carlo can deliver with any semblance of accuracy). Luckily, there are better ways of allocating computational resources. Which way is the best is a highly application-specific question and, moreover, the answer will likely change in time since this is a very active area of research. Of course, we will keep in mind the methods presented in Chap. 4 as standards for accuracy (to match as best as possible) and efficiency (to surpass as much as possible). Those methods solve the 3D RT equation while here we solve different transport equations, or even bypass equation solving altogether if physical intuition suggests that route.

Inasmuch as they can be written and implemented differently than limited run-time versions of methods from Chap. 4, truncated order-of-scattering expansions are not of direct interest to us either because they are adapted to optically thin media. For the same basic reason, we will also forgo the small scattering angle or "SSA" approximation[3] although it should not be overlooked for the optically thin regions of 3D cloudy media. These elements of the 3D RT approximation portfolio at large may prove important in future research into adaptive hybrid methods that combine approximation methods with restrictions to regions where they work well.

At the other end of the accuracy and efficiency scales that originate with Chap. 4's methods, we find the (local) ICA: 1D RT on a column-by-column basis. This is certainly expedient and can be implemented for efficiency as well with analytical two-stream approximations for the 1D RT and, as needed, ordering the computations by variable parameter value rather than by position on the horizontal spatial grid to be parsed. This *local* ICA is not to be confused with the *global* ICA, namely, the domain-average of the local ICA which can be very accurate, especially for fluxes, even though the local assumption can be catastrophically wrong. As stated in the Introduction, the prime application for the (global) ICA is GCM parameterization development and we differ further discussion of it to Sect. 5.4, the next chapter, and a few others in the volume. By contrast, we are interested here in small-scale variations of the radiation fields driven by specific variations in the optical properties. So we will focus on resolutely "post-ICA" models for 3D RT in cloudy atmospheres. These models will be more accurate than the local ICA by definition but they will necessarily consume at least a few more CPU cycles.

The emphasis will be on the computational aspects of 3D RT but that part of the discussion will be neither technical (in the sense of numerical analysis methods) nor quantitative (speed-up achieved versus cost in accuracy). Rather, we have strived to cover the diversity of the methods without losing track of how these methods relate to – and sometimes support – each other. We thus underscore the "work-in-progress" status of this active research area.

5.2.2 3D Diffusion Theory

This model goes back at least to Eddington's (1916) investigations into the equilibrium of stars based on the prior formulation of the RT equation by Schwarzschild (1906). One could argue that all two-stream theories, starting as far as we know with Schuster's(1905) seminal study of scattering atmospheres, are mathematically equivalent to a 1D diffusion theory. We can derive the diffusion equation from the RT equation in a variety of ways, and learn much about its limitations in the process. The most recent derivation is given in Sect. 5.2.3. The most illuminating derivation is through RT asymptotics, and we refer to Pomraning (1989) for the mechanics. The most straightforward derivation starts with the postulate that

$$I(x, \Omega) = \frac{1}{4\pi}[J(x) + 3\Omega \cdot F(x)] , \qquad (5.1)$$

---

[3] This technique is known in the particle transport literature as the Fokker-Planck approximation. Mathematically, it has the flavor of a diffusion theory, but one that operates in direction- or Ω-space (tangential on the unit sphere to the collimated source direction Ω_0). The original reference seems to be Fermi (1941).

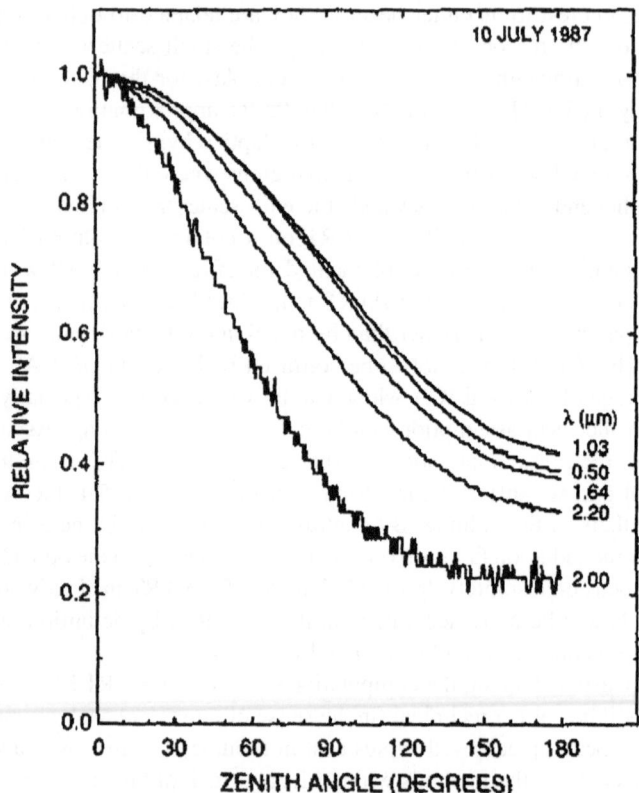

Fig. 5.2. Spectral radiance ratio $I_\lambda(x, \theta, \phi)/I_\lambda(x, 0, \phi)$ for the indicated values of λ where x is the position of the U. of Washington's Convair C-131A at 09:37 PDT along its flight inside a marine stratocumulus layer on July 10, 1987 (during FIRE) while ϕ defines the Cloud Absorption Radiometer's scan plane, perpendicular to the line-of-flight, reproduced with permission from King et al. (1990) (their Fig. 6). For all but $\lambda = 2$ μm we have a clear diffusion-domain dependence in $\cos\theta$ as prescribed in (5.1). For $\lambda = 2$ μm a CO_2 absorption feature leads to less light (increased noise) and the requirement of higher-order spherical harmonics to model the radiance. For more on the CAR instrument, see the URL http://car.gsfc.nasa.gov/

where the scalar and vector are defined as

$$\begin{Bmatrix} J(x) \\ F(x) \end{Bmatrix} = \int_{4\pi} \begin{Bmatrix} 1 \\ \Omega \end{Bmatrix} I(x, \Omega) d\Omega \tag{5.2}$$

and discussed in detail in Chap. 3. The radiance field, viewed as a distribution in direction-space, is thus limited to an isotropic term proportional to J and a dipole term oriented along F and proportional to $||F||$. Both quantities are expected to be functions of the position vector x. Figure 5.2 shows relative radiance measurements by King et al. (1990) inside a marine stratocumulus layer. We clearly see the characteristic cosine dependence with zenith angle for all but the noisy data at 2 μm where there is significant CO_2 absorption.

All higher-order spherical harmonics are neglected, making the postulate in (5.1) equivalent to a truncation in spherical harmonics (hence to a "closure" in angular moments). Accordingly, the scattering phase function in diffusion theory is effectively reduced to just two terms in its Legendre expansion,

$$P(\boldsymbol{x}, \theta_s) = 1 + 3g(\boldsymbol{x}) \cos \theta_s, \tag{5.3}$$

where θ_s is the scattering angle and $g(\boldsymbol{x})$ is, possibly position dependent, asymmetry factor from Chap. 3 (namely, the mean of $\cos \theta_s$ over $P(\boldsymbol{x}, \theta_s)/4\pi$). There are no higher-order terms in the radiance field to interact with after through higher-order Legendre coefficients. Note that the proposed phase function become negative if $|g| > 1/3$ and this immediately causes problems for diffusion theory at small optical depths (i.e., when phase function details make a difference).

Substitution of (5.1) and (5.3) into the 3D RT equation (in the absence of anisotropic volume sources) leads to an important hallmark of diffusion theory that is Fick's law:[4]

$$\boldsymbol{F} = -D'(\boldsymbol{x}) \nabla J, \tag{5.4}$$

where

$$D'(\boldsymbol{x}) = \ell_t(\boldsymbol{x})/3 \tag{5.5}$$

is photon "diffusivity" for steady-state 3D problems.[5] A fundamental scale of diffusion theory that appears in (5.5) is the *transport* mean-free-path (MFP)

$$\ell_t = \frac{1}{\sigma - \sigma_s g} = \frac{1}{(1 - \varpi_0 g)\sigma} \tag{5.6}$$

which is simply the usual photon MFP $1/\sigma$ divided by the famous Eddington/van de Hulst scaling factor $(1 - \varpi_0 g)$. We recall here that the notion of a MFP cannot be interpreted literally in 3D RT because only in case where $\sigma(\boldsymbol{x}) \equiv$ constant can we compute a priori what it is (cf. Chap. 3). So (5.6) is just a definition, and an indication of the value of the transport MFP to be expected.

In the frequently used formulation where un-collided (directly-transmitted) radiation is treated separately, and thus *not* included in J and $\boldsymbol{F}$, and if the scattering is not symmetric in $\cos \theta_s$ (hence if $g \neq 0$), then there is an additional term on the right-hand side of Fick's law in (5.4) that we will write as $3D'(\boldsymbol{x}) q_F(\boldsymbol{x})$. In particular, this term captures to some extent the directionality of the solar-beam injection source induced by the forward-peaked phase function. For this important kind of source, one should also apply the widely-used "δ-Eddington" rescaling. Joseph et al. (1976) introduced this improvement (originally for plane-parallel 2-stream theory) by noting that a better model than (5.3) for real phase functions uses two parameters:

---

[4] If (1) anisotropic volume sources are present and (2) the requested flux fields are for the diffuse component only in a solar RT problem, then there is an extra term in Fick's law.

[5] For time-dependent diffusion theory in d spatial dimensions, diffusivity is defined as $D = c\ell_t/d = cD'$ where c is the particle velocity, here, the speed of light.

$$P(\mu_s) = 2f\delta(1 - \mu_s) + (1 - f)(1 + 3g'\mu_s) , \qquad (5.7)$$

where $\mu_s = \cos\theta_s$. All the local optical quantities relevant to 3D diffusion theory are then modified:

$$\begin{aligned}
\sigma' &= (1 - \varpi_0 f)\sigma, \\
\varpi_0' &= \varpi_0(1 - f)/(1 - \varpi_0 f), \qquad (5.8)\\
g' &= (g - f)/(1 - f).
\end{aligned}$$

The choice of f is not obvious but an approach made popular by Joseph et al. is to fit the first two moments of the Henyey-Greenstein phase function, thus leading to $f = g^2$. For observed cloud phase functions (which do not resemble Henyey-Greenstein models at all, cf. Fig. 3.9 in Chap. 3), we have $g \approx 0.75-0.85$ hence $f \approx 0.56-0.72$ and $g' = g/(1+g) \approx 0.43-0.46$. This does not resolve the small optical depth issue noted earlier but helps. At any rate, the δ-rescaling makes no difference for thermal emission, the other important atmospheric source of radiation, which is isotropic, nor for isotropic boundary sources (as is the case for the computation of Lambertian surface albedo or surface thermal emission effects).[6]

The reason we talk about "diffusion" theory, rather than P_1 theory (a first-order spherical harmonic truncation in the terminology of Chap. 4), is because of the associated time-dependent theory. Although the time-dependent formulation of the problem is out of our present scope, it is helpful to bear in mind the idea of photons leaving their source and executing a long sequence of small ($\approx$MFP) steps, from one scattering event to the next, before being absorbed or escaping the medium. In the course of this random walk (which is precisely what is implemented in Monte Carlo schemes) the first thing that happens is loss of directional memory. It can be shown (Davis and Marshak, 1997) that this takes $\approx 1/(1 - g)$ steps if $\varpi_0 \approx 1$. By that time, it is as if the photon has made one larger step (≈ 1 transport MFP), and then scattered isotropically. Diffusion theory becomes increasingly accurate as the photon population becomes dominated by individuals that have performed one or more of these effective steps and isotropic scatterings. It is now easy to see when diffusion will work well. At least qualitatively, we can say that the more scattering (less absorption), the more isotropic the scattering, the further (a couple of ℓ_t at least) from strongly directional sources and from absorbing boundaries (hence quite large optical thicknesses), then the better for diffusion theory. What kind of variability diffusion can sustain without losing accuracy remains an open question. A priori, all we need to exclude are regions of very small extinction (making the local MFP too large).

To continue formulating mathematically the diffusion problem, one substitutes its defining Fickian law (5.4), with an extra source term denoted $3D'\boldsymbol{q}_F$ on the right-hand side if J and $\boldsymbol{F}$ model only the diffuse field, into the (exact) expression for

---

[6] To see this, note that $(1 - \varpi_0 g)\sigma$ hence the transport MFP, and $(1 - \varpi_0)(1 - \varpi_0 g)\sigma^2$ hence the diffusion scale, are left invariant by the proposed rescaling. The ratio of these quantities, the similarity factor $S(\varpi_0, g)$ in (5.16), is therefore also invariant. In the absence of anisotropic sources, these are indeed the only combinations that appear in the 3D diffusion equation and its boundary conditions.

photon conservation

$$\nabla \cdot \boldsymbol{F} = -\sigma_a(\boldsymbol{x})J(\boldsymbol{x}) + q_J(\boldsymbol{x}) \tag{5.9}$$

from Chap. 3, where $q_J(\boldsymbol{x})$ is the local rate of isotropic photon (equivalently, radiant energy) creation in $m^{-3}s^{-1}$ (or W/m^3). We obtain a standard 2nd-order partial differential equation (PDE) with variable coefficients:

$$-\nabla \cdot [D'(\boldsymbol{x})\nabla J] + \sigma_a(\boldsymbol{x})J = q(\boldsymbol{x}) \;, \tag{5.10}$$

where

$$q(\boldsymbol{x}) = q_J(\boldsymbol{x}) - 3\nabla \cdot (D'\boldsymbol{q}_F) \;, \tag{5.11}$$

where $\boldsymbol{q}_F = \int_{4\pi} \boldsymbol{\Omega} Q(\boldsymbol{x}, \boldsymbol{\Omega}) d\boldsymbol{\Omega}$ and $Q(\boldsymbol{x}, \boldsymbol{\Omega})$ is the source term in the 3D RT equation from Chap. 3.

As an important example, isotropic thermal sources are given by

$$q(\boldsymbol{x}) = q_J(\boldsymbol{x}) = \sigma_a(\boldsymbol{x})B[T(\boldsymbol{x})] \;, \tag{5.12}$$

where $T(\boldsymbol{x})$ is the temperature field (and $\boldsymbol{q}_F(\boldsymbol{x}) \equiv 0$). For solar illumination problems using the recommended direct+diffuse decomposition, the uncollided flux in the presence of forward-scattering is determined by

$$\begin{aligned} q_J(\boldsymbol{x}) &= F_0\sigma_s(\boldsymbol{x})\exp[-\tau_0(\boldsymbol{x};\boldsymbol{\Omega}_0)] \;, \\ \boldsymbol{q}_F(\boldsymbol{x}) &= 3F_0\sigma_s(\boldsymbol{x})g\boldsymbol{\Omega}_0\exp[-\tau_0(\boldsymbol{x};\boldsymbol{\Omega}_0)] \;, \end{aligned} \tag{5.13}$$

where F_0 is the solar constant and $\tau_0(\boldsymbol{x})$ is optical distance (integral of $\sigma(\boldsymbol{x})$) from point $\boldsymbol{x}$ to the upper boundary in the direction $-\boldsymbol{\Omega}_0$ of the sun:

$$\tau_0(\boldsymbol{x}, \boldsymbol{\Omega}_0) = \int_0^{s_0(\boldsymbol{x},\boldsymbol{\Omega}_0)} \sigma(\boldsymbol{x} - s\boldsymbol{\Omega}_0)ds \;, \tag{5.14}$$

where $s_0(\boldsymbol{x}, \boldsymbol{\Omega}_0)$ determines the unique interception point of the ray $\{\boldsymbol{x}, -\boldsymbol{\Omega}_0\}$ with the (convex) boundary ∂M. For a plane-parallel slab medium ($0 < z < h$), we have $s_0(\boldsymbol{x}, \boldsymbol{\Omega}_0) \equiv s_0(z, \mu_0) = (h - z)/\mu_0$ and $\tau_0(\boldsymbol{x}, \boldsymbol{\Omega}_0)$ will generally depend on the internal variability.

With the same caveat as for the meaning of the MFP in 3D media, we can see another important physical quantity appearing here, the so-called "diffusion" scale:

$$L_d = \sqrt{\frac{D'}{\sigma_a}} = \frac{1}{\sqrt{3(1 - \varpi_0)(1 - \varpi_0 g)}\sigma} \;. \tag{5.15}$$

This is the scale that we would use to render dimensionless the differential operator on the left-hand side of (5.10). The scale ratio

$$S(\varpi_0, g) = \frac{\ell_t}{L_d} = \sqrt{\frac{3(1 - \varpi_0)}{1 - \varpi_0 g}} \tag{5.16}$$

is known as the "similarity factor." It measures the relative importance of absorption versus scattering in the photon transport.

Mathematically equivalent elliptical PDE problems as in (5.10) are solved routinely by designers and diagnosticians in electrical, mechanical, thermal, and nuclear engineering as well as in subsurface hydrology and geological sciences. In most of these fields however, the 2nd-order PDE is considered as the *exact* formulation of the physical problem in steady-state conductivity, electro- or magneto-statics, thermal conductivity, neutron transport, flow through porous substrates, and so on. Any problem with a "continuity" equation like (5.9) and a "constitutive" equation like (5.4) will end in this kind of PDE problem. Yet another research area, presently very active, that uses photon diffusion per se is non-invasive medical diagnostics in soft tissue, specifically by diffusing photon spectroscopy/imaging. See, e.g., Yodh and Chance (1995) for an overview of this rapidly growing literature where photon diffusion is often called photon "migration."

Apart from the approximate-versus-exact difference in modeling perspective from their pervasive engineering and medical counterparts, radiation applications of diffusion theory have distinctive boundary conditions (BCs). They are necessarily of the mixed or Robin kind, rather than the more standard kinds: Dirichlet (given "density" $J(x)$), or von Neumann (given "current" across the boundary $n(x) \cdot F$), where $x \in \partial M$ (a standard notation for the outer boundary of a 3D domain M) and $n(x)$ is the normal to ∂M at x. That is because *incoming* flux can not be expressed with $J(x)$ nor with $n(x) \cdot F(x)$. Nor can any combination of these two available boundary quantities exactly translate a RT equation BC as simple as no incoming radiance, not even no incoming flux. However, approximations are possible and combinations are necessary to express them (Case and Zweifel, 1967). For non-reflective boundaries, we therefore impose

$$\frac{1}{2} \left[1 - \chi \ell_t(x) \, n(x) \cdot \nabla \right] J = f(x) , \tag{5.17}$$

where $f(x)$ is the local rate of photon (equivalently, radiant energy) creation at the boundary expressed in $m^{-2}s^{-1}$ (or W/m^2) and assumed isotropic. For instance, we take $f(x) \equiv 0$ to describe an absorbing surface – also called "vacuum" BCs – while the boundary-source Green function results for $f(x) = \delta(x)$.

The non-dimensional quantity χ determines the optimal mixture (at $\alpha(x) \equiv 0$) of boundary values for photon density $J(x)$ and photon current $-D'(x) \, n(x) \cdot \nabla J$. $\chi \ell_t$ is known as the "extrapolation length" and it has been the object of many classic studies, from which we retain:[7]

---

[7] It is not obvious that χ should always be constant along the boundary, but that is the conventional assumption.

$$\chi = \begin{cases} 1/3, & \text{Fick} \\ 1/\sqrt{3}, & \text{Marshak} \\ 2/3, & \text{Eddington} \\ 0.7104\cdots, & \text{Milne-Davison} \\ 1, & \text{6-flux theory} \\ 4/3, & \text{optically thin limit} \end{cases} . \tag{5.18}$$

The first and last values in the above list are primarily of academic interest. The remaining values have all proven useful in the experience of the present authors. "Utility" here is defined as better reproducing the solution of the RT equation in a validation exercise for the diffusion approximation where the reference computation is by a technique from Chap. 4. In particular, the Eddington value of $\chi = 2/3$ is the consistent choice for everywhere vacuum BCs, and therefore non-vanishing $q(x)$ from $q_J(x)$ and (optionally) $q_F(x)$ in (5.11).

This completes the mathematical formulation of the 3D photon diffusion problem at least for non-reflective BCs. For an illustration of the solution of the steady-state 1D plane-parallel diffusion problem, see Chap. 12 where the application is in fact to analytically estimate the mean number of scatterings (normally, a time-dependent problem). In numerical analysis, a 2nd-order elliptical PDE, even with position-dependent coefficients, is a much simpler problem to solve than the full RT equation. Because of its widespread applicability, there are many numerical schemes for solving the PDE in (5.10) with BCs in (5.17). Multi-grid and sparse-matrix methods are computationally optimal – algorithmic complexity is $O(N)$ – and can be made unconditionally stable. If only values at a few points are sought, then there are backward Monte Carlo solutions that bear very little resemblance to those used for the RT equation in Chap. 4. When "shopping" for numerical diffusion equation solvers, a radiation transport modeler should be aware of the a priori accuracy of the method. It can indeed be excessive – and therefore a waste of CPU time – since in 3D RT, unlike in its many engineering applications, the diffusion equation is itself an approximation to the actual problem at hand.

Certain results from the diffusion theoretical computation are of particular interest in atmospheric radiation: out-going 2D boundary flux fields, and the local 3D radiant energy absorption rate. Outgoing boundary fluxes are obtained simply by changing the sign of the second (photon current) term in (5.17).

Solar-type problems can be solved by setting $f(x) = |n(x) \cdot \Omega_0| F_0$ in the non-reflective BC in (5.17). For a geometrically plane-parallel medium, this translates simply to $f(x) \equiv \mu_0 F_0$ and $q_J(x) \equiv 0$. In this important case, the local albedo field is given by

$$R(x,y) = \frac{1}{2\mu_0 F_0} \left[1 + \chi \ell_t(x,y,h) \left(\frac{\partial}{\partial z} \right) \right] J \bigg|_{z=h} = \frac{J(x,y,h)}{\mu_0 F_0} - 1, \tag{5.19}$$

while transmittance is given by

$$T(x,y) = \frac{1}{2\mu_0 F_0} \left[1 - \chi \ell_t(x,y,0) \left(\frac{\partial}{\partial z} \right) \right] J \bigg|_{z=0} = \frac{J(x,y,0)}{\mu_0 F_0} . \tag{5.20}$$

Local solar absorption rate is given by

$$-\nabla \cdot \boldsymbol{F} = \sigma_a(\boldsymbol{x})J(\boldsymbol{x}) \tag{5.21}$$

a direct consequence of the energy conservation law in (5.9) when $q_J(\boldsymbol{x}) \equiv 0$.

Solar problems are however best treated using vacuum BCs and the anisotropic internal source terms in (5.13) rather than using an isotropic boundary source term in (5.17). In this case, we take $f(\boldsymbol{x}) \equiv 0$ and the albedo field is

$$R(x, y) = \frac{1}{2\mu_0 F_0}\left[1 + \frac{2\ell_t(x, y, h)}{3}\left(\frac{\partial}{\partial z}\right)\right]J\bigg|_{z=h} = \frac{J(x, y, h)}{\mu_0 F_0}, \tag{5.22}$$

while the transmittance fields are given by

$$T_{\text{dif}}(x, y) = \frac{1}{2\mu_0 F_0}\left[1 - \frac{2\ell_t(x)}{3}\left(\frac{\partial}{\partial z}\right)\right]J\bigg|_{z=0} = \frac{J(x, y, 0)}{\mu_0 F_0}, \tag{5.23}$$

$$T_{\text{dir}}(x, y) = \exp[-\tau_0(x, y, 0; -\boldsymbol{\Omega}_0)], \tag{5.24}$$

$$T(x, y) = T_{\text{dir}}(x, y) + T_{\text{dif}}(x, y) \tag{5.25}$$

corresponding respectively to the diffuse, direct and total transmissions. Local solar absorption is given in this context by the right-hand side of (5.21) with an extra term for uncollided flux, namely, $\sigma_a(\boldsymbol{x})F_0 \exp[-\tau_0(\boldsymbol{x}; -\boldsymbol{\Omega}_0)]$.

At a Lambertian reflective surface (cf. Chap. 3), we are simply sending back into the medium a fraction $\alpha_s(\boldsymbol{x}) \leq 1$ (the local albedo of the surface) of the flux. This leads to a simple modification of the BC in (5.17):

$$\frac{1}{2}\{[1 - \alpha_s(\boldsymbol{x})] - [1 + \alpha_s(\boldsymbol{x})]\chi\ell_t(\boldsymbol{x})\,\boldsymbol{n}(\boldsymbol{x}) \cdot \nabla\}J = f(\boldsymbol{x}) \tag{5.26}$$

where $\boldsymbol{x} \in \partial M$ with $\boldsymbol{n}(\boldsymbol{x}) \cdot \boldsymbol{\Omega}_0 < 0$. Note that for a purely reflective boundary, the only constraint in (5.26) is on the photon current (i.e., normal flux) at the interface. In the case where J is only for the diffuse radiation (and uncollided fluxes are treated separately), we can identify the given boundary source $f(\boldsymbol{x})$ on the right-hand side of (5.26) with the effective source created by reflecting the uncollided stream of photons hitting the reflective but otherwise non-illuminated part of the boundary:

$$f(\boldsymbol{x}) = \alpha_s(\boldsymbol{x})\boldsymbol{n}(\boldsymbol{x}) \cdot \boldsymbol{\Omega}_0 F_0 \exp[-\tau_0(\boldsymbol{x}; -\boldsymbol{\Omega}_0)]$$

where $\boldsymbol{x} \in \partial M$ with $\boldsymbol{n}(\boldsymbol{x}) \cdot \boldsymbol{\Omega}_0 > 0$.

We now describe three examples of diffusion theory application, one numerical and two analytical. We finish with a discussion of real-world observations of diffusion transport signatures.

Example 1: The I3RC Square-Wave Cloud Case

Figures 5.3–5.4 show comparisons of the outcomes of computational diffusion results and counterparts for an exact RT equation solver using Case 1 from the Intercomparison of 3D Radiation Codes project (I3RC; Cahalan et al., 2005). This is a cyclical alternation of tenuous and dense regions with Heaviside jumps in extinction along the x-axis characterized by the following structural parameters:[8] $\langle\tau\rangle = 10$, $\delta\tau = 8$, $h = 0.25$ km, and aspect ratio $L/h = 2$. Solar illumination and optical parameters are: $\mu_0 = 1$ and $1/2$ (collimated beam, with injection through a volume source term) and $\varpi_0 = 1$ and 0.99, Henyey–Greenstein phase function with $g = 0.85$. The ED3D code (Qu, 1999) was used to obtain the 3D diffusion results (provided graciously by Z. Qu) while TWODANT (Alcouffe et al., 1997) was used for the benchmark 3D RT equation solution (provided graciously by D. Kornreich). Benchmarks from 1D RT are the ICA estimates for each region, by 1D RT (DISORT) and by its well-known (analytical) diffusion approximation. We note the excellent performance of diffusion in 1D RT.

For the non-absorbing case ($\varpi_0 = 1$) we are looking at in Fig. 5.3, the variations of $R(x)$, $T(x)$ and the so-called horizontal flux field

$$H(x) = 1 - A(x) - R(x) - T(x) , \tag{5.27}$$

where column absorption $A(x) \equiv 0$ in this instance. This is, more precisely, the column-integrated and normalized horizontal divergence rate

$$H(x) = \frac{1}{\mu_0 F_0} \int_0^h \nabla_{\mathrm{h}} \bullet \boldsymbol{F} \mathrm{d}z = \frac{1}{\mu_0 F_0} \int_0^h \left(-\frac{\partial F_z}{\partial z} \right) \mathrm{d}z , \tag{5.28}$$

where $F_0 = 1$ in the present simulation. In the above, we have simply grouped and vertically integrated terms from the total (diffuse+direct) radiant energy conservation (5.9). The right-hand side indeed yields $[1-R(x)]-[T(x)-0]$. In the ICA, $H(x) \equiv 0$ and in post-ICA 3D photon transport $H(x)$ (in general, non-vanishing horizontal gradients) have been described justifiably as "pseudo-source/sink" terms from a vertical 1D RT perspective. For overhead sun, we note the transmittance in excess of unity in the middle of the tenuous region: a characteristic 3D RT effect caused by the channeling of radiation flow away from the neighboring dense regions. Another remarkable finding (captured by both exact and approximate theories) at $\mu_0 = 1$ is the powerful smoothing (reduced variance and no more ICA-induced jump) in reflectance by 3D RT. Opposite effects are found at oblique illumination: although the ICA jump is smoothed, the amplitude of the variability in reflectance is not reduced while the transmittance field is highly smoothed (compared to both the ICA and the overhead sun case). The panels showing $H(x)$ show the strong local violations of energy conservation when horizontal transport is ignored. For overhead sun, they quantify the channeling phenomenon (excess radiation in tenuous regions, deficit in dense ones). For the more complex scenario with slant illumination, the direct correlation of $H(x)$ with $\tau(x)$ is reversed.[9]

---

[8] Modulo a trivial quarter-wave phase shift, this is a square-wave equivalent of the sine-wave cloud model used further on, with a variable L/h.

[9] This does not mean that the same channeling phenomenon as observed for $\theta_0 = 0$ is not happening when $\theta_0 = 60°$, rather that we are looking at it from an awkward angle. See Fig. 5.12 and its discussion further on.

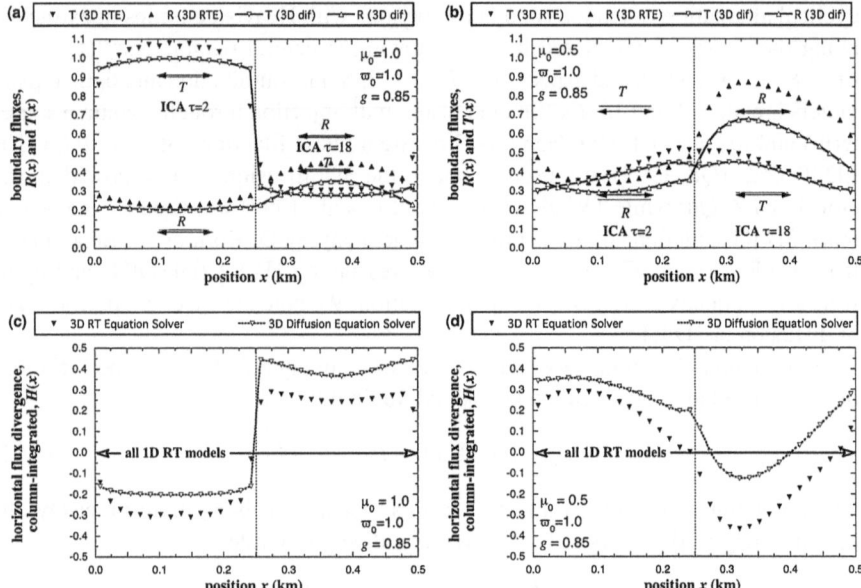

Fig. 5.3. Flux fields for the I3RC "Case 1" square-wave cloud (see text) in the conservative ($\varpi_0 = 1$) case. Benchmarks for comparison with the 3D diffusion theoretical results (ED3D code) are: a full 3D RT equation solution (TWODANT code), and the ICA (using both the 1D RT equation and the analytical diffusion solution). Two solar illumination angles and two boundary fluxes $\{R(x), T(x)\}$ are considered along with the "horizontal fluxes" (or apparent absorption) $H(x) = 1 - R(x) - T(x)$: (a) $R(x), T(x), \theta_0 = 0°$, (b) $R(x), T(x), \theta_0 = 60°$ (c) $H(x), \theta_0 = 0°$, and (d) $H(x), \theta_0 = 60°$. In spite of the mirror symmetry of cloud structure around the vertical planes at $x = 0.125$ and 0.375 km, the uniform $\mu_0 = 1$ illumination and the angularly-integrated response, we note a minor asymmetry in the results from TWODANT. That is because 3D RT equation solvers based on a grid proceed by "sweeps" in a given direction and iterations (see Chap. 4). This gives an indication of the residual numerical error

For the absorbing case ($\varpi_0 = 0.99$), we are looking in Fig. 5.4 precisely at the local variations of column absorption field $A(x)$ because it is closely related to solar heating rate; it is precisely the absorption rate in (5.21), column-integrated and normalized:

$$A(x) = \frac{\sigma_a}{\mu_0 F_0} \int_0^h J(x, z) dz . \tag{5.29}$$

The conditions are far from ideal for diffusion here: $(1 - \varpi_0 g)\tau = 2.9$ on the dense/right-hand side, but only 0.32 on the tenuous/left-hand side. Nonetheless, we see that the predicted radiation fields, in this case column absorption $A(x)$ values, compare quite well with the full 3D RT solutions even though they were obtained at a small fraction of the computer time. Again, the smoother transition around the jump in extinction is captured, to the extent that the output grid ($\Delta x = L/16$) can

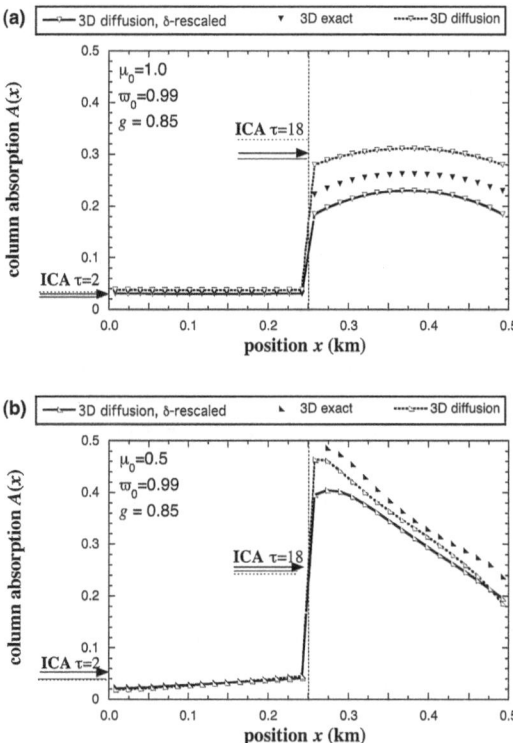

Fig. 5.4. Same as Fig. 5.3 but for column absorption (when $\varpi_0 = 0.99$). Here, the 1D and 3D diffusion results distinguish between with (thin solid line) and without (thin dashed line) the δ-rescaling. Two solar illumination angles are again considered: (**a**) $A(x), \theta_0 = 0°$, and (**b**) $A(x), \theta_0 = 60°$, both with and without δ-rescaling (same visual encodings but thick lines). In the former case, δ-rescaling helps, not in the latter. To appreciate the potential dynamical effect of the bias caused by the ICA assumption, we note that the local solar heating rate can be off by as much as a factor of 2. This happens near the strong gradients when the illumination is significantly off-zenith. By comparison, the error induced by the diffusion approximation is less than ≈10%

show it; this is something the ICA is completely blind to. The ICA adequately estimates the radiation field of interest but only far from the jumps, at high sun, and in the dense portion of the cloud. Otherwise, the ICA bias in absorption is significantly larger than the difference between the exact 3D RT result and the 3D diffusion theoretical result. For the slant illumination case in Fig. 5.4b, there is stronger absorption at the edge of the dense portion that is facing the sun because the abundant directly transmitted (or forward-scattered) sunlight is reaching deep into the medium.

Every approximation theory has its limitations. As mentioned above, it is fair to say that photon diffusion abhors an optical void because, far from executing convoluted random walks, photons will stream forward in ballistic trajectories. So 3D diffusion is well-suited for unbroken stratiform cloud layers but, even there, boundaries will be problematic. How to transition correctly from an interior diffusion solution to an exterior streaming solution across a well-defined radiative boundary layer remains an open question. An approach that has been exercised in 1D at least is to think of diffusion as a theory for computing the multiple scattering source function inside the cloudy portions of the atmospheric medium (see Chap. 3, Sect. 3.8.2),

$$S(\boldsymbol{x}, \boldsymbol{\Omega}) = \sigma_s(\boldsymbol{x}) \int_{4\pi} P(\boldsymbol{\Omega} \cdot \boldsymbol{\Omega}') I(\boldsymbol{x}, \boldsymbol{\Omega}') \frac{d\Omega'}{4\pi} = \sigma_s(\boldsymbol{x})[J(\boldsymbol{x}) + 3g\boldsymbol{\Omega} \cdot \boldsymbol{F}(\boldsymbol{x})] , \quad (5.30)$$

then the formal solution of the 3D RT equation (a.k.a. "long characteristics") can be applied to compute the radiance field at and outside the cloud boundary.

Some approximation theories have advantages that far outweigh their limitations, at least in specific applications. Diffusion is one such theory. A small but representative number of 3D RT problems are indeed tractable analytically in diffusion theory: simple, but not uniform, source distributions; simple, but not plane-parallel, outer structure; a general case of internal variability, but only for the domain average. The following Examples 2–3 illustrate the two former items while the last is mentioned in the Discussion and used in Sect. 5.4.

Example 2: The Steady-State Green Function

As another illustration of diffusion theory, consider the 3D RT problem of Green-function estimation (cf. Chap. 3) for a homogeneous conservatively scattering plane-parallel slab illuminated isotropically from an internal or boundary point. This leads to (5.10) with constant $D' \equiv 1/3(1 - g)\sigma$ and $f(\boldsymbol{x}) = \delta(x)\delta(y)\delta(z - z_0)$, $0 < z_0 < h$, a problem that can be reduced to the classic 2nd-order differential equation in 2-stream theory with a non-vanishing absorption coefficient by applying a 2D horizontal Fourier/Hankel transform,[10] $J(x, y, z) \equiv J(\rho, z) \mapsto J(k, z)$ where $\rho = \sqrt{x^2 + y^2}$ and k is its Fourier conjugate. Davis and Marshak (2002) take a special interest in the Green function at $z = 0$ when $z_0 = h$, the transmission case. Figure 5.5 shows the family of transmitted Green functions in Fourier space while Fig. 5.6 shows a couple of these in physical space. To produce Fig. 5.5, we define the transport (or rescaled) optical depth

$$\tau_t = h/\ell_t = (1 - g)\tau$$

and use $G_T(kh; \tau_t) = J(k, z = h) = T(k\ell_t, \tau_t)/T(0, \tau_t)$ where

$$T(S, \tau_t) = \frac{4\chi S}{2[1 + (\chi S)^2] \sinh(S\tau_t) + 4\chi S \cosh(S\tau_t)} . \quad (5.31)$$

Wavenumber k is identified here with $1/L_d$ in the definition (5.16) of S, hence

$$S = k\ell_t = kh/\tau_t .$$

---

[10] Whether we are discussing a function or its Fourier transform will by made clear by the argument; we will not need to change the symbol for the function itself.

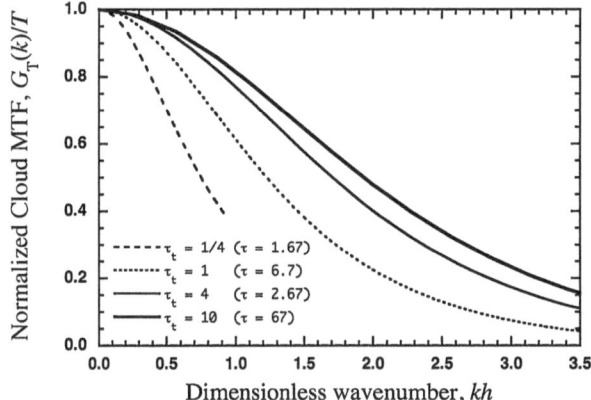

Fig. 5.5. Normalized Green function of a uniform plane-parallel cloud for transmittance in Fourier space. This quantity can be interpreted as the cloud's MTF (Modulation Transfer Function) and it has a closed-form analytical expression from (5.31)–(5.32) in diffusion theory. A number of rescaled optical depths $\tau_t = h/\ell_t$ are used (corresponding to the actual optical depths for $g = 0.85$ indicated in parentheses). We used $\chi = 0.71$ from (5.18); see Davis and Marshak (2002) for more details. This Fourier space representation of the Green function is used in those 3D RT approximation techniques were convolutions arise, namely, adjoint perturbation theory (Sect. 5.2.4) and the Nonlocal Independent Pixel Approximation or "NIPA" (Sects. 5.2.5 and 12.6).

The limit S or $k \to 0$ (required for the normalization) is singular but application of L'Hôpital's rule leads to

$$T(0, \tau_t) = \frac{1}{1 + \tau_t/2\chi} \, . \tag{5.32}$$

In Fig. 5.6, we present a direct comparison of a Monte Carlo solution (of the 3D RT equation with pencil beam source) and the *numerical* inverse Fourier transform of the analytic diffusion-based expression for the Green function in Fourier space. Agreement is excellent for all but the shortest distances from the collimated pencil beam in the forward scattering case.

The reflected Green function $G_R(kh; \tau_t) = J(k, z = 0) - 1 = R(k\ell_t, \tau_t)/R(0, \tau_t)$, is also of interest in many applications; it follows in closed-form from

$$R(S, \tau_t) = \frac{2[1 - (\chi S)^2] \sinh(S\tau_t)}{2[1 + (\chi S)^2] \sinh(S\tau_t) + 4\chi S \cosh(S\tau_t)} \tag{5.33}$$

with

$$R(0, \tau_t) = 1 - T(0, \tau_t) = \frac{\tau_t}{2\chi + \tau_t}$$

from (5.32) or from (5.33) using L'Hôpital's rule.

More recently, Polonsky and Davis (2004) used 3D diffusion theory to study Green functions using the more simply expressed "extended" BCs

$$J(\rho, -\chi\ell_t) = J(\rho, h + \chi\ell_t) = 0 \tag{5.34}$$

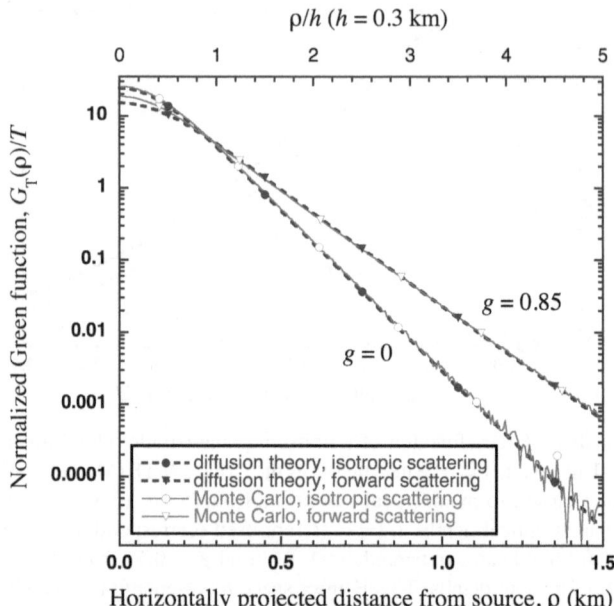

Fig. 5.6. Normalized Green function of a uniform plane-parallel cloud for transmittance in physical space. This quantity can be interpreted as the cloud's PSF (Point Spread Function). We compare here analytical results using both the 3D diffusion equation and numerical counterparts for the 3D RT equation. More precisely an analytical curve from the diffusion theoretical family plotted in Fig. 5.5 was inverse Fourier transformed numerically. For the RT computation, a Monte Carlo scheme was used in a slab of thickness $h = 0.3$ km where illumination was isotropic at a point and the conservative scattering was modeled by a Henyey-Greenstein phase function with the indicated value of g. The actual (not rescaled) optical depth τ is 16. We notice the inherent Monte Carlo noise for the isotropic scattering case which has a very low overall transmittance $T \approx 0.077$ from (5.32)

for a problem that in fact lives on $\{x \in \mathbb{R}^3; 0 < z < h\}$. This approximate expression of the diffusion BCs simplifies analytical computations. In this case, the inverse Fourier transforms can be taken analytically. This leads to a closed-form expression for the characteristic radius in the exponential tail observed in Fig. 5.6:

$$\left| \frac{\partial}{\partial \rho} \ln G_F(\rho; \tau_t) \right|^{-1} = \frac{G_F}{|\partial G_F / \partial \rho|} \approx \frac{h}{\pi R(0, \tau_t)}$$

for $\rho \gtrsim h$ and $F = R, T$. The same problem (with approximate BCs) for linear or power-law profiles in extinction is amenable to modified Bessel functions of arbitrary order.

Example 3: Finite/Isolated Clouds Under Solar Illumination

As a final example, the solar problem for homogeneous scattering/absorbing perpendicular parallelepipeds is tractable in diffusion theory (Davies, 1978), but their sharp edges lead to BC matching conditions that in turn lead to transcendental equations for the eigenvalues. By contrast, homogeneous non-absorbing spheres have no such problems and, rather than a sum of eigenfunctions, one obtains the following closed-form expression for the normalized boundary fluxes over the illuminated hemisphere (R) and the opposite hemisphere (T):

$$\frac{R}{T} = \frac{\tau_t}{2\chi} = \frac{(1-g)\tau}{2\chi} , \tag{5.35}$$

where τ is the optical diameter of the sphere. It is assumed here that the collimated solar beam used to modulate the intensity of an isotropic boundary source, rather than determine a volume source term. This amounts to making the assumption that the medium is opaque enough to randomize the directionality of the source in a vanishingly small boundary layer.

Given that $R + T = 1$, we can solve for T or R and, interestingly, we find the same expression as in (5.32). In Chap. 3 (Fig. 3.16), we show a validation[11] of (5.35) based on Monte Carlo simulations in spherical scattering media with Henyey-Greenstein phase functions over several orders of magnitude in τ. At large τ, we find $\chi \approx 0.71$ while, for small τ, we find $\chi \approx 4/3$; this is as expected from (5.18) and the ensuing discussion. The present authors have since generalized these analytical and numerical results to homogeneous ellipsoids with any semi-axis combination. These media lead to expansions in elliptical harmonic functions (Morse and Feshbach, 1953) as long as the solar beam is parallel to one axis.

Discussion

The closed-form solutions described in above Examples 2–3 are not just of academic interest. They have indeed shed light on challenging practical problems in cloud remote sensing. For instance,

- the Green-function problem is of interest in off-beam cloud lidar signal processing (Davis et al., 1999) as well as in adjoint perturbation theory for 3D RT effects (see Sect. 5.2.4);
- the ellipsoidal medium problem is useful in fully 3D remote sensing determinations of the optical properties of broken/isolated clouds at solar wavelengths (Davis, 2002) and associated cloud adjacency problems in surface remote sensing (see Chap. 3).

---

[11] "Validation" is concerned with matching model predictions and real-world observations or, in the case of an approximation theory, far-better-but-much-more-expensive modeling results.

Finally, one can obtain completely general results in 3D RT within the diffusion approximation; a formula for the domain-average impact of channeling derived by Davis and Marshak (2001) is given in (5.70)–(5.71) further on. This exact 3D diffusion result can be used in code verification.[12]

Validation

For an approximation model such as diffusion in radiation transport, good agreement with "exact" model results (i.e., solutions of the 3D RT equation) is a criterion for validity. Notwithstanding, we close our discussion of diffusion theory by noting that several empirical studies have demonstrated directly or indirectly that this simple transport modality dominates bulk photon flow in the VIS/NIR spectrum at least in dense cloud masses. Using the solar source illuminating stratus layers, we can mention:

- King et al. (1990), who observed the defining angular radiance pattern for diffusion in (5.1) *inside* marine boundary-layer clouds with the Cloud Absorption Radiometer (CAR), an airborne scanning multi-channel detector (cf. Fig. 5.2);
- Davis et al. (1997), who observe the signature diffusion phenomenon of radiative smoothing from *above* cloud with LandSat's imaging detector (cf. Fig. 12.2 in Chap. 12);
- Savigny et al. (1999), who observe radiative smoothing from *below* cloud in the time-series of a fixed zenith-pointing radiometer (cf. Fig. 12.24 in Chap. 12).

Still studying stratus layers, Davis et al. (1999) show a characteristic diffusion pattern in reflected radiance using an (effectively) steady laser source.

Although time-dependence is out of our present scope, other authors have found temporal diffusion signatures in detected photons emanating from pulsed lasers illuminating stratiform clouds:

- Love et al. (2001), looking up *from ground* using a wide-angle imaging lidar system;
- Davis et al. (2001), looking down *from space* using the large foot-print of the Lidar-In-space Technology Experiment (LITE);
- although not yet deployed in real clouds, Evans et al. (2003) use analytical and numerical results to show that their "in-situ" cloud lidar will measure diffusion-type temporal signals *from within* clouds.

In Chap. 13 we will see that time-dependence and pulsed sources are equivalent to steady sources and variable absorption by a well-mixed gas. Accordingly, Min and Harrison (1999) obtain using ground-based oxygen A-band spectrometry the linear diffusion-theoretical relation (Chap. 12) between cloud optical depth and the mean

---

[12] Verification is making sure that the computer code faithfully implements the conceptual model, asking the question *Are we solving the equations right?* Validation mentioned just above asks the next important question *Are we solving the right equations?* These defining questions are borrowed from the overview of "V&V" by Roache (1998).

photon pathlength in transmission for overcast skies with a single dense cloud layer (cf. Fig. 13.9).

Finally, turning to cumulus-type clouds, Davis (2002) finds reasonable agreement between the diffusion-based predictions of relative reflected and transmitted radiance levels in (5.35) and those observed in high-resolution multi-spectral satellite imagery. See Chap. 3, Fig. 3.12, for a geometrical explanation of how one can observe transmitted photons from space for finite (broken/isolated) clouds.

5.2.3 3D Discrete-Angle (or "6-Flux") RT Theory

Diffusion theory is equivalent to P_1 theory, the simplest possible spherical harmonic truncation with net transport.[13] In the same terminology from Chap. 4, Discrete-Angle (DA) RT theory is an S_1 theory in 3D.[14] In lieu of (5.1), we assume:

$$I(x, \Omega) = \sum_{\hat{i} \in \{\hat{x}, \hat{y}, \hat{z}\}} \sum_{\{\pm\}} F_{\pm\hat{i}}(x) \delta(\Omega \mp \hat{i}) . \tag{5.36}$$

This is basically a poor-man's discrete ordinate approach: it replaces the severe angular *smoothing* in diffusion theory by the minimal *sampling* in angle space that still enables 3D transport. In sharp contrast with diffusion, the DA model for 3D RT in (5.36) is clearly not at any risk of being validated by *direct* observations. But this does not mean that it is not of any value to predict radiation fields of interest in the applications, some of which may in turn be used to derive observables. This would lead to interesting validation questions.

It is easy to see why this was originally called "6-flux" theory (Chu and Churchill, 1955; Siddal and Selçuk, 1979); we note that the new spatially varying fields that multiply the directional δ's indeed have the same physical dimensions as fluxes. In fact, we will equate the $F_{\pm\hat{i}}(x)$, $\hat{i} \in \{\hat{x}, \hat{y}, \hat{z}\}$, to hemispherical fluxes with respect to all six directions along the axes of the natural ($\hat{z} = $ "up") Cartesian coordinate system. In this 6-flux transport theory, the self-consistent choice of scattering phase function is given by

$$\varpi_0 P(x, \theta_s, \phi_s)/4\pi = f(x) \times \delta(\theta_s) + b(x) \times \delta(\theta_s - \pi)$$
$$+ s(x) \times \delta(\theta_s - \pi/2) \sum_{n=0}^{3} \delta(\phi_s - n\pi/2) , \tag{5.37}$$

where $(\theta_s, \phi_s) = (\pi/2, 0)$ designates any Cartesian axis direction at a right angle with $\theta_s = 0, \pi$ (forward and backward scattering). We require $f + b + 4s = \varpi_0$ to ensure conservation then, using its usual definition, we see that the asymmetry factor is given by $g = (f - b)/\varpi_0$.

Lovejoy et al. (1990) present DA RT – the traditional 6-flux model and beyond – as a special case of standard (continuous-angle) RT with the above choice of phase

---

[13] Photons would have no net motion in a hypothetical P_0 theory.
[14] In 1D RT, S_1 theory is mathematically equivalent to the 2-stream/diffusion approximation, only the coefficients of the 2 coupled ODEs change (Meador and Weaver, 1980).

function. We already know that large-scale angularly-integrated radiative properties are not very sensitive to the precise choice of the phase function. Indeed, diffusion with its sole parameter g – and maybe f, from the δ-rescaling – does a reasonable job of reproducing continuous-angle fluxes. So DA RT may turn out to be a viable predictor of these coarse properties.

The goal in a DA RT problem is to determine everywhere the formal 6-vector made of all the fluxes in (5.36) given the usual extinction and scattering coefficients everywhere, as well as the required boundary conditions and source terms. In this representation of the model, the scattering phase function is reduced to a 6-by-6 matrix that couples the 6-vector of fluxes. As seen in (5.37), this scattering phase function has two free parameters, one more than in (5.3) after accounting for normalization to ϖ_0. Using a natural "first-±-then-$\{x, y, z\}$" ordering of the 6-vector of fluxes, we indeed have along the 2-by-2 block diagonal scattering matrix alternating factors of $0 \leq f \leq 1$ and $0 \leq b \leq 1 - f$ that denote respectively the probabilities of forward and backward scattering while the rest of the matrix is filled with the sideways-scattering probability value $0 \leq s = (1 - f - b)/4 \leq 1/4$ for transitioning to any of the remaining 4 Cartesian directions. By substitution of (5.36)–(5.37) into the (continuous-angle) RT equation, we obtain

$$
\begin{pmatrix}
+\partial_x\, F_{+\hat{x}} \\
-\partial_x\, F_{-\hat{x}} \\
+\partial_y\, F_{+\hat{y}} \\
-\partial_y\, F_{-\hat{y}} \\
+\partial_z\, F_{+\hat{z}} \\
-\partial_z\, F_{-\hat{z}}
\end{pmatrix}
= \sigma
\begin{pmatrix}
f-1 & b & s & s & s & s \\
b & f-1 & s & s & s & s \\
s & s & f-1 & b & s & s \\
s & s & b & f-1 & s & s \\
s & s & s & s & f-1 & b \\
s & s & s & s & b & f-1
\end{pmatrix}
\begin{pmatrix}
F_{+\hat{x}} \\
F_{-\hat{x}} \\
F_{+\hat{y}} \\
F_{-\hat{y}} \\
F_{+\hat{z}} \\
F_{-\hat{z}}
\end{pmatrix}
+
\begin{pmatrix}
q_{+\hat{x}} \\
q_{-\hat{x}} \\
q_{+\hat{y}} \\
q_{-\hat{y}} \\
q_{+\hat{z}} \\
q_{-\hat{z}}
\end{pmatrix},
$$
$$(5.38)$$

where extinction σ (and possibly all the phase-function coefficients) depend on x, and the last vector represents internal sources which will also generally depend on x. We see immediately that, simply by setting $s = 0$, we decouple the beams propagating back and forth along the x, y, z-axes. BCs are easy to express as long as the boundary ∂M is made of facets that follow the Cartesian planes. For instance, an internally variable but geometrically plane-parallel medium is $M = \{x \in \mathbb{R}^3; 0 < z < h\}$; in this case, no incoming flux at cloud base ($z = 0$) would appear as $F_{+\hat{z}}(x, y, 0) \equiv 0$ while a unit of incoming flux at cloud top ($z = h$) reads as $F_{-\hat{z}}(x, y, h) \equiv 1$.

As for continuous-angle RT, simplification will come from an eigenanalysis of the coupling term (which models scattering). The eigenvalues of the scattering-extinction matrix in (5.38) are:

$$\alpha = 1 - f - b - 4s \text{ (once)}, \quad \beta = 1 - f - b + 2s \text{ (twice)}, \quad \gamma = 1 - f + b \text{ (thrice)}. \quad (5.39)$$

The first and last are recognizable from diffusion theory,

$$\alpha = 1 - \varpi_0 \text{ and } \gamma = 1 - \varpi_0 g ; \qquad (5.40)$$

their respective eigenvectors are also well-known: scalar flux J, sum of all the F's; and vector flux $\boldsymbol{F}$, its components being

$$F_i = F_{+\hat{i}} - F_{-\hat{i}}, \ i \in \{x, y, z\} \ . \tag{5.41}$$

The second eigenvalue in (5.39) is less familiar. Its eigenvectors can be formed from the *even* combinations

$$J_i = F_{+\hat{i}} + F_{-\hat{i}}, \ i \in \{x, y, z\} \ , \tag{5.42}$$

where the corresponding *odd* combinations in (5.41) are the components of $\boldsymbol{F}$; a valid choice for the second set of eigenvectors is $J_x - J_z$ and $J_y - J_z$. We note that $J = \sum_i J_i$. We will return to this second eigenvalue and eigenspace to give them a physical meaning momentarily.

After straightforward manipulations of (5.38) guided by the eigenanalysis, we obtain three Fick-like relations,

$$-(1 - f + b)\sigma(\boldsymbol{x})F_i = \partial_i J_i - q_{i-}, \ i \in \{x, y, z\} \ , \tag{5.43}$$

reminiscent of (5.4) with the anticipated extra term for anisotropic sources, $q_{i-} = q_{+\hat{i}} - q_{-\hat{i}}$. Letting $q_{i+} = q_{+\hat{i}} + q_{-\hat{i}}$, we also obtain these three equations:

$$\partial_x F_x = \sigma(\boldsymbol{x}) \left[-(1 - f - b)J_x + 2s(J_y + J_z) \right] + q_{x+} \ ; \tag{5.44}$$

and permutations on y and z. Adding up these last three equations yields the usual conservation law in (5.9) since J and q are just the sums of all of their $\{x, y, z\}$-wise components. By combining (5.43) and (5.44), we arrive at three coupled 2nd-order PDEs that look like (5.10), plus coupling terms dependent on $s = (\beta - \alpha)/6$. Specifically, we get

$$\left[-\partial_i \left(\frac{1}{\gamma\sigma(\boldsymbol{x})} \right) \partial_i + \beta\sigma(\boldsymbol{x}) \right] J_i = (\beta - \alpha)\,\sigma(\boldsymbol{x})J/3 + q_i \ , \tag{5.45}$$

for $i \in \{x, y, z\}$, where we have used the eigenvalues in (5.39) and defined

$$q_i = q_{i+} + \partial_i(q_{i+}/\gamma\sigma) \ , \tag{5.46}$$

the DA counterpart of (5.11).

BCs in the dependent variables J_i of the coupled PDEs are mixed, as for the diffusion problem but with $\chi = 1$ in (5.17). Indeed, BCs in slab geometry are expressions for $F_{\pm\hat{z}} = (J_z \pm F_z)/2$; using (5.43) without the internal source terms, this yields

$$\frac{1}{2} \left[1 \mp \frac{1}{\gamma\sigma(\boldsymbol{x})}\partial_x \right] J_z = f(\boldsymbol{x}) \tag{5.47}$$

for the new boundary source function.

In summary, practically any of the standard numerical solutions of the diffusion PDE problem can be used to solve the discrete-angle problem in this (5.45)–(5.47) formulation since most methods apply equally well to single 2nd-order elliptical PDEs and to coupled systems of such PDEs. Formally, this is just like a PDE for a vector field $\mathbf{J} = (J_x, J_y, J_z)^{\mathrm{T}}$.

It is not hard to show that 3D diffusion theory is recovered identically from 6-flux theory in an intriguing (since non-physical) asymptotic limit of the phase-function parameters. This is new and unusual because traditional derivations of the diffusion equation act on the radiance field; along this road from RT per se to diffusion only the phase-function is manipulated, first going from continuous angles to discrete angles, then the asymptotic limit. Following Lovejoy et al. (1990) who reckoned in 2D, we remark that taking $\beta \rightarrow \infty$ in (5.45),[15] we have to require $J_i = J/3$ for $i \in \{x, y, z\}$ to balance both sides. In other words, the DA photons are partitioned equally along all three Cartesian axes (side-scattering will certainly promote this equipartition). In this case, we can sum up all the remaining terms in the 3 PDEs in (5.45) and, using (5.40), this amounts exactly to the 3D diffusion equation in (5.10) for J. This insight gives us a meaningful interpretation of the eigenvalue β and its associated eigenplane spanned by $(J_x - J_z)$ and $(J_y - J_z)$. A less than infinite value of β means a certain deviation from diffusion theory, and any non-vanishing projection of DA fluxes on the $\{J_x - J_z, J_y - J_z\}$ plane is a measure of "non-diffusive" transport. If we are seeking to improve upon diffusion theory, then 2nd-order DA PDEs with a small but finite value of $1/\beta$ (for fixed α and γ) might be an avenue to explore.

Another remarkable limit is obtained by setting $s = 0$ which is physically realizable, although probably not very realistic. We then have three decoupled two-stream models, each represented by a single second-order ordinary differential equation (ODE); basically, we are back to the two-stream ICA where propagation along z is allowed, but not along x or y (unless there are designated sources on those axes too). So, allowing for unphysical phase-function parameters, DA theory is intermediate between 2-stream ICA theory (decoupled 1D RT/diffusion equations) and 3D diffusion theory.

Because it is a minimalist discrete-ordinate approach, the above *continuous space* formulation suffers from a particularly severe form of the "ray effect" (as described in Chap. 4), at least for the physically accessible range of non-negative phase-function parameters that has been explored so far. Figures in the paper by Davis et al. (1993) illustrate this ray effect. More interesting for the applications is a *discrete space* formulation illustrated in Fig. 5.7. This formulation is reminiscent of a radiosity problem where *finite* elements (normally surfaces) redistribute the radiation they receive. However, while in radiosity it is usually quite full, the interaction matrix here is very sparse.

Assume that a unit of flux is received on one facet of one cell centered at $(x, y, z)^{\mathrm{T}} + (\Delta_x/2, \Delta_y/2, \Delta_z/2)^{\mathrm{T}}$ but registered with one of its corners, at a grid point

$$
\boldsymbol{m} = \begin{pmatrix} i \\ j \\ k \end{pmatrix} = \begin{pmatrix} x/\Delta_x \\ y/\Delta_y \\ z/\Delta_z \end{pmatrix} \in \{0, \dots, M_x - 1\} \otimes \{0, \dots, M_y - 1\} \otimes \{0, \dots, M_z - 1\}
$$

$$(5.48)$$

---

[15] Physically, the largest that β can achieve is 3/2; this happens when $f = b = 0$ and $s = 1/4$. So the idea in this formal limit is to emphasize side-scattering.

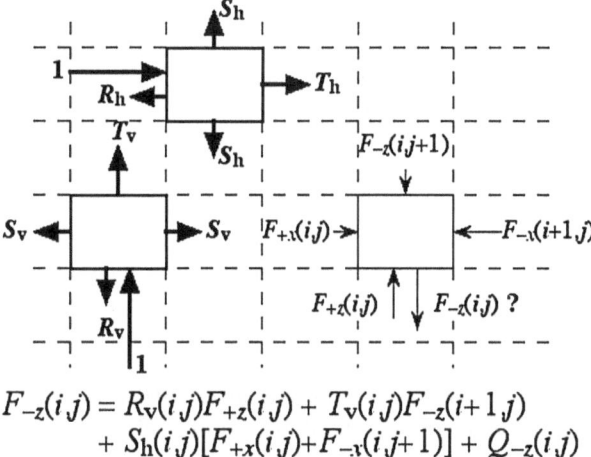

$$F_{-z}(i,j) = R_v(i,j)F_{+z}(i,j) + T_v(i,j)F_{-z}(i+1,j)$$
$$+ S_h(i,j)[F_{+x}(i,j)+F_{-x}(i,j+1)] + Q_{-z}(i,j)$$

Fig. 5.7. The orthogonal Discrete-Angle/Discrete-Space radiation transport model in 2D on a rectangular grid. The 2 cells on the left illustrate the unitary responses for the 2 possible directions of incidence. On the right, cell (i,j) receives radiation on all facets and we ask about the radiation escaping through the bottom facet, including photons created inside the cell. The balance equation for that cell is written below the schematic. See Gabriel et al. (1990) for an interesting analytical mean-field investigation of RT through a fractal structure using this 2D model based on real-space renormalization. Although the idea is far from optimal in computational DA RT, we can invoke a simple relaxation scheme where we take $F_{-z}^{(0)} = Q$ to initialize the process and the update equation for $F_{-z}^{(n+1)}$ would look like the balance equation using $F_{\pm z}^{(n)}$ and $F_{\pm x}^{(n)}$ on the right-hand side but without the source term. Generalization of this 4-flux model in 2D to the 6-flux scenario in 3D is straightforward (see text)

along the three axes where $\Delta_i, i \in \{x, y, z\}$, are the grid constants in the cuboidal medium with sides $L_i = M_i\Delta_i$. This unit of flux is redistributed according to a transmission T_m to the facet on the opposite side of the cell, a reflection R_m back to same facet (but the opposite direction), and a fraction A_m is absorbed inside the cell while sideways deflection towards all the adjacent facets is given by $S_m = (1 - A_m - R_m - T_m)/4$. Thus, part of the system of equations that describes radiative balance of this cell will look like

$$F_{-\hat{z}}(\boldsymbol{m}) = R_m \times F_{+\hat{z}}(\boldsymbol{m}) + T_m \times F_{-\hat{z}}(\boldsymbol{m} - (0,0,1)^T) + S_m$$
$$\times [F_{+\hat{y}}(\boldsymbol{m}) + F_{-\hat{y}}(\boldsymbol{m} + (0,1,0)^T) + F_{+\hat{x}}(\boldsymbol{m}) + F_{-\hat{x}}(\boldsymbol{m} + (1,0,0)^T)]$$
$$(5.49)$$

in the absence of internal sources (for simplicity). It expresses that what comes out of the bottom of the cell has to come from one of the 5 other faces. There are five more equations like this one for each node. Together, they express that what goes in to the cell has to come out, or stay inside if there is any absorption ($A_m > 0$). When a facet of the cell touches upper, lower, or lateral boundaries, then the appropriate BC must be applied. When internal sources are present, more terms appear that account for thermal emission or injection of solar photons. At the cell level, the expressions are in fact exact if the $F_{\pm i}(\boldsymbol{m})$, $i \in \{x, y, z\}$, are interpreted as total fluxes across a facet (cf. Chap. 3). The approximation here consists in assuming that all the 6-fold collections of equations for all the cells are coupled, in other words, that the interaction between the cells is only through the discrete angles. This is indeed a gross simplification, especially if there are optically thin cells in the medium.

Generalization to situations where the grid constant is different in one or all three directions is straightforward. This formulation makes no assumption about the optical thickness τ_m of the cell. It can be quite large, and then the question of how to obtain $\{T, R, A\}_m$ arises. This is not a hard question because, even if this is has to be done by Monte Carlo, these "expensive" mesoscopic coefficients can be stored in pre-computed look-up tables. Reasonably accurate diffusion solutions for rectangular parallelepiped media are also available (Davies, 1978).

As an example, we reconsider the I3RC "Case 1" cloud we already investigated with diffusion theory when $\varpi_0 = 0.99$, $g = 0.85$, and $\mu_0 = 1$. We can describe this as a very simple discrete-space DA transport problem with $M_x = 2$, $M_y = M_z = 1$ and $\Delta_x = \Delta_z = 0.25$ km, $\Delta_y = \infty$. First, we need to obtain $\{T, R, S\}$ for both the tenuous and the dense cells with square sections. Straightforward Monte Carlo simulations (with the prescribed Henyey-Greenstein phase functions and collimated illumination) yield $\{T, R, S\} = \{0.7148, 0.0336, 0.1154\}$ ($A = 0.0209$) for $\tau = 2$ and $\{T, R, S\} = \{0.112, 0.246, 0.227\}$ ($A = 0.184$) for $\tau = 18$. We can then write 8 coupled linear equations like (5.49) relating the 8 unknown fluxes coming out of the 2 cells in all 4 directions in these equations (where BCs determine the in-going fluxes in the z-direction). We find absorptions of 0.0363 and 0.3284 respectively for the $\tau = 2$ and 18 cells. This is quite close to the values found in Fig. 5.4a with 3D RT and 3D diffusion. Other optical and illumination parameter choices however do not work so well, which is not too surprising given how coarse the model is.

This discrete-angle/discrete-space way of modeling 3D radiation transport is particularly suitable for cloud-system models where cells are typically cloudy or not because, when they are, they can be quite opaque. In the end, there is a very large but very sparse matrix problem to solve. Again, there are optimally efficient/stable ways of doing this (e.g., multi-grid). Note also that, as for the diffusion PDE problem, if a time-sequence of 3D solutions is sought in a dynamically evolving cloud the previous time-step is probably a very good initial guess for the updated radiation field. As for the first step, the standard ICA is obtained by setting $S_m = 0$ and using 1D plane-parallel theory to compute $\{T, R, A\}_m$, in the 2-stream approximation.

If there are large optical voids in the cloud system then the picture of radiation propagating only along the $\{x, y, z\}$-axes is not much more realistic than the ICA picture we are trying to improve. In that model, radiation travels only along the z axis (cf. Fig. 5.1). If such optical voids exist, then a combination of radiosity (Siegel and Howell, 1981) and discrete-angle/discrete-space RT theory may be in order.

5.2.4 3D Adjoint Perturbation Theory

It was shown in Chap. 3 that the full 3D RT equation can be written as simply as

$$\mathcal{L}I = \mathcal{Q} , \tag{5.50}$$

where $I(\boldsymbol{x}, \boldsymbol{\Omega})$ is the unknown radiance field, $\mathcal{L}$ is the known integro-differential linear transport operator (with advection, extinction and scattering terms), and $\mathcal{Q}(\boldsymbol{x}, \boldsymbol{\Omega})$ is the known distribution of internal sources. Assuming I is just the diffuse component (uncollided solar photons are treated separately), the BCs in solar problems express respectively no incoming radiance at the upper level and either the same or, at the most, a ground reflection (which is actually just a surface scattering). In a typical RT problem, we are generally not interested in the 5-dimensional scalar field I but in some weighted integral[16]

$$E = \langle I, \mathcal{R} \rangle = \langle \mathcal{R}, I \rangle = \int_{4\pi} \int_{M} \mathcal{R}(\boldsymbol{x}, \boldsymbol{\Omega}) I(\boldsymbol{x}, \boldsymbol{\Omega}) \mathrm{d}\boldsymbol{x}\mathrm{d}\boldsymbol{\Omega} , \tag{5.51}$$

where $\mathcal{R}(\boldsymbol{x}, \boldsymbol{\Omega})$ can be viewed as the angular response of an instrument and/or a spatial averaging over a volume or a surface belonging to the optical medium M. We can think of E as an "effect," namely, the result of applying $\mathcal{R}$ to I. Examples of response functions are:

$$\mathcal{R}_T = |\mu|\Theta(-\mu)\delta(z)/(L_x L_y) , \tag{5.52}$$

for overall transmission $(T = E/\mu_0 F_0)$, where $\Theta(\cdot)$ is the Heaviside step function and $\{L_x, L_y\}$ are the horizontal dimensions of the domain (assumed rectangular); for overall albedo $(R = E/\mu_0 F_0)$, we use

$$\mathcal{R}_R = \mu\Theta(\mu)\delta(h - z)/(L_x L_y) ; \tag{5.53}$$

for nadir radiance at cloud top $(z = h)$ from position (x_o, y_o), we take

$$\mathcal{R}_{I\uparrow} = \delta(\boldsymbol{\Omega} - \hat{z})\delta(z - h)\delta(x - x_o)\delta(y - y_o) ; \tag{5.54}$$

and for zenith radiance out of cloud base $(z = 0)$ at the same position,

$$\mathcal{R}_{I\downarrow} = \delta(\boldsymbol{\Omega} + \hat{z})\delta(z)\delta(x - x_o)\delta(y - y_o) . \tag{5.55}$$

Going back to the definitions in Chap. 3, Sect. 3.10.2, we recall that the adjoint radiance field $I^+(\boldsymbol{x}, \boldsymbol{\Omega})$ obeys the adjoint RT equation

$$\mathcal{L}^+ I^+ = \mathcal{R} \tag{5.56}$$

where the adjoint transport operator $\mathcal{L}^+$ describes how adjoint photons propagate backwards from the receivers (adjoint sources) to the sources (adjoint sinks). The

---

[16] The $\langle \cdot, \cdot \rangle$ notation used here is the natural definition of the scalar product in a space of functions, viewed as an infinite-dimensional vector space. In such a vector space, each position-direction pair $\{\boldsymbol{x}, \boldsymbol{\Omega}\}$ is associated with a dimension, and the real number $f(\boldsymbol{x}, \boldsymbol{\Omega})$ is the associated component or projection of the "vector" f.

characteristic BCs for the adjoint RT equation express that there are no escaping adjoint photons. It can be shown that

$$E = \langle I, \mathcal{R} \rangle = \langle I^+, \mathcal{Q} \rangle . \tag{5.57}$$

The physical interpretation of $I^+(x, \Omega)$ is as a measure of the relative "importance" of position x and direction Ω for a given observation represented formally by the response function $\mathcal{R}(x, \Omega)$ in (5.56). So, not too surprisingly, the outcome E of a radiometric observation described by $\mathcal{R}$ can be estimated either by a direct integration of the radiance function I (as determined by $\mathcal{Q}$ and) weighted by $\mathcal{R}$, or through an integral over the (real) photon sources $\mathcal{Q}$ weighted by this importance function or adjoint radiance I^+ (as determined by $\mathcal{R}$). This is essentially the philosophy of backward Monte Carlo computation (cf. Chap. 4).

Keeping sources $\mathcal{Q}$ and receivers $\mathcal{R}$ constant, we can represent 3D variability (generically in the extinction σ and scattering σ_s coefficients, as well as in the phase function), as a perturbation on a base state that can be taken as a plane-parallel 1D medium for simplicity. So we write $\mathcal{L} = \mathcal{L}_b + \Delta\mathcal{L}$, and similarly for $\mathcal{L}^+$ where the subscript "b" designates a 1D uniform or vertically stratified medium. Just like the operators, we expand the radiance fields, $I = I_b + \Delta I$, and similarly for I^+. Finally, we expand the scalar quantity $E = E_b + \Delta E$. Since $\mathcal{L}_b I_b = \mathcal{Q}$ and $\mathcal{L}_b^+ I_b^+ = \mathcal{R}$, and exploiting linearity in (5.57), we obtain

$$\Delta E = E - E_b = -\langle I_b^+, \Delta\mathcal{L} I_b \rangle + \langle \Delta I^+, \mathcal{L}\Delta I \rangle + \cdots , \tag{5.58}$$

showing the first two terms of an infinite expansion that we denote respectively ΔE_1 and ΔE_2. The base-case transport operator $\mathcal{L}_b$ can be defined by horizontally averaging the 3D operator $\mathcal{L}$.

Suppose we are interested in a radiant energy budget problem in climate or in an effect of *unresolved* (sub-pixel) variability in remote sensing. Then $\mathcal{R}$ (which determines I_b^+) mandates a horizontal domain averaging. It is then easy to show that the 1st-order term

$$\Delta E_1 = -\langle I_b^+, \Delta\mathcal{L} I_b \rangle = -\langle \Delta\mathcal{L}^+ I_b^+, I_b \rangle \tag{5.59}$$

in (5.58) vanishes identically. Indeed, being solutions of 1D RT equations, both I_b and I_b^+ are invariant by horizontal translation and we know (by definition) that the horizontal mean of $\Delta\mathcal{L}$ is zero (i.e., the null operator). So domain-average 3D radiative transfer effects start at 2nd order.[17] To obtain an explicit algorithm from perturbation theory for domain averages, we now need to somehow replace the unknown ΔI and ΔI^+ with their known (subscript-"b") counterparts and, accordingly, introduce $\Delta\mathcal{L}$ and $\Delta\mathcal{L}^+$ in lieu of $\mathcal{L}$ in

$$\Delta E \equiv \Delta E_2 = \langle \Delta I^+, \mathcal{L}\Delta I \rangle . \tag{5.60}$$

---

[17] This is also made clear from the exact diffusion theoretical result described further on, in (5.71).

Remarking that $E = \langle \mathcal{R}, G * \mathcal{Q} \rangle = \langle G^+ * \mathcal{R}, \mathcal{Q} \rangle$ where G and G^+ are the Green function (GF) and its adjoint counterpart (with "*" denoting a convolution product over the 5-dimensional $\{x, \Omega\}$-space), Box et al. (1988, 1989) show that (5.60) can be estimated from quantities obtained entirely from the base case. The final expression for the 2nd-order perturbation term is

$$\Delta E_2 = \langle \Delta \mathcal{L}^+ I_\mathrm{b}^+, G_\mathrm{b} * \Delta \mathcal{L} I_\mathrm{b} \rangle = \langle G_\mathrm{b}^+ * \Delta \mathcal{L}^+ I_\mathrm{b}^+, \Delta \mathcal{L} I_\mathrm{b} \rangle . \tag{5.61}$$

The GFs are inherently 3D quantities here, even for the uniform base case because of their defining source distributions (cf. Chap. 3); they are indeed the solutions of $\mathcal{L}_\mathrm{b} G_\mathrm{b} = \delta(x - x')\delta(\Omega - \Omega')$ and $\mathcal{L}_\mathrm{b}^+ G_\mathrm{b}^+ = \delta(x' - x)\delta(\Omega' - \Omega)$. So, by virtue of source superposition, $I_\mathrm{b} = G_\mathrm{b} * \mathcal{Q}$ and $I_\mathrm{b}^+ = G_\mathrm{b}^+ * \mathcal{R}$ (even though $\mathcal{Q}$ and $\mathcal{R}$ are not, in the present case, dependent on the horizontal coordinates). These GFs will therefore interact with the (x, y)-dependence of $\Delta \mathcal{L}$ ($\Delta \mathcal{L}^+$) even if I_b (I_b^+) does not.[18]

There is the potential here for considerable computational complexity. GFs for the uniform base case have 10-dimensional arguments $(x, \Omega; x', \Omega')$ but depend only on relative values in the horizontal $(x - x', y - y', z, z')$. As soon as a GF appears in perturbation equations, the primed arguments are understood to be involved in a 5-dimensional integral. The scalar product in (5.61) is itself a 5-dimensional integral, and there are 2-dimensional integrals over angle-space in each of the transport operator perturbations. So (5.61) is in fact shorthand for a 14-dimensional integral. But this is somewhat naïve book-keeping. In reality, a judicious representation of operators and radiation fields in spherical harmonics can be used to do all of the 8 angular integrals analytically. Further simplifications are possible because of translational invariance in (x, y); see Box et al. (2003). More good news is that base-case ingredients in (5.61) only need to be computed once and can be then applied to arbitrary variability.

Now what if $\mathcal{R}$ describes the observation of a radiance- or flux-value at a point? Instead of averaging over space, we are then sampling points and, accordingly, $\mathcal{R}$ contains a spatial δ-function. This could be a way of modeling high-resolution remote sensing data and 3D effects across pixel boundaries. Since $\mathcal{R}$ is the source of the adjoint RT equation, the base-case adjoint radiance field I_b^+ in (5.59) is then de facto a spatial GF. So, in its broadest definition, adjoint perturbation theory is entirely based on GFs.

Figure 5.8, adapted from Polonsky et al. (2003), shows results from (5.59) for a sine-wave cloud model:

$$\sigma(x) = \sigma_\mathrm{b} + \delta\sigma \cos(2\pi x/L) , \tag{5.62}$$

for any $0 < z < h$. Structural parameters are $h = 1$ (nominally), $L/h = 2\pi \approx 6.3$, and $\varepsilon = \delta\sigma/\sigma_\mathrm{b} = 0.1$. The sources are uniform ($F_0 = 1$) and collimated ($\mu_0 = 1.0, 0.3$) along the top boundary ($z = h$). Optical properties are given by

---

[18] The expressions we provide for ΔE_1 and ΔE_2 are general, not limited to the case of uniform $\mathcal{Q}$ and $\mathcal{R}$, used here primarily for motivation of the higher-order theory.

$\tau_b = \sigma_b h = 10$; scattering is conservative ($\varpi_0 = 1$), and the phase function is Henyey–Greenstein with $g = 0.5$. In summary, we have

$$Q = F_0 \sigma_b \varpi_0 P_{HG}(\theta_s; g) \exp[-\sigma_b(h - z)/\mu_0] \qquad (5.63)$$

for the "solar photon injection" term used in Chap. 3 when the radiance I models only the once or more scattered light; alternatively, one could use an upper-boundary source term

$$Q = F_0 \delta(\mathbf{\Omega} - \mathbf{\Omega}_0) \delta(h - z) \qquad (5.64)$$

in which case I models the total (direct+diffuse) radiance field. The response we seek is nadir radiance at a point along the cloud top. So the $\mathcal{R}$ of interest is made explicit in (5.54). We plot ΔE_1 as a function of $\sigma_b x_o$ over a couple of cycles obtained by perturbation theory and with SHDOM (Chap. 4) for the two solar zenith angles. The agreement is quite good. Furthermore, we know that the 3D RT effect estimated here at 1st order is proportional to the extinction perturbation parameter ε and that it is sinusoidal, but it is not in phase with the extinction wave as soon as $\mu_0 < 1$. This distinguishes the perturbation method immediately from the local ICA.

The adjoint perturbation method is a closed-form approach to 3D RT. It also has the advantage of enabling the user to define (through $\mathcal{R}$) just how much or how little spatial and angular detail is wanted in 3D position space and 2D direction space. These features distinguish it from the two previously described methods which involve the numerical solution of a PDE or sparse-matrix problem. So the advantage of perturbation over alternate methods is not a simple issue. We can however anticipate that its accuracy will break down as the amplitude of the fluctuating coefficients increases with respect to their mean values. Again the break-even point in loss of accuracy, given all the advantages of the method, will be application dependent. Furthermore, there is no obligation in perturbation theory to take the base case, $\mathcal{L}_b I_b = Q$ (plus BCs), as homogeneous. It could be the ICA solution, and we seek the first- or higher-order corrections to it. It could be the 3D diffusion solution, and we seek to improve it with respect to the 3D RT equation. This is all about defining the perturbation $\Delta \mathcal{L}$ of the transport operator.

Finally, we note that in the above we have followed the formalism of Box et al. (1989) because of its relative simplicity, but there are other approaches. In fact, Li et al. (1994) combine a spatial perturbation technique with the 3D diffusion approximation and even apply it to a 2D sine-wave cloud model. Galinsky and Ramanathan (1998) furthermore examine the case where extinction is constant but the physical cloud thickness is modulated by a sine-wave.[19] Since the diffusion transport operator in (5.10) is based on 2nd-order spatial derivatives, it is *self*-adjoint and there is no need to bring in the concept of adjoint transport. On the other hand, the 2nd-order perturbation expansion, still in the diffusion limit, by Li et al. (1995) rivals in algebraic complexity the arbitrary-order adjoint RT equation approach by Box et al. (2003) which, to be fair, has yet to be encoded and assessed.

---

[19] Galinsky (2000) generalize their perturbation model from diffusion to a discrete-ordinate approach.

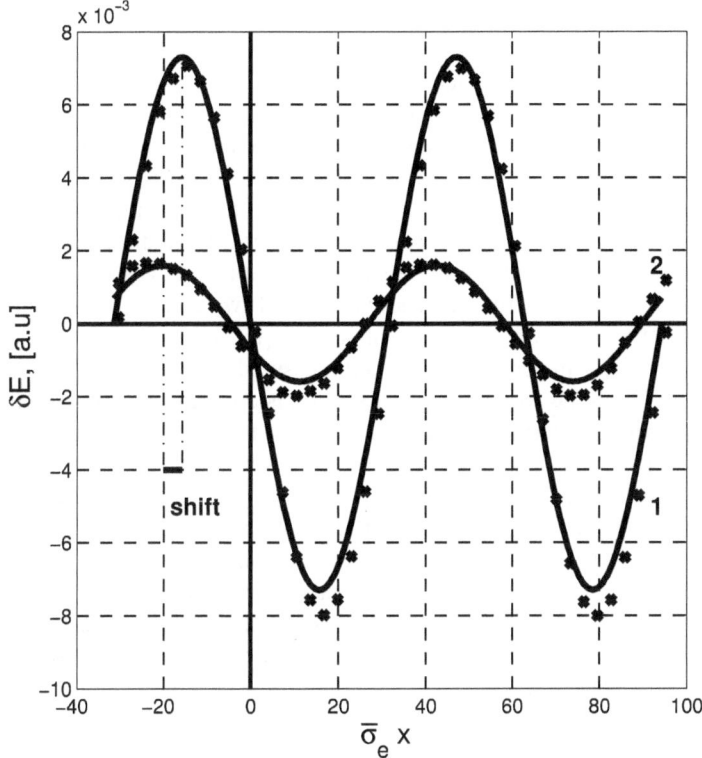

Fig. 5.8. Example of an adjoint perturbation prediction (solid lines) of 3D RT effects compared to exact SHDOM results (symbols) for a sine-wave cloud under two solar illumination angles, $\mu_0 = 1$ (#1) and 0.3 (#2), noting the shift of the 3D effect at slant illumination (details in the main text)

In conclusion, application of perturbation theory to radiation transport is a promising avenue to cope at least with weakly variable media, and especially when several realizations of the spatial variability around a given mean state are to be examined.

5.2.5 Hybrid and Semi-Empirical Methods

So far we have described three strikingly different approximation methods. But they all derive from the RT equation, precisely what is being approximated here. Given this connection, one can generally anticipate the conditions under which an approximation will work or not. Now we turn to methods which are predicated on other considerations: ad hoc hybridization, phenomenology, and borrowing across disciplinary boundaries. Representative examples follow.

- As a representative of a *hybrid* method applicable to solar problems, consider the straightforward idea of "feeding" a 1D-based ICA computation with a fully

3D source term. For a two-stream implementation, one would use the expressions in (5.13) computed (say) at the middle of the upper boundary of a cell of size $\delta x \times \delta y \times \delta z$ positioned at $\boldsymbol{x} = (x, y, z)^{\mathrm{T}}$ where x and y have now become fixed parameters while z scans the column as a (discretized) independent variable; in short, the solar constant F_0 in the analytical single-layer 2-stream model becomes $F_0(\boldsymbol{x}) = F_0 \exp[-\tau_0(x + \delta x/2, y + \delta y/2, z + \delta z)]$ obtained using (5.14). This collimated flux is then projected onto the z-axis, propagated into the cell, and partitioned between up- an down-ward directions. For the popular δ-Eddington version of the 2-stream model (actually just 1D diffusion theory), the source terms for the classic system of two coupled 1st-order ODEs are $F_0(\boldsymbol{x})\varpi_0(\boldsymbol{x})(1/2 \mp 3g(\boldsymbol{x})\mu_0/4) \exp[-\sigma(\boldsymbol{x})\zeta/\mu_0]$, respectively in the $\pm z$ directions, where ζ is the depth into the cell measured from the top (i.e., the independent variable in the two-stream ODEs). In this coupled ODE model, the forward and backward scattering coefficients (constant for a given cell) are similarly obtained: e.g., $[7 - \varpi_0(\boldsymbol{x})(4 + 3g(\boldsymbol{x}))]/4$ and $[1 - \varpi_0(\boldsymbol{x})(4 - 3g(\boldsymbol{x}))]/4$ respectively, using the diffusion-theory phase function in (5.3). Considering the cell as a homogeneous plane-parallel layer in a vertical column, the 2-stream theory (Meador and Weaver, 1980) uses the cell's optical coefficients $\tau(\boldsymbol{x}) = \sigma(\boldsymbol{x})\delta z$, $\varpi_0(\boldsymbol{x})$, and $g(\boldsymbol{x})$, all δ-scaled (5.9), to deliver albedo, diffuse- and direct transmissions $\{R(\boldsymbol{x}), T_{\mathrm{dif}}(\boldsymbol{x}), T_{\mathrm{dir}}(\boldsymbol{x})\}$. Then standard sparse matrix techniques are used to account for the radiative coupling of the layers throughout the column. As soon as the sun is off-zenith ($\mu_0 < 1$), results will differ from the standard ICA two-stream model where even the source term is computed with single-column quantities, and possibly quite dramatically different if the gridded optical medium is contains large regions of clear air. In particular, geometrical shadowing is accounted for exactly. Gabriel and Evan (1996) found this approximation to be a good compromise between efficiency and accuracy for domain-average flux profiles compared to several other techniques. We see no reason to not use it for the full 3D grid, as needed. This solution is certainly a reasonable first-guess for the 3D radiation field in a successive order of approximation technique such as the adjoint perturbation series.

- To illustrate an approximation grounded in phenomenology, we invoke the Nonlocal Independent Pixel Approximation (NIPA). The IPA is strictly local in the horizontal plane (cf. Chap. 12), so the chosen name here is somewhat self-contradictory by design. This approximation method was proposed by Marshak et al. (1998) as a way of improving the IPA/ICA, viewed as a prediction for radiance fields (not just the domain-average) when it was realized that the high-resolution IPA radiance fields inherited the characteristic roughness[20] of fractal stratocumulus cloud models (cf. Appendix) while observations and Monte Carlo simulations showed a remarkable degree of smoothness at the smallest scales (Marshak et al., 1995). This was soon understood as the natural averaging ef-

---

[20] This refers to the non-differentiability of the function assigning optical depth to horizontal position, not to the cloud boundaries which is in fact generally considered flat (i.e., the stratus-type cloud models are *geometrically* plane-parallel).

fect of net horizontal flux divergences that arise when pixels become optical thin in the transverse direction, i.e., when $\sigma(\boldsymbol{x})\Delta_h \approx 1$ where Δ_h is the horizontal dimension of the pixel/cell (as defined by the imaging radiometer or by the computational grid). For more on this "radiative smoothing" phenomenon, we refer to Chap. 12 (and references therein). Physical intuition tells us that a step in the right direction is to take the IPA field and to apply a low-pass spatial filter with a profile inspired by 3D RT experiments using pencil beams. The width of the filter thus captures parametrically the dependence of radiative smoothing on the relevant cloud properties, namely, thickness and mean optical depth. A reasonable choice for the smoothing kernel is in fact the boundary-source/boundary-sampled Green function, i.e., the reflected or transmitted radiance field excited by a δ-source at a boundary.[21] An example of a smoothing kernel that would be appropriate for a NIPA estimation of the transmitted flux field can be seen in Fig. 5.6, conveniently expressed in closed-form in the Fourier domain (where convolutions are the most efficiently performed) in Fig. 5.5. For a representation in physical space, see Polonsky and Davis (2004) who improve upon the empirical estimate of the exponential tail of the Green function used by Marshak et al. (1998). Since we end up using a closed-form convolution equation to predict the true 3D RT field starting from the IPA estimate, we can also attempt the opposite. The intention here is to use the IPA field inferred from actual 3D RT radiances (i.e., remote-sensing data) to back out cloud properties, starting with local optical depth. This is called the "inverse NIPA" (Marshak et al., 1998). As one might expect, this is a numerically unstable operation but amenable to standard regularization techniques in Fourier space (Tikhonov, 1977).

- To exemplify what can be gained by borrowing a technique from another discipline, in this case, computer science, we take Mapping Neural Networks (MNNs). MMNs are statistically trained algorithms developed primarily for complex problems in nonlinear function approximation using one or more layers of "links" and "nodes/neurons" that connect an input field to an output field. The links are essentially a series of weights used to collect the input into sums, then into intermediate "hidden-layer" values, and finally to redistribute these towards the output. The nodes nonlinearly transform the intermediate values. The values of the weights are determined iteratively by examining a "training" set of *given* input and *given* output, continuously adjusting the multi-layer weight collection to better approximate the output. So a "cost-function," say, the sum of (predicted minus actual) differences squared, is being minimized here. A good reference on this popular topic in computer science is Caudill and Butler (1992) and high-quality software encoded by subject-matter experts is freely-distributed over the internet. The MNN technique was first applied to 3D RT by Faure et al. (2001) and recently brought to almost operational functionality in cloud remote sensing by Cornet et al. (2004). The input is the IPA field, typically a reflected radiance, and we want to use it to

---

[21] This choice can be justified a posteriori by the 1st-order perturbation formula in (5.59) where I_b is the 1D RT equation solution (say for reflected radiance), $\Delta\mathcal{L}$ is (formally) the application of the IPA, and I_b^+ is the relevant Green function.

predict the true 3D RT field, or vice-versa in a remote-sensing application (Faure et al., 2002; Cornet et al., 2004). These are truly the same problems that are addressed with the NIPA and inverse NIPA respectively. The only difference is that MNNs are used in lieu of postulating the form of a smoothing kernel, applying it, and comparing the forecast with pre-computed 3D RT (using Monte Carlo or SHDOM). The cost-function is a natural measure of accuracy. Another measure of performance in machine learning is to consider input/output data that were not used for the training, so-called "generalization" data. A well-trained MNN will perform well for this new "out-of-sample" data.

These are just three examples, outcomes from creatively thinking about 3D RT with a practical problem-solving mindset. In particular, there is no attempt here to compute internal or transmitted radiance fields if reflected fields are the programmatic focus. There are other such methods, in particular, further variations on the IPA theme (e.g., Várnai and Davies, 1999) and also importation into deterministic remote-sensing problems of methods used for the statistical large-scale problems (e.g., Kokhanovsky, 2003).[22] We anticipate many more of these pragmatic approaches to be developed as the demand for 3D RT increases in specific situations. In particular, the tremendous advances in machine learning since the advent of MNNs (genetic programming, support vector machines, etc.) will likely be exploited.

5.3 Efficient Computation of Detailed 3D Radiation Fields: Discussion and Outlook

In the previous section we described approximation methods for computing detailed 3D radiation fields using elliptical PDE solvers, sparse-matrix equations, closed-form multi-dimensional integrals, smoothing kernels, and machine-learning modalities. Our goal was not to be exhaustive. More approaches are being developed as we go to press and will appear in the literature on a regular basis. Rather, we wanted to sample the diversity of current methods which reflects the magnitude and multi-faceted nature of the computational challenge at hand.

We did not discuss efficiency (speed-up with respect to the numerical 3D RT methods of Chap. 4) in quantitative terms, and we paid little attention to the associated tradeoff in accuracy. It seemed to us more important at this conjecture and in this venue to survey the available portfolio of methods because this is truly about work in progress. Even in the absence of a detailed study of efficiency/accuracy tradeoff, we already know that there is no "silver bullet" in 3D RT approximation. The method of choice will ultimately depend on the application. That is the main reason for maintaining the diversity of approaches.

We close with a brief discussion of the primary stakeholders that stand to benefit from efficient-but-approximate 3D radiation transport codes when they become

---

[22] See Chap. 6, Sect. 6.2, for a description of several such methods targeting domain-average fluxes.

available. In some case, we will allow ourselves to speculate on the approximation technique of choice . . . and be quite happy to be proven wrong.

5.3.1 Cloud and Cloud-System Modeling

Climate prediction is a capability we need as a society, to make sound decisions in energy policy that can have a huge impact the economy as well as on diplomacy. This project relies almost entirely on the forecasting skill of GCMs. This is a tremendous burden on predictive climate science, and a potential vulnerability in the greater scheme. Climate modelers are therefore constantly looking for ways of improving their representations of climatically important processes. This is especially true when these processes are not resolved by the GCM dynamics where grid-scales are on the order of several hundred km.[23] One of these processes is radiative transfer, and how to parameterize it in the presence of clouds was covered in the first sections of this Chapter as well as in several other parts in this volume. An arguably even more challenging sub-grid problem in GCMs is cloud physics and dynamics.

Cloud Process Modeling with CSRMs

RT subject-matter experts have 3D RT codes to investigate in great detail how 3D cloud structure may or may not impact the atmospheric and surface radiation budgets at GCM grid-scales. Their goal is to distill all the 3D information into a simple parameterization of boundary fluxes and heating-rate profiles, given a very coarse description of the state of the atmosphere and surface. Similarly, cloud experts have cloud-system resolving models or CSRMs which can cover a GCM grid-box yet resolve structures at a km or so in scale, although typically by reverting to 2D dynamics (only one horizontal and one vertical coordinate). These CSRMs were originally designed to be numerical laboratories where new parameterizations of cloud-scale processes can be developed and tested, and serve this purpose very well. However, a recent trend has been to short-circuit the parameterization step altogether and to insert a CSRM into each GCM grid-cell – this is called "super-parameterization" (Khairoutdinov and Randall, 2001). There are of course radiative processes at work at the scales resolved by the CSRM and they are 3D. However, because of their legacy, CSRMs use simple column-by-column RT computations in the 2-stream approximation. It is of course inconsistent to have 3D (or even 2D) dynamics and 1D radiation. So the CSRM community would surely welcome more realistic radiation codes with some 3D capability . . . as long as it does not cost too much in computer time. At a km or more across, CSRM grid-cells can be optically thick (clouds in their own right) as well as optically thin. This is not an easy 3D medium to model

---

[23] Japan's Earth Simulator (http://www.es.jamstec.go.jp/esc/eng/) is a dedicated super-computer that resolves horizontally details which, at about 10 km, are commensurate with the atmospheric scale-height. At this remarkable threshold-crossing in computational resolution, we will eventually need to reconsider the design of physical parameterizations, certainly radiation (cf. Fig. 5.1a).

in detailed 3D RT whether using the exact RT equation or an approximation thereof. We suspect a hybrid model is required here: discrete-angle/discrete-space 3D RT for interactions inside a multi-cell cloud mass (since it is a straightforward generalization of the current 2-stream model) and possibly a radiosity-based approach for interactions between such cloud masses.

Cloud Process Modeling with LESs

On smaller grids (cells not exceeding a 100 m or so), we have LES models that consider at most a few clouds at a time in a domain of a few tens of km, but they generally have better microphysics than CSRMs. At present, LES models use the same column-by-column 1D RT parameterizations as CSRMs and GCMs even though the width of these columns is now on the order of the local value of the photon mean-free-path, and often much less. In this situation, a 3D "remedy" to the unrealistic 1D RT is likely to be 3D diffusion theory, especially in the denser and more embedded portions of the cloud. This improved treatment of radiation has in fact already been implemented and tested in the UCLA cirrus cloud model by Gu and Liou (2001). For nocturnal (IR forcing only) boundary-layer stratocumulus, benchmark results have been obtained by running coupled dynamics and 3D RT using the exact-but-slow SHDOM model called by the U. Oklahoma LES (D. Mechem, private comm.). Having more boundary per unit of volume and being separated by quasi-vacuum, broken cloud fields are even more challenging.

5.3.2 Remote Sensing of Cloud and Surface Properties

Operational remote sensing of the Earth's environment is another major application area for atmospheric RT where computer time (measured in CPU cycles consumed per pixel) needs to be carefully managed. Indeed, space agencies worldwide – and NASA in particular – have considerable amounts of sophisticated instrumentation on orbiting satellite platforms that produce increasingly huge numbers of pixels to process, whether they are passive imagers or active profilers. In both of these observational modalities, spatial complexity across the horizontal boundaries separating the smaller pixels as well as inside the larger pixels is almost always ignored while spectral or temporal information layers gets all the attention. In other words, horizontal uniformity assumptions are the general rule in the RT modeling that supports operational remote-sensing observations, no matter how big or small the pixels are and whether or not it is justified. How much longer can this unrealistic modeling assumption be enforced on the questionable basis that we have nothing else than 1D RT to apply? Probably not very long. Is it really true that 1D RT is all we can afford to do anyway? Probably not. In the not-so-distant future, we will see real-time/high-resolution/whole-Earth observation from lunar and LaGrange stations. Thanks to advances in nuclear space reactors and broadband communications, there will be increasingly powerful imaging and/or profiling radiometers onboard planetary probes, fully calibrated and therefore capable of quantitative exploitation just like any Earth-bound instrument.

This vision puts tremendous pressure on the atmospheric RT community to develop physically realistic methods for extracting the inherent physical information from pixel-space as well as from the spectral dimension of remotely-sensed data ... without costing much more than a 1D RT computation. At this point, it is important to recall that clouds are the main cause of horizontal divergences and convergences in the photon flow, hence blatant violations of the operational uniformity assumption. We must therefore distinguish between the problems of remote sensing of cloud properties and of surface properties in the presence of clouds.

Cloud Remote Sensing Using Large Pixels

At resolutions worse than a few km per pixel, there is clearly more of a problem of unresolved variability than of cross-pixel radiation transport inside clouds. The obvious first-cut solution here is to use one of the effective-medium/mean-field theories discussed above even though they were developed with climate studies in mind; they can possibly remove some remote sensing biases at virtually no computational cost because they only change the optical parameters used in the retrievals based on 1D RT. As needed, another class of solutions will call on mean-field theories, again developed a priori for large-scale domain averages, that lead to new transport equations to solve. These equations predict the radiation fields from cloud optical properties, and in this application will need to be inverted. Either way, one can foresee multi-spectral – and multi-angular, multi-polarization, and eventually multi-pixel – retrievals of all the usual cloud properties of interest (optical thickness, droplet radius, etc.) with more accurate means. For an after-the-fact study looking towards future 3D retrievals, see Rossow et al. (2002). For a "recycling" of a climate-driven variability model, the Gamma-weighted two-stream approximation (GWTSA) described in the next Chapter, see Kokhanovsky (2003). As a matter of course, we will also see new retrievals to investigate in their own right: effective small-scale variability parameters.

Cloud Remote Sensing Using Small Pixels

Instruments in this class (resolutions of a couple of 100 m or better) are used, in particular, for detailed sub-pixel looks that support their counterparts we just discussed and that have poorer spatial resolution (e.g., ASTER for MODIS, on the same Terra platform). They thus provide validation data for improved remote-sensing algorithms that acknowledge unresolved variability. They are also of interest in their own right because they open up a new kind of 3D RT problem. Indeed, in high-resolution images of overcast scenes, pixel adjacency effects dominate since their size is comparable or smaller than the photon mean-free-path, as small as it already is inside clouds (typically less than 100 or so meters). Here, tricks like the above-described inverse NIPA and MNNs have already proven useful by attempting to restore the radiance that the strict (local) IPA would yield and, from there, cloud optical properties using standard inversions based on 1D RT. There are others, such as Marshak et al.'s

(2000) Normalized Difference Cloud Index (NDCI), also covered in Chap. 14. Conceivably, one could develop 3D tomographic techniques based on iterations of an efficient forward RT model. Diffusion and adjoint perturbation theories could be put to task here.

Broken/Isolated Cumulus Clouds

Small cumulus are problematic at any resolution because their RT is inherently 3D and IPA techniques are thus biased, unless informed somehow by the knowledge of 3D RT effects. Alternate techniques using radically non-plane-parallel cloud geometry are required. Davis (2002) has proposed a multi-pixel "sunny/shady ratio" technique for deriving broken cloud optical depth based on a diffusion solution for the RT in spheroidal cloud shapes. This is an avenue worth exploring for imagery with high-enough resolution to clearly distinguish the illuminated and non-illuminated portions of that clouds surface and analysis of 5-m resolution data from DOE's Multispectral Thermal Imager (MTI) VIS/NIR channels supports this proposal. Still assuming high-enough resolution, the contrast between radiance from ground pixels that are sunlight and in cloud shadows can also be used to estimate broken cloud optical depth (P. Chýlek, pers. comm.) In low-resolution images, up-welling radiances are a mixture of sun-light cloud surfaces, non-illuminated cloud surfaces, and similarly for the ground (surfaces that receive direct sunlight, and that are in cloud shadows). Furthermore, this is not a linear mixture because all these elements are in view of each other and therefore exchanging photons. Nonetheless, linear combinations can be used as a first cut, and line-of-sight arguments (a.k.a. "view-factors" in radiosity) can be used to go to the next level.

Cloud Contamination in Surface or Clear-Air Property Remote Sensing

In remote sensing of surface properties at high spatial resolutions, we already struggle with nonlinear mixtures of spectral signatures mediated by the molecular/aerosol atmosphere because the photon mean-free-path can be considerable, but not long enough to neglect the scattering altogether. This is known as the "adjacency" effect (Lyapustin and Kaufman, 2001). We can also become interested in determining ground or atmospheric properties, including surface fluxes, in the presence of broken clouds. What are their adjacency effects? What is the cloud impact on the illumination of a nearby surface element, hence on surface property retrievals? What is the impact of a dense cloud on the atmospheric "path radiance" contribution to the total radiance from a clear pixel near by? How does that influence clear-air property retrievals such as aerosol load or column water vapor? Intuition, supported by forward Monte Carlo simulations (Kobayashi et al., 2000; Wyser et al., 2002, 2005) or empirical assessments (Cahalan et al., 2001; Wen et al., 2001), suggests that at close enough range to the cloud these effects can be considerable (i.e., detectable, and thus damaging if neglected). To err on the safe side, industry providers of high-resolution satellite imagery such as DigitalGlobe do not consider their products of commercial value unless they are less than ≈20% cloudy. In contrast, Chýlek et al.

(2004) show empirically using MTI data that the effect of broken clouds on path radiance in water-vapor retrievals based on near-IR band ratios with 20-m resolution is often, but not always, negligible.[24] In the same paper, the effect of unresolved small clouds in imagery at lower resolution is simulated and found to be considerable; the same conclusion holds for the impact of unresolved clouds on aerosol optical depth retrievals (Henderson and Chýlek, 2005).

To estimate the magnitude and sign of these cloud contamination effects in real data, we need at least rough estimates of cloud properties. Geometrical parameters (height, thickness, extent, aspect ratio, etc.) can often be estimated using in-scene triangulation techniques, and improved estimates will be obtained if multiple viewing angles are acquired. Then optical properties must be determined by one or another of the above sketched techniques. This extension of atmospheric correction from purely aerosol/molecular atmospheres up to the edge of broken clouds may sound somewhat compulsive in many land-surface remote sensing applications in environmental science where it is easier to just wait for a more cloud-free collect. However, in the surveillance- or reconnaissance-driven remote sensing performed by the intelligence community, one may not have that luxury.

From Mitigation to Exploitation of 3D Radiative Transfer

So far, we have hinted at ways of compensating for 3D RT effects in and around clouds using rough approximations. This is not only for the sake of efficiency but because there is an error budget to balance: models have errors and so do the data. Beyond mitigation, we find deliberate exploitation of 3D RT effects in entirely new remote sensing technologies. One example, described in detail in Chap. 13 (and references therein), is to work with solar photon pathlength statistics obtained from differential absorption spectroscopy in well mixed gases, often oxygen bands or (at high-enough spectral resolution) oxygen lines. Another example, closely related to O_2 spectroscopy on the theoretical plane, is "large foot-print" lidar probing of clouds, surfaces or vegetation canopies. In both the passive and active modalities, 3D spatial information is packed into the time-domain via multiple scatterings and/or surface reflections. Illumination of a dense cloud at relatively close range by a narrow laser beam is a 3D RT problem that opens the path to off-beam/multiple-scattering cloud lidar (Davis et al., 1999) where the new closed-form lidar equation is essentially the (boundary-source) Green function sampled at a boundary point. The non-stationary version of the 3D diffusion equation from Sect. 5.2.2 is an accurate approximation for this signal at large enough times and distances (Polonsky and Davis, 2004). These recent developments mark the beginning of a new era in remote sensing science and technology where spatial and temporal relations between pixels bring as much if not more than what can be extracted from the multi- or even hyper-spectral layers of the image cube.

---

[24] This is however traced for the most part to the compensating effect of cloud droplet absorption in one of the reference (out-of-band) channels used by MTI to infer column water vapor. So the reassuring conclusion may not carry over to other instruments.

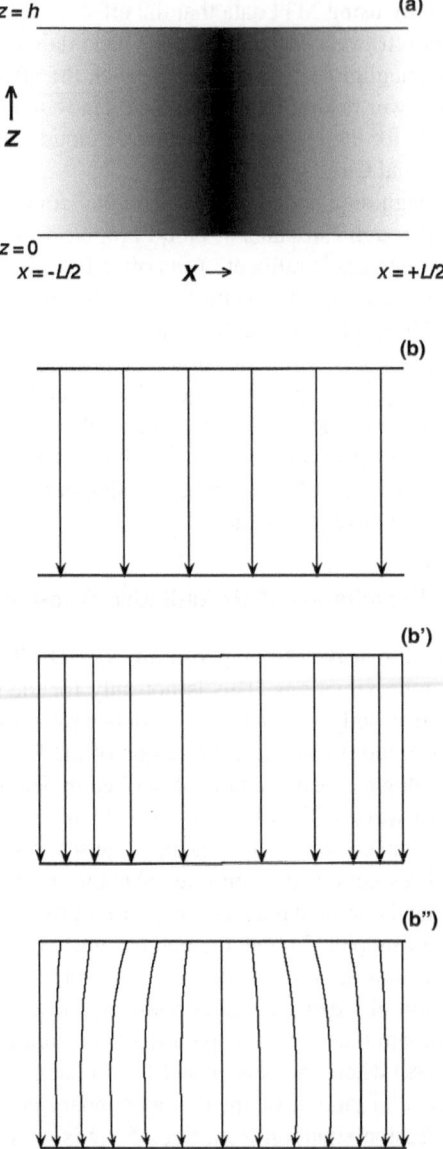

Fig. 5.9. Schematic of (**a**) the extinction field inside a sine-wave cloud model and (**b, b′, b″**) three approximations for the radiative transfer (photon flow) represented by vector-flux lines

5.4 Three-Dimensional Radiation Transport Phenomenology: A Case Study

Consider the schematic in Fig. 5.9. Panel (a) illustrates an extinction field with horizontal variability in the x-direction only in the form of a sine-wave oscillating around a mean value $\langle\sigma\rangle$ with an amplitude $0 \leq \delta\sigma \leq \langle\sigma\rangle$; see (5.62). The thickness is h and the wavelength (period) is L. Therefore $\tau(x) = \sigma(x)h$ varies in sync with extinction and its parameters are $\langle\tau\rangle = \langle\sigma\rangle h$ and $\delta\tau = h\delta\sigma$. This is a highly idealized cloud that has been considered over the decades in many theoretical studies using almost as many different RT techniques, starting with Giovanelli (1959) and ending (as far as we know) with Polonsky et al. (2003). Notable milestones in this ongoing series of 3D RT investigations using the idealized "sine-wave" cloud are the papers (known to us) by Weinman and Swartztrauber (1968); Cannon (1970); Romanova (1975); Stephens (1986); Li et al. (1994); Gabriel and Evan (1996); Galinsky and Ramanathan (1998); Galinsky (2000)[25] and Davis and Marshak (2001). This simple two-dimensional medium is understandably a preferred test-bed for new 3D RT equation solution methods as well as 3D RT approximation techniques, several of which are mentioned in the following Section.

To keep the following discussion as simple as possible, the medium will be assumed conservative ($\sigma_a = 0, \sigma = \sigma_s, \varpi_0 = 1$). Photon sources are furthermore concentrated at the top ($z = h$) of the cloud (the illumination pattern is assumed axisymmetric) and uniform in x (like sun at zenith or hemi-spherically averaged); the lower boundary ($z = 0$) is purely absorbing (no sources, nor reflection). This defines the 3D RT problem, up to the precise choice of scattering phase function and of the incident radiance pattern $I_{in}(\theta)$ with $\theta \in (\pi/2, \pi]$. We will assume that the incident flux, $F_0 = 2\pi \int_{\pi/2}^{\pi} I_{in}(\theta)|\cos\theta|d\theta$, is unity. Under these circumstances, the photon flow inside the medium, as represented by flux-lines, is two-dimensional and symmetric around the crests and troughs of the sine-wave extinction field. By definition, flux-lines are everywhere tangent to $F(x, z)$; they all start at cloud top (where the sources are located) and invariably end at cloud base (where the dominant sinks are located). In a very real sense, this photon flow is driven by an excess of "pressure" at cloud top but must negotiate with the variable extinction, i.e., resistance to flow. The resulting value of transmission $T(x)$, a net flux along the vertical axis across the lower boundary, is proportional to the *local* density of flux-lines. At cloud top, the same can be said of the net flux $1 - R(x)$ associated with the albedo field $R(x)$.

The next three panels illustrate schematically three levels of approximation to the 3D RT problem. Each approximation has a following in the atmospheric RT community but, for the moment at least, the number of practitioners decreases drastically as we progress down the sequence. This trend is in spite of the fact that accuracy increases at a computational cost that is not as forbidding as many imagine. The overarching goal of approximation theory in numerical 3D RT is to reverse this trend without making it difficult for the majority now using the 1D RT approximation, irrespective of the circumstances. We now visit each approximation in turn assuming we are after the normalized domain-average net flux

---

[25] In these two last papers, the authors examined a homogeneous cloud with a sine-wave varying thickness as well as the usual inhomogeneous cloud case.

$$\langle T \rangle = \int\limits_{-L/2}^{+L/2} T(x) \frac{\mathrm{d}x}{L} , \tag{5.65}$$

with $T(x)$ computed with one RT model or another, but we also have an interest in details of the RT inside the medium.

5.4.1 1D RT (Fig. 5.9b)

Here we make the simple assumption that $\delta\sigma = 0$, i.e., we ignore the unresolved variability in order to use classic plane-parallel/homogeneous theory which we have denote all along as "1D" or "PPH." This approximation imposes the maximum degree of translational symmetry on the underlying photon flow. Albedo and transmission fields are of course uniform. Although this volume is about 3D RT, 1D theory is not always a bad approximation depending on cloud type, scale and tolerance to error (cf. Chaps. 11–12). Furthermore, there are ways of informing the 1D theory, via parameter change, on how to give a more accurate answer in some necessarily limited sense. But that limitation may be good enough for some applications (cf. some of the methods discussed in previous Sections targeting domain-averages, as well as Chap. 8).

5.4.2 ICA/IPA (Fig. 5.9b′)

Here we make the assumption that $F_x(x, z) \equiv 0$, i.e., we ignore the *net* horizontal fluxes everywhere. This is commonly known as the Independent Column/Pixel Approximation (ICA/IPA) as long as that is understood as a model for *local* radiation fields, not just the domain-average that often follows. We are now in presence of a lesser degree of symmetry in the flow. In the present case of x-only variability, all happens in 3D as if the extinction sampled at cloud top by the incoming photon controls its whole trajectory, ending in reflection or in transmission. In other words, a zoom into any region at cloud top or bottom will look exactly like the 1D solution in panel (b). In a nutshell, this is NASA's operational strategy for generating cloud and surface remote sensing "products" starting from spectrally-sampled radiances from each pixel. Many studies have researched the validity of this assumption (cf. Chaps. 8–12 in this volume, and references therein). Most of the techniques targeting domain-averages assume that whatever errors the IPA/ICA incurs with respect to full 3D RT cancel sufficiently well for the purposes of radiation parameterization schemes in GCMs. But this does not exonerate the local IPA assumption used in remote sensing. Notice that the total number of flux-lines has increased from panels (b) to (b′). This is a direct consequence of Jensen's inequality,

$$T_{\mathrm{ICA}} = \langle T_{1\mathrm{D}}(\tau) \rangle > T_{1\mathrm{D}}(\langle \tau \rangle) , \tag{5.66}$$

for domain-average transmittance, simply because of the newly acknowledged variability. To see how inequality (5.66) arises, we note that 1D RT predictions for transmittance are concave functions of the horizontally varying optical thickness τ. As an example, consider (5.32) with $\tau_t = (1 - g)\tau$, thus leading to

$$T_{1D}(\tau) = \frac{1}{1 + (1 - g)\tau/2\chi} \,, \tag{5.67}$$

as was first obtained by Schuster (1905), where χ is an $O(1)$ numerical constant already discussed in connection with BCs in diffusion theory.[26] So total transmission across the domain has increased and albedo has, in this photon-conserving case, decreased by the same amount.

5.4.3 3D RT (Fig. 5.9b″)

Here we relax all assumptions about the internal variability, such as the radical one we made in connection with panel (b), or about the associated RT, such as the practical one we made in connection with panel (b′). So here we are doing the RT modeling at least accurately enough to capture trends in $F_x(x, z)$, the sign of which dictates how the flux-lines are deflected away from the vertical; as we will show further on, this need not be a full-blown solution of the 3D RT equation. The local translational symmetry in (b′) is now broken in all but $x = 0$ and $x = \pm L/2$ around which the extinction field is itself mirror-symmetric. Flux-line density is proportional to the magnitude of $F(x, z)$ everywhere, not just at the boundaries. This basic consequence of vector calculus shows us that this minimally symmetric vector flux field has evolved from (b′) in a completely predictable way. Indeed, the flux-line pattern tells us that photons will tend to flow around the dense region ($x \approx 0$), where flux is accordingly reduced, and into the more tenuous ones ($|x| \approx L/2$), where we see that flux is enhanced. In the schematic, we have predicted even more flux-lines than in (b′) which translates to even more overall transmission than in the IPA/ICA. As we now show numerically, this occurs if the angular illumination distribution is sufficiently concentrated around the zenith direction (e.g., $\Omega_0 \approx -\hat{z}$ if monodirectional). For illumination patterns with more slant incoming rays (not illustrated), we find less overall transmission than the ICA predicts but still more than predicted by 1D theory.

5.4.4 Quantitative Analysis of Fig. 5.9

We have used the simple sine-wave cloud as a framework to guide us from the scenario where variability is unresolved to one where we are informed about everything there is to know about the cloud. In the former case, an "ignorance is bliss" strategy can be implemented: use 1D RT and stop worrying. This is an often exercised option that may or may not be justified. At any rate, it is simply an approximation. In the later case, we know all the spatial detail and are at least compelled to go beyond 1D RT theory applied globally (PPH/1D) or locally (IPA/ICA). We continue with a short numerical investigation of the sine-wave cloud.

The next chapter relies heavily on the ICA and we will see a variety of assumptions about the 1-point variability of optical depth τ: Bernoulli (2-state), Gamma

---

[26] This model belongs to the broader class known as "Model 1" from Coakley and Chýlek (1975) with $\beta(\mu_0) = \mu_0(1 - g)/2\chi$, a special case where T_{1D} does not depend on the solar zenith angle.

and quasi-lognormal hypotheses. Here we have a special case of the beta distribution referred to colloquially as the "smile" PDF. From (5.62) using τ variables and parameters, we obtain

$$p(\tau; \langle\tau\rangle, \delta\tau) = \frac{1}{\pi\sqrt{\delta\tau^2 - (\tau - \langle\tau\rangle)^2}}, \tag{5.68}$$

if $\langle\tau\rangle - \delta\tau < \tau < \langle\tau\rangle + \delta\tau$ and 0 otherwise. Notably, this PDF is independent of L and h, as it should in an ICA. The graph of $p(\tau; \langle\tau\rangle, \delta\tau)$ has vertical asymptotes at $\tau = \langle\tau\rangle \pm \delta\tau$ and is symmetric around a minimum value of $1/\pi\delta\tau$ reached at $\tau = \langle\tau\rangle$. We now apply this PDF to Schuster's simple "diffusion in a slab" formula in (5.67). Figure 5.10a shows both $T_{1D}(\tau)$ and $p(\tau; \langle\tau\rangle, \delta\tau)$ for $\langle\tau\rangle = \delta\tau = 50$, a situation of interest further on.

The domain-average ICA transmittance can be computed with (5.65) using $T(x) = T_{1D}(\tau(x))$, or by

$$T_{ICA}(\langle\tau\rangle, \delta\tau) = \langle T_{1D}(\tau)\rangle = \int_0^\infty T_{1D}(\tau)p(\tau; \langle\tau\rangle, \delta\tau)d\tau \tag{5.69}$$

This yields the results in Fig. 5.10b where we have plotted the PPH bias for transmittance $\langle T_{1D}(\tau)\rangle - T_{1D}(\langle\tau\rangle) \geq 0$ versus the basic 1D prediction $T_{1D}(\langle\tau\rangle)$. All combinations of $\{\langle\tau\rangle, \delta\tau\}$ are explored. The important message here is that the PPH bias relative to the *observed* value, $[\langle T_{1D}(\tau)\rangle - T_{1D}(\langle\tau\rangle)]/\langle T_{1D}(\tau)\rangle$, can reach almost 50% for $\delta\tau = \langle\tau\rangle \approx 5.5 \times 2\chi/(1 - g)$ where we find $T_{1D}(\langle\tau\rangle)) \approx 0.15$ and $\langle T_{1D}(\tau)\rangle \approx 0.28$. For the canonical values of the optical parameters ($g = 0.85$, $\chi = 2/3$), This translates to $\langle\tau\rangle \approx 50$, hence a range $0 \leq \tau \lesssim 100$ ($1 \geq T_{1D}(\tau) \gtrsim 1/12$) for the PDF in (5.68). This is indeed the perfect situation to get the most out of Jensen's inequality in (5.66): the curvature of $T_{1D}(\tau)$ is effective enough on this support to go all the way from the linear regime at $\tau \to 0$ to the flat asymptotic regime at $\tau \to \infty$. Such apparently extreme variability is in fact typical of a partially cloudy scenario: practically all the optical depth is from the clouds but practically all the transmitted light is from the clear sky.

Figure 5.11 uses the same variability model as in the ICA exercise but now we set $\delta\tau = \langle\tau\rangle = 15$, $g = 0.85$, and assume a simple Henyey–Greenstein phase function (see Chap. 3) in a straightforward Monte Carlo scheme based on the maximum cross-section trick (see Chap. 4) with 10^5 histories. We then vary the sine-wave cloud's aspect ratio L/h from 0 (the PPH case) to ∞ (the ICA model) using two different illuminations, overhead sun ($\mu_0 = 1$) and isotropic illumination (μ_0 uniformly distributed on $(0, 1]$, each carrying a weight μ_0), and we plot the domain-average transmission $\langle T_{3D}\rangle$. We note that

- $\langle T_{3D}\rangle \approx T_{ICA} = \langle T_{1D}\rangle$ at $L \gg h$ because the fluctuations are *so slow* that the ICA becomes accurate locally, hence globally;

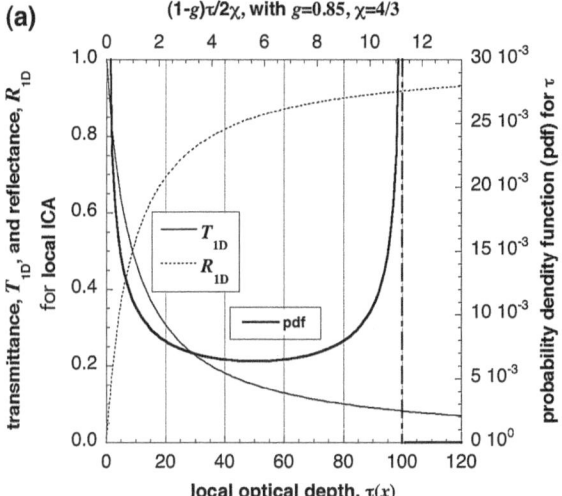

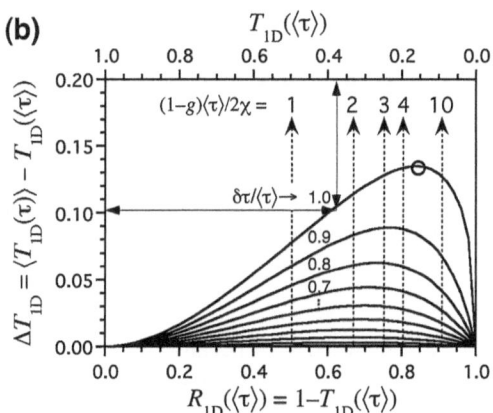

Fig. 5.10. (a) Simple diffusion prediction for $T_{1D}(\tau)$ in (5.67) and the "smile" PDF in (5.68) for $\langle\tau\rangle = \delta\tau = 50$. (b) Domain-average PPH bias, i.e., computed within the ICA, for transmittance is plotted versus the 1D RT result for selected variability parameters. For the maximum possible variability amplitude ($\delta\tau = \langle\tau\rangle$), the maximum bias is reached for the scaled mean optical thickness $(1-g)\langle\tau\rangle/2\chi \approx 5.5$, hence $\langle\tau\rangle = \delta\tau \approx 50$ if we take usual values $g = 0.85$ and $\chi = 2/3$. This situation is indicated by a bold circle in panel (b) and determines the choice of parameters in panel (a). The PPH bias for the parameters used in Fig. 5.11 (lower curve) are indicated by the double arrows. We notice the excellent agreement of the diffusion-based formulation used here and the MC technique used there in spite of a non-negligible contribution of small τ's

- $\langle T_{3D}\rangle \approx T_{1D}(\langle\tau\rangle)$ at $L \ll h$ because the fluctuations are *so fast* that all the radiation "feels" is the mean extinction;[27]

---

[27] Strictly speaking, this is true only for the diffuse source. For the overhead sun case, there is a finite jump between small but finite L/h and the PPH case since the quasi-holes between

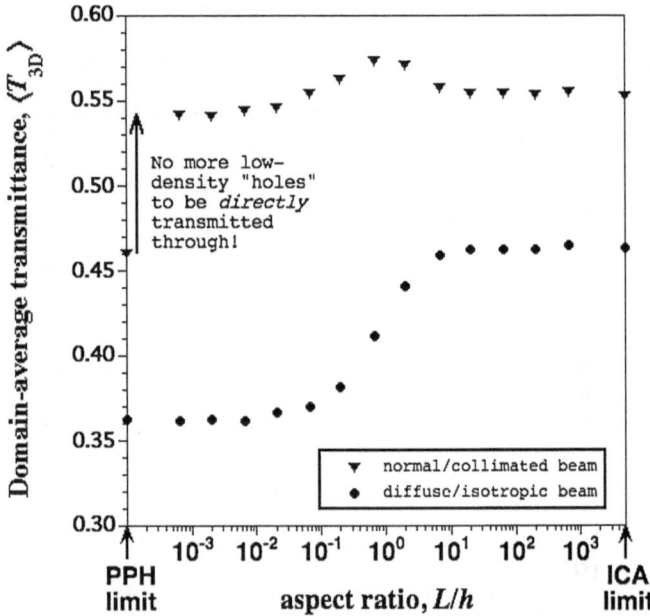

Fig. 5.11. Domain-average transmission for a non-absorbing Henyey–Greenstein ($g = 0.85$) sine-wave cloud with $\delta\tau = \langle\tau\rangle = 15$ under two illumination patterns, collimated/normal and isotropic. The aspect ratio L/h is varied along the abscissa and the most interesting region is where it is $O(1)$. See main text for a detailed discussion. Notice that the PPH limit on the left-hand side is singular for normal/collimated illumination because an angular alignment between incoming light and the striated medium is destroyed

- the interesting action happens for $L \approx h$ because we have an internal variability rhythm that matches ("resonates" with) the fixed outer scale of the medium, its thickness h.

For the overhead-sun illumination scenario (upper curve in Fig. 5.11), we find that a maximum in domain-average transmission occurs when 3D RT effects come into play *beyond the 1D and ICA approximations*. As was anticipated in the course of our discussion of Fig. 5.9b″, more overall flux is observed for all but the shortest values of L ($\lesssim 0.04h$). This is however not the case for diffuse illumination (lower curve in Fig. 5.11) where we see that $\langle T_{3D}\rangle$ is bracketed by T_{1D} and T_{ICA}. To see why, consider the schematic in Fig. 5.12 where we see a dense region under a typical oblique illumination. Photons will clearly be deflected towards both reflection and transmission while the geometrical shadow on the anti-solar side of the cloud reduces transmission as well as reflection. In other words, much cancellation is going on in the 3D scenarios (finite L/h) and the ICA bias in Fig. 5.10 dominates the overall mean to the point where the further 3D effects only control the details of the smooth

---

the clouds (i.e., the dense phase of the sine wave) are shut. This is an unimportant angular resonance phenomenon for this particular variability model.

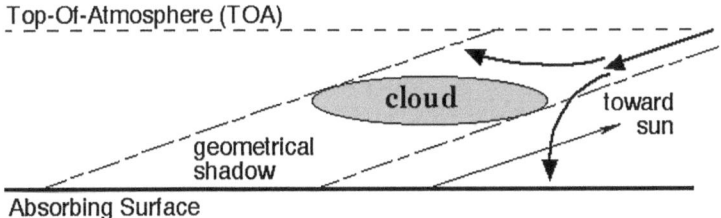

Fig. 5.12. Schematic showing a typical 3D atmospheric scenario under oblique sun

and monotonic transition from T_{1D} to T_{ICA} as L/h increases through the critical region where $L \approx h$.

5.4.5 When and Where Does 3D RT Matter?

In this section, chapter, volume, and the literature, 3D predictions have been compared to 1D predictions, usually based on the mean optical properties. Sometimes, the 1D benchmark is corrected for obvious variability-induced biases, e.g., the ICA is used as a baseline. All of this is using simulations of course. In spite of the plethora of predictions, 3D RT has not become a "predictive" science per se because we tend to first obtain our results (the radiative quantities) and then find interpretations for their changes, one way or another, caused by 3D effects. With few exceptions, not only discovery of any observed trend is obtained computationally rather than theoretically, but also its rationalization. A truly predictive science works only with the given quantities (cloud optical properties) and it anticipates what kind of phenomenon will occur and, hopefully, with what intensity and, ideally, completely new ones. Surely we need to establish that we have at least the beginning of such predictive capability to build approximations that we can trust. In this subsection, we take a few small steps in that direction.

First we remark that the conservatively scattering sine-wave cloud gives us

$$\Delta T = \langle T_{3D} \rangle - T_{1D}(\langle \tau \rangle) > 0 \tag{5.70}$$

under all the variability and illuminations circumstances that were investigated. Davis and Marshak (2001) prove rigorously – but using diffusion theory and diffuse boundary sources – that in arbitrary 3D optical media:

$$\frac{\Delta T}{1 - T_{1D}(\langle \tau \rangle)} = \frac{-\Delta R}{R_{1D}(\langle \tau \rangle)} = 3\chi \frac{\langle \Delta F_z \Delta \sigma \rangle}{F_0 \langle \sigma \rangle} , \tag{5.71}$$

where ΔF_z and $\Delta \sigma$ are the full 3D fields of 3D-to-1D differences and F_0 is the incoming flux. For ΔF_z, we assume the z-axis and the illumination are oriented as in Fig. 5.9, so $F_z < 0$ everywhere and $\Delta F_z < 0$ means a local increase in $|F_z|$. The extinction fluctuation field $\Delta \sigma$ integrates spatially to 0 by definition but ΔF_z does not. Conservation requires that its integral over a horizontal plane ($z =$ constant) be independent of z and, after normalization by F_0, equal to $-\Delta T$. The authors go on to show that $\Delta T > 0$, thus proving (5.70) from first principles. They show that the physical mechanism that leads to this inequality is the channeling[28] of photons into the more tenuous regions by the denser ones. The channeling determines the sign of the spatial cross-correlation term in (5.71). This is a fundamental result in solar 3D RT based on the classic diffusion approximation, just as the classic ICA obeys Jensen's fundamental inequality in (5.66).

Is this predictive capability? Not really. Channeling is nothing new in general-purpose transport theory and exploratory computations in 3D RT gave us a clue of what is going on. Still, it is reassuring to be able to trace an essential 3D RT effect all the way back to its primitive equations. To be facetious, we can say that we *had* the predictive capability, but we used it in a forensic analysis instead.

A major benefit of this hindsight is a progressive partitioning of ΔT in (5.70). First, we have the so-called "plane-parallel bias,"

$$\Delta T_{1D} = T_{ICA} - T_{1D}(\langle \tau \rangle) > 0 , \qquad (5.72)$$

which is the primary concern in the next chapter and has a definite sign (as indicated). The next important question is about the sign and magnitude of the so-called "ICA/IPA bias"

$$\Delta T_{ICA} = \Delta T - \Delta T_{1D} = \langle T_{3D} \rangle - T_{ICA} \qquad (5.73)$$

for domain-averages.[29] We retain from our sine-wave cloud study that the increment in domain-average transmission going from the 1D approximation to the ICA in (5.72) is larger in magnitude than what is gained or lost by going from the ICA to full 3D RT in (5.73). *However, this is only true for horizontal domain-average fields.* The horizontal domain is well-defined in computational 3D RT or in GCM parameterization work. But what is a "domain" in nature? Based on what the generic cyclical BCs applied laterally do for the numerical or analytical computations, we are driven to define a reasonable domain with a scale such that net transport through lateral boundaries can be neglected in the overall budget. Only then can we say that the variability that matters is *inside* the domain of interest.

So much for domain-average fluxes. Local flux values can be dramatically altered by 3D RT effects not captured by the ICA, including apparent violations of energy conservation such as transmittances or albedos in excess of unity (cf. Chaps. 3 and 12). And that is only for point-wise hemispherical fluxes (still averaging over the angular variability) at the boundaries. With radiance fields – all that can be observed remotely – the situation gets even worse for the local ICA. The local ICA/IPA is nonetheless used in routine pixel-scale cloud property retrievals from satellite data performed by NASA and space agencies worldwide, irrespective of the size of the pixel. If the pixels are large enough to qualify as a "domain" in the above sense,

---

[28] As far as we know, the first use of this descriptive language for photon flow seems to be by Cannon (1970) while explaining his astrophysical 3D RT computations.

[29] Empirical evidence, from the numerical experimentation presented in Fig. 5.11 for non-absorbing sine-wave media, tells us that this increment is relatively small compared to (5.72) and typically negative, the exception being overhead-sun ($\theta_0 = 0$) and not too fast variations ($L/h \gtrsim 0.04$).

then internal variability is neglected. If the pixels are too small to be considered radiatively independent domains, then photon cross-talk between neighboring pixels is neglected. Under these conditions, any cloud remote-sensing "product" such as optical depth, liquid water column or effective droplet radius based on 1D RT should be considered as a radiometric quantity expressed in convenient physical units. These numbers are at best loosely connected to the actual physical quantities. We refer to Chaps. 11–12 for an overview of how much these 1D RT retrievals can be off target, and also how to filter them down to some very narrow conditions (on spatial and angular sampling as well as apparent and actual variability) where they can be considered reasonably accurate on some statistical basis.

Returning to our quest for predictive capability at least in hindsight, we return to the abovementioned fact that the "interesting action" happens in Fig. 5.11 when the aspect ratio L/h of the sine-wave cloud is $O(1)$. This numerical finding deserves more attention because it can inform us about the circumstances under which 3D RT models (beyond the ICA) are required for domain average fluxes, and even more so for small pixel-scale radiance values. We have tacitly fulfilled a few conditions beyond $L \approx h$ in this case study, namely, (*i*) that the amplitude of the internal variability be as large as possible, (*ii*) that the mean cloud properties are such that the mean transmittance is in its "sweet-spot" for variability around 1/2, and (*iii*) that there is a certain balance of tenuous and dense regions in the cloudy medium.

In the case of the sine-wave cloud model, extra condition (*i*) mandates that $\delta\tau \approx \langle\tau\rangle$, and similarly for the extinction σ variation in (5.62). Anything less will clearly reduce spatial variability effects in the RT. To meet extra condition (*ii*), we have tuned $\langle\tau\rangle$, given the asymmetry factor $g = 0.85$ of the Henyey-Greenstein scattering phase function we used, so that the values of $\langle T_{3D}\rangle$ in Fig. 5.11 straddle 0.5. Bearing in mind Schuster's 1D RT expression for transmission in (5.67), this calls for $(1 - g)\langle\tau\rangle \approx 2\chi = 4/3$ for the standard (Eddington) value of χ. With $\langle\tau\rangle = 15$ and $g = 0.85$, we find that $(1 - g)\langle\tau\rangle = 2.25$ while $(1 - g)\tau(x)$ varies from 0 to 4.5. We have seen the $(1 - g)$ rescaling of τ before, in connection with diffusion theory (Sect. 5.2.2). So the key non-dimensional quantity in RT in the absence of absorption is the "transport" optical depth $\tau_t = (1 - g)\tau$, i.e., the ratio of cloud thickness h to rescaled or transport MFP $\ell_t = 1/(1 - g)\sigma$. The physical interpretation of the extension by $1/(1 - g)$ of the MFP is that it wipes out the directional memory effect of the forward scattering tendency for real cloud droplet size distributions at visible-through-thermal wavelengths: after a *transport* MFP or so, the scattering might as well have been isotropic. The simple RT consequence is that, if $\tau_t \ll 1$ ($\tau \lesssim 1 \ll 1/(1 - g)$ for $g = 0.85$), then ballistic and quasi-ballistic trajectories dominate, while if $\tau_t \gtrsim 1$ ($\tau \gtrsim 7 \approx 1/(1-g)$ for $g = 0.85$), then the photon transport is dominated by short steps and long random walks. At maximum amplitude, the sine-wave medium with $\langle\tau\rangle = 15$ and $g = 0.85$ spans these two extremes and, with its quadratic leveling at $\tau(x) - 0$ and $2\langle\tau\rangle$, it leaves plenty of space for both transport modes to be established. Finally, we turn to extra condition (*iii*). By using the sine-wave model in a way that satisfies (*i*) and (*ii*), we have effectively mandated that the inter-cloud distance is comparable to the cloud size. In other words, the effective cloud fraction A_c is not too small nor too close to 1. Otherwise, the 3D RT effects (at least for the domain-average) will vanish respectively for lack of cloud altogether or because ICA and maybe even 1D RT models become reasonably accurate for 100% cloud cover. In the real atmosphere and in many models, the definition of A_c is somewhat arbitrary. The sine-wave cloud model is no exception. That said, we could arguably draw the line between cloudy and clear portions at $\tau(x) = \tau_{cut}$. Thus, from (5.62), we find $A_c = 0\%$ if $\langle \tau \rangle + \delta\tau \leq \tau_{cut}$, $A_c = 100\%$ if $\langle \tau \rangle - \delta\tau \geq \tau_{cut}$, and otherwise

$$A_c(\langle \tau \rangle, \delta\tau; \tau_{cut}) = \frac{100}{\pi} \times \cos^{-1}\left(\frac{\tau_{cut} - \langle \tau \rangle}{\delta\tau}\right) \qquad (5.74)$$

in %. For the parameter values used in the numerical study that yielded Fig. 5.11, namely, $\langle \tau \rangle = \delta\tau = 15$ and $g = 0.85$, we find $A_c \approx 88\%$ for $\tau_{cut} = 1$ ($\approx 54\%$ for $\tau_{cut} = 2$). Since L is the spatial repeat cycle of the system, cloud size per se is $A_c L$ while inter-cloud distance is $(1 - A_c)L$.

In summary, to make interesting 3D RT processes happen, we have made

$$v = \frac{\delta\sigma}{\langle\sigma\rangle} \approx O(1), \langle \tau_t \rangle = \left\langle \frac{h}{\ell_t} \right\rangle \approx O(1), \frac{A_c}{1 - A_c} \approx O(1), a = \frac{A_c L}{h} \approx O(1) ,$$
$$(5.75)$$

where we need not interpret all the $\approx O(1)$ relations equally. Nor should we take the symbols' meanings strictly from the sine-wave model: v is just a dimensionless 1-point variability amplitude; $\langle \tau_t \rangle$ is a mean transport optical depth over a relevant domain, from which we can estimate mean *cloud* optical depth, $\approx \langle \tau_t \rangle / A_c$ (using the third item and the fact that the clear-sky optical depth is small by any definition); L is the size of this horizontal domain; and h is the thickness of the cloud layer or cloud system. As we have noted that, for the sine-wave model at least, the third item is basically a consequence of the first two. For illustration, take the square-wave cloud used in Sect. 5.2.2 (Figs. 5.3–5.4); we have $v = \delta\tau/\langle\tau\rangle = 8/10$, $\langle\tau_t\rangle = (1-g)\langle\tau\rangle = 1.5$, $A_c/(1 - A_c) = 1$ (if we interpret by extension the $\tau = 2$ region as "clear"), hence $A_c = 1/2$.

These conditions are not hard to come by in nature, even if we make "≈ 1" very specific, e.g., $\in [1/3, 3]$ where this makes physical sense. Indeed, we often see at least moderately opaque clouds ($\langle\tau_t\rangle/A_c \gtrsim 1$) along with comparatively very clear air ($v \approx 1$, $A_c \approx 1/2$); at the same time, the clouds are not infrequently of comparable thickness and horizontal extent ($A_c L \approx h$) and these dimensions are commensurate with their distance from each other ($A_c/(1 - A_c) \approx 1$). Convective and orographic cloud systems are good examples; stratiform systems are not for a couple of reasons ($A_c \approx 1$ and $A_c L \gg h$), neither are essentially isolated clouds ($A_c \approx 0$).

We now put these approximate but remarkable equalities together and ask a fundamental question about how the resulting variability will affect the flow of radiant energy.

First we acknowledge that the domain-average ICA is our main benchmark in 3D RT (beyond the 1D RT approximation based on mean cloud properties) and that the ICA has already capitalized on the first two (hence three) conditions in (5.75). The PPH bias in Fig. 5.10 is therefore already near its maximum for this particular model when we start using our Monte Carlo code to go from the left to right extremes in Fig. 5.11.

Now we ask about the value of the local extinction $\sigma(x)$ as a predictor of the *actual* MFP which will in general 3D media depend on both position and direction through the kind of variability to be found in that direction starting from that position. In the absence of other information, the a priori prediction for the MFP is indeed $1/\sigma(x)$. However, there is little chance of this prediction materializing unless $\Omega \cdot \nabla \sigma = 0$ for all x' to be found in direction Ω. In turn this can not happen for all Ω unless the medium is uniform. So, to further our predictive 3D RT agenda, we propose to use the field $\nabla \sigma$ or, better still, $\nabla \sigma^{-1}$, which is an appropriately nondimensional quantity. Furthermore, we should rescale extinction for forward scattering effects (i.e., use $\sigma_t = (1 - g)\sigma$) and evaluate the gradient (denoted $\nabla_\perp$) at right angles to the mean photon flow which we will generally assume to be along the vertical. For the sine-wave cloud, we find

$$\nabla_\perp \frac{1}{\sigma_t} = \frac{\mathrm{d}}{\mathrm{d}x}\left(\frac{1}{\sigma_t(x)}\right) = \frac{v \sin 2\pi x/L}{(1 + v \cos 2\pi x/L)^2} \times \frac{2\pi/L}{\langle \sigma_t \rangle} . \tag{5.76}$$

So, apart from a purely numerical term, we have

$$\left|\nabla_\perp \frac{1}{\sigma_t}\right| \approx \frac{v}{\langle \sigma_t \rangle L} = \frac{v}{a\langle \tau_t \rangle} . \tag{5.77}$$

From (5.75), we will therefore also have

$$\left|\nabla_\perp \frac{1}{\sigma_t}\right| \approx O(1) ; \tag{5.78}$$

In other words $(1/\sigma_t) \times \nabla_\perp \ln(\sigma) \approx O(1)$, therefore σ changes significantly over a transport MFP or so.

So, beyond the standard (vertically-measured) transport optical *thickness* of the cloud layer, the key non-dimensional quantity to watch in 3D RT is the (horizontally-measured) transport optical *wideness* of the clouds. Going back to out assessment of the sine-wave model in Fig. 5.11, we see that $|\nabla_\perp 1/\sigma_t| \propto h/L$ can be:

- so small that the ICA becomes accurate (for slow spatial variations), or
- so large that 1D RT becomes a reasonable model (i.e., only the mean optical properties matter because almost every lateral propagation sees both clear and cloudy air in their nominal amounts), or
- just right ("resonant") for 3D RT to have a significant impact in the case where (5.78) prevails.

In the course of this investigation of the basic phenomenology of 3D RT we have transitioned from small-scale/deterministic 3D RT (and approximations thereof) to large-scale/statistical approximations, the prime focus of the following Chap. 6. Our main message is that the elementary processes illustrated in Figs. 5.9b'', 5.11 and 5.12 are direct analytical consequences of the first principles, not just notable simulation results. Since they only use the given quantities, the two first criteria in (5.75) are predictive at the ICA level, as is (5.78) at the next level, i.e., "post-ICA" 3D RT. For more discussion, see Davis and Marshak (2001, 2004) as well as Várnai (2000) and Várnai and Marshak (2003).

5.5 Concluding Remarks

We have surveyed – all too briefly in the interest of space – some of the better-known approximation methods for computational 3D photon transport. We also covered the basic phenomenology of 3D radiative transfer which we see as completely entangled with the approximation program. This is because simple transport models (like diffusion) are needed to advance the theory (via analytical results) and, vice-versa, improved understanding (such as the radiative smoothing phenomenon) lead to new approximations (such as NIPA).

We consider this brevity-by-necessity a good sign. The extensive list of references at the end of the chapter will provide more information on a need-to-know basis. Beware however that approximations can be technically quite intricate, at par with the 3D RT equation solutions covered in Chap. 4. They can also be extraordinarily simple. We encourage the reader to go to the web-page of the Intercomparison of 3D Radiation Codes (I3RC) project (http://i3rc.gsfc.nasa.gov/). This is an ongoing comparison study, on the one hand, between different kinds of 3D RT equation solvers and, on the other hand, between these "exact" benchmarks (i.e., based on the 3D RT equation per se) and the outcome of different approximation techniques. At the time this volume is going to press, it is clear that the latter "Approximations" study group is far less advanced than the one-method-exact-to-another component (which is itself flooded with Monte Carlo models, SHDOM being the only alternative represented in the I3RC). This is (1) because the atmospheric RT community is relatively new to the art of approximation in computational 3D transport theory, and (2) because the challenges are considerable while the allocated resources are still meager. In time, this will change because many important applications demand 3D approximation techniques, no more and no less. In the interim, 3D radiative transfer modeling is precariously under-diversified.

The Manhattan project ushered us into the atomic era.[30] It also marks the beginning of the era of large-scale scientific computation with multi-purpose programmable electronic devices. Since then, science progresses within a qualitatively new paradigm. We used to have the theory–observation duality. Now we have the theory–computation–observation triad. A priori, computation is subordinated to theory (and some might argue that it is the modern substitute for theory). However, computational models are used as virtual laboratories where empiricism now roams freely, exploring often very complex systems. So, we see computation as a welcome third component of the scientific enterprise that is changing the way we theorize about and

---

[30] It is noteworthy that it is the diffusion approximation for neutron transport, with $\sigma_a < 0$ (i.e., $\varpi_0 > 1$) to model multiplication, that enabled the breakthrough computation of the critical mass of fissile material for a functional nuclear weapon (Serber and Rhodes, 1992).

observe nature's ways. We also raise the critically important and non-trivial question of computational model "validation" which ultimately involves detailed comparisons of model-predicted and observed data.[31]

While advances in information and sensor technologies (i.e., hardware) support directly the computational and observational pillars of the scientific edifice, theory remains the main source of the hypotheses that are tested in computational and/or observational procedures. Often breakthroughs use all three components and typically suggest new observational or computational strategies. So all three elements are intimately connected. We devoted the last few sections of this chapter to the phenomenology of 3D radiative transfer as an effort to invigorate the theoretical component of our subject area where, not infrequently, "theory" is equated with numerical simulation. In the present context, the contribution to phenomenology amounts to paying special attention to non-dimensional numbers and their physical meaning.

References

Alcouffe, R.E., R.S. Baker, F.W. Brinkley, D.R. Marr, R.D. O'Dell, and W.F. Walters (1997). *DANTSYS: A Diffusion Accelerated Neutral Particle Transport*, (UC-705), issued 06/95, revised 03/97. Los Alamos National Laboratory, Los Alamos (NM).

Box, M.A., M. Keevers, and B.H.J. McKellar (1988). On the perturbation series for radiative effects. *J. Quant. Spect. Radiat. Trans.*, **39**, 219–223.

Box, M.A., S.A.W. Gerstl, and C. Simmer (1989). Computation of atmospheric radiative effects via perturbation theory. *Beitr. Phys. Atmosph.*, **62**, 193–199.

Box, M.A., I.N. Polonsky, and A.B. Davis (2003). Higher-order perturbation theory applied to radiative transfer in non-plane-parallel media. *J. Quant. Spectrosc. Radiat. Transfer*, **78**, 105–118.

Cahalan, R.F., L. Oreopoulos, G. Wen, A. Marshak, S.-C. Tsay, and T. DeFelice (2001). Cloud characterization and clear-sky correction from Landsat-7. *Remote Sens. Environ.*, **78**, 89–98.

Cahalan, R.F., L. Oreopoulos, A. Marshak, K.F. Evans, A.B. Davis, R. Pincus, K. Yetzer, B. Mayer, R. Davies, T.P. Ackerman, H.W. Barker, E.E. Clothiaux, R.G. Ellingson, M.J. Garay, E. Kassianov, S. Kinne, A. Macke, W. O'Hirok, P.T. Partain, S.M. Prigarin, A.N. Rublev, G.L. Stephens, F. Szczap, E.E. Takara, T. Várnai,

---

[31] This comparison is not to be confused with the "calibration" of models where input parameters are determined by comparing model output to observational/experimental data. In approximation theory, the golden standard of real-world observations can arguably be replaced with theoretical results known to be very accurate. For instance, 3D photon diffusion theory can be both calibrated and validated with high-accuracy 3D Monte Carlo simulations. As needed, we fine-tune the free parameter χ in the boundary conditions using one set of Monte Carlo simulations (model calibration), and then compare diffusion predictions with no more free parameters to another set of Monte Carlo results (model validation).

G. Wen, and T.B. Zhuravleva (2005). The international Intercomparison of 3D Radiation Codes (I3RC): Bringing together the most advanced radiative transfer tools for cloudy atmospheres. *Bull. Amer. Meteor. Soc.*, to appear in Sept 2005 issue.

Cannon, C.J. (1970). Line transfer in two dimensions. *Astrophys. J.*, **161**, 255–264.

Case, K.M. and P.F. Zweifel (1967). *Linear Transport Theory*. Addison-Wesley, Reading (MA).

Caudill, M. and C. Butler (1992). *Understanding Neural Networks: Computer Explorations*. MIT Press, Cambridge (MA), 2nd edition.

Chu, M.C. and W.S. Churchill (1955). Numerical solution of problems in multiple scattering of electromagnetic radiation. *J. Chem. Phys.*, **59**, 855–863.

Chýlek, P., C. Borel, A.B. Davis, S. Bender, J. Augustine, and G. Hodges (2004). Effect of broken clouds on satellite based columnar water vapor retrieval. *IEEE Geosci. and Remote Sens. Lett.*, **1**, 175–179.

Coakley, J.A. and P. Chýlek (1975). The two-stream approximation in radiative transfer: Including the angle of the incident radiation. *J. Atmos. Sci.*, **32**, 409–418.

Cornet, C., H. Isaka, B. Guillemet, and F. Szczap (2004). Neural network retrieval of cloud parameters of inhomogeneous clouds from multispectral and multiscale radiance data: Feasibility study. *J. Geophys. Res.*, **109**, D12203, doi:10.1029/2003JD004186.

Davies, R. (1978). The effect of finite geometry on the three-dimensional transfer of solar irradiance in clouds. *J. Atmos. Sci.*, **35**, 1712–1725.

Davis, A.B. (2002). Cloud remote sensing with sideways-looks: Theory and first results using Multispectral Thermal Imager (MTI) data. SPIE Pro. Vol. 4725, pp. 397–405.

Davis, A. and A. Marshak (1997). Lévy kinetics in slab geometry: Scaling of transmission probability. In *Fractal Frontiers*. M.M. Novak and T.G. Dewey (eds.). World Scientific, Singapore, pp. 63–72.

Davis, A.B. and A. Marshak (2001). Multiple scattering in clouds: Insights from three-dimensional diffusion/P_1 theory. *Nuclear Sci. and Engin.*, **137**, 251–280.

Davis, A.B. and A. Marshak (2002). Space-time characteristics of light transmitted through dense clouds: A Green function analysis. *J. Atmos. Sci.*, **59**, 2714–2728.

Davis, A.B. and A. Marshak (2004). Photon propagation in heterogeneous optical media with spatial correlations: Enhanced mean-free-paths and wider-than-exponential free-path distributions. *J. Quant. Spectrosc. Radiat. Transfer*, **84**, 3–34.

Davis, A., S. Lovejoy, and D. Schertzer (1993). Supercomputer simulation of radiative transfer in multifractal cloud models.In *IRS'92: Current Problems in Atmospheric Radiation*. S. Keevallik and O. Kärner (eds.). Deepak Publ., Hampton (VA), pp. 112–115.

Davis, A., A. Marshak, R.F. Cahalan, and W.J. Wiscombe (1997). The LANDSAT scale-break in stratocumulus as a three-dimensional radiative transfer effect, Implications for cloud remote sensing. *J. Atmos. Sci.*, **54**, 241–260.

Davis, A.B., R.F. Cahalan, J.D. Spinhirne, M.J. McGill, and S.P. Love (1999). Off-beam lidar: An emerging technique in cloud remote sensing based on radiative

Green-function theory in the diffusion domain. *Phys. Chem. Earth (B)*, **24**, pp. 177–185 (Erratum 757–765).

Davis, A.B., D.M. Winker, and M.A. Vaughan (2001). First retrievals of dense cloud properties from off-beam/multiple-scattering lidar data collected in space. In *Laser Remote Sensing of the Atmosphere, Selected Papers from the 20th International Conference on Laser Radar*. A. Dabas, C. Loth and J. Pelon (eds.). Editions de l'École Polytechnique, Palaiseau (France), pp. 35–38.

Eddington, A.S. (1916). On the radiative equilibrium of stars. *Mon. Not. Roy. Ast. Soc.*, **77**, 16–35.

Evans, K.F., R.P. Lawson, P. Zmarzly, D. O'Connor, and W.J. Wiscombe (2003). In situ cloud sensing with multiple scattering lidar: Simulations and demonstration. *J. Atmos. and Oceanic Tech.*, **20**, 1505–1522.

Faure, T., H. Isaka, and B. Guillemet (2001). Neural network analysis of the radiative interaction between neighboring pixels in inhomogeneous clouds. *J. Geophys. Res.*, **106**, 14,465–14,484.

Faure, T., H. Isaka, and B. Guillemet (2002). Neural network retrieval of clouds parameters from high-resolution multi-spectral radiometric data. *Remote Sens. Environ.*, **80**, 285–296.

Fermi, E. (1941). Cosmic ray theory. *Rev. Modern Phys.*, **13**, 240.

Gabriel, P.M. and K.F. Evans (1996). Simple radiative transfer methods for calculating domain-averaged solar fluxes in inhomogeneous clouds. *J. Atmos. Sci.*, **53**, 858–877.

Gabriel, P.M., S.M. Lovejoy, A. Davis, D. Schertzer, and G.L. Austin (1990). Discrete angle radiative transfer II: Renormalization approach for homogeneous and fractal clouds. *J. Geophys. Res.*, **95**, 11,717–11,728.

Galinsky, V.L. (2000). 3D radiative transfer in weakly inhomogeneous media, Part II: Discrete ordinate method and effective algorithm for its inversion. *J. Atmos. Sci.*, **57**, 1635–1645.

Galinsky, V.L. and V. Ramanathan (1998). Three-dimensional radiative-transfer in weakly inhomogeneous media, Part I: Diffusive approximation. *J. Atmos. Sci.*, **55**, 2946–2959.

Giovanelli, R.G. (1959). Radiative transfer in non-uniform media. *Aust. J. Phys.*, **12**, 164–170.

Gu, Y. and K.-N. Liou (2001). Radiation parameterization for three-dimensional inhomogeneous cirrus clouds: Application to climate models. *J. Climate*, **14**, 2443–2457.

Henderson, B.G. and P. Chýlek (2005). The effect of spatial resolution on satellite aerosol optical depth retrieval. *IEEE Trans. Geosci. and Remote Sens.*, in press.

Joseph, J.H., W.J. Wiscombe, and J.A. Weinman (1976). The delta-Eddington approximation for radiative flux transfer. *J. Atmos. Sci.*, **33**, 2452–2459.

Khairoutdinov, M.F. and D.A. Randall (2001). A cloud-resolving model as a cloud parameterization in the NCAR Community Climate System Model: Preliminary results. *Geophys. Res. Lett.*, **28**, 3617–3620.

King, M.D., L.F. Radke, and P.V. Hobbs (1990). Determination of the spectral absorption of solar radiation by marine Streatocumulus clouds from airborne measurements within clouds. *J. Atmos. Sc.*, **47**, 894–907.

Kobayashi, T., K. Masuda, M. Sasaki, and J.-P. Mueller (2000). Monte Carlo simulations of enhanced visible radiance in clear-air satellite fields of view near clouds. *J. Geophys. Res.*, **105**, 26569–26576.

Kokhanovsky, A.A. (2003). The influence of horizontal inhomogeneity on radiative characteristics of clouds: An aysmptotic case study. *IEEE Trans. Geosci. and Remote Sens.*, **41**, 817–825.

Li, J., J.W. Geldart, and P. Chýlek (1994). Perturbation solution for 3D radiative transfer in horizontally periodic inhomogeneous cloud field. *J. Atmos. Sci.*, **51**, 2110–2122.

Li, J., J.W. Geldart, and P. Chýlek (1995). Second order perturbation solution for radiative transfer in clouds with a horizontally arbitrary periodic inhomogeniety. *J. Quant. Spectrosc. Radiat. Transfer*, **53**, 445–456.

Love, S.P., A.B. Davis, C. Ho, and C.A. Rohde (2001). Remote sensing of cloud thickness and liquid water content with Wide-Angle Imaging Lidar (WAIL). *Atmos. Res.*, **59-60**, 295–312.

Lovejoy, S., A. Davis, P. Gabriel, D. Schertzer, and G. Austin (1990). Discrete angle radiative transfer I: Scaling and similarity, universality and diffusion. *J. Geophys. Res.*, **95**, 11,699–11,715.

Lyapustin, A. and Y.J. Kaufman (2001). Role of adjacency effect in the remote sensing of aerosol. *J. Geophys. Res.*, **106**, 11909–11916.

Marshak, A., A. Davis, W.J. Wiscombe, and R.F. Cahalan (1995). Radiative smoothing in fractal clouds. *J. Geophys. Res.*, **100**, 26,247–26,261.

Marshak, A., A. Davis, R.F. Cahalan, and W.J. Wiscombe (1998). Nonlocal Independent Pixel Approximation: Direct and Inverse Problems. *IEEE Trans. Geosc. and Remote Sens.*, **36**, 192–205.

Marshak, A., Yu. Knyazikhin, A.B. Davis, W.J. Wiscombe, and P. Pilewskie (2000). Cloud - vegetation interaction: Use of normalized difference cloud index for estimation of cloud optical thickness. *Geophys. Res. Lett.*, **27**, 1695–1698.

Meador, W.E. and W.R. Weaver (1980). Two-stream approximations to radiative transfer in planetary atmospheres: A unified description of existing methods and a new improvement. *J. Atmos. Sci.*, **37**, 630–643.

Min, Q.-L. and L.C. Harrison (1999). Joint statistics of photon pathlength and cloud optical depth. *Geophys. Res. Lett.*, **26**, 1425–1428.

Morse, P.M. and H. Feshbach (1953). *Methods of Theoretical Physics,* 2 vols. McGraw-Hill, New York (NY).

Polonsky, I.N. and A.B. Davis (2004). Lateral photon transport in dense scattering and weakly-absorbing media of finite thickness: Asymptotic analysis of the Green functions. *J. Opt. Soc. Amer. A*, **21**, 1018–1025.

Polonsky, I.N., M.A. Box, and A.B. Davis (2003). Radiative transfer through inhomogeneous turbid media: Implementation of the adjoint perturbation approach at the first-order. *J. Quant. Spectrosc. Radiat. Transfer*, **78**, 85–98.

Pomraning, G.C. (1989). Diffusion theory via asymptotics. *Transp. Theory and Stat. Phys.*, **18**, 383–428.

Qu, Z. (1999). *On the Transmission of Ultraviolet Radiation in Horizontally Inhomogeneous Atmospheres: A Three-Dimensional Approach Based on the Delta-Eddington's Approximation*, Ph.D. Thesis. University of Chicago, Department of Geophysical Sciences, Chicago (IL).

Roache, P.J. (1998). *Verification and Validation in Computational Science and Engineering*. Hermosa, Albuquerque (NM).

Romanova, L.M. (1975). Radiative transfer in a horizontally inhomogeneous scattering medium. *Izv. Acad. Sci. USSR Atmos. Oceanic Phys.*, **11**, 509–513.

Rossow, W.B., C. Delo, and B. Cairns (2002). Implications of the observed mesoscale variations of clouds for the earth's radiation budget. *J. Climate*, **15**, 557–585.

Savigny, C. von, O. Funk, U. Platt, and K. Pfeilsticker (1999). Radiative smoothing in zenith-scattered sky light transmitted through clouds to the ground. *Geophys. Res. Lett.*, **26**, 2949–2952.

Schuster, A. (1905). Radiation through a foggy atmosphere. *Astrophys. J.*, **21**, 1–22.

Schwarzschild, K. (1906). Über das Gleichgewicht der Sonnenatmosphere (in German, English title: On the equilibrium of the Sun's atmosphere). *Gottingen Nachrichten*, **41**, 1–24.

Serber, R. and R. Rhodes (1992). *The Los Alamos Primer: The First Lectures on How to Build an Atomic Bomb*, annotated by Robert Serber, edited with an introduction by Richard Rhodes. University of California Press, Berkeley (CA).

Siddal, R.G. and N. Selçuk (1979). Evaluation of a new six-flux model for radiative transfer in recangular enclosures. *Trans. Inst. Chem. Eng.*, **57**, 163–169.

Siegel, R. and J.R. Howell (1981). *Thermal Radiation Heat Transfer*. McGraw-Hill, New York (NY), 2nd edition.

Stephens, G.L. (1986). Radiative transfer in spatially heterogeneous, two-dimensional anisotropically scattering media. *J. Quant. Spectrosc. Radiat. Transfer*, **36**, 51–67.

Tikhonov, A.N. (1977). *Solutions of Ill-Posed Problems*. Winston, New York (NY).

Várnai, T. (2000). Influence of three-dimensional radiative effects on the spatial distribution of shortwave cloud reflection. *J. Atmos. Sci.*, **57**, 216–229.

Várnai, T. and R. Davies (1999). Effects of cloud heterogeneities on shortwave radiation: Comparison of cloud-top variability and internal heterogeneity. *J. Atmos. Sci.*, **56**, 4206–4224.

Várnai, T. and A. Marshak (2003). A method for analyzing how various parts of clouds influence each other's brightness. *J. Geophys. Res.*, **108**, 4706, doi:10.1029/2003JD003561.

Weinman, J.A. and P.N. Swartztrauber (1968). Albedo of a striated medium of isotropically scattering particles. *J. Atmos. Sci.*, **34**, 642–650.

Wen, G., R.F. Cahalan, S.-C. Tsay, and L. Oreopoulos (2001). Imapct of cumulus cloud spacing on Landsat atmospheric correction and aerosol retrieval. *J. Geophys. Res.*, **106**, 12,129–12,138.

Wyser, K., W. O'Hirok, C. Gautier, and C. Jones (2002). Remote sensing of surface solar irradiance with corrections for 3-D cloud effects. *Remote Sens. Environ.*, **80**, 272–284.

Wyser, K., W. O'Hirok, and C. Gautier (2005). A simple method for removing 3-D radiative effects in satellite retrievals of surface irradiance. *Remote Sens. Environ.*, **94**, 335–342.

Yodh, A. and B. Chance (1995). Spectroscopy and imaging with diffusing light. *Phys. Today*, **48**, 34–40.

Part III

Climate

6

Approximation Methods in Atmospheric 3D Radiative Transfer Part 2: Unresolved Variability and Climate Applications

H.W. Barker and A.B. Davis

6.1   Introduction 343

6.2   **Analytical Models for Domain-Average Radiation Budgets** 345

6.3   **Computational Models for Domain-Average Radiation Budgets** 369

6.4   **Efficient Large-Scale Radiation Budget Estimation: Discussion and Outlook** 375

6.5   **Summary** 377

References 379

6.1 Introduction

Radiative transfer (RT) plays a crucial role within the atmospheric sciences. To first-order approximation, the most pressing climatic change scenarios, such as enhanced greenhouse gas emissions, injections of volcanic and anthropogenic aerosols, and land-use changes, are initiated by perturbations, or forcings, to Earth's radiative budget. Once forced, feedback mechanisms wihtin the Earth-atmosphere-ocean system take over as the system strives to achieve a new dynamic equilibrium. These feedbacks, or cause-and-effect relations, are governed much by the radiative sensitivity of Earth to changes in internal variables. In depth study of these forcings and feedbacks is made possible by numerical global climate models (GCMs). At this stage, GCMs are the primary tool used to study and predict past and future climates. Critical policy decisions depend on the verisimilitude of these virtual laboratories where, unlike the real world, climate change experiments can be executed safely and repeatedly. As GCMs are our most sound solution method to the question of climatic change, it is essential, given the central role of radiation to climate, that atmospheric radiation scientists minimize radiative-based errors in GCMs.

It is recognized generally that dubious representation of interactions between clouds and radiation in GCMs are responsible for much of the uncertainty surrounding predictions of (near-future) climatic change stemming from anthropogenic modification of Earth. Moreover, clouds are an inexorable part of the climate system that, unlike some greenhouse gases and aerosols, are not *directly* affected by human activity. Thus, our predictions of climate change will depend as much on our understanding of how clouds connect to other parts of the Earth system as on the natural variability of clouds in space and time. Admittedly, we appear to be far from representing clouds from first principles inside models as their nature depends intimately on other difficult subjects such as: nucleation; moist thermodynamics; fluid instability; turbulence; convection; precipitation processes, and radiative transfer.

The previous two paragraphs summarize the somewhat intimidating scientific and societal situation that atmospheric radiation scientists face as they attempt to uphold their end of the climate prediction bargain with society via various national governments: development of accurate and efficient methods of representing RT in GCMs. One ray of hope is that the primary radiative requirement of GCMs is broadband (spectrally-integrated) flux profiles (angular integrals of the height varying radiance field) averaged over large horizontal areas (a GCM grid-cell typically exceeds one hundred km on a side). The bad news is that GCMs provide very little information about the 3D structure of cloudiness at unresolved scales. In particular, methods from previous chapters do no apply and would cost too much computer time anyway. So there is much guessing to do, working with reasonable statistical hypotheses, and many computational corners to cut through, sacrificing as little accuracy as possible.

To this day, GCMs typically apply one of the most radical approximations one can make in 3D RT: subgrid-scale variability is treated as a simple linear mixture of "clear" and "cloudy" non-interacting plane-parallel, homogeneous regions in each layer along with layers linked vertically according to highly idealized rules. This approach is an elementary incarnation of the independent column approximation (ICA). There is, however, a plethora of approaches that attempt to address higher-order subgrid-scale fluctuations (primarily in cloud structure). Yet, like their reigning antecedents, they too almost always capitalize on plane-parallel RT and are even manifested as variants of the classic two-stream approximation.

Interestingly, the wider atmospheric modeling community has perceived radiation as a "messy," but grudgingly important, heating term and climatic forcing agent. For the spatial dimension of RT parameterization in GCMs, this disdain has led to an almost axiomatic, long-standing complacence with plane-parallel modeling. Coincident with this was much in-depth study of the spectral dimension of atmospheric RT. Efficient, yet accurate, methods of representing spectral information, such as correlated k-distributions and exponential sum fits, have thus been developed and tested against exact line-by-line spectral integrations. Even though this spectral work has unfolded entirely in the frame of 1D RT, it is not lost for 3D RT because the source-linearity of RT physically decouples the spatial and spectral aspects of the problem; so what's good for 1D is good for 3D.

In this chapter, we address RT problems that are driven by the needs of GCMs where only domain-average flux profiles are sought for columns with aspect ratios of vertical to horizontal sizes that are typically very large. Moreover, details of spatial variability are unresolved for these columns thereby rendering the need for *probabilistic* descriptions of optical-radiative fluctuations over wide ranges of scales.

Tradition has it that the most logical and satisfying solutions be analytical and based on insights into 3D RT processes. These solutions are discussed in Sect. 6.2. Recently however, there has been a push for straightforward computational answers and these are presented in Sect. 6.3. These two disparate philosophies are contrasted in Sect. 6.4 along with a look into the future of this very active area of research. Section 6.5 offers some concluding remarks.

6.2 Analytical Models for Domain-Average Radiation Budgets

This section is constructed on the assumption that domain-average radiative fluxes are required, first and foremost, in GCMs. The ultimate purpose of a RT model in a GCM is to compute surface fluxes and atmospheric layer heating rates. This involves integration over a wide swath of the electromagnetic spectrum, and broadband fluxes are discussed in detail in Chap. 9. Since broadband radiation calculations are integrations of monochromatic solutions of the 3D RT equation (see Chap. 3), it suffices to discuss computation of domain-average in monochromatic terms, for any model that fails there is sure to fail in the broadband. Furthermore, we will consider single cloud layers; consideration of layer-to-layer correlations come later.

6.2.1 Exact 3D and Independent Column Approximation (ICA) Solutions

If one is provided with a specific domain $\mathcal{D}$, such as from a CSRM, domain-average albedo $\langle R \rangle$ (or transmittance, or absorption, or fluxes in general) can be computed with an exact solution of the 3D RT equation (see Chap. 4 for numerical techniques) such that

$$\langle R \rangle = \iint_{\mathcal{D}} R_{3D}(x, y) \, dx \, dy \Big/ \iint_{\mathcal{D}} dx \, dy \, , \tag{6.1}$$

where R_{3D} is albedo from a solution that accounts for the flow of radiation in three dimensions. On several occasions the independent column approximation (ICA) has performed well for a variety of cloud regimes (Cahalan et al., 1994b; Chambers et al., 1997; Barker et al., 1999), so (6.1) can be approximated by

$$\langle R \rangle \approx \iint_{\mathcal{D}} R_{1D}(x, y) \, dx \, dy \Big/ \iint_{\mathcal{D}} dx \, dy \, , \tag{6.2}$$

where $R_{1D}(\cdot)$ is computed using a 1D RT model. As a rather typical example, the spatial dependence is inherited entirely from optical depth $\tau(x, y)$ but other optical parameters can change, even vertically in the column located at (x, y). The 1D RT computation can range from a simple two-stream approximation, the basis of all

GCM radiation codes, to an accurate numerical solution of the 1D RT equation (S_n P_n, adding/doubling, etc.).

Algorithmically, (6.2) is necessarily expressed as

$$\langle R \rangle \approx \frac{1}{N} \sum_{n=1}^{N} R_{1D}(n) , \tag{6.3}$$

where $\mathcal{D}$ now consists of N discrete columns. It might strike one as odd, but if N is very large, an arbitrarily accurate estimate of $\langle R \rangle$ with (6.1) can be obtained faster than with (6.2) or even (6.3), especially when broadband fluxes are sought.[1] Where (6.1) and (6.2) tend to differ most is in the solar spectrum at large solar zenith angles when energy input is relatively small. In fact, the ICA tends to become an increasingly better approximation of (6.1) as $\mathcal{D}$ increases in size for transport theoretical reasons articulated in Chap. 5, and lengthening temporal integration helps too (Benner and Evans, 2001). Thus, the ICA is a reasonable standard for less rigorous models to aim for in the context of relatively large $\mathcal{D}$'s, reasonably large dynamical time-scales, and little to no information describing horizontal correlations in optical fluctuations.

Contrary to being provided with a detailed description of the structure of $\mathcal{D}$, GCMs have only limited amounts of information available to describe the structure of unresolved fluctuations in optical properties. The current paradigm is to apply 1D codes based on the two-stream approximation in which unresolved variability of the surface-atmosphere system is either: (i) reduced to fractional coverage of homogeneous clouds that overlap according to extremely idealized configurations; or (ii) incorporated directly into 1D transport solvers. Some of the better known attempts to achieve the later are reviewed in the following subsection.

6.2.2 Solutions Using 1D Radiative Transfer Theory

To date, all 1D models that attempt to account for subgrid-scale fluctuations of cloud have been designed assuming that domain-average albedo can be computed as

$$\langle R \rangle = (1 - A_c) \langle R_{clr} \rangle + A_c \langle R_{cld} \rangle , \tag{6.4}$$

where A_c is layer cloud fraction, and $\langle R_{clr} \rangle$ and $\langle R_{cld} \rangle$ are mean albedos associated with the clear and cloudy portions of the layer, respectively. Of immediate concern is definition of A_c. Cloud fraction is a rather peculiar atmospheric quantity. It is discussed ad nauseam yet its definition, both from theoretical and practical perspectives, is invariably fraught with confusion and misinterpretation. In this volume alone, discussions of cloud fraction can be found in Chaps. 1, 2, 6 (here), and 7–11.

---

[1] To achieve this, one would exploit the ease with which straightforward 3D Monte Carlo schemes can estimate domain averages like $\langle R \rangle$. Indeed, only $n_R \approx 10^4$ histories ending in reflection ($N \approx 10^4/\langle R \rangle$ histories in all) will yield $\approx \sqrt{1 - \langle R \rangle} < 1$ error on $\langle R \rangle$ expressed in % (cf. Chap. 4). However, this advantage is not of any help in GCM-type problems since the detailed spatial information is simply not available.

What the models discussed here have focused on is computation of $\langle R_{\mathrm{cld}} \rangle$. Several models based on 1D RT yet approximate $\langle R_{\mathrm{cld}} \rangle$ are now reviewed. All are prime examples of the current paradigm for they attempt to fold descriptions of cloud structure, or cloud-radiation interactions, directly into the solution of the RT equation.

ICA-Based Approaches

The starting point of these models is to recast the Riemann integral in (6.2) into its Lebesgue equivalent which can be expressed as

$$\langle R_{\mathrm{cld}} \rangle = \int_0^\infty p(\tau) R_{1D}(\tau) \, d\tau \tag{6.5}$$

where $p(\tau)$ is a probability density function (PDF) that describes variations in τ over some domain (Ronnholm et al., 1980; Cahalan, 1989; Stephens et al., 1991). Equation (6.5) is simply a rearrangement of the infinitesimal terms in (6.2) into infinitesimal elements of probability of occurrence for a given value of τ. As a first example, note that (6.5) correctly yields $\langle R_{\mathrm{cld}} \rangle = R_{1D}(\bar{\tau})$ if $p(\tau) = \delta(\tau - \bar{\tau})$, the "degenerate" PDF. As another example, (6.4) can be interpreted as the outcome of (6.5) for a two-valued Bernoulli distribution superimposed on distributions of clear- and cloudy-sky optical properties.

There are several ways to use (6.5) depending on the functional forms of $p(\tau)$ and $R_{1D}(\tau)$. Clearly, if the forms are intractable and require numerical integration, then (6.5) is not tenable for use in GCMs. Several studies using satellite-inferred values of τ and cloud-resolving model data have shown, however, that for domains the size of those used in typical GCMs, it is reasonable to represent $p(\tau)$ by a Gamma distribution

$$p_\Gamma(\tau) = \frac{1}{\Gamma(\nu)} \left(\frac{\nu}{\bar{\tau}}\right)^\nu \tau^{\nu-1} e^{-\nu\tau/\bar{\tau}} \tag{6.6}$$

where by the method of moments

$$\nu = \frac{1}{\overline{\tau^2}/\bar{\tau}^2 - 1}, \tag{6.7}$$

and $\Gamma(\nu)$ is Euler's Gamma function. Here we use the over-score notation interchangeably with $\langle \cdot \rangle$ to designate ensemble/spatial averages.

Perhaps the simplest, yet credible, version of R_{1D} is Coakley and Chýlek's (1975) "Model 1" two-stream approximation with $\varpi_0 = 1$ in which

$$R_{1D}(\tau) = 1 - T_{1D}(\tau) = \frac{\beta(\mu_0)\tau}{\mu_0 + \beta(\mu_0)\tau}, \tag{6.8}$$

where μ_0 is cosine of solar zenith angle and $\beta(\mu_0)$ is the zenith-angle dependent backscatter function which is defined as

$$\beta(\mu_0) = \frac{1}{2} \int_0^1 P(\mu', -\mu_0) d\mu' \tag{6.9}$$

where $P(\mu', \mu)$ is the azimuthally-averaged phase function (normalized to 4π). The original two-stream model by Schuster (1905) for the non-absorbing case of immediate interest here is retrieved using

$$\beta(\mu_0) = \mu_0 (1 - g)/2\chi , \tag{6.10}$$

where g is asymmetry factor of the phase function and χ is the extrapolation length (in transport mean-free-paths), typically taken to be $\approx 2/3$ (cf. Chap. 5). This, "coupled with the fact that g is ≈ 0.85 for liquid water clouds", immediately sets the magnitude of $\beta(\mu_0)$ in (6.9) at ≈ 0.2 (Wiscombe and Grams, 1976).

Substituting (6.6) and (6.8) into (6.5) and evaluating the integral yields

$$\langle T_{\text{cld}} \rangle = 1 - \langle R_{\text{cld}} \rangle = e^{\xi + v \ln \xi} \Gamma(1 - v, \xi) \tag{6.11}$$

where $\Gamma(1 - v, \xi)$ is the incomplete Gamma function, and

$$\xi = \frac{v\mu_0}{\beta(\mu_0)\overline{\tau}} .$$

Equation (6.11) represents the simplest form of the Gamma-weighted two-stream approximation (GWTSA).

This solution can be improved upon by using the generalized, non-conservative scattering, two-stream approximation for collimated irradiance (Meador and Weaver, 1980) in which layer albedo is given by

$$R_{1D}(\tau) = \frac{\varpi_0}{a} \left(\frac{+r_+ e^{+k\tau} - r_- e^{-k\tau} - r_0 e^{-\tau/\mu_0}}{e^{+k\tau} - b e^{-k\tau}} \right) , \tag{6.12}$$

where ϖ_0 is droplet single-scattering albedo, and transmittance by

$$T_{1D}(\tau) = e^{-\tau/\mu_0} \left[1 + \frac{\varpi_0}{a} \left(\frac{-t_+ e^{+k\tau} + t_- e^{-k\tau} + t_0 e^{+\tau/\mu_0}}{e^{+k\tau} - b e^{-k\tau}} \right) \right] , \tag{6.13}$$

where the two terms are for the direct and diffuse components. The new quantities are defined as follows:

$$k = \sqrt{\xi_1^2 - \xi_2^2} ; \ a = [1 - (k\mu_0)^2] (k + \xi_1) ; \ b = \frac{\xi_1 - k}{\xi_1 + k} ;$$
$$r_\pm = (1 \mp k\mu_0) (\xi_1\xi_3 - \xi_2\xi_4 \pm k\xi_3) ; \ r_0 = 2k [\xi_3 - (\xi_1\xi_3 - \xi_2\xi_4) \mu_0] ;$$
$$t_\pm = (1 \pm k\mu_0) (\xi_1\xi_4 - \xi_2\xi_3 \pm k\xi_4) ; \ t_0 = 2k [\xi_4 - (\xi_1\xi_4 - \xi_2\xi_3) \mu_0] .$$

The coefficients $\xi_1, \ldots, \xi_4$ depend on the choice of two-stream approximation, droplet single-scattering parameters, and μ_0 (Meador and Weaver, 1980). For instance, in Eddington's model

$$\xi_1 = (7 - \varpi_0(4 + 3g))/4 \,,$$
$$\xi_2 = (1 - \varpi_0(4 - 3g))/4 \,,$$
$$\xi_3 = (2 - 3\mu_0 g)/4 \,,$$
$$\xi_4 = 1 - \xi_3 \,, \tag{6.14}$$

and hence

$$k = \sqrt{3(1 - \varpi_0)(1 - \varpi_0 g)} \,. \tag{6.15}$$

Because scattering phase functions for atmospheric particles are strongly peaked in the forward direction, it is recommended that the "δ-rescaling" of two-stream theory quantities τ, ϖ_0, and g be applied (Joseph et al., 1976).[2]

Substituting (6.12) and (6.13). into (6.5) leads to the more familiar GWTSA (Barker et al., 1996) in which

$$\langle R_{\text{cld}} \rangle = \phi_1^\nu \frac{\varpi_0}{a} \left[r_+ \mathcal{F}(b, \nu, \phi_1) - r_- \mathcal{F}(b, \nu, \phi_2) - r_0 \mathcal{F}(b, \nu, \phi_3) \right] \tag{6.16}$$

and

$$\langle T_{\text{cld}} \rangle = \left(\frac{\nu}{\nu + \bar\tau/\mu_0} \right)^\nu$$
$$+ \phi_1^\nu \frac{\varpi_0}{a} \left[-t_+ \mathcal{F}(b, \nu, \phi_4) + t_- \mathcal{F}(b, \nu, \phi_5) + t_0 \mathcal{F}(b, \nu, \phi_6) \right] \tag{6.17}$$

where again the two terms are for the mean direct and diffuse components. We have defined

$$\phi_1 = \frac{\nu}{2k\bar\tau} \quad ; \phi_4 = \phi_1 + \frac{1}{2k\mu_0} \,;$$
$$\phi_2 = \phi_1 + 1 \quad ; \phi_5 = \phi_4 + 1 \,;$$
$$\phi_3 = \phi_4 + \frac{1}{2} \,; \phi_6 = \phi_1 + \frac{1}{2} \,;$$

and

$$\mathcal{F}(b, \nu, \phi_i) = \sum_{n=0}^{\infty} \frac{b^n}{(\phi_i + n)^\nu} \,, \quad -1 \leq b < 1, \, \nu > 0, \, i = 1, \dots, 6 \,.$$

At $\varpi_0 = 1$ (hence $a = \xi_1 = \xi_2$, $k = 0$, and $b = 1$) there is a removable singularity in (6.12) and (6.13). Applying L'Hôpital's rule to (6.12) or (6.13) as $\varpi_0 \to 1$, albedo and transmittance for the generalized, conservative scattering two-stream approximation are defined as

$$R_{1D}(\tau) = 1 - T_{1D}(\tau)$$
$$= \frac{\xi_1 \tau + (\xi_3 - \xi_1 \mu_0)\left(1 - e^{-\tau/\mu_0}\right)}{1 + \xi_1 \tau} \,. \tag{6.18}$$

---

[2] Note that rescaled or "transport" optical depth $\tau_t = (1 - \varpi_0 g)\tau$ and the so-called similarity factor $S = \sqrt{3(1 - \varpi_0)/(1 - \varpi_0 g)}$ are invariant under the δ-rescaling of the phase function, as is $k\tau = S\tau_t$ in (6.12) and (6.13). So this transformation does nothing for two-stream theories with isotropic internal or boundary sources, for instance, the energy-conservating "Model 1" in (6.8) with $\beta(\mu_0) \propto (1 - g)\mu_0$.

Therefore, substituting (6.18) and (6.6) into (6.5) yields the conservative scattering GWTSA as

$$
\begin{aligned}
\langle T_{\mathrm{cld}} \rangle &= 1 - \langle R_{\mathrm{cld}} \rangle \\
&= \left(\frac{\nu}{\xi_1 \bar{\tau}} \right)^{\nu} \left[(1 - \xi_3 + \xi_1 \mu_0) \, \mathcal{G} \left(1 - \nu, \ \frac{\nu}{\xi_1 \bar{\tau}} \right) \right. \\
&\qquad \left. + (\xi_3 - \xi_1 \mu_0) \, \mathcal{G} \left(1 - \nu, \ \frac{\nu \mu_0 + \bar{\tau}}{\xi_1 \mu_0 \bar{\tau}} \right) \right]
\end{aligned}
\tag{6.19}
$$

where

$$
\mathcal{G} \left(1 - \nu, x \right) = e^{x} \Gamma \left(1 - \nu, x \right).
$$

Naturally, other solutions are possible for different representations of $p(\tau)$ but they are even more complex. For instance, if $p(\tau)$ is approximated by a beta distribution, one ends up with rather intractable solutions for $\langle R_{\mathrm{cld}} \rangle$ and $\langle T_{\mathrm{cld}} \rangle$ involving hypergeometric functions. A solution using a lognormal distribution for τ cannot be expressed in closed-form with simple functions.

Barker and Wielicki (1997) extended the GWTSA to transmittance of isotropic *longwave* irradiance in the absence of scattering. Begin by writing mean cloud transmittance as

$$
\langle T_{\mathrm{cld}} \rangle = 2 \int_0^{\infty} \int_0^1 p(\tau \mid \mu) e^{-\tau/\mu} \mu \, d\mu \, d\tau
\tag{6.20}
$$

where $p(\tau \mid \mu)$ is a zenith-angle dependent distribution of τ that can be estimated by equating τ/μ to the random value of the slant integral of extinction through the cloud layer. If $p(\tau \mid \mu) = \delta(\tau - \bar{\tau})$, (6.20) reduces to the familiar homogeneous solution given by

$$
\langle T_{\mathrm{cld}} \rangle = 2 \int_0^1 e^{-\bar{\tau}/\mu} \mu \, d\mu = 2 E_3 (\bar{\tau})
\tag{6.21}
$$

where $E_3(\bar{\tau})$ is the third-order exponential integral (Charlock and Herman, 1976). Assuming that the dependence of $p(\tau \mid \mu)$ on μ is negligible, substituting (6.6) into (6.20) leads to

$$
\langle T_{\mathrm{cld}} \rangle = x^{\nu} \left\{ 1 - (1 - x) \left[\nu - (\nu + 1) \bar{\tau} \sum_{n=0}^{\infty} \frac{x^{n+1}}{\nu + 2 + n} \right] \right\}
\tag{6.22}
$$

where

$$
x = \frac{\nu}{\nu + \bar{\tau}}.
$$

Li and Barker (2002) took a different approach using a single quadrature point to represent hemispherical integration in (6.20) and then applying the average over the Gamma distribution. As can be seen by examining the impact of the direct transmission term in (6.17) for the solar problem, this leads to

$$\langle T_{\text{cld}} \rangle = \left(\frac{\nu}{\nu + \bar{\tau}/\mu_1} \right)^{\nu} \tag{6.23}$$

where $\mu_1 \approx 0.601$ (Li and Fu, 2000). For integer values of ν, at least (6.23) can be computed faster than an exponential (i.e., the homogeneous solution). Moreover, this approach facilitates scattering computation via Li's (2002) perturbation approach.

To conclude, the appeal of this approach is that as long as one is willing to accept a particular underlying distribution of τ and that the ICA is the most precise estimate that 1D models can be expected to achieve in practice, then this approach represents an exact ICA solution for single layers. Efforts are currently under way to use local GCM variables to predict ν as well as $\bar{\tau}$.

Scaling of Mean Optical Depth

Several models have attempted to approximate $\langle R_{\text{cld}} \rangle$ by performing simple rescalings of mean cloud optical depth $\bar{\tau}$. Perhaps the earliest offering was from the real-space renormalization theory of Gabriel et al. (1990) and the numerical results of Davis et al. (1990) for very heterogeneous 2D fractal cloud models based on a singular cascade model[3] on large grids with, notably, one vertical and one horizontal dimension. These studies independently showed that $\langle R_{\text{cld}} \rangle$ computed within the frame of conservative discrete-angle RT (cf. Chap. 5) could be approximated as

$$\langle R_{\text{cld}} \rangle \approx R_{1D}(\bar{\tau}^{\delta}) \tag{6.24}$$

where δ was referred to as a "co-packing" exponent. The authors determined δ respectively using analytical arguments enabled (1) by the simplified "4-flux" RT theory and (2) by the deterministic monofractal nature of the cascade model (Gabriel et al., 1990), and using discrete- and continuous-angle 2D Monte Carlo simulations (Davis et al., 1990). This form of parameterization is highly desirable because it utilizes the standard 1D plane-parallel homogeneous (PPH) solution of the RT equation, which executes rapidly, with a minor adjustment to input data; the above ICA-based solutions generally require more CPU time than PPH models. This parameterization actually found its way into at least one operational model (McFarlane et al., 1992). In general, one expects

$$\delta \leq 1 \tag{6.25}$$

for cloudy cells with $\bar{\tau} > 1$ (and $\delta \geq 1$ if $\bar{\tau} < 1$ which would correspond to a relatively clear cell). This produces the desired effect by reducing the value of $R_{1D}(\bar{\tau})$. Although a rigorous foundation for this approach is lacking at present, it is consistent with Davis and Marshak's (1997) anomalous photon diffusion model, as shown further on. Notwithstanding, results can be expected at best to be approximate and limited in scope.

---

[3] In the present volume's Appendix, 1D examples of such singular cascades are illustrated; these can be likened to transects through the 2D media used in this project.

Alternatively, one could write

$$\langle R_{\rm cld} \rangle \approx R_{\rm 1D}(\eta\bar{\tau}) \tag{6.26}$$

where $\eta \leq 1$ could again be parameterized based on unresolved cloud structure. Cahalan et al. (1994b) advanced the potential applicability of this approach they call the "effective thickness approximation" (ETA), presented in more detail in Chap. 8. In comparison with (6.24), we are now operating on the prefactor and not the exponent of $\bar{\tau}$, a less radical change in view of its potentially large range of variation in all possible cloud scenes. This is consistent with the fact that Cahalan et al. (1994b) based their work on *bounded*, as opposed to *singular*, cascades where the heterogeneity is weaker in the sense of clumping (cf. Appendix). Bounded cascades are designed specifically to mimic the observed (Cahalan and Snider, 1989) horizontal variability of single stratocumulus layers. In retrospect, the singular cascades are more like cloud systems. Another key difference for the radiative transfer is that the fractal cloud structure generated by the bounded cascade model was assumed to unfold only horizontally in 1D or in 2D. This makes no difference for the ICA analysis, nor does any assumption about vertical uniformity or variability (such as a linear increase in extinction). Just as the numerical simulations of Davis et al. (1990) followed the analytical investigation of Gabriel et al. (1990) for singular cascade cloud models, Cahalan et al. (1994a) numerically substantiated the analytical results of Cahalan et al. (1994b) for bounded cascade cloud models using a conservative Monte Carlo scheme. The former pair of studies obtain only qualitative agreement (for reasons they explain), while the latter pair achieve excellent quantitative agreement (at least for the domain-averages of immediate interest). We will now examine in further depth the more-developed ETA rescaling model in (6.26) and contrast it with the GWTSA.

Cahalan et al. (1994b) noted that expansion of $R_{\rm 1D}$ in a Taylor series about $\overline{\log_{10}\tau}$ and ensemble averaging[4] yields

$$\langle R_{\rm cld} \rangle = R_{\rm 1D}\left(\eta\bar{\tau}\right) + \sum_{n=1}^{\infty} M_{2n} \frac{\partial^{2n} R_{\rm 1D}\left(\eta\bar{\tau}\right)}{\partial\left(\log_{10}\tau\right)^{2n}} \tag{6.27}$$

where M_{2n} are related to the assumed (quasi-lognormal) variability of τ, and

$$\eta = \frac{10^{\overline{\log_{10}\tau}}}{\bar{\tau}} . \tag{6.28}$$

We can assume, for tractability only, that the underlying distribution of τ is again the Gamma distribution $p_\Gamma(\tau)$ as defined in (6.6). It can be shown that use of $p_\Gamma(\tau)$ in (6.27) leads to

---

[4] To be precise, the average is over *spatial* fluctuations of the bounded cascade model described in the Appendix of this volume. By construction, every realization of this stochastic model has the same PDF so ensemble- or spatial-averaging yields the same result. At least in this 1-point statistical respect, the model is said to be "ergodic."

$$\langle R_{\text{cld}} \rangle = R_{1D}(\eta \bar{\tau}) + \left[\frac{1}{2 \ln 10} \frac{\partial \psi(v)}{\partial v} \right] \frac{\partial^2 R_{1D}(\eta \bar{\tau})}{\partial (\ln \tau)^2} + \cdots \qquad (6.29)$$

where

$$\eta = \frac{e^{\psi(v)}}{v}, \qquad (6.30)$$

and

$$\psi(v) = \frac{d}{dv} \ln \Gamma(v)$$

is Euler's ψ function.[5] Since $\psi(v)$ and $\psi'(v)$ are easy to compute, (6.29) is useful, but higher-order derivatives of R_{1D} are tedious in general. Using again Coakley and Chýlek's (1975) "Model 1" two-stream approximation with $\varpi_0 = 1$ to represent $R_{1D}(\bar{\tau})$, retention of just two terms in (6.29) gives

$$\langle R_{\text{cld}} \rangle \approx R_{1D}(\eta \bar{\tau}) \left\{ 1 + \frac{\mu_0 [\mu_0 - \beta(\mu_0)\eta \bar{\tau}]}{2 \ln 10 [\mu_0 + \beta(\mu_0)\eta \bar{\tau}]^2} \frac{\partial \psi(v)}{\partial v} \right\}. \qquad (6.31)$$

The third term is already too complicated to be of any use. The appeal of the ETA is that for visible radiation (at essentially non-absorbing wavelengths) (6.31) is well approximated by (6.26) either when the M_{2n} are all small (i.e., variability is weak), or when $\bar{\tau} \approx \mu_0/\beta(\mu_0)\eta$ (hence $\bar{\tau} \approx 15\mu_0$ for typical values of $\beta(\mu_0) \approx 0.1$ and $\eta \approx 0.7$).

Figure 6.1 shows plots of albedo predicted by the standard homogeneous solution $R_{1D}(\bar{\tau})$, the simple GWTSA given by (6.11), the relevant ETA using (6.30), and (6.31). Since in these cases, the GWTSA represents the full ICA, differences between it and the R_{1D} represent the PPH albedo bias with or without optical depth rescaling. In no way does this imply that the GWTSA is superior to the ETA approach; if another less convenient distribution were to be assumed, a curve for the GWTSA would appear explicitly. From these plots it is clear that the second term of the expansion in (6.31) is quite beneficial. The true ETA, however, encounters notable difficulties especially when η is small. In that case, for the important large values of μ_0 and for most anticipated values of $\bar{\tau}$, the ETA overestimates the PPH albedo bias. However, this is only a reminder that the ETA was never intended to be used with small values of η that correspond to strong variability, e.g., overwhelmingly frequent low values of τ associated with $v \le 1$ in (6.6).

At present, there is confusion regarding how to best estimate subgrid-scale descriptors like v and η (or δ) diagnostically within GCMs. Moreover, the range of realistic values of η for boundary-layer clouds is still unclear. Using microwave radiometer data, Cahalan et al. (1994b, 1995) estimated η to be roughly 0.6 to 0.7 for marine boundary layer clouds off the coast of California and on Porto Santo Island in the Açores. Later analyses of satellite data (Barker et al., 1996; Pincus et al., 1999; Rossow et al., 2002), however, indicate that most overcast marine boundary layer clouds yield $\eta \approx 0.9$, in which case the ETA, and likely all other approximations,

---

[5] Note that equating (6.28) and (6.30) leads to the maximum likelihood estimation of v which is commonly expressed as $\psi(v) + \ln(\bar{\tau}/v) - \overline{\ln \tau} = 0$.

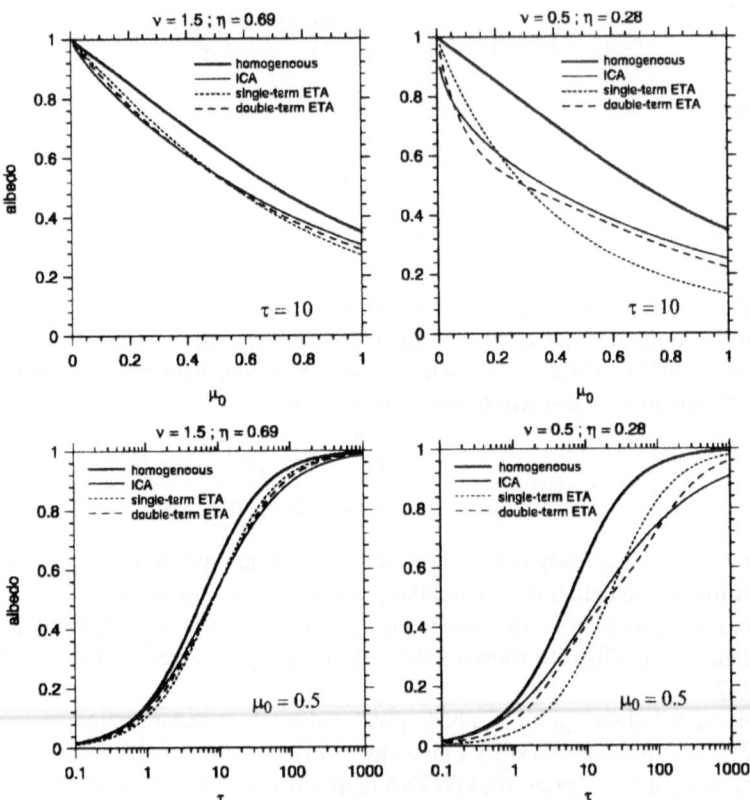

Fig. 6.1. The Gamma ICA for conservative scattering and 3 approximations. Top two plots show albedo of a cloud layer with $\bar{\tau} = 10$ as a function of cosine of solar zenith angle μ_0. Lower two plots show albedo of a cloud layer at $\mu_0 = 1/2$ as a function of $\bar{\tau}$. Results are shown for two inhomogeneous clouds for a given v, and associated η from (6.30), as listed at the top of each plot, and for their homogeneous counterpart (corresponding to $v = \infty$ and $\eta = 1$). Left-hand plots are for relatively large v and η where the 1-term ETA is expected to work well. Right-hand plots are for quasi-broken cloud scenarios with many tenuous pixels (relatively small values of v and η); the 1-term ETA does not do well but the second term helps greatly

will perform very well (even better than in Fig. 6.1 for $\eta \approx 0.7$). Nevertheless, like (6.24), (6.26) has also found its way into an operational GCM; specifically, Tiedtke (1996) applied the ETA globally to all clouds using $\eta \approx 0.7$. In both case studies, this demonstrates the readiness of the GCM community to do something about unresolved cloud variability effects. While the utility of, and interest in, such simplified experiments is widely recognized, care should be exercised in drawing too many conclusions too early.

In the same spirit as the ETA, Barker (1992) had suggested that the azimuthally-averaged 1D RT equation (see (4.5) with $m = 0$) could be modified as

$$\mu \frac{\partial}{\partial \tau} I(\tau, \mu) = -I(\tau, \mu) + \frac{\varpi_0}{2} \int\limits_{-1}^{1} P(\mu', \mu) I(\tau, \mu') \, d\mu' + F_0 \frac{\varpi_0}{4\pi} P(-\mu_0, \mu) e^{-a\tau/\mu_0}$$

(6.32)

where $I(\tau, \mu)$ is (the 0th azimuthal Fourier mode of) radiance, F_0 is incident solar irradiance, ϖ_0 is single-scattering albedo, and P is the azimuthally-averaged scattering phase function (see Wiscombe and Grams (1976) or Sect. 4.1.1 for its expression in spherical harmonics). The only difference between (6.32) and the regular transfer equation in 1D is the presence of $a < 1$ which allows the directly-transmitted solar beam to penetrate the cloud with greater ease relative to a homogeneous cloud. Once scattered, the photon transport proceeds as though the cloud were homogeneous. For instance, the two-stream solution that would result from (6.32) will resemble (6.12)–(6.13) with $\mu_0 \mapsto \mu_0/a$ in the solar forcing exponential term, but not when μ_0 appears in the scattering coefficients $\xi_{3,4}$ which result from $P(-\mu_0, \mu)$. Clearly, through parameterization of a, the solution for this model could be made to match more detailed ICA or 3D RT results. One could start by equating a with η in (6.28). Alternatively, the direct-beam transmittance term in (6.32) could be based on some assumed distribution of τ (cf. Gabriel and Evans, 1996). For example, assuming $p_\Gamma(\tau)$, as defined in (6.6), the scaled version of the exponential source term in (6.32) would be replaced by

$$\langle T_{\text{dir}} \rangle = \int\limits_{0}^{\infty} p_\Gamma(\tau) e^{-\tau/\mu_0} \, d\tau = \frac{1}{(1 + \bar{\tau}/\nu\mu_0)^\nu}$$

(6.33)

and the resulting 1D transport equation would be recast with $\bar{\tau}$ as the independent variable.

Further Examples and General Discussion

The 4 models described so far are interestingly different. The GWTSA starts with 1D RT and the ICA and ends with new formulas that can not be equated with 1D RT counterparts (since they are nonlinear averages of said 1D RT results). The Barker (1992) proposal to use the 1D RT equation with a modified but still exponentially decaying source term will lead to the usual 1D RT formulas with a scaled value for μ_0. The co-packing exponent model hails from a mean-field theory (real-space renormalization) – as such, it is a harbinger of models to come – but ends with a simple recipe for using 1D RT at non-absorbing wavelengths. The ETA ends the same way, but with a different and arguably even simpler recipe for rescaling the optical depth. Generalization of the last two fractal-based models to absorbing wavelengths was left by their originators as an open question.

That key question and others were addressed in a series of three computation-intensive papers co-authored by Prof. Harumi Isaka of the Laboratoire de Météorologie Physique (LaMP) in Clermont-Ferrand, France. Borde and Isaka (1996) revisit the 2D (vertical–horizontal) *singular* cascade model for cloud structure – this time, using a random (lognormal) multifractal instantiation – with and without absorption, but still in the frame of discrete-angle RT (with only 4 fluxes to determine at each grid-cell). On the one hand, they offer an alternative approach for finding the effective mean optical thickness in an equivalent 1D RT model. On the other hand, they find that in spite of the intense variability the domain-average absorption is not strongly affected; in particular, there is no compelling need to seek an effective single-scattering albedo (SSA) for most of their cloud modeling parameter space ... and the exception is better addressed in the second follow-on paper described next. Szczap et al. (2000a,b) revisit respectively non-absorbing and absorbing clouds configured by *bounded* cascade models in the frame of standard (continuous-angle) 3D RT using a Monte Carlo scheme. The Equivalent Homogeneous Cloud Approximation (EHCA) proposed by Szczap et al. (2000b) generalizes the ETA in (6.26) for averaging-scale and solar zenith-angle (SZA) effects. Szczap et al. (2000a) retains the EHCA for absorbing wavelengths but introduces an effective SSA to drive 1D RT models to the answer given by the 3D RT solution. The implicit parameterizations in these last two studies are empirical in nature, basically look-up tables populated with costly numerical simulation results.

One is left, in one way or the other, with a sense that methods for scaling mean optical depth (and possibly other optical properties) presented so far are simply too approximate and inadequate for operational GCM consumption. In particular, they are highly case-specific. Their significant advantage remains the fact that, in the end, they utilize simple PPH solutions of the 1D RT equation and therefore, assuming their variability parameter is given, they cost practically nothing to implement beyond the current RT investment in GCMs.

6.2.3 Solutions Using a Mean Field Theory for 3D Radiative Transfer

Here we cover another class of analytical methods for computing domain-average fluxes and heating rates in a cloudy atmospheric layer. Previously, we used closed-form solutions of the 1D RT problem and averaging over a given distribution of optical properties (typically, only the optical depth is varied). This has lead us to either new ways of using well-known 1D RT formulas (e.g., the ETA) or to new formulas to encode them (e.g., the GWTSA).

We now return to the 3D RT equation with spatially varying and, for all practical purposes, random coefficients. This *stochastic* RT equation is manipulated under a variety of assumptions until it yields an estimate of the domain-average quantities of interest. Borrowing from tradition in statistical physics, we call this a *mean field* RT theory. There are at least two distinct classes of mean field theory to consider:

- On the one hand, there are theories that consider no particular type of spatial variability but make a sequence of "reasonable" statistical assumptions about means and fluctuations and correlations driving towards a reduction of the 3D RT problem to its 1D counterpart, but with modified coefficients. This is also called "homogenization" and it is the mean-field counterpart of the ETA and other schemes that end with an application of 1D RT theory.

- On the other hand, there are theories where one first makes specific statistical assumptions about the spatial variability and then proceeds to solve the associated RT equation for the mean radiation field which can be quite different from the 1D RT equation. Both derivation and solution of the new equation typically call for problem-specific tricks. These are the mean-field counterparts of GWTSA-type solutions that may be based on 1D RT but end with non-standard formulas to apply.

In both categories, the assumptions are made in such a way as to obtain a certain outcome that enables analytical solutions, hence some insight is gained into the variability effects and/or the way they are modeled. Viable GCM parameterizations may even evolve from such theories.

We first discuss the general approaches. Whether explicitly or implicitly stated, the "reasonable" statistical assumptions they require eventually include a clear separation of scales between where the mean radiation field is estimated and where the variability actually occurs. In this case, the end result invariably reverts to a more-or-less standard RT equation for uniform but "effective" coefficients that will depend on both the means and on some parameter(s) introduced to describe the variability. So 1D RT solutions are invoked without further ado. Like the ETA described in the previous section, a single 1D RT computation is performed with modified optical parameters. Three examples come to mind here:

- Stephens (1988b) uses a simple closure scheme based on how unresolved variability affects the radiance in his previous (Stephens, 1988a) numerical experiments. By parameterizing the various correlations that appear in the mean-field equations, he arrives at a modified 2-stream approximation with new coefficients appearing.
- Cairns et al. (2000) use renormalization theory to address the 3D RT problem and they find fixed-point mappings of mean and variance parameters to effective quantities.
- Petty (2002) envisions "big scatterers" such as spherical clumps of scattering/absorbing material that are much smaller than the domain of interest but already one or more mean-free-paths in diameter. These clumps have a distribution of cross-sections and phase functions that are tabulated and they are assumed to be randomly distributed in space with some given density, and then used in a standard 2-stream approximation.

The former two studies make no specific assumption about the fluctuations of extinction and other optical properties, only on their spatial statistics (including the usual separation of the variability and averaging scales). They are discussed further on in some detail. The latter study assumes the 3D variability can be viewed as an assemblage of (say) spherical clumps, each of which is at once a multiple-scattering entity and an element of the whole medium. But the span of the size distribution of these big scatterers has to be small compared to their mean distance (according to their density) if they are to be considered as separate entities. So here again there is a tacit requirement of separation of scales.

The desired outcome of a mean field RT theory can also be a brand new RT equation (or system of equations) that needs to be solved from scratch. Again three examples come to mind, all of which apply to specific types of spatial variability:

- Chapter 7 by N. Byrne is entirely dedicated to "stochastic" RT in binary Markovian media where two coupled RT equations arise that can be solved by similar methods as for 1D RT (discrete ordinates and sweeps, or Monte Carlo). Recent and interesting developments are (*i*) to link the input cloud parameters to ground-based observations and compare predicted and observed fluxes (Lane et al., 2002) and (*ii*) to generalize to multiple layers with non-trivial overlap (vertical correlation) assumptions (Kassianov, 2003; Kassianov et al., 2003).
- Evans (1993) developed a non-Markovian Monte Carlo model for the domain-average fluxes that makes use of spatial/angular correlations between successive steps in the photon history. The method is an approximation because, in practice, it is limited to two-step correlations, not the full sequence of statistical memory effects along the photon path. The method is general but the correlation patterns need to be computed a priori from a standard Monte Carlo run in a specific kind of medium. So cloud structural properties determine the new Monte Carlo rules. Evans used a class of stochastic cloud models with realistic 2-point correlation structures.[6] Once this is done, the model is quite accurate and very efficient. However, it seems that the amount of overhead computation has limited the model's applicability. One can hope that in the future we will be able to parameterize the joint 2-step PDF for 3D optical media of interest in cloud optics and radiative energetics.
- Davis and Marshak (1997) assume that photon free paths (between all emission, scattering, absorption, and escape events) are not exponentially distributed but rather they have power-law tails, as can be expected in fractal-type optical media with no obvious characteristic scale. This ansatz leads to an "anomalous" diffusion process where the light particles trace Lévy walks which, in turn, call for new power-law kernels in the integral transport equation instead of the conventional exponential one (Buldyrev et al., 2001; Davis and Marshak, 2004). Here again, connections have recently been made with observed quantities (Pfeilsticker, 1999).

The latter model is discussed in some detail further on. The former model has a long and venerable history, going back at least to Avaste and Vainikko (1974). In particular, the two dedicatees of this volume, G. Titov and G. Pomraning, were fervent champions of stochastic RT in Markovian media. Assuming for simplicity a binary "cloudy/clear" medium, the enabling Markovian property is that there are two fixed probabilities per unit of length, one for each component, of crossing a cloudy/clear boundary. This is the equivalent of specifying a cloud fraction and a characteristic cloud size. The mean-field RT equations (two coupled RT equations arise) are valid over scales large enough to sample both the cloud size (distribution) and the cloud

---

[6] Specifically, these were 2D fractionally-integrated (power-law Fourier filtered) multiplicative cascades in the (x, z)-plane ending with a "$k^{-5/3}$" energy spectrum characteristic of turbulence, as is typically observed in stratocumulus; see Schertzer and Lovejoy (1987) or volume Appendix for more details.

fraction (distribution of inter-cloud distances), but not necessarily vast tracks of the medium. The stochastic RT model has an "effective medium" limiting case when the cloud size becomes tiny: alternations between cloudy and clear air are so fast that only the mean optical properties matter (this is know as the "atomistic mix"). It also has a relevant ICA or "linear-mixing" limit when cloud size becomes huge: interface effects can then be neglected and the linear mixture prediction in (6.4) becomes accurate.

We retain from this brief overview of mean-field RT theories that they come in a wide variety of flavors and that the research topic is by no means closed as we go to press with this chapter. Further on we discuss possible applications beyond parameterization of RT in GCMs. In the remainder of this section, we describe in more detail three selected mean-field RT theories: those of Stephens (1988b), Davis and Marshak (1997) and Cairns et al. (2000). The first and last are interestingly different in spite of the fact that they both lead back to familiar 1D RT with modified (or "effective") coefficients. In the earlier study, new optical properties appear that can be determined at least empirically; in the later one, the authors arrive at explicit formulas using means, variances, and correlation scales. In the third approach, a new class of 1D transport equations is obtained and certain important properties of their solutions are derived.

Radiance-Extinction Covariance and Statistical Closure

Following Stephens (1988b), we let $\widetilde{\mathcal{I}}$ represent the set of all horizontally decomposed Fourier components of the radiance field I. If $\mathcal{I}_0$ is the ensemble and/or domain average, the average equation of transfer for a sourceless medium can be expressed as

$$\left[\mu\frac{\partial}{\partial z} + \mathcal{S}_{0,0}\right]\mathcal{I}_0 = -\mathcal{S}_{0,>}\mathcal{I}_> \tag{6.34}$$

where

$$\mathcal{S}_{0,0} = \sigma_{0,0}\left[\cdot\right] - \int_{4\pi} S_{0,0}\left[\cdot\right]\,d\Omega \tag{6.35}$$

is the operator for the ensemble average radiance field in which $\sigma_{0,0}$ and $S_{0,0}$ are respectively the domain-average extinction and scattering kernel of the medium. More specifically, $S_{0,0}$ is the spatial mean of the product of the scattering coefficient σ_s and the phase function[7] $p(\Omega_{in}, \Omega)$. The pseudo-source/sink term in (6.34),

$$\mathcal{S}_{0,>}\mathcal{I}_> = \sum_{k>0} \mathcal{S}_{0,k}\mathcal{I}_k , \tag{6.36}$$

represents the impact of all smaller scales on the domain/ensemble average. Generally, there is a hierarchy of equations similar to (6.36) that describe mean fluctuations at scales $k = 1, 2, \ldots, K$ where some form of closure is invoked for all

---

[7] Here we normalize $p(\cdot)$ to 1 (rather than to 4π). This makes the physical interpretation of S as the *differential* scattering cross-section per unit of volume (cf. Chap. 3).

scales $k > K$, or else fluctuations are simply assumed to vanish (i.e., presumably "smoothed away" by multiple scattering).

The simplest form of (6.36), and the most tractable in terms of conceptualization and computation, beyond $\mathcal{S}_{0,>}\mathcal{I}_> = 0$ which reduces (6.34) to the regular equation of transfer for a homogeneous medium, is to invoke closure for all scales beyond the mean field (i.e., $K = 1$). Now let $I = \overline{I} + I'$ where the over-score designates the spatial/ensemble average, also the $k = 0$ Fourier component $\mathcal{I}_0$ in (6.34), and the primed quantity is the spatially varying fluctuation. We assume analogous decompositions for σ and S. Equation (6.34) can then be re-expressed as

$$\mu \frac{\partial \overline{I}}{\partial z} = \underbrace{-\overline{\sigma}\overline{I} + \int_{4\pi} \overline{S}(\mathbf{\Omega}_{in}, \mathbf{\Omega})\overline{I}\, d\mathbf{\Omega}_{in}}_{\substack{\text{regular transfer} \\ \text{equation: } \mathcal{S}_{0,0}\mathcal{I}_0}} \underbrace{-\overline{\sigma'I'} + \int_{4\pi} \overline{S'(\mathbf{\Omega}_{in}, \mathbf{\Omega})I'}\, d\mathbf{\Omega}_{in}}_{\substack{\text{1st-order closure} \\ \text{term: } \mathcal{S}_{0,>}\mathcal{I}_>}} \qquad (6.37)$$

where it is understood that all radiances have angular dependencies and that the dummy variable $\mathbf{\Omega}_{in}$ in the angular integrals is the incoming direction for the scattering event.

Stephens (1988b) offered a simple description of the mean fluctuation terms in (6.37) as

$$\begin{aligned} \overline{\sigma'I'} &= C_{\sigma I}\overline{\sigma}\overline{I} \\ \overline{S'I'} &= C_{SI}\overline{S}\,\overline{I} \end{aligned} \qquad (6.38)$$

where $C_{\sigma I}$ and C_{SI} are non-dimensional measures of correlation between the subscripted quantities. Note that this formulation rests on the tacit assumption that the medium has isotropic variability, i.e., it is statistically the same in all directions.

Substituting (6.38) into (6.37) gives

$$\mu \frac{\partial \overline{I}}{\partial z} = -(1 + C_{\sigma I})\,\overline{\sigma}\overline{I} + \int_{4\pi} (1 + C_{SI})\,\overline{S}(\mathbf{\Omega}_{in}, \mathbf{\Omega})\overline{I}\, d\mathbf{\Omega}_{in}\,. \qquad (6.39)$$

This resembles the 1D RT equation although a priori – and like for radiation transport through vegetation canopies (Chap. 14) – the extinction and scattering coefficients can depend on μ. If it is assumed that C_{SI} does not depend on the incoming direction $\mathbf{\Omega}_{in}$, the following transformed optical properties can then be used in any conventional 1D RT solution:

$$\begin{aligned} \widetilde{\sigma} &= (1 + C_{\sigma I})\,\overline{\sigma} \\ \widetilde{\varpi}_0 &= \left(\frac{1 + C_{SI}}{1 + C_{\sigma I}}\right) \varpi_0 \end{aligned} \qquad (6.40)$$

where $\overline{\sigma}$ and $\varpi_0 = \overline{\sigma}_s/\overline{\sigma}$ are domain-average extinction and the associated single-scattering albedo.

Stephens (1988b) proceeds to reformulate the Eddington 2-stream approximation for (6.39) for which he proposes the following parameterization

$$C_{\sigma I} = \tilde{C}_\sigma \mu$$
$$C_{SI} = \tilde{C}_S \mu \qquad (6.41)$$

based on an analysis of detailed numerical simulations by Stephens (1988a). This simple dependence on μ precludes the interpretation of (6.40) as a straightforward rescaling of optical properties.

Because of the linear dependence on μ in the parameterization proposed in (6.41), it adapts well to the 2-stream model based originally on

$$\bar{I}(z,\mu) = [J(z) + 3\mu F_n(z)]/4\pi \qquad (6.42)$$

and

$$p(\mu_{in}, \mu) = (1 + 3g\mu\mu_{in})/4\pi , \qquad (6.43)$$

in the so-called Eddington version (Meador and Weaver, 1980), a.k.a. 1D diffusion. Letting

$$F_\pm(z) = \int\limits_{\pm\mu>0} |\mu| \bar{I}(z,\Omega) d\Omega , \qquad (6.44)$$

and applying these definitions to (6.40) yields

$$\begin{pmatrix} +1 & 0 \\ 0 & -1 \end{pmatrix} \frac{d}{d\tau} \begin{pmatrix} F_+ \\ F_- \end{pmatrix} = \begin{pmatrix} -t_+ & +r_+ \\ +r_- & -t_- \end{pmatrix} \begin{pmatrix} F_+ \\ F_- \end{pmatrix} \qquad (6.45)$$

where $d\tau = \bar{\sigma} dz$. In the above coupled system of two ordinary differential equations (ODEs), we have used

$$t_\pm(z) = \bar{t} \pm t'$$
$$r_\pm(z) = \bar{r} \pm r' \qquad (6.46)$$

where (cf. (6.14), first 2 definitons)

$$\bar{t} = [7 - \varpi_0(4 + 3g)]/4$$
$$\bar{r} = [\varpi_0(4 - 3g) - 1]/4 \qquad (6.47)$$

and

$$t' = \tilde{C}_S \varpi_0(1 + g)/2 - \tilde{C}_\sigma$$
$$r' = \tilde{C}_S \varpi_0(1 - g)/2 \qquad (6.48)$$

We note an important difference with the classic 2-stream problem where the matrix of coefficients on the right-hand side of (6.45) has identical terms along the 2 diagonals, and here not necessarily. As usual in 2-stream theory (without internal source terms), the boundary conditions for the solar albedo problem are

$$F_-(\bar{\tau}) = F_0$$
$$F_+(0) = 0 \qquad (6.49)$$

respectively on the upper and lower cloud boundaries. Introducing

$$\kappa = \frac{\sqrt{(t_+ + t_-)^2 - 4r_+r_-}}{2} = \sqrt{\bar{t}^2 - \bar{r}^2 + r'^2} \tag{6.50}$$

where we recognize

$$\bar{t}^2 - \bar{r}^2 = 3(1 - \varpi_0)(1 - \varpi_0 g) ,$$

the calculus problem in (6.45) and (6.49) leads to

$$R = \frac{F_+(\bar{\tau})}{F_0} = \frac{\bar{r} - r'}{\bar{t} + \kappa \coth(\kappa\bar{\tau})} \tag{6.51}$$

for albedo,

$$T = \frac{F_-(0)}{F_0} = \frac{\exp(-t'\bar{\tau})}{\bar{t} \sinh(\kappa\bar{\tau})/\kappa + \cosh(\kappa\bar{\tau})} \tag{6.52}$$

for transmittance, and $A = 1 - R - T$ for absorptance (although we will soon see that radiant energy conservation is not a straightforward matter in this model).

While in the absence of fluctuations ($t' = r' = 0$), the regular Eddington 2-stream solution is recovered, there are some peculiarities about the new solution. For instance, albedo in (6.51) is independent of $\widetilde{C}_\sigma$ while transmittance in (6.52) depends on both $\widetilde{C}_\sigma$ and $\widetilde{C}_S$. Figure 6.2 shows albedo R, transmittance T, and absorptance A for a cloud with $\bar{\tau} = 10$, $\varpi_0 = 0.99$, and $g = 0.85$ as a function of $\widetilde{C}_\sigma$ and $\widetilde{C}_S$. If one allows for $\widetilde{C}_S < \widetilde{C}_\sigma$, absorptance is less than the homogeneous value (to the point of becoming negative for $\widetilde{C}_\sigma \approx 0.2$ and $\widetilde{C}_S \approx 0$); if $\widetilde{C}_S > \widetilde{C}_\sigma$, absorptance exceeds the homogeneous value. For such ϖ_0 close to unity, however, it is likely that this measure of inherent absorptivity is essentially independent of cloud structure, implying that $\widetilde{C}_\sigma$ should be $\approx \widetilde{C}_S$ in (6.40)–(6.41). In this also case (along the diagonals in Fig 6.2), albedo and transmittance change in the expected directions as fluctuations and correlations ($\widetilde{C}_\sigma \approx \widetilde{C}_S$) increase, but absorptance remains roughly constant.

To further discuss energy conservation in the new model, it is convenient to go back to the 1D diffusion quantities in (6.42). Direct application of the 2-stream definitions in (6.44) yields $F_\pm = J/4 \pm F_n/2$, hence

$$\begin{aligned} J &= 2(F_+ + F_-), \quad \text{scalar flux} \\ F_n &= F_+ - F_-, \quad\quad \text{net flux} \end{aligned} \tag{6.53}$$

Using (6.46)–(6.48), the ODE system in (6.45) then becomes

$$\frac{d}{d\tau}\begin{pmatrix} J \\ F_n \end{pmatrix} = \begin{pmatrix} \widetilde{C}_\sigma - \varpi_0\widetilde{C}_S & -(1 - \varpi_0) \\ -3(1 - \varpi_0 g) & \widetilde{C}_\sigma - \varpi_0 g\widetilde{C}_S \end{pmatrix}\begin{pmatrix} J \\ F_n \end{pmatrix} . \tag{6.54}$$

The cross-diagonal terms capture the standard model, specifically, Fick's (constituent) law and energy conservation (continuity equation) respectively. The on-diagonal terms contain the effects of the fluctuations and we see that, to enforce strict energy conservation (i.e., when $\varpi_0 = 1$), we need to mandate $\widetilde{C}_\sigma = \widetilde{C}_S$. Also the key parameter κ in (6.50) does not vanish when $\varpi_0 \to 1$ as it characteristically

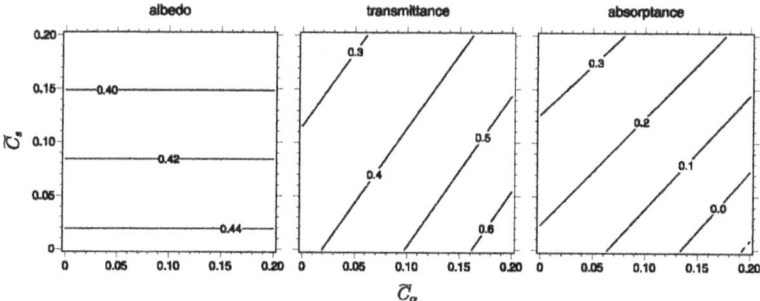

Fig. 6.2. Reflectance, transmittance, and absorptance for a cloudy layer as functions of parameters that describe correlations in fluctuations for radiance and extinction, $\widetilde{C}_\sigma$, and scattering extinction, $\widetilde{C}_S$. In this phenomenology, these are the new cloud properties, the traditional ones being $\overline{\tau} = 10$, $\varpi_0 = 0.99$, and $g = 0.85$. See text for more details on this modified 2-stream Eddington solution for an inhomogeneous plane-parallel slab. Values in lower left-hand corner of plots correspond to homogeneous conditions. Note that absorptance becomes negative for large $\widetilde{C}_\sigma$ and small $\widetilde{C}_S$

does in the standard 2-stream model, thus making (6.51) and (6.52) singular. Here, we find $\kappa = r' = \widetilde{C}_\sigma \varpi_0 (1 - g)/2$ and to obtain $R + T = 1$ in (6.51)–(6.52), we require $r' + t' = 0$, hence once again $\widetilde{C}_\sigma = \widetilde{C}_S$.

Finally, to underscore the fact that this closure model goes beyond the ICA, we note that it can be used to substantiate a general 3D diffusion-theoretical result presented in Chap. 5, (5.71) with $\chi = 2/3$, for diffuse illumination (or overhead sun) and conservative scattering:

$$\frac{R'}{\overline{R}} = 2\frac{\overline{\sigma'|F'_z|}}{\overline{\sigma}F_0} \leq 0 \tag{6.55}$$

in the present notations and sign conventions. We again denote domain-average albedo R, thus $\overline{R}$ is associated with the mean extinction and R' is its change due to 3D RT effects (which is always negative for the stated illumination conditions). This is a *deterministic* relation and associated inequality, i.e., valid for any realization of the variability with of course spatial averages in mind. The inequality follows directly from the physics of photon channeling in variable media (around the more opaque regions and into the more tenuous ones). See Davis and Marshak (2001) for a detailed derivation. Here, we can re-derive the inequality using Stephens' *statistical* parameterization of the extinction-radiance correlations in (6.38) and (6.41) where we normally have ensemble averages in mind. Indeed, we have

$$\overline{\sigma'|F'_z|} = -\int_{4\pi} \overline{\sigma' I'}(\mu)\mu d\Omega = -2\pi\overline{\sigma}\overline{I}\widetilde{C}_\sigma \int_{-1}^{+1} \mu^2 d\mu \leq 0 . \tag{6.56}$$

Furthermore, we obtain from (6.55) the following estimate for the non-dimensional correlation property

$$\tilde{C}_\sigma = \frac{3}{8}\left(\frac{F_0}{\pi\bar{I}}\right)\frac{|R'|}{R} \qquad (6.57)$$

where we can use[8] $\pi\bar{I} \approx F_0/4$. So $\tilde{C}_\sigma \approx \tilde{C}_S$ are roughly 3/2 of the relative 3D perturbation in albedo R. That is precisely what we see in the leftmost panel of Fig. 6.2, along the diagonal since we are in the conservative case.

In summary, this is an interesting model based on the largely unexplored concept of statistical closure in scale-space that is consistent with what we know from Chap. 5 about the general phenomenology of 3D radiation transport. As far as we know, it has not yet been used in a GCM yet, since it leads to a slightly modified 2-stream formalism that incorporates the impact of variability, it will cost almost nothing to implement. However, so far only the case of diffuse boundary sources has been examined, so collimated beams (for short wave RT) and isotropic internal sources (for long wave RT) need to be addressed. There is an important caveat to bear in mind before investing more effort in this model: if the parameters defining cloud fluctuations are not set carefully – and their values are by no means obvious – one can violate conservation of energy.

Renormalization Theory

Cairns et al. (2000) developed an approximate solution that is based on assumptions similar to those of Stephens (1988b), namely, a clear separation (hence a strong decoupling) of the averaging scale and the variability scales. But this requirement is brought on differently, so the outcome is also different. First, they remark that the number concentration of scatters can be described by

$$n(x) = \bar{n} + \delta n(x) \qquad (6.58)$$

where $\bar{n}$ is domain-average number concentration, $\delta n(x)$ is the local fluctuation at position x, and assume that these fluctuations are statistically isotropic in 3D space. They also focus on solar 3D RT problems where the source term can be confined to the boundary conditions and the 3D RT equation can be written more simply as

$$\mathbf{\Omega}\cdot\nabla I(x,\mathbf{\Omega}) + sn(x)\int_{4\pi} B(\mathbf{\Omega}'\cdot\mathbf{\Omega}')I(x,\mathbf{\Omega}')\,d\mathbf{\Omega}' = 0 \qquad (6.59)$$

where

$$B(\mathbf{\Omega}\cdot\mathbf{\Omega}') = \delta(\mathbf{\Omega}\cdot\mathbf{\Omega}' - 1) - \frac{\varpi_0}{4\pi}P(\mathbf{\Omega}\cdot\mathbf{\Omega}')$$

and s is the extinction cross-section (per cloud droplet); the fluctuating extinction is simply $\sigma(x) = sn(x)$. Averaging the above 3D RT equation over a large-enough volume leads them to domain-average, intensity $\overline{I(x,\mathbf{\Omega})}$ is given by

$$\mathbf{\Omega}\cdot\nabla\overline{I(x,\mathbf{\Omega})} + s\bar{n}\int_{4\pi} B(\mathbf{\Omega}'\cdot\mathbf{\Omega}')\overline{I(x,\mathbf{\Omega}')}\,d\mathbf{\Omega}'$$
$$= -s\int_{4\pi} B(\mathbf{\Omega}\cdot\mathbf{\Omega}')\overline{\delta n(x)I(x,\mathbf{\Omega}')}\,d\mathbf{\Omega}'. \qquad (6.60)$$

---

[8] This relation is exact for the conservative 2-stream model where $\bar{I} = \bar{J}/4\pi = J(\bar{\tau}/2)/4\pi$.

This looks formally like the usual integro-differential RT equation for the mean radiance $\overline{I(x, \Omega)}$ but, this time, with an non-vanishing source/sink term[9] on the right-hand side that is traced to the density fluctuations $\delta n(x)$ and how they interact with the 3D radiance field $I(x, \Omega)$.

In order to close (6.60), Cairns et al. (*i*) rewrite (6.60) as an integral equation involving the Green function (see Chap. 3), (*ii*) perform a formal perturbation expansion (akin to the von Neumann series described in Chap. 3), (*iii*) partially re-sum the series, and (*iv*) apply a classic 4th-order closure, (*v*) apply a nonlinear approximation (Rosenbaum, 1971) to the Green function. The last step improves the accuracy of the resulting mean-field integral transport equation. This procedure effectively decouples the *cross*-correlation term $\overline{\delta n(x) I(x, \Omega')}$ in (6.60) but at the cost of embedding the required $\overline{I(x, \Omega)}$ in a higher-dimensional integral over position and direction that involves the *auto*-correlation function $\overline{\delta n(x)\delta n(x')}$. Assuming the effects of fluctuations are local (this is critical), the authors are able to recover a tractable expression for the the right-hand side of (6.60) that can be grouped with the extinction-scattering term on the left-hand side. The mean-field RT equation can then be solved as though the medium was homogeneous with the following transformed domain-average optical properties:

$$\sigma' = (1 - \varepsilon)\overline{\sigma} \; ;$$

$$1 - \varpi_0' = (1 - \varpi_0) \left[1 - \varpi_0 \left(\frac{\varepsilon}{1 - \varepsilon} \right) \right] \; ; \qquad (6.61)$$

$$1 - \varpi_0'g' = (1 - \varpi_0 g') \left[1 - \varpi_0 \left(\frac{\varepsilon}{1 - \varepsilon} \right) \right] \; .$$

The key quantity that controls the above optical parameter renormalization induced by the variability effects is

$$\varepsilon = \frac{1}{2} \left(a - \sqrt{a^2 - 4V} \right) \; ,$$

where

$$a = \frac{1}{\overline{\sigma} l_c} + 1$$

and

$$V = \frac{\overline{n^2}}{\overline{n}^2} - 1$$

is the relative variance of n while l_c the effective correlation length of the variations. If particle density fluctuations follow lognormal-type distributions with 2-point correlations over a broad range, then we anticipate that sooner or later $\overline{\sigma} l_c \approx 1$. In this case, however, only moderate 1-point fluctuations are allowed, namely, $V < 1$.

From Cairns et al.'s (2000) initial formulation, it would appear that long-range fluctuations in $n(x)$ are neglected, thereby rendering the transformations in (6.61)

---

[9] The right-hand term in (6.60) balances the mean transport term on the left-hand side. Being dependent on the unknown 3D radiance field, this should more appropriately be called a *pseudo*-source/sink term.

applicable only to relatively small scales such as individual cells of a stratocumulus or individual cumuli. Cairns et al. allude to the ICA as being more suitable to describe the effects of fluctuations larger than the diffusion length, i.e., $1/\overline{\sigma}\sqrt{3(1-\varpi_0)(1-\varpi_0 g)}$.

Rossow et al. (2002) applied (6.61) to ISCCP (International Satellite Cloud Climatology Project) data, which have a horizontal resolution of ≈ 5 km, and redefined ε operationally as

$$\varepsilon = 1 - \frac{\widehat{\tau}}{\overline{\tau}} \tag{6.62}$$

where

$$\widehat{\tau} = R_{1D}^{-1}\left[\frac{1}{N}\sum_{n=1}^{N} R_{1D}\left(\tau_n\right)\right]$$

in which $R_{1D}^{-1}(R)$ is the solution of $R_{1D}(\tau) = R$, N is the number of satellite pixels in a domain, and τ_n is cloud optical depth inferred for the nth pixel.[10]

While Cairns et al.'s model does not suffer from the same ailment (potential for unphysical solutions) as Stephens', it appears from their initial presentation that it is meant to be applied at small scales; perhaps to be then superimposed onto another model designed to account for fluctuations at larger scales, such as the GWTSA.

Anomalous/Lévy Photon Diffusion

This model targets bulk radiation transport through the whole cloudy portion of the atmosphere, not necessarily a single cloud layer, starting with the empirical fact that this optical medium is variable on all observed scales. Furthermore, this 3D variability is shaped by complex thermodynamical processes in a highly dynamical environment of synoptic-scale geophysical turbulence as well as small-scale turbulence driven by shear- and buoyancy-driven instabilities. So we naturally tend to find robust turbulence-type variability laws such as the famous "$k^{-5/3}$" wavenumber spectrum (cf. Appendix). This characteristic scale-invariance implies long-range correlations and goes against the separation-of-scales rule for most common approaches to statistical 3D RT, i.e., that variability scales should be much smaller than the observation or computation scale. Motivated by these remarks, Davis and Marshak (1997) question the most basic premise of RT, namely, the exponential transport kernel that derives from Beer's classic law for direct transmission. Why should the mean-free-path (MFP) – the single parameter of the exponential free-path distribution – control every spatial aspect of the photon transport in such highly variable optical media? To explore a relevant alternative, Davis and Marshak (1997) assumed power-law free-path distributions. More precisely, they used symmetric Lévy-stable laws which are Gaussian-like for small steps but follow power-law tails for long ones. Letting s denote the random free path sampled by the photon population, Lévy-stable PDFs all have infinite variance: their power-law decay is indeed in $1/s^{1+\alpha}$ where $0 < \alpha < 2$.

---

[10] In the appendix to Rossow et al. (2002), it was shown that an accurate approximation relating v in (6.7) and ε in (6.62) is $v = 1/(\varepsilon - \ln(1-\varepsilon))$; see also footnote on p. 353.

In a later paper, Davis and Marshak (2004) showed that effective free-path distributions in spatially heterogeneous media are *always* sub-exponential and that the actual MFP is *always* larger than the inverse of the mean extinction.

To see more specifically why power-law distributions of photon free paths are effectively controlling the bulk radiative transfer in the Earth's cloudy atmosphere, we recall from Chap. 3 that the free-path distribution (equivalently, the effective transport kernel) derives from the law of direct transmission. Davis and Marshak (2004) argue that it is the *mean* direct transmission law averaged over the disorder in the medium that matters for the mean photon transport. In other words, to target large-scale fluxes one should average the transport kernel in isolation and thus derive a new equation to solve. In contrast, the GWTSA and ETA use the known solution of the classic (exponential kernel) equation. But why use kernels with power-law tails?

A mean law of direct transmission is obtained for other purposes in (6.33) for the reasonably accurate (Barker et al., 1996) Gamma distribution (6.6) for optical depths, implicitly for a fixed physical distance (namely, the thickness of the cloud layer). To be specific, we replace the slant optical path $\bar{\tau}/\mu_0$ by $\bar{\sigma}s$ where $\bar{\sigma}$ is the mean extinction (averaged over the Gamma PDF) and s is the (now random) step between scattering events; we thus obtain

$$\langle T_{\text{dir}}(s|\bar{\sigma})\rangle = \Pr\{\text{step} \geq s\} = \int\limits_0^\infty p_\Gamma(\sigma)e^{-\sigma s}\,d\sigma = \frac{1}{(1 + \bar{\sigma}s/v)^v}\,. \qquad (6.63)$$

As expected, the limit for $v \to \infty$ is the standard exponential law. In general, the photon MFP is $(1/\bar{\sigma}) \times v/(v - 1)$ which reverts properly to $1/\bar{\sigma}$ for $v \to \infty$ and diverges for $v \to 1$. This basically limits our interests to the regimes where $v > 1$. For the probability law in (6.63) free path variance is $(1/\bar{\sigma})^2 \times 2v^2/(v - 2)(v - 1)$ which is indeed divergent for $v < 2$. In the same way that all sums of random variables with *finite* variance converge to Gaussian laws (by virtue of the central limit theorem), sums of random variables with *infinite* variance, because of their power-law tails with exponent $1 + \alpha$ ($0 < \alpha < 2$), converge towards Lévy-stable (also called α-stable) laws; see, e.g., Samorodnitsky and Taqqu (1994). This class of PDFs include Gaussians in the limit $\alpha \to 2$. Since the multiple scattering is just a sum of random variables in all the coordinates, we expect the PDFs in (6.63) to behave like Lévy-Gauss PDFs with

$$\alpha = \min\{v, 2\} \qquad (6.64)$$

at least in transmission since it is dominated by photon paths with many scatterings. This argument applies strictly for isotropic scattering only because the random variables in the sums are presumably independent. However, Davis and Marshak (1997) extend their arguments to forward-scattering/non-absorbing scenarios by generalizing Eddington/van de Hulst rescaling, by $(1 - g)$, to situations where all we know is that the MFP exists (i.e., no need for the RT equation nor its diffusion counterpart).

Davis and Marshak's (1997) two key results for solar-type RT through cloudy atmospheres with asymptotically large optical depths are:

$$\langle T_{\text{cld}} \rangle \propto \frac{1}{[(1-g)\bar{\tau}]^{\alpha/2}} \tag{6.65}$$

where $\bar{\tau} = \bar{\sigma}h$ (h is the physical thickness of the medium) and, as an intermediate result,

$$\langle L \rangle_T \propto h \times [(1-g)\bar{\tau}]^{\alpha-1} \tag{6.66}$$

where $\langle L \rangle_T$ is the mean of the (total) path of the photons that escaped the medium in transmission.[11] The requirement for large $\bar{\tau}$ is simply to ensure that many scatterings photons in the terminology where "standard" diffusion is by small (Gaussian or exponential) steps. Starting with a very different statistical physics problem in mind, Buldyrev et al. (2001) derive a new integral transport equation for Lévy-distributed steps and solve it exactly. They retrieve the above asymptotic laws with precise prefactors and accurate transitions from the optically thin regime.

We note that the modified transmission law in (6.65) is equivalent to the rescaled reflectance law given in (6.24) for $\delta = \alpha/2 \le 1$ since $T_{1D} = 1 - R_{1D} \propto 1/(1-g)\tau$ in the asymptotic regime. As anticipated in (6.25), this expression for the co-packing exponent δ does not exceed unity and deviates from it more as the variability increases (α decreases from 2 to 1). We also note that, by multiplying the expression for pathlength in (6.66) by $\bar{\sigma}$ when $\alpha = 2$, we retrieve the classic statement that (the mean) order-of-scattering for transmitted light ($\approx \bar{\sigma}\langle L \rangle_T = \langle L \rangle_T/\text{MFP}$) is proportional to the square of the optical thickness, times $(1-g)$.

Using ground-based high-resolution O_2-line spectrometry to estimate $\langle L \rangle_T/h$, Pfeilsticker (1999) exploited (6.66) to show that the anomalous diffusion model captures the pathlength dimension (cf. Chap. 13) of solar photon transport in the whole atmospheric column far better than the exponential model under a wide variety of cloudy conditions (cf. Fig. 6.3). We can refer to Min and Harrison (1999) for a similar study optical depth was obtained directly from broad-band transmittance; their data follow the same kind of power laws as here with the same trend towards $\alpha \approx 2$ for single/continuous cloud layers and towards $1 < \alpha < 2$ for more complex cloud covers.

This direct observational validation is quite unique in 3D RT modeling at such large-scales. Other popular RT models for large-scale fluxes such as the GWTSA or the ETA are driven by parameters often determined from observations of cloud structure but their predictions for radiative fluxes are rarely compared with measurements because it is non-trivial to build up large-scale average fluxes from point-scale samples; from space the converse sampling problem occurs where much space is in view, but from a single direction; so a difficult conversion from radiance to flux is required. The appeal of the anomalous diffusion model for Pfeilsticker (1999) is that it addresses absorption by well-mixed gases directly (through the "equivalence theorem" discussed in Chap. 13); a major advantage is that pathlength studies using

---

[11] See Chap. 13 for an in-depth discussion and remote-sensing application of $\langle L \rangle_R$, the mean path for reflected photons.

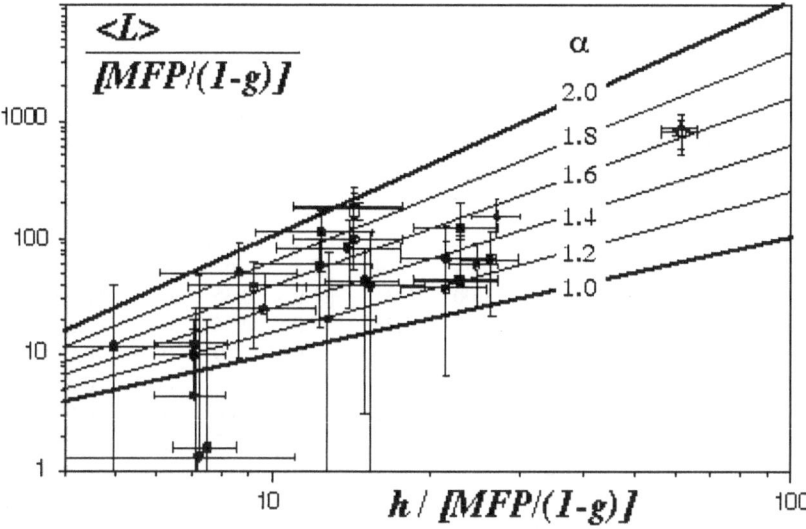

Fig. 6.3. Mean pathlength of transmitted solar light at 760 nm versus optical thickness: Oxygen A-band observations and Lévy transport model predictions. The observed pathlengths were corrected for pressure-weighting and expressed in units of "transport" MFPs, i.e., $h/\bar{\tau}$ scaled by $1/(1-g)$. This non-dimensional quantity is obtained as $\langle L \rangle_T/[h/(1-g)\bar{\tau}] \propto [(1-g)\bar{\tau}]^\alpha$ from (6.66). Optical depths were derived from the observed h (physical thickness of the whole cloud system) using commercial aircraft observations (the instrument was deployed near a major airport) and a climatological assumption about the MFP, then assigned generous error bars. The model predictions cover the data points with the proper association of single-layer un-broken clouds to $\alpha \approx 2$, sparse broken clouds to $\alpha \approx 1$, and complex multi-layered cloud scenes in between. The data were graciously provided by K. Pfeilsticker from Fig. 6 in his 1999 paper and re-plotted with a slightly improved representation of the effect of g in the theoretical curves

spectroscopy do not require absolute calibration, yet they show strong sensitivity to 3D structure.

In summary, the inability of the anomalous diffusion model to resolve layer-by-layer absorption is a liability in GCM parameterization development. However, its direct applicability to the emerging observational technology of oxygen A-band spectroscopy is an asset that begs for further exploitation.

6.3 Computational Models for Domain-Average Radiation Budgets

6.3.1 Horizontal Variability Aside From Water Content

All models discussed so far have been designed, and tested, with subgrid-scale fluctuations of cloud water content in mind. None of them address variations in particle

Table 6.1. Cloud microphysical data obtained on flights made by a Twin Otter (TW) and Convair-580 (CON) on dates given in dd/mm/yy. D signifies length of transect (km), A_c is cloud fraction, $\overline{\mathcal{L}}$ is mean liquid water content (g/m^3), $\overline{r_e}$ is droplet (optical) mean effective radius (μm), $v_{\mathcal{L}}$ and v_{r_e} are coefficients of variation (standard deviation/mean) for $\mathcal{L}$ and for r_e, ρ is the linear correlation coefficient between $\mathcal{L}$ and r_e, and b is the exponent in the nonlinear least-squares fit to $r_e = a\mathcal{L}^b$

Flight	Date	D	A_c	$\overline{\mathcal{L}}$	$\overline{r_e}$	$v_{\mathcal{L}}$	v_{r_e}	ρ	b
CON_02	16/08/95	82	0.94	1.44	12.3	0.62	0.22	0.74	0.30
TW_04B	21/08/95	38	1.00	0.31	11.9	0.35	0.13	0.55	0.22
CON_06	30/08/95	78	0.75	0.26	7.0	0.56	0.10	0.50	0.10
TW_11	30/08/95	68	0.74	0.10	5.9	0.49	0.15	0.25	0.06
TW_13C	01/09/95	45	1.00	0.23	6.7	0.30	0.14	0.53	0.25
TW_18A	08/09/95	82	0.76	0.05	9.4	0.60	0.17	0.59	0.18
CON_15	09/09/95	148	0.71	0.24	10.0	0.67	0.32	0.73	0.40
TW_19	09/09/95	40	1.00	0.13	7.9	0.42	0.08	0.78	0.16
TW_21	09/09/95	51	0.87	0.19	6.1	0.57	0.18	0.54	0.22
TW_24B	04/10/95	93	0.92	0.26	6.0	0.45	0.17	0.06	0.04

size for this entails horizontal variations in attenuation properties beyond extinction σ. This quantity of optical importance is related (in the large size-parameter limit) to two key parameters in cloud microphysics:

$$\sigma = \frac{3}{2}\frac{\mathcal{L}}{\rho_w r_e} \tag{6.67}$$

where $\mathcal{L}$ is liquid water content, ρ_w is density of water, and r_e is effective droplet radius. One can easily imagine simple scenarios where r_e is correlated with $\mathcal{L}$: large r_e and $\mathcal{L}$ in the cores of cells, and small r_e and $\mathcal{L}$ near edges where entrainment of unsaturated air occurs. Table 6.1 lists results from several aircraft flights made in 1995 during the Radiation and Cloud Experiment (RACE) through stratiform clouds over the Bay of Fundy (Räisänen et al., 2003). Here it is clear that r_e and $\mathcal{L}$ are usually correlated positively with correlation coefficients frequently exceeding 0.5. This reduces the impact of horizontal variability relative to when it is assumed that variations arise from $\mathcal{L}$ only: areas with small $\mathcal{L}$ tend to have small r_e which boosts extinction in (6.67); areas with large $\mathcal{L}$ tend to have large r_e which suppresses extinction relative to constant r_e. Räisänen et al. (2003) showed that this can mitigate the PPH bias by as much as 30%. Somewhat surprisingly, approximately 20% of this mitigation stems from fluctuations in phase function (hence the asymmetry parameter g); almost all of the remainder comes from modulations to extinction σ. Note the second-order correlation: regions with small r_e near cloud edges typically receive more irradiance than denser interiors with large r_e.

Since this situation is likely to be systematic, could we apply the same kind of nonlinear averaging in this multi-variate setting that was used previously for optical depth variability? In principle, yes, but in practice, it would appear to be close to intractable. Doing something about cloud microphysical variability could nonetheless

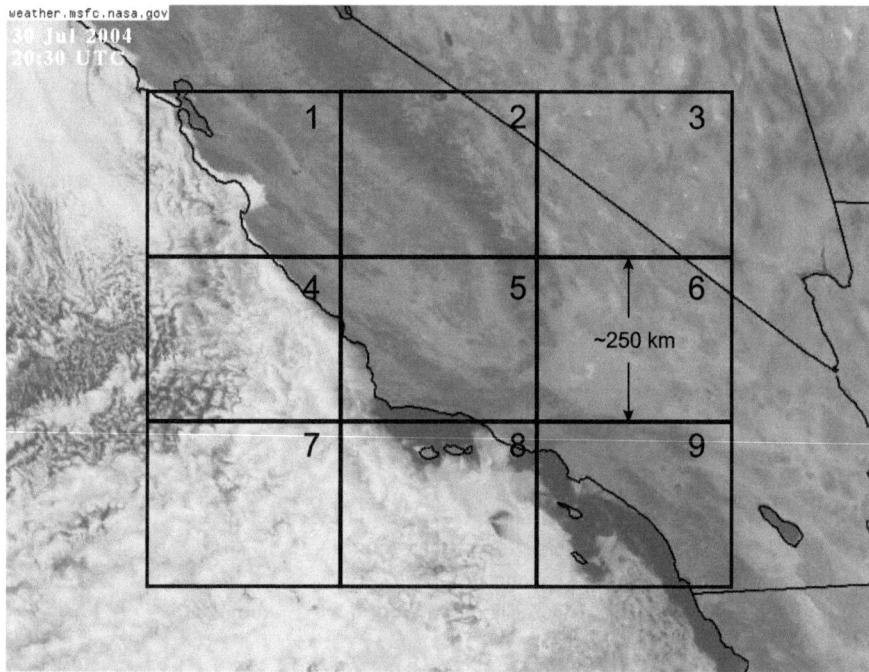

Fig. 6.4. This visible image is from NOAA's GOES-10 satellite on July 30, 2004 at 20:30 UTC (obtained from MSFC-NASA). The image is 1 km resolution and shows stratocumulus clouds off the SW coast of California. Nine cells corresponding to the size of a typical GCM grid-cell are shown. Cells 2, 3, and 6 are over land and are completely cloudless. Cell 7 is over ocean and is almost overcast. Cells 1, 4, 5, 8, and 9 straddle the coast and they exhibit almost cloudless conditions over land and mostly cloudy conditions over ocean

prove to be important for questions regarding indirect forcing by aerosols (Lohmann and Feichter, 1997) given that global average forcings are on the order of only $0.5\,\mathrm{Wm}^{-2}$.

Another issue that appears to be completely beyond the reach of the current 1D RT model paradigm involves variations in surface albedo that are correlated with clouds. This is a frequent occurrence for GCM cells that straddle continental coastal lines. Often, most notably in the large stratocumulus regions, cloud will be primarily over water and not over land. This is illustrated in Fig. 6.4. Current models simply construct an area average surface albedo and use that value as the lower boundary condition for the radiation model.

To demonstrate this effect, approximate domain-average albedo for diffuse irradiance as

$$\langle R \rangle = f_{\mathrm{w}} \left[\underbrace{(1 - A_{\mathrm{w}}) \left(R_{\mathrm{clr}} + \frac{T_{\mathrm{clr}}^2 \alpha_{\mathrm{w}}}{1 - \alpha_{\mathrm{w}} R_{\mathrm{clr}}} \right)}_{\text{clear-sky over water}} + \underbrace{A_{\mathrm{w}} \left(R_{\mathrm{cld}} + \frac{T_{\mathrm{cld}}^2 \alpha_{\mathrm{w}}}{1 - \alpha_{\mathrm{w}} R_{\mathrm{cld}}} \right)}_{\text{cloudy-sky over water}} \right]$$

$$\underbrace{\phantom{f_{\mathrm{w}} \left[(1 - A_{\mathrm{w}}) \left(R_{\mathrm{clr}} \right) + A_{\mathrm{w}} \left(R_{\mathrm{cld}} \right) \right]}}_{\text{contribution from water}}$$

$$+ (1 - f_{\mathrm{w}}) \left[\underbrace{(1 - A_{\mathrm{l}}) \left(R_{\mathrm{clr}} + \frac{T_{\mathrm{clr}}^2 \alpha_{\mathrm{l}}}{1 - \alpha_{\mathrm{l}} R_{\mathrm{clr}}} \right)}_{\text{clear-sky over land}} + \underbrace{A_{\mathrm{l}} \left(R_{\mathrm{cld}} + \frac{T_{\mathrm{cld}}^2 \alpha_{\mathrm{l}}}{1 - \alpha_{\mathrm{l}} R_{\mathrm{cld}}} \right)}_{\text{cloudy-sky over land}} \right]$$

$$\underbrace{\phantom{+ (1 - f_{\mathrm{w}}) \left[(1 - A_{\mathrm{l}}) \left(R_{\mathrm{clr}} \right) + A_{\mathrm{l}} \left(R_{\mathrm{cld}} \right) \right]}}_{\text{contribution from land}}$$

$$(6.68)$$

where f_{w} is fraction of domain that is water, A_{w} and A_{l} are cloud fractions over water and land, and α_{w} and α_{l} are albedos of water and land. To simplify matters, albedo and transmittance for clear-sky and cloud, R_{clr}, R_{cld}, T_{clr}, and T_{cld} are common to land and water. In contrast to (6.54), a conventional estimate of $\langle R \rangle$ would be

$$\langle R \rangle' = \left[(1 - \overline{A}) R_{\mathrm{clr}} + \overline{A} R_{\mathrm{cld}} \right] + \frac{\left[(1 - \overline{A}) T_{\mathrm{clr}} + \overline{A} T_{\mathrm{cld}} \right]^2 \overline{\alpha}}{1 - \overline{\alpha} \left[(1 - \overline{A}) R_{\mathrm{clr}} + \overline{A} R_{\mathrm{cld}} \right]} \qquad (6.69)$$

where

$$\overline{A} = f_{\mathrm{w}} A_{\mathrm{w}} + (1 - f_{\mathrm{w}}) A_{\mathrm{l}}$$

is overall cloud fraction, and

$$\overline{\alpha} = f_{\mathrm{w}} \alpha_{\mathrm{w}} + (1 - f_{\mathrm{w}}) \alpha_{\mathrm{l}}$$

is domain-average surface albedo. Figure 6.5 shows fractional differences between $\langle R \rangle'$ and $\langle R \rangle$ where it was assumed that $A_{\mathrm{l}} = 0$ and 1 (other parameter values are listed in the caption). Largest errors are committed for no cloud over land, very cloudy over ocean, for domains with approximately equal areas of land and water. The fact that differences can easily exceed 5% means that this effect often rivals that of the PPH bias for cloudy marine boundary layer clouds (Barker et al., 1996; Pincus et al., 1999).

6.3.2 Vertical Overlap of Clouds That Vary Horizontally

All the models discussed so far in this section were designed to yield domain-average albedo and transmittance for a single layer. It may happen, however, that a single-layer cloud occurs in an GCM, but the vertical layering of the GCM divides the cloud into several layers. When clouds are considered to be homogeneous, overlapping involves only cloud fraction. Implicit in overlapping of homogeneous clouds, however,

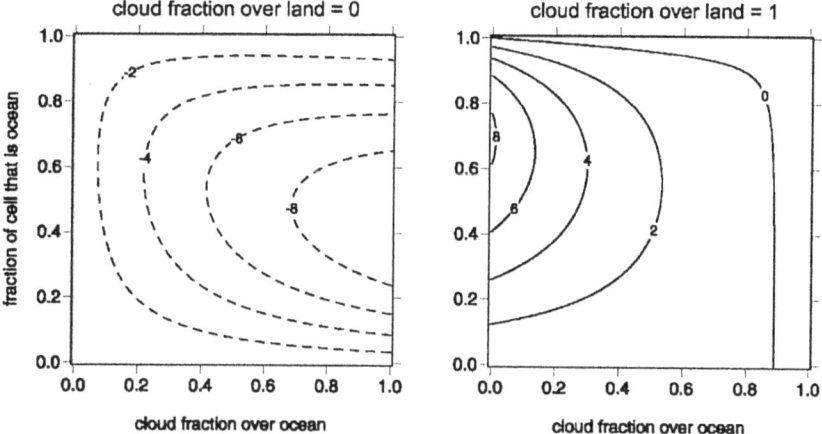

Fig. 6.5. Fractional error (in %) in domain-average albedo due to the use of (6.69) instead of the accurate (6.68) as functions of fraction of domain that is water f_w and cloud fraction over water A_w. Results are shown for two values of cloud fraction over land A_l. Values of the parameters in (6.68) and (6.69) are: $\alpha_w = 0.05$ and $\alpha_l = 0.3$, $R_{clr} = 0.1$ and $R_{cld} = 0.5$ (hence $T_{clr} = 0.9$ and $T_{cld} = 0.5$, assuming a conservative wavelength).

is a distribution of water path or optical depth, though they are generally unrealistic. When clouds vary horizontally, cloud fraction and condensate overlap. Hence, even for a cloud that is overcast in all its sub-layers, the problem of overlap remains.

Mean-field solutions that end up rescaling optical properties assume (at least implicitly) that fluctuations are isotropic but characterized by some correlation length. This assumption will likely be violated in multi-layer situations. With the GWTSA, one has to take explicit account of vertical correlations in condensates and allow for corresponding spatial variations in flux (Oreopoulos and Barker, 1999). This is a drawback to their method for it is difficult to see how this accounting can be anything other than very approximate (see Chap. 9).

The fact of the matter is that the overlap probabilities of cloud fraction and condensate appear to de-correlate at distinctly different rates and both influence the radiative transfer. Accounting for these fluctuations in the 1D models may be possible but will likely be complicated and certainly be rough approximations. Indeed, results from an inter-comparison study of 1D solar RT algorithms (Barker et al., 2003) indicate that even straightforward overlap of homogeneous clouds eludes perfect portrayal in the sense that different modelers arrive at surprisingly different results, even for similar overlap problems.

6.3.3 The Monte Carlo Independent Column Approximation (McICA)

For the foreseeable future, GCMs will continue to provide only domain-average profiles of conventional cloud information to their 1D radiation codes along with additional assumptions about cloud structure (e.g., overlap rates). With the ever increasing demand that radiation codes be more and more realistic and accurate, it is difficult to see how the current methods based on 1D monochromatic models, as discussed so far in this chapter, will meet this demand satisfactorily. Indeed, as their level of sophistication increases, so does their computational cost in general, and this is compounded with the already heavy burden of spectral integration. In stark contrast to this paradigm, Barker et al. (2002); Barker et al., (2004), Pincus et al. (2003), and Räisänen et al. (2005b) offer a simple compromise that solves many of the problems that bear down on the current paradigm. Their model, referred to as the Monte Carlo ICA (McICA), has the added advantages that it requires no more CPU resources than the fastest and simplest 1D solvers, it can facilitate any 1D solution of the RT equation, and it is unbiased with respect to the full ICA. This subsection outlines McICA which might prove to be a sufficient stopgap solution until full 3D simulations become routine in GCMs. This is achieved by using a random quadrature rule to sum simultaneously over the spectral domain and spatial variabilities in the GCM grid-cell.[12]

To begin, the most common means of accounting for absorption of broadband radiation in GCMs is the correlated-k distribution (CKD) method (Lacis and Oinas, 1991; Fu and Liou, 1992). With the CKD method, broadband fluxes are computed as

$$\mathcal{F} = \sum_{k=1}^{K} F(k) \tag{6.70}$$

where K is the number of monochromatic calculations and the $F(k)$ are fluxes. Therefore, returning to (6.4), we assume that a domain consists of M_{cld} cloudy sub-columns with parameters denoted generically as $\{s\}_m (m = 1, \ldots, M_{\mathrm{cld}})$. Using (6.3) gives the domain-average ICA flux as

$$\langle \mathcal{F} \rangle = (1 - A_{\mathrm{c}}) \sum_{k=1}^{K} F_{\mathrm{clr}}(k) + A_{\mathrm{c}} \left\{ \sum_{k=1}^{K} \left[\frac{1}{M_{\mathrm{cld}}} \sum_{m=1}^{M_{\mathrm{cld}}} F_{\mathrm{cld}}(\{s\}_m, k) \right] \right\} \tag{6.71}$$

$$= (1 - A_{\mathrm{c}}) \mathcal{F}_{\mathrm{clr}} + A_{\mathrm{c}} \langle \mathcal{F}_{\mathrm{cld}} \rangle .$$

Many GCM groups find it difficult to justify the computational requirements needed for $K \approx 30$ to 100 with $M_{\mathrm{cld}} = 1$ (i.e., a regular CKD application). Therefore, application of (6.71) with even a modest value of M_{cld} is untenable. The McICA solution (Räisänen and Barker, 2005; Räisänen et al., 2005b) to this problem is to randomly sample (generate) a subset of cloudy columns and approximate $\langle \mathcal{F}_{\mathrm{cld}} \rangle$ as

$$\langle \mathcal{F}_{\mathrm{cld}} \rangle \approx \sum_{k=1}^{K} \frac{1}{N(k)} \sum_{n=1}^{N(k)} F_{\mathrm{cld}}(\{s\}_n, k) \tag{6.72}$$

---

[12] Here "Monte Carlo" is used correctly but not in the usual sense for the atmospheric radiation community since here there is no 3D RT per se, only an ICA. Indeed, Monte Carlo is just a convenient (and never optimal) technique for estimating integrals numerically, in this case, those expressed in (6.2) or (6.5). However, we are now interested in cloud scenes with multiple layers where more than τ is varying.

where for each k, monochromatic fluxes are computed for $N(k)$ subgrid columns with randomly sampled properties $\{s\}_n$ and averaged. A total of

$$\mathcal{N} = \sum_{k=1}^{K} N(k)$$

sub-columns are used where ideally $\mathcal{N} \ll M_{\mathrm{cld}}K$. In practice, it is likely that $N(k) = 1$ would be considered the bare minimum, but in principle, there could be $N(k) = 0$ for some randomly selected values of k.

It is quite simple to show (Barker et al., 2002) that, even for $\mathcal{N} < K$, the expectation value of (6.72) is precisely the ICA. Thus, McICA is unbiased with respect to the exact ICA. So, in as much as the ICA accounts for all subgrid-scale fluctuations, so too does McICA. McICA does, however, produce additional random noise. Experiments using GCMs (Pincus et al., 2003; Räisänen et al., 2005b) indicate that the noise produced by McICA, assuming sufficiently large $\mathcal{N}$ (nominally > 30), is essentially consumed by the dynamics of the host model. This opens a new door regarding assessment of RT models: in addition to standalone tests, the RT model should be assessed actively in GCMs. Specifically, while McICA has a conditional random component, if GCMs are insensitive to that error, it can not be held against McICA. The McICA algorithm needs specific input from either explicit fields from which samples can be drawn or an adequate subgrid stochastic cloud generator (Räisänen et al., 2005a). If this pre-RT step is done correctly, McICA appears to be, for GCM applications at least, a satisfactory bridge between the plethora of analytical models based on 1D RT presented earlier in this chapter and full-3D computational models discussed in Chap. 4.

6.4 Efficient Large-Scale Radiation Budget Estimation: Discussion and Outlook

Up until the late 1980s, descriptions of clouds in GCMs left much to be desired and thus did not warrant overly sophisticated treatments of cloud-radiation interactions. The advent of detailed cloud parameterizations (e.g., Smith, 1990; Tompkins, 2002) and concern over seemingly small, but systematic, changes in cloud optical properties (e.g., Lohmann and Feichter, 1997) fueled the need for RT codes that consider interactions between unresolved clouds and radiation. Therefore, since GCMs require domain-average fluxes, the focus of the two previous sections was on methods of accounting for unresolved clouds in more-or-less standard 1D RT models or in somewhat more sophisticated approaches, namely, mean-field theories that do not reduce to *effective* local properties but rather to *new* transport equations.

There are two basic conditions that maintain the life of 1D/domain-average RT models in GCMs: (1) a paucity of information needed to initialize a full 3D solution; and (2) a paucity of computational resources needed to execute full 3D solutions even if sufficient information was available. As long as these conditions are met, development and use of 1D models that account for interactions with unresolved clouds (and other fluctuating media) is justified. Several methods that do just this have been proposed and developed over the past two decades and some were discussed here. The few results and the overarching message presented here are in step with the conclusions of a recent 1D RT model inter-comparison (Barker et al., 2003) and with recent advances in the representation of clouds in GCMs (e.g., Khairoutdinov and Randall, 2001): (1) no single 1D model in existence works well in all conditions and it is likely that none ever will; and (2) the conditions stated at the beginning of this paragraph are beginning to erode. Together, these trends seem to signal the beginning of the end for the current paradigm of 1D/domain-average RT models in GCMs.

The existing paradigm of weaving into 1D RT solutions increasingly sophisticated descriptions of unresolved optical property fluctuations may sometimes lead to elegant mathematical solutions but it also seems bound to endless approximation, limitation, and most crucially, biased[13] solutions. We should not lose sight of the fact that GCMs, the driving force behind 1D RT modeling, require unbiased estimates of radiative fluxes, not mathematically beautiful approximations the likes of which are expected in particle physics for example.

It seems likely that within the next decade or so, 1D solvers that attempt to account for unresolved interactions between radiation and optical fluctuations will be replaced by fully 3D solvers (see Chap. 4, Chap. 5, and Chap. 7). Some GCMs are already running CSRMs inside each of thousands of regular-sized columns (Randall et al., 2003) thereby providing radiation codes with distributions of hitherto unresolved clouds. While these "super-parameterized" GCMs currently run full, broadband ICA calculations, a limited number of simulations using true multi-dimensional RT have already occurred (J. Cole, priv. comm.). The point is that this methodology, which represents a significant step towards the endgame of modeling RT for cloudy atmospheres in GCMs, is knocking at the door and it will eventually enter. To some, this will seem like a homeland invasion, but to most it will be a welcome, and permanent, resident.

One must therefore ask: if conventional 1D RT methods are to be phased out of operational before too long, will it be worth the hefty intellectual effort of improving analytical 1D RT models, even modified for *some* 3D effects, to the point of meeting the demands of GCMs? Assuming that it will not be worth it and their obsolescence is coming, Barker et al. (2002) and Pincus et al. (2003) offered a simple compromise that appears to solve many significant problems facing current 1D methods. Their McICA model has been subjected to extensive testing and appears to be a reasonable and affordable stopgap solution.

---

[13] We recall that "bias" is not understood here as it is in the rest of the chapter and in much of this volume, to be with respect to 1D RT (with mean optical properties) nor with respect to full 3D RT (given detailed optical properties in 3D). Here, comparisons are between the *exact* ICA estimate for arbitrary 3D variability on one hand, while on the other, all 1D RT models using effective optical properties as well as all ICAs using specific PDFs for the optical depth distribution (typically assumed to be the *only* variable quantity). The bias discussed presently refers therefore to modeling errors resulting from ad hoc assumptions about the source and impact of variability inside a cloudy GCM layer. It also includes errors that follow from approximate ways of accounting for overlap effects.

In the above discussion we have pitted the vigorously progressing McICA project, whose widespread acceptance is still incomplete as this volume goes to press, against traditional 1D-based methods, whether they hail from analytical ICAs (e.g., GWTSA or ETA) or from mean-field theories (Stephens, 1988b; Cairns et al., 2000; Petty, 2002). Mean field theories that, by design, go beyond the ICA and effective medium theory – be it only for a specific variability model – are in a different class; members of this class discussed in this chapter are stochastic RT (see also Chap. 7) and anomalous photon diffusion (Davis and Marshak, 1997). They do not use the standard 1D RT equation nor any well-known 1D RT solutions. They can therefore be compared meaningfully only to full 3D numerical results, or to observations. In that sense, they are (like McICA) inherently more accurate than all 1D RT models, even using effective parameters (like the ETA) or parameterized ICAs (like the GWTSA). However, these analytical models do not have the flexibility of a purely numerical approach such as McICA. So we do not see them as serious contenders for operational GCM parameterizations. Rather, their relative simplicity will likely be used to gain understanding of large-scale 3D RT processes without limitation to single-layer clouds (where the GWTSA and ETA-type 1D models all work reasonably well).

Having strongly endorsed the computational McICA model for GCM parameterization, and assigned mean-field theories that go beyond the ICA to observational diagnostics, what is left for analytical ICA models and effective medium theories? We see their future in capturing the radiative effects of unresolved variability ... at the smallest conceivable scales in current climate-driven modeling, namely, inside cloudy CSRM cells. Indeed, our empirical knowledge of turbulence in clouds (cf. Chap. 2) tells us that it is unwise to assume homogeneity or even smoothness downwards from km-or-so scales. This final vocation may seem ironic for RT models that were designed originally to capture GCM grid-scale effects, but the cloudy atmosphere is variable at all scales from the Kolmogorov dissipation scale of a few mm to the planetary scale. Just as atmospheric dynamicists are learning how to adapt deterministic and stochastic phenomenologies across this spectrum, so too should atmospheric radiation experts – whether they choose to work with 1D or 3D, deterministic or stochastic methods.

6.5 Summary

We have surveyed from the literature both analytical and computational approaches to the estimation of large-scale radiative fluxes when little information about 3D cloud structure is provided.

The computational McICA (Monte Carlo Independent Column Approximation) model is making rapid advances in operational GCM parameterization of the full-spectrum RT when clouds are present. It offers more flexibility and robustness than both analytical counterparts that incorporate 3D variability effects, let alone the older 1D approaches that ignore variability altogether. The cost of the McICA upgrade for

GCMs is a tolerable increase to computer time and conditional noise that does not seem to affect the predicted climate.

There is however another emerging market for advances in analytical mean-field theory in 3D RT, and it is outside of traditional GCM parameterization work per se. There is indeed a pressing need for simple models that can be compared directly with real-world observations as well as "black-box" computations. For instance, intricate atmospheric/surface absorption and scattering processes in the presence of 3D clouds are captured by photon pathlength statistics obtained from O_2 A-band spectroscopy. Gaining physical insights from this new diagnostic calls for 3D RT models with statistically well-defined variability parameters (variances, correlation scales, overlap probabilities, etc.) that are typically found in analytical solutions, preferably with closed-form expressions. We foresee oxygen A-band products (photon path statistics), supported by mean-field modeling, and correlative data (optical depths, cloudiness structure, etc.) as an observational testing ground for GCM parameterization predictions at a deeper level than surface or TOA fluxes.

In spite of the considerable technical difficulties, the programmatic pressures, and the limited resources, approximation theory is a vibrant research area in 3D RT, in atmospheric science and elsewhere. As demonstrated in this chapter, it applies directly to the estimation of average fluxes and energy deposition in large domains, as used in GCMs, as well as to the observable effects of unresolved variability in remote sensing at moderate-to-coarse resolutions. As demonstrated in the previous chapter, it also applies to observational and computational problems that require detailed radiance fields everywhere in situations where there is intense radiative cross-talk between pixels or grid-cells. So there are many down-to-earth reasons to engage in this research, leading us to think that we are solving a practical engineering-type problem, intentionally cutting as many computational corners as we can get away with inside a given error budget and a given number of floating-point operations. Sometime we will indeed forge ahead with an algorithmically viable solution that does not draw on any of the analytical methods or physical insights used in their derivations. The current "McICA" solution for operational GCM radiation parameterization is a good example.

Having said this, we emphasize that we are pragmatic enough not to dismiss the analytical methods that seek to address aspects of RT that are beyond the scope and intent of McICA. As it turns out, the latter are a better match with new observations of photon pathlength distributions made available from advances in oxygen A-band spectroscopy, at least when the underpinning mean-field theory goes beyond the ICA and 1D "effective medium" solutions. As for mean-field theories that do lead to effective medium parameters, we must recall that in the end they only make a statement about local optical properties. In this sense, they may provide answers to the question of how unresolved variability impacts the smallest computationally resolved scales considered in Chaps. 4 and 5 when assuming homogeneity is not justified. As such, we begin to close the loop in the domain of spatial scales by, somewhat ironically, applying at the innermost scales of a CSRM a methodology that was designed originally to operate at the innermost scales of a GCM.

Paul Dirac is famous for pursuing exactness and formal elegance in the development of relativistic quantum physics. However, his early training was in electrical engineering, and he is on record for recognizing that:[14]

> *"The engineering course influenced me very strongly. [. . .] I've learned that, in the description of nature, one has to tolerate approximations, and that even work with approximations can be interesting and can sometimes be beautiful."*

References

Avaste, O.A. and G.M. Vainikko (1974). Solar radiative transfer in broken clouds. *Izv. Acad. Sci. USSR Atmos. Oceanic Phys.*, **10**, 1054–1061.

Barker, H.W. (1992). Solar radiative transfer for clouds possessing isotropic variable extinction coefficient. *Quart. J. Roy. Meteor. Soc.*, **118**, 1145–1162.

Barker, H.W. and B.A. Wielicki (1997). Parameterizing grid-averaged longwave fluxes for inhomogenous marine boundary layer clouds. *J. Atmos. Sci.*, **54**, 2785–2798.

Barker, H.W., B.A. Wielicki, and L. Parker (1996). A parameterization for computing grid-averaged solar fluxes for inhomogeneous marine boundary layer clouds – Part 2, Validation using satellite data. *J. Atmos. Sci.*, **53**, 2304–2316.

Barker, H.W., G.L. Stephens, and Q. Fu (1999). The sensitivity of domain-averaged solar fluxes to assumptions about cloud geometry. *Quart. J. Roy. Meteor. Soc.*, **125**, 2127–2152.

Barker, H.W., R. Pincus, and J.-J. Morcrette (2002). The Monte Carlo independent column approximation: Application within large-scale models. In *Proceedings from the GCSS Workshop*. Kananaskis, Alberta, Canada.

Barker, H.W., and Räisänen, P., (2004). Radiative sensitivities for cloud geometric properties that are unresolved by conventional GCMs. *Quart. J. Roy. Meteor. Soc.*, **130**, 2069–2086.

Barker, H.W., G.L. Stephens, P.T. Partain, J.W. Bergman, B. Bonnel, K. Campana, E.E. Clothiaux, S. Clough, S. Cusack, J. Delamere, J. Edwards, K.F. Evans, Y. Fouquart, S. Freidenreich, V. Galin, Y. Hou, S. Kato, J. Li, E. Mlawer, J.-J. Morcrette, W. O'Hirok, P. Räisänen, V. Ramaswamy, B. Ritter, E. Rozanov, M. Schlesinger, K. Shibata, P. Sporyshev, Z. Sun, M. Wendisch, N. Wood, and F. Yang (2003). Assessing 1D atmospheric solar radiative transfer models: Interpretation and handling of unresolved clouds. *J. Climate*, **16**, 2676–2699.

Benner, T.C. and K.F. Evans (2001). Three-dimensional solar radiative transfer in small tropical cumulus field derived from high-resolution imagery. *J. Geophys. Res.*, **106**, 14,975–14,984.

---

[14] We acknowledge Prof. F. Wilczek for finding this delectable quote in Dirac (1977). He used it to a similar end as we – but within a very different context – in "Analysis and Synthesis I: What Matters for Matter," the first installment of his insightful series that appeared in *Physics Today*, May 2003 issue.

Borde, R. and H. Isaka (1996). Radiative transfer in multifractal clouds. *J. Geophys. Res.*, **101**, 29,461–29,478.

Buldyrev, S.V., M. Gitterman, S. Havlin, A.Ya. Kazakov, M.G.E. da Luz, E.P. Raposo, H.E. Stanley, and G.M. Viswanathan (2001). Properties of Lévy flights on an interval with absorbing boundaries. *Physica A*, **302**, 148–161.

Cahalan, R.F. (1989). Overview of fractal clouds. In *Advances in Remote Sensing*. A. Deepak Publishing, Hampton, VA, pp. 371–388.

Cahalan, R.F. and J.B. Snider (1989). Marine stratocumulus structure during FIRE. *Remote Sens. Environ.*, **28**, 95–107.

Cahalan, R.F., W. Ridgway, W.J. Wiscombe, S. Gollmer, and Harshvardhan (1994a). Independent pixel and Monte Carlo estimates of stratocumulus albedo. *J. Atmos. Sci.*, **51**, 3776–3790.

Cahalan, R.F., W. Ridgway, W.J. Wiscombe, T.L. Bell, and J.B. Snider (1994b). The albedo of fractal stratocumulus clouds. *J. Atmos. Sci.*, **51**, 2434–2455.

Cahalan, R.F., D. Silberstein, and J. Snider (1995). Liquid water path and plane-parallel albedo bias during ASTEX. *J. Atmos. Sci.*, **52**, 3002–3012.

Cairns, B., A.A. Lacis, and B.E. Carlson (2000). Absorption within inhomogeneous clouds and its parameterization in general circulation models. *J. Atmos. Sci.*, **57**, 700–714.

Chambers, L.H., B.A. Wielicki, and K.F. Evans (1997). On the accuracy of the independent pixel approximation for satellite estimates of oceanic boundary layer cloud optical depth. *J. Geophys. Res.*, **102**, 1779–1794.

Charlock, T. and B.M. Herman (1976). Discussion of the Elsasser formulation for infrared fluxes. *J. Appl. Meteor.*, **15**, 657–661.

Coakley, J.A. and P. Chýlek (1975). The two-stream approximation in radiative transfer: Including the angle of the incident radiation. *J. Atmos. Sci.*, **32**, 409–418.

Davis, A. and A. Marshak (1997). Lévy kinetics in slab geometry: Scaling of transmission probability. In *Fractal Frontiers*. M.M. Novak and T.G. Dewey (eds.). World Scientific, Singapore, pp. 63–72.

Davis, A.B. and A. Marshak (2001). Multiple scattering in clouds: Insights from three-dimensional diffusion/P_1 theory. *Nuclear Sci. and Engin.*, **137**, 251–280.

Davis, A.B. and A. Marshak (2004). Photon propagation in heterogeneous optical media with spatial correlations: Enhanced mean-free-paths and wider-than-exponential free-path distributions. *J. Quant. Spectrosc. Radiat. Transfer*, **84**, 3–34.

Davis, A., P.M. Gabriel, S.M. Lovejoy, D. Schertzer, and G.L. Austin (1990). Discrete angle radiative transfer III: Numerical results and meteorological applications. *J. Geophys. Res.*, **95**, 11,729–11,742.

Dirac, P.A.M. (1977). In *History of Twentieth Century Physics*. C. Weiner (ed.). Academic Press, New York (NY).

Evans, K.F. (1993). A general solution for stochastic radiative transfer. *Geoph. Res. Lett.*, **20**, 2075–2078.

Fu, Q. and K.-N. Liou (1992). On the correlated-k distribution method for radiative transfer in inhomogeneous atmospheres. *J. Atmos. Sci.*, **49**, 2139–2156.

Gabriel, P.M. and K.F. Evans (1996). Simple radiative transfer methods for calculating domain-averaged solar fluxes in inhomogeneous clouds. *J. Atmos. Sci.*, **53**,
858–877.

Gabriel, P.M., S.M. Lovejoy, A. Davis, D. Schertzer, and G.L. Austin (1990). Discrete angle radiative transfer II: Renormalization approach for homogeneous and
fractal clouds. *J. Geophys. Res.*, **95**, 11,717–11,728.

Joseph, J.H., W.J. Wiscombe, and J.A. Weinman (1976). The delta-Eddington approximation for radiative flux transfer. *J. Atmos. Sci.*, **33**, 2452–2459.

Kassianov, E. (2003). Stochastic radiative transfer in multilayer broken clouds – Part
I: Markovian approach. *J. Quant. Spectrosc. Radiat. Transfer*, **77**, 373–393.

Kassianov, E., T.P. Ackerman, R. Marchand, and M. Ovtchinnikov (2003). Stochastic
radiative transfer in multilayer broken clouds – Part II: Validation tests. *J. Quant.
Spectrosc. Radiat. Transfer*, **77**, 395–416.

Khairoutdinov, M.F. and D.A. Randall (2001). A cloud-resolving model as a cloud
parameterization in the NCAR Community Climate System Model: Preliminary
results. *Geophys. Res. Lett.*, **28**, 3617–3620.

Lacis, A.A. and V. Oinas (1991). A description of the correlated-k method for modeling nongrey gaseous absorption, thermal emission, and multiple scattering in
vertically inhomogeneous atmospheres. *J. Geophys. Res.*, **96**, 9027–9063.

Lane, D.E., K. Goris, and R.C.J. Somerville (2002). Radiative transfer through broken cloud fields: Observations and model validation. *J. Climate*, **15**, 2921–2933.

Li, J. (2002). Accounting for unresolved clouds in a 1D infrared radiative transfer model. Part I: Solution for radiative transfer, scattering, and cloud overlap.
J. Atmos. Sci., **59**, 3302–3320.

Li, J. and H.W. Barker (2002). Accounting for unresolved clouds in a 1D infrared radiative transfer model. Part II: Horizontal variability of cloud water path.
J. Atmos. Sci., **59**, 3321–3339.

Li, J. and Q. Fu (2000). Absorption approximation with scattering effect for infrared
radiation. *J. Atmos. Sci.*, **57**, 2905–2914.

Lohmann, U. and J. Feichter (1997). Impact of sulfate aerosols on albedo and lifetime of clouds: A sensitivity study with the ECHAM4 GCM. *J. Geophys. Res.*,
102, 13,685–13,700.

McFarlane, N.A., G.J. Boer, J.-P. Blanchet, and M. Lazare (1992). The Canadian
Climate Centre second-generation general circulation model and its equilibrium
climate. *J. Climate*, **5**, 1013–1044.

Meador, W.E. and W.R. Weaver (1980). Two-stream approximations to radiative
transfer in planetary atmospheres: A unified description of existing methods and a
new improvement. *J. Atmos. Sci.*, **37**, 630–643.

Min, Q.-L. and L.C. Harrison (1999). Joint statistics of photon pathlength and cloud
optical depth. *Geophys. Res. Lett.*, **26**, 1425–1428.

Newman W.I., J.K. Lew, G.L. Siscoe, and R.G. Fovell (1995). Systematic effects of
randomness in radiative transfer. *J. Atmos. Sci.*, **52**, 427–435.

Oreopoulos, L. and H.W. Barker (1999). Accounting for subgrid-scale cloud variability in a multi-layer, 1D solar radiative transfer algorithm. *Quart. J. Roy.
Meteor. Soc.*, **125**, 301–330.

Petty, G.W. (2002). Area-average solar radiative transfer in three-dimensionally in-homogeneous clouds: The independently scattering cloudlets model. *J. Atmos. Sci.*, **59**, 2910–2929.

Pfeilsticker, K. (1999). First geometrical pathlengths probability density function derivation of the skylight from spectroscopically highly resolving oxygen A-band observations. 2. Derivation of the Lévy-index for the skylight transmitted by mid-latitude clouds. *J. Geophys. Res.*, **104**, 4101–4116.

Pincus, R., S.A. MacFarlane, and S.A. Klein (1999). Albedo bias and the horizontal variability of clouds in subtropical marine boundary layers: Observations from ships and satellites. *J. Geophys. Res.*, **104**, 6183–6191.

Pincus, R., H.W. Barker, and J.-J. Morcrette (2003). A new radiative transfer model for use in GCMs. *J. Geophys. Res.*, **108**, 4376–4379.

Räisänen, P. and H.W. Barker (2005). Evaluation and optimization of sampling errors for the Monte Carlo independent column approximation. *Quart. J. Roy. Meteor. Soc.*, in press.

Räisänen, P., G.A. Isaac, H.W. Barker, and I. Gultepe (2003). Solar radiative transfer for stratiform clouds with horizontal variations in liquid water path and droplet effective radius. *Quart. J. Roy. Meteor. Soc.*, **129**, 2135–2149.

Räisänen, P., H.W. Barker, M.F. Khairoutdinov, and D.A. Randall (2005a). Stochastic generation of subgrid-scale cloudy columns for large-scale models. *Quart. J. Roy. Meteor. Soc.*, **130**, 2047–2067.

Räisänen, P., H.W. Barker, and J.N.S. Cole (2005b). The Monte Carlo independent column approximation's conditional random noise: Impact on simulated climate. *J. Climate*, in press.

Randall, D., M. Khairoutdinov, A. Arakawa, and W. Grabowski (2003). Breaking the cloud parameterization deadlock. *Bull. Amer. Meteor. Soc.*, **84**, 1547–1564.

Ronnholm, K., M.B. Baker, and H. Harrison (1980). Radiation transfer through media with uncertain or random parameters. *J. Atmos. Sci.*, **37**, 1279–1290.

Rosenbaum, S. (1971). The mean Green's function: A nonlinear approximation. *Radio Sci.*, **6**, 379–386.

Rossow, W.B., C. Delo, and B. Cairns (2002). Implications of the observed mesoscale variations of clouds for the earth's radiation budget. *J. Climate*, **15**, 557–585.

Samorodnitsky, G. and M.S. Taqqu (1994). *Stable Non-Gaussian Random Processes.* Chapman and Hall, New York (NY).

Schertzer, D. and S. Lovejoy (1987). Physical modeling and analysis of rain and clouds by anisotropic scaling multiplicative processes. *J. Geophys. Res.*, **92**, 9693–9714.

Schuster, A. (1905). Radiation through a foggy atmosphere. *Astrophys. J.*, **21**, 1–22.

Smith, R.N.B. (1990). A scheme for predicting layer clouds and their water content in a GCM. *Quart. J. Roy. Meteor. Soc.*, **116**, 435–460.

Stephens, G.L. (1988a). Radiative transfer through arbitrary shaped optical media, I: A general method of solution. *J. Atmos. Sci.*, **45**, 1818–1836.

Stephens, G.L. (1988b). Radiative transfer through arbitrary shaped optical media, II: Group theory and simple closures. *J. Atmos. Sci.*, **45**, 1837–1848.

Stephens, G.L., P.M. Gabriel, and S.-C. Tsay (1991). Statistical radiative transfer in one-dimensional media and its application to the terrestrial atmosphere. *Transp. Theory and Stat. Phys.*, **20**, 139–175.

Szczap, F., H. Isaka, M. Saute, B. Guillemet, and A. Iotukhovski (2000a). Effective radiative properties of bounded cascade absorbing clouds: Definition of the equivalent homogeneous cloud approximation. *J. Geophys. Res.*, **105**, 20,617–20,634.

Szczap, F., H. Isaka, M. Saute, B. Guillemet, and A. Iotukhovski (2000b). Effective radiative properties of bounded cascade nonabsorbing clouds: Definition of an effective single-scattering albedo. *J. Geophys. Res.*, **105**, 20,635–20,638.

Tiedtke, M. (1996). An extension of cloud-radiation parameterization in the ECMWF model: The representation of subgridscale variations in optical depth. *Mon. Wea. Rev.*, **124**, 745–750.

Tompkins, A.M. (2002). A prognostic parameterization for the subgrid-scale variability of water vapor and clouds in large-scale models and its use to diagnose cloud cover. *J. Atmos. Sci.*, **59**, 1917–1942.

Wiscombe, W.J. and G.W. Grams (1976). The backscattered fraction in two-stream approximations. *J. Atmos. Sci.*, **33**, 2440–2451.

7

3D Radiative Transfer in Stochastic Media

N. Byrne

7.1 **Introduction** 385

7.2 **Line Statistics** 390

7.3 **Stochastic Transport Models** 400

7.4 **Validation** 413

7.5 **Conclusions** 420

References 421

Suggestions for Further Reading 424

7.1 Introduction

The purpose of this chapter is to provide a short introduction to the field of radiative transfer through media whose properties are known only probabilistically, to discuss its applications to problems of interest to atmospheric scientists, to present an introduction to the minimal amount of line statistics needed to describe the theory, to review the Levermore-Pomraning model, and to introduce two promising new studies.

7.1.1 Problem Description

Radiative transfer is a well-defined field, even apart from being the subject of this book. For this section we are going further, assuming that it is a solved problem, with analytic techniques, numerical schemes, and even general rules available to give us the predictions of the theory, with perhaps some tradeoff between desired accuracy and the cost of obtaining it. Climatology and a glance at the sky might serve if all we need is to see if it's sunny enough to go to the beach, a few-term parameterization may serve a crude global climate model in a developmental stage, while numerical codes such as as SHDOM (Evans, 1998) or Monte Carlo codes (e.g., O'Hirok and Gautier, 1998) can provide more accuracy. In certain cases where optical depths are everywhere large, analytic solutions of the diffusion equation can be the most accurate of all. We realize this is offensively crude to those who specialize in the problem of actually obtaining solutions but it will clarify this section if we assume it true for the moment.

The common term "stochastic radiation transfer" is, strictly, a misnomer. The situation we discuss is the problem of finding the average radiative properties of an "ensemble" (a generally infinite set) of problems, by means other than actually solving each problem and averaging the results. Each problem is assumed to be well defined with its properties and boundary conditions exactly specified. We ask that this set of problems be described by a reasonable statistical law, where reasonable can be defined more rigorously than we shall do herein. Since we assume that each problem has an obtainable solution, this average should exist. By a further restriction of the set of problems we will require sufficient continuity that the average is representative. This excludes for example sets for which the answers are all either one or zero so that the average of one half is unlike any particular solution.

7.1.2 Relevance

It has been observed (Cess et al., 1989) that the interaction of clouds with radiation is one of the dominant determinants of the global climate. This is true both for the source of energy (sunlight, shortwave) and for the losses thereof (thermal emission to space, longwave). An ordinary cloud can reflect up to 90% of the sunlight impinging on it, and can reduce the thermal losses through it by half.

A large volume of data is required for a detailed description of all the optical properties of the atmosphere, yet little is routinely available, although large efforts are been made to gather more in the US Department of Energy's Atmospheric Radiation Measurement (ARM) program. Predicting such properties in detail is a challenging if not impossible task, and in the context of an operational Atmospheric General Circulation Model (AGCM) it is unimaginable. We may hope for some cell-dependant information about liquid water content, ice, and dust, and their vertical distribution, but very little horizontal detail on a scale smaller than the grid size. There is a large ratio between the horizontal and vertical resolution of an AGCM cell and the typical distance between two nearby water droplet, which we take as a lower bound below which more detail does not much affect the solution.

Do we need that much accuracy and detail? It is useful to have in order to validate numerical models of 3D radiation transport, but our conjecture is that we do not need it for the humbler but practical purpose of improving our larger scale understanding of the climate. We assume that the purposes of an AGCM would be served if we could provide reasonable correlations between the available cell data and the resulting radiation field and its effects. Levermore et al. (1988) speculated that the predictions of stochastic radiation theory (for a certain class of statistics) may depend mainly on only two parameters – in their case the mean and variance of the chord length statistics. If true, we can hope that such simple data could be predicted by AGCMs on a cell-by-cell basis and would be useful.

Then we turn the large size of the AGCM cell to advantage by assuming that it can be subdivided into an infinite number of uncorrelated subcells, each populated by optical properties having the same statistics, those of the overall cell, or at least drawing upon its similarity to such a stochastic radiation transport problem.

An alternative approach is to regard each AGCM time step as comprising a large number of small time steps, in each of which the 3D configuration of cloud is different, and to argue that the time average of these is a stochastic problem.

These schemes will work if just the average albedo or transmissivity is required for a determination of the large scale dynamics of the Earth's atmospheric transport and energy balance, if there are enough subcells or subtimesteps, and if they are sufficiently uncorrelated. It may or may not improve the overall calculation in practice. A recent study (Lane, 2000) failed to demonstrate such improvement, but was inconclusive because in limitations in data gathering.

7.1.3 Historical Development

Exact Solutions

The idea of developing a radiation model that could find the average solution (at suitably low computational cost) to a problem whose details were only known statistically has been a goal for some time. For the case of light impinging on a source-free non-scattering Markovian stochastic set of problems, the exact result is that the average intensity $\hat{I}(s)$ declines with distance s into the medium as a sum of two exponentials:

$$\hat{I}(s) = \hat{I}(0) \left[f \, \exp(-s/l_1) + (1 - f) \, \exp(-s/l_2) \right] , \qquad (7.1)$$

where f, l_1, and l_2 are simple combinations of the four independent parameters of the problem:

- two *optical* parameters, the specific absorptivities of the two materials;
- two *structural* parameters, the mixing ratio and correlation scale of one of the two materials.

This result allows us to evaluate several other plausible approaches to the problem (that may apply in more generality) by comparing their predictions to the above.

Figure 7.1 shows a comparison of the exact solution with two other models. The atomic mix and fractional cloud cover models are of general interest. The former refers to the result of calculating the optical properties of a uniform medium made of the materials in the stochastic mix, each in their average quantities. FCC refers to the common stratagem in AGCMs of combining the results of two homogeneous problems according to the expected area fraction of each material in the actual stochastic mixture. In practice it is extended to acknowledge the effect of cloud overlap, as proposed for example by Geleyn and Hollingsworth (1979), although all other cloud-cloud interactions are ignored.

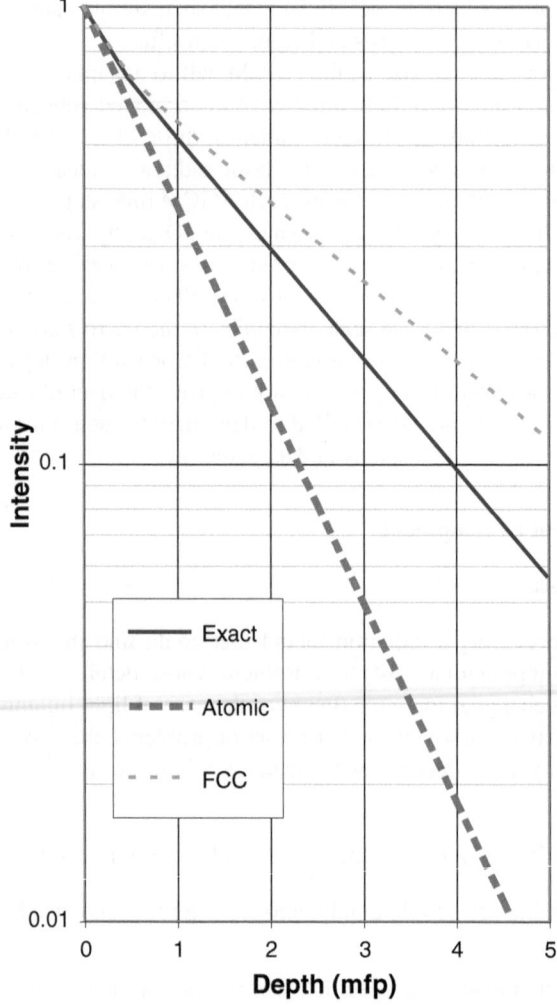

Fig. 7.1. Comparisons of various approximate solutions ("Atomic" refers to an atomic mix and "FCC" to fractional cloud cover) to an exact result

The graph presents the transmissivity as a function of (average) mean-free-path for a problem with a mixture of fairly opaque but small chunks in a sea of fairly transparent material. The point is that the availability of an exact solution, rare in this class of problems, provides a useful benchmark against which more general techniques can be measured. The actual numbers for the problem in hand are given in Table 7.1.

The results presented are very typical of the situation. The atomic mix model underpredicts transmission because there are no holes or thin spots; the FCC model

Table 7.1. Parameters used for Fig. 7.1. Note that not all are independent, as explained in Sect. 7.2.5

Material	#1	#2
Absorption	4	0.4
Fraction	17%	83%
Size	1	5

overpredicts it because it maximizes the effect of holes by aligning them. The actual result is between these rather wide limits.

Early Attempts

Many early treatments such as those of Vainikko (1973), Avaste et al. (1974) and Avaste and Vainikko (1974) considered a homogeneous Markovian distribution of clouds, which as we shall see is not a bad approximation, but then proceeded by assuming that the entire process was Markovian, which is not so in the presence of scattering.[1] Their results make use of the exact solution described above for the case without scattering.

Homogeneous mixtures are those for which statistics of the medium are everywhere the same, even though the actual realization (which material exists at which point in space) varies from point to point. Markovian statistics are those for which the probability of a transition out of one material as one moves along a line are independent of the position on the line. For the case here it implies an exponential distribution of cloud chord lengths and of cloud separations, perhaps with a different characteristic length from that of the chord distribution.

A General Formulation

The first mathematically rigorous treatment of the problem known to me is that of Levermore at Lawrence Livermore National Laboratory (and now U. of Maryland), Pomraning at UCLA, and their associates (Levermore et al., 1986). They were motivated to find an accurate transport description of a two-fluid turbulent mixture to support the laser fusion design codes of the inertially confined fusion program, but realized such a technique would have much wider application. These first results were limited to a homogeneous Markovian mixture of purely absorbing materials, but further progress would soon follow.

Pomraning's (1986) early review of the field gives an excellent description of the state of the art at that time, his later book on the subject (Pomraning, 1991b) brought the reader up to that date, and a final review article (Pomraning, 1998) updates the reader as to the state of the art in 1998. All three of these reviews are highly recommended to the reader.

---

[1] Titov (1990) later revisited and generalized their work.

7.2 Line Statistics

7.2.1 General Concepts

Much stochastic transport work uses the transport equation, although the diffusion model is often a better starting place for optically thick regions such as those often found in cloud. That said, a review of the statistics appropriate to an ensemble of lines is in order, since the radiation transport equation is a statement of balance along a line. There are good treatments of this subject by Sanchez et al. (1994) and in Pomraning's book (Pomraning, 1991b) .

Consider a set of realizations, each consisting of a random distribution of two materials over a line. The two materials are placed at random on the line in accordance with some specified probability density function. Every member of the set has one of these materials associated with every point on the line. For example, one such may have the interval (x, y) be all gray, in another this same interval may be white, in still another it may start out gray, become white, revert to gray, etc. Although everything in this section is applicable to any pair of materials, we will naturally be thinking of gray as cloud and white as clear sky. Figure 7.2 shows an example, a subgroup of 50 members of a set of realizations of a distribution of gray and white intervals. The statistics are such that white portions are four times as likely to occur as gray ones. The whole real line can be taken to be populated for each realization, but in order to save paper we show only the interval from zero to five.

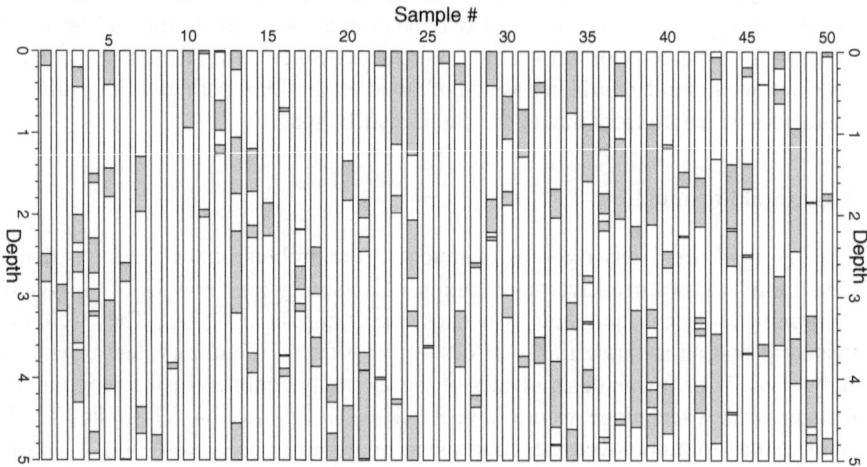

Fig. 7.2. 50 realizations of a given statistics

7.2.2 Two-Point Functions

The basic statistical properties of the gray material are illustrated by the cartoons in Fig. 7.3. The figure illustrates the section of the line between x and y, with additional

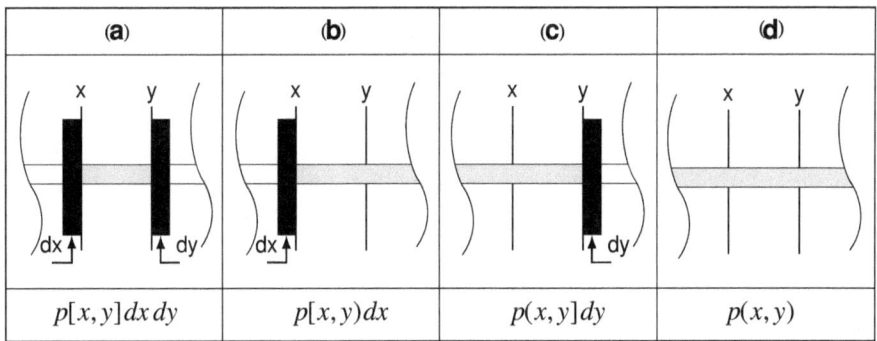

Fig. 7.3. Basic statistical properties for binary systems on a line

matter (of unspecified but positive lengths) to the left of x and right of y. Panel **(a)** shows a length of gray that definitely fills the entire interval between x and y. It starts somewhere in the ambiguous area of width dx to the left of x, and terminates in a similarly ambiguous area of width dy to the right of y. The probability of finding such an event depends upon x and y in general, and the further apart they are the less likely one is to find such an event. It is obviously proportional to the size of the areas in which the transitions occur: the probability of something happening exactly at a point is zero. With a little space to catch a transition we can expect to find some. For example, in Fig. 7.2 there are two samples (39 and 48) that fill the interval between depths 1 and 2. #39 drops out as we tighten dx to less than 0.11 because it drops over the left then. #48 drops out first as we tighten dy to less than 0.4, and by the time we shrink much more we don't have any samples left out of our initial 50. But of course 50 is not that large a number. It looks as though we were getting 2 out of 50 for dx and dy in the range of a few tenths. Had we looked at 5000 samples we might well have had 200 samples in the same net that caught #39 and #48 and we could have reduced dx and dy further and been able to find an estimate of the rate per unit dx and unit dy.

The name given to the probability density of finding an uninterrupted run of gray material that starts precisely at x and stops precisely at y is $p[x,y]$. The fraction of samples, out of a very large number, that is expected to be found meeting this criterion, given small transition regions dx and dy, is found by inspecting panel **(a)** of Fig. 7.3. In more mathematical terms, let:

$\Pr[x, y; \delta x, \delta y] =$ the probability that
    1) the material left of $x - \delta x$ is not gray
    2) AND the material at x, at y, and in between is gray
    3) AND the material beyond $y + \delta y$ is not gray.

Then:

$$p[x,y] = \lim_{\delta x, \delta y \to 0} \frac{\Pr[x, y; \delta x, \delta y]}{\delta x\, \delta y} . \tag{7.2}$$

For convenience we will restrict our discussion to sets where this probability is an integrable function and which contain no members with a finite number of transitions over an infinite range. Sanchez et al. (1994) is a bit freer with both of these restrictions and also treats an arbitrary number of materials.

Other interesting probabilities are related to the basic one. The probability density $p[x, y)$ of finding a strip of gray material that starts **at** x and continues **past** y is illustrated in panel **(b)** of Fig. 7.3 and defined by first setting:

$\Pr[x, y; \delta x) =$ the probability that
  1) the material to the left of $x - \delta x$ is not gray
  2) AND the material at x, at y, and in between is gray.

Then we find:

$$p[x, y) = \lim_{\delta x \to 0} \frac{\Pr[x, y; \delta x)}{\delta x} . \tag{7.3}$$

It is clear from the figure that this is the sum over all values of y of the stopping points; integration gives it in terms of our basic p:

$$p[x, y) = \int_y^\infty p[x, z] \, dz . \tag{7.4}$$

Here and in what follows we use a bracket " [" to indicate a point boundary ("starts **at**") and a parenthesis " (" to indicate continuation ("begins **before**"), and similarly for right designators.

There is a similar result involving a strip of gray starting **before** x and stopping **at** y, as indicated in panel **(c)** of Fig. 7.3. In this case we include all possible starting points:

$$p(x, y] = \int_{-\infty}^x p[w, y] \, dw . \tag{7.5}$$

If we don't care where it starts or stops, just that the interval (x, y) is all gray, then we have the situation pictured in panel **(d)** of Fig. 7.3, which includes all possible starting and stopping points:

$$p(x, y) = \int_{-\infty}^x \int_y^\infty p[w, z] \, dw \, dz , \tag{7.6a}$$

$$= \int_{-\infty}^x p[w, y) \, dw , \tag{7.6b}$$

$$= \int_y^\infty p(x, z] \, dz . \tag{7.6c}$$

The last two forms follow from the definitions of the respective integrands.

All the above four functions contain exactly the same information and can be transformed one into the other by differentiation and/or integration.

Changes

As can be seen by taking derivatives of the quantities of (7.6), the one-sided densities give the rates of change of the interval probability. The probability of finding an uninterrupted run of gray between x and y clearly increases as x increases (moves closer to y), because we include all the runs from the old x and add newly started ones. The formal statement of this is:

$$\partial_x p\,(x, y) = p[x, y]\,. \tag{7.7}$$

Of course, if we instead let y increase (move further from x) then p decreases because we start excluding runs as they drop out:

$$\partial_y p\,(x, y) = -p(x, y]\,. \tag{7.8}$$

The two sided, basic, density is the rate of change of either of the one-sided ones, to within a sign:

$$\partial_y p[x, y) = -p[x, y]\,, \tag{7.9a}$$
$$\partial_x p(x, y] = p[x, y]\,. \tag{7.9b}$$

7.2.3 One-Point Functions

The two-point probabilities can immediately be contracted to one-point versions by letting y approach x. The role of y was that the gray matter had to extend to it, at a minimum. Therefore setting y to x removes any constraint on the right.

Panel (**a**) of Fig. 7.4 illustrates the meaning of $p[x]$. The formal definition is:

$$p[x] = p[x, x]\,. \tag{7.10}$$

This gives us the creation rate of zero-length gray stuff. Such lengths are an anti-glue and prevent the coalescence of two abutting runs of white matter. In panel (**b**) we see the definition of $p[x)$:

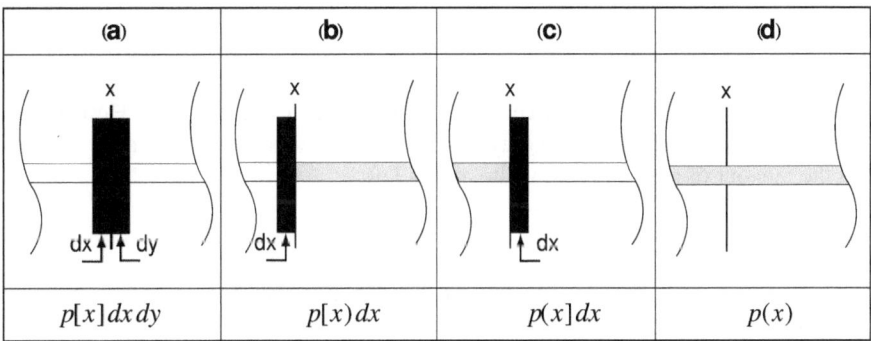

Fig. 7.4. One-point functions on a line

$$p[x] = p[x, x] \; . \tag{7.11}$$

This is the transition rate into gray **at** x without regard to what happens on the right: all lengths are accepted. We can relate this to the primitive function:

$$p[x] = \int_x^{\infty} p\,[x, z] \; \mathrm{d}z \; . \tag{7.12}$$

As shown in panel (**c**), we can also compute the transition rate out of gray **at** x:

$$p(x] = p(x, x] = \int_{-\infty}^{x} p\,[w, x] \; \mathrm{d}w \; . \tag{7.13}$$

Now we come to the simplest of all, as shown in panel (**d**). It is the probability that the point at x itself is gray:

$$p\,(x) = p\,(x, x) \; . \tag{7.14}$$

What About White?

We have so far only considered the statistics of the gray matter. There is a similar set for the white matter, too, which we will indicate by using q instead of p in our functions. Because there are only two materials we must have:

$$p\,(x) + q(x) = 1 \; . \tag{7.15}$$

Symmetry

If the statistics are not stationary (that is, if they differ from place to place) then in general for nonzero l the transitions into a run of l or better have no relation to the transitions out after a run of l or better:

$$p[x, x + l) \neq p(x - l, x] \tag{7.16a}$$
$$p[l) \qquad \neq p(l] \; , \tag{7.16b}$$

because there is no reason to expect symmetry about a general point. However, the transition rate out of white into gray, going to the right, is of course exactly the same as the rate out of gray and into white, going left:

$$q(x] = p[x) \; . \tag{7.17}$$

The same thing holds if the role of the two materials is reversed. p and q are independent, apart from the above constraints.

Changes

Even for $l = 0$ it is not necessary that the transition rates *into* a material balance those *out* of it, though. In fact it is this imbalance that gives rise to changes in the local probability $p(x)$: we could well be in a region where the probability of gray matter at x, $p(x)$, changes with x. In that case the entering and leaving rates will not be equal. The one-point analogue of (7.7) and (7.8) tells us exactly this:

$$\partial_x p(x) = p[x) - p(x] . \tag{7.18}$$

We can combine the two-point and one-point functions in a number of interesting ways. One such way is to recognize that the rate of transitions out of gray, going to the right, must surely be proportional to the probability that a point is gray in the first place (how else could it change?) so the ratio of these two is what is really significant. This ratio is the probability per unit length of a right-going gray stretch ending; the conditional probability that the gray stretch ends (per unit length) given that one is in gray:

rate of leaving gray $= p(x]$
probability of being in gray $= p(x)$

rate of leaving gray conditioned on being in gray $= \dfrac{p(x]}{p(x)} .$

It is convenient to define the scale length λ to be the inverse of this rate:

$$\lambda_p(x] \equiv \frac{p(x)}{p(x]} . \tag{7.19}$$

Since λ could depend upon just about anything we haven't learned a lot, but with it we can write fine-looking equations such as:

Probability of rightward leaving a gray stretch in $dx = p(x)\dfrac{dx}{\lambda_p(x]} . \tag{7.20}$

By substituting q for p in the above we get the probability of crossing out of a white stretch in length dx, headed right. This is of no small importance, since it is also the rate of entering gray stretches as one moves right. Thus we have found:

Probability of crossing into a gray stretch in $dx = q(x)\dfrac{dx}{\lambda_q(x]} . \tag{7.21}$

We could repeat all the above for the case of leftward motion, should we care.

Correlation Scale

In the general case the notion of the average length of a stretch of gray is not really well defined. We could refer to the average length of those stretches that begin at x, or are centered at x, or just about any other thing, and they could well all be different.

Say that for some reason we want the average length of stretches that end at x. We compute this by summing the length times the probability of said length, given that any length ends at x:

$$\ell_p(x) = \frac{\int_{-\infty}^{x} (x - w) \, p[w, x] \, dw}{\int_{-\infty}^{x} p[w, x] \, dw} . \tag{7.22}$$

Using the definitions in (7.4), (7.5), (7.6) and (7.19) and integrating by parts, we find that this is related to the scale length for changing out of gray at x, going right, plus another term which depends on the details of the distribution:

$$\ell_p(x) = \lambda_p(x) + \frac{\int_{-\infty}^{x} \{p(w, x) - p[w, x]\} \, dw}{p(x)} . \tag{7.23}$$

7.2.4 Homogeneity

Let us now consider homogeneous statistics,[2] a restricted branch of the general system above. All this means is that, on average, one position of the line is as good as another. One consequence is that we no longer need concern ourselves with constraints such as "right-going" or "left-going": homogeneity implies mirror symmetry.

Statistical properties can depend upon the separation of two or more points, however. Since we have already seen in the general case that just one function serves to determine the others we shall start with it and explicitly display the independence of position. We distinguish the homogeneous case by use of an "h" subscript, or in the short notation by the use of a capital letter. We suppose we are given the basic function P:

$$P[l] \equiv p_h[x, x + l] . \tag{7.24}$$

The other three two-point functions (which although functions of only one variable still involve two points, whose separation is l) follow by integration. The left-side starting density is the rate of finding a stretch of gray starting at x whose length is at least l:

$$p_h[x, x + l) = \int_{x+l}^{\infty} p_h[x, z] \, dz , \tag{7.25a}$$

$$= \int_{x+l}^{\infty} P[z - x] \, dz \tag{7.25b}$$

$$= \int_{l}^{\infty} P[z] \, dz . \tag{7.25c}$$

---

[2] "Homogeneity" is of course not to be confused here with uniformity (lack of spatial variability). It is the mathematically correct terminology to generalize for spatial statistics the concept of (statistical) "stationarity" used extensively in time-series analysis (a.k.a. stochastic processes theory), i.e., invariance of statistical properties by translation.

Naturally it does not depend on x at all.

With the reasonable notation:

$$P[l] \equiv \int_l^\infty P[z] \, dz \; , \qquad (7.26)$$

we have:

$$p_h[x, x+l] = P[l] \; . \qquad (7.27)$$

The right-side starting density works out to be the same, not the case in general as noted above:

$$p_h(y-l, y] = \int_{-\infty}^{y-l} p_h[w, y] \, dw \; , \qquad (7.28a)$$

$$= \int_{-\infty}^{y-l} P[y-w] \, dw \; , \qquad (7.28b)$$

$$= \int_l^\infty P[z] \, dz \; . \qquad (7.28c)$$

Merely for the sake of symmetry, we introduce $P(l]$:

$$P(l] \equiv \int_{-\infty}^l P[z] \, dz \qquad (7.29a)$$

$$= \int_l^\infty P[z] \, dz = P[l) \; . \qquad (7.29b)$$

This adds no new insight but lets us write:

$$p_h[x-l, x) = P(l] \; . \qquad (7.30)$$

The probability of finding a stretch of gray from x to $x+l$ is found by counting all possible starting places before x and stopping places beyond $x+l$:

$$p_h(x, x+l) = \int_{-\infty}^x \int_{x+l}^\infty P[z-w] \, dw \, dz$$

$$= \int_{-\infty}^x P[x+l-w) \, dw \qquad (7.31)$$

$$= \int_l^\infty P[w) \, dz \; .$$

Of course, we define:

$$P(l) \equiv \int_l^\infty P[w]\,dz\ .$$ (7.32)

And so finally:

$$p_h(x, x+l) = P(l)\ .$$ (7.33)

Again this does not depend upon x, as expected.

The corresponding one-point functions are found by evaluating the above functions at zero argument. As before, they are found by accepting all lengths greater than zero (which is of course all lengths). We have for the rate of generation of gray at x:

$$p_h[x] \equiv p_h[x, x]$$
$$= P[0]\ .$$ (7.34)

The rate of destruction of gray at x, similarly, is:

$$p_h(x] \equiv p_h(x, x]$$
$$= P(0]\ .$$ (7.35)

We note that the left-going and right-going transition rates out of gray are equal (they ought to be or else the gray level would change) and constant:

$$p_h[x] = p_h(x] = P[0] = P(0]\ .$$ (7.36)

The probability of finding gray at a point, also a constant, is:

$$p_h(x) = p_h(x, x)\ .$$ (7.37)

We have computed this above. It is:

$$p_h(x) = P(0)\ .$$ (7.38)

Correlation Scale

The situation is much simpler in the case of homogeneity since nothing depends upon the location of the stretch. We have in that case, by simple substitution:

$$\lambda_p[x] = \int_x^\infty (y - x)\, \frac{p_h[x, y]}{\int_x^\infty p_h[x, z]\,dz}\,dy\ ,$$ (7.39a)

$$= \int_0^\infty u\, \frac{P[u]}{\int_0^\infty P[v]\,dv}\,du\ .$$ (7.39b)

This shows that, as expected, the average length does not in fact depend on x at all. A similar exercise shows that it does not depend on left-right direction, either, so:

$$\lambda_p \left[x\right] = \lambda_p(x) = \lambda_p .$$
(7.40)

Integration by parts gives:

$$\lambda_p = \frac{\int_0^\infty P(u] \, du}{\int_0^\infty P[u] \, du} ,$$
(7.41a)

$$= \frac{P(0)}{P(0]} .$$
(7.41b)

Changes

Now recall (7.19) and (7.20), which tell us that this very ratio is the scale length for leaving gray in homogeneous systems. This holds true regardless of the underlying statistical law governing the system. This is a very attractive feature of homogeneous statistics. In the general case the scale length for leaving not only can vary from place to place, but is not simply related to any particular moment of the distribution. In the homogeneous case it is constant, independent of position.

7.2.5 Remarkable Fact

There is a further simplification with homogeneous statistics. Trivial manipulation gives the rate of leaving gray as:

$$P(0] = \frac{P(0)}{\lambda_p} ,$$
(7.42)

Recall (7.23) and note that it also follows (by virtue of the mirror symmetry of homogeneous statistics) that the argument of the integral on the right hand side vanishes. Thus the mean chord length is not only a constant but is equal to the scale length for changing:

$$\ell_p = \lambda_p ,$$
(7.43)

and this is true for any homogeneous statistics, not just Markovian.

Consider white space now. Define Q functions to be the analogue of the P functions, except pertaining to white rather than gray. We immediately have the perfectly symmetric result that the rate of leaving white is:

$$Q(0] = \frac{Q(0)}{\lambda_q} .$$
(7.44)

Taking advantage of the fact that the entering and leaving rates are the same for homogeneous statistics, this implies the rate for leaving white is also known:

$$Q[0) = \frac{Q(0)}{\lambda_q} .$$
(7.45)

Recall (7.17), which stated that the rate of entering white is the same as the rate for leaving gray. In this context this means:

$$\frac{P(0)}{\lambda_p} = \frac{Q(0)}{\lambda_q} \ .$$

(7.46)

$P(0)$ and $Q(0)$ sum to one, as mentioned in (7.15), so we see:

$$P(0) = \frac{\lambda_p}{\lambda_p + \lambda_q} \qquad Q(0) = \frac{\lambda_q}{\lambda_p + \lambda_q} \ .$$

(7.47)

If one knows any two of the three quantities $(P(0), \lambda_p, \lambda_q)$ then one also knows the other one. This remarkable result follows from homogeneity and is without regard to the underlying statistics, which could be quite complicated.

For illustration, we return to Table 7.1 and Fig. 7.1 where we can take Material #1 as grey (dense/cloudy) and Material #2 as white (tenuous/clear). Then $P(0) \approx 0.17$ and $Q(0) = 1 - P(0) \approx 0.83$ follow from $\lambda_p/\lambda_q \equiv \lambda_1/\lambda_2 = 5/1$.

7.2.6 Summary

There are some key results which we must retain from this discussion for use below. They are, first:

$$\text{Probability of leaving a gray stretch in } dx = p(x)\frac{dx}{\lambda_p(x)} \ ;$$

(7.48)

second:

$$\text{Probability of entering a gray stretch in } dx = q(x)\frac{dx}{\lambda_q(x)} \ ;$$

(7.49)

third, for homogeneous statistics:

$$\lambda_p(x) = \lambda_p[x) = \lambda_p = \text{constant} \ ,$$

(7.50a)

$$\lambda_q(x) = \lambda_q[x) = \lambda_q = \text{constant} \ ;$$

(7.50b)

and fourth, for homogeneous statistics:

$$P(0) = \frac{\lambda_p}{\lambda_p + \lambda_q} \qquad Q(0) = \frac{\lambda_q}{\lambda_p + \lambda_q} \ .$$

(7.51)

7.3 Stochastic Transport Models

7.3.1 Perturbative Approaches

Many treatments of the problem are based on some variation of mean field theory, in which the average solution is given by a transport-like equation with source terms involving ensemble average cross-correlations of the stochastic terms. Pomraning (1986) and Cairns et al. (2000), for example, have used this approach.

The transport equation for the nth realization in the ensemble is:

$$\mathbf{\Omega} \cdot \nabla I_n\left(\boldsymbol{x}, \mathbf{\Omega}\right) + \sigma_n\left(\boldsymbol{x}\right) I_n\left(\boldsymbol{x}, \mathbf{\Omega}\right) - \sigma_{sn}\left(\boldsymbol{x}\right) \int_{4\pi} \Phi_n\left(\boldsymbol{x}, \mathbf{\Omega} \cdot \mathbf{\Omega}'\right) I_n\left(\boldsymbol{x}, \mathbf{\Omega}'\right) \, d\mathbf{\Omega}'$$
$$= S_n\left(\boldsymbol{x}, \mathbf{\Omega}\right), \qquad (7.52)$$

where, for this realization, $\Phi_n\left(\boldsymbol{x}, \mathbf{\Omega} \cdot \mathbf{\Omega}'\right)$ is the scattering phase function,[3] $\sigma_{sn}(\boldsymbol{x})$ is the variable scattering coefficient, and $S_n\left(\boldsymbol{x}, \mathbf{\Omega}\right)$ is the source term. Introduce notation for the transport operator L_n and its ensemble average L:

$$L_n[G] \equiv \mathbf{\Omega} \cdot \nabla G + \sigma_n G - \sigma_{sn} \int_{4\pi} \Phi_n\left(\boldsymbol{x}, \mathbf{\Omega} \cdot \mathbf{\Omega}'\right) G \, d\mathbf{\Omega}' \qquad (7.53)$$

$$L[G] \equiv \mathbf{\Omega} \cdot \nabla G + \sigma G - \sigma_s \int_{4\pi} \Phi\left(\boldsymbol{x}, \mathbf{\Omega} \cdot \mathbf{\Omega}'\right) G \, d\mathbf{\Omega}' . \qquad (7.54)$$

Note that (7.54) is just the statement of the average operator, which is not the operator whose inverse will yield the average solution of (7.53). We must work harder to find that one.

Let δL_n, δI_n, and δS_n stand for the deviations from the average:

$$\delta L_n[G] \equiv L[G] - L_n[G]$$
$$\delta I_n \equiv I - I_n$$
$$\delta S_n \equiv S - S_n . \qquad (7.55)$$

Then the transport equation can be written as:

$$(L - \delta L_n) L^{-1} L[I_n] = (S - \delta S_n) . \qquad (7.56)$$

Assuming an inverse:

$$L[I_n] = \left\{1 - \delta L_n \times L^{-1}\right\}^{-1} [S - \delta S_n] . \qquad (7.57)$$

Expand the inverse operator:

$$L[I_n] = \{1 + \left(\delta L_n \times L^{-1}\right)$$
$$+ \left(\delta L_n \times L^{-1}\right) \times \left(\delta L_n \times L^{-1}\right) + \ldots\} [S - \delta S_n] . \qquad (7.58)$$

Now form the ensemble average of the last equation. Those terms first order in the deviations from the average all vanish by construction. If the deviations are in some sense small, we can neglect all but the lowest order, the second. Let $\langle \cdot \rangle$ denote the ensemble average:

---

[3] In the rest of this volume, the scattering phase function is denoted $p(\boldsymbol{x}, \mathbf{\Omega} \cdot \mathbf{\Omega}')$ or $P(\cdots)$ but in this chapter p's and P's are reserved for probabilities.

$$L[I] \approx S + \langle \delta L_n \times L^{-1} [S - \delta S_n] \rangle + \langle \delta L_n \times L^{-1} \times \delta L_n \rangle \times L^{-1} [S] \ . \quad (7.59)$$

If we neglect the second order terms, we have simply $L[I] = S$, so

$$I = L^{-1} [S] \ . \tag{7.60}$$

To the same order of accuracy as already exists in (7.59), we can use this approximation to find:

$$\{L - \langle \delta L_n \times L^{-1} \times \delta L_n \rangle\} [I] \approx S + \langle \delta L_n \times L^{-1} [S - \delta S_n] \rangle \ . \tag{7.61}$$

We have thrown away many complicated terms to get this far, and it looks rather simple, but it is still a formidable job to solve it for I. For one thing, the operators involve integrations over space as well as angle so we now have an explicitly nonlocal equation. We can solve it for only a few special cases, among them Markovian statistics for the mixture and no scattering. Even then the solution is not generally satisfactory because it is not robust: it does not always give reasonable results. Compare the solution labeled "Small" in Fig. 7.5 to that labeled "Exact" in Fig. 7.1; both are for the same parameters. The "Small" solution is not even within the bounds of "FCC" and "Atomic."

Cairns (1992) has taken this approach farther than anyone else. He developed a formalism using Feynman diagrams and the techniques of renormalization group theory to handle the higher order terms in the general solution. He stresses the need to seek and display the irreducible terms in the formal solution, and suggests closures based on neglecting these. That is, a four-point correlation $C[x_1, x_2, x_3, x_4]$ can be written as a sum of products of two-point correlations plus a cumulant, and it is this that is to be neglected.

7.3.2 Levermore-Pomraning Theory

Graphical Derivation

There are many ways of deriving the Levermore-Pomraning (L–P) equation. The initial treatment was based on the theory of alternating renewal processes, as discussed by Vanderhaegen (1986), Levermore et al. (1988), and Pomraning (1989), among others. The derivation below follows Adams et al. (1989) and is simple, general, and exact.

Figure 7.6 shows, at left, part of a scene looking up over Oklahoma and at right as a model might see it. Suppose that the atmosphere is composed of just two materials, cloud and clear air. Even so restricted, the right panel is oversimplified because it is presented as if the vertical height of cloud was infinite. The actual view from the bottom of a binary mixture would have a gradation of density (not shown here). With the scene extended to a large horizontal distance, up to the top of the atmosphere, and down to the Earth's surface, and with knowledge of the surface albedo and emissivity and the solar angle, we can in principal compute the radiation intensity anywhere. In particular we are for some reason interested in the *average* intensity of radiation going east (to the right in the figure) at the spot of given altitude, latitude, and longitude

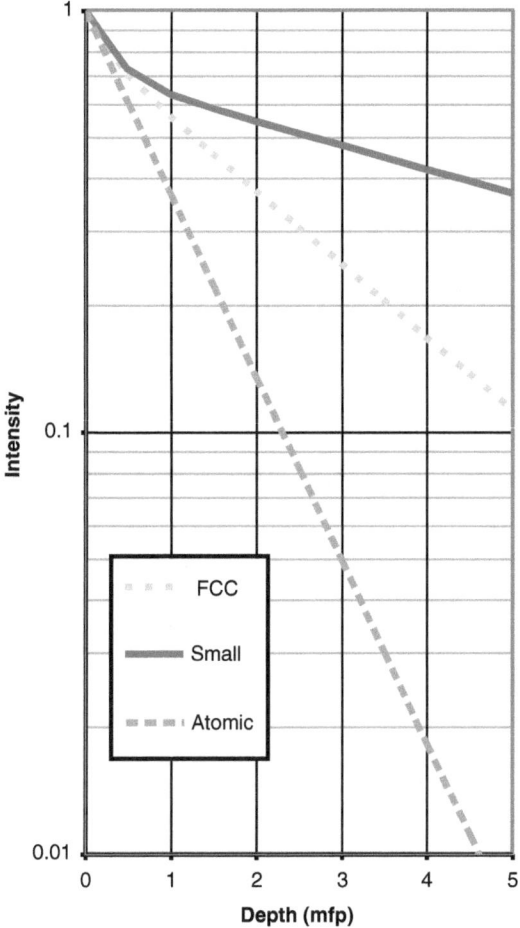

Fig. 7.5. "Small" is bad

indicated by the dot, which is to be understood as being of zero size. To simplify the discussion we will orient our coordinate system so that the origin is at the altitude, latitude, and longitude of the dot and the x axis is to the right in the figure. Then we can use this shorthand notation:

$$I(x) \equiv I(\mathbf{X} + x\mathbf{\Omega}, \mathbf{\Omega}) . \qquad (7.62)$$

Imagine we have data regarding the atmospheric, solar, and ground conditions going back many years, perhaps provided by a large government funded atmospheric radiation measurement program. This database is constructed from observations taken at a single spot and under conforming conditions (such as local liquid water content, solar angle, time of day, etc.). We will sharpen our predictive skill on this database so that we can use it to improve an AGCM model where the clouds occur

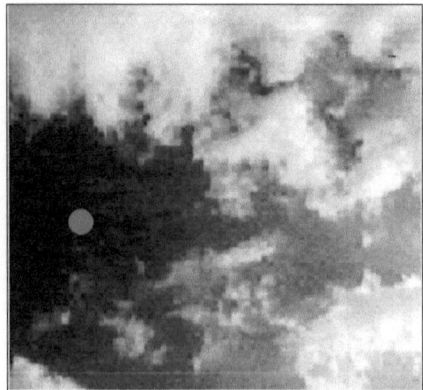 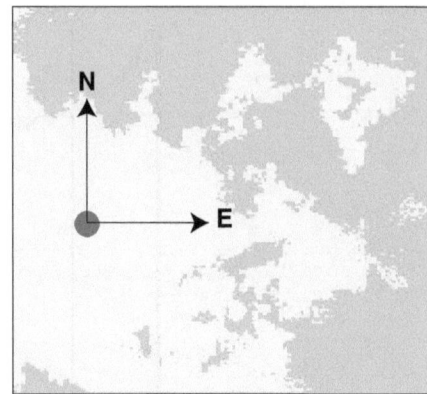

Fig. 7.6. A point in a clear region

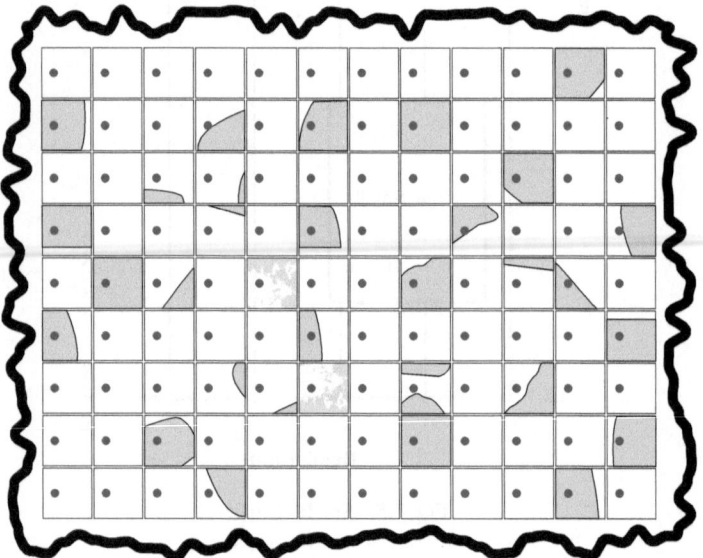

Fig. 7.7. Some of the many scenes in our database

in distributions known only statistically at best, yet we need an average answer as discussed above.

Our first task is to generate the appropriate line statistics from the database. This provides us with the statistics at every space point and for every direction.

Thus we have a (very) large number N of scenes, which we will number by n (going from 1 to N). Some of these are sketched in Fig. 7.7. We will have applied the appropriate boundary conditions and calculated or measured the intensity of radiation at x (the dot) in each sample. We can then estimate the average intensity of east-going (downstream) radiation at this point, $I(x)$, by adding up its value for each sample scene and dividing by the number of samples:

$$I(x) = \frac{1}{N} \sum_{1}^{N} I_n(x) .$$ (7.63)

It is convenient for the argument to consider separately the subset of scenes in the sum in which the point x (the dot) is in cloud, which we denote by $\{p\}$, and the subset in which it is not, denoted by $\{q\}$. Set $\{p\}$ is indicated in Fig. 7.8. Since $\{p\}$

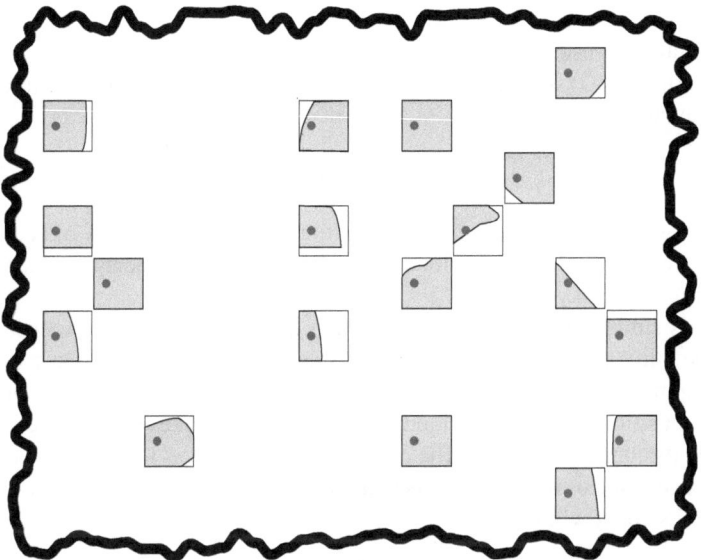

Fig. 7.8. Members of the set $\{p\}$ are in cloud at point x

and $\{q\}$ together make up the whole we have:

$$I(x) = \frac{1}{N} \left[\sum_{n \in \{p\}} I_n(x) + \sum_{n \in \{q\}} I_n(x) \right] .$$ (7.64)

Let $n_{\{p\}}$ be the number of scenes in which the point x is in cloud:

$$n_{\{p\}} = \sum_{n \in \{p\}} 1 ,$$ (7.65)

and notice that the average intensity, given that the point x is in cloud, $I_p(x)$, is just the average I in this set:

$$I_p(x) = \frac{1}{n_{\{p\}}} \sum_{n \in \{p\}} I_n(x) .$$ (7.66)

Do similarly for the set $\{q\}$, where the point x is *not* in cloud. Then the overall average intensity can be written:

$$I(x) = \frac{1}{N} \left[n_{\{p\}} I_p(x) + n_{\{q\}} I_q(x) \right] . \tag{7.67}$$

Assuming that N is large enough, the fraction of scenes for which the point x is in cloud can be replaced by the underlying statistical probability discussed above:

$$\lim_{N \to \infty} \frac{n_{\{p\}}}{N} = p(x) , \tag{7.68}$$

and similarly for $q(x)$.

Equation (7.67), in the limit of infinite N, thus splits the average intensity into its cloudy and clear portions:

$$I(x) = p(x) I_p(x) + q(x) I_q(x) . \tag{7.69}$$

The average intensity at a point (and direction $\mathbf{\Omega}$) is the probability of being in cloud times the average intensity in cloud, plus the probability of being in clear sky times the average clear sky intensity.

That's simple and obviously correct, but not yet informative. Proceed by doing the same thing at a new point x' a small distance dx downstream from x (to the right in the figures). We indicate this new position by the tip of an arrow extending from the start position. The new in-cloud subset would look like Fig. 7.9. We call this set $\{p'\}$ and its complement $\{q'\}$. The count of members of $\{p'\}$ will be called $n_{\{p'\}}$ and the average value of intensity for its members is $I_p(x')$. As in (7.68), the expectation value of its count will be:

$$\lim_{N \to \infty} \frac{n_{\{p'\}}}{N} = p(x') . \tag{7.70}$$

Most of the scenes in $\{p'\}$ appeared in our earlier subset $\{p\}$, because for small dx the point $x' = x + dx$ will likely still be in cloud if the point x was, but there are two little differences between the sets. Some of the scenes which were in cloud at x (and thus were in $\{p\}$) are in the clear at x' (and are in $\{q'\}$). Let the set of such be called $\{pq'\}$. Let $n_{\{pq'\}}$ be the count and $I_{\{pq'\}}(x)$ be the average intensity at point x in the set. There will also be the set for which the opposite is true, $\{qp'\}$. Its members are just those scenes which have cloud at x and clear sky at x'. Their count and average intensity will be denoted $n_{\{qp'\}}$ and $I_{\{qp'\}}(x)$.

We know how many scenes are to be expected in each of these new sets. It is just the number of transitions out of cloud in a distance dx, which we know from our discussion of line statistics in the previous section, (7.20) for example. Recall that our database has been mined to provide these statistics.

$$n_{\{pq'\}} \to p(x) \frac{dx}{\lambda_p(x)} \text{ as } N \to \infty . \tag{7.71}$$

The expected number in the other transition set is:

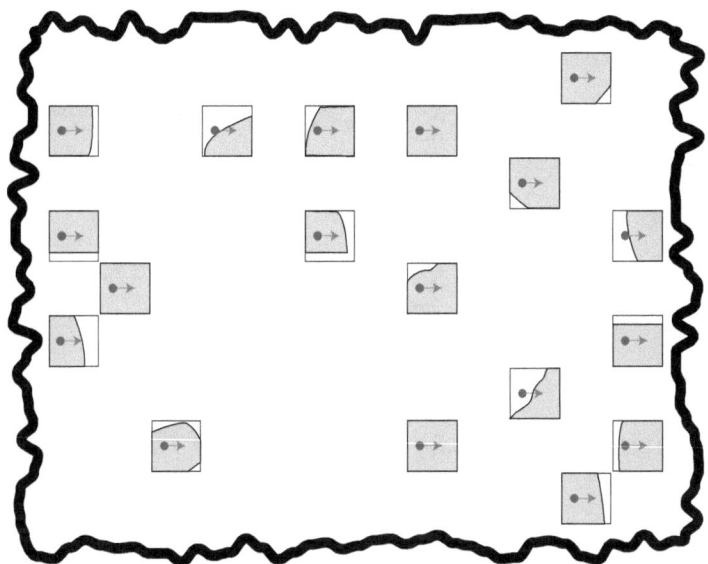

Fig. 7.9. Members of the set $\{p'\}$ are in cloud at point x'

$$n_{\{qp'\}} \rightarrow N\,q(x)\,\frac{\mathrm{d}x}{\lambda_q(x]} \text{ as } N \rightarrow \infty\,. \tag{7.72}$$

We choose $\mathrm{d}x$ small enough so that both of the above numbers are very small with respect to N, the total number of scenes in the database.

Define one last set $\{c\}$. It consists of all those scenes which are cloudy at *both* x and at x'. It is therefore the intersection of the two larger sets:

$$\{c\} = \{p\}\bigcap\{p'\}\,. \tag{7.73}$$

It will have its own count, $n_{\{c\}}$:

$$n_{\{c\}}(x) = \sum_{n\in\{c\}} 1\,. \tag{7.74}$$

For any x' the estimate of the two-point correlation between cloud at x and cloud at x' is $n_{\{c\}}/N$.

Now rewrite the formulas for the averages at the original point, x:

$$n_{\{p\}}\,I_p(x) = \sum_{n\in\{c\}} I_n(x) + \sum_{n\in\{pq'\}} I_n(x)$$

$$= \sum_{n\in\{c\}} I_n(x) + n_{pq'}\,I_{\{pq'\}}(x)$$

$$N\,p(x)\,I_p(x) = \sum_{n\in\{c\}} I_n(x) + N\,p(x)\,\frac{\mathrm{d}x}{\lambda_p(x]}\,I_{\{pq'\}}(x)\,. \tag{7.75}$$

The last equation follows by virtue of (7.68) and (7.71).

Do the same for the averages at x':

$$n_{\{p'\}} I_p(x') = \sum_{n\in\{c\}} I_n(x) + \sum_{n\in\{qp'\}} I_n(x')$$

$$= \sum_{n\in\{c\}} I_n(x') + n_{qp'} I_{\{qp'\}}(x)$$

$$N p(x') I_p(x') = \sum_{n\in\{c\}} I_n(x') + N q(x) \frac{dx}{\lambda_q(x)} I_{\{qp'\}}(x) . \qquad (7.76)$$

Subtract (7.75) from (7.76) and divide by N:

$$p(x') I_p(x') - p(x) I_p(x) = \frac{1}{N} \sum_{n\in\{c\}} [I_n(x') - I_n(x)]$$

$$+ q(x) I_{\{pq'\}}(x) / \lambda_q(x) \times \delta x$$

$$- p(x) I_{\{qp'\}}(x) / \lambda_p(x) \times \delta x . \qquad (7.77)$$

Recognize that $\delta x = |x' - x|$ is small; so for any $F(x)$:

$$F(x') - F(x) \approx \mathbf{\Omega} \cdot \nabla F(x) \times \delta x . \qquad (7.78)$$

Thus, taking the infinitessimal limit ($\delta x \mapsto dx$):

$$\mathbf{\Omega} \cdot \nabla (p(x) I_p(x)) \times dx = \frac{1}{N} \sum_{n\in\{c\}} \mathbf{\Omega} \cdot \nabla I_n(x) \times dx$$

$$+ q(x) I_{\{pq'\}}(x) / \lambda_q(x) \times dx$$

$$- p(x) I_{\{qp'\}}(x) / \lambda_p(x) \times dx . \qquad (7.79)$$

Look at the terms in the first sum on the right, which involve the small change in the intensity in a scene as the position changes slightly. Because of the way we have constructed the set $\{c\}$, all of its members are in cloud at both x and x'. Therefore the ordinary transport equation describes their local behavior, apart from a subset of members of $\{c\}$ which have a non-cloudy stretch between those points. For small enough dx the number of members in such a set approaches zero. That means that, for n in $\{c\}$, the local rate of change is given by:

$$\mathbf{\Omega} \cdot \nabla I_n(x) = -\sigma_n(x) I_n(x) + \sigma_{sn}(x) \int_{4\pi} \Phi_n I_n(x) \, d\mathbf{\Omega}' + S_n(x) . \qquad (7.80)$$

Assume that the optical properties of a scene depend upon position but have the same value in all the cloudy scenes at that position, and that the clear skies all share their own (different) common value at that point. The total cross-section, for example, is to be the same function of position in all cloudy scenes. This lets us collect the optical properties outside the first sum, since every term in the sum is in a cloudy scene:

$$\sum_{n\in\{c\}} \mathbf{\Omega}\cdot\nabla I_n(x) = -\,\sigma_p(x)\left[\sum_{n\in\{c\}} I_n(x)\right]$$

$$+\,\sigma_{sp}(x)\int_{4\pi}\Phi_p\left[\sum_{n\in\{c\}} I_n(x)\right]\mathrm{d}\Omega'$$

$$+\,S_p(x)\left[\sum_{n\in\{c\}} 1\right]. \tag{7.81}$$

We have written σ_p rather than $\sigma_{\{p\}}$ to emphasize the fact that $\sigma_n(x)$ is the same for all n in $\{p\}$; the same holds for all the optical properties.

All three sums on the right hand side of (7.81) over members of the set $\{c\}$ can be replaced by sums over the set $\{p\}$, since the two sets differ only by the fact that the latter includes a vanishingly small number of extra scenes and the summands are positive. After making this approximation and replacing the sums by the previously defined averages we have, to order $\mathrm{d}x$ and $1/N$:

$$\frac{1}{N}\sum_{n\in\{c\}} \mathbf{\Omega}\cdot\nabla I_n(x) = -\,\sigma_p(x)\,p(x)\,I_p(x)$$

$$+\,\sigma_{sp}(x)\int_{4\pi}\Phi_p\,p(x)\,I_p(x)\,\mathrm{d}\Omega'$$

$$+\,S_p(x)\,p(x). \tag{7.82}$$

Substitute this into (7.79) to find the L–P equation for the average intensity of radiation in cloud:

$$\mathbf{\Omega}\cdot\nabla\left[p(x)\,I_p(x)\right] = -\,\sigma_p(x)\,p(x)\,I_p(x) + p(x)\,S_p(x)$$

$$+\,\sigma_{sp}(x)\int_{4\pi}\Phi_p(x)\,p(x)\,I_p(x)\,\mathrm{d}\Omega'$$

$$+\,q(x)\,I_{\{pq'\}}(x)/\lambda_q(x)$$

$$-\,p(x)\,I_{\{qp'\}}(x)/\lambda_p(x). \tag{7.83}$$

Naturally there is also a completely symmetric equation for the average in clear sky with p and q interchanged.

These equations are exact, but they are incomplete; the last two average intensities on the right hand side are taken over different sets of scenes, and are unknown. They involve scenes in which the point in question is on the boundary of a cloud. We can proceed to write exact equations for these too, but they involve still more highly conditioned values. The series needs some sort of closure.

The simplest closure is to assume that the extra condition makes little difference. In Markov statistics this is true, because the basic transition probabilities are independent of current conditions. The resulting pair of equations is called the Levermore model; it is in fact an exact description of the situation in the case of Markov statistics without scattering and even with scattering if restricted to rod geometry (1D, only forward/backward directions). It is inexact elsewhere, but many benchmark calculations have demonstrated that it is a robust approximation to the more general 3D case.

Heuristic Extensions

Levermore et al. (1988) applied renewal theory to find the general solution to the case of decay of intensity as it penetrates a pure absorbing binary mixture of homogeneous statistical properties, with no sources and constant cross-sections. They found that the deep-in behavior could be captured by this simple prescription. Let a tilde denote the Laplace transform, so that:

$$\tilde{P}[s] \equiv \int_0^\infty P[l]\, e^{-s\,l} \mathrm{d}l$$

$$\tilde{Q}[s] \equiv \int_0^\infty Q[l]\, e^{-s\,l} \mathrm{d}l. \tag{7.84}$$

Form the correction terms:

$$I_p \equiv \frac{\lambda_p}{\sigma_p}\left[\frac{P(0)}{\tilde{P}[\sigma_p]} - 1\right]$$

$$I_q \equiv \frac{\lambda_q}{\sigma_q}\left[\frac{Q(0)}{\tilde{Q}[\sigma_q]} - 1\right]. \tag{7.85}$$

Then form the interface correction factor:

$$I_{\text{inter}} \equiv I_p + I_q - 1. \tag{7.86}$$

Now replace the interface-conditioned terms of the complete L–P equation as follows:

$$I_{\{qp'\}}(x) \rightarrow \frac{I_{\{q\}}(x)}{I_{\text{inter}}}$$

$$I_{\{pq'\}}(x) \rightarrow \frac{I_{\{p\}}(x)}{I_{\text{inter}}}. \tag{7.87}$$

Notice that for the case of Markov statistics the individual terms I_p and I_q are unity, so that if both the cloud and clear sky statistics are Markovian the overall correction factor I_{inter} is also unity. This is good, because this is known to be the correct result for this case.

Other closures have been proposed and evaluated by comparisons with the suite of 1D benchmark numerical calculations. Many of these have postulated algebraic relations, but Pomraning (1991a) proposed a new type of closure, adding a pair of transport-like equations to describe the interface intensity. This extends the system to four such equations, since it retains the L–P set and adds this one and (with a p, q interchange) its mate:

$$\mathbf{\Omega} \cdot \nabla \left(p \, I_{\{pq'\}} \right) + \sigma_p \, p \, I_{\{pq'\}} = \sigma_{sp} \left[\int_{2\pi+} \Phi_p \left(\mathbf{\Omega} \cdot \mathbf{\Omega}' \right) p \, I_{\{pq'\}} \, d\mathbf{\Omega}' \right.$$
$$\left. + \int_{2\pi-} \Phi_q \left(\mathbf{\Omega} \cdot \mathbf{\Omega}' \right) q \, I_{\{qp'\}} \, d\mathbf{\Omega}' \right]$$
$$+ p \, S_p + q \, I_{\{qp'\}} / \lambda_q - p \, I_{\{pq'\}} / \lambda_p \quad (7.88)$$

where $2\pi\pm$ designates the hemisphere going out of $(+)$ or in to $(-)$ the cloudy region.

The treatment of scattering in this interface model is very attractive in that it recognizes that light backscattered into the beam must have come from a beam which was itself conditioned on being at an interface, but on the other side and thus in the other material. Forward scattered contributions, similarly, must have come from interface-conditioned beams on the same side of the interface, that is, in the same material. This is illustrated in Fig. 7.10, wherein a ray in material α is supplemented by the scattering of another ray in material α, which forward scatters into the same beam. The blowup shows the interface as a layer of nonzero thickness δ; to get the probability density we count all such scenes and take the limit of that number as δ goes to zero. Clearly, if the inscattering ray had joined the main ray by scattering into the back hemisphere (not shown), it would have had to have originated in material β.

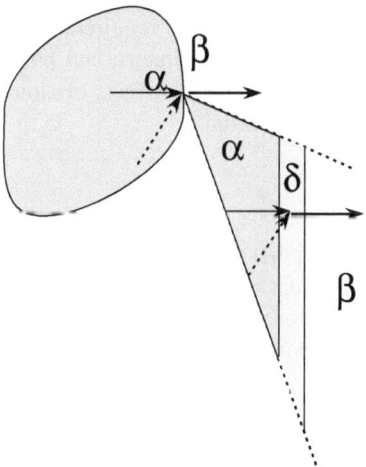

Fig. 7.10. In-scattering at an interface

Probably because of this, the interface model led to a great improvement in predicting transmission through scattering media. But in practice the interface model is not a success, because it proves not to be robust. In certain situations it predicts negative cross-sections, leading to (unphysical) growing amplitudes. This was pointed out by Zuchuat et al. (1994) and further analyzed by Su and Pomraning (1995), who also reviewed several algebraic closure models and suggested that the benefits of the interface model might be captured by a scheme for interpolating between it and the Levermore model.

It seems clear that this approach has the potential to produce a much better description of radiation in a stochastic medium, as can be seen from its derivation, and it does in some cases. I suspect that the trouble with the initial application lies with the very simple closure used, but until more work is done with closure models we cannot be sure that it can fulfill its promise.

7.3.3 Features of the Levermore-Pomraning Equations

Reduction of Dimension

Recall that if the underlying statistics is constant in space the transition rates are simply the average chord lengths, and they determine the fraction of the medium that is cloud (and clear). If the cloud statistics is independent of azimuth, varying only with altitude, then another very great simplification results. The ensemble average field of cloud then is also a one-dimensional system. Even though the presence of the solar beam turns the ensemble average into a three dimensional system, Malvagi et al. (1993) point out that one may expand the L–P equations in a Fourier series in the azimuthal angle. This reduces the system to an infinite set of one dimensional pairs of equations, no great advantage in and of itself. But the information required of the system for general use (the column absorption and transmission as functions of altitude, for example) are all given by the solution of the $m = 0$ mode, which is therefore the only one whose solution is required. It is not coupled to any higher mode, so we need solve only this one dimensional pair, for an enormous savings in computational effort. All the horizontal effects of cloud-clear interaction are still included exactly, as functions of altitude.

Interpolation

Byrne et al. (1996) pointed out that the Levermore-Pomraning equations reduce (correctly) to those of the atomic mix model in the limit of small chord lengths, and (also correctly) to those of the fractional cloud cover model in the opposite limit. They can be regarded as a way of interpolating between these two limits in the general case.

Pathlength

The 4π-integrated ($m = 0$ mode) intensity is also the total pathlength per unit of volume, and the L–P system therefore gives us immediately the pathlength separately for radiation in cloud and in clear sky. Byrne et al. (1996) use this feature to address questions about the origin of enhancement of absorption in fields of broken cloud noted by Cess et al. (1995) and Ramanathan et al. (1995).

7.3.4 Indicator Method

Titov (1990) proposed a different formalism for the binary problem. His approach was to restrict the stochastic behavior to an indicator field, a stochastic function of space that could take on the values zero or one. The underlying material could be assumed to be a uniform cloud; in any specific realization the clouds existed only when the indicator was one, elsewhere there was vacuum. He proposed explicit construction methods for such fields. One was to choose a particular geometric shape centered about an arbitrary point, then to populate space with a given density of points and locate the shape at each point. The overlaps could be handled by assuming a logical OR, which meant that any final realization could consist of quite complex indicator functions. If one used smaller and smaller shapes, increasing their number so that volume was preserved, the nature of the individual shape could be assumed to become irrelevant.

This approach has the distinct advantage over the Levermore-Pomraning one discussed above in that we are guaranteed that space is consistently defined everywhere for all of the realizations. The line statistics and correlation functions can in principle then be deduced from the shapes and the statistics of the distribution of centering points, whereas on the other hand, it is *not* certain that every arbitrary line statistics represents physical space. In the next section we discuss some recent work by Frank Graziani at Lawrence Livermore National Laboratory, which actually produces similar realizations computationally, as foreseen by Titov.

Once the line statistics of the indicator field is determined, the L–P machinery applies and can be used to solve the problem. In practice Titov's method is restricted to assuming Markovian correlation statistics, because otherwise it suffers from similar closure difficulties as the L–P model. In this case it is mathematically identical to L–P, as pointed out by Malvagi et al. (1993).

7.4 Validation

7.4.1 Direct Numerical Simulation (DNS)

Apart from purely absorbing Markovian systems it is rare to find analytic solutions to the problem of radiative transfer through binary systems. Therefore we turn to DNS for benchmarks against which to test our theories. The earliest set of these was published by Adams et al. (1989). They considered a one dimensional system with constant optical properties (different for the two materials), isotropic scattering, no internal sources of radiation, and Markov statistics. One set of problems was in rod geometry, where radiation travels in only two directions: straight ahead or straight back. This was extended by redoing the set in planar geometry where all directions of travel are possible, a much more challenging system to solve but still well within the means of modern computational power. Radiation was provided by imposing an isotropic field at one end of the system; the other end had zero incoming flux.

Adams et al. (1989) considered three different stochastic parameters, called A, B, and C in Table 7.2. The first two were the average optical thickness of a chunk of materials #1 and #2 ($\lambda_p \times \sigma_p$ and $\lambda_q \times \sigma_q$ respectively); the third was the ratio of the probability of finding the two materials, p/q. A was primarily optically thin material with a small amount of moderately opaque chunks in it (unit optical depth on the average), B was the same but at ten times the opacity for both materials, and C was an equal mixture of the fairly transparent material from A and the opaque material from B.

Table 7.2. Material sizes (in optical units) and probabilities

Set	Chunk #1	Chunk #2	p/q
A	1/10	1	9
B	1	10	9
C	1/10	10	1

For each of these they considered three sets of scattering properties, called I, II, and III in Table 7.3. These were then expressed in terms of the single-scattering albedo ϖ_0 (the ratio of the scattering to the total cross-sections) for each material.

Table 7.3. Scattering Properties

Set	ϖ_{0p}	ϖ_{0q}
I	0	1
II	1	0
III	0.9	0.9

This gave 3×3 sets of optical properties that together made a reasonable probe of interesting parameter space. Their decision to run each problem for 0.1, 1, and 10 (average) extinction lengths also strikes me as reasonable, for they explored the optically thin, moderate, and thick classes of problems.

For each of set of optical parameters 100,000 individual problem realizations were generated, each was solved numerically, and the average and the standard deviation of the intensity were computed from the results. In the course of this effort more transport problems were solved numerically than had been done in the preceding 10^{10} years.

This work was extended, in part, by Su and Pomraning (1993). They repeated the previous work for rod geometry, but instead of only Markov statistics they also included five other distributions. These were called step, ramp, tent, parabolic, and

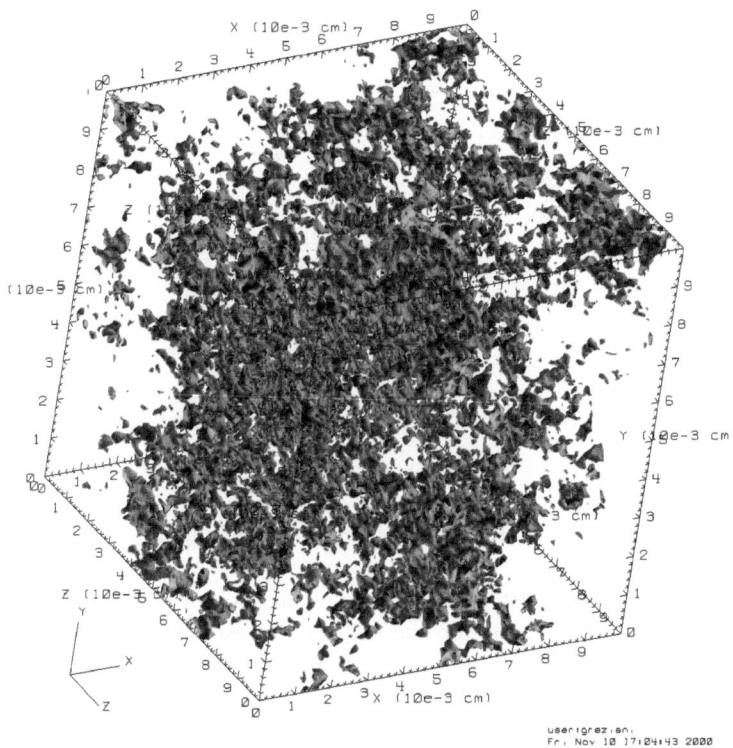

Fig. 7.11. View of a 3D population, from Graziani and Slone (2003) with permission

reverse ramp. Neither they nor Adams et al. (1989) compare their benchmark results to L–P predictions.

Graziani and Slone (2003) have recently begun a new benchmark series. They introduced a fundamental granularity into a finite 3D universe, which allows them to completely populate every point in the region in a finite number of operations. They have used Markov and Cauchy statistics on 100^3 grids, with a batch size of 100 complete realizations and two materials, each with their own constant optical properties. An example of one of their Markov volume realizations is shown in Fig. 7.11. The volume fraction of the dark material is 20%, and its length scale is 1% of the cube's width. As a test of the success of generating the desired statistics, Graziani's code projects a large number of rays through each realization and gathers statistics about what it finds. A sample of one such result, the frequency of occurrence of length, is presented in Fig. 7.12. Since for the case of Poisson statistics the result should be a straight line on a semi-log plot, the data show that the statistics are as intended. Data from a realization (not shown here) using Cauchy statistics are also displayed. The Cauchy (or Lorentzian) distribution exhibits a broader tail in the graph, since it falls off only like $1/x^2$. Along with gathering the run lengths, the raytracing code also

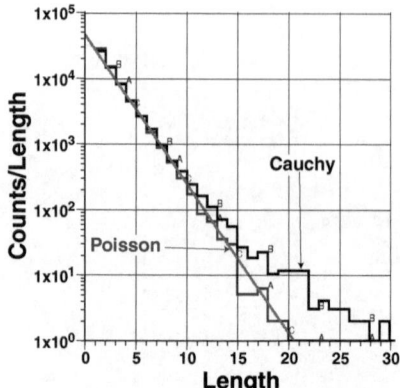

Fig. 7.12. Line statistics of one of F. Graziani's realizations

tabulates optical depths. The realizations discussed here use matter whose optical properties are constant, so this becomes a matter of multiplying the length times the opacity.

Although they have not yet addressed the question of scattering, they have already obtained useful results for one item of interest. This is the parameter ℓ_{eff}, which replaces the actual absorption length when the stochastic average is approximated by a single equation with renormalized extinction coefficients to account for the unresolved structure. Figure 7.13 shows results from a large number of runs of

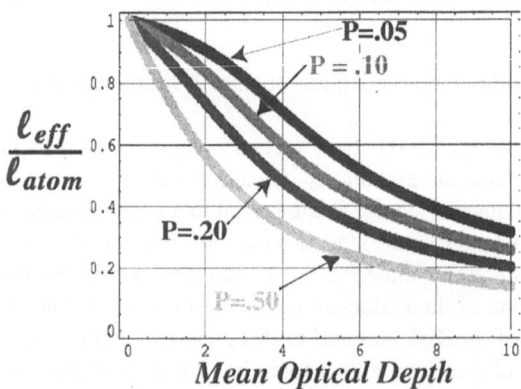

Fig. 7.13. Variation of renormalization parameter with mean optical thickness of chunks and fraction of the chunky material (through occurrence probability P). The effective photon mean-free-path in the variable medium is denoted by ℓ_{eff} while $\ell_{atom} = [P\sigma_1 + (1-P)\sigma_2]^{-1}$ is the mean-free-path based on the atomistic mixture of the two materials. The effective mean-free-path is defined by fitting an exponential to the numerically obtained distribution of physical penetration depths

batches of problems with varying parameters, in this case, the overall average optical depth per subcell and the fraction of absorbing material. This sort of program has been desired for years but has until now been too computationally expensive to carry out.

7.4.2 Experiments in the Cloudy Atmosphere

There are two tests a stochastic model must pass before being deemed useful. The DNS work mentioned above will eventually be available to guide the mathematical development toward some measurable and acceptable accuracy, but whether the model is useful to the atmospheric community also depends upon whether it can be applied to the problem of radiation transport through broken clouds, no matter how successful it may be on computer simulations.

Benner and Evans (2001) addressed exactly this issue. They reconstructed fields of small marine tropical cumulus (from visible and infrared imagery) and compared their Monte Carlo solutions, which may be taken as correct, to various approximate techniques (cf. Chaps. 6 and 8): Independent Pixel Approximation (IPA), "tilted" IPA, and conventional plane-parallel. For large-domain averages, they found only small errors introduced by the simpler methods, although some individual cases had significant ones.

Lane et al. (2002) also broke new ground on this issue in a recent study. They took data from many different sensors in the ARM testbed in Oklahoma and used them to generate statistics of the observed cloud fields. Then they compared the results of plane-parallel to L–P treatments of the reconstituted cloud fields. We will briefly sketch their technique, but the reader is referred to their article for details. One of the primary requirements of any analysis, stochastic or not, is the state of cloudiness. How much liquid water is there, and where? These data are desired on a 3D grid, and in this case we need enough samples to generate good statistics, yet it is not a standard ARM product. Lane et al. inferred information from a whole slew of sources of varying reliability, starting from hourly meteorological logs recorded by human observers and including various ceilometers, the Radio Acoustic Sounding System, radiosondes, the Time-Lapsed Video camera, the Whole Sky Imager, the Microwave Radiometer, the Radar Wind Profiler, and others. The Multi-Filter Rotating Shadowband Radiometer (MFRSR) played an especially important role. The MFRSR measures the direct and the diffuse solar radiation on the ground at a rate of up to four times per minute. Examples of MFRSR and MFRSR-like data on partially cloudy days are shown in Chap. 2.

Lane et al. (2002) inferred statistics of the size of the cloud by combining knowledge of cloud height, wind speed at cloud height, and the very reasonable assumption that the sudden dips in intensity of the direct beam measured by the MFRSR were due to the passage of a cloud between the instrument and the sun. The choice of intervals for which data were selected for this purpose was determined by a preset protocol designed to focus on periods where broken cloud fields dominated the insolation. The results of this analysis are shown in Fig. 7.14, which has separate panels for the cloudy and clear sky statistics. The fits $y = a \exp(-bD)$ (where D is the

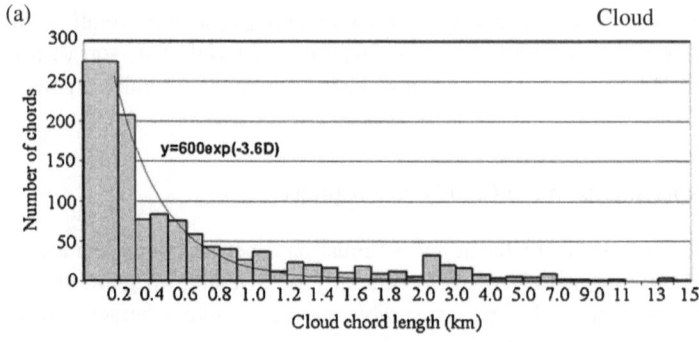

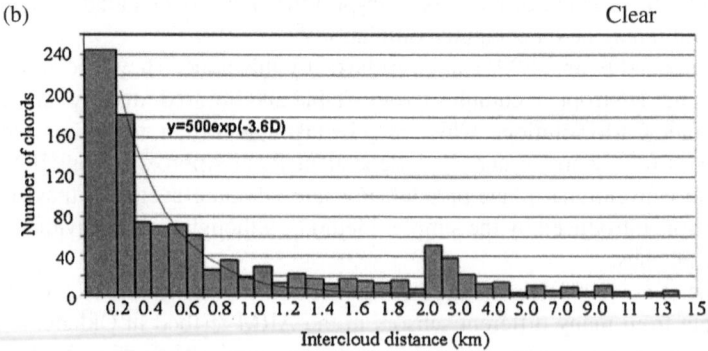

Fig. 7.14. (a) Cloud size histogram and (b) cloud spacing histogram for fair-weather cumulus using 6 MFRSRs: fits from Plank (1969) and figures from Lane et al. (2002), with permission

size of the cloud, a is the typical number of clouds, and b is empirically determined) that overlay the data are in terms of the horizontal axis in each panel; they represent Markovian distributions. Astin and Latter (1998) also pointed out that a Markovian distribution of cloud size is suggested by observations.

The averaging for the test of the stochastic model had two components, spatial and temporal. The spatial averaging was obtained by using data from the Oklahoma Mesonet, which provided downwelling shortwave radiation over four sites within 90 km of the central ARM facility, and six MFRSRs distributed over roughly the same area. The experimental data were collected frequently, but not at synchronized times from all the instruments. So each instrument's output was averaged over one-hour intervals, and these averages were used in making comparisons to the calculations. The findings are inconclusive as to which analysis technique is superior. The results for two days less than a month apart are given in Fig. 7.15 and Fig. 7.16. A superficial glance would indicate that the stochastic analysis was worse on the first day yet better on the second, but in fact the data are not adequate to discriminate between the two. (It was suggested that the main reason for these discrepancies is a "limited" ability of stochastic models to capture the vertical variability of clouds.)

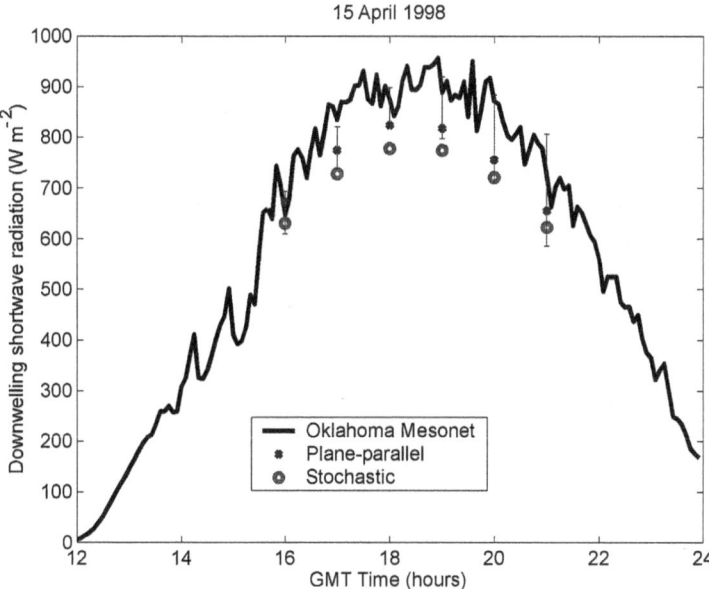

Fig. 7.15. Models results of downwelling shortwave radiation for 15 Apr 1998 compared with averaged observations from the Oklahoma Mesonet

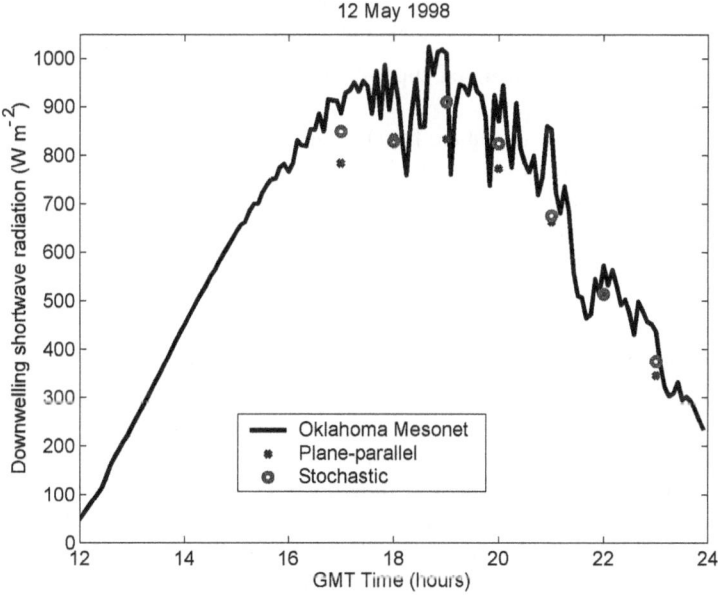

Fig. 7.16. Model results of downwelling shortwave radiation for 12 May 1998 compared with averaged observations from the Oklahoma Mesonet

One reason for this is indicated by the error bars on the figures, which represent estimates of the uncertainty in the radiation models due to the uncertainty in the liquid water path alone; there are many other uncertainties in the data as well, as discussed by Lane et al. (2002). The overall uncertainty is far too large to be tolerated for the purposes of climate prediction, for either method. What they have shown is that the principles of comparison with observation can be applied to this problem; they have developed new techniques for extracting the requisite information from the existing suite of ARM instruments; and they have found which types of data are not now adequately measured. We may expect future work to build on this pioneering study.

7.5 Conclusions

A pure absorber transmits light as $\exp(-\tau)$. A pure scatterer does so as τ^{-1} asymptotically. Fractal scattering media also does as a power, but not necessarily with an exponent of -1 (Gabriel et al., 1990; Davis et al., 1990). We have seen that a mixture of pure absorbers transmits light as a mixture of exponentials. We expect a mixture of pure scatterers will do so as a weighted average of $1/\tau$, and a mixture of a general material will somehow interpolate between these trends. The essence of the problem seems to be that the more transmissive materials regenerate light for the less transmissive ones. We have indicated some paths currently under investigation, especially the Levermore-Pomraning (L–P) stochastic model.

The shortcomings of the L–P system are clear:

1. L–P has not yet demonstrated superiority to plane-parallel (not to mention IPA) in practical use;
2. L–P offers the most advantage for statistically homogeneous systems, but these are not guaranteed to represent reality;
3. L–P is inexact with its present closure;
4. L–P, as presently applied, is developed for binary systems only.

However, just as many encouraging statements can be made:

1. The lack of demonstrated superiority is surely only temporary. The techniques mentioned herein can be developed to the point where decisions about model validity can be made.
2. Homogeneity is a relative term in spatial statistics. It is often observed that weather systems are very large, and there is even a World Meteorological Organization system of classification that breaks down all the world's clouds into a small number of types. Somewhat paradoxically the success of a thriving subculture finding common fractal character in cloud implies an underlying sameness in diversity. See Cahalan (1989) for an example.
3. The present closure is probably not the last word on the subject.
4. The restriction to binary systems is not fundamental, though the extensions have been to systems with countable, indeed small, number of submaterials. In any case

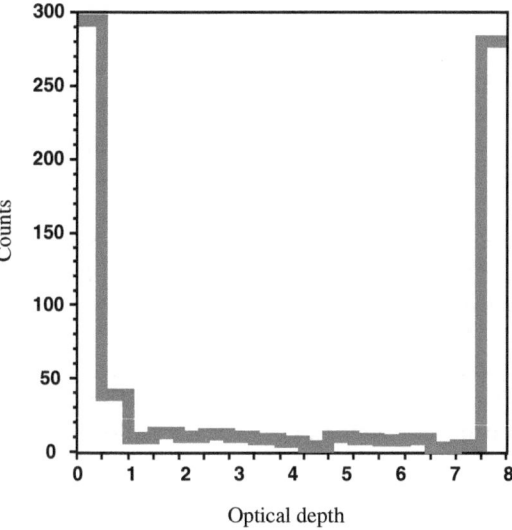

Fig. 7.17. Optical depths observed via MFRSR, ARM site, May 1993

there are many cases where a two material classification seems to be indicated observationally. For example, Fig. 7.17 shows a sample of optical depths retrieved over five minute intervals for a few days in May 1993 over the ARM site in Oklahoma. One is strongly tempted to divide things into two parts.

If it works, the promises are also clear:

1. L–P is simple and very efficient: it provides an answer to a complicated problem of vital interest, at a very low cost. It uses only a few statistical parameters and produces information About system-average answers without having to form answers for any individual realizations.
2. L–P provides useful diagnostics about details of the average radiation, separately in clouds and clear sky.
3. L–P looks very good against benchmarks, and it is robust.

References

Adams, M.L., E.W. Larsen, and G.C. Pomraning (1989). Benchmark results for particle transport in a binary Markov statistical medium. *J. Quant. Spectrosc. Radiat. Transfer*, **42**, 253–266.

Astin, I. and B.G. Latter (1998). A case for exponential cloud fields? *J. Appl. Meteor.*, **37**, 1375–1382.

Avaste, O.A. and G.M. Vainikko (1974). Solar radiative transfer in broken clouds. *Izv. Acad. Sci. USSR Atmos. Oceanic Phys.*, **10**, 1054–1061.

Avaste, O.A., Y.R. Mullamaa, K.Y. Niylisk, and M.A. Sulev (1974). On the coverage of sky by clouds. In *Heat Transfer in the Atmosphere*. Technical Report (Tech. Transl. TT-F-790), NASA, Washington (DC).

Benner, T.C. and K.F. Evans (2001). Three-dimensional solar radiative transfer in small tropical cumulus field derived from high-resolution imagery. *J. Geophys. Res.*, **106**, 14,975–14,984.

Byrne, R.N., R.C.J. Somerville, and B. Subasilar (1996). Broken-cloud enhancement of solar radiation absorption. *J. Atmos. Sci.*, **53**, 878–886.

Cahalan, R.F. (1989). Overview of fractal clouds. In *Advances in Remote Sensing*. A. Deepak Publishing, Hampton, VA, pp. 371–388.

Cairns, B. (1992). *An Investigation of Radiative Transfer and Multiple Scattering*. Ph.D. dissertation, University of Rochester, Rochester (NY).

Cairns, B., A.A. Lacis, and B.E. Carlson (2000). Absorption within inhomogeneous clouds and its parameterization in general circulation models. *J. Atmos. Sci.*, **57**, 700–714.

Cess, R.D., G.L. Potter, J.-P. Blanchet, G.J. Boer, S.J. Ghan, J.T. Kiehl, S.B.A. Liang, J.F.B. Mitchell, D.A. Randall, M.R. Riches, E. Roeckner, U. Schlese, A. Slingo, K.E. Taylor, W.M. Washington, R.T. Wetherald, and I. Yagai (1989). Interpretation of cloud-climate feedback as produced by 14 atmospheric general circulation models. *Science*, **245**, 513–516.

Cess, R.D., M.-H. Zhang, P. Minnis, L. Corsetti, E.G. Dutton, B.W. Forgan, D.P. Garber, W.L. Gates, J.J. Hack, E.F. Harrison, X. Jing, J.T. Kiehl, C.N. Long, J.-J Morcrette, G. L. Potter, V. Ramanathan, B. Subasilar, C.H. Whitlock, D.F. Young, and Y. Zhou (1995). Absorption of solar radiation by clouds: Observations versus models. *Science*, **267**, 496–499.

Davis, A., P.M. Gabriel, S.M. Lovejoy, D. Schertzer, and G.L. Austin (1990). Discrete angle radiative transfer III: Numerical results and meteorological applications. *J. Geophys. Res.*, **95**, 11,729–11,742.

Evans, K.F. (1998). The spherical harmonics discrete ordinate method for three-dimensional atmospheric radiative transfer. *J. Atmos. Sci.*, **55**, 429–446.

Gabriel, P.M., S.M. Lovejoy, A. Davis, D. Schertzer, and G.L. Austin (1990). Discrete angle radiative transfer II: Renormalization approach for homogeneous and fractal clouds. *J. Geophys. Res.*, **95**, 11,717–11,728.

Geleyn, J.-F. and A. Hollingsworth (1979). An economical analytical method for the computation of the interaction between scattering and line absorption of radiation. *Contrib. Atmos. Phys.*, **52**, 1–16.

Graziani, F. and D. Slone (2003). Radiation transport in 3D heterogeneous materials: Direct numerical simulation. In *Proceedings of the American Nuclear Society Winter Meeting, New Orleans, LA*. ANS, LaGrange Park, IL.

Lane, D.E. (2000). Ph.D. dissertation, University of California, San Diego (CA).

Lane, D.E., K. Goris, and R.C.J. Somerville (2002). Radiative transfer through broken cloud fields: Observations and model validation. *J. Climate*, **15**, 2921–2933.

Levermore, C.D., G.C. Pomraning, D.L Sanzo, and J. Wong (1986). Linear transport theory in a random medium. *J. Math. Phys.*, **27**, 2526–2536.

Levermore, C.D., J. Wong, and G.C. Pomraning (1988). Renewal theory for transport processes in binary statistical mixtures. *J. Math. Phys.*, **29**, 995–1004.

Malvagi, F., R.N. Byrne, G.C. Pomraning, and R.C.J. Somerville (1993). Stochastic radiative transfer in a partially cloudy atmosphere. *J. Atmos. Sci.*, **50**, 2146–2158.

O'Hirok, W. and C. Gautier (1998). A three-dimensional radiative transfer model to investigate the solar radiation within a cloudy atmosphere. Part I: Spatial effects. *J. Atmos. Sci.*, **55**, 2162–2179.

Plank, V.G. (1969). The size distribution of cumulus clouds in representative Florida populations. *J. Appl. Meteor.*, **8**, 46–67.

Pomraning, G.C. (1986). Transport and diffusion in a statistical medium. *Transp. Theory and Stat. Phys.*, **5**, 773–802.

Pomraning, G.C. (1989). Statistics, renewal theory and particle transport. *J. Quant. Spectrosc. Radiat. Transfer*, **42**, 279–293.

Pomraning, G.C. (1991a). A model for interface intesities in stochastic particle transport. *J. Quant. Spectrosc. Radiat. Transfer*, **46**, 221–236.

Pomraning, G.C. (1991b). Linear kinetic theory and particle transport in stochastic mixtures. In *Waves and Stability in Continuous Media*. S. Rionero (ed.). World Scientific Publishing, Singapore.

Pomraning, G.C. (1998). Radiative transfer and transport phenomena in stochastic media. *Int. J. of Eng. Sci.*, **36**, 1595–1621.

Ramanathan, V., B. Subasilar, G. Zhang, W. Conant, R. Cess, J. Kiehl, H. Grassl, and L. Shi (1995). Warm pool heat budget and shortwave cloud forcing: A missing physics? *Science*, **267**, 499–503.

Sanchez, R., O. Zuchuat, and F. Malvagi (1994). Symmetry and translations in multimaterial line statistics. *J. Quant. Spectrosc. Radiat. Transfer*, **51**, 801–812.

Su, B.J. and G.C. Pomraning (1993). Benchmark results for particle transport in binary non-Markovian mixtures. *J. Quant. Spectrosc. Radiat. Transfer*, **50**, 211–226.

Su, B.J. and G.C. Pomraning (1995). Modification to a previous higher order model for particle transport in binary stochastic media. *J. Quant. Spectrosc. Radiat. Transfer*, **54**, 779–801.

Titov, G.A. (1990). Statistical description of radiation transfer in clouds. *J. Atmos. Sci.*, **47**, 24–38.

Vainikko, G.M. (1973). Transfer approach to the mean intensity of radiation in non-continuous clouds. *Trudy MGK SSSR, Meteorological Investigations*, **21**, 28–37.

Vanderhaegen, D. (1986). Radiative transfer in statistically inhomogeneous mixtures. *J. Quant. Spectrosc. Radiat. Transfer*, **36**, 557–561.

Zuchuat, O., R. Sanchez, I. Zmijarevic, and F. Malvagi (1994). Transport in renewal statistical media: Benchmarking and comparison with models. *J. Quant. Spectrosc. Radiat. Transfer*, **51**, 689–722.

Suggestions for Further Reading

Audic, S. and H. Frisch (1993). Monte-Carlo simulation of a radiative transfer problem in a random medium: Application to a binary mixture. *J. Quant. Spectrosc. Radiat. Transf.*, **50**, 127–147.

Borovoi, A.G. (1984). Radiative transfer in inhomogeneous media. *Dok. Akad. Nauk SSSR*, **276**, 1374–1378. (English version: *Sov. Phys. Dokl.*, **29**, no 6.)

Davis, A. and A. Marshak (1997). Lévy kinetics in slab geometry: Scaling of transmission probability. In *Fractal Frontiers*. M.M. Novak and T.G. Dewey (eds.), World Scientific, Singapore, pp. 63–72.

Davis, A.B. and A. Marshak (2001). Multiple scattering in clouds: Insights from three-dimensional diffusion/P_1 theory. *Nucl. Sci. Eng.*, **137**, 251–288. (Special Issue "In Memory of Gerald C. Pomraning".)

Davis, A.B. and A. Marshak (2004). Photon propagation in heterogeneous optical media with spatial correlations: Enhanced mean-free-paths and wider-than-exponential free-path distributions. *J. Quant. Spectrosc. Rad. Transf.*, **84**, 3–34.

Evans, K.F. (1993). A general solution for stochastic radiative transfer. *Geophys. Res. Lett.*, **20**, 2075–2078.

Kassianov, E. (2003). Stochastic radiative transfer in multilayer broken clouds. Part I: Markovian approach. *J. Quant. Spectrosc. Radiat. Transf.*, **77**, 373–394.

Kassianov E., T.P. Ackerman, R. Marchand and M. Ovtchinnikov (2003). Stochastic radiative transfer in multilayer broken clouds. Part II: Validation test. *J. Quant. Spectrosc. Radiat. Transf.*, **77**, 395–416.

Kostinski, A.B. (2001). On the extinction of radiation by a homogeneous but spatially correlated random medium. *J. Opt. Soc. Am. A*, **18**, 1929–1933.

Lane-Veron, D.E. and R.C.J. Somerville (2004). Stochastic theory of radiative transfer through generalized cloud fields. *J. Geophys. Res.*, **109**, D18113, doi:10.1029/2004JD004524.

Lovejoy, S.M., A. Davis, P.M. Gabriel, D. Schertzer and G.L. Austin (1990). Discrete angle radiative transfer. Part 1: Scaling, similarity, universality and diffusion. *J. Geophys. Res.*, **95**, 11,699–11,715.

Newman W.I., J.K. Lew, G.L. Siscoe, and R.G. Fovell (1995). Systematic effects of randomness in radiative transfer. *J. Atmos. Sci.*, **52**, 427–435.

Peltoniemi, J. (1993). Radiative transfer in stochastically inhomogeneous media. *J. Quant. Spectrosc. Radiat. Transf.*, **50**, 655–672.

Pfeilsticker, K. (1999). First geometrical pathlength distribution measurements of skylight using the oxygen A-band absorption technique – Part II, Derivation of the Lévy-index for skylight transmitted by mid-latitude clouds. *J. Geophys. Res.*, **104**, 4101–4116.

Pomraning, G.C. (1996). Transport theory in discrete stochastic mixtures. *Adv. Nuclear Sci. Tech.*, **24**, 47–93.

Stephens, G.L. (1988). Radiative transfer through arbitrarily shaped media. Part 2: Group theory and closures. *J. Atmos. Sci.*, **45**, 1836–1848.

Titov G.A., T.B. Zhuravleva and V.E. Zuev (1997). Mean radiation fluxes in the near-IR spectral range: Algorithms for calculation. *J. Geophys. Res.*, **102**, 1819–1832.

8

Effective Cloud Properties for Large-Scale Models

R.F. Cahalan

8.1   **Introduction** 425

8.2   **Definitions** 428

8.3   **Independent Pixel Approximation** 431

8.4   **Diurnal Cycle** 435

8.5   **Effective Optical Thickness** 435

8.6   **Conclusions** 439

8.A   **Rescaling f Generates W Moments** 440

8.B   **Reduction Factor and Variance of $\log W$** 443

References 444

Suggested Reading 446

8.1 Introduction

The large-scale terrestrial climate is well-known to be sensitive to small changes in the average albedo of the earth-atmosphere system. Sensitivity estimates vary, but typically a 10% decrease in global albedo, with all other quantities held fixed, increases the global mean equilibrium surface temperature by 5°C, similar to the warming since the last ice age, or that expected from a doubling of CO_2 (e.g., Cahalan and Wiscombe, 1993). Yet not only is the global albedo of 0.31 only known to $\approx$10% accuracy[1] but current global climate models often do not predict the albedo in each gridbox from realistic cloud liquid water distributions; they normally tune the liquid until plane-parallel radiative computations produce what are believed to

---

[1] Estimates of global albedo range from 0.30 to 0.33, or 3 out of $31 \approx 10\%$, (e.g., Kiehl and Trenberth, 1997).

be typical observed albedos. The inability of global climate models to compute the albedo is due to their inability to predict the microphysical and macrophysical properties of cloud liquid water within each gridbox, and their reliance on plane-parallel radiative codes. As Stephens (1985) has emphasized, the mean albedo of each gridbox depends not only on the mean properties of clouds within each box, but also upon the variability of the clouds, which involves not only the fractional area covered by clouds, but also the cloud structure itself. During recent years many climate models began to carry liquid water as a prognostic variable, e.g., Sundqvist et al. (1989) and Tiedtke (1996). It is important to treat cloud radiation and cloud hydrology consistently, which requires that cloud parameterizations become dependent on the fractal structure of clouds. Radiative properties of singular multifractal clouds have been previously studied (e.g., Cahalan, 1989; Cahalan and Snider, 1989; Lovejoy et al., 1990; Gabriel et al., 1990; Davis et al., 1990). Here we shall show how radiative properties of marine stratocumulus boundary-layer clouds, and specifically area-average albedo of these clouds, depend on their structure. The central role of this cloud type in maintaining the current climate was clarified and quantified in Ramanathan et al. (1989).

The dependence of average albedo on cloud structure has been found to be especially important in the case of marine stratocumulus, a major contributor to net cloud radiative forcing. Computations based on observations of California stratocumulus during the First International Satellite Cloud Climatology Project (ISCCP) Regional Experiment (FIRE) have shown that stratocumulus have significant fractal structure, and that this "within-cloud" structure can have a greater impact on average albedo than cloud fraction (Cahalan and Snider, 1989; Cahalan et al., 1994a,b). These studies employed a "bounded cascade" model[2] to distribute the cloud liquid, defined in terms of two cascade or "fractal" parameters: f, the difference in cloud liquid fractions between two segments of the full cloudy domain being considered, and c, the difference of liquid fractions at the next smaller scale (within each segment) divided by f. Parameters c and f are empirically adjusted to fit the scaling exponent of the power spectrum of liquid water path (W), $\beta(c) \approx 5/3$, and the standard deviation of $\log W$, $\sigma(f, c)$, respectively. In order to isolate the effects of horizontal liquid water variations on cloud albedo, it is convenient to assume that the usual microphysical parameters are homogeneous, as is the geometrical cloud thickness. In order to simplify comparison with plane-parallel clouds, the area-averaged vertical optical depth is kept fixed at each step of the cascade. The albedo bias is then found as an analytic function of the fractal parameter, f, as well as the mean vertical optical thickness, τ_v, and sun angle, θ_0. For the diurnal mean of the values observed in FIRE ($f \approx 0.5$, $\tau_v \approx 15$, and $\theta_0 \approx 60°$) the absolute bias is approximately 0.09, nearly 15% of the plane-parallel albedo of 0.69. Diurnal and seasonal variations of cloud albedo bias

---

[2] Bounded cascades were first introduced in Cahalan et al. (1990), and their multi-scaling properties studied in Marshak et al. (1994). For a description of bounded cascades in terms of f and c, see the discussion following (8.2) below and the Appendix of the volume (using somewhat different notations).

have been determined from observations during the Atlantic Stratocumulus Transition Experiment (ASTEX) and compared to the FIRE results (Cahalan et al., 1995).

The goal of this chapter is to show how these results for the mean albedo of bounded cascade clouds, derived in the references cited above, may be applied to parameterizing the albedo of such clouds in terms of the plane-parallel albedo of a cloud having an "effective" optical thickness which is reduced from the mean thickness by a factor χ (or, equivalently, σ) which depends only on the fractal parameters, and not on the mean cloud properties. This Effective Thickness Approximation (ETA) is a special case of the more general Independent Pixel Approximation (IPA), sometimes referred to as the Independent Column Approximation (ICA) especially for gridded climate models. The key assumption of any IPA (or ICA) type approximation is the neglect of horizontal photon transport (see Chap. 12). In addition, it depends only on 1-point cloud probability distributions, not on the spatial arrangement or correlations of individual cloud elements. On the other hand, knowledge of the *accuracy* of any IPA depends on three-dimensional (3D) radiative transfer (i.e., with net horizontal fluxes) as well as on the spatial (typically fractal) cloud structure. In this chapter, though we compare the IPA/ICA with 3D radiative transfer as is done in other chapters, the primary purpose is to compare the IPA with the much simpler ETA. In particular, we use a simple fractal "bounded cascade" model to (1) motivate the ETA; and (2) determine the accuracy of the ETA by comparing it to the full IPA, using cloud parameters typical of marine stratocumulus. Moreover, some analytic results for bounded cascades are generalized and simplified in two appendices. In the "Further Readings" section at the end, we point the reader to simple alternatives to the ETA, each of which have particular advantages and points of view. We feel that each approximation is helpful insofar as it lends some insight into real clouds, which are far more complex than any of our mathematical idealizations, as anyone can discover who takes the opportunity to study the amazing variety of real cloud systems.

In the following, we first define some terms in Sect. 8.2. Then Sect. 8.3 shows that the IPA provides estimates of the plane-parallel albedo bias accurate to about 1% for bounded cascade clouds, and Sect. 8.4 applies the IPA to show that the total absolute bias reaches a maximum of about 0.10 during the morning hours, when the cloud fraction is nearly 100%. These two sections are primarily summaries of results from Cahalan et al. (1994a) and Cahalan et al. (1994b), although there a 1D cascade was employed, while a here a 2D cascade is applied. Section 8.5 gives the main result, that under certain commonly-observed conditions ($\beta \approx 5/3$, hence $c \approx 0.8$) the albedo is approximately the plane-parallel albedo at a reduced "effective optical thickness" $\tau_{\text{eff}} \equiv \chi \tau_{\text{v}}$, where the reduction factor χ decreases with σ, hence with f, approximately as $10^{-1.15\sigma^2}$ (see Fig. 8.6 and (8.B.11)), independently of the mean vertical optical depth, τ_{v}. The accuracy of this approximation is given as a function of both f and the mean thickness. The results are summarized and their limitations briefly discussed in Sect. 8.6. Appendix 8.A shows that all moments of a bounded cascade may be obtained by considering only the second moment as a function of the fractal parameter. This generalizes expressions for the second and third moments given in Cahalan et al. (1994b), and allows the lognormal behavior in the singular limit to be explicitly exhibited (see also Cahalan, 1994). Appendix 8.B gives expressions for $\chi(f, c)$ and $\sigma(f, c)$ as power series in f with coefficients depending on c, and evaluates the coefficients for the case of a $\beta(c) \approx 5/3$ $(c \approx 0.8)$ wavenumber spectrum.

8.2 Definitions

Many general circulation models (GCMs) are now predicting mean cloud liquid water in each gridbox, not merely diagnosing it from other quantities. The cloud albedo could potentially also be accurately predicted, if cloud liquid could be accurately distributed within each gridbox. Efforts are underway to improve the treatment of cloud distributions in global models, so that simulated clouds can respond more realistically to climate change. The hope is that average cloud liquid in each gridbox will be accurately predicted, and that the resulting cloud albedo will be correctly computed from this, and other average cloud parameters. It is important to recognize, however, that *mean cloud parameters are insufficient* to compute the mean albedo. The mean albedo also depends, at a minimum, on the deviations of the liquid water from the mean, for instance, on the mean *and standard deviation* of the logarithm of the liquid water. We demonstrate this here and in the next using the bounded cascade model.

The schematic in Fig. 8.1 shows three approaches to distributing a prescribed amount of liquid water in a given vertical level of a GCM gridbox. In (a) it is uniform over the whole area, and thus the albedo may be computed from plane-parallel theory, and depends only on the *average* optical thickness, effective particle radius, and so on. In (b) the cloud is assumed to cover only a fraction of the area, is somewhat thicker in order to contain the same total liquid, but is still assumed to be uniform on that so-called "cloud fraction." In this case the mean albedo of the gridbox is assumed to equal the area-weighted average of a "cloud albedo" and a "clear-sky" albedo. Finally, in (c) the cloud covers the same cloud fraction as in (b), with the same mean parameters, but is assumed to have a non-uniform structure which depends on one or more "fractal parameters." The cloud fraction and the fractal parameters are assumed to depend on geographic region, season, and time of day.

As a measure of the impact of cloud fraction and fractal parameters on the average albedo, we define the "absolute plane-parallel albedo bias" ΔR_{pp}, as the mean albedo computed in case (a) minus that in case (c). This may be expressed symbolically as:

$$\Delta R_{pp} = R_{pp} - [A_c R_f + (1 - A_c) R_s] , \qquad (8.1)$$

where R_{pp} is the plane-parallel reflectivity, R_f is the mean reflectivity of the fractal cloud, R_s is the mean clear-sky reflectivity, and the same total liquid water is used in all cases. The *relative* plane-parallel albedo bias is the absolute bias divided by R_{pp}. To avoid confusion, the absolute bias is always given as a fraction, while the relative bias is given in percent. Since the simple uniform cloud fraction model shown in Fig. 8.1b is currently widely employed, it is convenient to split the total plane-parallel bias into the difference between (a) and (b), plus the difference between (b) and (c). Symbolically:

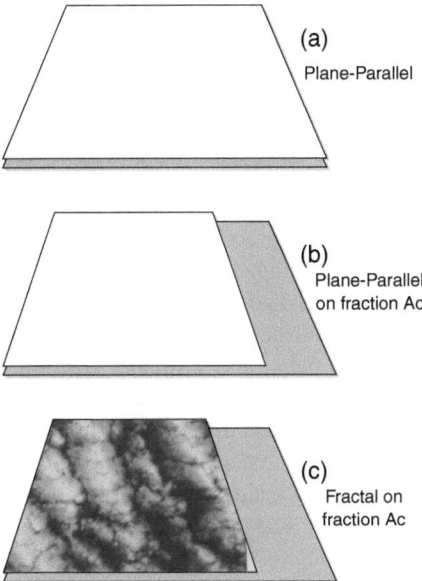

Fig. 8.1. Schematic showing three approaches to distributing the cloud liquid water in a GCM gridbox. In the top figure (**a**) the cloud has plane-parallel geometry, with cloud parameters such as vertical optical thickness, τ_v, uniform over the whole gridbox. In the middle figure (**b**) the parameters are uniform over a fraction A_c of the gridbox, with the same values as above, except that cloud vertical optical thickness increases to τ_v/A_c, thus preserving the total liquid, while the cloud thickness is zero on the remaining fraction $1 - A_c$. In the bottom figure (**c**) one has a fractal distribution of cloud parameters over the fraction A_c, with the same mean values as in the middle, and an identical clear fraction $1 - A_c$

$$\Delta R_{pp} = \{R_{pp} - [A_c R_{pp} + (1 - A_c)R_s]\}$$
$$+ \{[A_c R_{pp} + (1 - A_c)R_s]$$
$$- [A_c R_f + (1 - A_c)R_s]\} . \tag{8.2}$$

The first difference represents the bias due only to the reduction in cloud fraction from unity to A_c, and the corresponding increase in thickness, with no change in the plane-parallel assumption; the second difference is the *additional* bias due only to the within-cloud fractal structure, where again the same total liquid is employed in all cases. This section and the following considers the case of overcast clouds, having $A_c = 1$, so that the total bias depends only on the fractal parameters. Then Sect. 8.4 considers the case in which both the cloud fraction and the fractal parameters follow the diurnal variations observed in California marine stratocumulus. As we shall see, the $A_c = 1$ case produces the largest total bias, because of the sensitivity of the bias to the fractal structure, and the observed fact that in California stratocumulus the overcast cases have the greatest within-cloud variability.

In order to generate a bounded cascade cloud, we begin with a uniform cloud having a liquid water path of, e.g., $W = 100$ g/m^2, and corresponding vertical optical thickness of, e.g., $\tau_v = 15$ (assuming an effective drop radius of $r_e = 10\,\mu m$). We assume large but finite horizontal optical thicknesses in both horizontal directions, say $\tau_h = 1500$. This uniform distribution is then made non-uniform by a bounded multiplicative cascade process, in which the cloud is successively subdivided into smaller parts, and successively smaller fractions of liquid water are transferred among these parts, without changing the total.[3]

It is simplest to describe a one-dimensional (1D) bounded cascade, and we shall consider the simplest subdivision process: Divide the cloud in half along a north-south line. Flip a coin to select one half, and transfer a fraction, say $f_0 = f = 0.5$ from that half to the other one. The process is then iterated as follows: Each of the two halves is divided in half the same way, two coins are flipped to select one quarter from each of the two pairs, and a smaller fraction $f_1 = f \times c$, with say $c = 0.8$, so that $f_1 = 0.4$, is transferred from each chosen quarter to the other one. The resulting four quarters are in turn divided in half, four coins are flipped, and a fraction $f_2 = f_1 \times c = 0.32$ is transferred within the four pairs of eighths, and so on. The resulting distribution of liquid water path has a power spectrum behaving as $k^{-\beta}$, where $\beta \approx 5/3$ when $c = 2^{-1/3} \approx 0.8$, as observed (Cahalan and Snider, 1989), and an approximately lognormal probability distribution, with the standard deviation of $\log W, \sigma(f) \approx 0.39$ when $f = 0.5$, as is also observed (Cahalan et al. (1994b); see also Gage and Nastrom (1986) and Lilly (1989)).

A two-dimensional (2D) bounded cascade begins with the same initial cloud, which is then divided into quarters along both north-south and east-west lines, and liquid water fractions are then transferred among the quarters. One transfer method is as follows: The four quarters are divided into three pairs, aligned either north-south, east-west or diagonally, with equal probability for each of the three possible ways. One of the pairs is selected randomly, and a fraction $f_0 = f = 0.5$ is transferred within that pair, with either direction equally likely, while a fraction f_0' is transferred within the other pair. For simplicity we also take $f_0' = f$. The process is then repeated by quartering each quarter, transferring a fraction $f_1 = 0.8 \times f$, and so on. The set of optical depth values thus generated at steps $1, 2, 3, \ldots, N$ in the 2D cascade are identical to those generated at the same steps in the 1D cascade, except that each value appears twice in the first step, and 2^N times in the Nth step. The one-point probability distribution functions (the PDFs) of W and τ_v are identical in both 1D and 2D.

Table 8.1 summarizes the symbols and typical values of parameters in the bounded cascade cloud model. In addition to the bounded cascade, two additional assumptions are being made here. One is that the effective droplet radius is uniformly equal to 10 μm, so that the vertical optical thickness of each part of the cloud is linear in the liquid water:[4]

---

[3] If the fractions were kept the same at each step, the resulting distribution would be singular, and the power spectrum would have more small-scale variability than is observed in marine stratocumulus clouds.

[4] The proportionality constant $\tau_v/W = 3/2\rho_w r_e$, where ρ_w is the density of liquid water, in the limit of large size parameters in Mie scattering theory (Stephens, 1978); it equals 0.15 m^2/g if $r_e = 10\,\mu m$.

Table 8.1. Structural and optical parameters for bounded cascade cloud models

Parameter	Symbol	Typical Value
single-scattering albedo	ϖ_0	1
asymmetry	g	0.85
liquid water path	W	100 g/m^2
effective droplet radius	r_e	10 μm
vertical optical thickness	τ_v	15
solar zenith angle	θ_0	60°
scaling parameter	c	0.794
spectral exponent	$\beta(c)$	5/3
variance parameter	f	0.5
reduction factor	$\chi(f,c)$	0.7
effective optical thickness	τ_{eff}	10

$$\tau_v = 0.15\,W \qquad (8.3)$$

where W is expressed in g/m^2.

Second, we employ the Independent Pixel Approximation or IPA, which means that the reflectivity of each cloud pixel is assumed to depend only on its optical depth, $R = R(\tau)$, and not the optical depth of neighboring pixels. This is a strong assumption, and will be justified for the bounded model in the following section.

8.3 Independent Pixel Approximation

The grayscale map in Fig. 8.2a shows the reflectivity of 64×64 cloud cells as computed with a Monte Carlo method for a cloud generated by 6 cascade steps of a 2D bounded cascade with mean vertical optical thickness $\tau_v = 16$, $\theta_0 = 60°$, and fractal parameter $f = 0.5$. If there were no horizontal photon transport, the reflectivity of each of the $2^{12} = 4096$ cloud pixels would simply be determined by independent plane-parallel computations. The local differences between this "independent pixel approximation" (IPA) and the Monte Carlo reflectivities are shown by the grayscale map in Fig. 8.2b. The brighter areas of negative bias occur where the IPA underestimates the reflectivity of an optically thick region which lies on the sunward side of immediately adjacent thin regions and has an enhanced brightness due to photons escaping from those thin regions. Conversely, the darker positive regions occur where the IPA overestimates the brightness of a thin region which lies downstream of an adjacent thick region.[5] These local errors in the IPA can be quite large, with magnitudes exceeding the plane-parallel bias of about 0.1 and in one area even exceeding 0.25. However, the horizontal average of the IPA bias is an order of magnitude smaller than the plane-parallel bias, because the positive and negative regions tend to approximately cancel in the area average.

---

[5] Recall that the cloud has constant geometric thickness everywhere, so that the horizontal photon leakage is not simple geometrical shadowing. It occurs *within* the cloud.

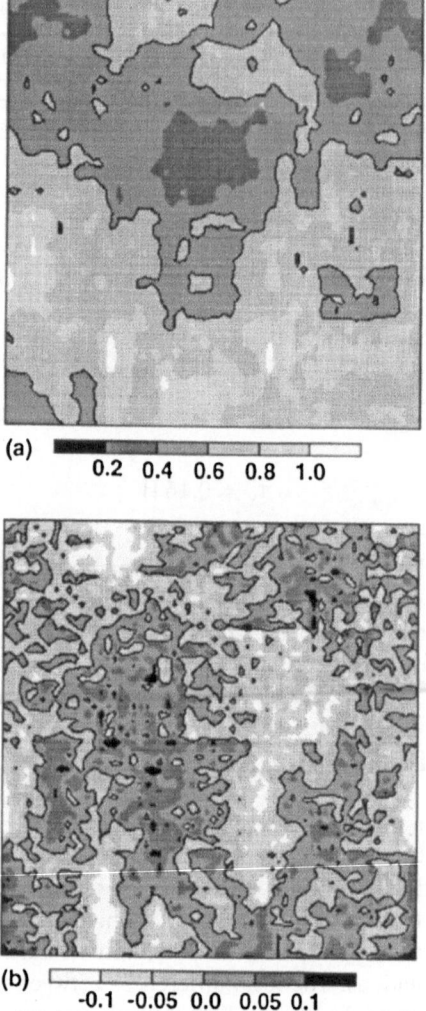

Fig. 8.2. (a) Contours of equal reflectance in a bounded cascade cloud with $A_c = 1$ and $f = 0.5$. Starting with a uniform cloud having mean vertical optical thickness $\bar{\tau} = \tau_v = 16$, 6 cascades were generated in each horizontal direction, giving $2^{12} = 4096$ uniform elements or "pixels". Reflectivities were computed by Monte Carlo with 10^7 photons. Microphysical properties are uniform, with single-scatter albedo $\varpi_0 = 1$ and asymmetry factor $g = 0.85$. The Henyey-Greenstein phase function was used, but essentially identical results are obtained from the fair weather cumulus phase function. The sun is $60°$ to the left of vertical. The black contour at 0.6 shows approximately where the reflectance equals the mean reflectance, with more reflective regions lighter, and less reflective regions darker. (b) Contours of equal "independent pixel bias" defined as the independent pixel reflectances (computed from the vertical optical thickness of each pixel) minus the Monte Carlo reflectances shown in (a). The average of these local algebraic biases is nearly an order of magnitude smaller than the "plane-parallel bias" namely the mean optical thickness minus the mean of the independent pixel reflectances, which is about 0.08

The IPA has a long history of use in remote sensing and was employed in a theoretical study by Ronnholm et al. (1980). But without any explicit model of the spatial structure, early studies could not examine the errors in the IPA. Here we find significant local errors in the IPA fluxes for the 2D bounded cascade, even though the model does not include geometrical cloud effects. The IPA is justified for the bounded cascade *only* for mesoscale-averaged fluxes, and even this simplification breaks down in the case of a singular cascade (Cahalan, 1989; Cahalan et al., 1994a).

When the sun is closer to the zenith than $\theta_0 \approx 60°$, the IPA errors tend to be of the same sign, but much smaller in magnitude. On the other hand, when the sun approaches the horizon, the reflectivity everywhere approaches unity, so all the biases are again smaller than at $\theta_0 \approx 60°$. As a result, the total IPA bias is maximum when the sun is near $60°$ (Cahalan et al., 1994b).

Since the horizontal average of the IPA errors is quite small, we may employ the IPA to estimate the average albedo, and compare it with the albedo of a uniform cloud having the same horizontal average optical depth. Thus we substitute this difference for the "plane-parallel albedo bias" defined in (8.1). It can be shown that the resulting plane-parallel bias is strictly positive as long as the reflection function is convex, unlike the IPA errors. (See Jensen (1906), also Sect. 12.3.) The plane-parallel albedo for the parameters used here is 0.69, while the average of Monte Carlo albedo (i.e., averaging over all pixels in a number of realizations such is the one in Fig. 8.2a) is 0.60. Thus the bias associated with using the area-average optical thickness is 0.09, which is 13% of the plane-parallel albedo.

As a result of the IPA , the mean albedo may be computed by simply transforming the optical depth of each pixel to reflectivity, and then averaging over all pixels. The results in the case of conservative scattering are shown in Fig. 8.3. The upper curve is the plane-parallel ($f = 0$) albedo as a function of mean liquid water path, and the lower curve is the fractal ($f = 0.5$) albedo. For a typical mean liquid water path of $W \approx 100$ g/m^2 ($\tau_v \approx 15$), Fig. 8.3 shows that the plane-parallel albedo of about 0.69 is reduced to about 0.60 by the fractal structure, implying a relative bias of approximately 15%. In order to obtain the correct albedo from a plane-parallel cloud, it is necessary to reduce the liquid water path, or optical thickness, by 30%. An explicit expression for this reduction is derived in Sect. 8.5.

Since in the IPA the reflectivity of a given pixel is a function of the local liquid water path, it may be expanded in a Taylor series as follows:

$$R(W) = R(\overline{W}) + (W - \overline{W})R'(\overline{W}) + \frac{1}{2}(W - \overline{W})^2 R''(\overline{W})$$
$$+ O((W - \overline{W})^3 R''') , \tag{8.4}$$

where $\overline{W}$ is the average liquid water path, and $R'(\overline{W})$, $R''(\overline{W})$ and $R'''(\overline{W})$ are the successive derivatives of R with respect to W, evaluated at $\overline{W}$. (We have suppressed for simplicity the dependence of R on the solar zenith angle.) Averaging both sides of (8.4) eliminates the linear term on the right side, and we obtain

$$\overline{R(W)} = R(\overline{W}) + \frac{1}{2}\mu_2(f)R''(\overline{W}) + O(\mu_3 R''') , \tag{8.5}$$

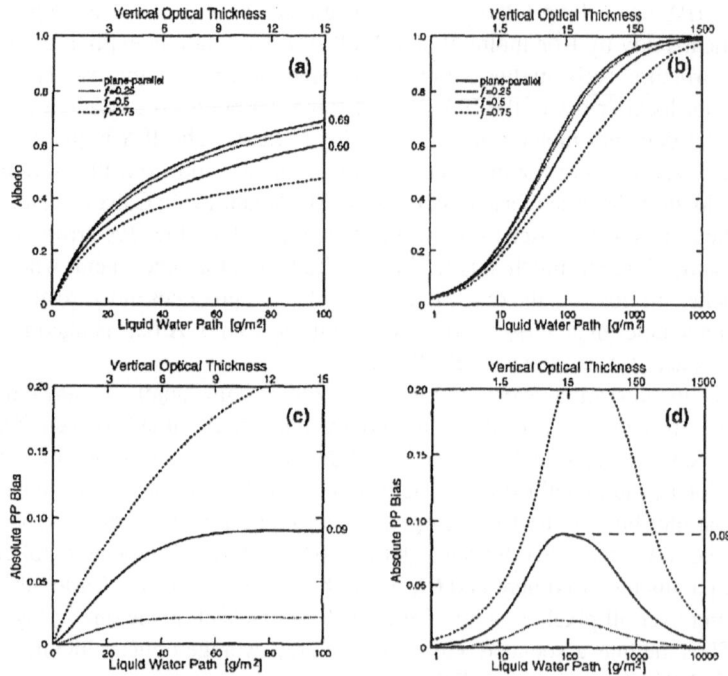

Fig. 8.3. (a) Albedo versus mean liquid water path [0–100 on lower axes], and vertical optical thickness [0–15 on upper axes], for the two approaches shown in Figs. 8.1a and 8.1c, where the fractal case is computed from the bounded model for $A_c = 1$ and $f = 0.25, 0.5, 0.75$, using the independent pixel approximation. (b) Same as (a) except plotted on log scales for liquid water [1–10,000 on lower axes], and optical thickness [0.15–1500 on upper axes]. (c) The "plane-parallel bias" obtained by subtracting the mean reflectance (the lower curves in (a) from the reflectance of the mean (the upper curve). (d) Same as (c) except plotted on the same log scales used in (b). Note from (a) that for the typical $f = 0.5$, when $\tau_v = 15$ the bias is $0.69 - 0.60 = 0.09$, or $0.09/0.69 \approx 15\%$ of the plane-parallel albedo. Drawing a horizontal line in (a) at the mean reflectance 0.60 at $f = 0.5$ shows that this reflectance is that of a plane-parallel cloud having 30% less liquid water, or an optical thickness $\tau_{eff} = 10$. The 30% reduction in cloud liquid corresponds to a value of the "reduction factor" of $\chi = 0.7$ (cf. (8.8) and (8.9)). Lack of significant curvature in (b) near $\tau_v = 15$, compared to the curves in (a), is the reason that an expansion in logs as in (8.6) is preferred over the ordinary Taylor expansion in (8.5)

where μ_2 and μ_3 are the second and the third centered moments, respectively, of the one-point distribution of W generated by the bounded cascade. Subtracting (8.5) from $R(\overline{W})$ gives the plane-parallel albedo bias. The lowest-order term is positive, since the curvature R'' is negative (i.e., Fig. 8.3a shows convex graphs). This term overestimates the bias, while inclusion of the μ_3 term underestimates, and so on (see Cahalan et al., 1994b). Appendix 8.A shows that all the moments of the bounded model may be obtained from μ_2 (as a function of f), thus formally determining all the coefficients in the above expansion. In Sect. 8.5 we consider an alternative expansion about $\overline{\log(W)}$ (see Fig. 8.3b), which leads to a simple expression for the effective liquid water path and effective thickness. First, however, we briefly review the dependence of the bias on cloud fraction, A_c, to show that the overcast case, $A_c = 1$, assumed in the above discussion, is associated with the largest plane-parallel albedo bias during the diurnal cycle of California marine stratocumulus.

8.4 Diurnal Cycle

The total plane-parallel albedo bias has two contributions, as described in (8.2): (1) that due only to cloud fraction, which is given by the albedo for Fig. 8.1a minus that of Fig. 8.1b, and (2) that due to the fractal structure, given by the difference between Fig. 8.1b and Fig. 8.1c. The fractal structure contribution is largest when the liquid water variance is largest, which in the case of California marine stratocumulus occurs during the morning hours, when the cloud fraction is nearly 100%, as shown in Cahalan et al. (1994b). Although the cloud fraction contribution to the bias is larger in the afternoon, when the cloud fraction drops to 60%, this is more than offset by the decrease in the liquid water variance, which reduces both the fractal contribution and the total bias. The fact that the cloud variance is largest when the cloud cover is largest leads to the surprising result that plane-parallel estimates are most in error when the usual "cloud fraction" corrections vanish!

In Cahalan et al. (1994b) the diurnal cycle of the albedo bias was estimated indirectly, by first computing the diurnal cycle of f, determined from hourly values of the variance of $\log W$. Here we compute the bias directly from the time series of W, by performing a plane-parallel computation of reflectance for each observation, and then compositing the results hourly. The direct results agree qualitatively with the earlier indirect approach, and are shown in Fig. 8.4. Here the lower curve is the usual correction due only to cloud fraction, and vanishes when the fraction reaches 100% around 10 am. The middle curve is the additional correction due to the fractal distribution of the cloud liquid water. The upper curve is the total albedo bias. Note that the cloud fraction correction is much smaller than the total, and is $180°$ out of phase with the total during most of the day (except when the sun is setting) when the total is dominated by the cloud fraction correction due to the neglect of the clear-sky albedo. The 0.09 albedo reduction needed when the clouds are overcast represents a major change in the average cloud albedo of 0.6.

8.5 Effective Optical Thickness

Since the largest albedo bias occurs for overcast cloudiness conditions, when $A_c = 1$, let us further consider that case, represented by the 15% increase in Fig. 8.3 of the albedo of a plane-parallel cloud over that of a fractal with the same total cloud water. As shown in Cahalan et al. (1994b), this bias may be estimated from a simple "effective thickness approximation" which is a lowest-order approximation to the

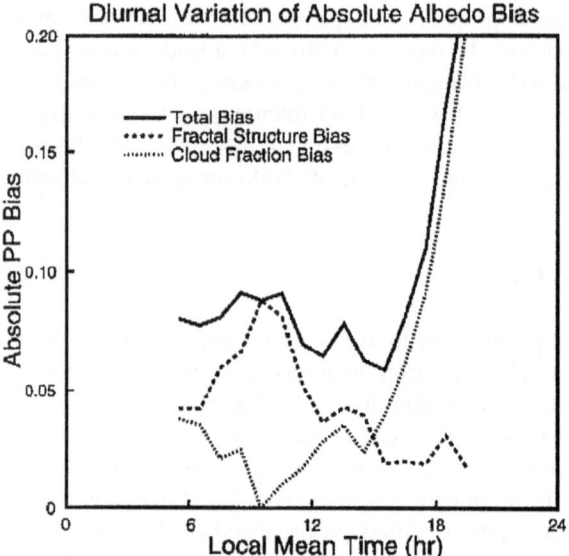

Fig. 8.4. Absolute plane-parallel albedo bias as a function of time-of-day for California marine stratocumulus, determined directly from microwave measurements of liquid water path during 18 days in June 1987, by computing an independent reflectivity from each measurement. The same computation using the bounded cascade model with diurnally varying f and A_c is given in Cahalan et al. (1994b), and is qualitatively similar. The upper *solid curve* is the total bias defined as in (8.1), while the *dotted* and *dashed* curves are the contributions due to cloud fraction and fractal structure, respectively, as defined in (8.2). Cloud fraction is defined as the fraction of values exceeding 10 g/m^2, and clear-sky albedo is taken to be zero

bias determined from the IPA. To derive it, consider an expansion similar to (8.5), except now the local reflectance is considered as a function of the *logarithm* of the local liquid water path, $\log W$, and is expanded in a Taylor series about the mean, $\overline{\log W}$. Taking the mean of the result gives the mean cloud reflectivity as:

$$\overline{R(\log W)} = R(\overline{\log W}) + \frac{1}{2}M_2(f)R''(\overline{\log W}) + O(M_4R'''') , \qquad (8.6)$$

where M_2 is the variance of $\log W$, given in Appendix 8.B, and R'' is the second derivative of R with respect to $\log W$ evaluated at the mean of $\log W$. As a function of $\log W$, the conservative reflection function has an inflection point, where the slope stops increasing with $\log W$ and begins to decrease, and the curvature goes through zero, as seen in Fig. 8.3b. This typically occurs near $\overline{\log W}$. Thus the second term in the preceding equation is small, so that the mean reflectivity is approximately given by the reflectivity evaluated at $\overline{\log W}$. In the bounded cascade model, the mean of $\log W$ is given by

$$\overline{\log W} = \log W_{\text{eff}} , \qquad (8.7)$$

where

$$W_{\text{eff}} = \overline{W} \times \chi(f, c) \, , \qquad (8.8)$$

and $\chi(f, c) < 1$ is the "reduction factor" given in Appendix 8.B of this chapter, and is approximately 0.7 when $f = 0.5$ and $c = 0.8$, the appropriate values for typical cloud liquid water distributions (see below).

Combining (8.8) with (8.3), allows us to define the "effective optical thickness":

$$\tau_{\text{eff}} = \tau_{\text{v}} \times \chi(f, c) \, , \qquad (8.9)$$

where τ_{v} is the mean vertical optical thickness. Taking only the first term in the expansion in (8.6), and using (8.3), it is clear that for a range of intermediate mean cloud thicknesses near the inflection point of the reflectivity, the mean albedo may be approximated by the plane-parallel albedo evaluated at the effective thickness, as follows:

$$\overline{R(\tau)} \approx R(\tau_{\text{eff}}) \, , \qquad (8.10)$$

An estimate of the plane-parallel albedo bias may be obtained by subtracting (8.10) from the plane-parallel albedo, $R(\tau_{\text{v}})$. The relative error in the bias estimate derived

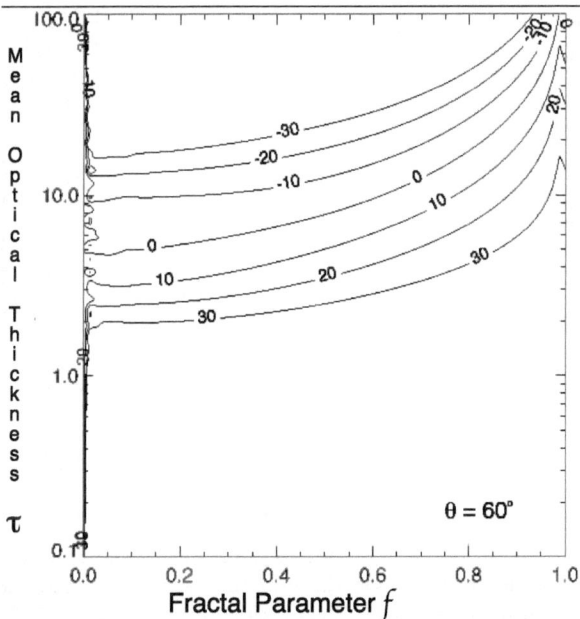

Fig. 8.5. Relative error in percent in the plane-parallel bias when the actual albedo is approximated by the plane-parallel albedo at a reduced "effective thickness" as a function of mean optical thickness τ_{v} and fractal parameter f. If the effective thickness gives an absolute bias of 0.10 near the −20 contour, for example, then the actual bias should be increased 20%, to 0.12, and similarly an estimate of 0.10 near then +20 contour should be decreased to 0.09. These same corrections can also be applied to the relative bias

from (8.10) is shown in Fig. 8.5 as a function of f and τ_v, for $c = 0.8$ and a solar zenith angle of $\theta_0 = 60°$, which is typical for stratocumulus. For the contours labeled ± 30, for example, the bias obtained from the simple effective thickness approximation should be multiplied by 1 ∓ 0.3. Since the bias itself is on the order of 0.1, this corresponds to corrections of $\approx \mp 0.03$. The correction is dominated by the M_2 term in (8.6), and thus changes sign near the inflection point of $R(\log W)$.

According to (8.9), the effective optical thickness depends on the fractal structure through χ, which is a known analytic function of fractal parameters f and c. The fractal parameter f is in turn adjusted to give the observed value of σ, also a known analytic function of f and c, while c is fixed by the exponent of the wavenumber spectrum. Thus τ_{eff} is parametrically determined as a function of σ by varying f. Details are given in Appendix 8.B, and results are shown in Fig. 8.6 for both $c = 2^{-1/3} \approx 0.8$ needed to give a $\beta = 5/3$ wavenumber spectrum, and for the singular limit $c \to 1$, for which χ is a simple exponential given in (8.B.11). The point labeled

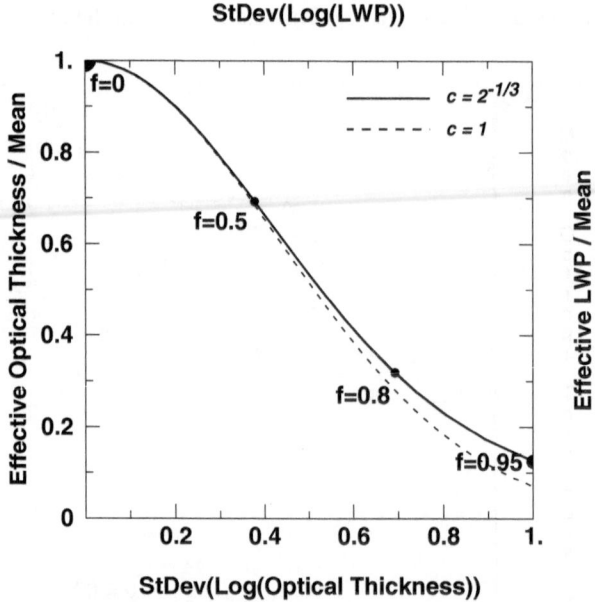

Fig. 8.6. Plot of χ, the reduction factor, versus σ, the standard deviation of $\log W$. Both the horizontal and vertical scales are independent of the number of cascade steps, and apply to either W or τ because of the simple linear relation expressed in (8.3). The *solid curve* is for the bounded model with $c \approx 0.8$, while the *dashed curve* is the singular limit given by the simple expression in (8.B.11). Labeled points apply only to the *upper curve*. The value of σ derived from observations of California marine stratocumulus is 0.39, corresponding to $\chi \approx 0.7$, which occurs at $f = 0.5$. (This is the diurnal mean in the summer, when f varies from about 0.6 in the morning to 0.3 in the afternoon.) The global reduction factor $\chi \approx 1/3$ discussed by Harshvardhan and Randall (1985) occurs at $f = 0.8$, and requires a global value of $\sigma \approx 0.7$

$f = 0.5$ in Fig. 8.6 corresponds to the diurnal average value of $\sigma = 0.39$, determined from the stratocumulus observations discussed in Sect. 8.4. In this case $\chi \approx 0.7$, so for example when $\tau = 15$ we have $\tau_{\text{eff}} \approx 10$, and $\overline{R} \approx 0.6$.

Harshvardhan and Randall (1985) found that the global average cloud liquid must be reduced by a factor of approximately 0.3 in order to obtain the correct global albedo. To obtain this value of the reduction factor, $\chi = 0.3$, for the bounded cascade model requires an increase in the fractal parameter to $f = 0.8$, and an increase in the standard deviation to $\sigma = 0.7$, as seen in Fig. 8.6. This in turn increases the plane-parallel albedo bias by a factor of 5. The fact that a much larger bias is found on a global basis is presumably due to the much wider variation in cloudiness over the globe, as compared to the relatively benign variation in marine stratocumulus. Davis et al. (1990) considered a related quantity, the "packing factor" the inverse of the reduction factor, and studied the thick cloud limit in a singular model, for which $\chi \to 0$, and the packing factor diverges. A similar singular model was studied in Cahalan (1989). The bounded model considered here is a relatively conservative extension of the plane-parallel idealization. More radical, and perhaps singular, models may be needed to better represent radiative processes in deep convective cloud systems.

8.6 Conclusions

A number of results on the mesoscale-average albedo of marine stratocumulus clouds, known to be a major contributor to cloud radiative forcing, have been reviewed in this chapter. A fractal cloud model which reproduces the observed power spectrum and low-order moments of the liquid water distribution in these clouds was studied by both 3D Monte Carlo and analytic methods. Local horizontal fluxes were determined from a 2D bounded cascade in Sect. 8.3, showing that errors in estimates of such fluxes by the Independent Pixel Approximation or IPA can be large in some regions, though still producing an area-averaged reflectivity accurate to about 1%. Section 8.4 discusses the diurnal cycle of the variability of marine stratocumulus, showing the plane-parallel biases are largest when cloud fraction is near 100%.

The results suggest a way of parameterizing the impact of such cloud variability on the large-scale albedo in terms of an "effective" liquid water path, W_{eff} (or, equivalently, an "effective" optical thickness, τ_{eff}), smaller than the mean by a factor which depends on the fractal cloud structure. Section 8.5 determined the accuracy of this Effective Thickness Approximation or ETA, as a function of the fractal parameter f and the mean liquid water path $\overline{W}$ (or equivalently the mean optical thickness τ_{v}). The ratio of $\chi = W_{\text{eff}}/\overline{W}$ (or $\tau = \tau_{\text{eff}}/\tau_{\text{v}}$) was determined as an analytic function of the fractal parameters, and as a parametric function of σ, the standard deviation of $\log W$ (or $\log \tau$), which may be estimated from observations.

For marine stratocumulus, we find $\sigma \approx 0.4$ and $\chi \approx 0.7$ giving a mean albedo approximately equal to that of a plane-parallel cloud having 30% less liquid water, or 15% less than the plane-parallel albedo of a cloud with the same liquid water amount. A surprising result is that the plane-parallel albedo requires the largest adjustment when the cloud fraction is nearly 100%, since that is when the largest variability is observed. Thus the largest correction occurs when the usual cloud fraction correction is small.

The bounded cascade model studied here represents a relatively conservative extension to plane-parallel clouds, since the cloud height and base are fixed. We also keep the microphysics uniform. Yet even this conservative model shows that the variability of liquid water in marine stratocumulus can have a larger impact on the mesoscale average albedo than the usual cloud fraction corrections. For cloud types not confined to a single vertical layer, such as those found in deep convective regions, geometrical fractal properties neglected here may also impact large-scale radiative properties, and may well require more radical departures from conventional plane-parallel ideas. Fractal models for various surface types, including topography, vegetation and sea ice, need to be combined with cloud models in order to fully understand effects of inhomogeneity on atmospheric radiative transfer (see, e.g., Rozwadowska and Cahalan, 2002). Further study of the structure and radiation of real clouds and surfaces in their full complexity will be needed in order to understand how Earth's climate is being regulated, and in order to consistently quantify the role played by Earth's cloud systems on the energy and hydrological cycles.

Appendices

8.A Rescaling f Generates W Moments

Here we derive expressions for the moments of a bounded cascade, as a function of the cascade parameters f and c. We show that the moments may all be obtained from the second moment considered as a function of f, by rescaling the values of f. We then consider the singular limit $c \to 1$, and show that all moments approach those of a lognormal.

It is convenient to first define two sets of nth-order polynomials:

$$P_n(x) \equiv \frac{(1 + \sqrt{x})^{2n} + (1 - \sqrt{x})^{2n}}{2} = \sum_{m=0}^{n} \binom{2n}{2m} x^m, \qquad (8.A.1)$$

and

$$Q_n(x) \equiv \frac{(1 + \sqrt{x})^{2n+1} + (1 - \sqrt{x})^{2n+1}}{2} = \sum_{m=0}^{n} \binom{2n+1}{2m} x^m. \qquad (8.A.2)$$

For example, the first four are given by:

n	$P_n(x)$	$Q_n(x)$	
1	$1 + x$	$1 + 3x$	
2	$1 + 6x + x^2$	$1 + 10x + 5x^2$	(8.A.3)
3	$1 + 15x + 15x^2 + x^3$	$1 + 21x + 35x^2 + 7x^3$	
4	$1 + 28x + 70x^2 + 28x^3 + x^4$	$1 + 36x + 126x^2 + 84x^3 + 9x^4$.	

Realizations of bounded cascade have the form

$$W = \prod_{k=0}^{\infty} \left(1 \pm f\, c^k\right) , \tag{8.A.4}$$

where $f, c \in (0, 1]$. After averaging over $\pm$, the moments of W depend only on $a = f^2$ and $s = c^2$, and can be written in terms of the above polynomials in the form:

$$\mu_{2n}(a, s) = \prod_{k=0}^{\infty} P_n\left(as^k\right) , \tag{8.A.5}$$

and

$$\mu_{2n+1}(a, s) = \prod_{k=0}^{\infty} Q_n\left(as^k\right) . \tag{8.A.6}$$

For example, when $n = 1$,

$$\mu_2(a, s) = \prod_{k=0}^{\infty} \left(1 + as^k\right) = 1 + \sum_{m=1}^{\infty} \left(\frac{s^{m(m+1)/2}}{\prod_{k=1}^{m}\left(1 - s^k\right)}\right) a^m , \tag{8.A.7}$$

and

$$\mu_3(a, s) = \prod_{k=0}^{\infty} \left(1 + 3as^k\right) = \mu_2(3a, s) . \tag{8.A.8}$$

The last expression for μ_2 in (8.A.7) was originally derived by Euler, as discussed by Hardy and Wright (1979), page 280. Taking the limit $s \to 1$, we can use the fact that

$$\lim_{s \to 1} \frac{1 - s^k}{1 - s} = k$$

to show that

$$\lim_{s \to 1} \left(\frac{\mu_2}{\exp(\frac{a}{1-s})}\right) = 1 , \tag{8.A.9}$$

which implies an essential singularity in μ_2. The third moment is also singular, since

$$\lim_{s \to 1} \left(\frac{\mu_3}{(\mu_2)^3}\right) = 1 . \tag{8.A.10}$$

We now generalize (8.A.8) and (8.A.10) to the remaining moments. By application of Sturm's theorem on polynomial roots localization, it can be shown that the roots of P_n, Q_n all lie on the negative real axis, so that we may write:

$$P_n(x) = \prod_{i=1}^{n} \left(1 + R_i^{(n)} x\right) , \tag{8.A.11}$$

and

$$Q_n(x) = \prod_{i=1}^{n} \left(1 + \tilde{R}_i^{(n)} x\right) \tag{8.A.12}$$

where the $R^{(n)}$, $\tilde{R}^{(n)}$ are sets of positive real numbers with n elements. The first four sets are:

n	$R^{(n)}$	$\tilde{R}^{(n)}$	
1	1	3	
2	$3 - \sqrt{8},\ 3 + \sqrt{8}$	$5 - \sqrt{20},\ 5 + \sqrt{20}$	(8.A.13)
3	$7 - \sqrt{48},\ 1,\ 7 + \sqrt{48}$	$0.232,\ 1.572,\ 19.196$	
4	$0.040,\ 0.446,\ 2.240,\ 25.274$	$0.132,\ 0.704,\ 3,\ 32.163$.	

Moments of order $2n$ and $2n + 1$ thus factor into n products:

$$\mu_{2n}(a, s) = \prod_{i=1}^{n} \mu_2 \left(R_i^{(n)} a, s\right) , \tag{8.A.14}$$

and

$$\mu_{2n+1}(a, s) = \prod_{i=1}^{n} \mu_2 \left(\tilde{R}_i^{(n)} a, s\right) , \tag{8.A.15}$$

so that all moments are determined by products of the second moment evaluated at various rescaled values of the fractal parameter $a = f^2$.

Combining (8.A.9) and (8.A.14), we find that in the singular limit,

$$\lim_{s \to 1} \left(\frac{\mu_{2n}}{(\mu_2)^{\sum_{i=1}^{n} R_i^{(n)}}}\right) = 1 , \tag{8.A.16}$$

with a similar expression for the odd moments with $R_i \to \tilde{R}_i$. The sum of the roots can be shown to equal the coefficient of the linear term in (8.A.1), so that:

$$\sum_{i=1}^{n} R_i^{(n)} = 2n(2n - 1)/2 , \tag{8.A.17}$$

and similarly for $\tilde{R}_i$. The limits for the even and odd moments can then be combined to yield:

$$\lim_{s \to 1} \left(\frac{\mu_n}{(\mu_2)^{n(n-1)/2}}\right) = 1 , \tag{8.A.18}$$

consistent with the behavior of moments of a lognormal.

Summary

By considering the roots of P_n, Q_n in (8.A.11)–(8.A.12), we have seen in this Appendix that

1. all the moments are explicitly determined by the second moment evaluated at rescaled values of f, cf. (8.A.14) and (8.A.15); and
2. that the moments progress in the same ratios as for a lognormal distribution.

Note that this second fact does *not* imply that the pdf is lognormal, since a lognormal is *not* uniquely determined by its moments, though it is determined by the *moments of the logarithm*, which are discussed in Appendix 8.B. However, having the same ratios as a lognormal does show that the higher moments diverge in the same way as a lognormal when $s = c^2 \to 1$, each with an essential singularity, just like the variance in (8.A.9), diverging like $\exp[a/(1 - s)]$.

8.B Reduction Factor and Variance of log W

Here we derive simple polynomial approximations for the reduction factor and the standard deviation of $\log W$, or equivalently $\log \tau$, as a function of the fractal parameter f with coefficients depending on c. For $c = 2^{-1/3}$ (i.e., for –5/3 wavenumber spectral exponent) these are well approximated by rational functions of f accurate for $f < 0.9$. Also, we show that the reduction factor is approximately given by $\chi \approx \exp^{-\sigma^2/2}$, and insensitive to c as seen in Fig. 8.6.

The "effective" optical thickness defined in Sect. 8.5 is based on the following result for the liquid water path W:

$$\overline{\log W} = \log\left(\overline{W}\,\chi(f,c)\right), \tag{8.B.1}$$

where the overbar signifies an area and ensemble average, and where the "reduction factor" is given by

$$\chi(f,c) = \left(\prod_{n=0}^{\infty}\left(1 - f^2 c^{2n}\right)\right)^{1/2}. \tag{8.B.2}$$

Here f varies diurnally, as discussed in Sec. 8.4, but c is assumed constant, given by $c = 2^{-1/3}$, or

$$c^2 = 0.630, \tag{8.B.3}$$

as required for a $k^{-5/3}$ wavenumber spectrum. Equations (8.B.1) and (8.B.2) were derived in Cahalan et al. (1994b) assuming the statistical distribution generated by the bounded cascade model. The reduction factor may also be expressed as

$$\chi(f,c) = 10^{-\Delta(f,c)}, \tag{8.B.4}$$

where

$$\Delta(f,c) \equiv \log \overline{W} - \overline{\log W}. \tag{8.B.5}$$

A polynomial expression for Δ is obtained by taking log of (8.B.2), changing to base e by multiplying by log e, and expanding in a power series in f, leading to

$$\Delta(f,c) = \frac{\log e}{2} \frac{f^2}{1-c^2} \left(1 + \frac{f^2}{1+c^2} + O\left(f^4\right)\right). \tag{8.B.6}$$

For the value of c^2 given in (8.B.3), a better fit than (8.B.6) is given by the rational approximant:

$$\Delta(f) = 0.594 \, f^2 \left(\frac{1 - 0.485 f^2}{1 - 0.739 \, f^2}\right), \tag{8.B.7}$$

which is accurate to 1% as long as $f < 0.9$.

The second moment of $\log W$ was derived in Cahalan et al. (1994b), and is given by

$$M_2(f) = \sum_{k=1}^{\infty} \left(\frac{1}{2} \log \left(\frac{1 + f c^k}{1 - f c^k}\right)\right)^2. \tag{8.B.8}$$

If we take the square root of (8.B.8), and expand the result in powers of f, we obtain the standard deviation of $\log W$ in the form:

$$\sigma(f,c) = \frac{f \log e}{\sqrt{1-c^2}} \left(1 + \frac{1}{3} \frac{f^2}{1+c^2} + O\left(f^4\right)\right). \tag{8.B.9}$$

The first term here agrees with the standard deviation obtained by taking the square root of the exponent of μ_2 in the singular limit in (8.A.9). For the value of c^2 in (8.B.3) a better fit is given by the approximant:

$$\sigma(f) = 0.718 \, f \left(\frac{1 - 0.556 \, f^2}{1 - 0.729 \, f^2}\right), \tag{8.B.10}$$

which is accurate to 1% as long as $f < 0.9$.

Solving for f in (8.B.9), and substituting the result in (8.B.6) allows us to write (8.B.4) to lowest order as:

$$\chi(\sigma) = 10^{-\sigma^2/2 \log e} \approx 10^{-1.15 \sigma^2}. \tag{8.B.11}$$

The leading term in the exponent in (8.B.11) is independent of c, and the correction terms are of order σ^4 and quite small as long as $\sigma < 0.8$. The insensitivity to c is verified in Fig. 8.6.

References

Cahalan, R.F. (1989). Overview of fractal clouds. In *Advances in Remote Sensing*. A. Deepak Publishing, Hampton, VA, pp. 371–388.

Cahalan, R.F. (1994). Bounded cascade clouds: Albedo and effective thickness. *Nonlinear Proc. Geophys.*, **1**, 156–167.

Cahalan, R.F. and J.B. Snider (1989). Marine stratocumulus structure during FIRE. *Remote Sens. Environ.*, **28**, 95–107.

Cahalan, R.F. and W.J. Wiscombe (1993). Impact of cloud structure on climate. In *Current Problems in Atmospheric Radiation*. A. Deepak Publishing, Hampton, VA, pp. 120–124.

Cahalan, R.F., M. Nestler, W. Ridgway, W.J. Wiscombe, and T. Bell (1990). Marine stratocumulus spatial structure. In *Proceedings of the Fourth International Conference on Statistical Climatology*, New Zealand Meteorological Service, Wellington, N.Z., J. Samson (ed.). pp. 28–32.

Cahalan, R.F., W. Ridgway, W.J. Wiscombe, S. Gollmer, and Harshvardhan (1994a). Independent pixel and Monte Carlo estimates of stratocumulus albedo. *J. Atmos. Sci.*, **51**, 3776–3790.

Cahalan, R.F., W. Ridgway, W.J. Wiscombe, T.L. Bell, and J.B. Snider (1994b). The albedo of fractal stratocumulus clouds. *J. Atmos. Sci.*, **51**, 2434–2455.

Cahalan, R.F., D. Silberstein, and J. Snider (1995). Liquid water path and plane-parallel albedo bias during ASTEX. *J. Atmos. Sci.*, **52**, 3002–3012.

Davis, A., P.M. Gabriel, S.M. Lovejoy, D. Schertzer, and G.L. Austin (1990). Discrete angle radiative transfer III: Numerical results and meteorological applications. *J. Geophys. Res.*, **95**, 11,729–11,742.

Gabriel, P.M., S.M. Lovejoy, A. Davis, D. Schertzer, and G.L. Austin (1990). Discrete angle radiative transfer II: Renormalization approach for homogeneous and fractal clouds. *J. Geophys. Res.*, **95**, 11,717–11,728.

Gage, K. and D. Nastrom (1986). Atmospheric wavenumber spectra of wind and temperature observed by comercial aircraft during GASP. *J. Atmos. Sci.*, **43**, 729–740.

Hardy, G.H. and E.M. Wright (1979). *An Introduction to the Theory of Numbers*. Oxford University Press, Oxford, UK.

Harshvardhan and D. Randall (1985). Comments on "The parameterization of radiation for numerical weather prediction and climate models". *Mon. Wea. Rev.*, **113**, 1832–1833.

Jensen, J.L.W.V. (1906). Sur les fonctions convexes et les inégalitiés entre les valeurs moyennes. *Acta Math.*, **30**, 175–193.

Kiehl, J.T. and K.E. Trenberth (1997). Earth's annual global mean energy budget. *Bull. Amer. Meteor. Soc.*, **78**, 197–208.

Lilly, D.K. (1989). Two-dimensional turbulence generated by energy sources at two scales. *J. Atmos. Sci.*, **46**, 2026–2030.

Lovejoy, S., A. Davis, P. Gabriel, D. Schertzer, and G. Austin (1990). Discrete angle radiative transfer I: Scaling and similarity, universality and diffusion. *J. Geophys. Res.*, **95**, 11,699–11,715.

Marshak, A., A. Davis, R.F. Cahalan, and W.J. Wiscombe (1994). Bounded cascade models as non-stationary multifractals. *Phys. Rev. E*, **49**, 55–69.

Ramanathan, V., R.D. Cess, E.F. Harrison, P. Minnis, B.R. Barkstrom, E. Ahmad, and D. Hartmann (1989). Cloud-radiative forcing and climate: Results from the Earth-radiation Budget Experiment. *Science*, **243**, 57–63.

Ronnholm, K., M.B. Baker, and H. Harrison (1980). Radiation transfer through media with uncertain or random parameters. *J. Atmos. Sci.*, **37**, 1279–1290.

Rozwadowska, A. and R.F. Cahalan (2002). Plane-parallel biases computed from inhomogeneous clouds and sea ice. *J. Geophys. Res.*, **107**, 4384–4401.

Stephens, G.L. (1978). Radiation profile in extended water clouds - Part 1, Theory. *J. Atmos. Sci.*, **35**, 12111–2122.

Stephens, G.L. (1985). Reply (to Harshvardhan and Randall). *Mon. Wea. Rev.*, **113**, 1834–1835.

Sundqvist, H., E. Berge, and J.E. Kristjansson (1989). Condensation and cloud parameterization studies with a mesoscale Numerical Weather Prediction model. *Mon. Wea. Rev.*, **117**, 1641–1657.

Tiedtke, M. (1996). An extension of cloud-radiation parameterization in the ECMWF model: The representation of subgridscale variations in optical depth. *Mon. Wea. Rev.*, **124**, 745–750.

Suggested Reading

1. For an early clear description of the cascading effect of dynamical processes on cloud structure, see:

Welander, P. (1955). General development of motion in a 2D ideal fluid. *Tellus*, **7**, 141–156.

A modern and popularized discussion of the evolution of cascades is in

Barabasi, A.-L. (2003). *Linked: The New Science of Networks*. 256 pp., Perseus Publishing, Boulder (CO).

For a useful general approach to treating clouds in large-scale models, see:

Tiedtke, M. (1993). Representation of clouds in large-scale models. *Mon. Wea. Rev.* **121**, 3040–3061.

2. There are many interesting alternatives to the "effective thickness approximation" (ETA). Listed here are a few that are particularly instructive. A "generalized ETA" is described in:

Szczap, F., H. Isaka, M. Saute, B. Guillemet and A. Iolthukhovski (2000). Effective radiative properties of bounded cascade non-absorbing clouds: Definition of the equivalent homogeneous cloud approximation. *J. Geophys. Res.*, **105**, 20,617–20,633.

A generalization of the IPA to account for sun angle effects is the "tilted IPA" or "TIPA" of:

Várnai, T. and Davies, R. (1999). Effects of cloud heterogeneities on shortwave radiation: comparison of cloud-top variability and internal heterogeneity. *J. Atmos. Sci.*, **56**, 4206–4224.

Especially useful in the context of the "two-stream approximation" often used in large-scale models is the "Gamma-weighted two-stream approximation" described for example in:

Barker, H.W. (1996). A parameterization for computing grid-averaged solar fluxes for inhomogeneous marine boundary layer clouds - Part 1, Methodology and homogeneous biases. *J. Atmos. Sci.,* **53**, 2289–2303.

Oreopoulos, L. and H.W. Barker (1999). Accounting for subgrid-scale cloud variability in a multi-layer 1D solar radiative transfer algorithm. *Quart. J. Roy. Meteor. Soc.,* **126**, 301–330.

An interesting approach to taking advantage of cloud scaling properties in parameterizing effective cloud properties is the "renormalization" method described in (see also Chap. 6):

Cairns, B., A.A. Lacis and B.E. Carlson (2000). Absorption within inhomogeneous clouds and its parameterization in general circulation models. *J. Atmos. Sci.* **57**, 700–714.

A model that treats a cloud distribution analogously to the drop distribution within a cloud is the "independent scattering cloudlets" model described in:

Petty, G.W. (2002). Area-average solar radiative transfer in three-dimensionally inhomogeneous clouds: the independently scattering cloudlets model. *J. Atmos. Sci.,* **59**, 2910–2929.

---

3. There are many generalizations of the simple bounded cascade. For example, see:

Gollmer, S., Harshvardhan, R.F. Cahalan and J.B. Snider (1995). Windowed and wavelet analysis of marine stratocumulus cloud inhomogeneity. *J. Atmos. Sci.,* **52**, 3013–3030.

There are also a multiplicity of cloud structure models that can be compared to the bounded cascade. We list here a few that are appropriate for various cloud types and applications. A model helpful in understanding large-scale cloud statistics, especially in the storm tracks, is the "cloud dot" model given in the appendix of:

Cahalan, R.F., D.A. Short and G.R. North (1982). Cloud fluctuation statistics. *Mon. Wea. Rev.,* **110**, 26–43.

Interesting cascades have been developed for other geophysical fields, such as rainfall and vegetation. See for example:

Waymire, E. and V.J. Gupta (1981). The mathematical structure of rainfall representations 1. A review of the stochastic rainfall models. *Water Resour. Res.,* **17**, 1261–1294.

As c approaches unity in the bounded cascade, we reach an "essential " singularity that prevents the analytic continuation of the cloud properties to the singular $c = 1$ case. The singular case can be treated directly, as in the so-called "p-model" of

Meneveau, C. and K.R. Sreenivasan (1987). Simple multifractal cascade model for fully developed turbulence. *Phys. Rev. Lett.,* **59**, 1424–1427.

Singularities in unbounded cascades can be tamed by integration, for example (see also the volume's Appendix at the end of the book) in the "fractionally integrated" model of:

Schertzer, D. and S. Lovejoy (1987). Physical modeling and analysis of rain and clouds by anisotropic scaling multiplicative processes. *J. Geophys. Res.,* **92**, 9693–9714.

The above is not meant to be an exhaustive bibliography, which would take many pages, but only to point the reader in some interesting directions that may be helpful in their own areas of interest. Simple models of complex structure share many common features, so that often qualitative behavior discovered in the context of one simple model such as the bounded cascade, or any other listed above, may have surprisingly broad applicability. A model has served its purpose well if it prompts the researcher to ask productive questions that lead to a better understanding.

9

Broadband Irradiances and Heating Rates for Cloudy Atmospheres

H.W. Barker

9.1   Introduction and Outline 449

9.2   Clouds, Radiation, and Climate 450

9.3   Characteristics of Clear-Sky Transmittance, Surface Albedo,
      and Cloud Optical Properties 454

9.4   Computing Broadband Irradiances
      using Monte Carlo Algorithms 458

9.5   Computing Broadband Irradiances using 1D Algorithms 462

9.6   1D versus 3D Radiative Transfer 470

9.7   Summary and Conclusions 478

References 479

Suggested Reading 482

9.1 Introduction and Outline

Arguably, much of what makes Earth's climate so complex, difficult to predict, and worthy of study arises from the four-dimensional interaction between broadband (BB) radiation and the three phases of water. The most elusive of these interactions almost certainly involves clouds. Indeed, it is widely recognized that interactions between clouds and radiation are central to Earth's climate and thus climatic change (e.g., Intergovernmental Panel on Climate Change, 1996). Though radiative transfer in narrow spectral ranges is often key to climatic sub systems (e.g., stratospheric dynamics, living organisms), anthropogenic climate forcings (e.g., increasing concentrations of CO_2 and CFCs), and remote sensing, Earth's hydrologic cycle is determined much by the BB radiative budget of the surface-lower atmosphere system which is, in turn, governed much by clouds. Thus, given the lofty climatic role of

BB radiation and the heavy restrictions on modelling clouds and radiative transfer in large-scale atmospheric models (LSAMs), it is essential that this book contain an extended discussion of BB radiative transfer for cloudy atmospheres.

As in most studies, BB irradiances are defined here for incoming solar radiation and terrestrial radiation emitted by the Earth-atmosphere system. Naturally these sources overlap but for the most part they can be considered as exclusive and to span wavelengths between $0.2 - 5$ μm and $5 - 50$ μm, respectively.

The second section of this chapter gives a brief overview of how radiation figures into the climate system and the important role that BB radiation plays in diagnosing both real and modelled climate. The third section briefly presents popular methods for computing BB irradiances as well as issues facing all BB radiation codes. The fourth and fifth sections discuss modelling BB radiative transfer using 3D Monte Carlo methods and 1D algorithms, respectively, with emphasis on realistic cloudy atmospheres. Here the focus is on solar radiation because: i) much more research has been done on solar transfer through realistic cloud fields than on longwave transfer; and ii) BB longwave transfer is addressed thoroughly in Chap. 10 of this book. In the sixth section results from 3D Monte Carlo algorithms and 1D models are compared and contrasted. The final section provides a summary and conclusion.

9.2 Clouds, Radiation, and Climate

The purpose of this section is to provide an overview of the climatic importance of clouds from a BB radiative standpoint. It discusses the impact of clouds on Earth's radiation budget and the prominent role of radiation in cloud-climate feedback processes.

9.2.1 The Impact on Clouds on Earth's Radiation Budget

Perhaps the most straightforward and simple diagnostic measure of the impact of clouds on Earth's radiation budget is *cloud radiative forcing*. Since cloud radiative forcing is not a *forcing* in the conventional sense of a perturbation to a system in (quasi-)equilibrium but rather the integrated radiative impact of clouds as they (and the rest of the system) evolved over a period of time, this has prompted some to use the more apt term *cloud radiative effect* (CRE). Since CRE is a more precise term, it is used in this chapter and defined as

$$CRE = F_{all} - F_{clr} , \qquad (9.1)$$

where F_{all} and F_{clr} are generally net BB irradiances for all-sky and cloudless conditions. Irradiances can be measured at any level, including the surface, or they can be replaced with net fluxes for some layer, including the entire atmosphere. More often than not, F_{clr} and F_{all} in (9.1) are discussed in terms of monthly means. Net fluxes at the top of the atmosphere (TOA) are used most commonly to define CRE. Not only is this because the TOA represents the ultimate energy budget level, but satellites are the only means of providing near-global coverage of CRE.

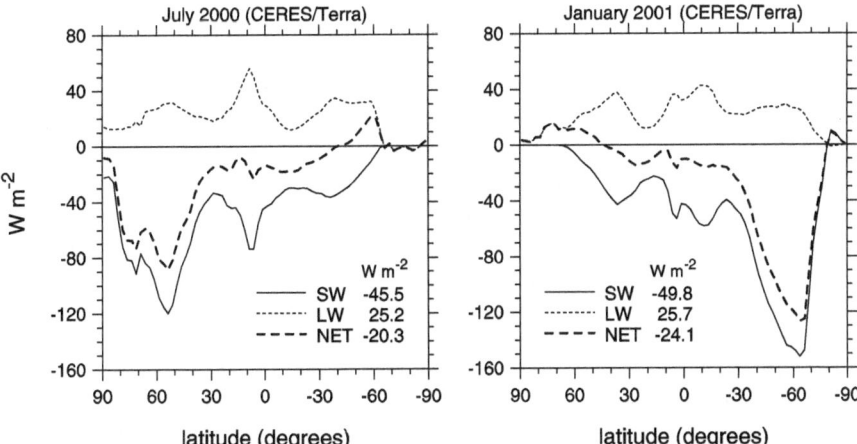

Fig. 9.1. Zonally averaged, monthly mean TOA CRE for shortwave (SW), longwave (LW), and net (NET) radiation as functions of latitude. These values were inferred from CERES satellite radiances using Earth Radiation Budget Experiment (ERBE) conversion algorithms. Global averages are listed on the plots (data courtesy of T. Wong and N. Loeb)

The attractiveness of CRE is that the integrated impact of clouds on Earth's radiation budget can be both measured and computed in a climate model fairly easily. Concerns have been raised over computation of F_{clr}. The point being that within an LSAM it is extremely convenient and simple to remove clouds and compute diagnostic clear-sky irradiances. With observations, however, measurements of F_{clr} require truly clear-skies whose atmospheric characteristics can be expected to differ systematically from mean all-sky conditions. Nevertheless, Fig. 9.1 shows zonally averaged, monthly mean CRE for shortwave, longwave, and net radiation as inferred from CERES satellite radiances using Earth Radiation Budget Experiment (ERBE) conversion algorithms. Shortwave $CREs$ are maximized ($>120\,\mathrm{Wm^{-2}}$) in the summer hemisphere at about latitude $60°$ where solar input is largest and low clouds are abundant, with a secondary maximum in the tropics. The reverse is true for longwave $CREs$ where near-ubiquitous high cool clouds in the tropics contrast sharply with warm oceans. Spatial breakdowns of these plots can be seen in Chap. 10. Globally averaged, net CRE is $-22\,\mathrm{Wm^{-2}}$ indicating that for present-day climate, clouds act to cool Earth.

9.2.2 Radiation and Cloud-Climate Feedbacks

The fundamental working hypothesis in analysis of climate feedbacks is that over a sufficiently long period of time T (>1 year), the Earth-atmosphere system (technically at the tropopause) is in radiative equilibrium such that

$$\int_T \left\{ \frac{F_\odot(t)}{4}\left[1 - \alpha_p(t)\right] - \mathcal{I}(t) \right\} \mathrm{d}t = 0 \,, \tag{9.2}$$

where $F_\odot$ is incoming normal solar irradiance (annual mean of $\approx 1368\,\mathrm{Wm^{-2}}$ as deduced from several satellites over the past 20 years), α_p is planetary albedo, and $\mathcal{I}$ is net longwave radiation at the tropopause. All of the quantities in (9.2) are spectrally-integrated. By virtue of external forcings such as fluctuations in $F_\odot$, evolution of Earth (including human impacts and volcanism), and the internal chaotic nature of Earth's climate, (9.2) is never satisfied perfectly. Conventionally, however, climate sensitivity has been assessed by assuming that (9.2) holds and that perturbations, either internal or external, to (9.2) by amounts ΔR are sudden and followed by restoration of equilibrium so that (9.2) holds again. In actuality, ΔR are always time-dependent thereby making forcing, restoration of equilibrium, and internal chaotic behaviour inexorably intertwined. Nevertheless, the force-restore paradigm of analyzing (9.2) is tractable, succinct, and can be expanded easily to address many detailed questions.

The common view of climatic change has been, and in many respects still is, how much Earth's mean annual surface temperature T_s changes given ΔR. Using linear feedback theory and (9.2), it can be shown (e.g., Schlesinger and Mitchell, 1987) that for a perturbation of ΔR, change in T_s can be approximated as

$$\Delta T_s \approx \frac{-\Delta R}{\underbrace{\frac{\partial \mathcal{I}}{\partial T_s} + \frac{F_\odot}{4}\frac{\partial \alpha_p}{\partial T_s}}_{\text{initial}} + \underbrace{\sum_i \left[\frac{\partial \mathcal{I}}{\partial x_i} + \frac{F_\odot}{4}\frac{\partial \alpha_p}{\partial x_i}\right]\frac{\mathrm{d}x_i}{\mathrm{d}T_s}}_{\text{feedbacks}}} , \qquad (9.3)$$

where x_i are climate variables including cloud-related properties such as cloud fraction, cloudtop altitude, and water path. The term labelled *initial* is simply the system response in the absence of internal adjustment. For example, a doubling of CO_2 concentration ($[CO_2]$) will increase the opacity of Earth's atmosphere and reduce net longwave radiation at the tropopause resulting in $\Delta R \approx -4\,\mathrm{Wm^{-2}}$ (Cess et al., 1993) and

$$\frac{\partial \mathcal{I}}{\partial T_s} \approx 4\varepsilon\sigma_B T_s^3 \approx 3.3\,\mathrm{W\,m^{-2}/K} \qquad (9.4a)$$

and

$$\varepsilon \approx \frac{\dfrac{F_\odot}{4}(1-\alpha_p)}{\sigma_B T_s^4} \approx 0.62 , \qquad (9.4b)$$

where σ_B is Stefan-Boltzmann's constant, $T_s \approx 287\,\mathrm{K}$, ε is mean atmospheric longwave transmittance for radiation emitted by the surface given a mean value for α_p of 0.3. Since changing $[CO_2]$ impacts α_p minimally, doubling $[CO_2]$ would result in roughly

$$\Delta T_s \approx 1.2\,\mathrm{K} \qquad (9.4c)$$

in the absence of feedbacks.

The term in (9.3) labelled *feedbacks* is, however, where much of the uncertainty about, and research into, climate prediction rests. Each feedback process consists of two parts:

$$\underbrace{\left[\frac{\partial \mathcal{I}}{\partial x_i} + \frac{F_\odot}{4}\frac{\partial \alpha_\text{p}}{\partial x_i}\right]}_{\text{radiative}} \times \underbrace{\frac{\mathrm{d}x_i}{\mathrm{d}T_\text{s}}}_{\text{dynamic}} \tag{9.5a}$$

and feedback processes are inexorably linked to each other as demonstrated by

$$\frac{\mathrm{d}x_i}{\mathrm{d}T_\text{s}} = \frac{\partial x_i}{\partial T_\text{s}} + \sum_j \frac{\partial x_i}{\partial x_j}\frac{\mathrm{d}x_j}{\mathrm{d}T_\text{s}} \ . \tag{9.5b}$$

As such, each feedback process has a clear impact on Earth's radiation budget which means direct reliance on a model's treatment of radiative transfer and optical properties.

Several studies have shown that the climate system appears to be sensitive to seemingly small but systematic changes in cloud properties and assumptions about clouds (Senior, 1999). Moreover, model intercomparison studies reveal that much of the uncertainty associated with climate sensitivity (i.e., the denominator in (9.3)) stems from clouds. Thus, when feedbacks are included, estimates of ΔT_s lie between 1°C and 5°C with representation of clouds being responsible for much of this range (Intergovernmental Panel on Climate Change, 1996). Radiative transfer model intercomparison studies reported by Fouquart et al. (1991) and Barker et al. (2003) indicate that when several different radiative transfer models act on identical clear and cloudy atmospheres, the range of responses can be surprisingly large. Thus, it is still not clear how much of the disparity among GCM-predicted cloud-radiative feedbacks is due to different treatments of clouds and their optical properties and how much is due to different treatments of radiative transfer (particularly for cloudy atmospheres).

A drawback to conventional feedback analysis is that climatic change is portrayed as changes to mean values only. For example, if changes to cloud fraction A_c occurred only at night, the solar portion of the feedback would be zero. Straight application of (9.3) would, however, yield a *phantom* solar feedback of

$$\frac{\partial \alpha_\text{p}}{\partial A_\text{c}}\frac{\mathrm{d}A_\text{c}}{\mathrm{d}T_\text{s}} \approx (\alpha_\text{cld} - \alpha_\text{clr})\frac{\Delta A_\text{c}}{\Delta T_\text{s}} \tag{9.6}$$

where α_cld and α_clr are cloudy-sky and clear-sky albedos, at the expense of what should have been ascribed entirely to the longwave. A more thorough feedback analysis would involve conditional distribution functions. This way, mean values of x_i may remain fixed but their distributions may flex (perfectly legitimate climatic change). Still, however, this would fall short of a truly satisfying method of analyzing climatic change. Take, for example, a situation where everything about clouds remains unchanged except the frequency of occurrence of multi-layered systems and their overlapping pattern. This would certainly impact Earth's radiation budget but it would be very difficult to tie the change to something other than the hazy notion of *total* cloud fraction for layer cloud fractions remaining fixed.

To summarize, representation of cloud-radiative feedbacks in LSAMs is crucial for confident prediction of climate, and this points directly to how well LSAMs account for BB cloud optical properties and radiative transfer for cloudy atmospheres. The following section discusses current procedures for modelling BB radiation.

9.3 Characteristics of Clear-Sky Transmittance, Surface Albedo, and Cloud Optical Properties

When performing BB simulations, clear-sky transmittances are required by all atmospheric radiative transfer models. Therefore it is essential to at least review popular techniques for simulating clear-sky, BB irradiances. Likewise, since all BB codes require estimates of spectral surface albedo and cloud optical properties, these are reviewed briefly as well.

9.3.1 Clear-Sky Transmittance

Obviously, the ideal way to compute BB irradiances is to perform extremely high spectral resolution line-by-line (LBL) simulations of the monochromatic radiative transfer equation and integrate results for individual lines (e.g., Mlawer et al., 2000). These models are based on detailed descriptions of spectral lines in conjunction with assumptions about line structure and interactions. Generally, broadening of lines are assumed to extend for about $30\,\mathrm{cm}^{-1}$. By most people's standards, LBLs are still computationally debilitating (even for perfectly clear-sky conditions) and are viewed by most as purveyors of valuable high spectral resolution benchmarks for well-defined atmospheres that can be used to assess other, more approximate, models. As yet, LBL simulations are performed for plane-parallel, homogeneous (PPH) atmospheres only. The two radiative transfer model intercomparison studies mentioned above have employed LBL calculations as industry benchmarks.

Figure 9.2 shows differences between measured and LBL modelled direct and diffuse spectral irradiances for a moderate solar slant path through a midlatitude summer-like water vapour burden. Only a small amount of aerosol was needed to account for direct irradiance, and reasonable aerosol optical properties ($\varpi_0 = 0.9$ and $g = 0.7$) were required to bring modelled diffuse irradiances into line with measured values. Examples such as this are important because they show no significant anomalous absorption and provide confidence that LBLs are benchmark-worthy. In this spirit, Fig. 9.3 shows results from Barker et al. (2003) for the tropical clear-sky atmosphere where the CHARTS model was used to assess numerous 1D models. Here it is clear that most 1D models (many of which see active service in operational LSAMs) underestimate absorption by water vapour by between $10\,\mathrm{Wm}^{-2}$ and $40\,\mathrm{Wm}^{-2}$ for overhead Sun in the tropics.

For most applications, LBL calculations are not tenable so simplifications must be made. For example, in an LSAM one can expect to execute only a relatively small number of multi-layer radiative transfer simulations for each atmospheric column. As such, sweeping integrations over spectral bands must be done. The biggest challenge is accounting for the rapid fluctuations of gaseous absorption in spectral space

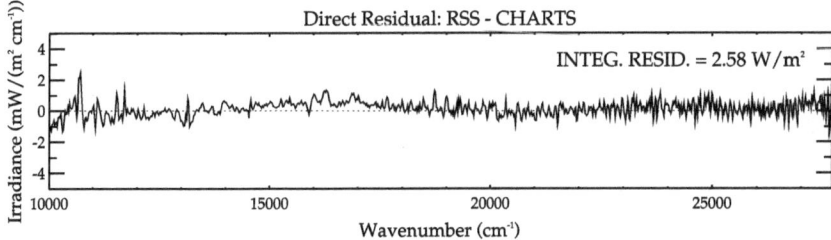

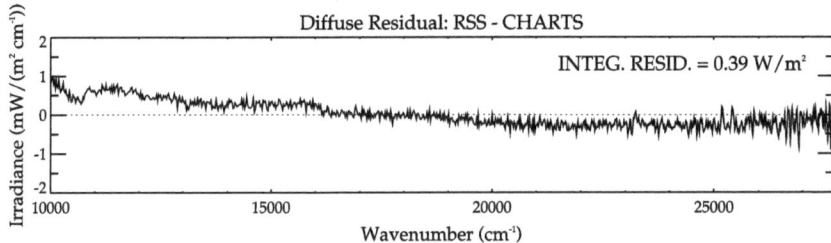

Fig. 9.2. Differences between measured and modelled spectral fluxes from 0.36 μm to 1 μm. Measurements were made on 03/20/2000 at 1226 LST with the Rotating Shadowband Spectrometer (RSS) at the Atmospheric Radiation Measurement (ARM) site in OK, USA. Model results were obtained from the CHARTS model maintained by Atmospheric Environment Research (AER). Solar zenith angle was 40° precipitable water amount was 1.16 cm, and assumed aerosol optical depth was 0.0344 (at 0.7 μm; Ångström exponent of 2.24). Integrated direct and diffuse measured irradiances were 707 Wm^{-2} and 56.2 Wm^{-2}. Integrated differences are listed on the plots (data courtesy of E. Mlawer)

and their spectrally-unresolved convolution with irradiances and cloud optical properties.

Though several methods exist for computing gaseous transmittances, it is probably safe to assume that operational BB atmospheric radiative transfer modelling is firmly ensconced in the correlated k-distribution (CKD) paradigm (Lacis et al., 1979; Lacis and Oinas, 1991), the essence of which rests in its parent k-distribution method.

Spectrally-averaged transmittance (for uniform spectral irradiance) along a straight line between two levels z_1 and z_2 would be computed via LBL as

$$T_{\Delta \nu} = \int\limits_{\Delta \nu} \exp\left[-\int\limits_{z_1}^{z_2} k\left(\nu; p(z), T(z)\right) \rho(z) \mathrm{d}z\right] \frac{\mathrm{d}\nu}{\Delta \nu}, \qquad (9.7)$$

where p, T, and ρ are pressure, temperature, and density. However, with the correlated k-distribution method, we have

$$T_{\Delta \nu} = \int\limits_{0}^{1} \exp\left[-\int\limits_{z_1}^{z_2} k\left(g; p(z), T(z)\right) \rho(z) \mathrm{d}z\right] \mathrm{d}g, \qquad (9.8)$$

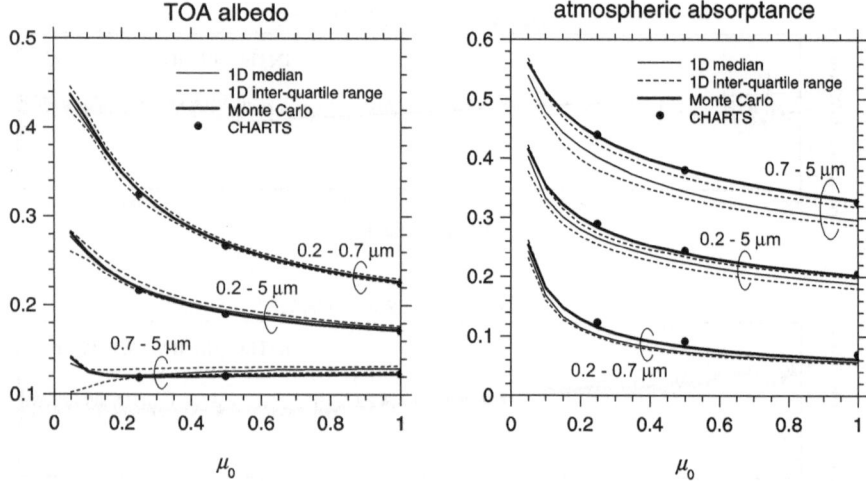

Fig. 9.3. TOA albedo and atmospheric absorptance for three spectral ranges (wavelengths listed on the plots) as functions of μ_0 for the clear-sky tropical standard atmosphere with broadband surface albedo 0.2. Results are shown for an LBL model (CHARTS) and a Monte Carlo algorithm that accounted for gaseous absorption with a CKD model. Also shown are means and quartiles for 18 1D solar transfer codes. Many of these 1D models were employed at the time in operational weather prediction or global climate models

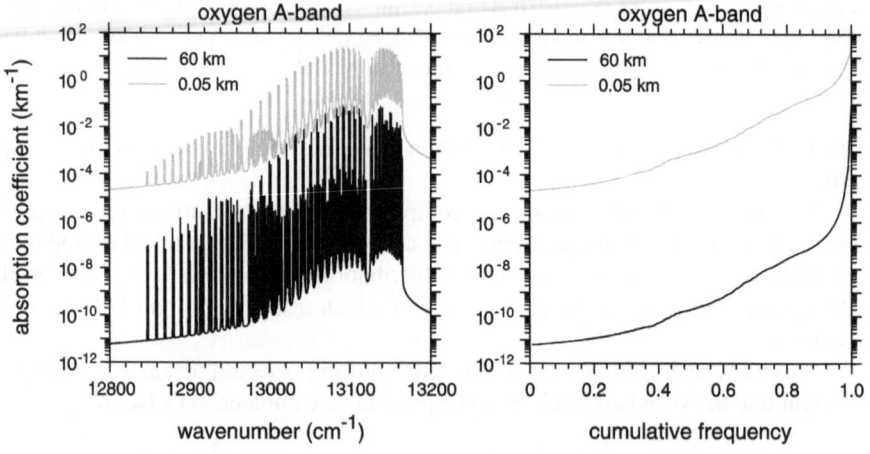

Fig. 9.4. Plot on the left shows absorption coefficients k_ν across the O_2 A-band at two altitudes as functions of wavenumber. Plot on the right shows normalized cumulative distributions for the same k_ν as plotted on the left

where k has been transformed from wavenumber space into cumulative probability-space where g is the cumulative distribution of k over the interval $\Delta\nu$. Figure 9.4 shows $k(\nu)$ and $k(g)$ over the oxygen A-band at two altitudes. Clearly, integration of $k(g)$ can be achieved with relatively few judiciously chosen quadrature points while numerical integration of $k(\nu)$ is formidable. The CKD method builds on the k-distribution method by harvesting obvious strong correlations between spectral signatures at different conditions (see Fig. 9.4). For practical issues involving spectral-weighting and overlap of different gases, see Fu and Liou (1992).

9.3.2 Surface Albedo

Spectral surface albedos $\alpha_s(\lambda)$ are used as lower boundary conditions for atmospheric radiative transfer models. What is subsumed under the convenient blanket of surface albedo is actually the result of an immensely complex radiative transfer process. This is elaborated on in Chap. 14 for vegetated surfaces. For now it suffices to say that like gases, some surfaces exhibit complex spectral albedos that are complicated by changing internal conditions, such as soil moisture and snow cover, and dependencies on illumination conditions.

Figure 9.5 shows visible and near-IR $\alpha_s(\lambda)$ for various surface types at solar zenith angles θ_0 near 60°. Clearly, atmospheric radiative transfer models face the same problem with surface albedo as they do with gases. That is, BB models require spectrally-weighted surface albedos, but the necessary spectral irradiances needed to compute mean albedo are not available. Generally, simple uniform weightings are applied and surfaces are assumed to be perfectly Lambertian. There is, however, the possibility of assigning spectral subgrid values of $\alpha_s(\lambda)$ to each k-value in a CKD algorithm. This is at best a long way off as most applications draw only a gross distinction between albedo in a few bands across the solar spectrum (e.g., $<0.7\,\mu$m and

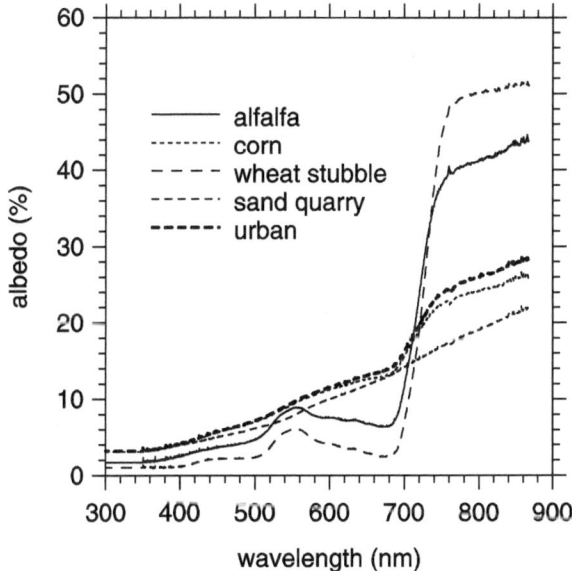

Fig. 9.5. Examples of spectral surface albedos for various surface types (data, courtesy of Z. Li, collected during spring in OK, USA)

>0.7 μm). Likewise, dependencies on θ_0 are generally crude and often do not distinguish between direct and diffuse albedos (but again, this partition actually requires solution of the radiative transfer algorithm).

9.3.3 Cloud Optical Properties

For liquid clouds it is sufficient to assume that droplets are spherical and pure water. Cloud optical properties can be computed using Mie theory as functions of refractive indices, wavelength of radiation, and droplet radius r.

For both LSAMs and Monte Carlo simulations, it is neither efficient nor necessary to perform on-line Mie calculations. Instead, σ, ϖ_0 , and g are parametrized as functions of droplet effective radius which is defined as (Hansen and Travis, 1974)

$$r_e = \frac{\int_0^\infty r^3 n(r)\mathrm{d}r}{\int_0^\infty r^2 n(r)\mathrm{d}r} \, , \tag{9.9}$$

where $n(r)$ is droplet size distribution. The most popular cloud optical property parametrizations for solar wavelengths are those of Hu and Stamnes (1993) and Slingo (1989); both allow up to 24 bands across the solar spectrum. Hu and Stamnes's covers a wider range of r_e than does Slingo's and it also covers the terrestrial spectrum (allowing up to 50 bands). While both parametrizations agree well for σ and ϖ_0, Fig. 9.6 shows that at $r_e = 10\,\mu m$ they disagree for g with Hu and Stamnes in better agreement with Mie values but their parametrization exhibits discontinuities at certain values of r_e. The size distributions used here were Gamma distributions with effective variances of 0.1. Effective variance is defined as

$$v_e = \frac{\int_0^\infty (r - r_e)^2 r^2 n(r)\mathrm{d}r}{r_e^2 \int_0^\infty r^2 n(r)\mathrm{d}r} \, . \tag{9.10}$$

Though the discrepancy may appear to be small, to bring Hu and Stamnes's estimates down to Slingo's requires $r_e = 6\,\mu m$. This change in r_e exceeds even the largest changes to mean r_e expected to have taken place over the industrial era due to increased condensation of cloud condensation nuclei.

For clouds consisting of ice, be they mixed phase or entirely ice, specification of optical properties is difficult and results are always questionable. This is because of the myriad of forms that ice crystals can assume (e.g., Liou, 1992). The relative rarity of optical phenomena like halos coupled with evidence from particle images suggests, however, that from a climatological standpoint, the habit of most ice crystals is highly irregular. If so, this makes specification of their optical properties for climate studies somewhat more tractable and certain.

9.4 Computing Broadband Irradiances using Monte Carlo Algorithms

While the Monte Carlo method *can be* the most accurate solution of the radiative transfer equation, it can, under certain conditions, also be the *fastest* way to compute

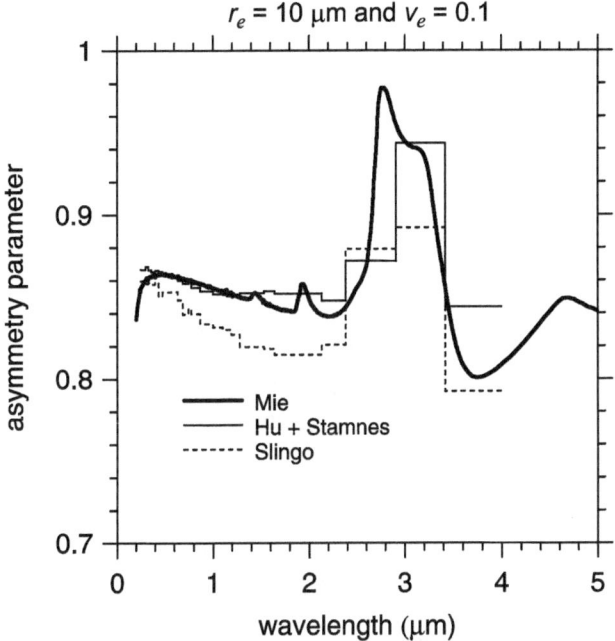

Fig. 9.6. Asymmetry parameter g predicted by Mie theory for a cloud droplet size distribution described by a Gamma distribution with $r_e = 10\,\mu m$ and effective variance $v_e = 0.1$. Also shown are broadband values of g as given by the parametrizations of Hu and Stamnes (1993) and Slingo (1989)

BB solar irradiances for 3D cloudy atmospheres. Hence, issues of particular concern to computation of BB solar irradiances using Monte Carlo algorithms are considered here, thereby augmenting part of Chap. 4 (see also Chap. 10 for a discussion of Monte Carlo LW codes).

The primary advantage that Monte Carlo codes have over analytic codes when performing BB calculations is that spectral integration can be performed via the Monte Carlo method. In general, the intention is to compute integrated irradiance ($\mathrm{Wm^{-2}}$) defined as

$$\mathcal{F} = \int_{\lambda_1}^{\lambda_2} S(\lambda)T(\lambda)\mathrm{d}\lambda \,, \tag{9.11}$$

where $S(\lambda)$ is a spectral source ($\mathrm{Wm^{-2}\,\mu m^{-1}}$), and $T(\lambda)$ is a fractional irradiance (essentially transmittance). Since $S(\lambda)$ is generally non-analytic, assume that it is tabulated for J bands across the interval $[\lambda_1, \lambda_2]$ and that the fractional amount of energy in the jth band is

$$s_j = \frac{\int_{\lambda_j}^{\lambda_{j+1}} S(\lambda)\mathrm{d}\lambda}{\int_{\lambda_1}^{\lambda_2} S(\lambda)\mathrm{d}\lambda} = \frac{S_j}{\int_{\lambda_1}^{\lambda_2} S(\lambda)\mathrm{d}\lambda} \,. \tag{9.12}$$

Though sophisticated strategies exist for optimizing integration of (9.11), the simplest way to proceed is to approximate $\mathcal{F}$ by

$$\mathcal{F} \approx \sum_{j=1}^{J} S_j T_j , \qquad (9.13a)$$

where T_j are obtained with the Monte Carlo transport algorithm using

$$n_j = s_j N_p , \qquad (9.13b)$$

photons where N_p is the total number of photons for the simulation. With this approach, time required to compute $\mathcal{F}$ is independent of J and so as more spectral information is included, time required to perform radiative transfer calculations does not change. Only the time required to compute atmospheric optical properties increases. For a given N_p, however, the statistical significance of irradiances in each band decreases as J increases. An analytic model, on the other hand, requires the same optical properties as the Monte Carlo, but it must run through full computations for each band. So as J increases, time to compute $\mathcal{F}$ increases too.

As with spectral weighting, temporally-averaged quantities can be obtained rapidly via Monte Carlo integration over time. When mean values are sought over a time period, one has to: i) specify latitude θ_L and day number N_d; ii); set the start t_1 and stop t_2 times (LST); and iii) inject photons at cosine of zenith angles

$$\mu_0 = \sin \delta(N_d) \sin \theta_L + \cos \delta(N_d) \cos \theta_L \cos(\alpha \theta_h) , \qquad (9.14a)$$

where $\delta(N_d)$ is solar declination, α a uniform random number between 0 and 1, and θ_h is hour angle between sunrise and noon. When diurnal means are required, photons are injected at

$$\begin{aligned} \mu_0 = &\sin \delta(N_d) \sin \theta_L \\ &+ \cos \delta(N_d) \cos \theta_L \cos \left\{ \alpha \cos^{-1} \left[-\tan \delta(N_d) \tan \theta_L \right] \right\} . \end{aligned} \qquad (9.14b)$$

This procedure affects a uniform, random sampling of time. In these cases, initial weights of photons are not 1 but μ_0, and accumulated radiative quantities are not normalized by N_p, but rather by

$$\sum_{i=1}^{N_p} \mu_0(i) , \qquad (9.15)$$

where $\mu_0(i)$ is the incident cosine of zenith angle for the ith photon.

Since irradiances are of concern, details of scattering phase functions are usually irrelevant thereby rendering analytic approximations sufficient. The Henyey-Greenstein phase function is the most common analytic phase function and it is expressed as

$$P_{\text{HG}}(\mu_s) = \frac{1 - g^2}{(1 + g^2 - 2g\mu_s)^{3/2}} , \qquad (9.16a)$$

where μ_s is cosine of scattering angle, and g is asymmetry parameter defined as

$$g = \frac{1}{2} \int_{-1}^{1} P_{HG}(\mu_s)\mu_s d\mu_s \ . \tag{9.16b}$$

Random cosine of scattering angles can be determined analytically as

$$\mu_s = \begin{cases} \dfrac{1}{2g}\left[1 + g^2 - \left(\dfrac{1-g^2}{1+g-2g\alpha}\right)^2\right]; & |g| \le 1, g \ne 0 \\ 1 - 2\alpha & ; \quad g = 0 \end{cases}, \tag{9.16c}$$

where α is again a uniform random number between 0 and 1.

Imagine that each cell in a 3D Monte Carlo simulation has M constituents with optical properties $\left(\sigma^{(m)}, \varpi_0^{(m)}, g^{(m)}\right)$ for a particular spectral band. These could be any number of gases, aerosols, and cloud droplet size ranges. The most common way to perform the simulation is to create a single set of optical properties for each grid-cell by summing according to

$$\sigma = \sum_{m=1}^{M} \sigma^{(m)}$$

$$\varpi_0 = \frac{\displaystyle\sum_{m=1}^{M} \sigma^{(m)}\varpi_0^{(m)}}{\sigma} \tag{9.17}$$

$$g = \frac{\displaystyle\sum_{m=1}^{M} \sigma^{(m)}\varpi_0^{(m)}g^{(m)}}{\sigma\varpi_0} \ .$$

Alternatively, one could carry M fields of optical properties and have the algorithm *decide* at each photon-matter interaction what constituent M' has been encountered by solving

$$\sum_{m=0}^{M'-1} \sigma^{(m)} < \alpha\sigma \le \sum_{m=1}^{M'} \sigma^{(m)} \qquad \left[\sigma^{(0)} = 0, M' = 1,\ldots,M\right], \tag{9.18}$$

where σ is defined in (9.17) and α a uniform random number between 0 and 1. This method becomes almost essential if Mie phase functions are desired for it circumvents the need to assign a detailed Mie phase function (and cumulative phase function) to each cell; only the cumulative distribution of σ need be carried.

Once M' is determined, the probability of surviving the encounter is $\varpi_0^{(M')}$. If it survives, the appropriate phase function (be it P_{HG} or a full Mie function) is used to determine the scattering direction. This way, 3D distributions of atmospheric absorption by each constituent can be computed easily.

9.5 Computing Broadband Irradiances using 1D Algorithms

For several decades now, LSAMs have employed multi-layer 1D codes for computation of heating and cooling rate profiles. Up until the 1990s, descriptions of clouds
by LSAMs left much to be desired and thus did not warrant sophisticated treatments
of cloud-radiation interactions. The advent of detailed cloud parametrizations (e.g.,
Smith, 1990) has, however, fuelled the need for radiative transfer codes that consider
interactions between unresolved clouds and radiation fields. The focus of this section is therefore on methods of accounting for unresolved clouds in multi-layer 1D
radiative transfer models.

9.5.1 Cloud Fraction

The most basic issue involving clouds for a 1D code is cloud fraction A_c. Matrix
methods for solving the 1D radiative transfer equation had been used widely up until
attempts to account explicitly for layers filled partially by cloud. Partial cloudiness
(which can be taken as the rule at typical large-scale model horizontal grid-spacings;
Rossow (1989)) is handled better by adding solutions that ensure continuity of fluxes
at levels using layer reflectance and transmittances for collimated and diffuse irradiance (e.g., Liou, 1992).

Typically, a 1D radiative transfer algorithm computes mean irradiances emerging
from a layer as

$$F_{all} = (1 - A_c) F_{clr} + A_c F_{cld} , \qquad (9.19)$$

where F_{cld} is irradiance associated with the cloudy portion of the layer. There are,
however, some implicit assumptions behind (9.19) as expressed by Stephens (1988).
First is the definition of A_c itself. Clearly it rests on imposition of a threshold at some
spatial scale. The more fundamental definition would involve the actual cloud itself,
and this is known as *intrinsic* cloud fraction. Here one could imagine cloud water
content $\mathcal{L}$ varying across a cell and defined with a probability density function $p(\mathcal{L})$
such that

$$A_c \equiv \int_{\mathcal{L}^*}^{\infty} p(\mathcal{L}) d\mathcal{L} , \qquad (9.20a)$$

where $\mathcal{L}^* \geq 0$ is some inner-scale-dependent threshold. Likewise, an *extrinsic* cloud
fraction would be defined by a less direct measure such as shortwave reflectance R:

$$A_c \equiv \int_{R^*}^{\infty} p(R) dR , \qquad (9.20b)$$

where again $R^* \geq 0$ is a scale-dependent threshold above which cloud is deemed to
be present and $p(R)$ is a density function describing the distribution of reflectance
over a domain. This definition of A_c is actually more in line with an intrinsic definition that uses cloud water path integrated through a layer and so would correspond
to the more common interpretation of A_c as a vertically-projected fractional area

(rather than a volume). What Stephens (1988) showed was that (9.19) assumes there are no correlations between fluctuations in the radiation and cloud fields. In a similar vain, Barker and Wielicki (1997) considered domain-averaged diffuse transmittance through a partially cloudy layer to be defined as

$$T_{\text{all}} = 1 - \widehat{A}_c + 2 \int_0^1 \int_0^\infty A_c\,(\mu)\,e^{-\tau/\mu} p\,(\tau \mid \mu)\,\mu d\tau d\mu\,, \tag{9.21}$$

where $\widehat{A}_c$ is cloud fraction presented to a diffuse beam, $A_c\,(\mu)$ is cloud fraction as a function of zenith angle, $e^{-\tau/\mu}$ is direct-beam transmittance, and $p\,(\tau \mid \mu)$ is the distribution of cloud optical depth τ conditional on μ. The conditionality on μ arises due to horizontal smoothing along lines of sight as zenith angle increases. This, in conjunction with Stephens's elucidations, demonstrates that decoupling cloud fraction and radiation must always be approximate when considering domain-averaged computations. To add to this difficulty, values of A_c as supplied by the cloud routine in an LSAM may not be the ones that the radiation, or some other physics routine, is expecting (cf. volume to area definitions). This is further confounded by the fact that what cloud atlases (e.g., Warren et al., 1986) supply may differ from what an LSAM produces and/or uses in its radiation code (though it is not possible to label one as more realistic than the other given the highly subjective nature of A_c).

9.5.2 Multi-Layer Radiation Codes and Cloud Overlap

Ideally, we would like to represent clouds in a radiative transfer model the same way we tend to think of them: as entities. For large domains, however, this simple notion clearly breaks down. Moreover, *if* a cloud could be represented successfully as an entity (e.g., marine boundary layer clouds), there is nothing stopping the cloud from being *sliced* into arbitrary layers as dictated by the LSAM. This leads immediately to the question of how to convey information to the radiative transfer model about the vertical overlapping of partially cloudy layers.

The simplest overlap configuration is to assume that clouds overlap randomly. This could be achieved two ways. Consider the simple case of transparent clear-sky, non-absorbing clouds, and diffuse irradiance. The first way is to specify the excepted overlapping pattern such that diffuse albedo for a two-layer system is described as

$$\overline{R}_{12} = \underbrace{A_{c1}\,(1 - A_{c2})\,R(\tau_1)}_{\substack{\text{non-overlapping} \\ \text{upper cloud}}} + \underbrace{A_{c2}\,(1 - A_{c1})\,R(\tau_2)}_{\substack{\text{non-overlapping} \\ \text{lower cloud}}} + \underbrace{A_{c1}A_{c2}R(\tau_1 + \tau_2)}_{\substack{\text{overlapping} \\ \text{clouds}}} \tag{9.22a}$$

where A_{c1} and A_{c2} are cloud fractions in the top and bottom layers, R is albedo, τ_1 and τ_2 are cloud optical depths for the two layers, and

$$R(\tau_1 + \tau_2) = R(\tau_1) + \frac{[1 - R(\tau_1)]^2\,R(\tau_2)}{1 - R(\tau_1)R(\tau_2)} \tag{9.22b}$$

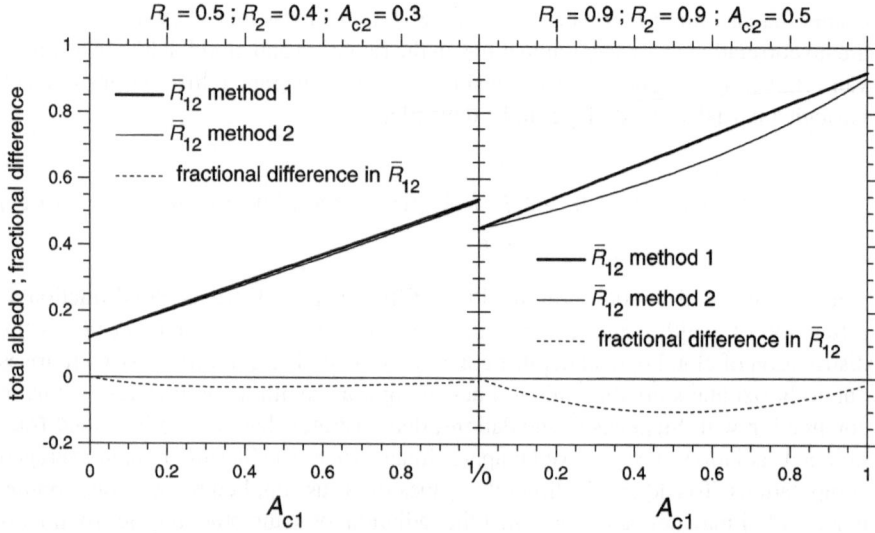

Fig. 9.7. Total albedo $\overline{r}_{12}$ of a simple two-layer cloud system as a function of cloud fraction in the *upper layer* A_{c1} computed by two methods of accounting for random overlap of cloud: method 1 is for (9.22); and method 2 is for (9.23). Also shown are fractional differences between $\overline{r}_{12}$ as predicted by these methods. Values in the titles are: albedo for upper layer clouds r_1; albedo for *lower layer* clouds r_2; and cloud fraction in lower layer A_{c2}

is albedo of the combined layers for the cloudy portion. This combinatorial approach is employed at times but can become overwhelming when many layers are involved. Alternatively, albedo for the two layers could be defined as

$$\overline{R}_{12} = A_{c1}R(\tau_1) + \frac{[1 - A_{c1} + A_{c1}T(\tau_1)]^2 A_{c2}R(\tau_2)}{1 - A_{c1}R(\tau_1)A_{c2}R(\tau_2)} , \qquad (9.23)$$

where T is transmittance, and it has been assumed that there is no correlation between radiation emerging from the top layer and optical properties of the lower layer. Figure 9.7 shows $\overline{R}_{12}$ computed by both (9.22) and (9.23). For the realistic case on the left, differences (even at $A_{c1} = 1$) between the methods are minor, but the extreme case produces differences in excess of 10%. Under most conditions having two distinct cloudy layers, (9.23) is the better approximation (Barker et al., 1999).

The other extreme overlap approximation that is used with increasingly regularity follows from the assumption that clouds in contiguous layers are maximally overlapped and randomly overlapped when separated by a clear layer (Geleyn and Hollingsworth, 1979). While maximum overlap of contiguous layers becomes an increasingly better assumption as layer thicknesses approach 0, for large grid cells, one can expect a mixture of random and near-maximum overlap; due simply to entrainment, shearing, and a continuum of cloud development (Barker et al., 1999; Hogan and Illingworth, 2000). In response to this, several models are beginning to take a more general approach to overlap where the fractional overlap between clouds in any layer can take on any possible value (e.g., Li, 2002).

9.5.3 Horizontal Variability of Overlapping Cloud

While there are several ways to account for horizontal variations in cloud water path, only the Gamma-weighted two-stream approximation (GWTSA) method (Barker et al., 1996) is considered here. This is because other methods are documented in Chap. 6, the multi-layer GWTSA for both shortwave and longwave radiation has been assessed extensively in the literature (e.g., Oreopoulos and Barker, 1999; Barker and Fu, 2000; Li and Barker, 2002), and some pertinent issues facing 1D models that attempt to account for unresolved cloud effects have been investigated only through the GWTSA. Thus, in this subsection, the GWTSA will be reviewed, first for shortwave and then for longwave, and general issues facing all 1D codes that attempt to address unresolved clouds will be discussed.

Shortwave

To recapitulate part of Chap. 6, the GWTSA is a brute-force example of a stochastic Independent Column Approximation (ICA) where mean transmittance and albedo for direct solar irradiance are expressed as

$$\left\{ \begin{array}{c} T_\Gamma \\ R_\Gamma \end{array} \right\} = \int_0^\infty p_\Gamma\left(\tau\right) \left\{ \begin{array}{c} T_{pp}(\tau) \\ R_{pp}(\tau) \end{array} \right\} d\tau , \tag{9.24}$$

where T_{pp} and R_{pp} are transmittance and albedo as predicted by a standard two-stream model,

$$p_\Gamma\left(\tau\right) = \frac{1}{\Gamma(\nu)} \left(\frac{\nu}{\overline{\tau}}\right)^\nu \tau^{\nu-1} e^{-\nu\tau/\overline{\tau}} , \tag{9.25}$$

for $\tau > 0$, is the Gamma distribution that approximates the distribution of τ over an LSAM's grid-cell, $\Gamma(\nu)$ is the Gamma function, $\overline{\tau} > 0$ is mean optical depth, and $\nu > 0$ depends on the variance $\overline{\tau^2} - \overline{\tau}^2$ of τ and can be solved for by either the method of moments ($1/\nu = \overline{\tau^2}/\overline{\tau}^2 - 1$) or by maximum likelihood estimation. Note that the neither the GWTSA nor any other 1D radiative transfer models account for the truly geometric aspect of 3D radiative transfer. As such, the natural standard against which 1D parametrizations should be judged is the explicit ICA (affected in a Monte Carlo algorithm by setting horizontal grid-spacing arbitrarily large). Substituting generalized two-stream expressions for T_{pp} and R_{pp} for $\varpi_0 < 1$ (Meador and Weaver, 1980) along with (9.25) into (9.24) leads to

$$T_\Gamma = \left(\frac{\nu}{\nu + \overline{\tau}/\mu_0}\right)^\nu - \phi_1^\nu \frac{\varpi_0}{a} \left[t_+ \mathcal{F}\left(\beta, \nu, \phi_4\right) - t_- \mathcal{F}\left(\beta, \nu, \phi_5\right) - t\mathcal{F}\left(\beta, \nu, \phi_6\right)\right] \tag{9.26a}$$

$$R_\Gamma = \phi_1^\nu \frac{\varpi_0}{a} \left[r_+ \mathcal{F}\left(\beta, \nu, \phi_1\right) - r_- \mathcal{F}\left(\beta, \nu, \phi_2\right) - r\mathcal{F}\left(\beta, \nu, \phi_3\right)\right] , \tag{9.26b}$$

where

$$\mathcal{F}(\beta, v, \phi) = \sum_{n=0}^{\infty} \frac{\beta^n}{(\phi + n)^v} , \quad \text{with } |\beta| \leq 1 \ (\beta \neq 1) \text{ and } v > 0 , \tag{9.26c}$$

and the parameters $\phi_{1,\dots,6}$, β, a, r, t, $t_{\pm}$, and $r_{\pm}$ depend on either the choice of two-stream approximation, cloud optical properties, or μ_0. The removable singularity in the two-stream at $\varpi_0 = 1$ $(\beta = 1)$ leads to different expressions for T_{Γ} and R_{Γ}. Corresponding expressions for diffuse irradiance can be found in Oreopoulos and Barker (1999).

While derivation and application of (9.26) is straightforward, matters become much more complex when one attempts to use (9.26) in a multi-layer code. This goes for any model that addresses subgrid-scale variable cloud-radiation interactions. The essence of the issue is the fact that, unlike a 1D homogeneous overcast, irradiances associated with cloudy regions are no longer unique but rather they too vary horizontally. This is easiest to appreciate with normal direct-beam irradiance. Consider a cloud that is *vertically homogeneous* but has horizontal variability defined by $p_{\Gamma}(\tau)$. Following from (9.24) and (9.25), mean direct-beam transmittance is

$$T_{\text{dir}} = \int_0^{\infty} p_{\Gamma}(\tau) e^{-\tau} d\tau$$
$$= \left(\frac{v}{v + \bar{\tau}}\right)^v . \tag{9.27}$$

Assume now that this cloud is partitioned into n equal layers, each having $p_{\Gamma}(\tau)$ defined with $(\bar{\tau}/n, v)$. The conventional method for computing overall transmittance would be to simply multiply layer transmittances thereby giving

$$T'_{\text{dir}} = \prod_{i=1}^{n} \left(\frac{v}{v + \bar{\tau}/n}\right)^v$$
$$= \left(\frac{vn}{vn + \bar{\tau}}\right)^{vn} \leq \left(\frac{v}{v + \bar{\tau}}\right)^v = T_{\text{dir}} . \tag{9.28}$$

The important point is that the number of layers n and the variance parameter v are commutable, so in the limit as either n or v goes to ∞, the latter representing homogeneity,

$$T'_{\text{dir}} = \lim_{(n \text{ or } v) \to \infty} \left(\frac{vn}{vn + \bar{\tau}}\right)^{vn} = e^{-\bar{\tau}} \leq T_{\text{dir}} . \tag{9.29}$$

Thus, each time the original layer is partitioned and the *correct* Gamma-weighted layer mean transmittances are multiplied together, a discrete step is affected back towards homogeneity; recovering Beer's law (9.29) in the limit of $n \to \infty$.

Oreopoulos and Barker (1999) showed that the standard homogeneous multiplication approach can be maintained and the correct overall transmittance preserved

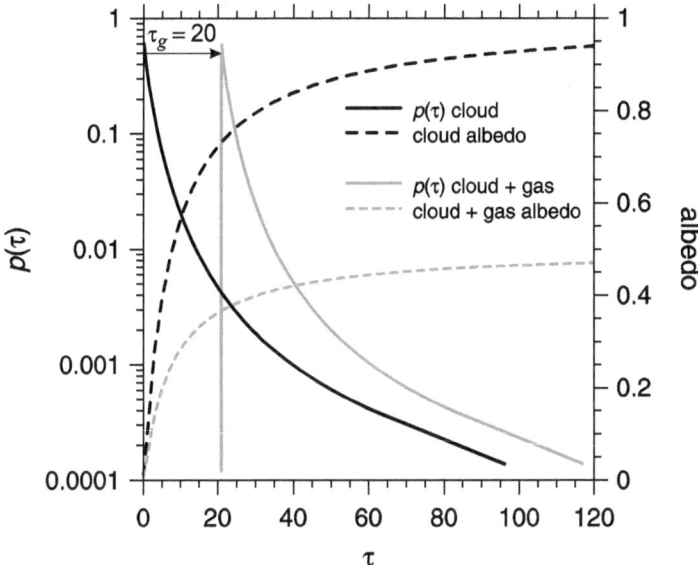

Fig. 9.8. Distribution of cloud optical depth $p(\tau)$ and 1D cloud (layer) albedo as functions of optical depth τ at a wavelength where gaseous (water vapour) optical depth is zero. Also shown is $p(\tau)$ and 1D (layer) albedo at another wavelength where water vapour optical depth is 20

by renormalizing $p_\Gamma(\tau)$ for each layer in a block of contiguous layers containing cloud (basically reducing $\bar{\tau}$ for layers beneath the uppermost layer). The problem is significantly more complicated for the scattered radiation field. Nevertheless, this simple illustration points to the concept of generalized cloud overlap where vertical correlation in extinction must be addressed; even for completely overcast skies.

It is worth pointing out that for bands where gaseous absorption is strong, the effects of variable cloud are diminished. For example, Fig. 9.8 shows a distribution of τ for cloud and for cloud plus a homogeneous gas of optical depth $\tau_g = 20$. It also shows $R_{pp}(\tau)$ for non-absorbing cloud and for a mixture of cloud and gas. For cloud-only, the highly non-linear portion of $R_{pp}(\tau)$ cuts across densely populated values of τ and so $R_\Gamma \ll R_{pp}(\bar{\tau})$. For cloud plus gas, however, most of $p(\tau + \tau_g)$ occurs where $R_{pp}(\tau)$ has saturated and so $R_\Gamma \approx R_{pp}(\bar{\tau})$. This squelching effect is even more prevalent in the longwave (see Chap. 11).

Longwave

In the formulation of the GWTSA for longwave radiative transfer (Li and Barker, 2002), cloud overlap and vertical correlation of extinction are treated differently than in the solar. To begin, for isothermal homogeneous layers (Li, 2002), downwelling radiance in direction μ at level $j + 1$ is defined as

$$I_{j+1}^-(\mu) = I_j^-(\mu)e^{-\langle \kappa \rangle_j/\mu} + B_{j+\frac{1}{2}}\left(1 - e^{-\langle \kappa \rangle_j/\mu}\right), \tag{9.30a}$$

where

$$\langle \kappa \rangle_j = (1 - \varpi_0)\,\tau_j \,, \tag{9.30b}$$

is absorption optical depth for the jth layer, and $B_{j+\frac{1}{2}}$ is the Planck function evaluated at the temperature of layer j. Therefore, downwelling irradiance is

$$F_{j+1}^- = 2\pi \int_0^1 I_{j+1}^-(\mu)\mu\,\mathrm{d}\mu \tag{9.31a}$$

$$= F_j^- e^{-\langle \kappa \rangle_j/\mu_1} + \tilde{B}_{j+\frac{1}{2}}\left(1 - e^{-\langle \kappa \rangle_j/\mu_1}\right),$$

where

$$\tilde{B}_{j+\frac{1}{2}} = \pi B_{j+\frac{1}{2}} \,, \tag{9.31b}$$

and $\mu_1 \approx 0.601$.

Now, define the amounts of radiation emerging out the base of the clear and cloudy portions of partially cloudy layer i as

$$\begin{cases} \mathcal{M}_{i+1}^- = (F_i^- - \tilde{B}_{i+\frac{1}{2}})e^{-\tau_i/\mu_1} + \tilde{B}_{i+\frac{1}{2}}\,; & \text{cloudless} \\ \mathcal{N}_{i+1}^- = (F_i^- - \tilde{B}_{i+\frac{1}{2}})e^{-\langle \kappa \rangle_i/\mu_1} + \tilde{B}_{i+\frac{1}{2}}\,; & \text{cloudy} \end{cases}, \tag{9.32a}$$

where τ_i and $\langle \kappa \rangle_i$ are clear-sky and clear+cloud optical depths. Therefore, mean downward irradiance is

$$F_{i+1}^- = (1 - A_{\text{ci}})\mathcal{M}_{i+1}^- + A_{\text{ci}}\mathcal{N}_{i+1}^- \,, \tag{9.32b}$$

where A_{ci} is cloud fraction in the ith layer. If clouds are separated by a clear layer, the next layer is irradiated uniformly with F_{i+1}^- and the next partially cloudy layer to be encountered is also irradiated uniformly.

For contiguous cloudy layers, components of irradiance at level $i + 2$ are

$$\mathcal{M}_{i+2}^- = \left[\frac{\overbrace{(A_{\text{ci}} - o_{i,i+1})\mathcal{N}_{i+1}^-}^{1} + \overbrace{(1 - A_{\text{ci}} - A_{\text{ci+1}} + o_{i,i+1})\mathcal{M}_{i+1}^-}^{2}}{1 - A_{\text{ci+1}}} - \tilde{B}_{i+1+\frac{1}{2}} \right]$$

$$\times\, e^{-\tau_{i+1}/\mu_1} + \tilde{B}_{i+1+\frac{1}{2}} \tag{9.33a}$$

$$\mathcal{N}_{i+2}^- = \left[\frac{\overbrace{o_{i,i+1}\mathcal{N}_{i+1}^-}^{3} + \overbrace{(A_{\text{ci+1}} - o_{i,i+1})\mathcal{M}_{i+1}^-}^{4}}{A_{\text{ci+1}}} - \tilde{B}_{i+1+\frac{1}{2}} \right]$$

$$\times\, e^{-\langle \kappa \rangle_{i+1}/\mu_1} + \tilde{B}_{i+1+\frac{1}{2}} \,, \tag{9.33b}$$

where the overlapping fraction $o_{i,i+1}$ is

$$\max\{0, A_{\mathrm{c}i} + A_{\mathrm{c}i+1} - 1\} \le o_{i,i+1} \le \max\{A_{\mathrm{c}i}, A_{\mathrm{c}i+1}\} . \tag{9.33c}$$

The denumerated terms in (9.33a,b) are shown graphically in Fig. 9.9.

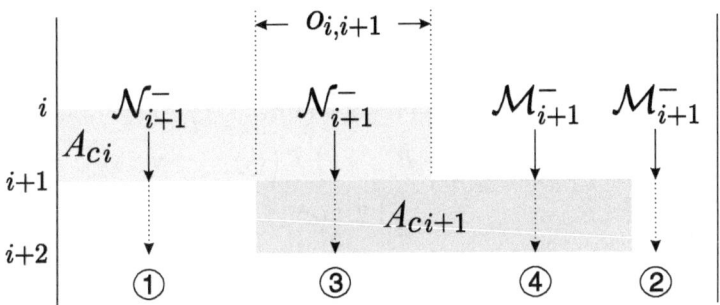

Fig. 9.9. Graphical illustration of the radiative components of a generalized cloud overlap scheme for longwave radiation (Li, 2002). Quantities $\mathcal{N}^-$ and $\mathcal{M}^-$ are downwelling irradiances associated with either cloudy or clear portions of the upper layer, respectively. Clouds in these layers overlap by an amount $o_{i,i+1}$

In an attempt to deal with vertical correlation of extinction, explicit account of contributions to F_j^- from all contiguous cloudy layers must be considered so (9.31a) is expanded as

$$
\begin{aligned}
F_j^- &= \left(F_{j-1}^- - \tilde{B}_{j-1+\frac{1}{2}}\right) \mathrm{e}^{-\langle\kappa\rangle_{j-1}/\mu_1} + \tilde{B}_{j-1+\frac{1}{2}} \\
&= \left(F_{j-2}^- - \tilde{B}_{j-2+\frac{1}{2}}\right) \mathrm{e}^{-(\langle\kappa\rangle_{j-2}+\langle\kappa\rangle_{j-1})/\mu_1} \\
&\quad + \left(\tilde{B}_{j-2+\frac{1}{2}} - \tilde{B}_{j-1+\frac{1}{2}}\right) \mathrm{e}^{-\langle\kappa\rangle_{j-1}/\mu_1} \\
&\quad + \tilde{B}_{j-1+\frac{1}{2}} \\
&\ \ \vdots \\
&= \left(F_{i-n}^- - \tilde{B}_{i-n+\frac{1}{2}}\right) \mathrm{e}^{-\langle\kappa\rangle_{i\ n,j}/\mu_1} \\
&\quad + \left(\tilde{B}_{i-n+\frac{1}{2}} - \tilde{B}_{i-n+1+\frac{1}{2}}\right) \mathrm{e}^{-\langle\kappa\rangle_{i-n+1,j}/\mu_1} \\
&\quad + \ldots + \left(\tilde{B}_{j-3+\frac{1}{2}} - \tilde{B}_{j-2+\frac{1}{2}}\right) \mathrm{e}^{-\langle\kappa\rangle_{j-2,j}/\mu_1} \\
&\quad + \left(\tilde{B}_{j-2+\frac{1}{2}} - \tilde{B}_{j-1+\frac{1}{2}}\right) \mathrm{e}^{-\langle\kappa\rangle_{j-1,j}/\mu_1} \\
&\quad + \tilde{B}_{j-1+\frac{1}{2}} ,
\end{aligned}
\tag{9.34a}
$$

where

$$\langle \kappa \rangle_{k,j} = \sum_{m=j-1}^{k} \langle \kappa \rangle_m \tag{9.34b}$$

is mean optical depth between levels k and j. Then, each component in (9.34a) is operated on as in (9.27) yielding

$$
\begin{aligned}
\mathcal{F}_j^- = {}& \left(F_{i-n}^- - \tilde{B}_{i-n+\frac{1}{2}} \right) \mathcal{T} \left(\langle \kappa \rangle_{i-n,j}, \nu_{i-n,j} \right) \\
& + \left(\tilde{B}_{i-n+\frac{1}{2}} - \tilde{B}_{i-n+1+\frac{1}{2}} \right) \mathcal{T} \left(\langle \kappa \rangle_{i-n+1,j}, \nu_{i-n+1,j} \right) \\
& + \ldots + \left(\tilde{B}_{j-3+\frac{1}{2}} - \tilde{B}_{j-2+\frac{1}{2}} \right) \mathcal{T} \left(\langle \kappa \rangle_{j-2,j}, \nu_{j-2,j} \right) \\
& + \left(\tilde{B}_{j-2+\frac{1}{2}} - \tilde{B}_{j-1+\frac{1}{2}} \right) \mathcal{T} \left(\langle \kappa \rangle_{j-1,j}, \nu_{j-1,j} \right) \\
& + \tilde{B}_{j-1+\frac{1}{2}},
\end{aligned}
\tag{9.35a}
$$

where

$$
\mathcal{T}(\langle \kappa \rangle_{k,j}, \nu_{k,j}) = \int_0^\infty p_\Gamma(\kappa \mid \langle \kappa \rangle_{k,j}, \nu_{k,j}) e^{-\kappa/\mu_1} \mathrm{d}\kappa
$$

$$
\tag{9.35b}
$$

$$
= \left(\frac{\nu_{k,j}}{\nu_{k,j} + \langle \kappa \rangle_{k,j}/\mu_1} \right)^{\nu_{k,j}},
$$

and $p_\Gamma(\kappa \mid \langle \kappa \rangle_{k,j}, \nu_{k,j})$ is as in (9.25) except it applies to the collection of layers between levels k and j whose variance parameter is described by $\nu_{k,j}$.

While this formulation is a more precise treatment than that presented for the solar, it requires not only estimates of ν for each layer, but also for each grouping of layers. The complexity of this is illustrated in Fig. 9.10 for a $(50\,\mathrm{km})^2$ domain of convective clouds simulated by a cloud-resolving model (CRM). While ν for each layer is ≈ 1, it is 0.67 for the field integrated through the entire depth of cloud, and reaches as low as 0.38 for the collection of layers between $\approx 4.5\,\mathrm{km}$ and the lowest cloudy layer at $\approx 0.6\,\mathrm{km}$. Such a low value of $\nu_{k,j}$ is due to a combination of well correlated shafts of convection in conjunction with substantial near-random overlap of less dense clouds (and regions of entrained cloud near edges of convective cores). Modelling this is a substantial challenge but is likely to have a secondary effect relative to errors in A_c, $\langle \kappa \rangle$, ν, and overlap on a per layer basis.

9.6 1D versus 3D Radiative Transfer

The purpose of this section is to demonstrate some differences between BB irradiances computed with 1D (and enhancements such as GWTSA) and fully 3D radiative transfer models. Focus is on both domain-average profiles of heating and spatial distributions of heating. Results pertain to solar radiation.

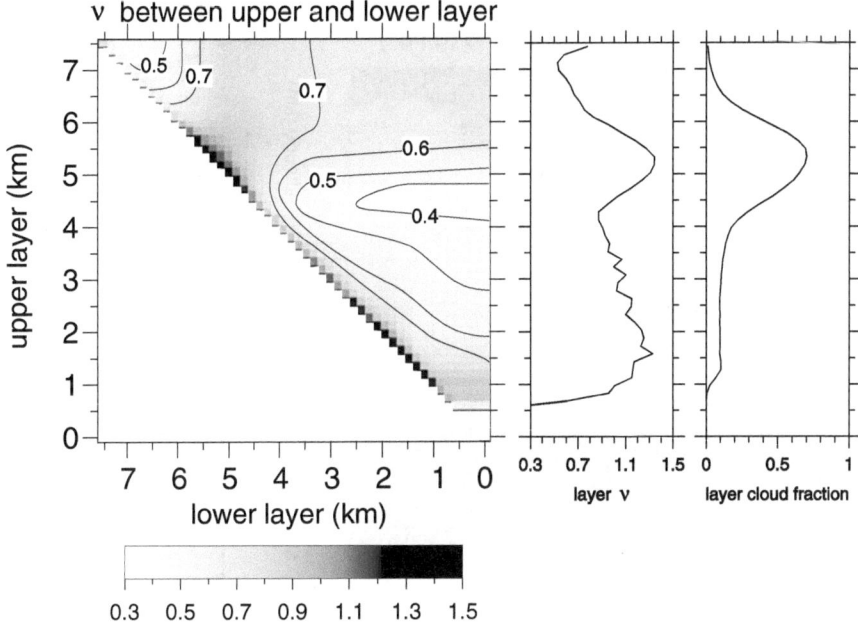

Fig. 9.10. Plots on the right show profiles of layer ν (see 9.25) and cloud fraction for a cloud field derived from a cloud-resolving model. Panel on the left shows ν for collections of layers bounded between lower altitude listed along the abscissa and upper altitude listed on the ordinate (cloud data courtesy of W. Grabowski and V. Grubišić)

9.6.1 High Resolution Differences

Relatively little has been reported on high resolution distributions of radiative heating. This goes for both spatial and spectral distributions. One of the more thorough studies is that of O'Hirok and Gautier (1998a,b) where they simulated spatial distributions of solar heating with a Monte Carlo algorithm acting on a couple of large convective clouds inferred partly from satellite imagery with rather ad hoc alterations. They also assessed some aspects of spectral structure of solar absorption.

To augment their findings, results are presented here for part of a shallow, nonprecipitating, cumulus cloud field simulated by a 2D CRM. Horizontal and vertical grid-spacings were 50 m and droplet size distribution was carried for each cell (all drops had radii <20 μm). Figure 9.11 shows cross-sections of water vapour mixing ratio q_v, liquid water content LWC, and droplet effective radius r_e for a 6.25 km long stretch of the field along with three sample profiles. The panel for q_v shows complete convective plumes and the other panels show near-constant lifting condensation levels with LWC and r_e generally increasing with height. Figure 9.12 shows that while these clouds are shallow, $\bar{\tau}$ is large corresponding to a mean cloud extinction coefficient of ≈85 km^{-1}.

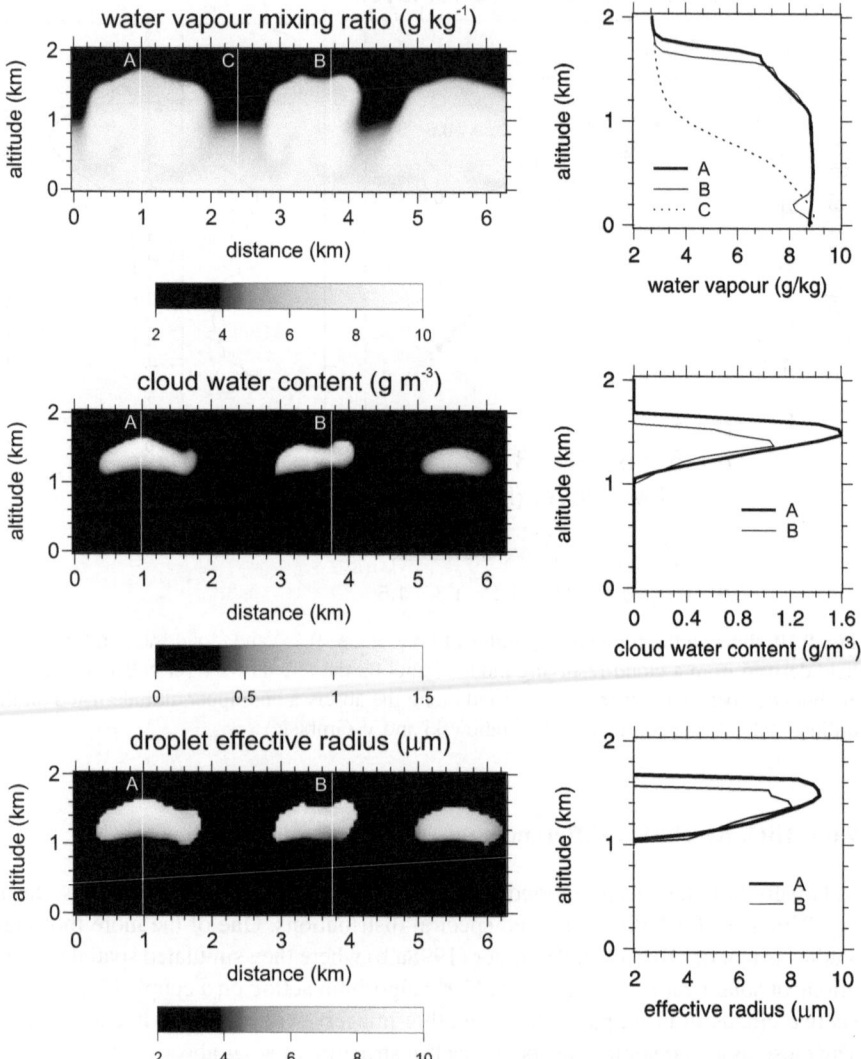

Fig. 9.11. Top panel shows a profile of water vapour mixing ratio as a function of distance for a 6.25 km stretch (of a 100 km domain) of atmosphere as simulated by a cloud-resolving model. Middle and lower panels show corresponding cloud water content and droplet effective radius. Line plots on the right show profiles of respective quantities for profiles labelled on the cross-section panels (data courtesy of J.-P. Blanchet)

Figure 9.13 shows distributions of heating rate for the field shown in Fig. 9.11 as computed by a BB Monte Carlo algorithm that resolves the solar spectrum at a resolution of 0.01 μm (Barker et al., 1998). A constant BB Lambertian surface albedo of 0.06 was used and Slingo's (1989) 24-band parametrization for cloud optical

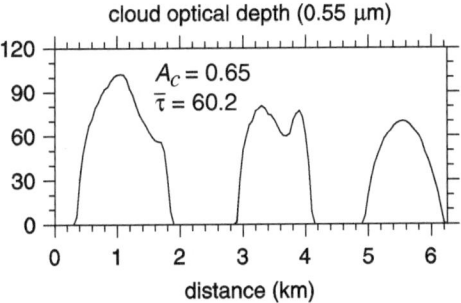

Fig. 9.12. Vertical cloud optical depth for the domain illustrated in Fig. 9.11

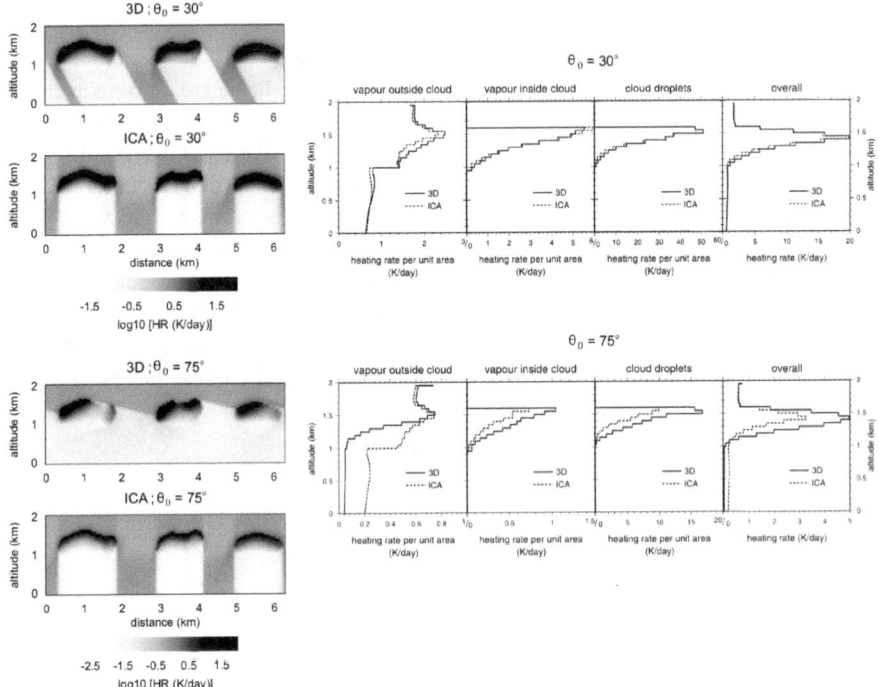

Fig. 9.13. Panels on the *bottom* show distribution of heating rates for two solar zenith angles θ_0 as predicted by two executions of a Monte Carlo algorithm: the first used horizontal grid-spacings $\Delta x = 0.05$ km (3D) while the second used Δx arbitrarily large (ICA). Line plots show corresponding domain-averaged heating profiles per unit area for water vapour inside and outside of cloud and droplets. Plots labelled "overall" show domain-averaged heating due to all constituents

properties was employed. Results are shown for both the full 3D and the ICA (which would be used in a CRM). Here the vertically-aligned pattern of the ICA (regardless of θ_0) is obvious but so too are radical differences in surface and cloud heating patterns (more so at large θ_0). It remains to be seen whether these differences are important for cloud simulations as full 3D radiation codes have as yet not been run within a CRM simulation.

The domain average profiles in Fig. 9.13 show interesting differences between 3D and ICA and echo those reported by others. At $\theta_0 = 30°$, differences are minor but at $\theta_0 = 75°$, the ICA significantly underestimates in-cloud absorption due to neglect of cloud side illumination, and overestimates out of cloud absorption by vapour from about mid-cloud level down to the surface. This is because the ICA allows passage of direct-beam between clouds while in the 3D case cloud fraction presented to the direct-beam is 100%. The plot showing overall heating, however, indicates that overestimation of heating by water vapour outside cloud is minor compared to underestimation of heating by droplets.

The left panel in Fig. 9.14 shows average dispositions of solar radiation as a function of μ_0 for the domain shown in Fig. 9.11. While these are domain averages, the domains are still small relative to a typical LSAM cell. Domain-average albedo α is reminiscent of results shown in many studies (e.g., Welch and Wielicki, 1985; Barker et al., 1998) where 3D clouds reflect less at high Sun (due to photon leakage out sides along predominantly downward trajectories) and more at low Sun (due to side illumination and increased effective cloud cover). The reverse is true for surface absorptance a_{sfc} while atmospheric absorptances a_{atm} for both methods are very similar. In fact, at $\mu_0 = 1$, a_{atm} for 3D and ICA are equal but as the middle panel in Fig. 9.14 shows, ratios $\mathcal{R}$ of CRE at the surface to that at the top of the model domain are systematically largest for the 3D simulation. Though $\mathcal{R}$ has been used to

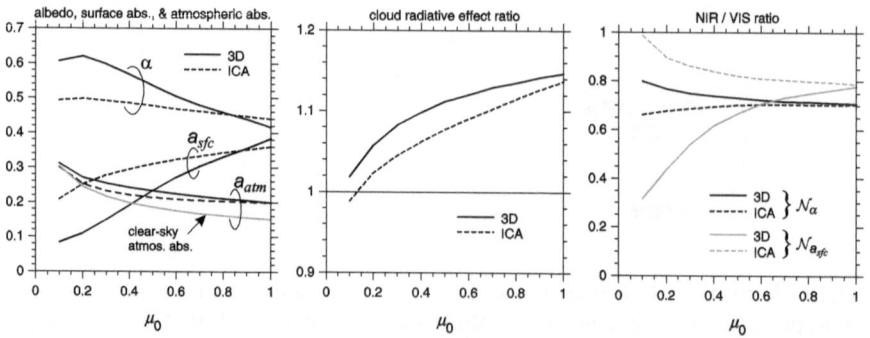

Fig. 9.14. *Left* plot shows domain averaged TOA albedo α, surface absorptance a_{atm}, and surface absorptance a_{sfc} as functions of μ_0 as predicted by the Monte Carlo in full 3D and ICA modes for the domain shown in Fig. 9.11. Also shown is clear-sky a_{atm}. Centre plot shows ratios $\mathcal{R}$ for CRE at the surface to that at the TOA for both modes. *Right* plot shows ratios of NIR to VIS values for albedo $\mathcal{N}_\alpha$ and surface absorptance $\mathcal{N}_{a_{sfc}}$ for both modes of the Monte Carlo algorithm

diagnose the impact of cloud on a_{atm} (e.g., Cess et al., 1995), in this case at overhead Sun its ambiguous nature shows. This is because, in addition to cloud absorption, $\mathcal{R}$ also responds to the partition of radiation by cloud into albedo and transmittance. Here, 3D clouds yield more transmittance thereby enhancing $\mathcal{R}$ relative to the ICA.

The right panel in Fig. 9.14 shows the ratios of near-IR ($>0.7\,\mu m$) to visible ($<0.7\,\mu m$) albedo $\mathcal{N}_\alpha$ and surface absorptance $\mathcal{N}_{a_{sfc}}$ for 3D and ICA estimates. Like $\mathcal{R}$, $\mathcal{N}_\alpha$ has been used to diagnose impacts of cloud on a_{atm}. $\mathcal{N}_\alpha$ is almost independent of both μ_0 and method of solution; 3D values are slightly larger than ICA values. On the other hand, $\mathcal{N}_{a_{sfc}}$ for 3D and ICA are almost equal at overhead Sun but as μ_0 decreases, 3D values do too though ICA estimates increase (both monotonically).

One can speculate about trends in $\mathcal{N}$ only so far; to comprehend them fully, detailed spectral information is needed. Figure 9.15 shows ratios of 3D/ICA for α, a_{sfc}, and a_{atm}. For overhead Sun, it is apparent that the near equality of both $\mathcal{N}_\alpha$ and $\mathcal{N}_{a_{sfc}}$ for 3D and ICA is actually indicative of the near equality of spectral ratios across the entire spectrum (see O'Hirok and Gautier, 1998b). For $\theta_0 = 75°$, however, the enhancement in $\mathcal{N}_\alpha$ for 3D conditions relative to ICA is due primarily to suppression of the 3D effect near the UV rather than enhanced reflectance expected due to cloudside illumination (as is clearly visible for $\lambda > 0.5\,\mu m$). The converse is true for $\mathcal{N}_{a_{sfc}}$. The middle panel of Fig. 9.15 shows 3D/ICA for a_{atm}. For $\theta_0 = 0°$ and $75°$, 3D absorbs respectively less than and more than ICA in the gaseous widows. This is due to cloudside leakage and fewer scattering events by droplets at $\theta_0 = 0°$ and more scattering events due to side illumination at $\theta_0 = 75°$ (see Fig. 9.13).

This case study demonstrates clearly that the essence of understanding the impact of clouds on BB atmospheric absorption should be explored using high spectral resolution measurements and modelling whenever possible.

9.6.2 Domain Average Differences

The study by Barker et al. (2003) intercompared domain-average BB irradiances for cloudy atmospheres as computed by 1D and 3D solar radiative transfer codes. The objective was to assess how well 1D codes interpret and handle unresolved clouds. Four BB Monte Carlo algorithms set benchmark irradiances and they in turn were assessed for clear-sky and homogeneous overcast skies against LBL results. Cloud fields produced by several different CRMs where used. The Monte Carlo runs acted on the full 3D fields while 1D codes operated on degenerate versions represented by vertical profiles of A_c, $\bar{\tau}$, $\overline{\ln \tau}$, cloud overlap rate, and $\bar{q}_v$.

In addition to full 3D benchmarks, this study also used *intermediate* benchmarks applicable to PPH clouds that follow: exact overlapping structure of cloud; maximum/random overlap; and random overlap. ICA estimates were given too. As such, modellers were able to see if their 1D code was doing what it was expected to do as well as gauge how far it was from the full 3D solutions.

Of the several cases analyzed, a particularly demanding one was selected for showing here: a $(400\,\text{km})^2$ domain containing towering convective clouds reaching from $\approx 1\,\text{km}$ to $\approx 15\,\text{km}$ high (Grabowski et al., 1998). Total cloud fraction was about

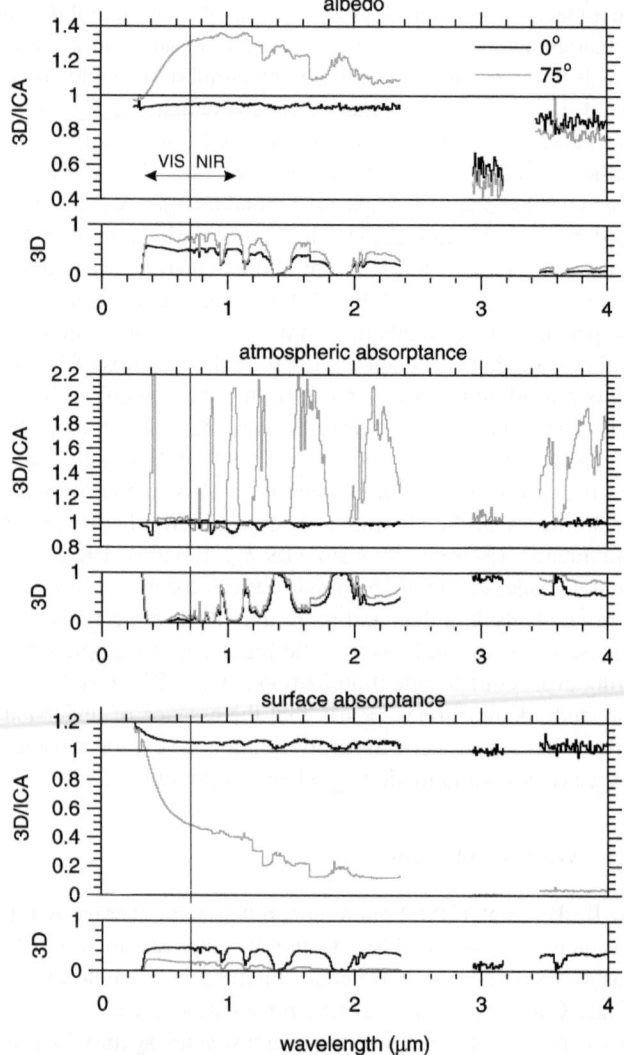

Fig. 9.15. Domain average TOA albedo, atmospheric absorptance, and surface absorptance as functions of wavelength for the cloud field shown in Fig. 9.11 at two values of θ_0 as indicated. Each grouping contains two plots: the small lower one shows estimates from the full 3D simulation while the large upper one shows ratios between 3D and ICA results. Blanked-out areas are where statistical noise, from the Monte Carlo simulation, was overwhelming

45% and mean visible $\bar{\tau} \approx 90$ for cloudy columns only. The 1D codes were partitioned into four categories:

- ICA: attempted to account for horizontal fluctuations in τ;
- exact overlap: used PPH clouds but attempted to utilize overlap information;

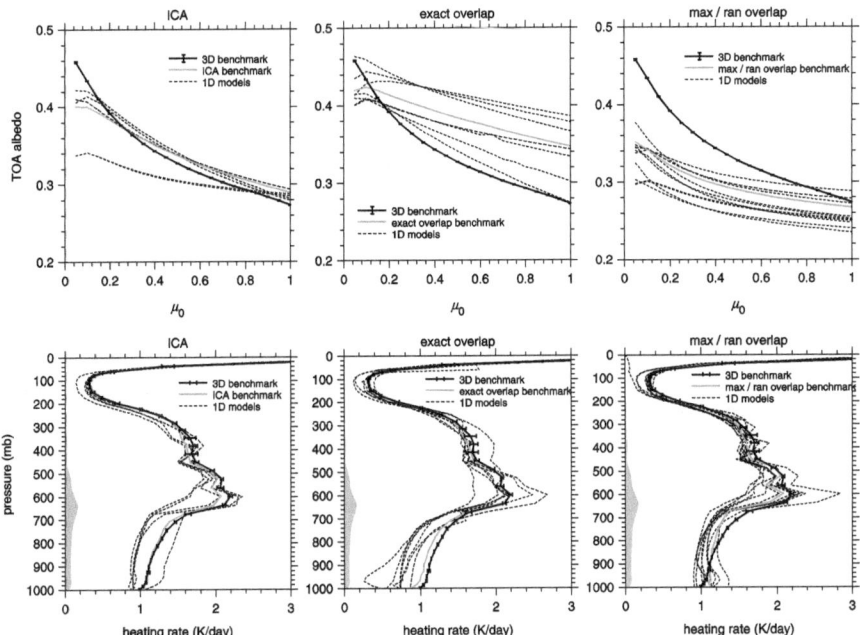

Fig. 9.16. Upper plots show broadband TOA albedos as functions of μ_0 and *lower* plots show heating rate profiles for $\mu_0 = 0.5$. Each plot shows the 3D Monte Carlo benchmark solution (mean of 4 models with standard deviation bars) along with a particular intermediate benchmark (set by one of the Monte Carlo codes) and results for 1D codes that address unresolved clouds differently. 1D codes in the ICA genre attempt to account for unresolved horizontal fluctuations while the other two use PPH clouds with different overlap assumptions. *Shaded* regions on heating rate plots indicate layer cloud fraction profile (maximized at $\approx$20%)

- max/ran overlap: used PPH clouds with maximum/random overlap assumption;
- random overlap: used PPH clouds with random overlap assumption.

The most populous category was max/ran overlap.

Figure 9.16 shows albedo as a function of μ_0 and heating rate profiles at $\mu_0 = 0.5$. Each plot shows mean values with standard deviation bars for full 3D solutions (based on four Monte Carlo codes). Clearly, these codes are in extremely good agreement. Also shown on each plot is the intermediate benchmark, simulated by one of the Monte Carlos (gray line), as well as the 1D results (dashed lines). Interestingly, those codes that acknowledge horizontal variability of cloud come very close to the ICA benchmark which in turn is very close to the full 3D solution. Both the exact and max/ran overlap genre of 1D codes appear to be *attracted* to their respective intermediate benchmark though exhibit a fair degree of variance among themselves. Moreover, their intermediate benchmarks deviate, significantly at times, from the full 3D solutions. The max/ran values are *less than* the full 3D solution despite use of PPH clouds. This is because the max/ran assumption is an extreme case that usually

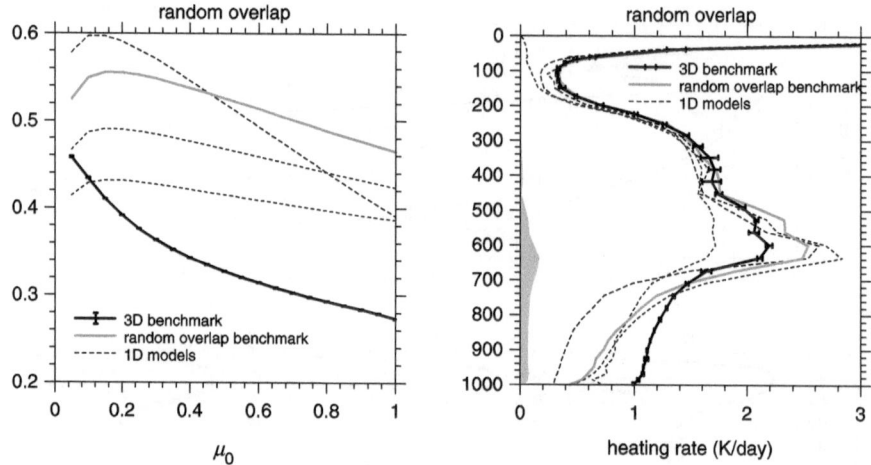

Fig. 9.17. As in Fig. 9.16 but showing intermediate benchmark and 1D results assuming that clouds in all layers overlapped randomly

underestimates cloud fraction; a first-order error. If the max/ran codes were to account for horizontal variability, their estimates of α would be slightly smaller still.

Surprisingly, Fig. 9.17 shows that the random overlap codes show little affinity for their benchmark. While this approximation is extreme and rarely used throughout a code, it is invoked when blocks of cloudy layers are separated by clear layers. Thus, this result is a bit disconcerting.

Errors in heating rates are less easy to document than those for α. Nevertheless, it is clear from Figs. 9.16 and 9.17 that those codes that overestimate α, underestimate heating in near-surface layers. Note also the tendency for PPH clouds of all overlap persuasions to slightly overestimate heating near 600 mb; where cloud mass presented to the direct-beam begins to be abundant. This is due to excessive numbers of scattering events by droplets as photons are not allowed to escape horizontally.

9.7 Summary and Conclusions

This chapter opened by reiterating the inexorable links between BB radiation, clouds, and climate. The importance of these links rests on:

(i)   BB radiation being at the core of climate feedback mechanisms;
(ii) seemingly small perturbations to cloud properties can modulate Earth's radiation budget significantly; and
(iii) our ability to represent cloud-radiation interactions in large-scale atmospheric models (LSAMs) is recognized generally as lacking. Thus, the issue here that is at the heart of this book is not just BB radiative transfer for 3D cloudy atmospheres, but the ability of 1D radiative transfer models, as used in LSAMs,

to capture spatially unresolved interactions between 3D clouds and BB radiation. As such, the main body of this chapter addressed, and compared results from, common techniques for modelling BB irradiances (and heating rates) using both Monte Carlo algorithms (see Chap. 4) and approximate 1D solutions (see Chaps. 6, 7, and 8).

At this point in time, several 3D Monte Carlo BB algorithms exist and some have been intercompared in detail. These are primarily solar codes though longwave ones are beginning to emerge and there are plans to intercompare them too. The consensus so far is that these codes are in excellent agreement and the majority of the minor discrepancies arise from specification of optical properties. With the proliferation of cloud model data from either cloud-resolving models (CRMs) or phenomenological models, we have now documented quite well the nature of BB radiative transfer and heating for realistic cloudy atmospheres. We are still, however, a fair distance from understanding the full ramifications of 3D BB transfer on the scale-dependent nature of atmospheric circulation. It is only a matter of time though before full 3D BB radiation codes will be used actively in CRMs. In fact, 1D parametrizations for BB radiative transfer that attempt to address key aspects of realistic 3D transfer are already being used in LSAMs, but again, confident assessment of their impact on climate has yet to be made.

What we have in hand now with 3D BB codes and cloud fields is the ability to rigorously assess the capabilities of multi-layer 1D BB models. Although differences between BB irradiances predicted by 3D and 1D codes can be large, particularly for convective clouds at large solar zenith angles, when the 1D codes are initialized with accurate information about cloud structure, these differences are often at a level that is likely to be acceptable for many purposes (i.e., plane-parallel, homogeneous biases are reduced markedly).

It would appear then that the major challenge facing LSAMs regarding BB radiation is not radiative transfer per se but rather parametrization of vertical profiles of unresolved cloud structure; namely (and roughly in order of importance) cloud fraction, mean optical thickness, overlap structure, and horizontal fluctuations. To a great extent the elements of this list merge and overlap with one another, but this is to be expected given that we are discussing ad hoc discretizations of the continuous atmosphere. Some may argue that some of the variables listed above are represented fairly well and that the order of importance might even be reversed to that as listed. Regardless, reaching the stage of confident representation of radiation-cloud feedbacks in LSAMs will require not only continued development of unresolved cloud and BB radiative transfer algorithms, but also extensive cloud and radiation data from long-term monitoring programs such as DOE's Atmospheric Radiation Measurement (ARM) program (Stokes and Schwartz, 1994).

References

Barker, H.W. and Q. Fu (2000). Assessment and optimization of the gamma-weighted two-stream approximation. *J. Atmos. Sci.*, **57**, 1181–1188.

Barker, H.W. and B.A. Wielicki (1997). Parameterizing grid-averaged longwave fluxes for inhomogenous marine boundary layer clouds. *J. Atmos. Sci.*, **54**, 2785–2798.

Barker, H.W., B.A. Wielicki, and L. Parker (1996). A parameterization for computing grid-averaged solar fluxes for inhomogeneous marine boundary layer clouds - Part 2, Validation using satellite data. *J. Atmos. Sci.*, **53**, 2304–2316.

Barker, H.W., J.-J. Morcrette, and G.D. Alexander (1998). Broadband solar fluxes and heating rates for atmospheres with 3D broken clouds. *Quart. J. Roy. Meteor. Soc.*, **124**, 1245–1271.

Barker, H.W., G.L. Stephens, and Q. Fu (1999). The sensitivity of domain-averaged solar fluxes to assumptions about cloud geometry. *Quart. J. Roy. Meteor. Soc.*, **125**, 2127–2152.

Barker, H.W., G.L. Stephens, P.T. Partain, J.W. Bergman, B. Bonnel, K. Campana, E.E. Clothiaux, S. Clough, S. Cusack, J. Delamere, J. Edwards, K.F. Evans, Y. Fouquart, S. Freidenreich, V. Galin, Y. Hou, S. Kato, J. Li, E. Mlawer, J.-J. Morcrette, W. O'Hirok, P. Räisänen, V. Ramaswamy, B. Ritter, E. Rozanov, M. Schlesinger, K. Shibata, P. Sporyshev, Z. Sun, M. Wendisch, N. Wood, and F. Yang (2003). Assessing 1D atmospheric solar radiative transfer models: Interpretation and handling of unresolved clouds. *J. Climate*, **16**, 2676–2699.

Cess, R.D., M.-H. Zhang, G.L. Potter, H.W. Barker, R.A. Colman, D.A. Dazlich, A.D. Del Genio, M. Esch, J.R. Fraser, V. Galin, W.L. Gates, J.J. Hack, W. Ingram, J.T. Kiehl, A.A. Lacis, H. Le Treut, Z.-X. Li, X.-Z. Liang, J.F. Mahfouf, B.J. McAvaney, V.P. Meleshko, J.-J. Morcrette, D.A. Randall, E. Roeckner, J.-F. Royer, A.P. Sokolov, P.V. Sporyshev, K.E. Taylor, W.-C. Wang, and R.T. Wetherald (1993). Intercomparison of CO_2 radiative forcing in atmospheric general circulation models. *Science*, **262**, 1252–1255.

Cess, R.D., M.-H. Zhang, P. Minnis, L. Corsetti, E.G. Dutton, B.W. Forgan, D.P. Garber, W.L. Gates, J.J. Hack, E.F. Harrison, X. Jing, J.T. Kiehl, C.N. Long, J.-J Morcrette, G. L. Potter, V. Ramanathan, B. Subasilar, C.H. Whitlock, D.F. Young, and Y. Zhou (1995). Absorption of solar radiation by clouds: Observations versus models. *Science*, **267**, 496–499.

Fouquart, Y., B. Bonnel, and V. Ramaswamy (1991). Intercomparing shortwave radiation codes for climate studies. *J. Geophys. Res.*, **96**, 8955–8968.

Fu, Q. and K.-N. Liou (1992). On the correlated-k distribution method for radiative transfer in inhomogeneous atmospheres. *J. Atmos. Sci.*, **49**, 2139–2156.

Geleyn, J.-F. and A. Hollingsworth (1979). An economical analytical method for the computation of the interaction between scattering and line absorption of radiation. *Contrib. Atmos. Phys.*, **52**, 1–16.

Grabowski, W.W., X. Wu, M.W. Moncrieff, and W.D. Hall (1998). Cloud-resolving modeling of cloud systems during phase III of GATE. Part II: Effects of resolution and the third spatial dimension. *J. Atmos. Sci.*, **55**, 3264–3282.

Hansen, J.E. and L.D. Travis (1974). Light scattering in planetary atmospheres. *Space Sci. Rev.*, **16**, 527–610.

Hogan, R.J. and A.J. Illingworth (2000). Derived cloud overlap statistics from radar. *Quart. J. Roy. Meteor. Soc.*, **126**, 2903–2909.

Hu, Y.-X. and K. Stamnes (1993). An accurate parameterization of the radiative properties of water clouds suitable for use in climate models. *J. Climate*, **6**, 728–742.

Intergovernmental Panel on Climate Change (1996). The Science of Climate Change: Report of the Intergovernmental Panel on Climate Change (IPCC), J.T. Houghton et al. (eds.). Cambridge University Press, New York (NY).

Lacis, A.A. and V. Oinas (1991). A description of the correlated-k method for modeling nongrey gaseous absorption, thermal emission, and multiple scattering in vertically inhomogeneous atmospheres. *J. Geophys. Res.*, **96**, 9027–9063.

Lacis, A.A., W.C. Wang, and J.E. Hansen (1979). Correlated-k method for radiative transfer in climate models: Application to the effect of cirrus clouds on climate. Technical Report NASA Conf. Publ. 2076, NASA.

Li, J. (2002). Accounting for unresolved clouds in a 1D infrared radiative transfer model. Part I: Solution for radiative transfer, scattering, and cloud overlap. *J. Atmos. Sci.*, **59**, 3302–3320.

Li, J. and H.W. Barker (2002). Accounting for unresolved clouds in a 1D infrared radiative transfer model. Part II: Horizontal variability of cloud water path. *J. Atmos. Sci.*, **59**, 3321–3339.

Liou, K.-N. (1992). *Radiation and Cloud Processes in the Atmosphere*. Oxford University Press, New York (NY).

Meador, W.E. and W.R. Weaver (1980). Two-stream approximations to radiative transfer in planetary atmospheres: A unified description of existing methods and a new improvement. *J. Atmos. Sci.*, **37**, 630–643.

Mlawer, E.J., P.D. Brown, S.A. Clough, L.C. Harrison, J.J. Michalsky, P.W. Kiedron, and T. Shippert (2000). Comparison of spectral direct and diffuse solar irradiance measurements and calculations for cloud-free conditions. *Geophys. Res. Lett.*, **27**, 2653–2656.

O'Hirok, W. and C. Gautier (1998a). A three-dimensional radiative transfer model to investigate the solar radiation within a cloudy atmosphere. Part I: Spatial effects. *J. Atmos. Sci.*, **55**, 2162–2179.

O'Hirok, W. and C. Gautier (1998b). A three-dimensional radiative transfer model to investigate the solar radiation within a cloudy atmosphere. Part II: Spectral effects. *J. Atmos. Sci.*, **55**, 3065–3075.

Oreopoulos, L. and H.W. Barker (1999). Accounting for subgrid-scale cloud variability in a multi-layer, 1D solar radiative transfer algorithm. *Quart. J. Roy. Meteor. Soc.*, **125**, 301–330.

Rossow, W.B. (1989). Measuring cloud properties from space: A review. *J. Climate*, **2**, 201–213.

Schlesinger, M.E. and J.F.B. Mitchell (1987). Climate model simulations of the equilibrium climatic response to increased carbon dioxide. *Review of Geophys.*, **4**, 760–798.

Senior, C.A. (1999). Comparison of mechanisms of cloud-climate feedbacks in GCMs. *J. Climate*, **12**, 1480–1489.

Slingo, A. (1989). A GCM parameterization for the shortwave radiative properties of water clouds. *J. Atmos. Sci.*, **46**, 1419–1427.

Smith, R.N.B. (1990). A scheme for predicting layer clouds and their water content in a GCM. *Quart. J. Roy. Meteor. Soc.*, **116**, 435–460.

Stephens, G.L. (1988). Radiative transfer through arbitrary shaped optical media, II: Group theory and simple closures. *J. Atmos. Sci.*, **45**, 1837–1848.

Stokes, G.M. and S.E. Schwartz (1994). The Atmospheric Radiation Measurement (ARM) program: Programmatic background and design of the cloud and radiation test bed. *Bull. Amer. Meteor. Soc.*, **75**, 1201–1221.

Warren, S.G., C.J. Hahn, J. London, R.M. Chervin, and R.L. Jenne (1986). Global distribution of total cloud cover and cloud type amounts over land. Technical Report TN-273+STR., NCAR, Boulder (CO).

Welch, R.M. and B.A. Wielicki (1985). A radiative parameterization of stratocumulus cloud fields. *J. Atmos. Sci.*, **42**, 2888–2897.

Suggested Reading

Clouds, Radiation, and Climate

Cess, R.D., M.-H. Zhang, W.J. Ingram, G.L. Potter, V. Alekseev, H.W. Barker, E. Cohen-Solal, R.A. Colman, D.A. Dazlich, A.D. Del Genio, M.R. Dix, M. Esch, L.D. Fowler, J.R. Fraser, V. Galin, W.L. Gates, J.J. Hack, J.T. Kiehl, H. Le Treut, K.K.-W. Lo, B.J. McAvaney, V.P. Meleshko, J.-J. Morcrette, D.A. Randall, E. Roeckner, J.-F. Royer, M.E. Schlesinger, P.V. Sporyshev, B. Timbal, E.M. Volodin, K.E. Taylor, W.C. Wang and R.T. Wetherald (1996). Cloud feedback in atmospheric general circulation models: An update. *J. Geophys. Res.*, **101**, 12,791–12,794.

Cess, R.D., M.-H. Zhang, G.L. Potter, V. Alekseev, H.W. Barker, S. Bony, R.A. Colman, D.A. Dazlich, A.D. Del Genio, M.Déqué, M.R. Dix, V. Dymnikov, M. Esch, L.D. Fowler, J.R. Fraser, V. Galin, W.L. Gates, J.J. Hack, W.J. Ingram, J.T. Kiehl, Y. Kim, H. Le Treut, X.Z. Liang, B.J. McAvaney, V.P. Meleshko, J.-J. Morcrette, D.A. Randall, E. Roeckner, M.E. Schlesinger, P.V. Sporyshev, K.E. Taylor, B. Timbal, E.M. Volodin, W. Wang, W.C. Wang and R.T. Wetherald (1997). Comparison of the seasonal change in cloud-radiative forcing from atmospheric general circulation models and satellite observations. *J. Geophys. Res.*, **102**, 16,593–16,603.

Fu, Q., S.K. Krueger and K.-N. Liou (1995). Interactions of radiation and convection in simulated tropical cloud clusters. *J. Atmos. Sci.*, **52**, 1310–1328.

Hansen, J.E., D. Russell, D. Rind, P. Stone, A. Lacis, L. Travis, S. Lebedeff and R. Ruedy (1983). Efficient three-dimensional global models for climate studies: Models I and II. *Mon. Wea. Rev.*, **111**, 609–662.

Liou, K.-N. (1992). *Radiation and cloud processes in the atmosphere*. Oxford University Press, New York, USA, 487pp.

Mitchell, J.F.B., R.A. Davis, W.J. Ingram and C.A. Senior (1995). On surface temperature, greenhouse gases, and aerosols: Models and observations. *J. Climate*, **8**, 2364–2386.

Potter, G.L., J.M. Slingo, J.-J. Morcrette and L. Corsetti (1992). A modelling perspective on cloud radiative forcing. *J. Geophys. Res.*, **97**, 20,507–20,518.

Ramanathan, V. (1987). The role of Earth radiation budget studies in climate and general circulation research. *J. Geophys. Res.*, **92**, 4075–4095.

Schlesinger, M.E. and J.F.B. Mitchell (1987). Climate model simulations of the equilibrium climatic response to increased carbon dioxide. *Rev. of Geophys.*, **4**, 760–798.

Clear Sky

Arking, A. and K. Grossman (1972). The influence of of line shape and band structure on temperatures in planetary atmospheres. *J. Atmos. Sci.*, **29**, 937–949.

Chou, M.D. and K.T. Lee (1996). Parameterizations for the absorption of solar radiation by water vapor and ozone, *J. Atmos. Sci.*, **53**, 1203–1208.

Clough, S.A., F.X. Kneizys and R.W. Davies (1989). Line shape and the water vapor continuum. *Atmos. Res.*, **23**, 229–241.

Fu, Q. and K.-N. Liou (1992). On the correlated-k distribution method for radiative transfer in inhomogeneous atmospheres. *J. Atmos. Sci.*, **49**, 2139–2156.

Kato, S, T.P. Ackerman, J.H. Mather and E.E. Clothiaux (1999). The k-distribution method and correlated-k approximation for a shortwave radiative transfer model. *J. Quant. Spectrosc. Radiat. Transfer*, **62**, 109–121.

Lacis, A.A. and V. Oinas (1991). A description of the correlated-k method for modeling nongrey gaseous absorption, thermal emission, and multiple scattering in vertically inhomogeneous atmospheres. *J. Geophys. Res.*, **96**, 9027–9063.

Wiscombe, W.J. and J.W. Evans (1976). Exponential-sum fitting of radiative transmission functions. *J. Comp. Phys.*, **24**, 416–444.

Ocean Albedo

Cox, C. and W. Munk (1956). Slopes of the sea surface deduced from photographs of the sun glitter. *Bull. Scripps Inst. Ocean.*, **6**, 401–488.

Payne, R.E. (1972). Albedo of the sea surface. *J. Atmos. Sci.*, **29**, 959–970.

Preisendorfer, R.W. and C.D. Mobley (1986). Albedos and glitter patterns of a wind-roughened sea surface. *J. Phy. Ocean.*, **16**, 1293–1316.

Cloud Optical Properties

Fu, Q. and K.-N. Liou (1993). Parameterization of the radiative properties of cirrus clouds. *J. Atmos. Sci.*, **50**, 2008–2025.

Hu, Y.-X. and K. Stamnes (1993). An accurate parameterization of the radiative properties of water clouds suitable for use in climate models. *J. Climate*, **6**, 728–742.

Li, J., S.M. Freidenreich and V. Ramaswamy (1997). Solar spectral weight at low cloud tops. *J. Geophys. Res.*, **102**, 11,139–11,143.

Macke, A., J. Mueller and E. Raschke (1996). Single scattering properties of atmospheric ice crystals. *J. Atmos. Sci.*, **53**, 2813–2825.

Räisänen, P. (1999). Parameterization of water and ice-cloud near-infrared single-scattering co-albedo in broadband radiation schemes. *J. Atmos. Sci.*, **56**, 626–641.

Slingo, A. (1989). A GCM parameterization for the shortwave radiative properties of water clouds. *J. Atmos. Sci.*, **46**, 1419–1427.

Stephens, G.L. and S.-C. Tsay (1990). On the cloud absorption anomaly. *Quart. J. Roy. Meteo. Soc.*; **116**, 671–704.

Sun, Z. and K.P. Shine (1994). Studies of the radiative properties of ice and mixed-phase clouds. *Quart. J. Roy. Meteo. Soc.*; **120**, 111–137.

1D Models

Edwards, J.M. and A. Slingo (1996). Studies with a flexible new radiation code. I: Choosing a configuration for a large-scale model. *Quart. J. Roy. Meteo. Soc.*; **122**, 689–719.

Fouquart, Y. and B. Bonnel (1980). Computations of solar heating of the Earth's atmosphere: A new parameterization. *Cont. Atmos. Phys.*, **53**, 35–62.

Li., J. (2002). Accounting for unresolved clouds in a 1D infrared radiative transfer model. Part I: Solution for radiative transfer, scattering, and cloud overlap. *J. Atmos. Sci.*, **59**, 3302–3320.

Ramaswamy, V. and S. Freidenreich (1991). Solar radiative line-by-line determination of water vapor absorption and water cloud extinction in inhomogeneous atmospheres. *J. Geophys. Res.*, **96**, 9133–9157.

Ritter, B. and J.-F. Geleyn (1992). A comprehensive radiation scheme for numerical weather prediction models with potential applications in climate simulations. *Mon. Wea. Rev.*, **120**, 303–325.

1D Models and Unresolved Clouds

Barker, H.W. and B.A. Wielicki (1997). Parameterizing grid-averaged longwave fluxes for inhomogeneous marine boundary layer clouds. *J. Atmos. Sci.*, **54**, 2785–2798.

Cairns, B., A.A. Lacis and B.E. Carlson (2000). Absorption within inhomogeneous clouds and its parameterization in general circulation models. *J. Atmos. Sci.*, **57**, 700–714.

Hogan, R.J. and A.J. Illingworth (2000). Derived cloud overlap statistics from radar. *Quart. J. Roy. Meteo Soc.*, **126**, 2903–2909.

Li, J. and H.W. Barker (2002). Accounting for unresolved clouds in a 1D infrared radiative transfer model. Part II: Horizontal variability of cloud water path. *J. Atmos. Sci.*, **59**, 3321–3339.

Morcrette, J.-J. and Y. Fouquart (1986). The overlapping of cloud layers in shortwave radiation parameterizations. *J. Atmos. Sci.*, **43**, 321–328.

Oreopoulos, L. and H.W. Barker (1999). Accounting for subgrid-scale cloud variability in a multi-layer, 1D solar radiative transfer algorithm. *Quart. J. Roy. Meteo. Soc.*, **125**, 301–330.

Stubenrauch, C.J., A.D. Del Genio and W.B. Rossow (1997). Implementation of subgrid cloud vertical structure inside a GCM and its effects on the radiation budget. *J. Climate*, **10**, 273–287.

Tian, L. and J.A. Curry (1989). Cloud overlap statistics. *J. Geophys. Res.*, **94**, 9925–9935.

3D Simulations

Barker, H.W., J.-J. Morcrette and G.D. Alexander (1998). Broadband solar fluxes and heating rates for atmospheres with 3D broken clouds. *Quart. J. Roy. Meteo. Soc.*, **124**, 1245–1271.

Barker, H.W., G.L. Stephens and Q. Fu (1999). The sensitivity of domain-averaged solar fluxes to assumptions about cloud geometry. *Quart. J. Roy. Meteo. Soc.*, **125**, 2127–2152.

O'Hirok, W. and C. Gautier (1998). A three-dimensional radiative transfer model to investigate the solar radiation within a cloudy atmosphere. Part I: Spatial effects. *J. Atmos. Sci.*, **55**, 2162–2179.

O'Hirok, W. and C. Gautier (1998). A three-dimensional radiative transfer model to investigate the solar radiation within a cloudy atmosphere. Part II: Spectral effects. *J. Atmos. Sci.*, **55**, 3065–3075.

Model Intercomparisons

Barker, H.W., G.L. Stephens, P.T. Partain, J.W. Bergman, B. Bonnel, K. Campana, E.E. Clothiaux, S.A. Clough, S. Cusack, J. Delamere, J. Edwards, K.F. Evans, Y. Fouquart, Freidenreich, S., Galin, V., Hou, Y., Kato, S., Li, J., E. Mlawer, J.-J. Morcrette, W. O'Hirok, P. Räisänen, V. Ramaswamy, B. Ritter, E. Rozanov, M. Schlesinger, K. Shibata, P. Sporyshev, Z. Sun, M. Wendisch, N. Wood and F. Yang (2003). Assessing 1D atmospheric solar radiative transfer models: Interpretation and handling of unresolved clouds. *J. Climate*, **16**, 2676–2699.

Boucher, O., S.E. Schwartz, T.P. Ackerman, T.L. Anderson, B. Bergstrom, B. Bonnel, P. Chylek, A. Dahlback, Y. Fouquart, Q. Fu, R.N. Halthore, J.M. Haywood, T. Iverson, S. Kato, S. Kinne, A. Kirkevag, K.R. Knapp, A. Lacis, I. Laszlo, M.I. Mishchenko, S. Nemesure, V. Ramaswamy, D. L. Roberts, P. Russell, M.E. Schlesinger, G.L. Stephens, R. Wagener, M. Wang, J. Wong and F. Yang (1998). Intercomparison of models representing direct shortwave radiative forcing by sulfate aerosols. *J. Geophys. Res.*, **103**, 16,979–16,998.

Cess, R.D., M.-H. Zhang, G.L. Potter,d H.W. Barker, R.A. Colman, D.A. Dazlich, A.D. Del Genio, M. Esch, J.R. Fraser, V. Galin, W.L. Gates, J.J. Hack, W. Ingram, J.T. Kiehl, A.A. Lacis, H. Le Treut, Z.-X. Li, X.-Z. Liang, J.F. Mahfouf,

B.J. McAvaney, V.P. Meleshko, J.-J. Morcrette, D.A. Randall, E. Roeckner, J.-F. Royer, A.P. Sokolov, P.V. Sporyshev, K.E. Taylor, W.-C. Wang and R.T. Wetherald (1993). Intercomparison of CO_2 radiative forcing in atmospheric general circulation models. *Science*, **262**, 1252–1255.

Ellingson, R.G. and Y. Fouquart (1991). The intercomparison of radiation codes in climate models (ICRCCM): An overview. *J. Geophys. Res.*, **96**, 8926–8929.

Fouquart, Y., B. Bonnel and V. Ramaswamy (1991). Intercomparing shortwave radiation codes for climate studies. *J. Geophys. Res.*, **96**, 8955–8968.

Kinne, S., R. Bergstrom, O.B. Toon, E. Dutton and M. Shiobara (1998). Clear-sky atmospheric solar transmission: An analysis based on FIRE 1991 field experiment data. *J. Geophys. Res.*, **103**, 19,709–19,720.

10

Longwave Radiative Transfer
in Inhomogeneous Cloud Layers

R.G. Ellingson and E.E. Takara

10.1 Introduction 487

10.2 Models for Thermal Radiative Transfer Calculations 491

10.3 Overcast/Clear Linear Mixing 493

10.4 Analytical Results for 3D Clouds 499

10.5 Monte Carlo Calculations for 3D Clouds 508

10.6 Summary 517

References 518

Suggested Further Reading 519

10.1 Introduction

The solar energy absorbed by the Earth-atmosphere system is balanced in the long
term by radiant loss of energy by the system to space in the thermal infrared. The
manner by which this loss to space occurs involves absorption, scattering and thermal emission by the surface and the gaseous and suspended matter (i.e., clouds
and aerosols) within the atmosphere. These complex processes in the thermal infrared comprise the "atmospheric greenhouse effect," which makes the Earth's surface warmer than it would be if the atmosphere were not present and leads to a
complex vertical temperature profile.

The Earth-atmosphere thermal radiative processes are non-linear in atmospheric
properties, and there are complex radiation-climate feed back mechanisms. Furthermore, unlike solar radiation, thermal radiative processes occur continuously in time.
Thus, realistic modeling of the climate system, as well as remote sensing techniques
to accurately infer properties of the atmosphere, require accurate models of the various radiative processes that occur in the thermal infrared.

When absorption, thermal emission and scattering occur, the equation of radiative transfer under local thermodynamic equilibrium may be written as (see Chap. 3)

$$\boldsymbol{\Omega} \cdot \nabla I = - \sigma_e(\boldsymbol{x}) I(\boldsymbol{x}, \boldsymbol{\Omega})$$
$$+ \sigma_s(\boldsymbol{x}) \int_{4\pi} p(\boldsymbol{x}, \boldsymbol{\Omega}' \to \boldsymbol{\Omega}) I(\boldsymbol{x}, \boldsymbol{\Omega}') \mathrm{d}\boldsymbol{\Omega}' + \sigma_a(\boldsymbol{x}) B_v(T(\boldsymbol{x})) \qquad (10.1)$$

where σ_e, σ_s, and σ_a are the transport coefficients for extinction, scattering, and absorption respectively. Here, $\sigma_e = \sigma_s + \sigma_a$ and p is scattering phase function (normalized in such a way that its integral over 4π steradians is unity); ϖ_0, the albedo of single scattering, is defined as the ratio of σ_s to σ_e. Finally, B_v is the Planck function depending on local temperature T, an isotropic source term. Neglecting incident solar radiation in the longwave portion of the spectrum (wavelength $\lambda \gtrsim 3\,\mu\mathrm{m}$), solutions of this equation for the entire atmosphere are usually sought by assuming as boundary conditions zero incident radiation at the top of the atmosphere (TOA), and thermal emission and reflected incident radiation at the base. The effects of the solar longwave radiation, if important, are typically calculated separately, by means discussed in previous chapters. For downward flux considerations, this term is typically the order of $10\,\mathrm{Wm}^{-2}$, on the order of a few percent of the thermal longwave flux incident on the surface.

When only gases are considered, ϖ_0 is essentially 0 in the thermal infrared due to the λ^{-4} decrease of the molecular scattering coefficient. For this case, the radiation field depends only on the absorption properties, amounts and distributions of the active gases, the temperature distribution, and the emission properties of the underlying surface. Due to the relative opacity of the atmosphere and the generally slow horizontal variation of temperature and absorbing gases, clear-sky radiation calculations are generally performed assuming a horizontally homogeneous atmosphere.

The main difference in the treatment of shortwave and longwave radiation is due to the spectral absorption of atmospheric gases. Figure 10.1 is a low-resolution depiction of the major absorption features of the dominant active gases and the approximate spectral distributions of incoming solar energy and terrestrial radiation emitted by the atmosphere. In the solar portion of the spectrum ($\lambda \lesssim 4\,\mu\mathrm{m}$), there is little gaseous absorption across large regions of the spectrum, particularly the region from 0.3 to 1 $\mu\mathrm{m}$, a region containing more than 50% of the incident solar radiation. This region is particularly sensitive to the presence of clouds since ϖ_0 for cloud particles is close to 1, and thermal emission by the gases is practically 0.

In the longwave, the atmosphere as a whole is nearly opaque to energy incident on its boundaries due to the strong vibration-rotation bands of H_2O, CO_2, O_3, CH_4 and N_2O. The major exception is the interval from 8 to 12 $\mu\mathrm{m}$ or 1250 to 833 cm^{-1} (wavenumber in cm^{-1} is 10^4 times the reciprocal of wavelength λ in $\mu\mathrm{m}$ and vice-versa). Spectral intervals of significant transmission (lower absorption) are called windows. The primary window is the 8–12 $\mu\mathrm{m}$ interval; this is where clouds and 3D radiative transfer have their largest effects. There is also a "dirty" window centered near 500 cm^{-1} (20 $\mu\mathrm{m}$) which becomes significant in dry atmospheres.

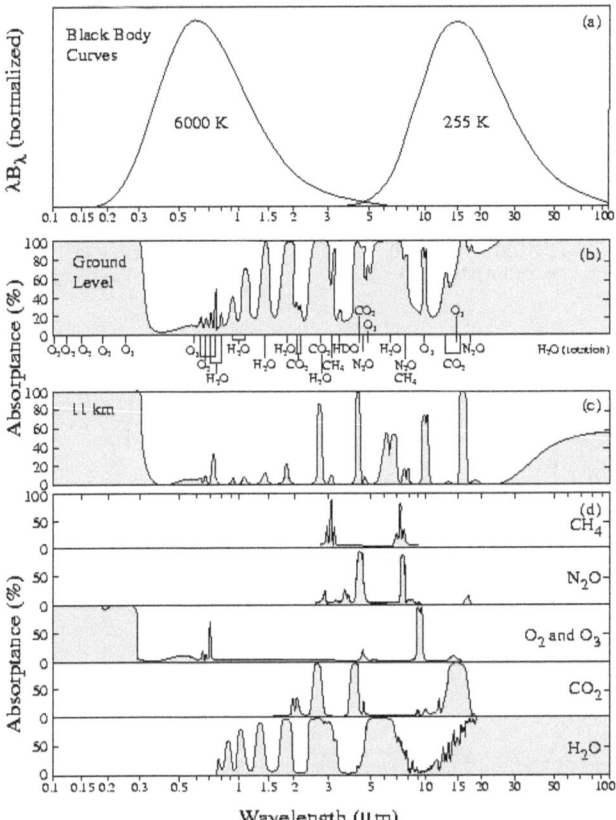

Fig. 10.1. (a) Planck function curves for approximate solar and terrestrial temperatures. (b) Absorption spectra for the entire vertical extent of the atmosphere. (c) Absorption spectra for the atmosphere above 11 km. (d) Atmospheric absorption spectra for the major active gases. From Thomas and Stamnes (1999), reprinted with the permission of Cambridge University Press

The temperature variation in the atmosphere is relatively small, so there is little contrast between the incident and emitted radiation. Scattering effects are most pronounced when the emission source is at a much higher temperature than the scattering medium as in the shortwave. Therefore, scattering is much less significant in the longwave than the shortwave. Because the gas concentrations, temperature and pressure vary with altitude, not every part of the atmosphere is opaque, and radiative transfer calculations are complicated by the structure of molecular line absorption.

The longwave optical properties of water clouds, shown in Fig. 10.2, are not nearly as spectrally detailed as those of the gases. ϖ_0 and extinction cross-sections vary strongly with particle size, but for typical cloud particle sizes, $\varpi_0 \approx 0.5$. However, the ϖ_0 to be used in calculations is that for the cloud-air mixture. Although cloud particles have non-zero scattering albedo across the longwave spectrum, the

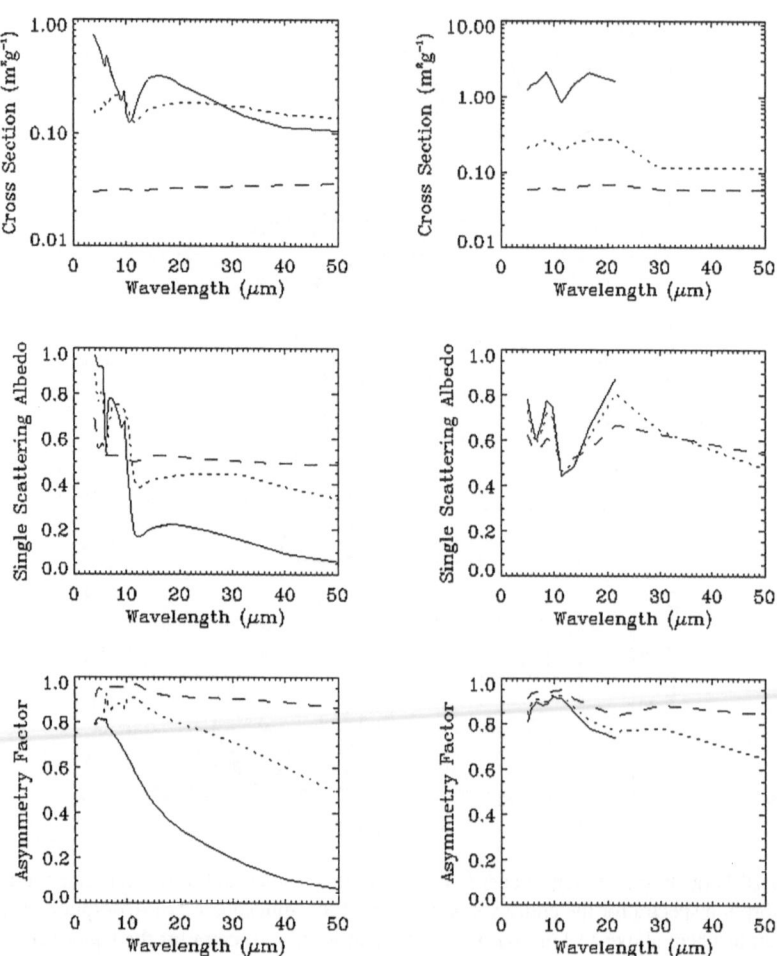

Fig. 10.2. Dependence of the cloud particle longwave optical properties on wavelength and effective radius, from Thomas and Stamnes (1999), reprinted with the permission of Cambridge University Press. The three *left* panels pertain to water clouds and the *right* panels to ice clouds. The curves in the left panels pertain to different effective cloud droplet radii; solid 3 μm; dotted 10 μm; dashed line 50 μm, The right panel curves correspond to different ice particle effective diameters; solid 6 μm; dotted 25 μm; dashed 100 μm

strong absorption by water vapor and other atmospheric gases reduces the effectiveness of scattering by cloud particles in the longwave compared to the shortwave. This is true even in the 8–12 μm window region, where the H_2O continuum absorbs a significant amount in humid atmospheres. Longwave 3D cloud scattering effects are reduced in those portions of the atmosphere where there is strong gaseous absorption. As a result, 3D scattering effects are typically unimportant in climate models. The exception to this is for conditions typical of some cirrus clouds – thin cold clouds with large particles at low pressures where the concentrations of water vapor and the pressure broadening effects on line absorption are relatively small compared to the lower troposphere. Also, 3D scattering effects may be important for more detailed models such as cloud resolving models (CRMs) and large eddy simulations (LESs). Even in those cases, the spectroscopy of atmospheric gases plays a very large role in longwave 3D radiative transfer.

Although 3D cloud effects are potentially important for a variety of problems in remote sensing and climate studies, the discussion in this chapter is focused toward climate applications. In the material that follows we summarize information on the calculation of radiation quantities for a two-dimensional, horizontally homogeneous atmosphere. Next, longwave Monte Carlo calculations are introduced with some results expanding on the two-dimensional calculations and results for a 3D calculation.

10.2 Models for Thermal Radiative Transfer Calculations

It is virtually impossible to describe the calculation details necessary to account for the spectroscopy of the longwave active gases and cloud particles in a review chapter such as this. Instead, only a verbal summary of the techniques is provided. Readers are urged to consult any of the referenced texts and journal articles for the important details.

Climate modelers are only interested in the spectrally integrated fluxes and heating rates. Unfortunately, the mathematical formalism developed with the monochromatic solutions is not transferable to frequency-averaged radiation. Furthermore, modelers are generally interested in effects of quantities over large spatial domains. The challenge of developing models for climate applications is to find solutions to (10.1) that adequately account for frequency variations as well as the spatial variations of the radiative fields.

The major obstacle for frequency integration of (10.1) is the spectral detail of gaseous absorption. Absorption features are composed of many discrete quantized events that are broadened due to collisions and/or thermal motions of the molecules. The discontinuous nature of these absorption features prohibits analytical solutions without approximations. As noted in Fig. 10.2, cloud particles and other aerosols have relatively slowly varying spectral absorption features. Outside of homogeneous spherical particles, however, our knowledge of cloud and aerosol radiative properties is still in its infancy. This is especially true for suspended ice particles. Furthermore, the horizontal and vertical distributions of clouds fall into a seemingly infinite range of possibilities, thereby precluding easy solutions.

Most climate models tend to determine fluxes and cooling rates by transforming the general three-dimensional radiative transfer problem into one-dimension. This is usually done by taking the average of clear and overcast quantities weighted by cloud amount(s). For example, the upward or downward flux of radiation F at a given level above or below a cloud layer is often calculated in climate models as

$$F = (1 - A_c)F_{clear} + A_c F_{overcast} \qquad (10.2)$$

where F_{clear} and F_{overcast} are the fluxes calculated assuming homogeneous clear and overcast conditions respectively, and A_c is the cloud fraction. A more detailed discussion of (10.2) is provided in Sect. 10.4. Note that F_{clear} includes the effects of absorption and thermal emission by the atmospheric gases and non-cloud aerosols. F_{overcast} includes the simultaneous effects of clouds and gases.

Techniques for calculating radiative transfer are most easily classified according to the methods employed to perform the spectral integration of the gaseous absorption features, namely line-by-line (LBL) models and band models. Calculations in LBL models are performed monochromatically, and they often include all of the detailed physics, including multiple scattering. Generally, these techniques offer the possibility of performing all of the numerics with high accuracy. They are particularly useful for checking the approximations made in less detailed models.

However, LBL models require substantial computing resources. For example, the thermal infrared portion of the spectrum may require the order of 10^7 intervals for monochromatic computations. Such computations were prohibitively expensive until the advent of supercomputers, but they are now relatively easy to do on desktop workstations. Furthermore, it is now possible to perform such calculations for a wide variety of atmospheric gases because of the wide distribution of tabulated values of the line strengths and half-widths important for line absorption. These models are very useful for remote sensing applications, but they consume too much computer time to be used in climate models.

Models for spectrally integrated gaseous absorption (i.e., band models) offer the possibility of rapid, but accurate, evaluation of the integral over detailed gas absorption spectra. The common approach used in band models is to make assumptions concerning the spectral distribution of absorption features for a homogeneous path that allow a solution in terms of analytical functions with a few adjustable parameters. The values of the adjustable parameters are specified either through fitting to more detailed model calculations or to asymptotic limits expressed in terms of line strengths and half-widths. Such models are applied to the inhomogeneous atmospheric path through a variety of scaling approximations developed from asymptotic solutions or linear expansions (e.g., the Curtis-Godson approximation). Excellent discussion of such techniques may be found in Goody and Yung (1989) and Liou (1980).

The analytical models take on a variety of forms and spectral resolution. Typically, those with spectral resolution less than about $50\,\text{cm}^{-1}$ are called narrow band models, and those for larger intervals being labeled as wide band models. There has been a general assumption that the smaller the spectral interval, the more accurate the result. However, the results of the Intercomparison of Radiation Codes used in Climate Models (ICRCCM; Ellingson and Fouquart, 1991) do not substantiate that position.

In an attempt to surmount the problem of arbitrary assumptions concerning the properties of analytic models, there have been several attempts to develop tabular values from LBL calculations. Such approaches increase numerical accuracy and speed at the expense of storage allocation, which is not a crucial consideration for current computers. The tabulations can be used to improve band models; but band models have a serious limitation: it is difficult to use them for problems involving multiple scattering. In monochromatic radiation the transmittance between two levels can be computed by multiplying the transmittances of sub-intervals. But this does not hold for band transmittances of the same gas, which are averaged over spectral intervals.

To avoid this problem, modelers have developed what is commonly called the k-distribution technique (see also Chap. 9). This technique transforms the integral over frequency to one over absorption coefficient. This is based on the observation that the intensity is the same when the absorption coefficient is the same for a narrow spectral interval. Thus, the frequency integration is reduced to a sum of monochromatic problems. This approach requires the probability or cumulative probability distribution of absorption coefficients that may be determined numerically from LBL calculations or analytically from some band models. The k-distribution technique is not without limitations, because approximations must be made for application to the inhomogeneous atmosphere, and one may be required to keep large numbers of k's for applications at low pressures (i.e., above about 20 km).

In summary, it must be remembered that band models are approximations and unwanted errors may occur unexpectedly. Some general statements concerning such models are: their absolute accuracy is only as good as that of the more detailed models or data on which they are based; fits based on asymptotic limits may be inaccurate for atmospheric calculations; accounting for overlapping absorption by two or more gases may be difficult for large band areas; application of scaling approximations are inaccurate in some situations; and except for the k-distribution, they are difficult to use in problems involving multiple scattering.

Parameterizations of the absorption by cloud particles typically express the wavenumber averaged transmittance through a cloud for a given direction in terms of the average cloud optical depth τ for the interval under consideration and the cosine of the local zenith angle μ as

$$\mathcal{T}(\tau; \mu) = e^{-\tau/\mu} .$$

For a given spectral interval, τ is the product of an effective radius dependent mass absorption coefficient (in $m^2 \, g^{-1}$) and liquid water path. For the $10 \, \mu m$ window region, the upper panel of Fig. 10.2 shows mass absorption coefficients are about 0.15 and 0.1 $m^2 \, g^{-1}$ for typical water and ice clouds, respectively. For liquid water contents of 0.1 $g \, m^{-3}$, typical of cumulus clouds, a 300 m deep cloud has an optical depth of about 4.5 – opaque although not black. As discussed below, cloud opacity extends over a wide range, with cirrus clouds tending to be the most transparent, and cumulus clouds the most opaque.

10.3 Overcast/Clear Linear Mixing

To illustrate the bulk spectral effects of gaseous and cloud absorption and thermal emission, Fig. 10.3 shows the spectral distribution of the nadir (a) and zenith (b) radiances at the top and bottom of the atmosphere, respectively, with spectrally black

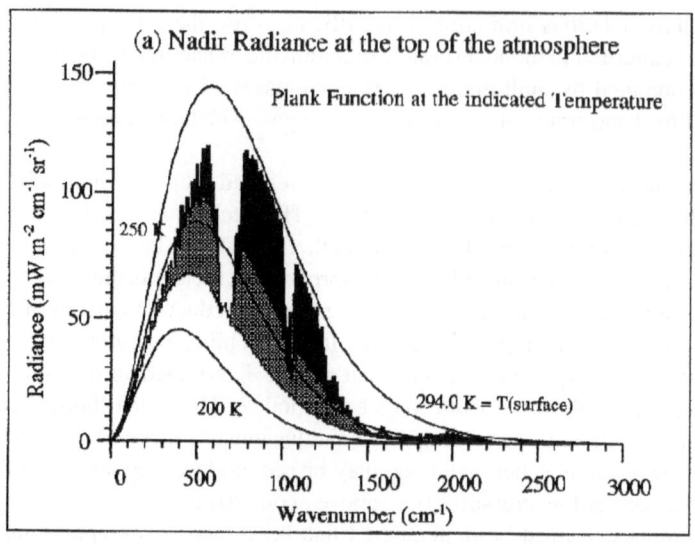

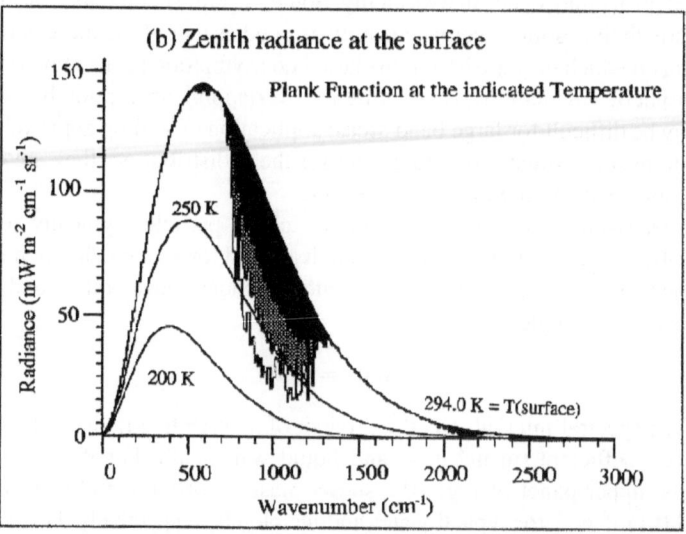

Fig. 10.3. Nadir (**a**) and zenith (**b**) radiances at the top and bottom of the atmosphere, respectively, for the Air Force Geophysical Laboratory (AFGL) mid-latitude summer (MLS) atmosphere calculated with the model of Warner and Ellingson (2000). The solid and hashed regions correspond to the change of the respective radiances resulting from increasing the altitude of a black plane-parallel cloud from the surface to 4 km, and from 4 to 11 km, respectively. Note that the uppermost curve in (**a**) and the lowermost curve in (**b**) correspond to the clear-sky radiances

clouds at different levels for mid-latitude summer (MLS) conditions calculated with a detailed band model. When the cloud level is at the surface (294 K), the downwelling radiance at the surface is the same as the Planck function for the temperature of the surface, whereas the upwelling radiance at the top is the same as for clear skies with a black surface. The TOA clear-sky upwelling radiance shows that outside of the 833–$1250\,\mathrm{cm}^{-1}$ window, the radiance originates at different levels of the atmosphere depending upon the opacity of the given spectral region. As the cloud level increases from the surface to 4 km, the TOA window region radiance follows the Planck function of the cloud temperature (250 K), whereas the remainder of the spectrum changes little, except for the $500\,\mathrm{cm}^{-1}$ "dirty" window This occurs because the absorption features of the atmospheric gases outside of the window regions are large enough, even with the decreased absorption path, to block emission from the 4 km level. When the cloud reaches 11 km, the atmosphere above the clouds is sufficiently transparent for the radiance to closely follow the Planck function of the assumed black cloud, except for the strongest regions of the $667\,\mathrm{cm}^{-1}$ ($15\,\mu\mathrm{m}$) CO_2 and $1042\,\mathrm{cm}^{-1}$ ($9.6\,\mu\mathrm{m}$) O_3 bands.

Except for the very driest atmospheres, the atmosphere is very opaque across the longwave spectrum, so the clear-sky downwelling radiance at the surface resembles the Planck function at the surface temperature. As stated in Sect. 10.1, the primary window is at 833–$1250\,\mathrm{cm}^{-1}$; within this window surface emission can transmit directly to space. For drier atmospheres the $500\,\mathrm{cm}^{-1}$ "dirty window" can also transmit to space. Clouds effectively block these windows. As the cloud's base altitude increases, two factors decrease its effect on downwelling at the surface. First, the cloud becomes colder. Second, the distance between the cloud and surface increases, increasing the amount of the atmosphere that can absorb emission from the cloud. When the cloud reaches 11 km, the overcast radiance distribution resembles the clear-sky radiance distribution.

In general, cloud effects on the downwelling radiance primarily occur in the 833–$1250\,\mathrm{cm}^{-1}$ window, whereas for the TOA radiance, clouds dominate the window only when they are above about 4 km. When the clouds are above 4 km, the TOA radiance is affected by cloud absorption and scattering properties across the entire spectrum because of the decreased water vapor concentration. This shows quite dramatically that ice clouds can influence a large portion of the spectrum. Recall that scattering effects are most important when there is a significant temperature difference between the emission source and the scattering medium, as in the shortwave. Since ice clouds have the greatest temperature difference with the surface, their scattering can be important when they are located in colder regions of the atmosphere.

As previously discussed, cloud forcing is an important concept in climate studies. It is informative to examine it in the longwave. Cloud forcing (denoted CF) is defined as $CF = F - F_{\mathrm{clear}}$, at the TOA and the surface. From (10.2), we obtain $CF = A_c(F_{\mathrm{overcast}} - F_{\mathrm{clear}})$. Figure 10.4 shows CF/A_c at the TOA and the surface for overcast black clouds at different levels. Note that the effects of non-blackness may be calculated for plane parallel clouds by multiplying A_c by the plane parallel cloud flux emissivity, $\varepsilon_{\mathrm{cpp}}$ given as

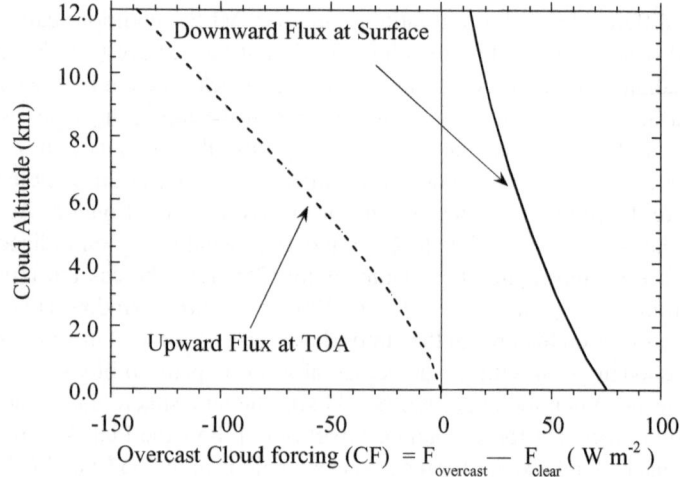

Fig. 10.4. Longwave black cloud forcing CF for overcast conditions at the top of the atmosphere (*dashed*) and surface (*solid*) as a function of cloud altitude for MLS conditions as calculated with the model of Warner and Ellingson (2000)

$$\varepsilon_{\mathrm{cpp}} = 2 \int_0^1 \left(1 - e^{-\tau/\mu}\right) \mu d\mu = 1 - 2E_3(\tau) \qquad (10.3)$$

where E_3 is the third-order exponential integral. Letting

$$A_{\mathrm{c}}^* = A_{\mathrm{c}}\varepsilon_{\mathrm{cpp}} \ ,$$

we note that the effect of the variations δA_{c}^* is simply $(F_{\mathrm{overcast}} - F_{\mathrm{clear}})\delta A_{\mathrm{c}}^*$.

For $A_{\mathrm{c}}^* = 1$, the magnitude of TOA CF increases as the cloud altitude increases, whereas that for the surface decreases, both in accord with the expectations from Fig. 10.3. As clouds become more transparent, the magnitude of CF decreases for all cloud altitudes. Note however, that relatively small uncertainties in A_{c}^* yield large errors in CF. At the surface, a 15% error in A_{c}^* for a cloud at 0.5 km yields about a $10\,\mathrm{Wm}^{-2}$ uncertainty in the surface CF, whereas there is little change at the TOA. Such systematic, potentially global, 3D cloud effects are missing in current climate model simulations, leaving the possibility of significant climatic importance.

Cloud effects on the radiation field away from the boundaries of the atmosphere are more difficult to quantify in a simple fashion. Instead of looking at the flux field, it is more informative to examine the altitude and spectral distributions of the radiative cooling rate – the divergence of the net flux (Fig. 10.5). The clear-sky tropospheric distributions occur because cooling to space primarily controls the cooling rate. At a given wavenumber the maximum longwave cooling to space in an atmosphere with temperature decreasing with altitude occurs at optical depth 1. If the optical depth is less than 1, the maximum occurs in the layer closest to the surface. This occurs because of the near exponential decreases with height of the absorbing/emitting gas

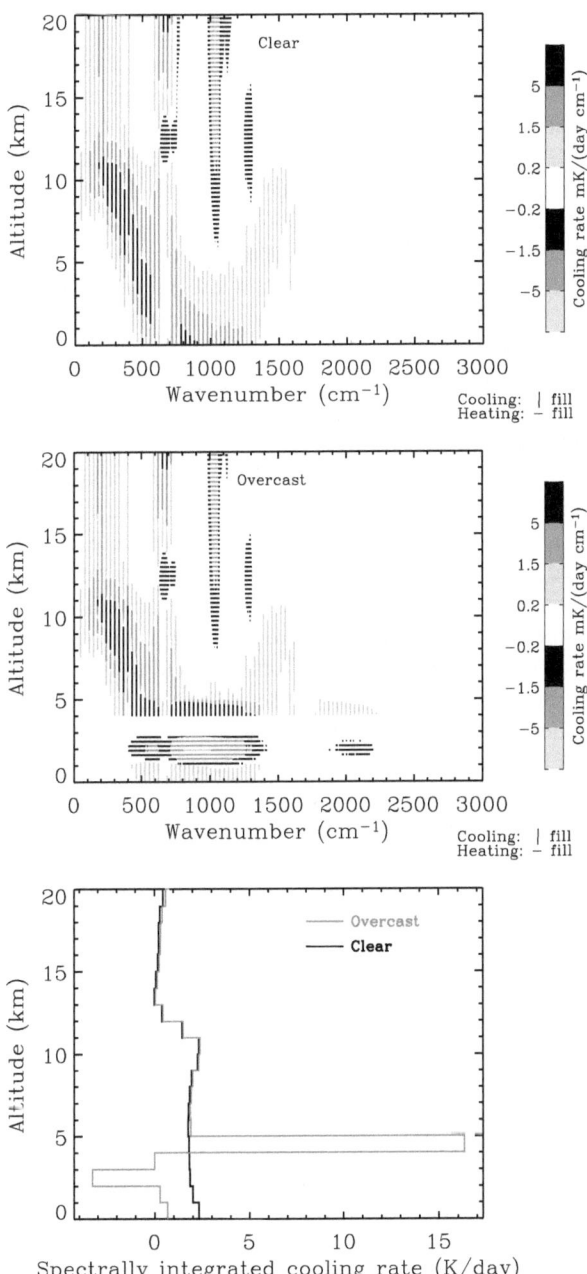

Fig. 10.5. Spectral distributions of longwave clear-sky (*upper*) and black overcast (*lower*) cooling rates for MLS conditions as calculated with the model of Warner and Ellingson (2000). The spectrally integrated cooling rates for these conditions are noted on the inserted middle figure. Clouds are assumed for the layer at 3–4 km

concentrations. At low altitudes the emission is high, but there is large attenuation. At high altitudes, the density is low, and there is less emission and attenuation. Thus, for clear-sky conditions, cooling in the lower troposphere is dominated by the atmospheric window where the optical depth is less than 1. As altitude increases, the very strong H_2O pure rotation band controls cooling. Since the optical depth varies with wavenumber, the altitude of maximum cooling shifts to the altitude where the optical depth equals 1.

The temperature gradient changes sign at the tropopause (near 13 km for the MLS sounding) and the cooling to space scenario is altered. In this sounding the heating in the $667\,cm^{-1}$ region near 13 km is a result of the thermal emission by the very strong CO_2 band at the cold tropopause being less than the absorption from warmer layers nearby. The heating in the 10-20 km range in the 1000 to $1100\,cm^{-1}$ region is due to the emission by O_3 being less than the absorption of relatively un-attenuated radiation from the lower troposphere.

When a cloud layer is added to the atmosphere (here for simplicity a black cloud in the 3-4 km layer), the cooling rate distribution changes dramatically in the vicinity of the cloud layer. Heating occurs immediately below the cloud layer due to absorption by the cloud of radiation from the more transparent regions. As expected, there is very little change at altitudes where the gaseous opacity is large. No cooling occurs in the assumed black cloud layer. There is strong cooling in the immediate layer above the clouds in the window regions, again related to the strong gradient in the cloud opacity. Within a couple of km of the cloud top, the cooling rate profile becomes almost identical to the clear-sky one as the local distribution of atmospheric opacity governs the cooling. Note that there is less heating by ozone due to the decreased radiation incident on the stratosphere when clouds are present. Also note that the higher a cloud is placed in the atmosphere, the larger the spectral range over which it is effective.

The earlier discussion for the cloud forcing may also be applied to the cooling rate. That is, uncertainties in the cloud fraction A_c^* lead to uncertainties in the cooling rate proportional to δA_c^* and the difference between the overcast and clear-sky cooling rates. For the clouds shown in Fig. 10.5, a 0.15 uncertainty in A_c^* results in about a $2.4°K\ day^{-1}$ error in the cooling rate in the 1 km layer above the cloud, a number larger than the clear-sky cooling rate. The magnitude of the effects depends primarily upon the cloud location. As shown below, some longwave 3D effects yield uncertainties in A_c^* that will result in very large cooling rate errors. Since the general effect of longwave cooling is to destabilize the atmosphere by heating below the clouds and cooling above, uncertainties caused by neglecting 3D cloud effects in dynamical models will change the dynamics by virtue of a modified heating rate profile.

10.4 Analytical Results for 3D Clouds

Atmospheric modelers have typically reduced the 3D longwave cloud problem to a solution based on one-dimensional models by a series of assumptions concerning cloud properties. Consider the following assumptions:

- the molecular atmosphere is plane-parallel and horizontally homogeneous;
- the cloud field for a given area is statistically homogeneous and isotropic.

Now determine the average flux for an area large compared to the individual cloud elements by summing separately the contributions of those solid angles that are completely clear and those that have cloud contributions. With these assumptions and with no scattering or ground reflection, the equation for the area-averaged downward flux at an altitude z below the clouds may be written as

$$F(z) = 2\pi \int_0^{\pi/2} I_{\text{clear}}(z, \theta) P(\theta) \cos\theta \sin\theta d\theta$$

$$+ 2\pi \int_0^{\pi/2} I_{\text{cloudy}}(z, \theta)[1 - P(\theta)] \cos\theta \sin\theta d\theta \qquad (10.4)$$

where $P(\theta)$ is the azimuthally-averaged probability of a clear line of sight (PCL) through the clouds at zenith angle θ, and I_{clear} and I_{cloudy} are the average radiant intensities from the clear and cloudy areas of the sky, respectively. Here it is assumed that the PCL $P(z, \theta) \equiv P(\theta)$ which is true at least for cloudy layers with vertical cloud boundaries.

Note that the above assumptions are only simplifying if it is possible to specify $P(\theta)$ and the area-averaged clear and cloudy sky radiances. For the clear-sky component, this is not a major concern barring frontal boundaries and orography, since the clear-sky radiative properties are quasi-horizontally homogeneous over large length scales. Clouds present more of a challenge because cloud field and radiance properties require major simplifications before $P(\theta)$ and I_{cloudy} can be determined. Furthermore, there are sparse observations or calculations with which to test the approximations.

Figure 10.6 illustrates the overall geometrical effects of broken cloudiness. For illustration purposes, the clouds shown have the same horizontal lengths, but they have different vertical dimensions and irregular spacing. In this 2D transect, each cloud projects the same length at the level of cloud base from all view angles if only the cloud horizontal dimension is considered. Note that for very thin, randomly dispersed thin plates, $P(\theta)$ is $(1 - A_c)$, where A_c is the absolute cloud fraction. Assuming the thin plate clouds have identical radiative properties, (10.4) reverts to the standard approximation (10.2) used in most climate models. Thus for all z,

$$F_{\text{thin}}(z) = (1 - A_c)F_{\text{clear}}(z) + A_c F_{\text{overcast}}(z) . \qquad (10.5)$$

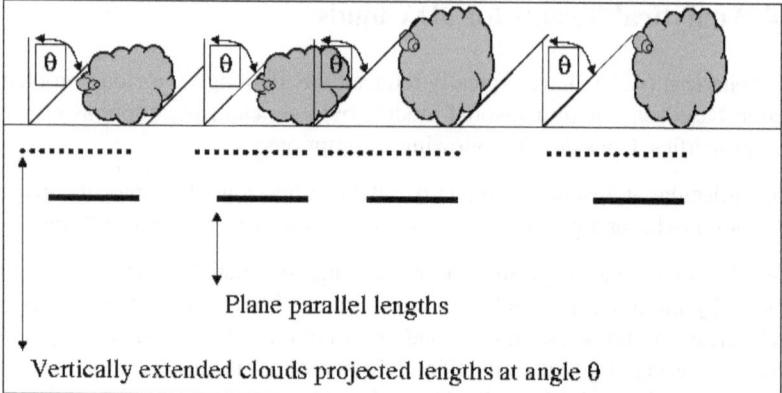

Plane parallel lengths

Vertically extended clouds projected lengths at angle θ

Fig. 10.6. Two dimensional view of an array of vertically extended clouds each with the same width. When viewed at angle θ, the clouds project as the dotted lines. When projected vertically downward, the cloud project as the solid horizontal lines (displaced here to coincide with the start of the *dashed lines*). The plane-parallel cloud assumption, the *solid lines*, underestimates cloud cover at all view angles >0

That is, the flux is simply the cloud amount weighted average between the clear and overcast fluxes, each of which can be calculated with plane-parallel techniques.

Now consider the effects of the vertical dimensions of the clouds by examining the projections of the vertical faces of the clouds on the horizontal plane when viewed at an angle θ. As shown, when clouds are broken and have vertical extent, they project a larger area than their bases, thereby obscuring a larger fraction of the sky at a given angle than do flat plates with the same horizontal dimensions. One might anticipate that the effects of the vertical dimension to become larger as the clouds become tall relative to their bases. As will be seen below, when the cloud fraction is small, the clouds tend not to obscure one another. However, as fractional coverage of the bases increases, cloud sides become less important as mutual obscuration increases, a feature also connected with the distribution of cloud spacing. Additionally, since clouds are not isothermal, temperature variations along their sides must be accounted for in calculations.

Note that if the clouds have any depth, one cannot define an appropriate cloud fraction without also requiring the average fluxes to be dependent on the cloud properties. For example, if we define a hemispherically averaged cloud fraction A_{cH} as

$$A_{cH} = 2 \int_0^{\pi/2} [1 - P(\theta)] \sin\theta \cos\theta d\theta , \qquad (10.6)$$

we write the average flux as

$$\bar{F}(z) = (1 - A_{cH})\bar{F}_{clear}(z) + A_{cH}\bar{F}_{cloudy}(z) . \qquad (10.7)$$

This will require the average fluxes to be defined as

$$\bar{F}_{clear}(z) = \frac{2\pi}{(1 - A_{cH})} \int_0^{\pi/2} P(\theta) I_{clear}(z, \theta) \sin\theta \cos\theta d\theta \qquad (10.8)$$

and

$$\bar{F}_{cloudy}(z) = \frac{2\pi}{A_{cH}} \int_0^{\pi/2} [1 - P(\theta)] I_{cloudy}(z, \theta) \sin\theta \cos\theta d\theta . \qquad (10.9)$$

That is, the effective cloud fraction and the radiation field cannot be determined independently of each other.

10.4.1 Effective Cloud Fraction and Probability of Clear Line of Sight

The suggested reading contains references to many attempts that have been made to determine the form of $P(\theta)$ by making assumptions concerning the cloud field and the geometry of the individual cloud elements. In general, the approaches allow the equations for the upward and downward area-averaged fluxes to be written as

$$F = (1 - A_{ce})F_{clear} + A_{ce}F_{overcast} \qquad (10.10)$$

where A_{ce} is an effective cloud fraction that depends upon the assumptions made concerning the cloud field, F_{clear} is the clear-sky flux calculated with the domain averaged clear-air radiative properties, $F_{overcast}$ is the flux calculated for overcast conditions with assumed cloud radiative properties. The advantage of this approach is that it allows climate codes to use their radiation models with new parameterizations of A_{ce}. However, it must be remembered that none of these attempts have included the effects of scattering between broken cloud elements or have been shown conclusively to account adequately for all three-dimensional cloud effects.

Many of the bulk geometrical effects of clouds on $P(\theta)$ may be appreciated by considering an array of identical right-circular cylinders of thickness h, radius R, and cloud positions distributed according to a Poisson distribution with areal density parameter γ. For such an array, $P(\theta)$ may be written as (Avaste et al., 1974)

$$P(\theta) = \exp[-\gamma(\pi R^2 + 2hR\tan\theta)] . \qquad (10.11)$$

Note that the terms in parentheses are simply the areas of a cloud base and the side, respectively, projected onto the base level. $P(\theta)$ is a maximum when looking directly overhead ($\theta = 0$) and goes to 0 when looking at the horizon ($\theta = \pi/2$).

Defining the absolute cloud amount as $1 - P(0)$, allows $P(\theta)$ to be written as

$$P(\theta) = (1 - A_c) \exp(b\tan\theta) \qquad (10.12)$$

where

$$b = 2\beta\ln(1 - A_c)/\pi ,$$

and β, the aspect ratio, is h/R.

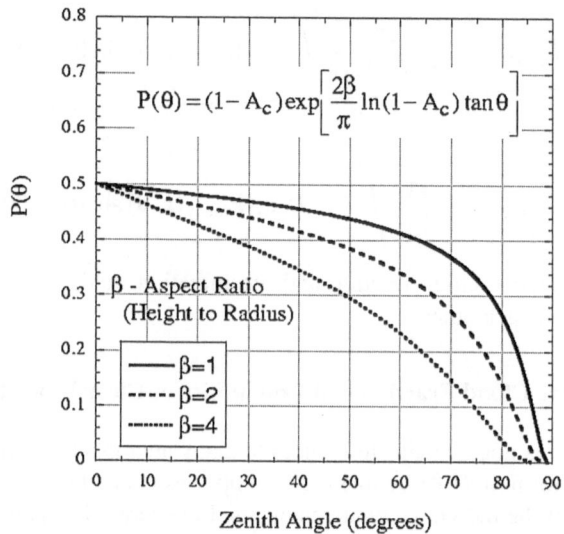

Fig. 10.7. Probability of a clear line of sight $P(\theta)$ for a random cylindrical cloud field as a function of zenith angle for 50% absolute cloud cover A_c and different aspect ratios β

Figure 10.7 illustrates $P(\theta)$ as a function of θ for a few different aspect ratios for 50% absolute cloud cover. $P(\theta)$ decreases with increasing θ and β, since more cloud side areas are seen at a given θ. Note that the horizon is not visible for these type clouds.

Note that even with the simple form of (10.12), it is easily seen that A_{ce} cannot be separated from the radiance fields without further assumptions. However, due to the quasi-isotropic nature of the radiance fields, the angular integration poses no major computational problem. For an isotropic radiance assumption and for isothermal black clouds, A_{ce} is easily shown to be a function of A_c and β given by

$$A_{ce} = A_c + (1 - A_c)\left[1 - 2\int_0^{\pi/2} \exp(b\tan\theta)\sin\theta\cos\theta d\theta\right] . \qquad (10.13)$$

Thus, A_{ce} is always $> A_c$, and varies with A_c in a complex fashion dependent upon β.

10.4.2 Geometrical Effects of Broken "Black" Clouds

Despite some of the shortcomings of such simplifications for the properties of the cloud and radiance fields, it is illustrative to examine the results from at least one model to glean the overall bulk 3D geometrical effects, sans scattering. The results shown herein draw heavily from the work of Han and Ellingson (1999) that assumes truncated square pyramids, a critical nearest neighbor spacing of clouds, and power law distributions for both cloud size and spatial distributions. Clouds are assumed

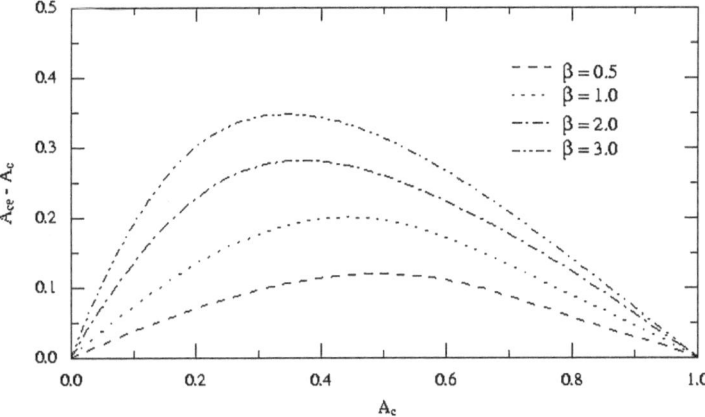

Fig. 10.8. $A_{ce} - A_c$ as a function of A_c and β on A_{ce} for isothermal cylindrical clouds

to be spectrally black, a good assumption for cumulus clouds with optical thickness greater than 3 – a few hundred meters thick. Since we wish to highlight the effects relative to the thin plate approximation, results are shown as $A_{ce} - A_c$. This combined with the results shown in Fig. 10.4 provides a useful method for estimating the possible flux errors inherent in neglecting cloud geometry.

Figure 10.8 shows $A_{ce} - A_c$ as a function of A_c for isothermal cylindrical clouds for different aspect ratios, assuming realistic values for cloud spacing and sizes. Note that the larger β, the larger the cloud side area relative to base area. When A_c is small, mutual shading is not significant and cloud sides make a large contribution to A_{ce}.

The error in neglecting cloud geometry can be estimated by combining the results shown in Figs. 10.4 and 10.8. Since the cumulus clouds depths are approximately the same as their widths; $\beta \approx 2$. From Fig. 10.8 the error for $\beta = 2$ peaks at $A_c \approx 0.35$; $A_{ce} - A_c \approx 0.275$. From Fig. 10.4 the overcast cloud forcing at the surface for cloud altitude 0.5 km is approximately $70\,\mathrm{Wm}^{-2}$. To get the forcing for fractional cloudiness, multiply by A_{ce} to account for cloud geometry. Multiplying by A_c neglects cloud geometry. In this case, $(A_{ce} - A_c)CF \approx 20\,\mathrm{Wm}^{-2}$; neglecting cloud geometry will underestimate the downward flux at the surface by $20\,\mathrm{Wm}^{-2}$.

Non-isothermal clouds in an atmosphere with temperature decreasing with increasing altitude have smaller bulk geometry effects, because the cloud sides are closer to the clear-sky radiance field. If the clouds are tall enough so that the flux from top to bottom increased by a factor of two, the maximum effect would decrease by about 30%.

The effects of varying the exponent v of an assumed power law distribution of cloud sizes are shown in Fig. 10.9. A larger v implies a larger number of clouds of the same cloud size. For low A_c, the spatial distances remain large. More clouds tend to offer a greater area of cloud sides and a greater A_{ce}. Nonetheless, it is quite clear that the geometrical effects can be reduced by about 50% and moved to larger cloud

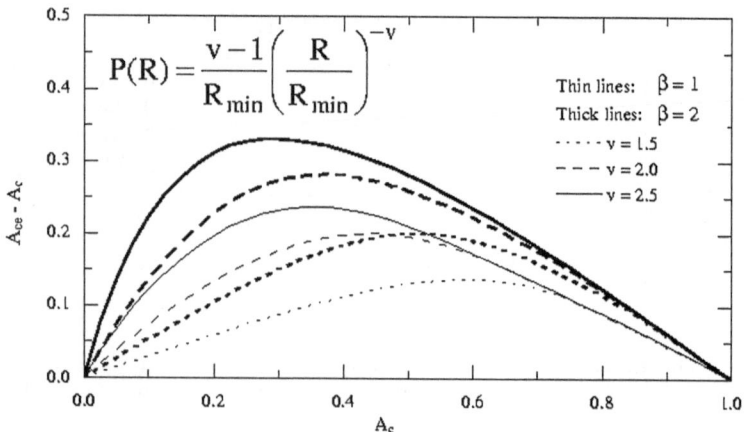

Fig. 10.9. The effects of size distribution exponent v on A_{ce} for different β. $P(R)dR$ is the probability of a cloud of size R and $R + dR$. A larger v implies a larger number of clouds of the same cloud size. For low A_c, the spatial distances remain large. More clouds tend to offer a greater area of cloud sides and a greater A_{ce}

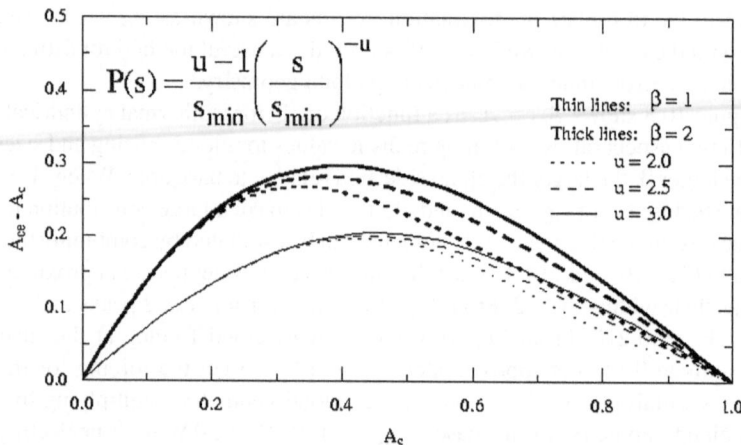

Fig. 10.10. The effects of spatial distribution exponent u on A_{ce}. $P(s)ds$ is the probability of spacing between s and $s + ds$; $u > 1$. When u is large, clouds are more sparsely distributed and cloud sides tend to be obscured less and generate a larger A_{ce}

fractions by having a wider spread in the sizes of the clouds. Note that v is not a routinely observed quantity or one that is predicted by climate models.

The effects of varying the exponent u of an assumed power law distribution of cloud spacing are shown in Fig. 10.10. When u is large, clouds are more sparsely distributed and cloud sides tend to be obscured less and generate a larger A_{ce}. Note that the major effects of spacing become more apparent for A_c between 0.4 and 0.7,

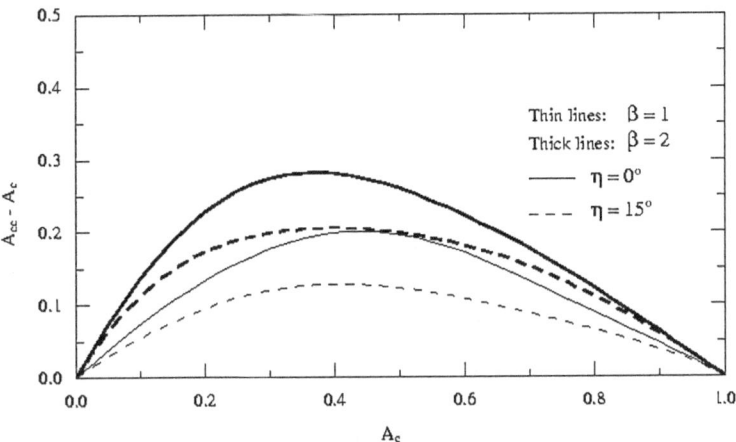

Fig. 10.11. The effect of side inclination η on the effective cloud fraction for different β

and tend to be about 25% of the effects caused by realistic variations in the cloud size distributions.

The inclination of cloud sides is also important, as this reduces the area of the projected sizes (see Fig. 10.11). A modest change in the side inclination ($\eta = 15°$) reduces the maximum geometric effect by about 33% over a broad range of A_c. Although not shown, the effects of different cloud shapes are realized primarily through the inclination factor.

It is not at all clear which, if any, of the geometrical models represent real clouds, because few observational studies have been performed, and there are conflicting results from those that have been performed. Figure 10.12 show the results of an attempt to compare observed with model calculated values of $A_{ce} - A_c$ for relatively thick, single layer cumulus clouds. In general, the observations show the same general trend of different models, but there are few observations in the range of parameter space where the models are most sensitive.

10.4.3 Heterogeneous Non Black Clouds

As noted in Sect. 10.3, for non-black plates of uniform optical thickness, A_c in (10.2) is replaced by the product of A_c and the plane-parallel cloud flux emissivity,

$$\varepsilon_{cpp}(\tau) = 1 - 2E_3(\tau) ,$$

where τ is the cloud optical depth, E_3 is the third-order exponential integral, and $F_{overcast}$ is replaced by $F_{overcast}(black)$, the flux resulting from a black cloud; that is,

$$F = (1 - A_c\varepsilon_{cpp}(\tau))F_{clear} + A_c\varepsilon_{cpp}(\tau)F_{overcast}(black) . \tag{10.14}$$

There is large body of literature on analysis of observations that find a wide range of optical depths for liquid water clouds, even for marine stratocumulus cloud fields

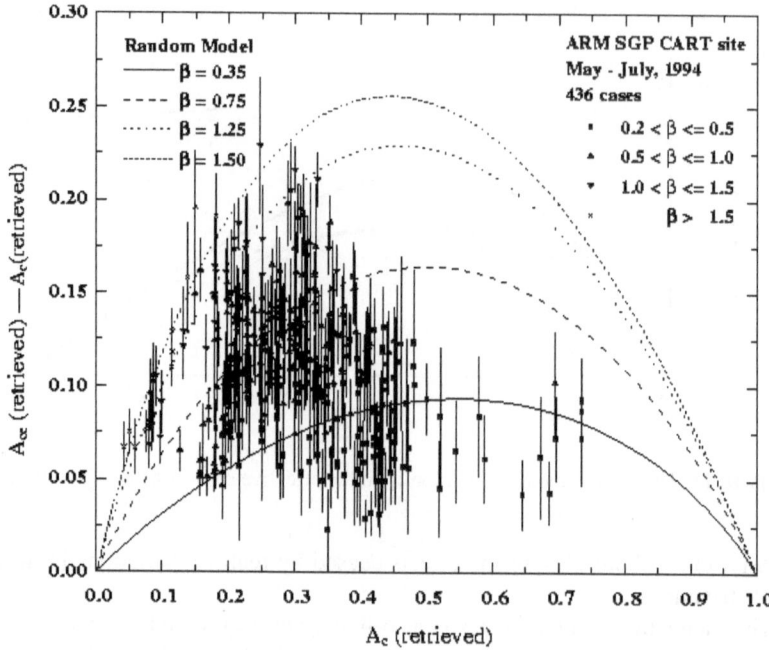

Fig. 10.12. The distribution of $A_{ce} - A_c$ as functions of A_c and β from 436 cases of single-layer cumulus clouds observed in 1994 at the ARM site in Oklahoma, USA. Half of each vertical bar indicates the standard deviation of uncertainty in the retrieved A_{ce}. The curves are from a model assuming random cylinders. Data are from Han (1999)

that appear to be horizontally homogeneous. Due to the highly non-linear dependence of the cloud emissivity on τ for moderate τ, it is important to consider the appropriate average ε rather than ε at the average τ.

To illustrate the potential effects of the horizontal distribution of optical depth, assume that in our domain, the cloud radiance at a given level and zenith angle, $I_{cloudy}(z, \theta)$, at any position in the domain under consideration varies from other positions due only to the cloud absorption optical depth τ viewed at θ. We'll neglect the possibility that a given view actually intersects more than one cloud. Consider probability distributions of τ for lines of sight along θ, and denote these probability densities of τ, conditional on θ, as $P_c(\tau | \theta)$.

Assuming $P(\theta)$ (line-of-sight variability) and $P_c(\tau | \theta)$ (optical depth variability) are independent allows us to write the domain averaged cloudy sky radiance $\bar{I}_{cloudy}(z, \theta)$ as

$$\bar{I}_{cloudy}(z, \theta) = \int_0^\infty P(\tau | \theta) I_{cloudy}(z, \theta, \tau) d\tau . \qquad (10.15)$$

Assume now that the clouds are isothermal and that the monochromatic cloud emission $I_{cloudy}(z, \theta, \tau)$ be approximated as $[1 - e^{-\tau \sec \theta}] B_{vc}$ where B_{vc} is the Planck function for the temperature of the cloud. For illustration purposes, the appropriate domain averaged monochromatic emissivity ε_{ca} for vertically extended clouds is then given as

$$\varepsilon_{ca} = 2 \int_0^\infty d\tau \int_0^{\pi/2} P(\tau|\theta)[1 - e^{-\tau \sec\theta}] \cos\theta \sin\theta d\theta . \qquad (10.16)$$

Recent observational studies of marine stratocumulus with high resolution Landsat data by Barker and Wielicki (1997) have shown that the $1000\,\mathrm{cm}^{-1}$ ($10\,\mu\mathrm{m}$) window region optical depth is represented well by the Gamma distribution written as

$$P_c(\tau|\theta) = \frac{1}{\Gamma(\nu)} \left(\frac{\nu}{\bar{\tau}}\right)^\nu \tau^{\nu-1} e^{-\nu\tau/\bar{\tau}}; \tau > 0, \nu > 0 \qquad (10.17)$$

where $\Gamma(\nu)$ is the Gamma function, irrespective of θ. Parameter $\nu = (\bar{\tau}/\sigma)^2$, where σ is the standard deviation of the distribution of τ with mean $\bar{\tau}$.

With this definition of $p(\tau)$, Barker and Wielicki derive a cloud emissivity $\varepsilon_{c\Gamma}$ using (10.16) and (10.17) to obtain

$$\varepsilon_{c\Gamma}(\bar{\tau}) = 1 - \frac{1}{\Gamma(\nu)} \left(\frac{\nu}{\bar{\tau}}\right)^\nu \left\{ \left(\frac{\bar{\tau}}{\nu+\bar{\tau}}\right)^\nu \left[\Gamma(\nu) - \frac{\bar{\tau}}{\nu+\bar{\tau}} \Gamma(\nu+1)\right] \right.$$
$$\left. + \int_0^\infty E_1(\tau)\tau^{\nu+1} e^{-\nu\tau/\bar{\tau}} d\tau \right\} \qquad (10.18)$$

where E_1 is the first-order exponential integral.

It should be noted that (10.18) holds strictly for monochromatic radiation. However, since the liquid water optical properties vary slowly with wavenumber in the thermal infrared, this form might be used for the spectrally averaged flux transmittance.

Figure 10.13 shows $\varepsilon_{c\Gamma}$ and the ε_{cpp} for a range of ν and τ. For $\nu \lesssim 2$, $\varepsilon_{c\Gamma}$ depends markedly on ν, and $\varepsilon_{c\Gamma} < \varepsilon_{cpp}$ for all τ. Note that for $\tau > 3$, $\varepsilon_{cpp} > \varepsilon_{c\Gamma}$ by more than a factor of 2 for $\nu = 2$, an observed set of parameters. For $\tau \approx 1$, a value representative of relatively thin stratocumulus, $\varepsilon_{cpp} \approx 0.8$, whereas $\varepsilon_{c\Gamma}$ varies between about 0.40 for $\nu = 0.2$ to 0.75 for $\nu = 2$. Clearly, horizontal variations of τ are important.

Note that the effects of variations in τ on the downward flux at the surface are dramatic for low-level clouds. For overcast conditions typical of stratocumulus, the plane-parallel black cloud approximation would be about $35\,\mathrm{Wm}^{-2}$ greater than values computed of $\varepsilon_{c\Gamma}(\tau = 3, \nu = 0.2)$. For 50% cloud cover with cumuli of aspect ratios ≈ 1, the flat plate approximation would be $20\,\mathrm{Wm}^{-2}$ greater than the combined effects of geometry and optical depth variability. Incorporating the cloud geometry alone exceeds the combined effects by $35\,\mathrm{Wm}^{-2}$. Clearly, neglecting optical depth variability can lead to significant flux errors.

The relative sensitivity of the flux to the combined effects of geometry and optical depth variability, with respect to the flat plate approximation, may be written as

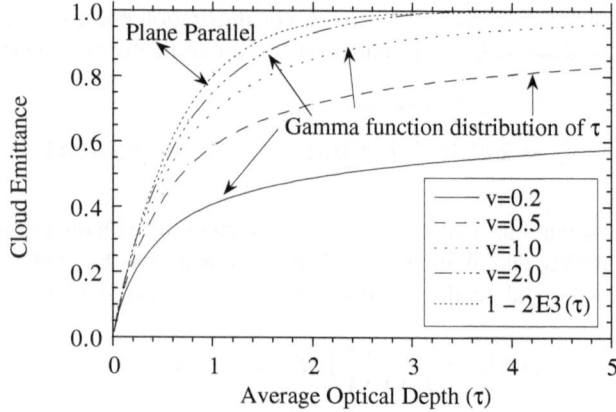

Fig. 10.13. Comparison of $\varepsilon_{\mathrm{cpp}}$ (*dotted*) and $\varepsilon_{\mathrm{c\Gamma}}$ as a function of average cloud optical depth for different Gamma distributions of optical depth as embodied in parameter v

$$\frac{\delta F}{F_{\mathrm{overcast}} - F_{\mathrm{clear}}} \approx \frac{A_{\mathrm{ce}} - A_{\mathrm{c}}}{A_{\mathrm{ce}}} + \frac{\varepsilon_{\mathrm{c\Gamma}} - \varepsilon_{\mathrm{cpp}}}{\varepsilon_{\mathrm{c\Gamma}}} . \tag{10.19}$$

Assuming the $P(\theta)$ and $p(\tau|\theta)$ given herein are approximately correct when $v \gtrsim 2$, cloud geometrical effects dominate for all τ. Note that non blackness is an important consideration for those cases with $\tau \lesssim 3$. For $v \lesssim 0.5$ and all τ, variations in τ will dominate. For other values of τ and v, the effects of cloud geometry and optical depth variability may well be of similar magnitude but of opposite sign, depending upon the values of various parameters. Additional observations and/or numerical modeling are necessary to ascertain the atmospheric conditions controlling the importance of geometry and optical depth variability.

10.5 Monte Carlo Calculations for 3D Clouds

In the longwave, the Monte Carlo method is nearly identical with that in the shortwave. Tracking beams, or photon "bundles," through probabilistic transmission, absorption, and scattering events simulates the radiative transfer. The only difference is the source of the photon bundles – in the longwave, the surface, cloud elements, and the atmospheric gases are emitters. Any simulation must account for these emission sources.

For a diffuse surface, the emission zenith angle is

$$\theta = \cos^{-1}\sqrt{\alpha} \tag{10.20a}$$

while for a volumetric emitter, such as a gaseous or cloud element,

$$\theta = \cos^{-1}(1 - 2\alpha) ; \tag{10.20b}$$

in both cases, the emission azimuth angle is

$$\phi = 2\pi\alpha \ . \tag{10.20c}$$

Here, α denotes independently generated random numbers uniformly distributed between 0 and 1. The above simulation rules follow directly from the simulation of continuous random variables with a given probability density function described in Sect. 4.2.2. For example, for diffuse surface, probability density function is $P(\theta) = \sin 2\theta$, $0 \le \theta < \pi/2$; thus (10.20a) follows directly from (4.35)–(4.36).

10.5.1 Forward vs. Backward Monte Carlo

As described in Sect. 4.2, forward and backward Monte Carlo algorithms have advantages and disadvantages. These are more pronounced in the longwave since every element of the atmosphere and the surface emits in the longwave.

Because the emitting temperatures are almost the same, it is common to emit the same number of photons from each element and normalize the fractions according to the temperature and optical properties. A more sophisticated method is to use the temperature and optical properties to partition the photon bundles among the emitting elements. The fractional probability of being emitted by the ith surface element, f_{si} at temperature T_i and a wavenumber interval centered on v is

$$f_{si} = \frac{\pi \int_{A_i} \varepsilon_{vi} B_v(T_i) \mathrm{d}A_i}{U} \ ; \tag{10.21a}$$

ε_{vi} is the emissivity of surface element i; A_i is the area of element i; B_v is the Planck function at wavenumber v. Similarly, the fractional probability of being emitted by the ith volumetric (gaseous and cloud) element, f_{vi} at temperature T_i and a wavenumber interval centered on v is

$$f_{vi} = \frac{\pi \int_{V_i} a_{vi} B_v(T_i) \mathrm{d}V_i}{U} \ ; \tag{10.21b}$$

a_{vi} is the absorptivity of volumetric element i; V_i is the volume of element i. U is the sum of all emitted energy at wavenumber v over the M surface elements and N volumetric elements:

$$U = \pi \sum_{i=1}^{M} \int_{A_i} \varepsilon_{vi} B_v(T_i) \mathrm{d}A_i + \pi \sum_{i=1}^{N} \int_{V_i} a_{vi} B_v(T_i) \mathrm{d}V_i \ . \tag{10.21c}$$

The primary disadvantage of forward Monte Carlo arises when only a limited number of quantities need to be computed and there are a large number of emitting elements, such as for remote sensing. In that case a large number of the tracked photon bundles will not contribute to the solutions; energy partitioning as in (10.21a-c) will reduce this problem.

Backwards Monte Carlo is advantageous when the number of computed quantities is small compared to the number of emitting elements. When the computed quantities are roughly equal in number and location as the emitting elements, the methods are equivalent. If there are more computed quantities than emitting elements, forward Monte Carlo is more efficient.

Parallel computing is an obvious path to improve the speed of Monte Carlo calculations. Since each photon bundle is independent, Monte Carlo is readily adapted to parallel computing. Backwards Monte Carlo can be directly implemented; the backward tracking origin points can be spread among the processors and the results gathered at the end. This requires minimal communication between processors, a single number (the answer) can be passed. Forward Monte Carlo requires more communication. Like the backward calculation, the bundle tracking from emission point can be spread among the processors. To compute the quantities the statistics from each emission point must be gathered together. Gathering the statistics requires passing arrays instead of single numbers. Unlike the serial case, it is not clear which method has the advantage in a given situation.

10.5.2 Quasi-3D Example

In Sect. 10.4.2 we used analytical solutions for broken black clouds. Non-black and scattering clouds were not examined because they greatly increased the complexity. However, the Monte Carlo method can easily use the probability of clear line sight expressions and account for cloud transmission and scattering. For example, a single layer cloud field composed of homogenous cylinders with random horizontal overlap as in Takara and Ellingson (2000). To emphasize the cloud geometry and scattering, consider only the $833–1250\,\mathrm{cm}^{-1}$ atmospheric window.

The upward and downward fluxes are found through a spectral and angular integration of radiances,

$$F^{\uparrow\downarrow} = \int_{0}^{\pi/2} I^{\uparrow\downarrow}(\theta)\cos(\theta)d\theta = \sum_{j=1}^{3} w_j I^{\uparrow\downarrow}(\theta_j)\,. \tag{10.22}$$

The w_j are the Gaussian weights for the Gaussian angles in the interval $0 < \theta < \pi/2$. Here a 3-point quadrature is used. The upward and downward radiances, $I^{\uparrow\downarrow}$, are computed by summing over the wavenumber intervals:

$$I^{\uparrow\downarrow} = \int_{833\,\mathrm{cm}^{-1}}^{1250\,\mathrm{cm}^{-1}} I_\nu^{\uparrow\downarrow}d\nu = \sum_{i=1}^{6} I_i^{\uparrow\downarrow}\Delta\nu_i\,. \tag{10.23}$$

Here, $I_i^{\uparrow\downarrow}$ is the spectral radiance or specific intensity of the ith wavenumber interval. The intervals (in cm^{-1}) are: $833 < \nu_1 < 909 < \nu_2 < 1000 < \nu_3 < 1081 < \nu_4 < 1143 < \nu_5 < 1212 < \nu_6 < 1250$.

Outside the cloud layer, the radiances are computed using the transmission and emission calculated using a line-by-line radiative transfer model (or "LBLRTM") assuming a black surface. Backward Monte Carlo simulations are used to find the radiances emerging from the cloud layer; to find the emerging radiance at Gaussian angle θ_j, bundles are emitted into the layer at $(\pi-\theta_j)$. Using the PCL it is determined whether or not bundles intersected a cloud element. For random number α,

$$\begin{cases} \alpha \leq \text{PCL: bundle does not intersect clouds} \\ \alpha > \text{PCL: bundle intersects clouds} \end{cases} . \qquad (10.24)$$

The transmissivity is computed throughout the cloud layer at 25-meter intervals. Since transmissivities range between 0 and 1 they can be used directly to model transmission or absorption probabilities. For a bundle emitted at z_j traveling upward to z_i and random number α,

$$\begin{cases} \alpha \leq T(z_i, z_j, \theta): & \text{bundle is transmitted} \\ \alpha > T(z_i, z_j, \theta): & \text{bundle is extinguished} \end{cases} . \qquad (10.25)$$

The bundle extinction location, z_{ext}, is found by equating the ratios of T and z:

$$\frac{\alpha - 1}{T(z_i, z_j, \theta) - 1} = \frac{z_{\text{ext}} - z_j}{z_i - z_j} \qquad (10.26a)$$

where $T = 1$ corresponds to the emission point at z_j and $T(z_i, z_j, \theta)$ to the end of the path at z_i. This linear approximation assumes that each layer is optically thin. Solving for z_{ext} gives:

$$z_{\text{ext}} = z_j + \frac{z_i - z_j}{T(z_i, z_j, \theta) - 1}(\alpha - 1) . \qquad (10.26b)$$

The transmission from z_j traveling upward to z_i through the cloud is the product of gaseous and cloud transmission.

$$T(z_i, z_j, \theta) = T_{\text{gas}}(z_i, z_j, \theta) T_{\text{cloud}}(z_i, z_j, \theta) \qquad (10.27a)$$

$$T_{\text{cloud}}(z_i, z_j, \theta) = \exp\left(-\frac{\tau}{\cos\theta}\right) = \exp\left[-\sigma_e\left(\frac{z_i - z_j}{\cos\theta}\right)\right] \qquad (10.27b)$$

$$T(z_i, z_j, \theta) = T_{\text{gas}}(z_i, z_j, \theta) \exp\left[-\sigma_e\left(\frac{z_i - z_j}{\cos\theta}\right)\right] \qquad (10.27c)$$

where σ_e is the extinction coefficient.

If the photon bundle is extinguished then it is either scattered or absorbed. The albedo of single scattering, ϖ_0, determines which occurs. For random number α,

$$\begin{cases} \alpha \leq \varpi_0: & \text{bundle is scattered} \\ \alpha > \varpi_0: & \text{bundle is absorbed} \end{cases} . \qquad (10.28)$$

There is no scattering for a bundle with a clear line of sight, $\varpi_0 = 0$, so the bundle is always absorbed. If the bundle is scattered, a new direction is assigned using the phase function through the same process described in Sect. 4.2.

Results are computed for the McClatchey tropical (TRP) and sub-arctic winter (SAW) soundings. A single layer of water clouds is inserted at three different cloud base altitudes (z_b), 0.5, 2, and 4 km. Within the clouds, an effective radius (r_e) of 5 μm is assumed for the droplets. The clouds are given two different geometries, large and small. For the large clouds, the aspect ratio (β) is 2 and the cloud diameter (D) is 1 km; the cloud thickness is 1 km. For the small clouds, $\beta = 1$ and $D = 0.25$ km; the cloud thickness is 0.125 km. The liquid water content (LWC) is 0.1 gm^{-3} for SAW and 1 gm^{-3} for TRP. For these LWC, the SAW small water cloud optical thickness is approximately 3 in the vertical direction, and 30 for the TRP.

To determine the relative importance of scattering and geometry for water clouds the results are compared to Monte Carlo computations with black clouds – the black cloud approximation. The bias error for the black cloud approximation is

$$\delta X = X(\text{approximation}) - X; \quad X = F^\uparrow, F^\downarrow . \tag{10.29}$$

In Fig. 10.14a (TRP) the absolute values of the errors are less than 0.9 Wm^{-2}, quite small. For water clouds the black cloud approximation works because the downward emission by the water clouds is augmented by the downward reflection of upward flux at the cloud bottom. This can be seen by noting that the error is negative at $A_c = 1$. Since the apparent cloud emission is greater than the blackbody emission, the apparent emissivity of the water cloud is larger than one. The high temperature and water vapor concentration in the first kilometer masks the scattering effect at the surface. This agrees with observational studies of cumulus cloud fields.

In Fig. 10.14b (SAW), the errors are less than 2 Wm^{-2} for the large clouds, and approach 11 Wm^{-2} for small clouds. The small cloud errors decrease as cloud height increases, peaking at $A_c = 0.7$. Because the small clouds are not opaque, there is direct transmission from the surface to the upper atmosphere. The surface "sees" the cold upper atmosphere. Though the cloud bottom continues to reflect, a good deal of the surface radiation is transmitted through the clouds. The reflection from the cloud bottom is not large enough to compensate for the reduced emission, unlike Fig. 10.14a. As a result, the black cloud approximation fails since the small water cloud vertical optical thickness is approximately 3; this agrees with the results in Sect. 10.4.2 and results by Harshvardhan (1982).

The errors for the upward flux at 15 km are shown in Figs. 10.15a,b. In Fig. 10.15a (TRP), the absolute value of the error is less than 4 Wm^{-2}. The error increases almost linearly with A_c; the larger clouds having more error than the smaller clouds. In this case, the black cloud approximation overestimates the flux above the clouds. The clouds are too opaque to allow transmission from the lower atmosphere. Scattering clouds trap their emissions within themselves; so the outward emission by the clouds is reduced. Unlike for surface fluxes, there is no reflection from the cloud top to compensate for the lower emission by the cloud. This effect is more noticeable above the thicker water clouds where atmospheric emission is reduced. Fig. 10.15b (SAW) also shows a similar pattern, except for the small water cloud at 4 km. From Fig. 10.14b, the small clouds are partially transparent and the large clouds are opaque. It might be expected that the small cloud errors would cluster together as in Fig. 10.14b. But only the small water cloud at 4 km stands apart. For the other clouds, the black cloud approximation overestimates the flux and the error increases almost linearly with

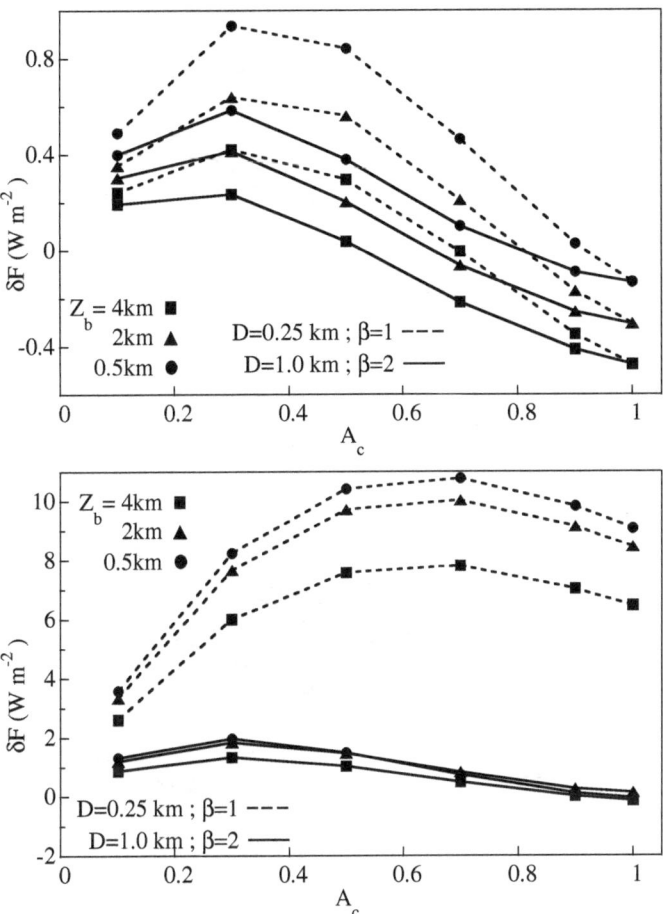

Fig. 10.14. (a) TRP downward flux error at surface for large and small water clouds at various z_b: LWC = 1 g m^{-3}, $r_e = 5\,\mu$m. **(b)** SAW downward flux error at surface for large and small water clouds at various z_b: LWC = 0.1 g m^{-3}, $r_e = 5\,\mu$m. Notice the vastly different ordinate scales

A_c. Since the optical thickness of the small clouds is low enough to allow transmission while the large clouds are opaque, it is curious to see similar errors for large and small clouds at 0.5 and 2 km. The explanation is in the temperature profile. In this sounding, the temperature increases slightly with altitude for the first kilometer, drops back down at 2 km and remains almost the same up to 3 km. As a result, 0.5 and 2 km clouds (both large and small) are at almost the same temperature as the atmosphere below. Only the 4 km clouds are significantly colder than the lower atmosphere. Since the small cloud is partially transparent, the energy from the lower levels radiates through the cloud layer. This leads to an underestimation of the flux by the black cloud approximation.

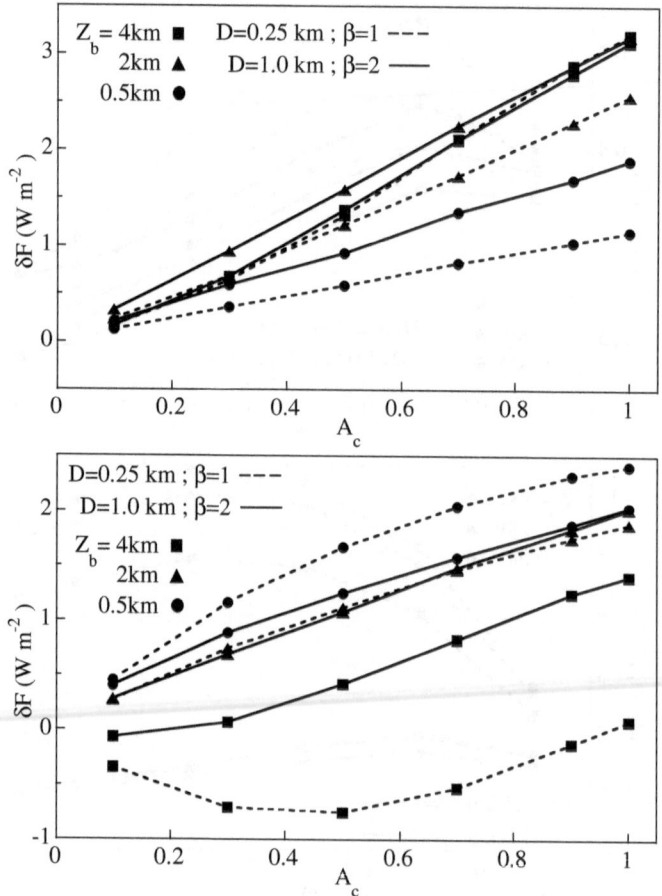

Fig. 10.15. (a) TRP upward flux error at 15 km for large and small water clouds at various z_b: LWC = $1 \, \mathrm{g \, m^{-3}}$, $r_e = 5 \, \mu\mathrm{m}$. (b) SAW upward flux error at 15 km for large and small water clouds at various z_b: LWC = $0.1 \, \mathrm{g \, m^{-3}}$, $r_e = 5 \, \mu\mathrm{m}$

In general the black cloud approximation worked well for opaque water clouds with flux errors less than $5 \, \mathrm{Wm^{-2}}$. When the cloud optical thickness is reduced so that the cloud is no longer opaque, the black cloud approximation fails. The results indicate that individual cloud geometry is of primary importance for opaque water clouds. The errors due to neglecting cloud scattering are largest close to the cloud layer and decrease as the distance from the cloud layer increases. This reduction occurs most rapidly below 3 km. Since the effects of water cloud longwave scattering are effectively muted in the 833–1250 $\mathrm{cm^{-1}}$ window, scattering effects from optically thick water clouds should not be significant over the longwave spectrum.

10.5.3 Fully-3D Example

The Intercomparison of Three-Dimensional Radiation Codes (I3RC[1]) has provided excellent examples of three-dimensional cloud problems. The most computationally intensive is from an LES-based prediction of a cumulus field featuring a 100×100 grid in the horizontal with a vertical grid of 30 layers and periodic boundary conditions (see Fig. 10.16a and also Fig. 4.6).

Note that the fourth quadrant ($x > 3$ km, $y < 3$ km) is almost cloud free. The I3RC specified several problems for this simulated field, one of which is to find the surface downward flux at $1000 \, \text{cm}^{-1}$ for each grid point (10,000 points in all). The backward Monte Carlo computation used here assumed cells centered on each grid point with constant properties. Quantities are calculated at the midpoint or center point of each zone. The bundles are tracked from the middle or center of each cell.

Bundle tracking proceeded as in Sect. 10.5.2, i.e., (10.25) for transmission and (10.28) for scattering are directly applied. Since it is assumed that there is no horizontal variation in temperature, it is not necessary to tabulate the horizontal components of the absorption locations, only the vertical locations are necessary. Once the bundle tracking is completed, the fractions accumulated in each vertical location can be used to compute the downward flux at the surface ($z = 0$) $F^{\downarrow}$:

$$F^{\downarrow} = \pi f_{\text{s}} B_{\text{v}}(T_{\text{s}}) + \pi \sum_{k=1}^{30} f_{\text{v}k} B_{\text{v}}(T_k) \tag{10.30a}$$

where subindex "s" stands for surface while "v" stands for volume. Here f_{s} is the ratio of the number of bundles that reach and are absorbed by the surface, N_{s}, to the total number of bundles emitted at the computation point, N:

$$f_{\text{s}} = \frac{N_{\text{s}}}{N} \tag{10.30b}$$

and $f_{\text{v}k}$ is the same ratio for the number of bundles that reach and are absorbed in layer k, $N_{\text{v}k}$:

$$f_{\text{v}k} = \frac{N_{\text{v}k}}{N} \, . \tag{10.30c}$$

Note that the number of bundles that reach the TOA (set at $z = 30$ km) is not tabulated since there is no downward emission from space (other than the longwave portion of the solar spectrum).

To compute the upward flux at TOA $F^{\uparrow}$, photons are tracked backwards from the starting points at the top of the atmosphere

$$F^{\uparrow} = \pi f_{\text{s}} B_{\text{v}}(T_{\text{s}}) + \pi \sum_{k-1}^{30} f_{\text{v}k} B_{\text{v}}(T_k) \, . \tag{10.31}$$

To compute q, the heating rate per unit volume at a particular point $(x, y, z)^{\text{T}}$, photons are tracked backwards from that point:

---

[1] http://i3rc.gsfc.nasa.gov/

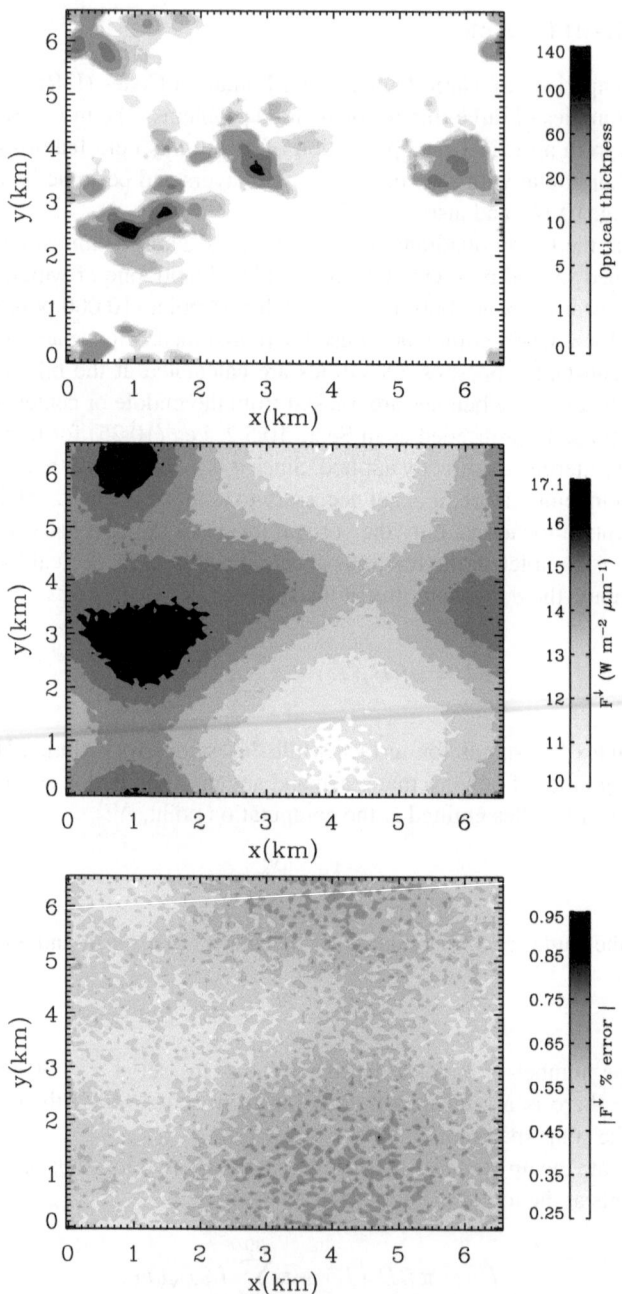

Fig. 10.16. (a) Column integrated optical thickness for I3RC cumulus field at $1000\,\mathrm{cm}^{-1}$. **(b)** $1000\,\mathrm{cm}^{-1}$ downward flux at surface for I3RC cumulus field. **(c)** Percent Monte Carlo error estimate (*absolute value*) for $1000\,\mathrm{cm}^{-1}$ downward flux at surface for I3RC cumulus field

$$q = 4\pi\sigma_a \left[f_s B_v(T_s) + \sum_{k=1}^{30} f_{vk} B_v(T_k) \right] - 4\pi\sigma_a B_v(T) \qquad (10.32)$$

where σ_a is the absorption coefficient and T is the temperature at position $(x, y, z)^T$, typically only a function of z. The first term is the absorbed energy per unit volume; $4\pi\sigma_a B_v(T)$ is the energy emitted per unit volume.

Fig. 10.16b is the downward surface spectral $10\,\mu m$ flux for the 100×100 horizontal grid. The fluxes at each point are computed by backward tracking 10,000 bundles. Approximately 12 hours are needed to compute the 10,000 fluxes on an 800 Mhz Linux workstation. The three-dimensional effects of the clouds will be most apparent for this spectral flux since it is near the center of the $833–1250\,cm^{-1}$ window region. In Fig. 10.16b, the clear fourth quadrant from Fig. 10.16a is a region of low downward flux. The areas below the clouds have high downward flux. The estimated percentage numerical error (absolute value) estimate is shown Fig. 10.16c. The estimated error is under one percent throughout, indicating that 10,000 bundles from each surface mid-point will give a good estimate for the surface flux over the entire region. Note that the areas of lower error are below the clouds (high flux) and the higher errors are in the clear areas (low flux). The time required to compute these fluxes is a drawback of the Monte Carlo method and emphasizes the importance of carefully implementing a Monte Carlo calculation.

10.6 Summary

Within this chapter basic information on gas and cloud spectroscopy and an introduction to longwave calculations with results are presented. This is followed by analytical results for 3D clouds and then a discussion of longwave Monte Carlo with some results.

The discussion of spectroscopy in Section 10.1 shows that atmospheric gases are quite opaque. For clear skies, there are two spectral intervals with significant surface to TOA transmission: the primary $833–1250\,cm^{-1}$ ($8–12\,\mu m$) window and the "dirty" window $500\,cm^{-1}$ ($20\,\mu m$), which becomes significant in dry atmospheres. In comparison to the gaseous absorption which varies greatly within a small spectral interval the cloud radiative properties vary quite slowly. The introduction to longwave calculations in Sect. 10.2 and the results shown in Sect. 10.3 show the blocking effect of clouds in the primary $833–1250\,cm^{-1}$ window, which yields higher downwelling at the surface and lower upwelling at the TOA. Clouds affect the atmospheric cooling rate through the optical depth gradients at their boundaries. This results in heating below the cloud layer and cooling above. In partially cloudy cases, a 15% error in plane parallel cloud fraction leads to very large cooling rate errors.

The analytical results for 3D clouds in Sect. 10.4 use the concept of the probability of a clear line of sight to generate analytical solutions for broken cloud fields of various size and space distributions. From this the effect of cloud geometry is determined. For the simple isothermal cylindrical cloud model, neglecting cloud geometry can lead to underestimating the downwelling flux at the surface by as much as

20 Wm^{-2}. A more complex model featuring truncated square pyramids with power law distributions for cloud size and spacing generally agrees with measurements. The comparison of the effect of cloud non blackness to geometry shows that for clouds with large optical depth and low variability, cloud geometry is most important. For clouds with high variability in optical depth, the variability and non-blackness is most important. Sect. 10.5 describes aspects of the Monte Carlo method in the longwave; expressions for longwave emission and the relative merits of forward and backward Monte Carlo are described. Results from an extension of the work on analytical results for 3D clouds that includes scattering and transmission are presented. They show that in very humid atmospheres with opaque clouds, cloud scattering has very little effect on surface flux compared to the black cloud approximation. The high humidity masks out the scattering effect. Scattering does have an effect for the upward flux at 15 km; it lowers the emission from the cloud top. Lastly, surface fluxes for a full-blown monochromatic 3D Monte Carlo model are shown. The computation is quite lengthy for 10,000 bundles from each of the 10,000 computation points. Even with the vast increases in computer speed, long computational times are a problem for Monte Carlo calculations.

References

Avaste, O.A., Y.R. Mullamaa, K.Y. Niylisk, and M.A. Sulev (1974). On the coverage of sky by clouds. In *Heat Transfer in the Atmosphere*. Technical Report (Tech. Transl. TT-F-790), NASA, Washington (DC).

Barker, H.W. and B.A. Wielicki (1997). Parameterizing grid-averaged longwave fluxes for inhomogenous marine boundary layer clouds. *J. Atmos. Sci.*, **54**, 2785–2798.

Ellingson, R.G. and Y. Fouquart (1991). The intercomparison of radiation codes in climate models (ICRCCM): An overview. *J. Geophys. Res.*, **96**, 8925–8927.

Goody, R.M. and Y.L. Yung (1989). *Atmospheric Radiation Theoretical Basis*. Oxford University Press, New York (NY).

Han, D. (1999). *Studies of Longwave Radiative Transfer Under Broken Cloud Conditions: Cloud Parameterizations and Validations*. Ph.D. dissertation, University of Maryland, College Park, College Park (MD).

Han, D. and R.G. Ellingson (1999). Cumulus cloud formulations for longwave radiation calculations. *J. Atmos. Sci.*, **56**, 837–851.

Harshvardhan (1982). The effect of brokenness on cloud-climate sensitivity. *J. Atmos. Sci.*, **39**, 1853–1861.

Liou, K.-N. (1980). *An Introduction to Atmospheric Radiation*. Harcourt Brace Jovanovich, San Diego (CA).

Takara, E.E. and R.G. Ellingson (2000). Broken cloud field longwave scattering effects. *J. Atmos. Sci.*, **57**, 1298–1310.

Thomas, G. and K. Stamnes (1999). *Radiative Transfer in the Atmosphere and Ocean*. Cambridge University Press, New York (NY).

Warner, J. and R. Ellingson (2000). A new narrowband radiation model for water vapor absorption. *J. Atmos. Sci.*, **57**, 1481–1496.

Suggested Further Reading

Benassi, A., F. Szczap, A.B. Davis, M. Masbou, C. Cornet and P. Bleuyard (2004). Thermal radiative fluxes through inhomogeneous cloud fields: A sensitivity study using a new stochastic cloud generator. *Atmos. Res.*, **72**, 291–315.

Cahalan, R.F. and J.H. Joseph (1989). Fractal statistics of cloud fields. *Mon. Wea. Rev.*, **117**, 261–272.

Clough, S.A., F.X. Kneizys and R.W. Davies (1989). Line shape and the water vapor continuum. *Atmos. Res.*, **23**, 229–241.

Clough, S.A., M.J. Iacono and J.L. Moncet (1992). Line-by-line calculations of atmospheric fluxes and cooling rates: Application to water vapor. *J. Geophys. Res.*, **97**, 15,761–15,785.

Ebert, E.E. and J.A. Curry (1992). A parameterization of ice cloud optical properties for climate models. *J. Geophys. Res.*, **97**, 3831–3836.

Ellingson, R.G. and J.C. Gille (1978). An infrared transfer model. Part I: Model description and comparison of observations with calculations. *J. Atmos. Sci.*, **35**, 523–545.

Ellingson, R.G. (1982). On the effects of cumulus dimensions on longwave irradiance and heating rates. *J. Atmos. Sci.*, **39**, 886–896.

Ellingson, R.G., J. Ellis and S. Fels (1991). The intercomparison of radiation codes in climate models (ICRCCM): Longwave results. *J. Geophys. Res.*, **96**, 8929–8953.

Fu, Q. and K.N. Liou (1993). Parameterization of the radiative properties of cirrus clouds. *J. Atmos. Sci.*, **50**, 2008–2025.

Hu, Y.X. and K. Stamnes (1993). An accurate parameterization of the radiative properties of water clouds suitable for use in climate models. *J. Climate*, **6**, 728–742.

Kauth, R.J. and J.L. Penquite (1967). The probability of clear lines of sight through a cloudy atmosphere. *J. Appl. Meteor.*, **6**, 1005–1017.

Killen, R. and R.G. Ellingson (1994). The effects of shape and spatial distribution of cumulus clouds on longwave irradiance. *J. Atmos. Sci.*, **51**, 2123–2136.

Niylisk, K.Y. (1968). Atmospheric thermal radiation in partly cloudy regions. *Izv. Acad. Sci. USSR, Atmos. Oceanic Phys.*, **4**, 383–396.

Plank, V.G. (1969). The size distribution of cumulus clouds in representative Florida populations. *J. Appl. Meteor.*, **8**, 46–67.

Takara, E.E. and R.G. Ellingson (1996). Scattering effects on longwave fluxes in broken cloud fields. *J. Atmos. Sci.*, **53**, 1464–1476.

Zhu, T., J. Lee, R.C. Weger and R.M. Welch (1992). Clustering, randomness, and regularity in cloud fields: 2. Cumulus cloud fields. *J. Geophys. Res.*, **97**, 20,537–20,558.

Part IV

Remote Sensing

Remote Sensing

11

3D Radiative Transfer in Satellite Remote Sensing of Cloud Properties

R. Davies

11.1 Introduction 523

11.2 The Relation Between Cloud Heterogeneity
and 3D Radiative Transfer in Global Remote Sensing 525

11.3 Evidence of Cloud Heterogeneity 526

11.4 Consequences of Cloud Heterogeneity 529

11.5 Some Current Uses of Heterogeneity 533

11.6 Summary and Outlook 536

References 537

Suggested Further Reading 538

11.1 Introduction

The effects of three-dimensional radiative transfer (3DRT) are evident in a variety of satellite remote sensing applications. They are most frequently associated with the remote sensing of cloud optical properties such as optical depth, but may also be relevant to other applications, such as the remote sensing of aerosol properties over heterogeneous land (Diner and Martonchik, 1984), and even the retrieval of rainfall rates. One of the earliest applications of a 3DRT model to remote sensing was in fact a study of the effect of cloud heterogeneity on the retrieval of rainfall rates from passive microwave radiances by Weinman and Davies (1978), who used both analytical and Monte Carlo 3DRT models to obtain ambiguities of the order of a factor of 2 in the inferred rainfall rate when the rain did not uniformly fill the radiometer's field of view. On the other hand, many other types of remote sensing, notably those that involve line-of-sight absorption and emission with little atmospheric scattering, such as temperature and humidity sounding or the retrieval of surface features, remain essentially one-dimensional radiative transfer (1DRT) problems for which the complexity of 3DRT is not needed.

Because the earliest practical uses of radiative transfer in satellite remote sensing were entirely one-dimensional, the initial work with 3DRT tended to involve theoretical studies that examined the potential limitations of the one-dimensional retrievals. In the case of finite clouds it was relatively trivial to show large differences in retrieved properties compared with the one-dimensional counterpart. Even for the example of cloud top temperatures inferred from 10 μm radiances, the inclusion of three-dimensional scattering effects gave differences of several degrees from the one-dimensional approach for finite clouds, leading to possible 'superblack' cloud tops, as shown by Harshvardhan et al. (1981). Further, and progressively more sophisticated, theoretical studies have continued to explore the limitations of one-dimensional retrievals with the goal of establishing a basis for the application (largely unrealized, as yet) of three-dimensional retrieval techniques.

Specific case studies of finite clouds clearly show glaring differences between the one- and three-dimensional approaches, but may not necessarily be typical of global cloudiness. For this and other reasons, the global remote sensing community initially appeared to be unresponsive to such findings and continued to accept the results of one-dimensional retrievals with remarkably little concern. It took what might be called a second wave of studies to place the consequences of 3DRT squarely in a global context that cannot be ignored. These studies have been statistical in nature, summarizing satellite measurements in ways that illustrate dependencies that are inconsistent with the assumptions of 1DRT (e.g., the breakdown of directional reciprocity), and that lead to clearly undesirable consequences (e.g., cloud optical depths that increase with solar zenith angle). As new satellites are deployed to measure the Earth-leaving radiation with increasingly higher spectral and spatial resolutions, and at more viewing angles, statistical studies continue to provide new quantitative insights into the global nature and consequences of cloud heterogeneity, and consequently of 3DRT. The key fact that clouds are predominantly heterogeneous on a global scale, however, already appears inescapable.

This brings us to what might be called the third wave of 3DRT studies, still very much in its infancy. This refers to the use of remote sensing techniques that exploit cloud heterogeneity to retrieve new properties (as in the multi-angle remote sensing of cloud-top heights using stereo), and to modify or optimize the use of 1DRT (as in identifying cases where one-dimensional retrievals can be safely made).

In this chapter we include examples of each of these three types of study. The second type receives the most attention, given its relative maturity and significance. Our emphasis is on the remote sensing of cloud properties using satellite measurements of shortwave spectral radiances. While this limitation reflects the prejudice of the author's research interests, it is also the type of remote sensing that is most dependent on the need for a 3DRT perspective. Shortwave cloud properties and, notably, the relationship between cloud albedo and microphysical content are also of great interest in studies of equilibrium climate and climate change, given the huge uncertainties still evident in assessing cloud-radiative feedback processes. Much of the remote sensing challenge is therefore directed at obtaining accurate cloud albedos, and in relating these albedos to cloud optical depth, which is frequently taken as a proxy measure of cloud water amount. In the interests of space and time, however, we simply scratch the surface of possible examples, and the content is illustrative rather than comprehensive (but see the suggested reading for additional topics).

The chapter opens with a discussion of cloud heterogeneity in a global context, and presents some evidence for the global nature of cloud heterogeneity. Cloud heterogeneity is a nuisance in that it sabotages the retrieval of cloud properties using one-dimensional techniques, and the impact of this on cloud optical depth retrieval is discussed at some length, followed by the impact on cloud albedo estimation. Cloud heterogeneity also has its positive uses, and can be exploited by multi-angle remote sensing techniques to obtain cloud-top heights and height-resolved winds, as discussed in subsequent sections.

11.2 The Relation Between Cloud Heterogeneity and 3D Radiative Transfer in Global Remote Sensing

Clearly, if clouds were truly homogeneous horizontally, there would be no horizontal gradients in their water concentration, and the one-dimensional plane-parallel paradigm would apply. Given uniform solar illumination and sufficiently thick cloud (or thin cloud and a sufficiently homogeneous underlying surface), then the reflected radiances must also be horizontally uniform. On occasion, when satellite-measured radiances at high spatial resolution differ by less than a percent or two over several hundred kilometers, this appears in fact to be the case. Such occasions, however, are few and far between, and the vast bulk of the measured radiances exhibit much greater differences, and on scales often less than one kilometer.

This should come as no surprise given the dynamic nature of cloud formation and dissipation. Clouds are formed by rising air motion. Rising air of convective origin is localized in cells surrounded at least partially by sinking air. Stratiform uplift on the other hand tends to be more uniform and horizontally widespread, but is also slow, allowing ample time for longwave cooling to erode the cloud-top surface (upside-down convection, if you like). At one extreme, we have heterogeneous clouds that are small, localized cloud elements surrounded by clear air, as in the case of fair weather cumulus. At the other extreme, we have continuously overcast stratiform cloud with an irregular (i.e., bumpy) cloud-top structure. In between these extremes there are many heterogeneous examples, often with degrees of self-organization that provides a quasi-periodic horizontal variability of both cloud water and reflected radiance that may have a cellular or striated appearance. The case of deep convection, in the form of cumulonimbus or cumulus congestus, deserves special mention as here the shapes are most obviously three-dimensional and the water contents the highest. Cirrus cloud may also be significantly heterogeneous, and often extends vertically over several kilometers, encountering a wind shear that may stretch the falling ice crystals into intricate three-dimensional patterns. However, the anvil tops of some supercell thunderstorms (removed from any overshoot region or edges) give the opposite appearance, and represent the most frequent examples that approximate the ideal of a locally homogeneous cloud.

Radiative transfer models show that the presence of cloud heterogeneity requires 3DRT in order to accurately represent the external radiation field of the cloud. Conversely, departures of remotely measured radiance from the expectations of 1DRT usually indicate the existence of some measure of cloud heterogeneity. We therefore start with some global evidence of cloud heterogeneity based on remotely measured radiances.

11.3 Evidence of Cloud Heterogeneity

11.3.1 A Spatial Test of Heterogeneity

The simplest and clearest measure of cloud heterogeneity is the spatial change in scattered radiance with distance across a cloud top. Assuming uniform solar irradiance incident on the cloud, and either a uniform underlying surface (such as the ocean), or sufficiently thick cloud so that any changes in the underlying surface do not matter, then a spatial change in reflected radiance is directly caused by the presence of cloud heterogeneity. Individual clouds may occasionally be remarkably uniform over large horizontal distances, but more commonly show significant variation in their reflected radiances over distances short enough (e.g., 1–8 km) to affect typical satellite remote sensing.

A quantitative measure of this type of spatial heterogeneity was defined by Genkova and Davies (2003), who set a threshold, T, as the maximum percent variability in reflected radiance allowed across a cloudy region of specified size, measured at some specified higher resolution, to be regarded as being spatially homogeneous. They then applied this threshold to a global distribution of clouds, measured from space using the Multiangle Imaging SpectroRadiometer (MISR) on NASA's Terra satellite. Sufficient data (58 different orbits) were analyzed to reach statistically representative results. Excluding thin clouds with nadir visible reflectivity less than 20%, and high latitudes that are dominated by underlying snow and ice surfaces, they found only a very small percentage of the remaining clouds were typically classified as being spatially homogeneous over scales of 1-8 km. Figure 11.1 summarizes the homogeneity pass rates for $T = 10\%$, for cloudy regions of different size and for two measurement resolutions (275 m and 1100 m). The error bars are mainly an indication of the uncertainty in the mean due to natural variability. Note that because the Terra orbit is sun-synchronous with a 10:45 a.m. equator crossing time these results are not necessarily representative of the full diurnal behavior of global cloudiness.

In terms of this spatial homogeneity test, it appears that only about 11% of globally distributed clouds appear to be spatially homogeneous (at the 10% level) over a scale of 1.1 km. This number drops quickly with region size, to a pass rate of less than 2% for regions larger than 8.8 km. These low numbers call into question the applicability of 1DRT approaches to cloud remote sensing, but as we will see in the next section, this may be too conservative of a test to answer such questions.

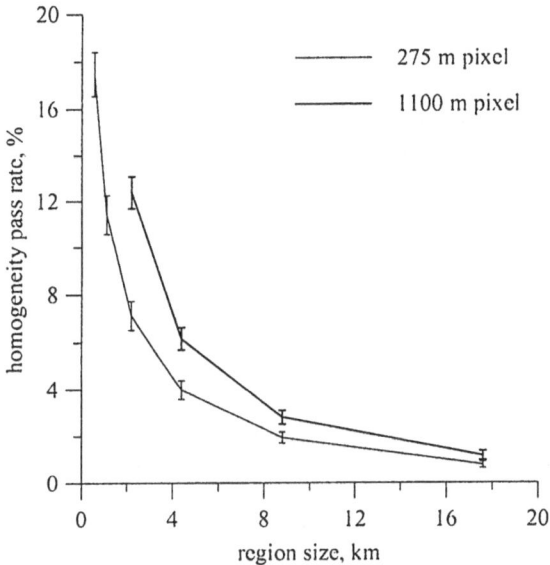

Fig. 11.1. Dependence of the spatial homogeneity pass rate on region size and resolution for 14 orbits on 9 April 2001, from Genkova and Davies (2003) with permission

11.3.2 An Angular Test of Heterogeneity

In a similar vein to the preceding section, Horváth and Davies (2004) defined an angular test of cloud homogeneity. Here the reflected radiances from globally distributed cloud tops measured by the nine viewing directions of MISR are examined individually to retrieve nine collocated measures of cloud optical depth, as retrieved using 1DRT theory. If an optical depth exists for which the 1DRT radiances agree to within 10% of the nine measured values, then the cloud is deemed to pass the angular test for homogeneity, otherwise it is classified as heterogeneous.

The MISR instrument has a nadir view, complemented by four pairs of oblique cameras positioned at nominal viewing zenith angles of 26.1°, 45.6°, 60°, and 70.5°. Each oblique pair consists of one camera looking forward and one looking backward with respect to the flight direction. Since the time interval between the two most oblique observations is only 7 min, the instrument allows the almost instantaneous sampling of the bi-directional reflectance field. The cross-track resolution is 275 m, while the along-track resolution increases with view angle, from 214 m at nadir to 707 m at the most oblique angle. The along-track sample spacing, however, remains at 275 m.

To minimize ambiguities due to other factors, Horváth and Davies (2004) included only maritime clouds between 60°N and 60°S, for regions devoid of sea ice. Ice and mixed phase clouds, as identified by the Moderate-Resolution Imaging Spectroradiometer (MODIS) cloud phase product, were also excluded. In applying the 1DRT algorithm to retrieve cloud optical depth consideration was given to the

effects of Rayleigh scatter and ozone absorption that become increasingly important for the oblique views. Uncertainties in the anisotropy of ocean surface reflection
were also reduced by excluding thin clouds with optical depths less than 3. The
one-dimensional retrieval also assumed standard liquid water microphysics, with an
effective radius of 8 µm. Other values of effective radius were examined but gave
lower passing rates.

A major difficulty with applying this technique is the need to accurately co-
register the multiangle views to the same point on the cloud top, allowing for parallax
effects due to height, and displacement effects due to wind. The operational products
apply a stereo retrieval technique to accomplish this over 2.2 km regions, the results of which were used by Horváth and Davies (2004) to assess the uncertainty in
their angular test, as the co-registration for their study was based on the mean cloud
height and wind over 70.4 km regions. The effect of cloud height variations within a
70.4 km region on co-registration produced an uncertainty in the pass rates of about
3%, which is added to the uncertainty due to natural variability, giving an overall
uncertainty of about 5% in the pass rates.

As shown in Fig. 11.2, the pass rates for the angular test are higher than for the
previous spatial test. At an effective resolution of 1.1 km, the passing rate is now
about 23%. The pass rate increases as the resolution is degraded, to a maximum
value of about 37%. At the highest resolution of 275 m the homogeneity pass rate

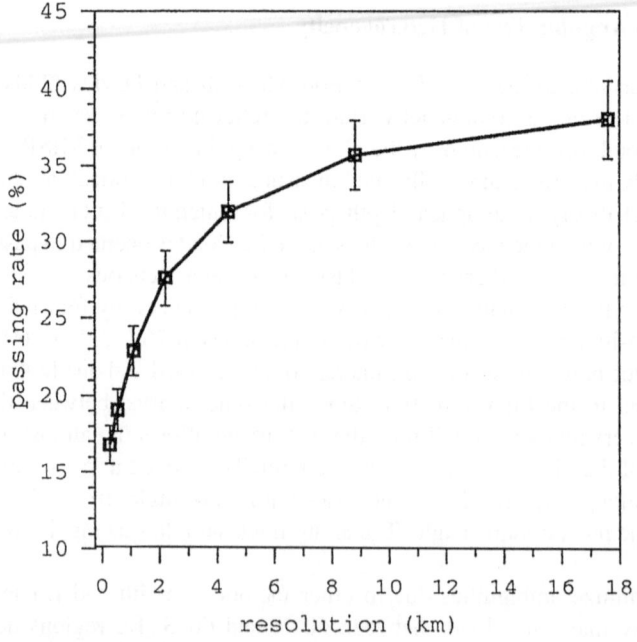

Fig. 11.2. Angular test passing rate vs. pixel resolution for $r_e = 8$ µm, from Horváth and
Davies (2004) with permission

drops to about 17%. While the effects of co-registration make this value somewhat less certain, the angular pass rate at 275 m is roughly comparable to the spatial pass rate for a 550 m region.

The fact that the spatial homogeneity pass rate decreases as the region considered expands, whereas the angular pass rate increases as the resolution is degraded has several implications. Angular modeling, needed in the context of estimating cloud albedo from satellite-measured radiances, is evidently easier at coarser resolution. A consistent retrieval of cloud properties at coarse resolution with view angle, however, does not appear to be a guarantee of accuracy. It may simply be a nonlinear averaging effect. This theme is continued in the following section. Conversely, when the spatial and angular tests are both passed for a given region, irrespective of its size, we may have much more confidence in the applicability of 1DRT and the cloud properties retrieved.

11.4 Consequences of Cloud Heterogeneity

11.4.1 Optical Depth Biases

The practical effects of cloud heterogeneity on the global distribution of cloud optical depths retrieved using 1DRT were first demonstrated by Loeb and Davies (1996). Using a year's worth of Earth Radiation Budget Satellite (ERBS) shortwave observations at nadir over ocean between 30°N and 30°S, they found that the observed nadir reflectances increased with solar zenith angle. Matching the observations to 1DRT on a pixel-by-pixel basis to retrieve cloud optical depth produced optical depths that also increased with solar zenith angle. However, stratification of the data into morning and afternoon measurements gave virtually identical results at the same solar zenith angles, indicating a true diurnal dependence for this data set did not appear to be significant.

Figure 11.3 compares the observed nadir reflectances with those calculated assuming a distribution of optical depths that is invariant with solar zenith angle. The optical depth distribution was based on 1DRT retrieval at high sun, and separated into classes either side of its median value. Relatively good agreement is obtained for the thinner cloud class for solar zenith angles less than about 70°, but the thicker cloud class shows almost immediate divergence as the solar zenith angle increases.

The optical depths retrieved using 1DRT thus have a spurious dependence on solar zenith angle that causes them to be biased high at large solar zenith angles. This bias is greater for thicker clouds. Loeb and Davies (1997) showed that this bias is also greater for nadir views, compared with oblique views. ERBS has a relatively coarse resolution ($31 \times 47 \, km^2$ at nadir) so that some 40% of the clouds would likely pass the angular homogeneity test, with the bias being due to the remaining 60%. Loeb and Coakley (1998) showed consistent results even when the analysis was restricted to marine stratocumulus using the higher spatial resolution ($\approx 4 \times 4 \, km^2$) of Advanced Very High Resolution Radiometer (AVHRR) measurements.

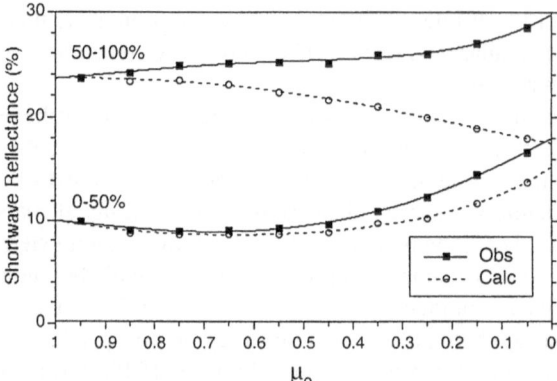

Fig. 11.3. Average reflectance versus μ_0 for the observations and for calculations derived using cloud optical depths and cloud fractions inferred from observations for μ_0 between 0.9 and 1.0, separately for reflectances lying below the 50th percentile (0–50%) and for pixels lying between the 50th and 100th percentiles (50–100%), from Loeb and Davies (1996) with permission

An explanation for how cloud heterogeneity produces this bias has been provided by Loeb et al. (1997). Using a 3DRT Monte Carlo model, they noted that the increase in nadir radiance with solar zenith was, in the case of overcast clouds, produced by cloud top texture, with the local slopes of the cloud tops scattering more radiation vertically than their flat-topped one-dimensional counterparts. In the case of broken clouds, illumination of cloud sides also added to this effect.

11.4.2 Violation of Directional Reciprocity

One of the classic tests of a radiative transfer model, comparable to checking for energy conservation, is to check whether it satisfies the principle of reciprocity. For one-dimensional atmospheres this reduces to checking the directional relationship, $\mu_A I_A = \mu_B I_B$, where A and B refer to two reciprocal directions, I the radiance, and μ the cosine of the zenith angle. In other words, I_A is the radiance in direction A due to incident irradiance from direction B, and I_B is the radiance in direction B due to incident irradiance from direction A. A necessary condition for a cloudy scene to be modeled with 1DRT is therefore that the directional principle be satisfied for that scene. Conversely, violation of directional reciprocity for a given scene implies that it is not horizontally homogeneous, and satisfies only the general principle of reciprocity (which also includes consideration of horizontal displacement).

In a study using spatial autocorrelation functions derived from ERBS data, Davies (1994) showed that the directional principle is clearly violated for scenes measured at the ERBS spatial resolution. Figure 11.4 shows an example of this violation for the reciprocal pair – nadir view with 60° solar zenith angle and 60° viewing zenith with overhead sun. Di Girolamo et al. (1998) explored this breakdown further, by investigating the spatial scales over which directional reciprocity might

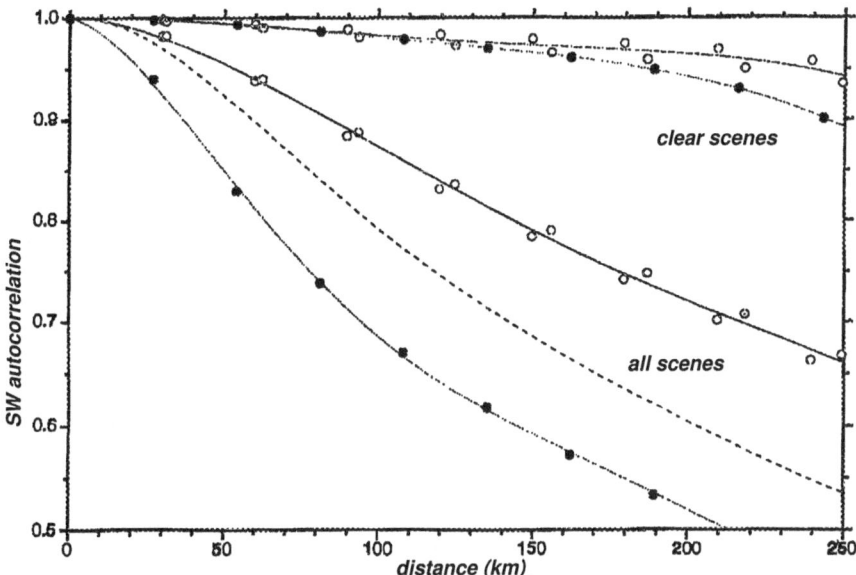

Fig. 11.4. Autocorrelation functions for reciprocal pairs of reflected shortwave radiance measured by the ERBS scanner, April-July 1985 within 30° of the equator. *Solid lines (open circles)* correspond to a view angle of 60° and a solar zenith angle of 0°. *Dotted lines (solid circles)* correspond to a view angle of 0° and a solar zenith angle of 60°, at the resolution of the nadir footprint. The *dashed line* corresponds to a view angle of 0° and a solar zenith angle of 60°, integrated over a pixel area corresponding to a 60° view angle

best be applied for different scenes. They noted scales in excess of 100 km were typically required for cloudy scenes, but scales of only a few meters to a kilometer were required for clear scenes.

Other than demonstrating the breakdown of 1DRT, there is a very practical application of this result. Albedo estimation from satellite radiances requires the application of angular models of scene reflectivity in order to account for the effects of anisotropy. In building such models, there is often a shortage of data in certain directions, and a temptation to fill in the values for missing directions by applying directional reciprocity. Recognizing that this would be erroneous, the angular models used by the Clouds and Earth's Radiant Energy System (CERES) instrument, which measures radiances on variable spatial scales with a minimum of 10 km, make use of directional reciprocity relations for clear scenes only (Loeb et al., 2003).

11.4.3 Plane-Parallel Albedo Biases

As noted elsewhere in this book, the relationship between cloud optical properties and reflected radiation is decidedly nonlinear. Heterogeneous cloud fields with varying optical depths thus create a biased relationship between the area-averaged radiation and the area-averaged optical depth that depends on the size of the region being averaged, the variation of the properties within the region and average brightness of the scene. From a remote sensing perspective, applying 1DRT to area-averaged radiation tends to produce an optical depth that is biased low compared with the true average. Similarly, from a modeling perspective, use of the correct average optical depth tends to produce an area-averaged albedo that is biased high.

The plane-parallel albedo bias for a given region containing N cloudy pixels is defined as

$$B = R(\bar{\tau}, \theta_0) - \frac{1}{N} \sum_{n=1}^{N} R(\tau_n, \theta_0) ,$$

where $\bar{\tau}$ is the mean optical depth of the region averaged over the cloudy pixels, R is the plane-parallel albedo, and θ_0 is the solar zenith angle. Note that the second term is just the average cloud albedo calculated using the independent pixel approximation. Oreopoulos and Davies (1998) analyzed a large set of AVHRR data over the North Atlantic to assess the magnitude of B, and its dependence on region size. Figure 11.5 is adapted from their results, showing average biases in the visible albedo of about 0.08 for $(55\,\mathrm{km})^2$ regions. This value rises as the region size, typical of that used in General Circulation Models, is increased further. Because the plane-parallel albedo bias is defined in a 1DRT context, its value also suffers from the optical depth bias mentioned above, and Oreopoulos and Davies (1998) noted that this causes B to rise with solar zenith angle also. This has perhaps the greatest practical application to the construction and interpretation of global climate models. Such models require an accurate parameterization of albedo in order to correctly assess the absorbed solar energy available to power the climate system. The bias shown here has therefore to be removed by appropriately adapting the apparent thickness or amount of clouds. Approximate corrections are certainly possible, but we note here the difficulty implied by the dependence on region size and the dependence on solar zenith angle.

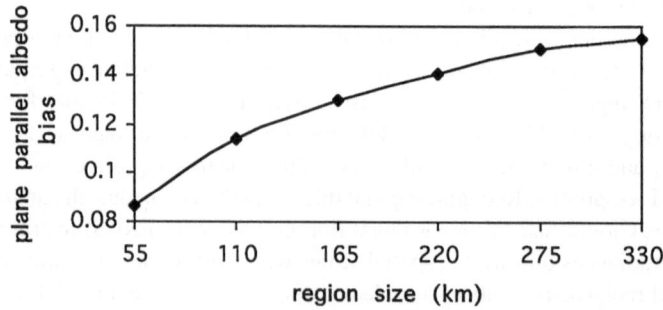

Fig. 11.5. Plane-parallel visible albedo bias of North Atlantic clouds as function of region size, based on AVHRR local area coverage (LAC) data from August–October, 1993, adapted from Oreopoulos and Davies (1998)

11.5 Some Current Uses of Heterogeneity

11.5.1 Stereo Heights and Winds

Now that we have examined some of the more obvious negative consequences of cloud heterogeneity to remote sensing, let us turn our attention to some positive applications, wherein the presence of cloud heterogeneity is put to productive use to retrieve properties that would be unattainable from homogeneous clouds. These applications have only recently been implemented and depend more on geometrical and statistical analyses than on a detailed knowledge of 3DRT.

We use the MISR data products as examples. MISR measures the reflected radiance from a cloud at 275 m resolution from nine different directions as it passes over the cloud. If the cloud is heterogeneous, possessing detectable features that change horizontally, these create reflectivity patterns that can be matched from the different viewing directions to determine the parallax attributable to cloud height or wind displacement. As described by Moroney et al. (2002), this leads to the operational retrieval of cloud-top heights on a 1.1 km horizontal grid, with a nominal accuracy of 550 m.

Parallaxes obtained by two pairs of views, with one viewing direction in common between two pairs, can also be analyzed to separate the effect of height from that of wind displacement, leading to the novel retrieval of height-resolved cloud tracked winds. As shown by Horváth and Davies (2001), such winds can be retrieved to nominal accuracy of 3 m/s with a height resolution of 400 m over 70 km × 70 km domains, raising the prospect of making this an operational remote sensing technique in the future for numerical weather prediction application. Figure 11.6 shows an example of the MISR height-resolved winds.

11.5.2 Cloud Detection and Classification

The conventional classification of cloud types from space, approximately consistent with that of a surface observer, is exemplified by the approach of the International Satellite Cloud Climatology Program (Rossow and Schiffer, 1999). This approach classifies clouds by their cloud-top altitude (low, middle, and high) and apparent optical thickness (based on visible reflectivity), providing a time-honored division of clouds into about 9 major classes. This type of cloud classification yields no direct insight on cloud heterogeneity, other than the qualitative expectation that cumuliform clouds are likely to be more heterogeneous than stratiform clouds. The effects of heterogeneity on spatial and angular variability can be exploited, however, to provide additional quantitative information about certain types.

For example, the conventional detection of polar clouds over bright, cold surfaces is adversely affected by a lack of contrast in either reflectivity or brightness temperature. A measure of the anisotropy of visible reflectivity from three different viewing angles, expressed simply as a false color composite, using red for the oblique forward angle, green for nadir, and blue for the oblique backward angle, finds clouds that may be completely undetectable in single-angle true color imagery. Isotropic

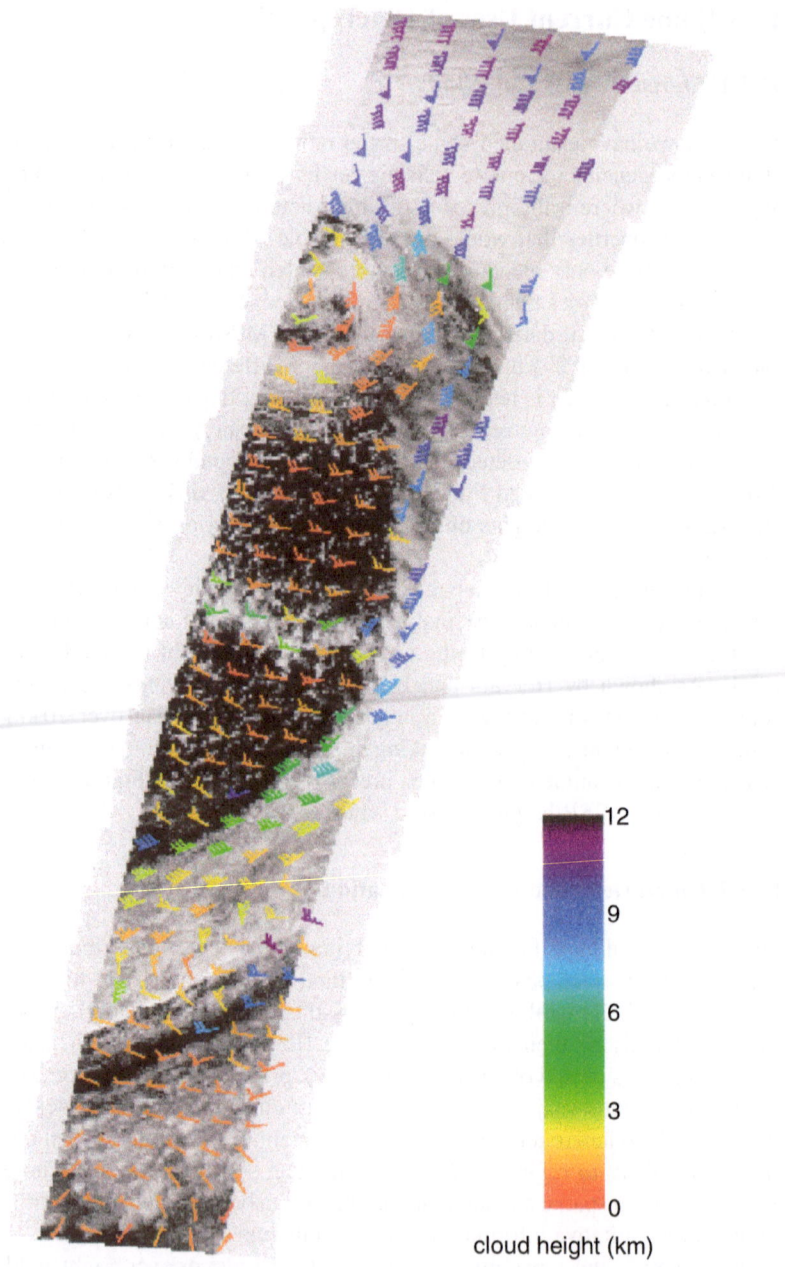

Fig. 11.6. Height-resolved cloud tracked winds (m/s) from MISR, adapted from Horváth and Davies (2001), from Terra orbit # 1900 on 26 April 2000, approx. 24°–52°N, 128°–142°W

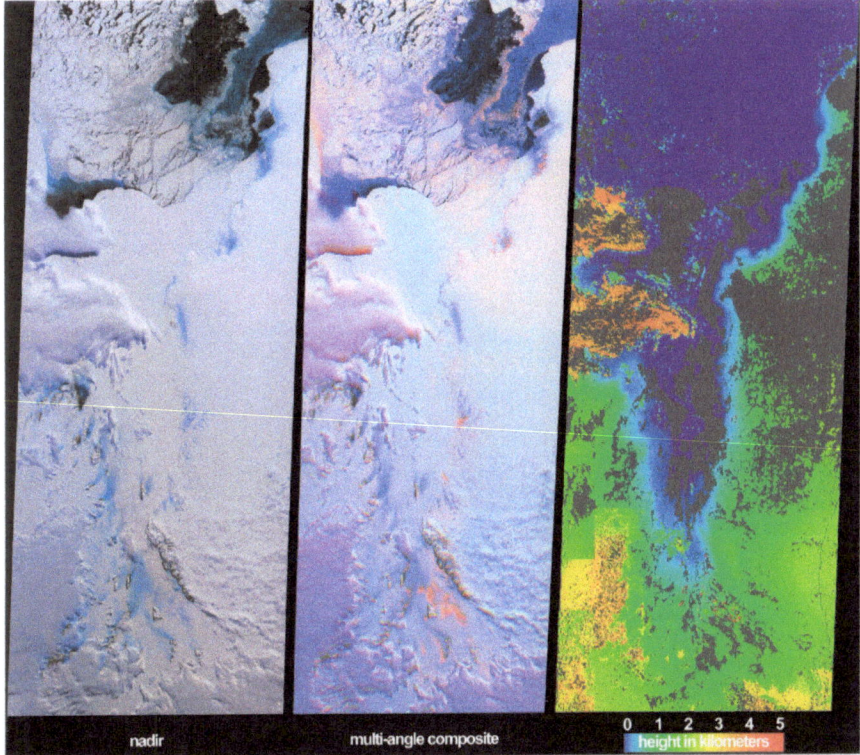

Fig. 11.7. Cloud detection using multi-angle views from MISR. The *left panel* is the natural color nadir view. Clouds show up in the false-color three-angle composite in the *middle panel*, and as indicated by the height color bar based on stereo retrievals in the *right hand panel*. The image is over the Amery Ice Shelf in East Antarctica, on October 25, 2002. Image credit: NASA's GSFC/LaRC/JPL MISR Team, http://www-misr.jpl.nasa.gov/gallery/

reflection appears white in such a scheme, characteristic of snow. Clouds have relatively less reflectivity in the nadir redirection, relatively more in oblique forward and back directions, resulting in a purple color, as shown in the example of Fig. 11.7 taken from the MISR image gallery.

A second example, Fig. 11.8, is the automatic determination of presence and fractional coverage of some types of marine stratocumulus using spatial coherence (Coakley and Bretherton, 1982). Here the spatial variability in brightness temperature is exploited by plotting a scatter diagram of the local variance of an 8×8 pixel array against its mean.

Clear regions have low variance and warm brightness temperature. Completely overcast regions similarly have low variance but distinctly different mean brightness temperature. The end points of the scatter diagram span the range from 0 to 100% cloud fraction. An absence of low variance data for intermediate mean values is required for the cloud field to be identified as a single layer system.

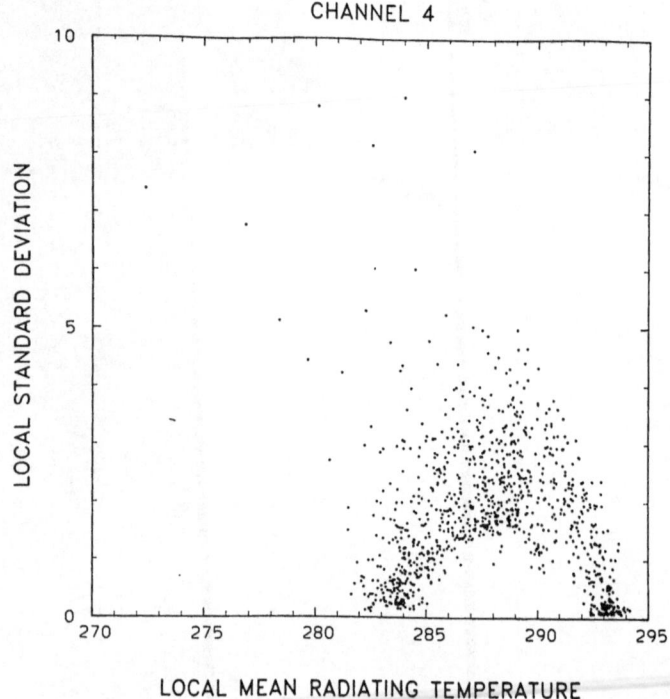

Fig. 11.8. The 11-micron local mean brightness temperature and local standard deviations for 8×8 arrays of global area coverage data, reproduced from Coakley and Bretherton (1982) with permission. The cluster of points near 293 K represents cloud-free scan spots; the cluster near 283.5 K represents cloud-cover scan plots. The points between these clusters represents partially covered fields of view

We note, however, that especially for marine stratocumulus, which cover large areas of the world's oceans, the spatial variability of clouds occurs on a variety of scales, and considerably more research is likely to be needed in order to implement an automatic classification scheme from space that usefully captures this variety. Proposed techniques for this include the use of Grey Level Difference Statistics (Chen et al., 1989) that measure the texture of the clouds, or direct cluster analysis to subdivide such clouds into their major subclasses.

11.6 Summary and Outlook

In summary, we note that cloud heterogeneity is a major factor that strongly affects the remote sensing of certain cloud properties. While 1DRT may be an appropriate tool for the small subset of global clouds that are sufficiently homogeneous to satisfy the plane-parallel paradigm, 3DRT is generally needed. This is especially relevant when retrieving properties that are most sensitive to the effects of heterogeneity, such as the cloud optical depth at non-absorbing wavelengths. Certain properties, such as cloud albedo, while still being sensitive to heterogeneity, can be estimated fairly well from empirically derived angular models without direct recourse to 3DRT. Similarly, geometrically based stereo retrievals of cloud heights or statistically based cloud classification techniques, while relying on cloud heterogeneity for their functionality, do not make direct use of 3DRT.

At the time of writing, direct application of 3DRT to the remote sensing of cloud properties remains substantially underdeveloped. A recent study by Zuidema et al. (2003) indicates one direction that future applications might take. They applied a 3DRT Monte Carlo model to the analysis of multi-angle radiances for a limited case study of heterogeneous clouds, and gleaned some insight regarding the vertical distribution of the volume extinction coefficient. It seems apparent that similar applications in the future must rely heavily on 3DRT as well as adequate geometrical descriptions of the cloud or cloud field.

References

Chen, N.C.L., S.K. Sengupta, and R.W. Welch (1989). Cloud field classification based upon high spatial resolution textural features 2. Simplified vector approaches. *J. Geophys. Res.*, **94**, 14,749–14,765.

Coakley, J.A. and F.P. Bretherton (1982). Cloud cover from high-resolution scanner data: Detecting and allowing for partially filled fields of view. *J. Geophys. Res.*, **87**, 4917–4932.

Davies, R. (1994). Spatial autocorrelation of radiation measured by the Earth Radiation Budget Experiment: Scene inhomogeneity and reciprocity violation. *J. Geophys. Res.*, **99**, 20,879–20,887.

Di Girolamo, L., T. Várnai, and R. Davies (1998). Apparent breakdown of reciprocity in reflected solar radiances. *J. Geophys. Res.*, **103**, 8795–8803.

Diner, D.J. and J.V. Martonchik (1984). Atmospheric transfer of radiation above an inhomogeneous non-lambertian reflective ground-II. Computational considerations and results. *J. Quant. Spectrosc. Radiat. Transfer*, **32**, 279–304.

Genkova, I. and R. Davies (2003). Spatial heterogeneity of reflected radiance from globally distributed clouds. *Geophys. Res. Lett.*, **30**, 2096–2099.

Harshvardhan, J.A. Weinman, and R. Davies (1981). Transport of infrared radiation in cuboidal clouds. *J. Atmos. Sci.*, **38**, 2500–2513.

Horváth, Á. and R. Davies (2001). Anisotropy of water cloud reflectance: a comparison of measurements and 1D theory. *Geophys. Res. Lett.*, **28**, 2915–2918.

Horváth, Á. and R. Davies (2004). Simultaneous retrieval of cloud motion and height from polar-orbiter multiangle measurements. *Geophys. Res. Lett.*, **31**, L01102, doi:10.1029/2003GL018386.

Loeb, N.G. and J.A. Coakley (1998). Inference of marine stratus cloud optical depths from satellite measurements: Does 1D theory apply? *J. Climate*, **11**, 215–233.

Loeb, N.G. and R. Davies (1996). Observational evidence of plane parallel model biases: The apparent dependence of cloud optical depth on solar zenith angle. *J. Geophys. Res.*, **101**, 1621–1634.

Loeb, N.G. and R. Davies (1997). Angular dependence of observed reflectances: A comparison with plane parallel theory. *J. Geophys. Res.*, **102**, 6865–6881.

Loeb, N.G., T. Várnai, and R. Davies (1997). Effect of cloud inhomogeneities on the solar zenith angle dependence of nadir reflectance. *J. Geophys. Res.*, **102**, 9387–9395.

Loeb, N.G., N. Manalo-Smith, S. Kato, W.F. Miller, S.K. Gupta, P. Minnis, and B.A. Wielicki (2003). Angular distribution models for top-of-atmosphere radiative flux estimation from the Clouds and the Earth's Radiation Energy System instrument on the Tropical Rainfall Measuring Mission Satellite. Part I: Methodology. *J. Appl. Meteor.*, **42**, 240–265.

Moroney, C., R. Davies, and J.-P. Muller (2002). Operational retrieval of cloud-top heights using MISR data. *IEEE Trans. Geosci. and Remote Sens.*, **40**, 1532–1540.

Oreopoulos, L. and R. Davies (1998). Plane parallel albedo biases from satellite observations. Part I: Dependence on resolution and other factors. *J. Climate*, **11**, 919–932.

Rossow, W.B. and R.A. Schiffer (1999). Advances in understanding clouds from ISCCP. *Bull. Amer. Meteor. Soc.*, **11**, 2261–2287.

Weinman, J.A. and R. Davies (1978). Thermal microwave radiances from horizontally finite clouds of hydrometeors. *J. Geophys. Res.*, **83**, 3099–3107.

Zuidema, P., R. Davies, and C. Moroney (2003). On the angular closure of tropical cumulus congestus clouds observed by the Multiangle Imaging Spectroradiometer. *J. Geophys. Res.*, **108**, 4626, doi:10.1029/2003JD003401.

Suggested Further Reading

The conventional approach to cloud remote sensing, without regard to cloud heterogeneity, is provided by:

Nakajima, T.Y. and M.D. King (1990). Determination of the optical thickness and effective particle radius of clouds from reflected solar radiation measurements – Part I, Theory. *J. Atmos. Sci.*, **47**, 1878–1893.

Nakajima, T.Y., M.D. King, J.D. Spinhirne and L. F. Radke (1991). Determination of the optical thickness and effective particle radius of clouds from reflected solar radiation measurements – Part II, Marine stratocumulus observations. *J. Atmos. Sci.*, **48**, 728–750.

Rossow, W.B. (1993). Clouds, In *Atlas of Satellite Observations Related to Global Change*. R.J. Gurney, J.L. Foster and C.L. Parkinson (eds.), Cambridge University Press, Cambridge (United Kingdom).

Stephens, G.L. (1994). *Remote Sensing of the Lower Atmosphere*. Oxford University Press, Oxford (United Kingdom).

Platnick, S., M.D. King, S.A. Ackerman, W.P. Menzel, B.A. Baum, J.C. Riedi and R.A. Frey (2003). The MODIS cloud products: Algorithms and examples from Terra. *IEEE Trans. Geosci. and Remote Sens.*, **41**, 459–473.

Additional recent references that address cloud heterogeneity effects on optical thickness retrievals include:

Faure, T., H. Isaka and B. Guillemet (2001). Neural network retrieval of cloud parameters of inhomogeneous and fractional clouds: Feasibility study. *Remote Sens. Environ.*, **77**, 123–138.

Várnai, T. and A. Marshak (2002). Observations of three-dimensional radiative effects that influence MODIS cloud optical thickness retrievals. *J. Atmos. Sci.*, **59**, 1607–1618.

Iwabuchi, H. and T. Hayasaka (2002). Effects of cloud horizontal inhomogeneity on the optical thickness retrieved from moderate-resolution satellite data. *J. Atmos. Sci.*, **59**, 2227–2242.

Davis, A.B. (2002). Cloud remote sensing with sideways-looks: Theory and first results using Multispectral Thermal Imager (MTI) data. *SPIE Proc.*, **4725**, 397–405.

Additional recent references related to 3D radiative effects in satellite observations include:

Oreopoulos, L., A. Marshak, R.F. Cahalan, and G. Wen (2000). Cloud three-dimensional effects evidenced in Landsat spatial power spectra and autocorrelation functions. *J. Geophys. Res.*, **105**, 14,777–14,788.

Kobayashi, T., K. Masuda, M. Sasaki and J.P. Mueller (2000). Monte Carlo simulations of enhanced visible radiance in clear-air satellite fields of view near clouds. *J. Geophys. Res.*, **105**, 26,569–26,576.

Cahalan, R.F., L. Oreopoulos, G. Wen, A. Marshak, S.-C. Tsay and T. DeFelice (2001). Cloud characterization and clear-sky correction from Landsat-7. *Remote Sens. Environ.*, **78**, 83–98.

Wen, G., R.F. Cahalan, S.-C. Tsay and L. Oreopoulos (2001). Imapct of cumulus cloud spacing on Landsat atmospheric correction and aerosol retrieval. *J. Geophys. Res.*, **106**, 12,129–12,138.

Buriez, J.-C., M. Doutriaux-Boucher, F. Parol and N.G. Loeb (2001). Angular variability of the liquid water cloud optical thickness retrieved from ADEOS-POLDER. *J. Atmos. Sci.*, **58**, 3007–3018.

Chýlek, P., C. Borel, A.B. Davis, S. Bender, J. Augustine and G. Hodges (2004). Effect of broken clouds on satellite based columnar water vapor retrieval. *IEEE Geosci. and Remote Sens. Lett.*, **1**, 175–179.

Additional references on the plane-parallel albedo bias include:

Kobayashi, T. (1993). Effects due to cloud geometry on biases in the cloud albedo derived from radiance measurements. *J. Climate*, **6**, 120–128.

Cahalan, R.F., D. Silberstein and J.B. Snider (1995). Liquid water path and plane-parallel albedo bias during ASTEX. *J. Atmos. Sci.*, **52**, 3002–3012.

Marshak, A., A. Davis, R.F. Cahalan and W.J. Wiscombe (1998). Nonlocal Independent Pixel Approximation: Direct and Inverse Problems. *IEEE Trans. Geosc. and Remote Sens.*, **36**, 192–205.

Wood, R. and J.P. Taylor (2001). Liquid water path variability in unbroken marine stratocumulus cloud. *Quart. J. Roy. Meteor. Soc.*, **127**, 2635–2662.

Kokhanovsky, A.A. (2003). The influence of horizontal inhomogeneity on radiative characteristics of clouds: An asymptotic case study. *IEEE Trans. Geosci. and Remote Sens.*, **41**, 817–825.

Additional references related to stereo retrievals of cloud height include:

Prata, A.J. and P.J. Turner (1997). Cloud-top height determination using ATSR data. *Remote Sens. Environ.*, **59**, 1–13.

Naud, C., J.P. Muller and E.E. Clothiaux (2002). Comparison of cloud top heights derived from MISR stereo and MODIS CO_2-slicing. *Geophys. Res. Lett.*, **29**, 42–45.

Wang, H.Q., S.H. Lu, Y. Zhang and Z.Y. Tao (2002). Determination of cloud-top height from stereoscopic observation. *Prog. in Natural Sci.*, **12**, 689–694.

Lancaster, R.S., J.D. Spinhirne and K.F. Manizade (2003). Combined infrared stereo and laser ranging cloud measurements from Shuttle mission STS-85. *J. Atmos. Ocean. Tech.*, **20**, 67–78.

Naud, C., J.P. Muller, M. Haeffelin, Y. Morille and A. Delaval (2002). Assessment of MISR and MODIS cloud top heights through intercomparison with a back-scattering lidar at SIRTA. *Geophys. Res. Lett.*, **31**, 4114–4117.

Additional references related to cloud classification based on spatial features include:

Lee, J., R.C. Weger, S.K. Sengupta and R.M. Welch (1990). A neural network approach to cloud classification. *IEEE Trans. Geosci. and Remote Sens.*, **28**, 846–855.

Lewis, H.G., S. Cote and A.R.L. Tatnall (1997). Determination of spatial and temporal characteristics as an aid to neural network cloud classification. *Int. J. Remote Sens.*, **18**, 899–915.

Saitwal, K., M.R. Azimi-Sadjadi and D. Reinke (2003). A multichannel temporally adaptive system for continuous cloud classification from satellite imagery. *IEEE Trans. Geosci. and Remote Sens.*, **41**, 1098–1104.

Christodoulou, C.I., S.C. Michaelides and C.S. Pattichis (2003). Multifeature texture analysis for the classification of clouds in satellite imagery *IEEE Trans. Geosci. and Remote Sens.*, **41**, 2662–2668.

Additional references that address multi-scale phenomena in cloud remote sensing include:

Cahalan, R.F., W. Ridgway, W.J. Wiscombe, T. L. Bell and J.B. Snider (1994). The albedo of fractal stratocumulus clouds. *J. Atmos. Sci.*, **51**, 2434–2455.

Marshak, A., A. Davis, W.J. Wiscombe and G. Titov (1995). The verisimilitude of the independent pixel approximation used in cloud remote sensing. *Remote Sens. Environ.*, **52**, 72–78.

Marshak, A., A. Davis, W.J. Wiscombe and R.F. Cahalan (1995). Radiative smoothing in fractal clouds. *J. Geophys. Res.*, **100**, 26247–26261.

Davis, A., A. Marshak, R.F. Cahalan and W.J. Wiscombe (1997). The LANDSAT scale-break in stratocumulus as a three-dimensional radiative transfer effect, Implications for cloud remote sensing. *J. Atmos. Sci.*, **54**, 241–260.

Cornet, C., H. Isaka, B. Guillemet and F. Szczap (2004). Neural network retrieval of cloud parameters of inhomogeneous clouds from multispectral and multiscale radiance data: Feasibility study. *J. Geophys. Res.*, **109**, D12203, doi:10.1029/2003JD004186.

12

Horizontal Fluxes and Radiative Smoothing

A. Marshak and A.B. Davis

12.1 Introduction 543

12.2 Horizontal Fluxes 546

12.3 Horizontal Photon Transport: Diffusion-Based Predictions 555

12.4 Radiative Smoothing and Roughening in Simulations 564

12.5 Radiative Smoothing and Roughening in Data 572

12.6 Nonlocal Independent Pixel Approximation (NIPA) 576

12.7 Summary 581

References 583

Suggestions for Further Reading 585

12.1 Introduction

We will start this chapter with an intricate image of marine stratocumulus (Sc) clouds captured by LandSat 5 during First International Satellite Cloud Climatology Project (ISCCP) Regional Experiment (FIRE) field program (Fig. 12.1). This is a $60 \times 60\,\mathrm{km}^2$ completely cloudy subscene over the eastern Pacific Ocean illuminated by the Sun with a zenith angle of about $30°$. The first thing our eyes pick up in the image is the large scale structure or so-called cloud "streets" oriented parallel to the wind. The width of the streets is typically 8 km. "The 8 km streets reveal considerable structure on smaller scales, the whole scene being covered by an intricate dark filigree," Cahalan and Snider wrote in their 1989 paper where they first analyzed statistically the structure of this scene. Plotting a wavenumber spectrum averaged over 10 scan lines of the scene vs. wavenumber, they uncovered a scale-invariant behavior for large scales (a straight line on a log-log plot) and a smoother structure (steeper slope) for small (less than 200 m) scales (Fig. 12.2). Comparing

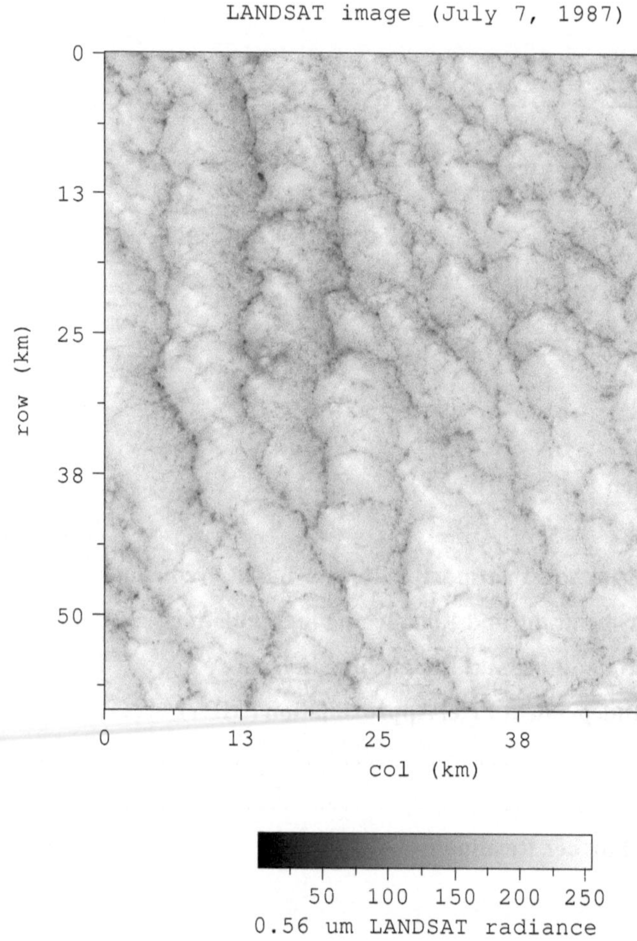

LANDSAT image (July 7, 1987)

Fig. 12.1. Portion of a Thematic Mapper (TM) LandSat image in channel 2 (0.52–0.69 μm). This 61 × 61 km² scene was captured at 28.5 m resolution on July 7, 1987, off the coast of California. This is a small portion of an extensive marine Sc deck. Note that the cloud "streets" are oriented parallel to the mean wind. Solar zenith angle is about 30°

the large-scale structure of LandSat radiances with the structure of cloud liquid water measured by ground-based microwave radiometer, they found them to be very similar; this validates the retrieval of cloud structure from high-resolution satellite measurements at these scales. However for small scales, fluctuations of cloud liquid water are much stronger than those for LandSat radiance; this prohibits small-scale retrievals using 1D (pixel-by-pixel) methods. Indeed, as long as fluctuations of cloud liquid water and LandSat radiance are statistically different at a given scale, it is impossible to reliably retrieve liquid water at this scale.

The transition to smoother behavior at small scales for horizontally inhomogeneous clouds and its implementation in remote sensing is the subject of this chapter.

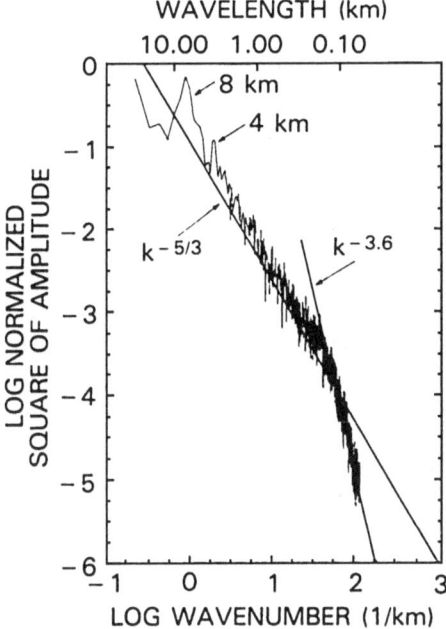

Fig. 12.2. Estimated wavenumber spectrum of the cloud reflectivity field in Fig. 12.1. The upper horizontal axis shows wavelength (or scale) r which is related to wavenumber k (the lower horizontal axis) as $k = 2\pi/r$. The peaks at $r = 4$ and 8 km correspond to cloud streets visible in Fig. 12.1. Two straight lines correspond to slopes (spectral exponents) of $-5/3$ (large scales) and -3.6 (small scales), respectively. Reprinted from Cahalan and Snider (1989) with permission from Elsevier

We will study the physical mechanism that is responsible for the transition and how it relates to cloud structure. We start with photon horizontal transport, determine horizontal fluxes and show how nonvanishing net horizontal fluxes effect the accuracy of the so called Independent Pixel Approximation or simply "IPA" (Cahalan et al., 1994) separately for reflected, transmitted or absorbed photons (Sect. 12.2). Then, based on the diffusion approximation, we derive the average distance that photons travel horizontally (Sect. 12.3). Though the derivation is done in a plane-parallel geometry, we discuss its generalization to inhomogeneous clouds. Next we define radiative smoothing as a radiative transfer process that smoothes the small-scale fluctuations of cloud structure and illustrate its signature as a scale-break on a log-log plot of wavenumber spectra or structure functions (Sects. 12.4–12.5). The scale break location is related to the distance reflected or transmitted photons travel horizontally. Finally, we use our new understanding of radiative smoothing to improve the IPA accounting for photon horizontal transport (Sect. 12.6).

12.2 Horizontal Fluxes

12.2.1 Energy Balance Equation

To determine photon horizontal transport, we start with the radiative transfer equation (see Sect. 3.7)

$$\mathbf{\Omega} \cdot \nabla I = -\sigma(\mathbf{x})I(\mathbf{x}, \mathbf{\Omega}) + \sigma(\mathbf{x})\varpi_0 \int_{4\pi} p(\mathbf{\Omega'} \to \mathbf{\Omega})I(\mathbf{x}, \mathbf{\Omega'})\mathrm{d}\mathbf{\Omega'} \qquad (12.1)$$

where $I(\mathbf{x}, \mathbf{\Omega})$ is radiance at position $\mathbf{x} = (x, y, z)^{\mathrm{T}}$ in the cloudy medium confined between $z = 0$ and $z = h$. Propagation direction is denoted by $\mathbf{\Omega} = (\Omega_x, \Omega_y, \Omega_z)^{\mathrm{T}}$; $p(\mathbf{\Omega'} \to \mathbf{\Omega})$ is the scattering phase function; $\sigma(\mathbf{x})$ is the extinction coefficient; and ϖ_0 is the single-scattering albedo. For simplicity we assume no reflection from surface. Let us first integrate (12.1) term-by-term with respect to $\mathbf{\Omega}$ (over 4π). Invoking normalization of phase function introduced in Sect. 3.4, we obtain

$$\nabla \cdot \mathbf{F}(\mathbf{x}) = -\sigma_a(\mathbf{x})J(\mathbf{x}) \qquad (12.2)$$

where the (net) photon flux vector is

$$\mathbf{F}(\mathbf{x}) = (F_x, F_y, F_z)^{\mathrm{T}} = \int_{4\pi} \mathbf{\Omega}I(\mathbf{x}, \mathbf{\Omega})\mathrm{d}\mathbf{\Omega} , \qquad (12.3)$$

the scalar flux is

$$J(\mathbf{x}) = \int_{4\pi} I(\mathbf{x}, \mathbf{\Omega})\mathrm{d}\mathbf{\Omega} , \qquad (12.4)$$

and $\sigma_a(\mathbf{x}) = (1 - \varpi_0)\sigma(\mathbf{x})$ is the absorption coefficient. We assume that a uniform solar flux

$$\mu_0 F_0 = \int_0^{2\pi} \int_{\mu<0} |\mu| \, I(x, y, h, \mu, \phi)\mathrm{d}\mu\mathrm{d}\phi \qquad (12.5)$$

is incident normally to the cloud top at $z = h$. Integrating (12.2) over the parallelepiped $C(r; x, y) = \{(x', y', z')^{\mathrm{T}} \in \mathbb{R}^3 : x \le x' \le x+r, \, y \le y' \le y+r, \, 0 \le z' \le h\}$, we get

$$\int_0^h \int_y^{y+r} \int_x^{x+r} \left(\frac{\partial F_x}{\partial x} + \frac{\partial F_y}{\partial y} + \frac{\partial F_z}{\partial z} \right) \mathrm{d}x'\mathrm{d}y'\mathrm{d}z'$$

$$= \int_0^h \int_y^{y+r} \int_x^{x+r} \sigma_a(x', y', z')J(x', y', z')\mathrm{d}x'\mathrm{d}y'\mathrm{d}z' . \qquad (12.6)$$

Using the divergence theorem (e.g., Korn and Korn, 1968) to replace the volume integral by a surface integral on the left side of (12.6) and dividing both sides by the total incoming radiation equal to $r^2\mu_0 F_0$, we obtain

$$[R(r; x, y) - 1] + [T(r; x, y) - 0] + H(r; x, y) = -A(r; x, y) . \tag{12.7}$$

Here

$$R(r; x, y) = \frac{1}{r^2 \mu_0 F_0} \int\limits_y^{y+r} \int\limits_x^{x+r} \int\limits_0^{2\pi} \int\limits_{\mu \geq 0} \mu I(x', y', h, \mu, \phi) \, d\mu \, d\phi \, dx' \, dy' \tag{12.8}$$

is the local reflectance (or albedo) and

$$T(r; x, y) = \frac{1}{r^2 \mu_0 F_0} \int\limits_y^{y+r} \int\limits_x^{x+r} \int\limits_0^{2\pi} \int\limits_{\mu \leq 0} |\mu| \, I(x', y', h, \mu, \phi) \, d\mu \, d\phi \, dx' \, dy' \tag{12.9}$$

is the local transmittance; further,

$$A(r; x, y) = \frac{1}{r^2 \mu_0 F_0} \int\limits_0^h \int\limits_y^{y+r} \int\limits_x^{x+r} \sigma_a(x', y', z') J(x', y', z') \, dx' \, dy' \, dz' \tag{12.10}$$

and

$$H(r; x, y) = \frac{1}{r^2 \mu_0 F_0} \int\limits_0^h \left\{ \int\limits_x^{x+r} [F_y(x', y' + r, z') - F_y(x', y', z')] dx' \right.$$
$$\left. + \int\limits_y^{y+r} [F_x(x' + r, y', z') - F_x(x', y', z')] dy' \right\} dz' \tag{12.11}$$

where F_x and F_y are the *net* fluxes in the x and y directions respectively from (12.3); e.g.,

$$F_x(x^*, y, z) = F_+(x^*, y, z) - F_-(x^*, y, z) \tag{12.12}$$

with $x^* = x$ or $x + r$ and F_+ (or F_-) is the flux through the pixel lateral side to positive (or negative) direction as can be seen by partitioning the angular integral in (12.3) into two hemispheres. We will call $H(r; x, y)$ the "horizontal flux term" although technically it is *vertically-integrated horizontal flux divergence*.

Rearranging the terms in (12.7) for easier interpretation, we obtain

$$H(r; x, y) = 1 - [R(r; x, y) + T(r; x, y) + A(r; x, y)] . \tag{12.13}$$

Equation (12.13) is a local energy balance for the parallelepiped $C(r; x, y)$. Indeed, it accounts for photons that are reflected from the top (R), transmitted through the bottom (T), absorbed (A) within the confines of C, as well as for those photons that left or entered C through the sides of the column (H). It is obvious that, if we integrate (12.2) over the whole domain, H vanishes; i.e.,

$$\lim_{r \to \infty} H(r; x, y) \equiv \langle H \rangle = 0 \tag{12.14a}$$

and

$$\langle R \rangle + \langle T \rangle + \langle A \rangle = 1 \qquad (12.14b)$$

where $\langle \cdot \rangle$ means domain-averaged. For computational purposes, using cyclical boundary conditions with the size of computational domain equal to $L_D \times L_D$, we also obtain $H(L_D; x, y) \equiv 0$, hence

$$R(L_D; x, y) + T(L_D; x, y) + A(L_D; x, y) = 1 . \qquad (12.14c)$$

It also follows from the energy balance (12.13) that the horizontal flux term can be determined as a difference between "true" column absorption $A(r; x, y)$ and its "apparent" counterpart, $1 - [R(r; x, y) + T(r; x, y)]$. For measuring column absorption of a cloud, or the whole atmosphere, only $R(r; x, y)$ and $T(r; x, y)$ can be estimated from data using radiometers above and below the system. It is clearly important to find the scale r at which the contribution of $H(r; x, y)$ to the radiation budget can justifiably be neglected; thus measurable apparent absorption estimates unknown true absorption (Marshak et al., 1997; Titov, 1998).

To illustrate the magnitude of the net horizontal fluxes excited by horizontal inhomogeneity, we set up the following numerical experiment. A homogeneous slab of 300 m depth is divided into two parts: one is set to the optical thickness 30 and the other to 5. A backward Monte Carlo (Chap. 4) was used to simulate separately vertical and horizontal fluxes at different levels. Sun was in zenith to eliminate any effect of geometrical shadowing. Figure 12.3 illustrates the results "measured" at 15 m from the discontinuity, in the more tenuous region. For the homogeneous case ($\tau_{\text{thin}} = \tau_{\text{thick}} = 5$), the net vertical flux (the straight dash line at 0.775) is computed using 1D radiative transfer. The net horizontal flux is of course directed towards the thinner region; its maximum is reached about 50 m from cloud top which is close to one transport mean-free-path in the dense region $h/(1 - g)\tau_{\text{thick}} \approx 67$ m. Near cloud bottom, the effect of horizontal fluxes is small. The most interesting fact is that the maximum value of net horizontal fluxes is about 30% of their vertical counterpart for the gradient between thick and thin regions of 25 which is not a rare event in horizontal variations in optical depth of real marine Sc (see Chap. 2).

Finally, to illustrate horizontal fluxes in a more realistic cloud simulated by a bounded cascade model (see volume Appendix) in Fig. 12.4 we plotted R, T and H defined by (12.8), (12.9) and (12.11) respectively at the pixel scale of 12.5 m, assuming that the extinction field varies only in the x-direction, i.e., $\sigma(\mathbf{x}) \equiv \sigma(x)$. In this case,

$$H(r; x) = 1 - [R(r; x) + T(r; x) + A(r; x)] . \qquad (12.15)$$

For simplicity, clouds are assumed to be purely scattering ($\varpi_0 = 1$), thus $A(r; x) \equiv 0.0$ and $H = 1 - (R + T)$. We see that, even for this mildly inhomogeneous cloud, the sum of albedo and transmittance on a pixel-by-pixel basis can be different from unity by up to 15%.

12.2.2 Accuracy of IPA on a Per-Pixel Basis

We will relate the horizontal fluxes $H(r; x, y)$ to pixel-by-pixel accuracy of the IPA which treats each pixel as an independent plane-parallel medium, neglecting any *net*

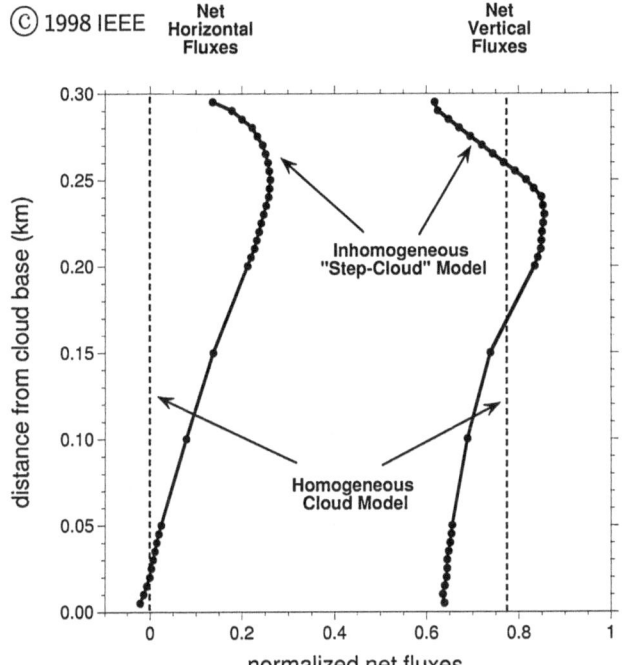

Fig. 12.3. Net horizontal (F_x) and vertical (F_z) fluxes. A homogeneous cloud layer 300 m thick was divided into two regions with optical thicknesses 30 and 5. Point-wise fluxes were calculated using backward Monte Carlo at 15 m from the discontinuity in extinction. Sun was in zenith and a Henyey–Greenstein phase function with $g = 0.85$ was used. Net fluxes are defined as the differences between down and up fluxes (net vertical, F_z), and right and left fluxes (net horizontal, F_x). The straight line at 0.775 is computed using the 1D DISORT code (Stamnes et al., 1988) and corresponds to the constant net vertical flux in a homogeneous medium with $\tau = 5$. Note that, near cloud top, net horizontal flux F_x exceeds 30% of net vertical flux F_z. Obviously, for homogeneous clouds and overhead Sun, $F_x = 0$. Reproduced from Marshak et al. (1998) with permission from IEEE

horizontal photon transport, i.e., $H_{\mathrm{IPA}}(r; x, y) \equiv 0$. If we replace unity in (12.13) by the sum $R_{\mathrm{IPA}}(r; x, y) + T_{\mathrm{IPA}}(r; x, y) + A_{\mathrm{IPA}}(r; x, y)$, we get (Marshak et al., 1999),

$$
\begin{aligned}
H(r; x, y) &= R_{\mathrm{IPA}}(r; x, y) - R(r; x, y) + T_{\mathrm{IPA}}(r; x, y) - T(r; x, y) \\
&\quad + A_{\mathrm{IPA}}(r; x, y) - A(r; x, y) \\
&= H_R(r; x, y) + H_T(r; x, y) + H_A(r; x, y) .
\end{aligned} \tag{12.16}
$$

Each component in (12.16) is a pixel-by-pixel IPA accuracy,

$$
H_F(r; x, y) = F_{\mathrm{IPA}}(r; x, y) - F(r; x, y), \; F = R, T \text{ and } A . \tag{12.17}
$$

We will call H_R, H_T and H_A horizontal fluxes for photons reflected from cloud top, transmitted to cloud base, or absorbed by cloud column, respectively.

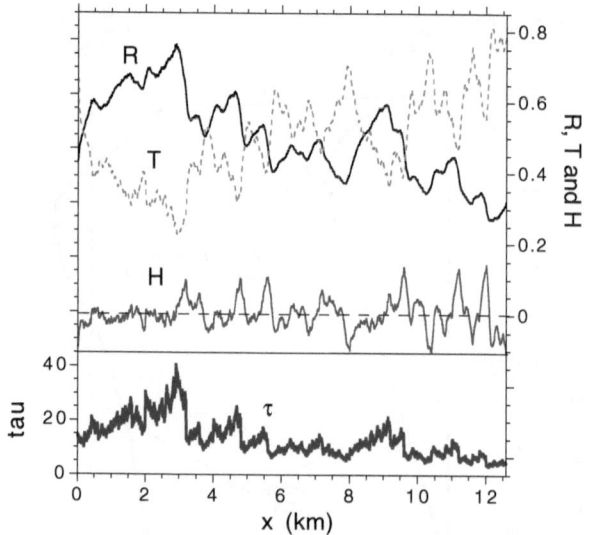

Fig. 12.4. Albedo, transmittance and horizontal flux. Monte Carlo reflectance $R(r; x)$ and transmittance $T(r; x)$ fields are plotted vs. position x for $r = 12.5$ m (pixel-scale); also plotted are the horizontal flux term $H(r; x) = 1 - R(r; x) - T(r; x)$ and optical depth $\tau(x)$. The 1D cloud model with mean optical depth $\langle \tau \rangle = 13$ was variable only in the x-direction. We used solar zenith angle 22.5°; other parameters are: physical thickness $h = 300$ m, Henyey–Greenstein phase function with $g = 0.85$ and $\varpi_0 = 1$

To measure the magnitude of horizontal fluxes, we will use the norm,

$$\|H_F(r)\| = \frac{1}{L_D^2} \left[\int_0^{L_D} \int_0^{L_D} [H_F(r; x, y)]^2 dx dy \right]^{\frac{1}{2}}. \tag{12.18}$$

Equation (12.16) and the norm definition (12.18) yield

$$\|H_F\| \leq \|H_R\| + \|H_T\| + \|H_A\|. \tag{12.19}$$

Note that IPA accuracy H_F are the average pixel-by-pixel absolute differences between the full 3D calculations of reflectance (or transmittance, or absorptance), through the solutions of (12.1), and 1D computations of the same quantities independently performed at each pixel. We will study here the dependence of each component in (12.16) on single-scattering albedo, ϖ_0, examining whether or not the magnitude of horizontal fluxes is related to IPA accuracy.

It is natural to expect that, as r increases, the IPA becomes more accurate and horizontal fluxes $\|H_F(r)\|$ get smaller. Figures 12.5a, b, and c show that this is true for H_R, H_T and H_A, i.e.,

$$\|H_F\| \to 0, r \to L_D, F = R, T, \quad \text{and } A. \tag{12.20}$$

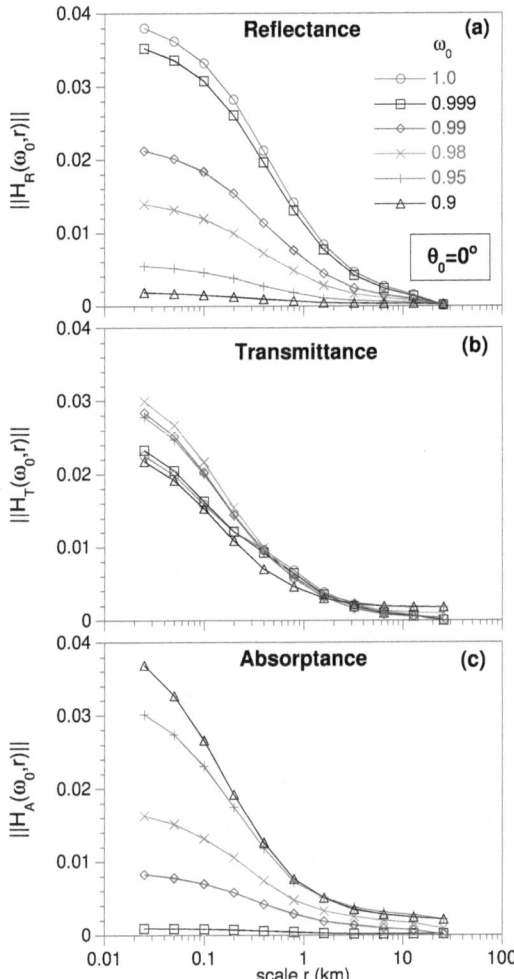

Fig. 12.5. Dependence of pixel-by-pixel IPA accuracy, $\|H_F\|(F = R, T \text{ or } A)$, on averaging scale r and single-scattering albedo ϖ_0, from Marshak et al. (1999). (a) Reflectance, $\|H_R\|$; (b) Transmittance, $\|H_T\|$; (c) Absorptance, $\|H_A\|$. Horizontal distribution of cloud optical depth is simulated with 10-step bounded cascades models (cf. Appendix) where $p = 0.3$, $H = 1/3$, $\langle \tau \rangle = 13$, pixel size $r = 25$ m. Flat cloud top and cloud base, geometrical thickness $h = 300$ m. Henyey–Greenstein phase functions with asymmetry parameter $g = 0.85$ is used. Surface is absorbing. The results are averaged over 10 independent realizations. Solar zenith angle $\theta_0 = 0°$, single-scattering albedo $\varpi_0 = 1.0, 0.999, 0.99, 0.98, 0.95, 0.9$.

It follows from inequality (12.19) that (12.20) is valid for H as well. The effect of single-scattering albedo ϖ_0 on horizontal fluxes is, however, different for reflected, transmitted and absorbed photons.

Figure 12.5a illustrates the dependence of H_R on both r and ϖ_0. We see that the more absorption the shorter photon horizontal transport for reflected photons.

As a result, the IPA pixel-by-pixel accuracy for reflectance improves with more absorption. Note that, for $\varpi_0 = 0.9$, IPA is almost accurate even on a per-pixel basis. This is expected since the contribution of multiple scattering to the reflectance field decreases and fewer photons travel between pixels.

The situation with transmitted photons (at least for high Sun) is surprisingly different (Fig. 12.5b): the accuracy of the IPA on a per-pixel basis first decreases (from $\varpi_0 = 1.0$ down to 0.98) and only for strongly absorbing wavelengths ($\varpi_0 > 0.98$) does it increase again reaching the accuracy level of the conservative scattering in case of $\varpi_0 = 0.9$. To explain this, we go back to a distribution of homogeneous clouds and calculate the standard deviation

$$
s_F = \left\{ \int_{\tau_{min}}^{\tau_{max}} [F(\tau)]^2 p(\tau) d\tau - \left[\int_{\tau_{min}}^{\tau_{max}} F(\tau) p(\tau) d\tau \right]^2 \right\}^{\frac{1}{2}}
\tag{12.21}
$$

for reflectance ($F = R$) and transmittance ($F = T$). In (12.21), $p(\tau)$ is the probability density of the optical depth distribution. As shown in Fig. 12.6a, for a log-normal type of $p(\tau)$, the range of reflectance, $R(\tau_{max}) - R(\tau_{min})$, sharply decreases with more absorption. At the same time, the range of transmittance, $T(\tau_{max}) - T(\tau_{min})$, increases, at least for weakly absorbing wavelengths due to finite sampling or truncation. As a result (see Fig. 12.6b), standard deviation s_T first increases (down to $\varpi_0 = 0.98$) and then decreases for strongly absorbing wavelengths, while s_R decreases monotonically. These results are almost independent of $p(\tau)$, unless it has an unrealistically long tail. Moreover, similar trends are found for all solar zenith angles within the 2-stream approximation (e.g. Meador and Weaver, 1980), even for a uniform (but truncated) distribution of τ. Note that the behavior of s_R and s_T is similar to what we see in Fig. 12.5a and b for the IPA accuracies $\|H_R\|$ and $\|H_T\|$, respectively, at the smallest scales. Indeed, Fig. 12.6c illustrates a surprisingly good linear correlation between s_F and $\|H_F\|$ for both $F = R$ and $F = T$. This completes the explanation of both Figs. 12.5a and b.

Finally, the increase of pixel-by-pixel IPA absorption errors, $\|H_A\|$, with more absorption (Fig. 12.5c) is easily understandable; it follows directly from both the natural increase of $A_{\mathrm{IPA}}(\tau)$ itself and its standard deviation with more absorption. Besides that, the magnitude of $\|H_A\|$ monotonically increases with stronger cloud variability and more oblique illumination for any $\varpi_0 < 1$.

Figures 12.7a and b illustrate the joint effect of all horizontal fluxes for both high ($\theta_0 = 0°$) and low ($\theta_0 = 60°$) Sun. The general tendencies of horizontal fluxes are similar for both solar angles (with more absorption $\|H_A\|$ sharply increases, $\|H_R\|$ monotonically decreases, and $\|H_T\|$ slowly decreases for $\theta_0 = 60°$ and first increases and then decreases for $\theta_0 = 0°$). However, the vertically integrated horizontal fluxes $\|H\|$ are much closer to the sum $\|H_A\| + \|H_T\| + \|H_R\|$ in case of slant illumination (Fig. 12.7b) than in case of high Sun (Fig. 12.7a).

To interpret this, note that for high Sun horizontal fluxes for reflected and transmitted photons are mostly anticorrelated (see Fig. 12.4), while for low Sun they are mostly correlated. This is a direct consequence of radiative "channeling" around

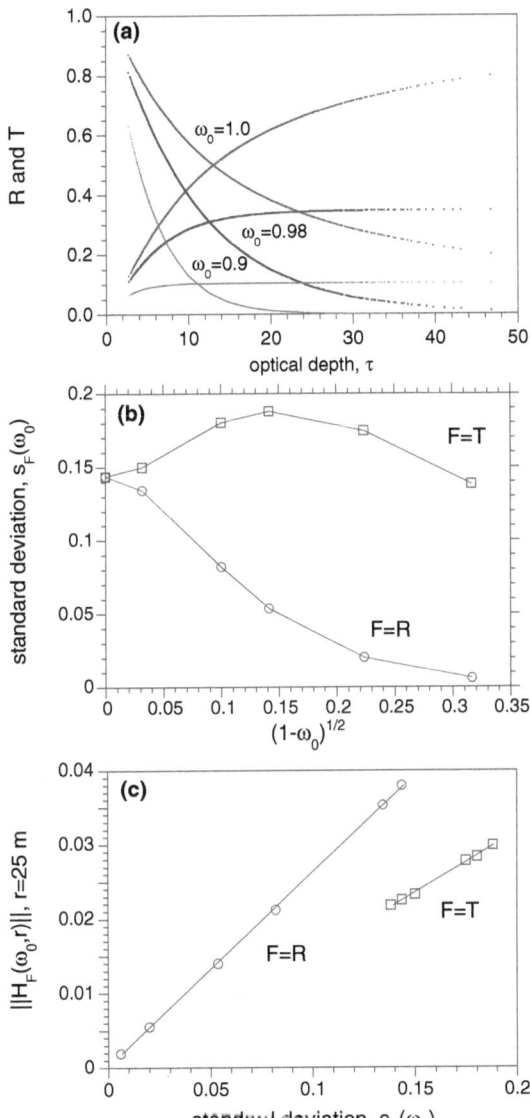

Fig. 12.6. Standard deviation s_F and pixel-by-pixel IPA accuracy for reflectance and transmittance, from Marshak et al. (1999). Illumination and scattering conditions are the same as in Fig. 12.5. **(a)** 1D reflectances (increasing curves) and transmittance (decreasing curves) calculated using DISORT for $\varpi_0 = 1.0, 0.98,$ and 0.9. Distribution of optical depth, $p(\tau)$, is defined by a bounded cascade model and is close to log-normal. The range of τ is from 3 to 47. **(b)** Standard deviations s_R and s_T defined in (12.21) versus $\sqrt{1-\varpi_0}$ for $\varpi_0 = 1.0, 0.999, 0.99, 0.98, 0.95,$ and 0.9. **(c)** Standard deviations s_R and s_T versus IPA accuracies $\|H_R\|$ and $\|H_T\|$ at the pixel scale $r = 25\,\mathrm{m}$

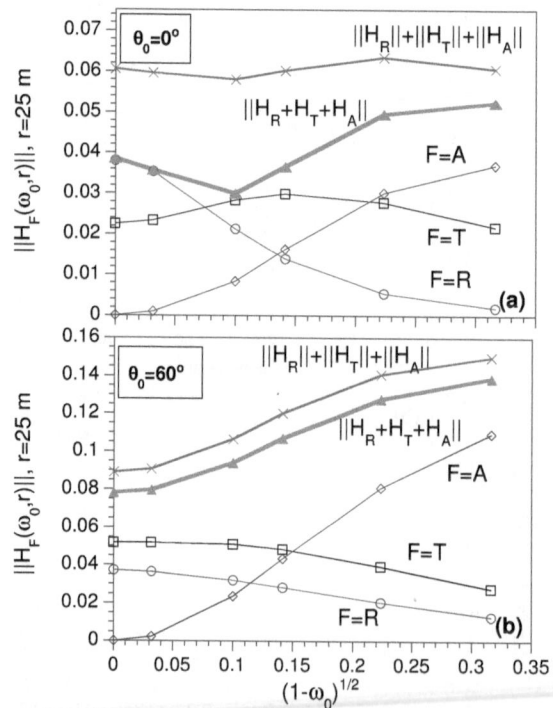

Fig. 12.7. Pixel-by-pixel IPA accuracy for reflectance, transmittance, absorptance, their sum, and the total horizontal fluxes H, from Marshak et al. (1999). Cloud model and scattering conditions are as in Fig. 12.5. (**a**) $\theta_0 = 0°$; (**b**) $\theta_0 = 60°$. Note the two-fold difference in vertical scale going from panels (a) to (b)

the dense regions into the tenuous ones (Davis and Marshak, 2001). Indeed, while reflected photons travel from optically thick (dense) to optically thin (tenuous) regions, H_R is positive for thicker pixels and negative for their thinner counterparts. In contrast, horizontal fluxes for transmitted photons, H_T, have the opposite tendency. As a result, for the majority of pixels, H_R and H_T have opposite signs if $\theta_0 = 0°$ and the same sign if $\theta_0 = 60°$. Thus

$$\|H\| \equiv \|H_R + H_T + H_A\| \ll \|H_R\| + \|H_T\| + \|H_A\|, \quad \theta_0 = 0° ; \quad (12.22a)$$
$$\|H\| \equiv \|H_R + H_T + H_A\| \lesssim \|H_R\| + \|H_T\| + \|H_A\|, \quad \theta_0 = 60° , \quad (12.22b)$$

as we see in Figs. 12.7a and b. (Note that "$\ll$" is used only to emphasize the contrast between high and low Sun.)

Finally and most importantly, the increase in vertically integrated horizontal fluxes $\|H\|$ with the increase of absorption is entirely attributable to the increase of $\|H_A\|$. In the case of slant illumination, this is true for all single-scattering albedos and, in case of high Sun, only for strongly absorbing wavelengths. To conclude, horizontal flux H defined in (12.13) is not directly related to IPA accuracy for either reflectance or transmittance; for strongly absorbing wavelengths, the IPA errors

H_R and H_T can be sufficiently small but nevertheless horizontal fluxes H are large because of the IPA absorptance error H_A.

12.3 Horizontal Photon Transport: Diffusion-Based Predictions

12.3.1 Average Number of Scatterings

Using time-dependent radiative transfer equation in homogeneous medium and 3D diffusion enables us to obtain an analytical estimate of the mean photon pathlength from injection to escape (Davis and Marshak, 2001). It follows from there that the average number of scatterings suffered by a photon in an optically thick homogeneous medium, from injection to escape, can be estimated as follows,

$$\langle n \rangle \approx \frac{\sigma \int_0^h J(z)dz}{\int_0^h S(z)dz} \tag{12.23}$$

where $J(z)$ is a scalar flux defined in (12.4) and $S(z)$ is the diffusion source term. In the diffusion limit, $J(z)$ satisfies the following boundary-value problem (Case and Zweifel, 1967):

$$\begin{cases} -\frac{\ell_t}{3}\frac{d^2 J}{dz^2} + (1 - \varpi_0)\sigma J = S(z), 0 < z < h \\ \left[J - \chi\ell_t\frac{dJ}{dz}\right]_{z=0} = \left[J + \chi\ell_t\frac{dJ}{dz}\right]_{z=h} = 0 \end{cases}, \tag{12.24}$$

where

$$\ell_t = \frac{1}{(1 - \varpi_0 g)\sigma} \tag{12.25}$$

is the photon transport mean-free-path, $\chi\ell_t$ is the "extrapolation" length, and h is geometrical thickness of the homogeneous plane-parallel medium. If we set $S(z) = \delta(z - z^*), 0 \le z^* \le h$, then $J(z) = G(z, z^*)$ is the Green function of the boundary-value problem in (12.24), and it follows from (12.23) that

$$\langle n \rangle \approx \sigma \int_0^h G(z, z^*)dz \tag{12.26}$$

since the integral of $S(z)$ is unity.

We will start from the conservative scattering ($\varpi_0 = 1$) case. It is easy to verify (Kamke, 1959, p. 396) that the Green function for the boundary-value problem in (12.24) can be expressed as

$$G(z, z^*) = \frac{3}{\ell_t(h + 2\chi\ell_t)}\begin{cases} (\chi\ell_t + z)(h + \chi\ell_t - z^*), z \le z^* \\ (\chi\ell_t + z^*)(h + \chi\ell_t - z), z \ge z^* \end{cases}. \tag{12.27}$$

Substituting (12.27) into (12.26), we obtain

$$\langle n \rangle \approx \frac{3\chi}{2}\sigma h \left[1 + \frac{z^*}{h}\left(1 - \frac{z^*}{h}\right)\frac{h}{\chi \ell_t} \right]. \tag{12.28}$$

Recalling that $\sigma h = \tau$ (optical thickness of the slab) and $\ell_t = 1/[(1 - g)\sigma]$ for conservative scattering, we have

$$\langle n \rangle \approx \frac{3\chi}{2}\tau \left[1 + \frac{z^*}{h}\left(1 - \frac{z^*}{h}\right)\frac{(1 - g)\tau}{\chi} \right]. \tag{12.29}$$

We see that, if the source is located on a boundary ($z^* = 0$ or $z^* = h$), the second term vanishes. Multiplying the result by 2 to account for the loss of source strength in case of $z^* \to 0$ or $z^* \to h$, we have

$$\langle n_R \rangle \approx 3\chi\tau \tag{12.30}$$

which corresponds to reflected photons (thus subindex R). For a source deep inside the medium, by setting $z^* = h/2$, we obtain

$$\langle n_T \rangle \approx \frac{3}{8}(1 - g)\tau^2 + \frac{3\chi}{2}\tau \tag{12.31a}$$

which corresponds to transmitted photons (thus subindex T). Recently Davis and Marshak (2002), using a Fourier-Laplace transform approach, improved the representation (12.31a); they found

$$\langle n_T \rangle \approx \frac{1}{2}(1 - g)\tau^2 \left[1 + \frac{\varepsilon(4 + 3\varepsilon)}{2(1 + \varepsilon)} \right] \tag{12.31b}$$

where

$$\varepsilon = \frac{2\chi}{(1 - g)\tau}. \tag{12.31c}$$

Motivated by cloud lidar instrument development, Davis et al. (1999) obtained similar pre-asymptotic correction terms in ε and a more accurate prefactor for the simple expression in (12.30) pertaining to reflected photons.

Comparing (12.30) with (12.31) we note that $\langle n_T \rangle$ is proportional to τ^2 independent of boundary conditions (i.e., the extrapolation length factor χ) when $(1 - g)\tau \gg 4\chi$, while $\langle n_R \rangle$ is proportional to τ independent of scattering details (i.e., asymmetry factor g). The latter means that, while transmitted photons experience fewer scatterings for anisotropic ($g > 0$) than for isotropic ($g = 0$), the number of scatterings for reflected photons are independent of g even though, intuitively, reflected photons with isotropic scattering are more readily reversed. Figure 12.8 illustrates formulae (12.30) and (12.31a) with numerical calculations for optical depth τ from 2 to 128: the value $\chi \approx 0.5$ in (12.30) fits well the results of numerics for reflected photons, as shown for the $\varpi_0 = 1$ case in Figs. 12.9a,b.

For absorbing clouds the solution of (12.24) with $S(z) = \delta(z - z^*)$ has the explicit expression (Morse and Feshbach, 1953)

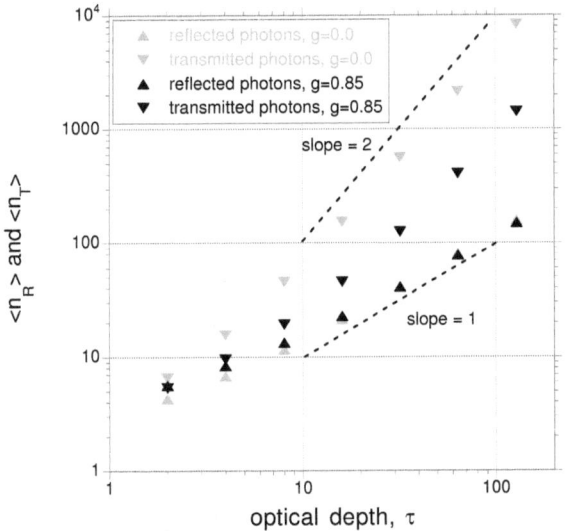

Fig. 12.8. Average number of scatterings for reflected from and transmitted through homogeneous slab photons (conservative scattering). The results of Monte Carlo simulations for diffusely illuminated homogeneous plane-parallel media with seven optical depths from 2 to 128, by powers of 2, and a Henyey-Greenstein phase function with $g = 0$ and $g = 0.85$

$$G(z, z^*) = \frac{3}{2\xi} \frac{1}{\kappa_+^2 e^{h/L_d} - \kappa_-^2 e^{-h/L_d}}$$

$$\times \begin{cases} (\kappa_+ e^{z/L_d} - \kappa_- e^{-z/L_d})(\kappa_+ e^{(h-z^*)/L_d} - \kappa_- e^{-(h-z^*)/L_d}), & z \leq z^* \\ (\kappa_+ e^{(h-z)/L_d} - \kappa_- e^{-(h-z)/L_d})(\kappa_+ e^{z^*/L_d} - \kappa_- e^{-z^*/L_d}), & z \geq z^* \end{cases}$$

$$(12.32)$$

where

$$L_d = \frac{1}{\sigma\sqrt{3(1 - \varpi_0)(1 - \varpi_0 g)}} \qquad (12.33a)$$

is the characteristic diffusion length scale,

$$\xi = \frac{\ell_t}{L_d} = \sqrt{\frac{3(1 - \varpi_0)}{(1 - \varpi_0 g)}}, \qquad (12.33b)$$

is the similarity factor and

$$\kappa_\pm = 1 \pm \chi\xi. \qquad (12.33c)$$

In the case of vanishing absorption ($\varpi_0 \to 1$, thus $L_d \to \infty$ and $\xi \to 0$), (12.32) converges to (12.27).

Setting $z^* = 0$ or h in (12.32) and then integrating the Green function from 0 to h (and multiplying by 2 as explained above), we obtain for reflected photons

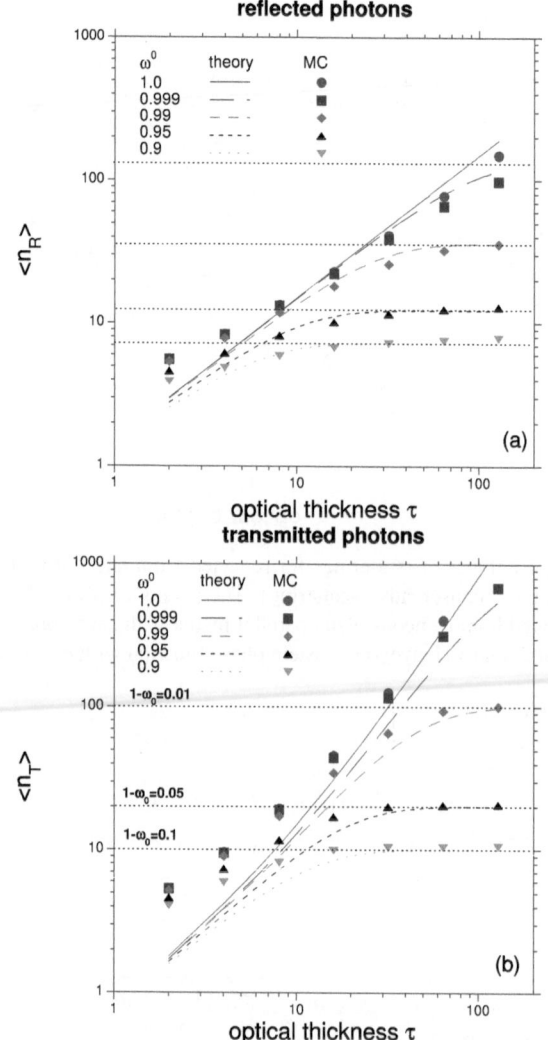

Fig. 12.9. Average number of scatterings in light reflected from and transmitted through homogeneous scattering/absorbing slabs. (a) Reflected photons; (b) Transmitted photons. We
compare the predictions of diffusion theory (curves) with results of a numerical solution of
the Radiative Transfer Equation (RTE) (symbols). The latter were obtained from Monte Carlo
simulations for diffusely illuminated homogeneous plane-parallel media with seven optical
depths from 2 to 128, by powers of 2, and a Henyey–Greenstein phase function with $g = 0.85$.
The absorption parameter $1 - \varpi_0$ ranges from 0 (conservative case) to the point where diffusion theory is expected to fail when $g = 0.85$. We used $\chi \approx 0.5$ in (12.33c) to fit the numerical
data in the diffusion regime ($\tau \gtrsim 10$). Note that for the conservative scattering and transmitted
photons, the prefactor 1/2 (as in (12.31b)) was used instead of 3/8 (as in (12.31a))

$$\langle n_R \rangle = \sigma L_{\mathrm{d}} \frac{6\chi}{\kappa_+^2 e^{h/L_{\mathrm{d}}} - \kappa_-^2 e^{-h/L_{\mathrm{d}}}} \left(\kappa_+ e^{h/L_{\mathrm{d}}} + \kappa_- e^{-h/L_{\mathrm{d}}} - 2 \right), \varpi_0 < 1 .$$
(12.34)

By analogy, for transmitted photons, we set $z^* = h/2$ in (12.32), and after integrating (12.26), we obtain

$$\langle n_T \rangle = \frac{\sigma L_{\mathrm{d}}}{\xi} \frac{3\chi}{\kappa_+^2 e^{h/L_{\mathrm{d}}} - \kappa_-^2 e^{-h/L_{\mathrm{d}}}}$$
$$\times \left[\kappa_+^2 e^{h/L_{\mathrm{d}}} - \kappa_-^2 e^{-h/L_{\mathrm{d}}} - 2 \left(\kappa_+ e^{h/2L_{\mathrm{d}}} - \kappa_- e^{-h/2L_{\mathrm{d}}} \right) \right], \varpi_0 < 1 .$$
(12.35)

Equations (12.34)–(12.35) generalize (12.30)–(12.31) for $\varpi_0 < 1$ noting however that, to retrieve (12.30) from (12.34) and (12.31) from (12.35) in the limit $\varpi_0 \to 1$ ($L_{\mathrm{d}} \to \infty$) at fixed τ, calls for a 2nd-order expansion of the exponentials in h/L_{d} then application of L'Hopital's rule.

Let us study the asymptotic behavior of $\langle n_R \rangle$ and $\langle n_T \rangle$, starting with reflected photons. We see from (12.34) that instead of increasing linearly with τ indefinitely, as in (12.30), $\langle n_R \rangle$ now crosses over to a flat asymptote at

$$\langle n_R \rangle \approx \frac{6\chi}{\kappa_+ \sqrt{3(1 - \varpi_0)(1 - \varpi_0 g)}}, \varpi_0 < 1, \tau \to \infty .$$
(12.36)

The accuracy of this approximation, that also derives from asymptotic transport theory, has been investigated by Kokhanovsky (2002). In the limit $\varpi_0 \to 1$ ($\xi \to 0$) and for very large τ, ignoring prefactors, we have (see also Uesugi and Irvine (1970))

$$\langle n_R \rangle \propto \frac{1}{\sqrt{1 - \varpi_0}}, \varpi_0 \to 1, \tau \to \infty .$$
(12.37)

For transmitted photons, from (12.35) we have a flat asymptote at

$$\langle n_T \rangle \propto \frac{1}{1 - \varpi_0}, \varpi_0 < 1, \tau \to \infty$$
(12.38)

independent of phase function. Note that the last asymptote corresponds to the sum $1 + \varpi_0 + \varpi_0^2 + \varpi_0^3 + \dots$. In both cases, the asymptote is reached when τ exceeds $\sigma L_{\mathrm{d}} = [3(1 - \varpi_0)(1 - \varpi_0 g)]^{-1/2}$.

Figure 12.9 shows both $\langle n_R \rangle$ and $\langle n_T \rangle$ as functions of τ for $g = 0.85$ and selected values of ϖ_0 : 1.0, 0.999, 0.99, 0.95, and 0.9. The formulas in (12.34)–(12.35) with $\chi \approx 0.5$ follow quite closely the numerical results obtained by Monte Carlo solution of the RTE, at least at optical thickness large enough ($\tau \gtrsim 10$) for the diffusion model to apply. This is true at all the levels of absorption considered.

12.3.2 Extension to Horizontal Photon Transport

The characteristic order-of-scattering can be mapped to characteristic spatial scale that describes the area of the slab-cloud that is "explored" by photons during their diffusive random walks from injection $(x_{\mathrm{in}}, y_{\mathrm{in}}, z_{\mathrm{in}})^{\mathrm{T}}$ to escape $(x_{\mathrm{out}}, y_{\mathrm{out}}, z_{\mathrm{out}})^{\mathrm{T}}$. Here photons start their trajectory on the upper boundary, i.e., $z_{in} = 0$, and escape at $z_{out} = 0$ (reflected photons) or at $z_{out} = h$ (transmitted photons). To this effect, we use Einstein's relation for Brownian motion:

$$\langle \boldsymbol{x}^2(t) \rangle = Dt \tag{12.39}$$

where $\boldsymbol{x}(t) = (x(t), y(t), z(t))^{\mathrm{T}}$ is the random vector-position of the particle at time t after leaving the origin $(x_{in}, y_{in}, 0)$ at $t = 0$, and the constant

$$D = \frac{c\ell_t}{3} = \frac{c}{3(1 - \varpi_0 g)\sigma}$$

is photon diffusivity (where c is the velocity of light). Time is now mapped to the number of scatterings *at escape* by

$$t \approx \frac{\langle L \rangle}{c} \approx \frac{\ell \langle n \rangle}{c} = \frac{\langle n \rangle}{c\sigma} \tag{12.40}$$

where $\langle L \rangle$ is the total in-cloud photon path and $\ell = 1/\sigma$ is the photon mean-free-path. The first approximation in (12.40) is the replacement of the parameter t in the boundary-free diffusion described by (12.39) by the average $\langle L \rangle / c$ in the bounded diffusion problem of interest here. The second approximation – already used in (12.23) – is the replacement of the continuous variable σL (optical path) by the discrete quantity n.

   We are interested here in the lateral transport distance,

$$\rho = \sqrt{(x_{out} - x_{in})^2 + (y_{out} - y_{in})^2} \tag{12.41}$$

which is the (x, y)-projection of the displacement $\boldsymbol{x}$ in 3D space, and its variance $\langle \rho^2 \rangle$. Since $\boldsymbol{x}$ is statistically isotropic, $\langle \rho^2 \rangle \propto \langle \boldsymbol{x}^2 \rangle$ and it follows from (12.39) and (12.40) that

$$\langle \rho^2 \rangle \propto \frac{\langle n \rangle}{(1 - \varpi_0 g)\sigma^2} . \tag{12.42}$$

   Using (12.30)–(12.31) that estimate the average scattering order $\langle n_F \rangle (F = R, T)$, we obtain the root-mean-square (rms) of the horizontal displacement for conservative scattering,

$$\sqrt{\langle \rho_R^2 \rangle} \propto \frac{h}{\sqrt{(1 - g)\tau}} = \sqrt{\ell_t h}, \quad \varpi_0 = 1 \tag{12.43}$$

and

$$\sqrt{\langle \rho_T^2 \rangle} \propto h, \quad \varpi_0 = 1 . \tag{12.44}$$

We deliberately used here the "proportional to" instead of "approximately equal to." The important issue of prefactors, that are of order 1, is beyond the scope of this chapter. Interested readers can find detailed discussions of prefactors in Davis and Marshak (2001, 2002). Here we only state that numerical results (below) together with theoretical explanations by Davis and Marshak support prefactors of order unity.

Note that for transmitted photons, (12.44), the horizontal displacement is proportional to the geometrical thickness of the cloud. This has a simple geometrical explanation: a "wave" of diffusing photons is emanating from a point on the upper boundary and propagating "slowly" but isotropically, therefore ρ at cloud bottom is independent of g and τ. For reflected photons, (12.43), the horizontal displacement is roughly equal to the harmonic mean of transport mean-free-path ℓ_t and geometrical cloud thickness h; these two are the main characteristic scales in radiative transfer in a plane-parallel slab in the absence of absorption.

For absorbing wavelengths, we derive the rms of the horizontal displacement for reflected photons:

$$\sqrt{\langle \rho_R^2 \rangle} \propto \frac{h}{\tau} \sqrt{\frac{\langle n_R \rangle}{(1 - \varpi_0 g)}} \tag{12.45}$$

where $\langle n_R \rangle$ is defined in (12.34). Note that the numerator in (12.45) ceases to increase with τ for $\tau > \sigma L_d = [3(1 - \varpi_0)(1 - \varpi_0 g)]^{-1/2}$ with $\varpi_0 < 1$. Thus the behavior of $\langle \rho_R^2 \rangle^{1/2}$ changes from decreasing with $h/\tau^{1/2}$ as in (12.43) for $\varpi_0 = 1$ to a steeper decrease that follows h/τ, for given g and ϖ_0. Even more importantly, $\langle \rho_R^2 \rangle^{1/2}$ is a strong function of $1 - \varpi_0$ which varies far more than g in clouds (with wavelength across the solar spectrum). Indeed, we see that in the limit $\varpi_0 \to 1$, (12.36) and (12.45) yield

$$\sqrt{\langle \rho_R^2 \rangle} \propto \frac{h}{\tau} \sqrt[4]{\frac{1 - \varpi_0 g}{1 - \varpi_0}} \tag{12.46}$$

as long as $\tau > \sigma L_d = [3(1 - \varpi_0)(1 - \varpi_0 g)]^{-1/2}$ which is itself increasing as $\varpi_0 \to 1$.

Figure 12.10 shows our diffusion-based prediction for the dependence on τ of the rms horizontal displacement in (12.45) using (12.30) or (12.34), respectively for $\varpi_0 = 1$ and $\varpi_0 = 0.99, 0.95,$ or 0.90. Note that the analytical results of (12.45) again compare well with the corresponding numerical results obtained by Monte Carlo solutions of the RTE.

12.3.3 Application to Inhomogeneous Clouds

Analytical expressions derived in the previous section are for plane-parallel homogeneous clouds. The natural question is, of course, how well do they represent "real" clouds? The answer to this question depends on the degree of inhomogeneity of the real clouds. To describe inhomogeneity, we will use parameter v as a squared ratio of mean and standard deviation of optical depth τ. For 18 LandSat images, Barker et al. (1996) observed v between 1.5 and 22.5 with a typical value of 4 to 5 for marine stratocumulus clouds. Below we will compare the total photon path and the horizontal displacement of light reflected from scattering and absorbing slabs for horizontally homogeneous and inhomogeneous media prescribing $v = 4.5$ for inhomogeneous clouds.

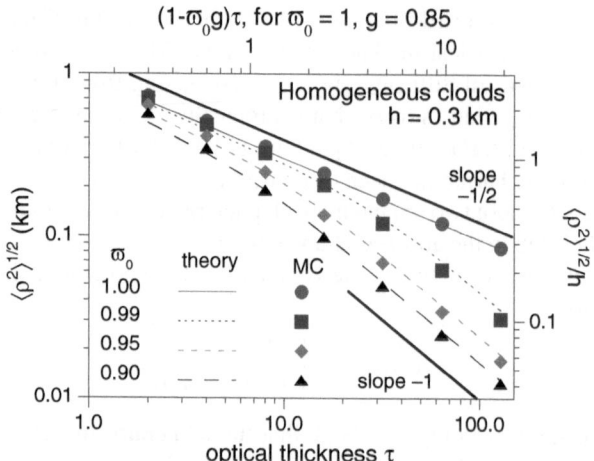

Fig. 12.10. The rms horizontal displacement of light reflected from scattering/absorbing slabs. Homogeneous plane-parallel media with $h = 0.3$ km. The point-source at the upper boundary is isotropic. Four single-scattering albedos ϖ_0 are used: 1.00, 0.99, 0.95, and 0.90. The optical depth sequence and phase function are as in Fig. 12.9; the reference lines are $\propto \tau^{-1/2}$ corresponding to $\varpi_0 = 1$ in (12.43) and $\propto \tau^{-1}$ for $\varpi_0 < 1$ and $\tau \gg 1$ in (12.45)

For conservative scattering, the total photon path $\langle L \rangle \approx \langle n \rangle / \sigma$ for homogeneous and inhomogeneous media is almost the same ($\langle L \rangle \sim h$), while for absorbing wavelengths the total path for inhomogeneous medium is longer than for homogeneous. The difference increases with absorption. This is consistent with Jensen's inequality $[f(\frac{x_1+x_2}{2}) \leq \frac{f(x_1)+f(x_2)}{2}$ for any concave ($f''(x) \geq 0$) function] since the total photon path $\langle L \rangle$ is concave as a function of τ. Figure 12.11 illustrates this behavior for $\varpi_0 = 1.0, 0.99, 0.95,$ and 0.90 using a bounded cascade model with parameters compatible with $v = 4.5$.

Next we compare the rms value of ρ, the horizontal distance between the photon's entry and exit points in (12.41). Figure 12.12 shows this statistic for reflected and transmitted photons for both homogeneous and inhomogeneous models with conservative scattering. We see that $\langle \rho_T^2 \rangle^{1/2}$ shows a trend towards constancy as τ (or $\langle \tau \rangle$) becomes large, as predicted by the diffusion-based results in (12.44). Plotted on double-log axes, $\langle \rho_R^2 \rangle^{1/2}$ closely follows the $-1/2$ slope for both models, as predicted by (12.43).

To illustrate the horizontal distance a photon travels in media that is homogeneous and/or inhomogeneous, composite Fig. 12.13 shows logarithmically spaced isophotes for the 2D albedo and transmittance fields obtained from two normally illuminated cloud models: homogeneous and fractal with the same optical depth on average (Davis et al., 1997). In the homogeneous case, responses are described by a series of concentric almost equidistant circles that implies an exponential decay explained recently by Polonsky and Davis (2004) for both reflection and transmission. Response for the fractal case shows a degree of azimuthal anisotropy traceable to the

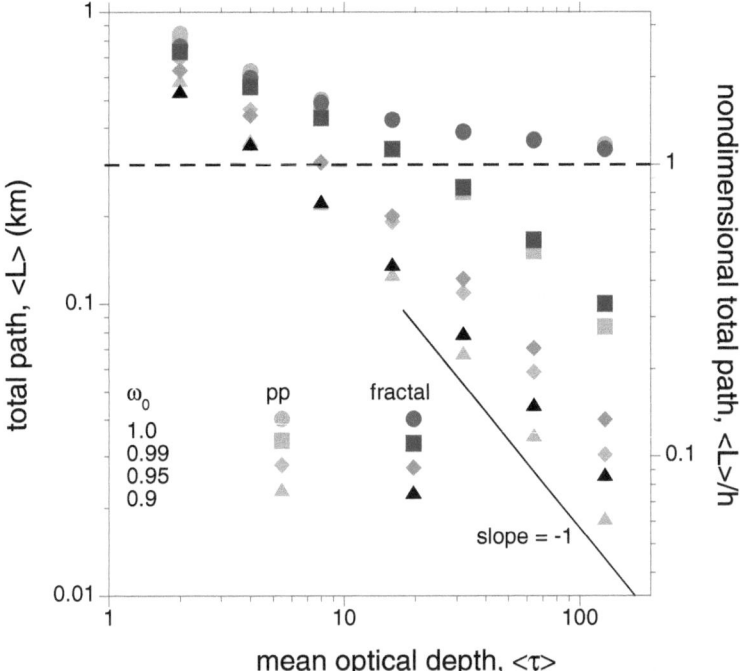

Fig. 12.11. Total photon path for horizontally homogeneous (pp) and inhomogeneous (fractal) media. Horizontal distribution of cloud optical depth is simulated with 10-step bounded cascades models (cf. Appendix) where $p = 0.35$, $H = 1/3$ (this gives $\nu \approx 4.5$), pixel size $r = 50$ m, geometrical thickness $h = 300$ m. Henyey–Greenstein phase functions with asymmetry parameter $g = 0.85$ is used. Surface is absorbing. The point-source at the upper boundary is isotropic. Four single-scattering albedos ϖ_0 are used: 1.0, 0.99, 0.95, and 0.9. A dash line $\langle L \rangle = h$ is added for reference, as well as a line corresponding to $\langle L \rangle \sim 1/\langle \tau \rangle$.

variability in optical depth. Also $\langle \rho^2 \rangle^{1/2}$ is larger here than in homogeneous case in reflection, as expected from Jensen's inequality for concave functions.

Finally, as a counterpart to Fig. 12.10, in Fig. 12.14 we plotted $\langle \rho_R^2 \rangle^{1/2}$ for inhomogeneous clouds. We see that for this relatively mild level of inhomogeneity $(\nu - 4.5)$ the agreement with theoretical predictions would be acceptable; however the theory deteriorates as absorption increases. Note that $\langle \rho_R^2 \rangle^{1/2}$ exceeds its homogeneous counterpart which is again consistent with Jensen's inequality.

To conclude, though the theoretical predictions obtained for plane-parallel homogeneous slabs for average number of scatterings, total photon path and horizontal displacement deviate from their inhomogeneous numerical results, the general behavior remains the same. Thus the above formulae are also valid (with slightly different prefactors) for inhomogeneous clouds as long as the degree of inhomogeneity is relatively mild as is the case for boundary layer stratiform clouds (Barker et al., 1996). The sign of deviation from the homogeneous medium is always compatible with Jensen's inequalities for concave/convex functions.

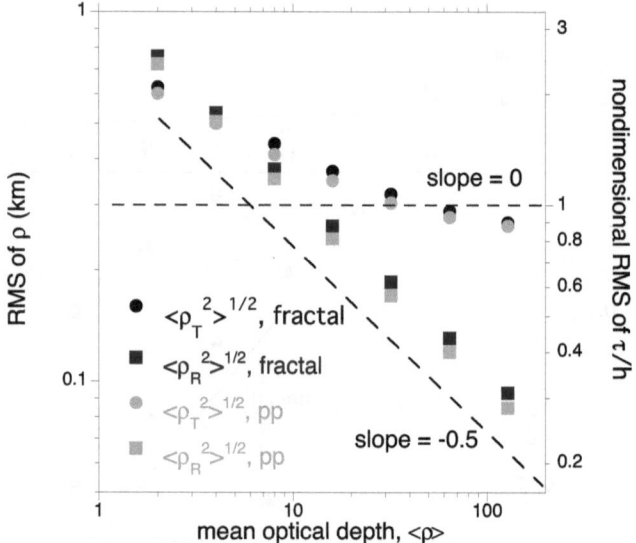

Fig. 12.12. The rms of the horizontal displacement for conservative scattering and horizontally homogeneous (pp) and inhomogeneous (fractal) media. Clouds and illumination conditions are the same as in Fig. 12.11. Slopes 0 and −0.5 illustrate asymptotic behavior and follow from the diffusion theoretical prediction in (12.43)–(12.44)

12.4 Radiative Smoothing and Roughening in Simulations

12.4.1 Conservative Scattering

Stephens (1988) noticed that "multiple scattering tends to filter out the fine structure in the radiation field" thus smoothing its small-scale structure. Marshak et al. (1995) defined radiative smoothing as a radiative transfer process whereby radiation does not follow the small-scale fluctuations of cloud structure producing much smoother radiation fields. Panel (a) in Fig. 12.15 illustrates this phenomenon. It shows both the optical depth and the IPA and Monte Carlo (MC) nadir radiance fields plotted against horizontal distance x. While both fields have almost the same domain average of 0.5, there are large differences in individual pixels. First of all, there is a direct one-to-one relation between the IPA radiance and local optical depth; the horizontal fluctuations of $I_{\text{IPA}}(x)$ follow those of optical depth, showing no smoothing whatsoever. By contrast, $I_{\text{MC}}(x)$ shows considerable smoothing. To better illustrate the smoothness of $I_{\text{MC}}(x)$, panel (b) of Fig. 12.15 shows a 2 km fragment of panel (a), consisting of 160 pixels, each 12.5 m wide; the smoothness of $I_{\text{MC}}(x)$ is now apparent. The smoothness of albedo is even more pronounced. To illustrate this, we plotted the albedo field in panel (c) for the same 2 km fragment as in panel (b). Because of a hemispherical field of view the albedo field is much smoother than its radiance counterpart.

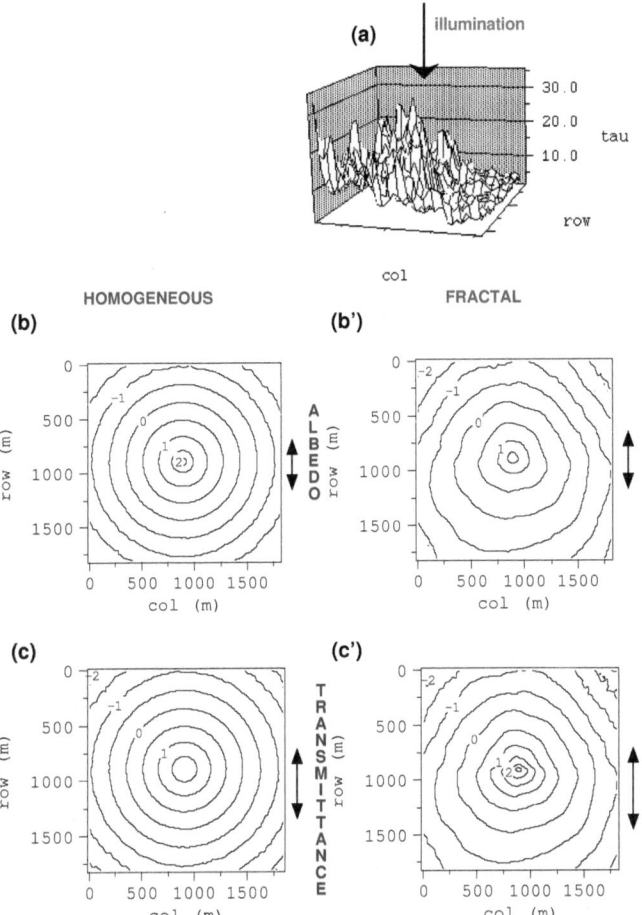

Fig. 12.13. "Spots" for 2D homogeneous and fractal clouds , from Davis et al. (1997) with permission. (**a**) Central 2 km by 2 km portion of a cloud model based on a 2D bounded cascade model (cf. Appendix) with $\langle \tau \rangle = 13$, on a 128×128 grid. The phase function was Henyey–Greenstein with $g = 0.85$ and the illumination from zenith. (**b**) Reflected spot, using flux (local albedo), for a homogeneous cloud with $\tau = 13$. (**b'**) Same as (**b**) for the fractal model in panel (**a**), overall transmittance $T \approx 0.5$. (**c, c'**) Same as (**b, b'**) for spots in transmittance. Isophotes are traced for integer values on a $\log_{10}$ scale; the flux units are arbitrary but uniform, so the higher values and slower spread in panel (**c'**) compared to (**c**) lead to a substantially larger overall transmittance $T' \approx 0.7$. The rms horizontal displacement $\langle \rho^2 \rangle^{1/2}$ is shown by a double-ended arrow for each plot

What is the signature of radiative smoothing in a wavenumber spectrum? In Fig. 12.16 we plotted wavenumber spectra $E(k)$ for the three fields: optical depth, and two nadir radiance fields calculated by IPA and MC. For our present purposes, it suffices to assume that the two-point autocorrelation properties of cloud optical

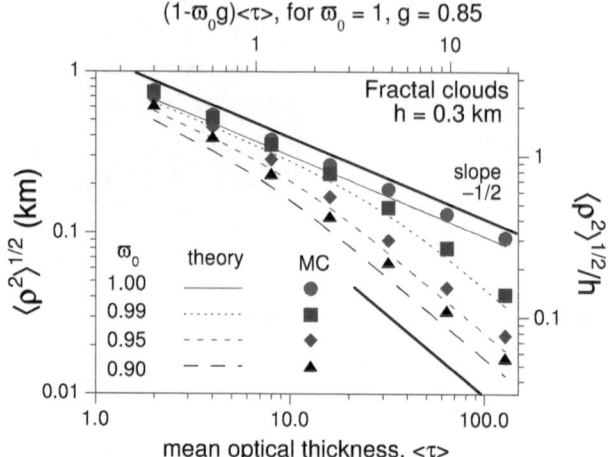

Fig. 12.14. RMS horizontal displacement of light reflected from inhomogeneous clouds. Same outer geometry and optical properties as in Fig. 12.10, but extinction is horizontally variable, as in Fig. 12.11; in this case, the point-sources are uniformly distributed over the upper boundary. The analytical diffusion theory is the same as in Fig. 12.10

depth are described by a power-law wavenumber spectrum:

$$E_\tau(k) \propto k^{-\beta} \tag{12.47}$$

with $\beta \approx 5/3$ (see volume's Appendix) for scales $r = 1/k$ ranging from at least tens of kilometers down to only tens of meters. Plotted in log-log axes, the Fourier power is a straight line with a negative slope β. The IPA radiance field associated with this optical depth has a spectrum that follows a similar power-law as the optical depth (only the prefactor changes). In contrast, the spectrum of the numerically calculated 3D radiance fields follow a similar power-law, but only down to a few hundred meters. At the smallest scales (below this scale break), there is a significant deficit of variance, hence the term radiative smoothing. The special scale at which this scale break occurs (envision the intersection of two lines on a log-log plot of $E(k)$ versus k) is denoted by η. Can the value of η be predicted?

Extensive numerical experimentation showed that in the case of conservative scattering for reflected photons the scale $\eta = \eta_R$ has the same dependence on average cloud geometrical thickness $\langle h \rangle$, average cloud optical depth $\langle \tau \rangle$ and asymmetry factor g as $\langle \rho_R^2 \rangle^{1/2}$ in (12.43), i.e.,

$$\eta_R \propto \frac{\langle h \rangle}{\sqrt{(1-g)\langle \tau \rangle}}, \tag{12.48}$$

with a prefactor 2 ± 1 depending primarily on exactly how η_R is defined. So the diffusion-based theory of horizontal photon transport in homogeneous slab clouds captures the basic phenomenology of radiative smoothing in (stratiform internally

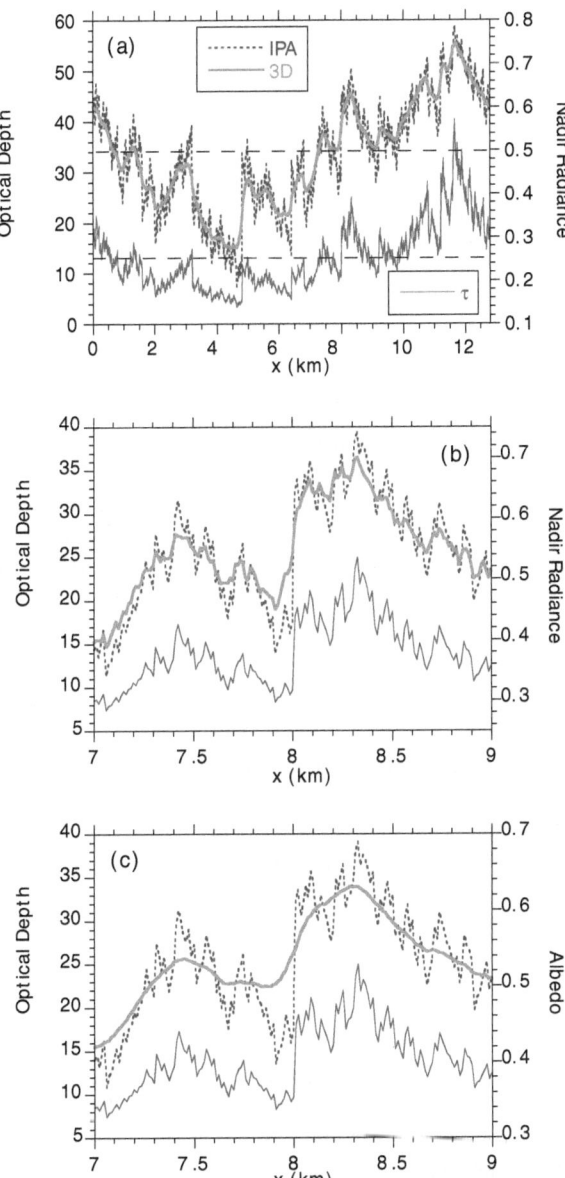

Fig. 12.15. Comparison of 3D (Monte Carlo) and IPA reflectivity fields. The optical depth τ field as in Fig. 12.4, Sun at 22.5° and scattering determined by a Deirmendjian's C1 phase function. (**a**) Nadir radiance for the entire 12.8 km computational domain, dashed lines indicating domain averages; (**b**) 2 km zoom from panel (a); (**c**) Same 2 km zoom but for albedo

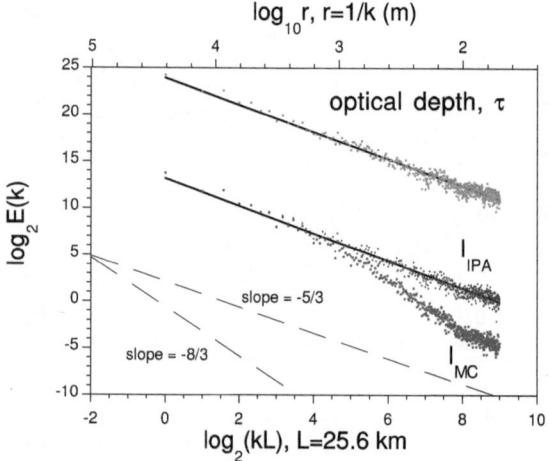

Fig. 12.16. Wavenumber spectra of optical depth, 3D (MC) and IPA nadir radiance fields. Conservative scattering. The optical depth field as in Fig. 12.4 but with pixel size 25 m and spectra are averaged over 10 realizations. Sun is in zenith and scattering was determined by a Henyey–Greenstein phase function with asymmetry parameter $g = 0.85$. Slopes of $-5/3$ and $-8/3$ are added for reference

variable) fractal clouds. Figure 12.16 with $\langle h \rangle = 300\,\text{m}, g = 0.85$ and $\langle \tau \rangle = 13$ shows $\eta_R \approx 400\,\text{m}$.

For radiation transmitted through dense clouds, diffusion theory predicts that horizontal transport is proportional to cloud geometrical thickness and is independent of its optical depth and scattering "details" (see (12.44)). In this case, the radiative smoothing scale $\eta = \eta_T$ can be expressed as

$$\eta_T \propto \langle h \rangle . \tag{12.49}$$

Thus, for the cloud fields as in Fig. 12.15 we expect a scale-break to occur at $\approx$ 300 m. This is confirmed for $\varpi_0 \approx 1$ by Fig. 12.18, which will be discussed in the next subsection.

Note that radiative smoothing scale η is the critical value where IPA effectively breaks down: for scales smaller than η, real radiation fields are much smoother than their IPA counterparts (see Fig. 12.15b) for the same cloud structure. Since the radiative smoothing scale characterizes horizontal photon transport, it is also important to know how η is related to the magnitude of horizontal fluxes H and to the pixel-by-pixel IPA accuracy H_R, H_T and H_A discussed in Sect. 12.2. To answer these questions we start with the effect of absorption on radiative smoothing scale η.

12.4.2 Nonconservative Scattering

In addition to wavenumber spectra of optical depth and radiances for conservative scattering plotted in Fig. 12.16, Fig. 12.17 shows IPA and 3D wavenumber spectra

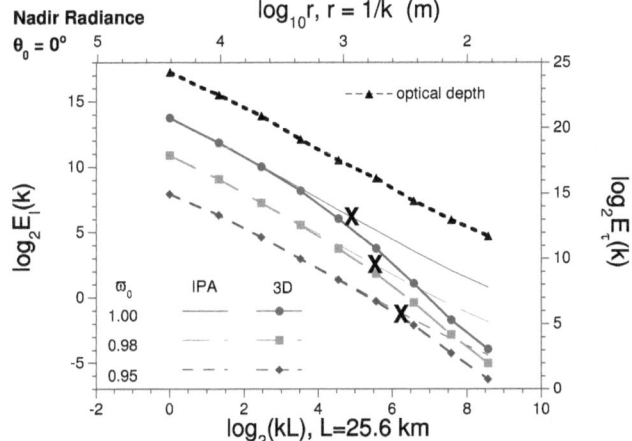

Fig. 12.17. Wavenumber spectra of optical depth, 3D and IPA nadir radiance fields: same data as in Fig. 12.16 (now octave-averaged) with two absorbing wavelengths with $\varpi_0 = 0.98$ and 0.95 added

for two absorbing wavelengths with single scattering albedos $\varpi_0 = 0.98$ and 0.95, respectively. First, we see that all three IPA nadir radiance fields have scale-invariant spectra with a slope similar to the one of cloud optical depth field described by (12.47). In contrast, all three 3D fields exhibit scale breaks determined by the radiative smoothing scale η_R that is a direct indication of photon horizontal transport. We see that the scale break moves towards smaller scales with the increase of absorption, i.e.,

$$\eta_R(\varpi_0) > \eta_R(\varpi_0'), \text{if } \varpi_0 > \varpi_0' . \tag{12.50}$$

This trend is expected and η_R decreases rapidly with ϖ_0, as does $\langle\rho_R^2\rangle^{1/2}$ in (12.45)–(12.46). Furthermore, the variance is less reduced relative to the IPA prediction as ϖ_0 decreases because more absorption means less scattering, hence less smoothing. The last statement is in good agreement with Fig. 12.5a: the pixel-by-pixel IPA error in the reflected radiation H_R monotonically decreases with ϖ_0.

Now we focus on transmitted radiation. Figure 12.18 illustrates its wavenumber spectra. As we see, the inequality (12.50) is not valid for transmitted photons. Indeed, comparing Fig. 12.18 with Fig. 12.5b, we find the scale break (thus the radiative smoothing scale η_T) is determined by the horizontal fluxes for transmitted photons which do not show much variability with respect to ϖ_0 (see Figs. 12.5b and 12.7a). Another indication of weak dependence of η_T on ϖ_0 is (12.49): it depends only on mean cloud geometrical thickness $\langle h\rangle$ and does not depend on cloud optical properties. Figure 12.18 confirms the scale break around $\langle h\rangle = 300$ m.

To conclude, both reflected and transmitted radiation fields exhibit scale breaks which are characterized by their respective radiative smoothing scales $\eta_F(F = R, T)$; it is directly related to the pixel-by-pixel IPA accuracy H_R and H_T for reflected and transmitted fields, respectively. In contrast, the vertically-integrated horizontal fluxes H in (12.16) increase with more absorption while the H_R decreases

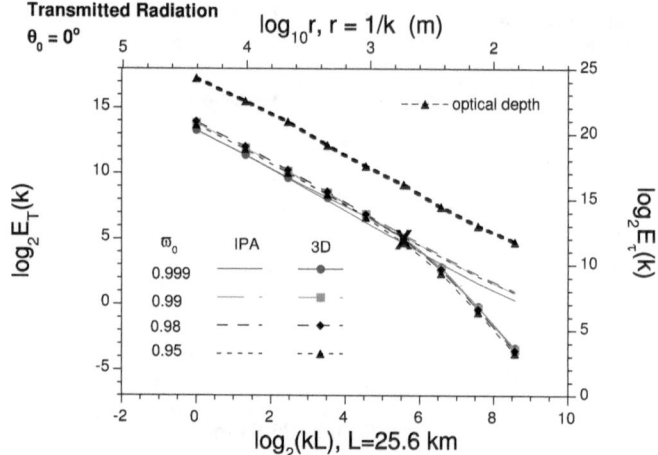

Fig. 12.18. Wavenumber spectra of optical depth, 3D and IPA for radiation transmitted through clouds. The same as in Fig. 12.17 but for four absorbing wavelengths with $\varpi_0 = 0.999, 0.99,$ 0.98 and 0.95. Wavenumber spectrum of optical depth is the same as in Figs. 12.16 and 12.17. Note that due to finite sampling, the overall variance, the integral of $E(k)$, does not vary much (compare with Fig. 12.6b for standard deviation)

and H_T remains almost unchanged (Fig. 12.7b). Therefore, the vertically-integrated horizontal fluxes H for absorbing wavelengths are determined by the IPA absorption error H_A and are not directly related to the radiative smoothing scale η_F, at least for near-zenith Sun.

12.4.3 Oblique Illumination and Radiative Roughening

The situation with oblique illumination (Fig. 12.19) is more complex. As a direct consequence of shadowing, the 3D reflection fields have much more fluctuations at intermediate scales than their IPA counterparts (Zuidema and Evans, 1998; Várnai, 2000; Oreopoulos et al., 2000). The small-scale behavior is still governed by radiative smoothing, but it is not as clearly seen in the wavenumber spectra. In general, there are two competing processes that determine the wavenumber spectra: roughening by shadowing and brightening (via side illumination of clumps) and smoothing by diffusion processes. While roughening flattens the spectra, smoothing steepens it; however roughening is more pronounced for intermediate scales while smoothing always dominates at small scales.

For high Sun (Fig. 12.19a) radiative smoothing dominates, especially for conservative scattering where photons bounce systematically from the dense regions to the tenuous ones. In these cases, the wavenumber spectrum consists of only two regimes: IPA regime for large scales and smoothing for small scales. With more absorption, even for high Sun, low-order scattered photons dominate and we begin to see a signature of shadowing and side illumination – a bump at intermediate scales. This bump

Nadir
Radiance

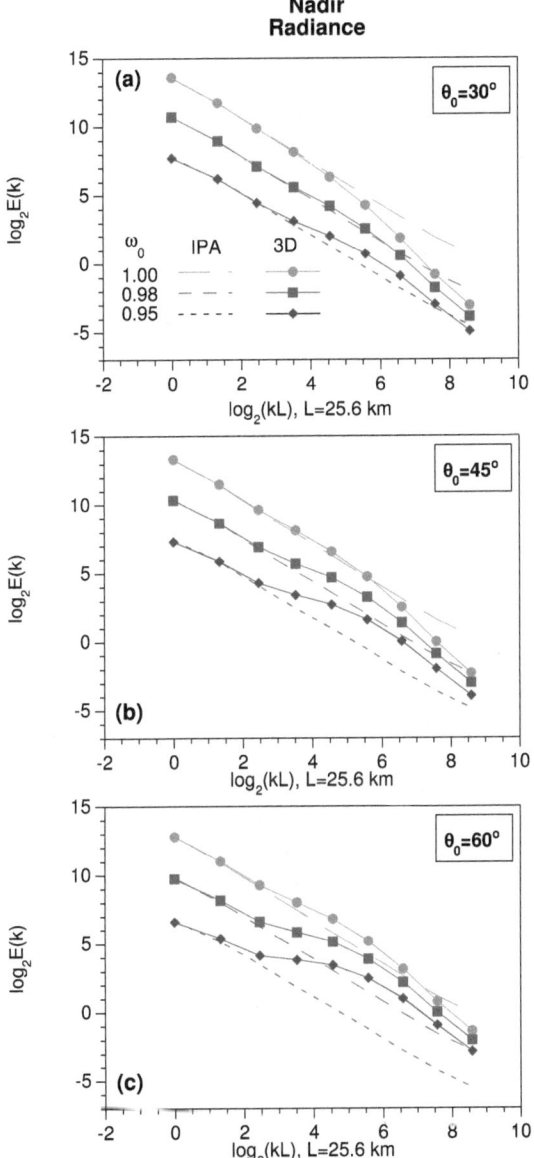

Fig. 12.19. Wavenumber spectra of 3D and IPA nadir radiance fields: same as in Fig. 12.17 ($\varpi_0 = 1.00$, 0.98 and 0.95) but for **(a)** $\theta_0 = 30°$, **(b)** $\theta_0 = 45°$, and **(c)** $\theta_0 = 60°$

becomes more pronounced as absorption and solar zenith angle increase (Fig. 12.19b and c). In these cases, roughening dominates over smoothing.

In summary, nadir reflectivity wavenumber spectra for oblique illumination are characterized by three distinct regimes:

1. large scales where reflectivity follows the fluctuations of optical depth;
2. intermediate scales, where radiative roughening makes reflectivity fluctuations stronger than those of optical depth; and
3. small scales where radiative smoothing makes reflectivity fluctuations weaker.

This is schematically illustrated in Fig. 12.20 for scale-by-scale variance instead of the wavenumber spectrum. By variance we can understand here a counterpart of wavenumber spectrum in physical space such as the 2nd-order structure function (as used in the volume's Appendix). Scale r is best measured in units of cloud geometrical thickness h. So we see that large scales are for r/h greater than 15 to 20; intermediate scales are for r/h between 3 and 15; and, finally, small scales are for r/h less than 1 to 3. Note that roughening is actually controlled by $h \tan \theta_0$ – the horizontal length of the direct beam through a cloud.

For transmitted radiation (Fig. 12.21), the situation is different. First, as for overhead illumination (Fig. 12.18), wavenumber spectra for low Sun do not exhibit much ϖ_0 dependence. Because of the longer photon paths, the small-scale behavior of transmitted light for oblique illumination is even smoother than in case of zenith Sun. A steeper small-scale slope is a clear indication of smoother behavior. Also, the smoothing for oblique illumination is observed for larger scales, i.e.,

$$\eta_T(\theta_0) < \eta_T(\theta_0'), 0° \le \theta_0 < \theta_0' < 90° . \tag{12.51}$$

Finally, in contrast to the up-welling radiation (Fig. 12.19), there is no indication of roughening: the shadowing effect is well seen only for broken clouds with a strong side illumination.

In summary, smoothing, as a radiative transfer process, always happens independent of the illumination conditions; however, it is superseded by radiative roughening in up-welling radiation at low Sun. Thus, using only the wavenumber spectrum it is not always possible to estimate a smoothing effect of horizontal fluxes and IPA accuracy. This is true for both up- and down-welling radiation.

12.5 Radiative Smoothing and Roughening in Data

12.5.1 LandSat Images and Wavenumber Spectra

Thematic Mapper (TM) radiance observations from LandSat with pixels only 30 m wide provide an excellent opportunity to study real-world cloud morphology. One of the most remarkable properties of LandSat cloud scenes is their statistical scale-invariance: wavenumber spectra dependent on a scale parameter r follow power laws over a large range of values r. These are the properties heavily exploited in the present chapter to understand and characterize 3D radiative effects in LandSat images.

Based on the theory of radiative smoothing detailed above, we are now able to explain the scale break in the LandSat wavenumber spectrum in Fig. 12.2. Indeed, a strictly scale-invariant behavior from 8 km to 200 m in LandSat radiances

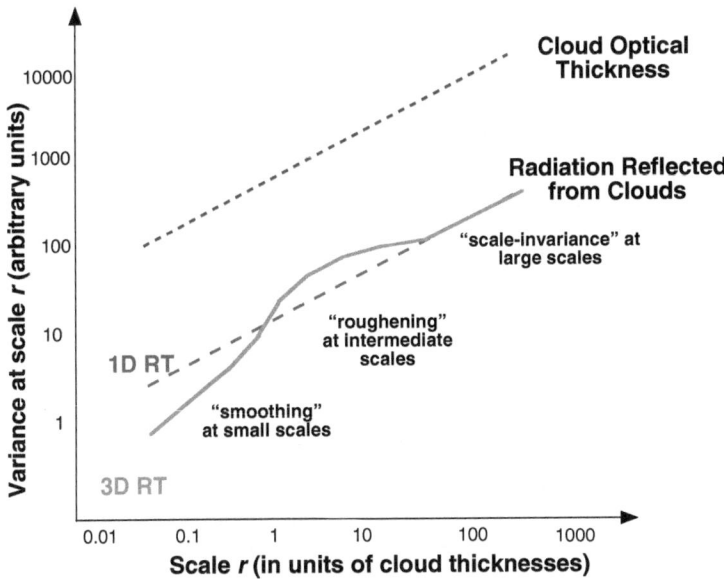

Fig. 12.20. A schematic log-log plot of variance of reflected radiation fields at a given scale versus scale expressed in the units of cloud thickness. 1D radiative transfer corresponds to the IPA. Its slope is parallel to the slope corresponding to cloud structure which is about 2/3 (cf. Appendix)

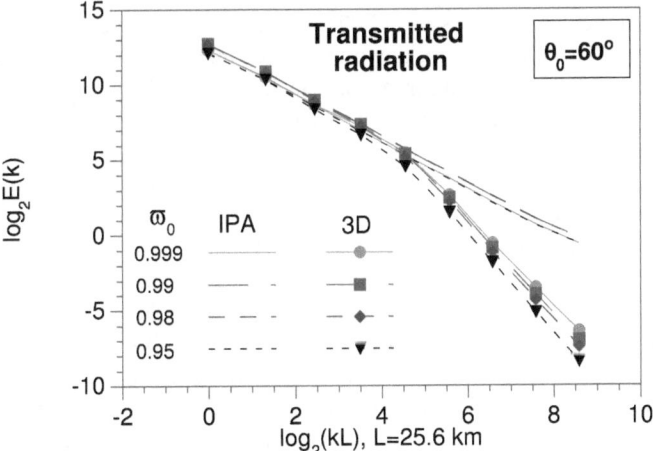

Fig. 12.21. Wavenumber spectra of 3D and IPA radiation transmitted through clouds: same as in Fig. 12.18 but for $\theta_0 = 60°$

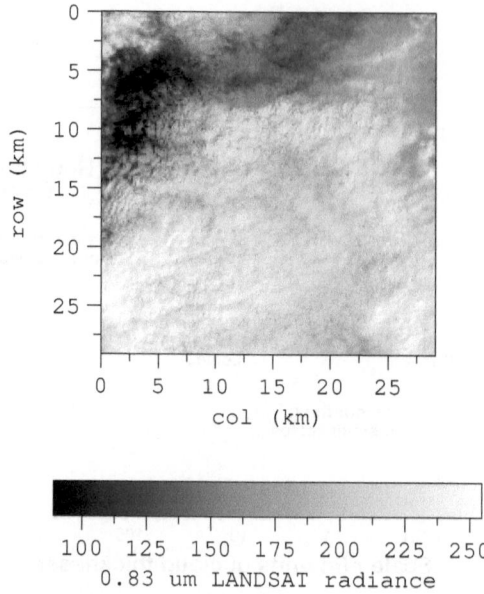

Fig. 12.22. Portion of TM LandSat scene in channel 4 centered around 0.83 μm taken on September 24, 1996, over Oklahoma. The dimensions are 30×30 km², solar zenith angle is ≈45°, and the mean reflectivity at 0.83 μm is 0.68

(inside cloud "streets" in Fig. 12.1), suggests that cloud liquid water has a power-law wavenumber spectrum as in (12.47). We know (see Appendix) that power law behavior goes far below 200 m, down to tens of meters. Thus, as it is described in Sects. 12.4.1–12.4.3, a photon transport mechanism mediated by horizontal fluxes would be able to smooth a scale-invariant cloud liquid water field and bend its wavenumber spectrum. With high Sun ($\theta_0 \lesssim 30°$), the location of a scale break around 200 m is in agreement with (12.48) and Fig. 12.19a where $\langle h \rangle$ and $\langle \tau \rangle$ have typical marine Sc values of 250–300 m and 10–20, respectively.

Generally, LandSat cloudy images are much more complex than the one shown in Fig. 12.1. In Fig. 12.22 we plotted a visible cloudy LandSat image captured over Oklahoma on September 24, 1996. With solar zenith angle at 45° and partially broken clouds, this image represents a challenge for the analysis of cloud morphology. Wavenumber spectra for both conservative scattering (band 4 centered at 0.83 μm) and absorbing media (band 7 centered at 2.2 μm) for this 1024 × 1024 pixel region are shown in Fig. 12.23. We see that both wavenumber spectra have all 3 distinct regimes illustrated schematically in Fig. 12.20. Radiative smoothing is observed for scales smaller than 300–400 m while radiative roughening (a bump in the spectrum) is seen for intermediate scales around 0.5–2.0 km. This is in good agreement with Fig. 12.19b for the same solar angle of 45°. We also see that spectra of absorbing wavelengths (band 7) show less smoothing than nonabsorbing ones (band 4); this is natural since absorbing channels have less scattering thus less smoothing. Finally,

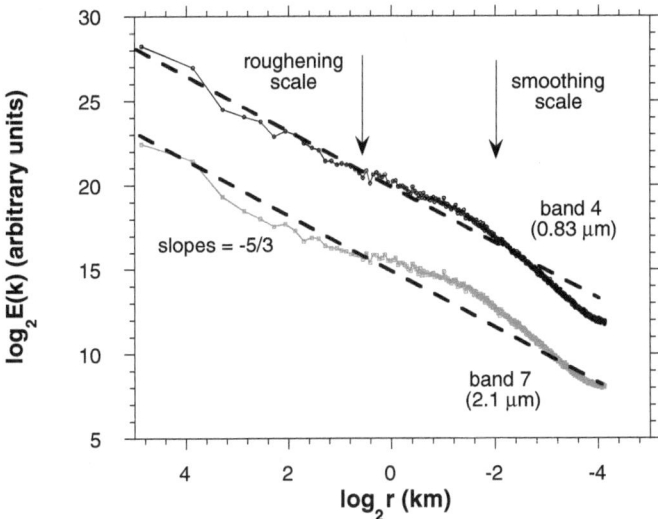

Fig. 12.23. Wavenumber spectra of the LandSat image from Fig. 12.22. Two LandSat TM channels are used: channel 4 centered around 0.83 μm and channel 7 centered around 2.2 μm. For clarity, channel 7 wavenumber spectra are shifted down by 5. The *dash lines* are a slope of −5/3 added for visual comparison. Arrows approximately indicate roughening and smoothing scales that are 1.4 km and 300 m, respectively

bumps are stronger for absorbing channels which is again in good agreement with Fig. 12.19b and Sect. 12.4.3.

Although the wavenumber spectra of LandSat cloud images exhibit a great variety of different shapes and forms depending on cloud top variability, surface reflectance, cloud fraction, illumination conditions, etc. (Oreopoulos et al., 2000), there are several major features common to all images and have a definite signature in wavenumber spectra. These are governed by two radiative transfer processes: smoothing and roughening; both of them are defined as violations of the scale-invariance of the liquid water field. The first process decreases variance at small scales while the second one increases it at intermediate scales. As a result, both 3D radiative transfer processes make retrieval of cloud liquid water from high resolution satellite images an indeterminate or multi-valued problem that causes inaccurate results.

12.5.2 Zenith Radiance Time-Series and Structure Functions

High resolution satellite imagery such as LandSat illustrates radiative smoothing in reflection. As shown in (12.44) and Fig. 12.12, there is also horizontal transport, hence smoothing, by photon diffusion in the transmitted light field. Savigny et al. (1999) set out to observe this smoothing in long time-series of narrow-band zenith radiance at 0.77 μm, where there is no significant absorption, coming from dense unbroken clouds. Their data were collected during the 1998 Cloud Lidar And Radar

Experiment (CLARE) campaign at Chilbolton Observatory (UK). Figure 12.24a shows a merger of 1-minute averages from 3 datasets covering most of the daytime on Oct. 16, 1998. This signal is clearly quite rough, but Fig. 12.24b shows a random zoom into the raw 2 Hz data that illustrates the smooth variations of zenith radiance.

To quantify the smoothing, Savigny et al. (1999) performed (2nd-order) structure function analyses, namely, the variance of increments in radiance over different time-lags, and they looked for scaling regimes:

$$\langle [I(t + \Delta t) - I(t)]^2 \rangle \propto \Delta t^{\zeta(2)} \tag{12.52}$$

where $0 \le \zeta(2) \le 2$. It can be shown that, in theory (i.e., with perfect continuous sampling), $\beta = \zeta(2) + 1$ in (12.47) as long as $1 \le \beta \le 3$, and that $\zeta(2) = 0$ for $\beta < 1$; more details in the Appendix at the end of the volume. Figure 12.24c shows the authors' results for the Oct. 16 data. They invariably found scale breaks separating a smoothed regime, with $\zeta(2)$ approaching 2, and a rough one with an exponent $\zeta(2)$ around $2/3$ which is the canonical value for a turbulent field ($\beta \approx 5/3$). Moreover these breaks are always on the scale of 0.1 km using the mean wind to convert time to space in a "frozen turbulence" hypothesis. As expected from diffusion theory, these scales are commensurate with cloud thickness which was measured independently during CLARE. So this finding is in every respect the empirical structure-function analog of the scale-break in $E(k)$ for simulated data in Fig. 12.18. Finally, we note that agreement with the theoretical estimate of the scale-break based on $\sqrt{\langle \rho_T^2 \rangle}$ improves when accounting for the use of transmitted radiance rather than flux on the one hand, and the accurate prefactor for (12.44) obtained by Davis and Marshak (2002) on the other hand.

Looking at the handful of day-long datasets from the CLARE campaign in October 1998, Savigny et al. (2002) found another scale-break at much larger scales (over 10 km) where the IPA is an accurate description of the transport, see Fig. 12.24c. This time $\zeta(2)$ goes from $\approx 2/3$ to almost 0 which, for all practical purposes, is a roughening. This transition to stationarity is attributed to physical impossibility of radiance increments to grow much beyond the natural unit of zenith or nadir radiances (namely, $\mu_0 F_0 / \pi$), and this value is reached at about 10 km. Optical depth values and increments can in principle grow without bounds (at least until conversion of droplets to precipitation, starting with drizzles, dominates cloud evolution) but albedo, transmittance, and radiances are inherently bounded.

12.6 Nonlocal Independent Pixel Approximation (NIPA)

Let us come back to Fig. 12.15b that shows a 2 km fragment of cloud optical depth field and two fields of cloud reflectivities corresponding to it. The first one was calculated using 1D radiative transfer theory, namely the IPA-plane-parallel theory on a pixel-by-pixel basis that neglects the net horizontal fluxes excited by the spatial variability. In contrast, the second field was calculated using 3D radiative transfer theory represented here by MC, a robust but rather costly numerical technique that

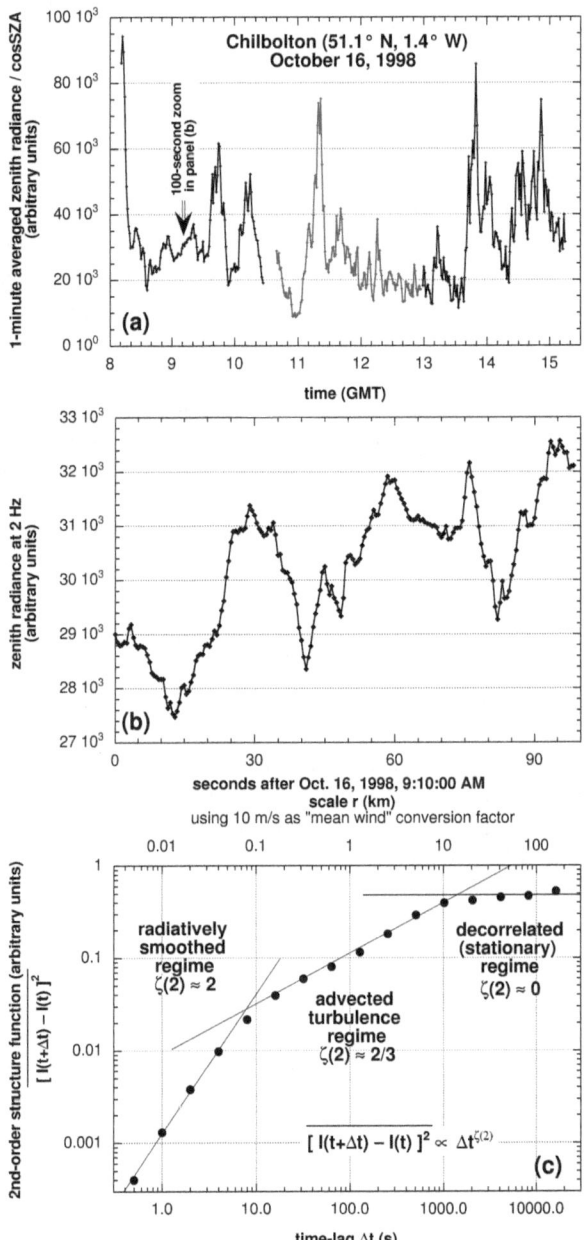

Fig. 12.24. Radiative smoothing in transmission. (**a**) A day-long time-series of 1-minute averages of zenith radiance in a narrow band around 770 nm. (**b**) A "zoom" into the area indicated in panel (a) to show the smoothness of the raw (2 Hz) data. (**c**) Structure function analysis of the radiance data across almost 5 orders of magnitude; see main text for details. The data and statistical results presented here are courtesy of C. von Savigny

uses photon trajectories with given probability densities defined by the integral radiative transfer equation (Chap. 4). We see that because of radiative smoothing, the 3D field is remarkably smoother than the IPA one that follows the fluctuations in cloud structure. While relatively accurate for large enough scales (Chap. 9), there are dramatic errors in individual pixel radiances calculated by IPA. This section will show how to adjust the IPA to substantially decrease these errors.

As illustrated in Fig. 12.16, radiation reflected from clouds and calculated by the IPA is scale-invariant and the scaling of its wavenumber spectrum is therefore defined by just one exponent β over the full range of scales. In contrast, to describe the scaling properties of 3D radiative transfer, one needs at least two more parameters: α which determines small-scale behavior in Fig. 12.16 (and/or Figs. 12.17, 12.20), and the characteristic scale η which defines the scale-break location. Thus, to "simulate" the realistic scaling properties starting with the IPA field (in absence of roughening), we can convolve it with a two-parameter smoothing kernel. This kernel is associated with the radiative transfer Green's function – the radiative response to a point-wise source.

Studying numerically the profile of the "spot" (Fig. 12.13) of reflected light from a point-wise source it was found (Marshak et al., 1995) that a two-parameter Gamma distribution

$$G(\alpha, \eta; |x|) = \frac{1}{2\Gamma(\alpha)\eta^\alpha} |x|^{\alpha-1} \exp\left(-\frac{|x|}{\eta}\right), \qquad (12.53)$$

where $\Gamma(\alpha)$ is Euler's Gamma function, is a good candidate to approximate cloud's Green function if $0 < \alpha < 1$. There are other possibilities, especially with convenient closed-form expressions in Fourier space, resulting from the diffusion-based theory of off-beam cloud lidar signals (Davis et al., 1999). Expressions in physical space are harder to obtain but the asymptotic analysis by Polonsky and Davis (2004) yields the same form as (12.53) with $\alpha = 1/2$ and

$$\eta = \frac{h}{\pi R(\tau)}$$

where $1/R(\tau) = 1 + \varepsilon(\tau)$ is from (12.31c), i.e., the simplest diffusion-theoretical estimate of cloud albedo $R(\tau)$.

With an emphasis on the scaling properties, we represent symbolically by $\text{IPA}(\beta; x)$ a *one*-parameter family of radiation fields reflected from clouds (flux or radiance) and calculated using pixel-by-pixel 1D radiative transfer. The 3D radiative effects will accounted by a *three*-parameter family

$$\text{NIPA}(\alpha, \eta, \beta; x) = G(\alpha, \eta; |x|) * \text{IPA}(\beta; x) \qquad (12.54)$$

where $*$ is a convolution product. We call this whole operation the "Nonlocal Independent Pixel Approximation" (NIPA).

Let I_{IPA} and I_{NIPA} be nadir radiances calculated using IPA and NIPA, respectively. The convolution product (12.54) is best done in Fourier space

$$\tilde{I}_{\text{NIPA}}(\alpha, \eta, \beta; k) = \tilde{G}(\alpha, \eta; k)\tilde{I}_{\text{IPA}}(\beta; k) \qquad (12.55)$$

with (Gradshteyn and Ryzhik, 1980)

$$\tilde{G}(\alpha, \eta; k) = 2 \int_0^\infty G(\alpha, \eta; x) \cos(kx) \mathrm{d}x = \frac{\cos\left[\alpha \tan^{-1}\left(\frac{\eta k}{\alpha}\right)\right]}{\left[1 + \left(\frac{\eta k}{\alpha}\right)^2\right]^{\alpha/2}}. \tag{12.56}$$

The convolution does not effect the domain average $(k = 0)$ in any way, since integral of $G(\alpha, \eta; x)$ is equal to 1. The wavenumber spectrum of the NIPA field is thus given by

$$E_{\mathrm{NIPA}}(\alpha, \eta, \beta; k) = E_{\mathrm{IPA}}(\beta; k) \frac{\cos^2\left[\alpha \tan^{-1}\left(\frac{\eta k}{\alpha}\right)\right]}{\left[1 + \left(\frac{\eta k}{\alpha}\right)^2\right]^{\alpha}} \tag{12.57}$$

where we can assume that $E_{\mathrm{IPA}}(k) \sim k^{-5/3}$. Special cases of interest are $\alpha = 1$ where (12.53) becomes an exponential smoothing kernel:

$$E_{\mathrm{NIPA}}(\alpha, \eta, \beta; k) = E_{\mathrm{IPA}}(\beta; k)\tilde{G}(1, \eta; k)^2 \sim \frac{k^{-5/3}}{[1 + (\eta k)^2]^2} \sim k^{-\frac{17}{3}} \text{ as } k \to \infty;$$
$$\tag{12.58}$$

and $\alpha = 0.5$:

$$E_{\mathrm{NIPA}}(\alpha, \eta, \beta; k) = E_{\mathrm{IPA}}(\beta; k)\tilde{G}(0.5, \eta; k)^2$$
$$\sim k^{-\frac{5}{3}}\left[\frac{1}{\sqrt{1 + (2\eta k)^2}} + \frac{1}{1 + (2\eta k)^2}\right] \sim k^{-\frac{8}{3}} \text{ as } k \to \infty. \tag{12.59}$$

In general, a little algebra shows that the small scale (large k) behavior is $E_{\mathrm{NIPA}}(k) \sim k^{-5/3-2\alpha}$ for all $\alpha < 1$ (asymptotic approach from above) and $E_{\mathrm{NIPA}}(k) \sim k^{-17/3}$ for $\alpha = 1$ (asymptotic approach from below). Figure 12.25 illustrates $E_{\mathrm{NIPA}}(k)$ for $\alpha = 0, 0.25, 0.5, 0.75$, and 1.

We see from (12.56) and Fig. 12.25 that for large scales r (thus small wavenumbers $k \propto 1/r$), both I_{NIPA} and I_{IPA} have the same spectrum which follows a $k^{-5/3}$ power law, while for small scales r (large k) the behavior is quite different. Being scale-invariant, I_{IPA} has a $k^{-5/3}$ spectrum for all scales, while I_{NIPA} exhibits much smoother behavior for small scales.

Now we return to Figs. 12.15a and b and for the same cloud structure we compute I_{NIPA} as a convolution between I_{IPA} and G as prescribed by (12.54). For simplicity, we chose $\eta = 0.25$ km as follows from (12.48) and $\alpha = 0.5$ that corresponds to a small-scale slope of $-8/3$ (see Fig. 12.16). Figure 12.26a is an NIPA-counterpart of Fig. 12.15b; here for a 2 km zoom, in addition to IPA and MC fields, we plotted the NIPA field. Errors at the bottom of Fig. 12.26a demonstrate the improvement. Finally, both NIPA and IPA radiances are plotted versus MC radiance on a scatter plot (Fig. 12.26b) for each pixel. We see that NIPA points concentrated along the

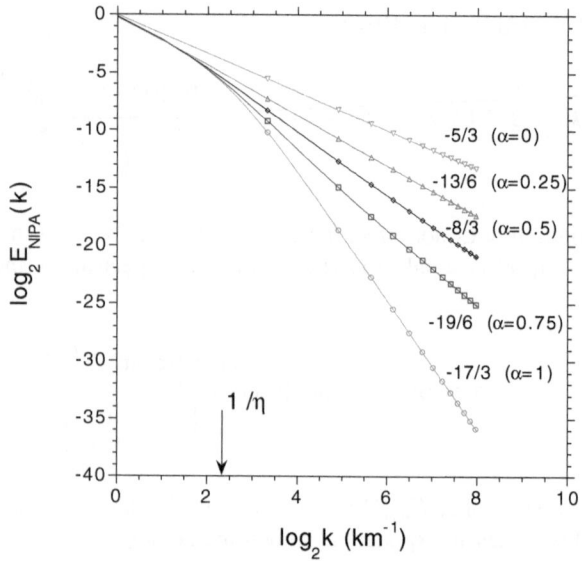

Fig. 12.25. Theoretical wavenumber spectra for NIPA fields of scale invariant clouds. Equation (12.56) is plotted for an IPA that scales as $k^{-5/3}$; η is held constant at 0.2 km and $\alpha = 0, 0.25, 0.5, 0.75,$ and 1

diagonal show very good agreement with MC results. (Note that the computer time for calculation of NIPA radiances is several thousand times shorter than for MC ones.)

To summarize, ignoring net horizontal fluxes for horizontally variable clouds yields incorrect radiation fields for scales on the order of (or smaller than) the smoothing scale η. NIPA incorporates the effect of photon horizontal transport in a simple IPA-type radiative transfer method by using a convolution product of IPA with an approximate Green function for radiative transfer. The Green function can be approximated by a two-parameter Gamma distribution. The first parameter (η) indicates the scale from which the smoothing occurs, while the second parameter (α) defines the amount of smoothing. Both parameters can be estimated from a log-log plot of wavenumber spectrum of cloud reflectivity.

Finally, we note that the performance of NIPA deteriorates substantially with the increase of solar zenith angle. Indeed, the correction factor G as in (12.53) is independent of solar angle. All solar angle dependence of the NIPA is concentrated in the IPA, which is insensitive to the pixel-by-pixel correlation. As a result, NIPA is not designed to correct for shadowing/brightening that results in roughening (not smoothing!) of the radiation reflected from clouds (cf. Fig. 12.19b). Some techniques that deal with roughening by using both nonabsorbing and absorbing wavelengths are described by Oreopoulos et al. (2000) and Faure et al. (2001a).

Generalization of NIPA to two-dimensional fields is straightforward; it is described in Marshak et al. (1998). An inverse NIPA that corrects the retrieval of cloud

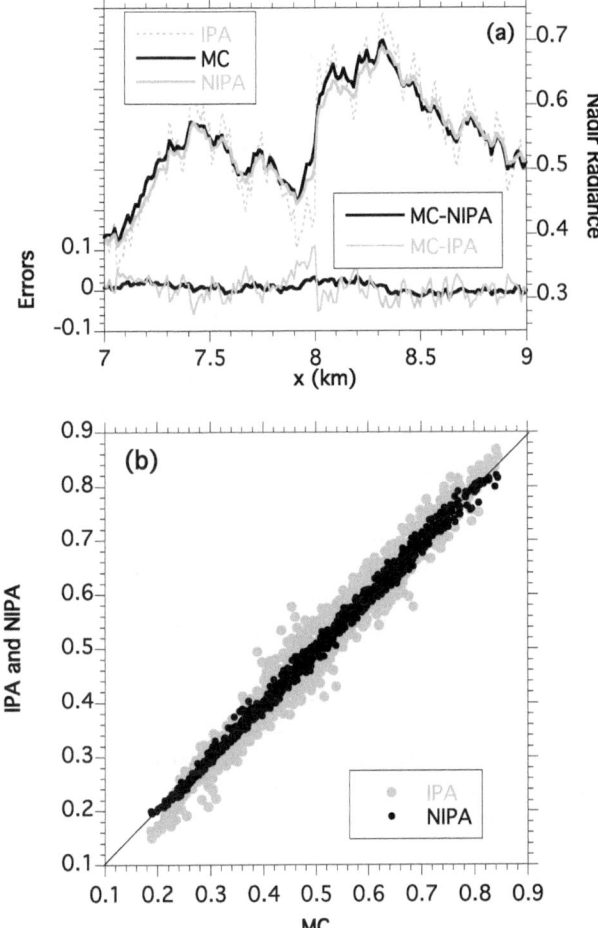

Fig. 12.26. IPA, MC and NIPA for cloud nadir radiances. (**a**) The same as in Fig. 12.15b but with NIPA for $\alpha = 0.5$, $\eta = 0.25$ km in 12.53. Errors MC-IPA and MC-NIPA are also plotted. (**b**) Scatter plot of IPA and NIPA radiances versus their MC counterpart

optical depth for radiative smoothing at small scales is a typical "ill-posed" problem and calls for a "regularization." Theoretical background for the inverse NIPA is given in Marshak et al. (1998) while Oreopoulos et al. (2000) apply this technique to LandSat radiance fields to retrieve cloud optical depths and Faure et al. (2001b) demonstrate a 3D retrieval methods based on neural networks.

12.7 Summary

According to in-situ and ground-based microwave probing of real clouds (Cahalan and Snider, 1989), the horizontal distribution of cloud liquid water shows a power law behavior in wavenumber spectra $\sim k^{-\beta}$ with spectral exponent $\beta > 1$; typically $\beta \approx 5/3$ for marine stratocumulus clouds. The range of scales goes from a few meters to tens of kilometers.

For horizontally inhomogeneous clouds, the Independent Pixel Approximation (IPA) calculates the radiation properties of each pixel by treating it as a homogeneous plane-parallel layer, ignoring net horizontal transport. It is routinely used in current remote sensing applications to infer cloud properties from measured radiance fields. Although the IPA is a nonlinear transformation, to a first approximation, it preserves the scaling properties of cloud liquid water. However, high resolution satellite cloud images show that while the fluctuations of the radiance field follow those of cloud liquid water at large scales, at small scales they exhibit much smoother behavior. The same is true for time-series of zenith radiances measured from ground. Hence, the radiance wavenumber spectrum has a scale break; meaning that there is a characteristic scale that separates two distinct scaling behaviors. As a result, IPA describes well large-scale fluctuations of radiances, but overestimates them at small scales.

The shortcoming of IPA for small scales comes from the fact that while using plane-parallel radiative transfer theory locally, it ignores net horizontal photon transport, i.e., horizontal fluxes, which are a direct consequence of the inhomogeneity in cloud structure. Horizontal fluxes smooth small-scale fluctuations of cloud liquid water; hence, they are responsible for the scale break in the radiance wavenumber spectrum.

In this chapter we studied the properties and magnitude of horizontal fluxes at both absorbing and nonabsorbing wavelengths. We showed the connection between pixel-by-pixel accuracy of the IPA and horizontal fluxes, from one side, and radiative smoothing, from another side. Based on the diffusion theory of horizontal photon transport in homogeneous slab clouds, we were able to capture the basic phenomenology of radiative smoothing in inhomogeneous clouds and predict the location of the wavenumber spectra scale break. In case of conservative scattering, the radiative smoothing scale η is proportional to the geometrical mean of cloud geometrical thickness h and photon transport mean-free-path ℓ_t for reflection, and to the geometrical thickness in transmission. For absorbing wavelengths and reflected photons, η decreases with ϖ_0, while for transmitted photons, η-dependence on ϖ_0 is insignificant.

Finally, we described a method that improves the small-scale performance of IPA without jeopardizing its computational efficiency. This method uses a convolution of IPA with the radiative transfer Green function that can be approximated by a Gamma distribution. We called this the "nonlocal IPA" or NIPA. Unlike IPA, NIPA takes into account net horizontal transport; the large-scale fluctuations of NIPA are similar to those of IPA while the small-scale behavior is smoother and reproduces statistical properties observed in LandSat cloud data and zenith radiance time-series.

In general, LandSat cloud images exhibit a variety of different shapes depending on cloud geometry, surface reflectance, and solar zenith angle. Smoothing is the main radiative transfer process that is responsible for small-scale behavior. For larger scales, there is also radiative roughening that is driven by shadowing and brightening by side illumination; obviously, shadowing is more pronounced for more oblique illumination. Both radiative transfer processes (smoothing and roughening) are a direct consequence of inhomogeneity in cloud structure. Ignoring them in processing satellite data causes inaccurate retrievals of cloud properties.

Ground-based measurements of down-welling radiance at a fixed location as a function of time (clouds are advected by) are also dominated by horizontal transport at small increments in time. At time intervals associated with scales larger than the cloud thickness, the IPA becomes a better approximation and the turbulent structure of clouds is apparent in the signal. At very large scales, more than 10 km (over 1/2 hour in time), the range of possible radiance values has been saturated and a transition to a statistically stationary regime is observed. Although not a 3D effect, this is another mechanism by which radiation hides the inherent variability of clouds.

References

Barker, H.W., B.A. Wielicki, and L. Parker (1996). A parameterization for computing grid-averaged solar fluxes for inhomogeneous marine boundary layer clouds - Part 2, Validation using satellite data. *J. Atmos. Sci.*, **53**, 2304–2316.

Cahalan, R.F. and J.B. Snider (1989). Marine stratocumulus structure during FIRE. *Remote Sens. Environ.*, **28**, 95–107.

Cahalan, R.F., W. Ridgway, W.J. Wiscombe, T.L. Bell, and J.B. Snider (1994). The albedo of fractal stratocumulus clouds. *J. Atmos. Sci.*, **51**, 2434–2455.

Case, K.M. and P.F. Zweifel (1967). *Linear Transport Theory*. Addison-Wesley, Reading (MA).

Davis, A.B. and A. Marshak (2001). Multiple scattering in clouds: Insights from three-dimensional diffusion/P$_1$ theory. *Nuclear Sci. and Engin.*, **137**, 251–280.

Davis, A.B. and A. Marshak (2002). Space-time characteristics of light transmitted through dense clouds: A Green function analysis. *J. Atmos. Sci.*, **59**, 2714–2728.

Davis, A., A. Marshak, R.F. Cahalan, and W.J. Wiscombe (1997). The LANDSAT scale-break in stratocumulus as a three-dimensional radiative transfer effect, Implications for cloud remote sensing. *J. Atmos. Sci.*, **54**, 241–260.

Davis, A.B., R.F. Cahalan, J.D. Spinhirne, M.J. McGill, and S.P. Love (1999). Off-beam lidar: An emerging technique in cloud remote sensing based on radiative Green-function theory in the diffusion domain. *Phys. Chem. Earth (B)*, **24**, 757–765.

Faure, T., H. Isaka, and B. Guillemet (2001a). Neural network analysis of the radiative interaction between neighboring pixels in inhomogeneous clouds. *J. Geophys. Res.*, **106**, 14,465–14,484.

Faure, T., H. Isaka, and B. Guillemet (2001b). Neural network retrieval of cloud parameters of inhomogeneous clouds. Feasibility test. *Remote Sens. Environ.*, **77**, 123–138.

Gradshteyn, I.S. and I.M. Ryzhik (1980). *Table of Integrals, Series and Products*. Academic Press, San Diego.

Kamke, E. (1959). *Differentialgleichungen, Losungsmethoden und Losungen, Vol. 1*. Chelsea Publishing Company, New York.

Kokhanovsky, A.A. (2002). Simple approximate formula for the reflection function of a homogeneous semi-infinite medium. *J. Opt. Soc. Am. A*, **19**, 957–960.

Korn, G.A. and T.A. Korn (1968). *Mathematical Handbook.* McGraw-Hill Book Company, New York (NY).

Marshak, A., A. Davis, W.J. Wiscombe, and R.F. Cahalan (1995). Radiative smoothing in fractal clouds. *J. Geophys. Res.*, **100**, 26,247–26,261.

Marshak, A., A. Davis, W.J. Wiscombe, and R.F. Cahalan (1997). Inhomogeneity effects on cloud shortwave absorption measurements: Two-aircraft simulations. *J. Geophys. Res.*, **102**, 16,619–16,637.

Marshak, A., A. Davis, R.F. Cahalan, and W.J. Wiscombe (1998). Nonlocal Independent Pixel Approximation: Direct and Inverse Problems. *IEEE Trans. Geosc. and Remote Sens.*, **36**, 192–205.

Marshak, A., L. Oreopoulos, A.B. Davis, W.J. Wiscombe, and R.F. Cahalan (1999). Horizontal radiative fluxes in clouds and accuracy of the Independent Pixel Approximation at absorbing wavelengths. *Geophys. Res. Lett.*, **11**, 1585–1588.

Meador, W.E. and W.R. Weaver (1980). Two-stream approximations to radiative transfer in planetary atmospheres: A unified description of existing methods and a new improvement. *J. Atmos. Sci.*, **37**, 630–643.

Morse, P.M. and H. Feshbach (1953). *Methods of Theoretical Physics,* 2 vols. McGraw-Hill, New York (NY).

Oreopoulos, L., A. Marshak, R.F. Cahalan, and G. Wen (2000). Cloud three-dimensional effects evidenced in Landsat spatial power spectra and autocorrelation function. *J. Geophys. Res.*, **105**, 14,777–14,788.

Polonsky, I.N. and A.B. Davis (2004). Lateral photon transport in dense scattering and weakly-absorbing media of finite thickness: Asymptotic analysis of the Green functions. *J. Opt. Soc. Amer. A*, **21**, 1018–1025.

Savigny, C. von, O. Funk, U. Platt, and K. Pfeilsticker (1999). Radiative smoothing in zenith-scattered sky light transmitted through clouds to the ground. *Geophys. Res. Lett.*, **26**, 2949–2952.

Savigny, C. von, A.B. Davis, O. Funk, and K. Pfeilsticker (2002). Time-series of zenith radiance and surface flux under cloudy skies: Radiative smoothing, optical thickness retrievals and large-scale stationarity. *Geophys. Res. Lett.*, **29**, 1825, doi:10.1029/2001GL014153.

Stamnes, K., S.-C. Tsay, W.J. Wiscombe, and K. Jayaweera (1988). Numerically stable algorithm for discrete-ordinate-method radiative transfer in multiple scattering and emitting layered media. *Appl. Opt.*, **27**, 2502–2509.

Stephens, G.L. (1988). Radiative transfer through arbitrary shaped optical media, I: A general method of solution. *J. Atmos. Sci.*, **45**, 1818–1836.

Titov, G.A. (1998). Radiative horizontal transport and absorption in stratocumulus clouds. *J. Atmos. Sci.*, **55**, 2549–2560.

Uesugi, A. and W.M. Irvine (1970). Multiple scattering in a plane-parallel atmosphere 1. Successive scattering in a semi-infinite medium. *Astrophys. J.*, **159**, 127–135.

Várnai, T. (2000). Influence of three-dimensional radiative effects on the spatial distribution of shortwave cloud reflection. *J. Atmos. Sci.*, **57**, 216–229.

Zuidema, P. and K.F. Evans (1998). On the validity of the Independent Pixel Approximation for the boundary layer clouds observed during ASTEX. *J. Geophys. Res.*, **103**, 6059–6074.

Suggestions for Further Reading

Ackerman, S.A. and S.K. Cox (1981). Aircraft observations of the shortwave fractional absorptance of non-homogeneous clouds. *J. Appl. Meteor.*, **20**, 1510–1515.

Antyufeev, V.S. (1996). Solution of the generalized transport equation with a peak-shaped indicatrix by the Monte Carlo method. *Russ. J. Numer. Anal. Mth. Modeling*, **11**, 113–137.

Barker, H.W. and J.A. Davies (1992). Cumulus cloud radiative properties and the characteristics of satellite radiance wavenumber spectra. *Remote Sens. Environ.*, **42**, 51–64.

Barker, H.W. (1995). A spectral analysis of albedo and bidirectional reflectances for inhomogeneous clouds. *Remote Sens. Environ.*, **54**, 113–120.

Barker, H. and D. Liu (1995). Inferring cloud optical depths from LANDSAT data. *J. Climate*, **8**, 2620–2630.

Boers, R., A. van Lammeren and A. Feijt (2000). Accuracy of optical depth retrieval from ground-based pyranometers. *J. Atmos. Ocean. Tech.*, **17**, 916–927.

Cahalan, R.F., W. Ridgway, W.J. Wiscombe, S. Gollmer and Harshvardhan (1994). Independent pixel and Monte Carlo estimates of stratocumulus albedo. *J. Atmos. Sci.*, **51**, 3776–3790.

Cahalan, R.F. (1994). Bounded cascade clouds: Albedo and effective thickness. *Nonlinear Proc. Geophys.*, **1**, 156–167.

Chambers, L., B. Wielicki and K.F. Evans (1997). On the accuracy of the independent pixel approximation for satellite estimates of oceanic boundary layer cloud optical depth. *J. Geophys. Res.*, **102**, 1779–1794.

Chambers, L., B. Wielicki and K.F. Evans (1997). Independent pixel and two-dimensional estimates of Landsat-derived cloud field albedo. *J. Atmos. Sci.*, **54**, 1525–1532.

Davis, A., A. Marshak, R.F. Cahalan and W.J. Wiscombe (1997). Interactions: Solar and laser beams in stratus clouds, fractals & multifractals in climate & remote-sensing studies. *Fractals*, **5** suppl., 129–166.

Ivanov, V.V. and S.D. Gutshabash (1974). Propagation of brightness wave in an optically thick atmosphere. *Physika atmosphery i okeana*, **10**, 851–863.

Loeb, N.G. and J.A. Coakley (1998). Inference of marine stratus cloud optical depths from satellite measurements: Does 1D theory apply? *J. Climate*, **11**, 215–233.

Marshak, A., A. Davis, W.J Wiscombe and G. Titov (1995). The verisimilitude of the independent pixel approximation used in cloud remote sensing. *Remote Sens. Environ.*, **52**, 72–78.

Marshak, A., A. Davis, W.J. Wiscombe and R.F. Cahalan (1998). Radiative effects of sub-mean free path liquid water variability observed in stratiform clouds. *J. Geophys. Res.*, **103**, 19557–19567.

Marshak, A., W.J. Wiscombe, A. Davis, L. Oreopoulos and R.F. Cahalan (1999). On the removal of the effect of horizontal fluxes in two-aircraft measurements of cloud absorption. *Quart. J. Roy. Meteor. Soc.*, **558**, 2153–2170.

O'Hirok, W. and C. Gautier (1998). A three-dimensional radiative transfer model to investigate the solar radiation within a cloudy atmosphere. Part I: Spatial effects. *J. Atmos. Sci.*, **55**, 2162–2179.

Oreopoulos, L. and R. Davies (1998). Plane parallel albedo biases from satellite observations. Part I: Dependence on resolution and other factors. *J. Climate*, **11**, 919–932.

Oreopoulos, L., R.F. Cahalan, A. Marshak and G. Wen (2000). A new normalized difference cloud retrieval technique applied to Landsat radiances over the Oklahoma ARM site. *J. Appl. Meteor.*, **39**, 2305–2321.

Platnick, S. (2001). Superposition technique for deriving photon scattering statistitics in plane-parallel cloudy atmospheres. *J. Quant. Spectrosc. Radiat. Transfer*, **68**, 57–73.

Platnick, S. (2001). Approximations for horizontal transport in cloud remote sensing problems. *J. Quant. Spectrosc. Radiat. Transfer*, **68**, 75–99.

Romanova, L.M. (2001). Narrow light beam propagation in a stratified cloud: Higher transverse moments. *Izv. Atmos. Oceanic Phys.*, **37**, 748–756.

Várnai, T. (1999). Effects of cloud heterogeneities on shortwave radiation: comparison of cloud-top variability and internal heterogeneity. *J. Atmos. Sci.*, **56**, 4206–4224.

Várnai, T. and A. Marshak (2001). Statistical analysis of the uncertainties in cloud optical depth retrievals caused by three-dimensional radiative effects. *J. Atmos. Sci.*, **58**, 1540–1548.

Várnai, T. and A. Marshak (2002). Observations and analysis of three-dimensional radiative effects that influence MODIS cloud optical thickness retrievals. *J. Atmos. Sci.*, **59**, 1607–1618.

13

Photon Paths and Cloud Heterogeneity: An Observational Strategy to Assess Effects of 3D Geometry on Radiative Transfer

G.L. Stephens, A.K. Heidinger and P.M. Gabriel

13.1 Introduction 587

13.2 Elementary Concepts 592

13.3 Photon Paths and Radiative Transfer 596

13.4 Analyses of Surface Transmission Measurements 601

13.5 Spatial Heterogeneity Effects on Radiances
     and Photon Paths.. 605

13.6 Implications for the Remote Sensing of Clouds................. 608

13.7 Summary and Concluding Comments 610

Notes and Further Reading....................................... 611

References... 614

13.1 Introduction

Two obstacles that limit both our understanding of climate change and our ability to predict this change are our lack of definition of climate forcing (Intergovernmental Panel on Climate Change, 2001) and our inability to quantify the response of the climate system to this forcing. At the core of these topics is the need to specify the radiative properties of the atmosphere and, in one way or another, knowledge of the optical properties of the particulate matter of the global atmosphere, either in the form of aerosol or clouds.

Developing knowledge about the radiative properties of clouds and aerosol requires progress on a number of fronts, notably in improving methods to observe these properties globally and in the translation of these global observations to radiative properties relevant to radiative transfer. Challenges confronting these observational efforts include better detection of thin layers of cloud and aerosol especially over bright land and snow covered surfaces, clarifying the ambiguity introduced by non-spherical particle scattering, complications introduced by 3D structures of the scattering layers (a problem that is not solved simply by increasing the spatial resolution of observations) and a critical lack of vertical resolution characteristic of current global observing systems.

The radiative properties of clouds that are so critical to many climate problems, in particular, are significantly influenced by the three dimensional organization of clouds. For example, the amount of IR radiation absorbed and emitted by clouds is primarily influenced by the vertical structure of clouds. This structure determines how much radiation is exchanged between the surface, between clouds at different levels in the atmosphere and between clouds and space (Stephens, 1999). Furthermore, the amount of solar radiation reflected by clouds is directly influenced by the extent of horizontal variability that dictates how much solar radiation flows horizontally between cloud elements and between these cloud elements, the surface and back to space.

Global satellite observational studies of clouds are commonly based on the interpretation of measurements of reflected sunlight in terms of column-averaged cloud properties. Unfortunately, there is a general lack of quantitative error analyses and other information that can be used to determine the reliability of the retrieved information. The inversion methods developed are, by necessity, based on crude assumptions about the vertical structure of clouds and furthermore that the radiative transfer within each pixel occurs independently of neighboring pixels (the so-called Independent Pixel Approximation, hereafter IPA).

At this time, our ability to observe the general 3D cloud structure from the ground is limited to scanning pulsed active sensors. For all practical purposes, we have no capability to observe 3D cloud structure from space although certain aspects of the horizontal variability are well captured by imaging instruments. Although several theoretical studies point to the general importance of 3D effects on the radiative properties of clouds, it is a challenge to bring direct observational evidence to support these studies, cf. Chaps. 11 and 12. Stephens and Greenwald (1990) provide the first global-scale indication of the effects of 3D structure on the radiation budget but this too is inconclusive owing to the lack of both resolution and coincidence of the data used in that study. Meaningful observational approaches to this problem are not yet firmly in hand and we are left with a challenge to develop observational techniques for determining and testing the significance of 3D effects on radiative transfer. This chapter describes preliminary thoughts on one such approach.

13.1.1 An Old Idea Revisited

The concepts described in this chapter are based on the idea of using measurements in a portion of the solar spectrum that contains an absorption band associated with a gas of known concentration and known distribution. As such we turn an old idea around, namely to use known absorption properties of a well mixed gas such as molecular oxygen as a vehicle to understand radiative transfer in a multiple scattering medium, rather than the more traditional application that exploits the scattering

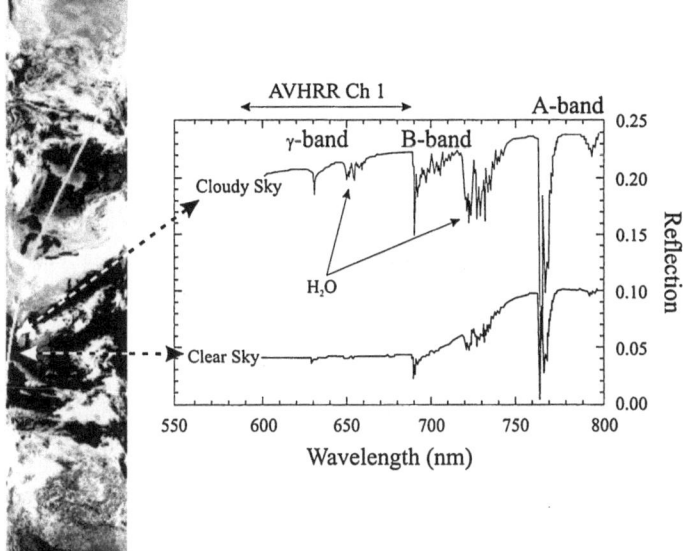

Fig. 13.1. Two spectra measured by GOME over the regions indicated from the overlay of the orbit path of GOME on a near-coincident AVHRR visible (channel 1) image of the atmosphere and surface below. The spectral width of the AVHRR channel 1 is shown for reference

of the atmosphere and surface to deduce the properties of the absorbing species. We will focus on measurements of oxygen absorption, either in the O_2 A-band region between 760–770 nm or in the weaker B-band between approximately 680–690 nm. Figure 13.1 is an example of spectra obtained over a cloudy and clear region measured by the Global Ozone Monitoring Experiment (GOME) instrument (Burrows et al., 1999). The two absorption bands of oxygen that are relevant to the topics of this chapter are noted. Inverting these types of measurements in the form of information about an absorbing gas is an old idea dating back at least to Yamamoto and Wark (1961) who proposed that oxygen absorption observed in measured reflection spectra from satellites could be used to infer column oxygen amounts thereby providing a way of estimating the measurements of surface (and cloud top) pressure. Upon closer scrutiny, however, the relatively small effects of atmospheric scattering that cannot be easily accounted for (scattering that arises for example from undetected thin cirrus and aerosol within the field of view of the instrument) introduces significant error in the retrieval of surface pressure.

Small amounts of atmospheric scattering can be characterized and thus corrected for provided the measurements of absorption are of sufficient spectral resolution. Measurements obtained with a spectral resolution of $1 \, cm^{-1}$ across the O_2 A-band provide a means for differentiating between the paths of photons that are reflected from the surface thus experiencing the entire column of oxygen from paths of photons that penetrate only to some level in the atmosphere before being scattered back to space. It is worth remarking that this resolution is significantly higher than the

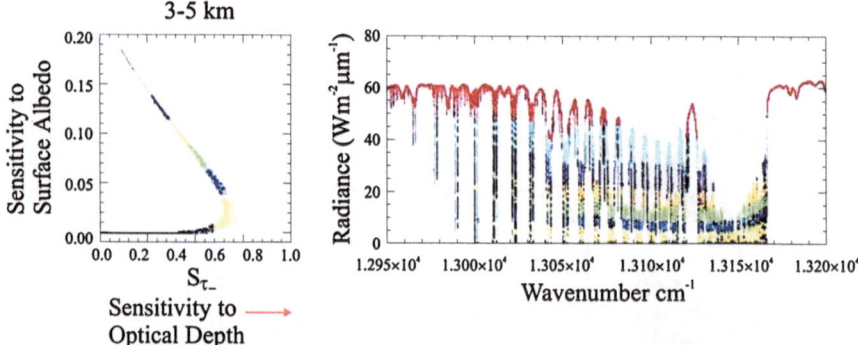

Fig. 13.2. A simulated example of an ultra-fine reflection spectrum for an aerosol layer over a reflecting surface of albedo 20% (*right*). The sensitivity to aerosol optical depth is mostly contained in the centers of absorbing lines (*yellow, left panel*) whereas the non-absorbing wavelengths are mostly sensitive to surface albedo changes (*red, left panel*)

resolution of GOME or of other instruments flown previously in space. Hereafter the terminology "ultra-fine resolution" is used as a convenient way of distinguishing measurements with resolutions of $1\,\mathrm{cm}^{-1}$ or better from measurements with characteristically coarser resolution more typical of present satellite and many surface instruments.

That ultra-fine spectral resolution measurements are capable of discriminating atmospheric scattering from surface scattering is demonstrated in the theoretical results shown in Fig. 13.2. The right-hand panel is a model-derived spectrum of reflected sunlight as it would be measured by a spectrometer on a satellite measuring reflected sunlight across the O_2 A-band at a resolution of $1\,\mathrm{cm}^{-1}$. The spectrum shown is for a case that corresponds to an optically thin aerosol layer over a moderately bright land surface (20% albedo). The left-hand panel is the sensitivity of the spectral radiance to surface albedo changes (ordinate) and changes to aerosol optical depth (abscissa). The portions of the spectrum corresponding to a given sensitivity are matched by the color code. An ideal aerosol observing system corresponds to a maximum sensitivity with respect to optical depth and minimum sensitivity with respect to surface albedo. The former is maximum in the centers of the absorbing lines whereas the latter exhibits most sensitivity in those non-absorbing regions of the spectrum which also have minimum sensitivity to optical depth. These non-absorbing wavelengths contain least information about aerosol and are unfortunately more typical of the channels adopted by current imaging instruments such as the AVHRR and MODIS imaging instruments. This example offers a perspective on the potential value of highly resolved spectral measurements for separating surface from atmospheric scattering and, as such, implies that these spectra also contain information about the vertical distribution of scattering particles in the atmosphere (e.g., Heidinger and Stephens, 2000).

Another advantage offered by spectrally resolved measurements of the absorption of uniformly mixed gases is the information these measurements provide on photon path distributions. The path a photon travels in an absorbing and scattering medium is a useful concept in the study of molecular absorption in a multiple scattering atmosphere. The equivalent width of a molecular absorption line grows with pathlength in well known ways (Chamberlain and Hunten, 1987). Multiple scattering prolongs the path of a photon through the absorbing gas thereby enhancing the absorption. As early as 1995, Stephens proposed to the Atmospheric Radiation Measurement (ARM) community that the photon path information in multiple scattering media might also be considered a fundamental property in understanding radiative transfer in such a medium and furthermore proposed that this information is particularly germane to 3D transport problems for at least two reasons:

i. The characteristics of this 3D transport, at least as they effects measurable radiation fields, are uniquely defined by these path distributions; and
ii. the path distribution information can also be extracted from observations with gross assumptions about the nature of the geometry of the medium.

13.1.2 Outline of this Chapter

The results of Fig. 13.2 point to the value of spectral absorption measurements for obtaining information about the scattering particles suspended in the atmosphere. By inverting the more classical problem as posed originally by Yamamoto and Wark (1961) to one that focuses on the scattering processes, we may be able to provide new insights into the radiative transfer processes in a multiple scattering atmosphere. This concept for studying clouds and aerosol optical properties led to a proposal in 1994 by the lead author of this chapter to fly ultra-fine A-band spectrometers on cloud and aerosol satellite experiments that were being formulated at that time, culminating in the CloudSat and CALIPSO mission concepts (Stephens et al., 2002). These missions were subsequently selected by NASA with anticipated launches in 2004 but unfortunately the spectrometers from both missions were descoped due to cost overruns. Since that time, supporting surface measurements have become more prevalent and analyses of these data further demonstrate the advantages of spectroscopic measurements applied to the study of clouds.

The remainder of this chapter constructs the arguments for the value of spectroscopic measurements of the reflection and transmission for studying radiation transfer through 3D clouds. The following section introduces the general notion of how these measurements might be analyzed based on actual aircraft measurements of reflection in the O_2 A-Band. Here the particular advantage of ultra-fine resolution data is demonstrated in a practical way. The concept of photon path is introduced in this section leading to a discussion of more formal concepts of photon path distribution described in Sect. 13.3. Examples of data analyzed from surface based spectrometer systems supporting the general theory introduced in Sect. 13.3 are described in Sect. 13.4. The effects of 3D geometry on measured spectra and photon pathlengths inferred from such measurements is presented in Sect. 13.5 followed by a discussion of the possible application of ultra-fine resolution data in the remote sensing of clouds.

13.2 Elementary Concepts

13.2.1 Ideal Case of Reflection from a Solid Slab

Consider the simple problem defined in Fig. 13.3a. Sunlight is transmitted through an upper layer of absorbing gas (in this case molecular oxygen) and then reflected from a solid, uniform surface of reflectance α_s. Suppose the reflected sunlight is measured at two wavelengths by an instrument above the gaseous layer (although we could equally pose the problem in terms of transmission measurements by an instrument below this layer as described later). These wavelengths are selected to lie in spectral regions of weak and strong absorption by the gas in question. Radiation at the wavelength of weak absorption, denoted I_1, passes through this gaseous layer largely unattenuated whereas the radiation corresponding to the region of strong absorption, I_2, is attenuated to a greater degree. We further suppose the wavelengths to be close enough that the reflection of the underlying surface and the incident solar radiation I_o are the same for each wavelength. For this simple case, the sunlight detected by an instrument observing the surface at an angle θ from nadir is

$$I_2 = I_o \alpha_s \exp(-\tau_2 m)$$
$$I_1 = I_o \alpha_s \exp(-\tau_1 m)$$

(13.1)

for each wavelength where τ_1 and τ_2 are the optical depths of the layer at the respective wavelengths. The airmass factor is

$$m = 1/\mu_0 + 1/\mu$$

where μ_0 is the cosine of the solar zenith angle θ_0 and μ is the cosine of the (nadir) observing angle θ. With the introduction of radiance ratio

$$X = \ln(I_2/I_1)$$

(13.2)

then after some re-arrangement, (13.1) becomes

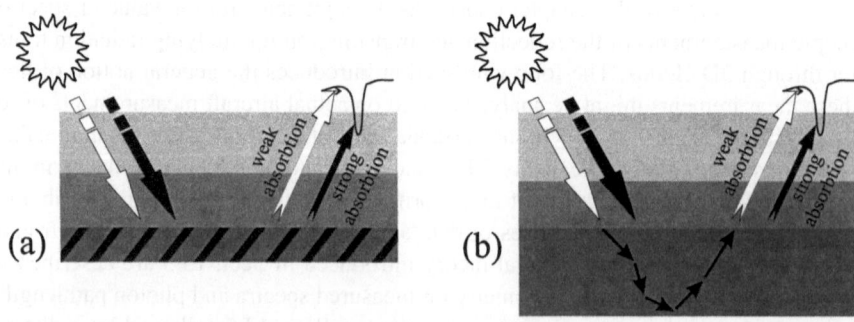

Fig. 13.3. (a) Geometry for the problem of absorption by a gaseous atmosphere overlying a reflecting surface; (b) as in (a) but for the case of a gaseous atmosphere overlaying and imbedded in a scattering layer

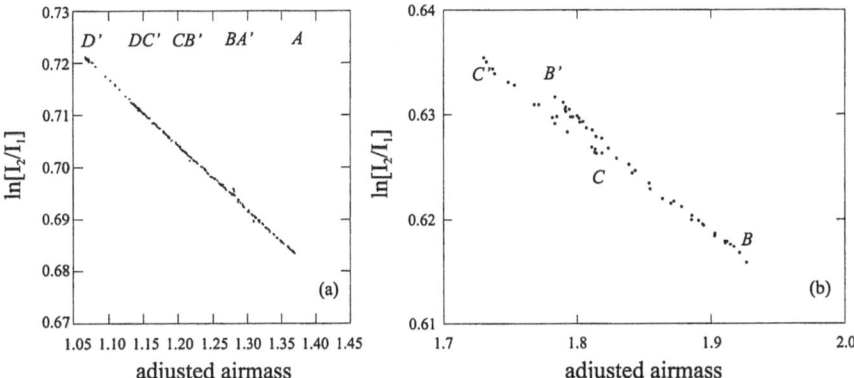

Fig. 13.4. (a) The measured relationship between the radiance ratio factor X (see text) and the (adjusted) airmass $\widetilde{m}$. The radiances were obtained by averaging spectral radiances measured by an airborne spectrograph. The data for this case refer to a very clean atmosphere. **(b)** As in (a) but for two other flight legs affected by diffuse radiation created by a thin aerosol and cirrus layer respectively below and above the aircraft. From O'Brien et al. (1999), with permission.

$$X = -(\tau_2 - \tau_1)m \qquad (13.3)$$

assuming that both α_s and I_o cancel. The resultant linear relation between X and m forms a Langley-like relationship wherein the measurements expressed in terms of X and their variation with m (as typically occurs with changing solar elevation) provides a way of retrieving the optical depth factor $(\tau_2 - \tau_1)$ of the gaseous layer. Through physical relationships, this optical depth difference relates to the gas path and thus amount of absorbing gas. This is the essential idea introduced by Yamamoto and Wark (1961).

Figure 13.4a confirms the predicted relationship between X and m obtained from the experimental airborne spectrograph data of O'Brien et al. (1999). These results correspond to the reflection measurements in the oxygen A-band portion of the solar spectrum using an airborne instrument flown above a very clean atmosphere overlying a dark ocean surface. The radiance data obtained from the spectrograph were averaged over two relatively broad spectral regions 3 nm wide to produce a single radiance ratio. Even at this relatively coarse resolution, which is similar to the resolution of GOME (see also Fig. 13.1), the predicted linear relation of (13.3) expressed in the form,

$$\widetilde{m} = cX + d \qquad (13.4)$$

accurately fits the data. This coarse-band 'model' is referred to as model A in the O'Brien et al. (1999) study and in subsequent discussion of the results. In this expression, $\widetilde{m}$ is an airmass factor adjusted for the pressure height of the aircraft

$$\widetilde{m} = [1 - (p/p_s)^2]m \qquad (13.5)$$

where p_s is the surface pressure, p is the pressure altitude of the aircraft. The quadratic form of this correction reflects the pressure dependence in the wings of oxygen lines. From the estimation of $\tilde{m}$ via model A (or via model B introduced later), inversion of (13.5) provides an estimate of p given p_{s}. This estimate can then be compared to the actual recorded pressure altitude of the aircraft and the closeness of these values is a form of validation of the given model.

13.2.2 Reflection from a Tenuous Medium

The simple analyses above not only provide a general background for the discussion to follow but also introduces the (reference) air mass m which can be thought of as a reference pathlength factor. With this background, we now consider the case of a layer for which reflection varies according to the distribution of scatterers within the layer and the strength of gaseous absorption superimposed on this scattering. Photons now travel over a variety of paths depending on the probabilities of the direction of scattering within the scattering layer. Unlike the simple example above, the path distributions differ according to the strength of the absorption by the interstitial gas. In spectral regions of strong absorption, photons detected at the top of the atmosphere travel relatively short paths within the scattering volume (they would not be observed otherwise) whereas photons from weak absorption regions penetrate further into the layer before re-emerging. We represent this situation in a simple way according to

$$I_2 = I_o \alpha_{\mathrm{s}} \exp(-\tau_2 m)$$
$$I_1 = I_o \alpha_{\mathrm{s}} \exp(-\Lambda \tau_1 m) \tag{13.6}$$

where as above I_2 corresponds to reflection from line centers where absorption is strong and photons are considered in this case to arise only from the top of the layer. The radiance I_1 corresponds to reflection in regions of line wings and the Λ factor is meant to represent the enhancements of the paths relative to the vertical paths due to the zigzag motions of the photons diffusely scattered within the layer (Fig. 13.3b) at these wavelengths. As above, the ratio of reflected radiances now becomes

$$X = -(\tau_2 - \Lambda \tau_1)m \tag{13.7}$$

where the slope of the relationship between X and m now differs from the slope factor contained in (13.3).

The experiments of O'Brien et al. (1999) also provide the opportunity to examine the effects of diffuse scattering on the relationships between radiances and airmass. Scattering by thin cirrus and aerosol was encountered during some of the flights reported by O'Brien et al. (1999). On these occasions, the relationship between X and $\tilde{m}$ responded in a manner anticipated by the simple analysis described above. The data are presented in Fig. 13.4b in the same way as in Fig. 13.4a with different flight legs indicated by the letter pairs. The flight leg labeled CC' and BB' occurred under a patchy layer of thin cirrus and above a very thin aerosol layer respectively. These layers create a source of diffuse radiation between the sun and the nadir viewing instrument.

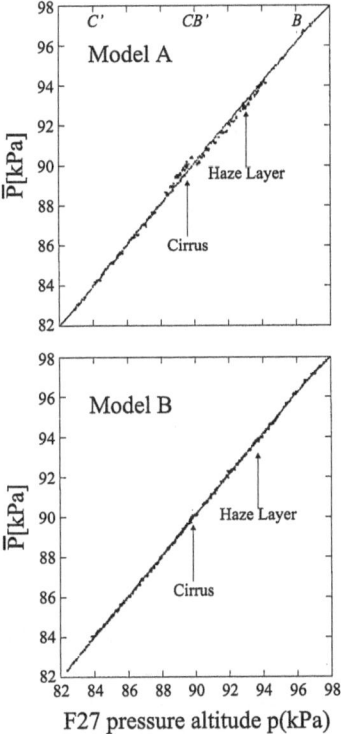

Fig. 13.5. The comparison between retrieved aircraft pressure altitude and directly measured pressure altitude according to model A (*upper*) and model B (*lower*) for the flight legs of Fig. 13.4b; from (O'Brien et al., 1999), with permission

Figure 13.5 (upper panel) indicates the relationship between the observed aircraft pressure altitude and the air-mass inferred pressure altitude from model A whereas the lower panel portrays results of the same comparison where the adjusted airmass is derived from the following model (hereafter model B):

$$\widetilde{m} = \sum_{j=1}^{N} f_j X_j + cX + d \qquad (13.8)$$

where the coefficients f_j, c and d are derived from empirical fits of $\widetilde{m}$ as functions of X and X_j. In this case the X is as before and the X_j's are the ratios derived from the individual radiances measured at the native resolution of the instrument (approximately 0.03 nm in contrast to the 3 nm radiances used to form X for model A). The ability to fit the data with the higher resolution information confirms the increased information content contained in higher spectral resolution measurements. These measurements contain specific information about the diffuse radiances and the scattering properties of the clouds and aerosol that create this diffuse field.

With the comparison of the results of Figs. 13.4a and 13.4b placed in the context of the simple analysis above, we note the following:

i.   The effect of scattering, when superimposed on absorption, introduces a form of ambiguity between reflected radiances and airmass. This ambiguity is directly related to the changing pathlengths of photons as they travel in an absorbing gas due to scattering by particles.
ii.   The degree of multiple scattering experienced by photons changes as the strength of the absorption changes. This implies that the pathlengths photons travel, among other factors, varies according to the strength of absorption.
iii. Ultra-fine spectral radiance data contains enough information to account for the variations of the pathlengths that result from the processes that create diffuse sunlight. These data presumably also contain relevant information about the scattering properties that characterize the multiple scattering process.

13.3 Photon Paths and Radiative Transfer

We now offer a way of connecting the spectrally resolved measurements of gaseous absorption like those of the O'Brien et al. (1999) study to optical properties of the scattering medium in which absorbing gases are imbedded. We return to the geometry shown in Fig. 13.3b. Photons emerging from the layer do so by traveling along different paths in the layer. Radiances $I_v(x, \Omega)$ emerging from the layer at point x in a direction Ω after traversing a *geometrical* path of length s from point x_0 may be expressed *in the absence of scattering* as

$$I_v(x, \Omega) = I_v(x_0, \Omega)e^{-\lambda_v} \tag{13.9}$$

where $x = x_0 + \Omega s$ and the exponential factor defines the attenuation within the medium governed by $\lambda_v = \sigma_{ev} s$ which is the *optical* path and σ_{ev} is the extinction coefficient, assumed constant along the beam. In general, *when multiple scattering is occurring*, more than one pathlength $\lambda_v = \sigma_{ev} L$, where the total geometrical path is denoted L, contributes to the measured radiance I_v. So the exponential attenuation term in (13.9) needs to be modulated by a probability density function $p(\lambda)$, leaving the v-dependence implicit, where

$$\int_0^\infty p(\lambda)d\lambda = 1 . \tag{13.10}$$

From there, the mean (optical) path follows as

$$\langle \lambda \rangle = \int_0^\infty \lambda p(\lambda)d\lambda . \tag{13.11}$$

The ratio of $\langle \lambda \rangle$ to cloud optical depth τ_c is a measure of how much the photon paths are enhanced through multiple scattering; $\langle \lambda \rangle$ is thus analogous to Λ in (13.7). It thus follows that the average distance $\langle L \rangle$ a photon travels in a uniform layer (i.e., for σ_e constant), is

$$\langle L \rangle = \int_0^\infty \frac{\lambda}{\sigma_e} p(\lambda) d\lambda = \frac{\langle \lambda \rangle}{\sigma_e} \tag{13.12}$$

where $\langle L \rangle$ is the mean photon *geometric* path, independent of ν.

It is worth distinguishing clearly between the *optical* path λ and *geometric* path L (beyond the important dependence of λ on ν). For a homogeneous medium defined by a constant extinction coefficient, these differ from one another only by this constant extinction coefficient. Any heterogeneity in extinction introduces a degree of de-correlation between $\langle \lambda \rangle$ and $\langle L \rangle$, as will be seen further on. (See Notes for the connection between λ statistics and those of the photon's order of scattering.)

13.3.1 The Equivalence Theorem

The equivalence theorem establishes a relation between radiances measured in a spectral region free of the molecular absorption to radiances measured within an absorption line.

Imagine a scattering layer in which the scattering particles are pervaded by a homogenous absorbing gas and that the scattering properties are constant across the spectral region in which the gas absorbs. (The scattering particles may also be partially absorbing as long this is also in a spectrally neutral way.) The dimensionless quantities that enter this problem are: (*i*) the particle single scatter albedo $\varpi_0 = \sigma_s/(\sigma_s + \sigma_a)$; (*ii*) the optical thickness outside the absorption band $\tau_c = (\sigma_s + \sigma_a)h$ (where h is the cloud's geometrical thickness); (*iii*) the optical path outside the band $\lambda = (\sigma_s + \sigma_a)L$ where L is the geometric path of the photon; and (*iv*) the optical path at a frequency ν inside the absorption line

$$\lambda_\nu = (\sigma_s + \sigma_a + \sigma_{a\nu})L = (1 + \gamma_\nu)\lambda \tag{13.13}$$

where $\sigma_{a\nu}$ is the absorption coefficient of the gas and

$$\gamma_\nu = \sigma_{a\nu}/(\sigma_s + \sigma_a) \tag{13.14}$$

is the ratio of the line absorption $\sigma_{a\nu}$ to particle extinction $\sigma_o = \sigma_s + \sigma_a$. Outside the absorption band $\gamma_\nu = 0$ and the probability that a photon contributes to $I(\gamma_\nu = 0, \Omega)$ traveling along an optical path between λ and $\lambda + d\lambda$ is denoted $p(\lambda)d\lambda$. The equivalence theorem is simply stated as

$$I(\gamma_\nu \neq 0, \Omega) = I(\gamma_\nu = 0, \Omega) \int_0^\infty \mathcal{T}_\nu(\lambda) p(\lambda, \Omega) d\lambda \tag{13.15}$$

where the transmission function

$$\mathcal{T}_\nu(\lambda) = e^{-\lambda_\nu} = \exp[-(1 + \gamma_\nu)\lambda]$$

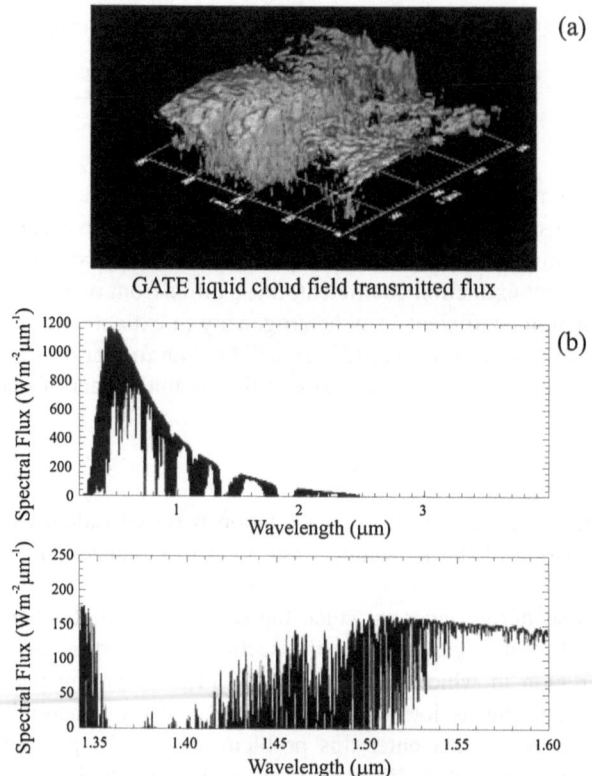

Fig. 13.6. (**a**) The cloud distributions derived from a 3D cloud resolving model simulation of a tropical squall line. (**b**) Downwelling spectral solar flux at the surface under this cloud system calculated using the equivalence theorem and photon path histories derived from a Monte Carlo simulation. The fluxes are an average over the entire domain containing the cloud system. The second panel of fluxes shows a portion of the spectrum of the upper panel centered on a water vapor band at a higher resolution. From Partain et al. (2000), with permission

depends on the optical pathlength λ_v in (13.13) that includes the effects of gaseous absorption. We note that, in homogeneous media, the constituents that do the scattering do not have to be also those that absorb; in either case the problem is the same as if all the absorption was concentrated at the scattering centers.

The equivalence theorem is valid for any cloud geometry and applies equally to other radiometric quantities although the photon distributions vary according to the quantity of interest. Stated differently, the pathlength statistics of transmission are not the same as they are for reflection. Partain et al. (2000) employ this principle to calculate spectral solar fluxes associated with sunlight scattered by a complex 3D cloud field. The cloud field in question, shown in Fig. 13.6a, is the output from a cloud resolving model. The continuum (i.e., particle) absorption and scattering properties were resolved explicitly in the model of Partain et al. (2000) by dividing the spectrum into a small number of spectral bands. The pathlength distributions were then calculated for each band free of gaseous absorption. Given this distribution of path lengths $p(\lambda)$, the equivalence theorem is then applied at a spectral resolution that resolves the molecular line absorption resulting in the line-by-line spectra shown in Fig. 13.6b.

In summary, the equivalence theorem states that

- if the properties of the radiation field along a given observing direction Ω are known outside the absorbing lines, i.e., if the radiance $I(\gamma_v = 0, \Omega)$ is known or inferred in some way, and
- if the corresponding pathlength distribution $p(\lambda, \Omega)$ is also known (we bring the directional dependence of this distribution to view as a reminder that the path distributions depend on the specified viewing geometry),

then the intensity along the same direction Ω follows immediately from (13.15).

13.3.2 Photon Path Properties Derived from Reflected Sunlight

If we introduce the ratio of intensities as

$$1 - s(\gamma_v) = \frac{I(\gamma_v, \Omega)}{I(\gamma_v = 0, \Omega)}$$

then it follows from (13.12) that the "line-profile" quantity

$$1 - s(\gamma_v) = \int_0^\infty T_v(\lambda)p(\lambda)d\lambda . \tag{13.16}$$

This quantity can be evaluated from measurements at different wavelengths representing different gaseous absorption strengths and thus a range of values of γ_v. An inverse Laplace Transform can then be applied to these measurements to retrieve the pathlength distribution. Stephens and Heidinger (2000) showed that the mean pathlength follows from (13.16) as

$$\langle \lambda_v \rangle = -\tau_c(1 + \gamma_v)\frac{\partial \ln s(\gamma_v)}{\partial \tau_{O_2}(\gamma_v)} . \tag{13.17}$$

Since the continuum optical depth τ_c is constant by design, γ_v is directly proportional to the optical depth of the absorbing gas τ_{O_2} since the ratio γ_v in (13.14) is simply $\tau_{O_2}(v)/\tau_c$.

The relationship in (13.17) is illustrated in Fig. 13.7. Panel (a) shows how the mean pathlength $\langle \lambda \rangle$ for $\gamma_v = 0$ varies almost linearly with τ_c (see Notes). In this example, the mean photon path was derived using (13.17) with the radiance quantities obtained from a doubling-adding model. Panel (b) shows the spectral character of $\langle \lambda_v \rangle$ as it varies with $\tau_{O_2}(v)$ (or equivalently with γ_v). This mean path is also compared to the mean photon path obtained independently using Monte Carlo simulations for $\gamma_v = 0$. Two properties of the mean photon path length highlighted in these figures are:

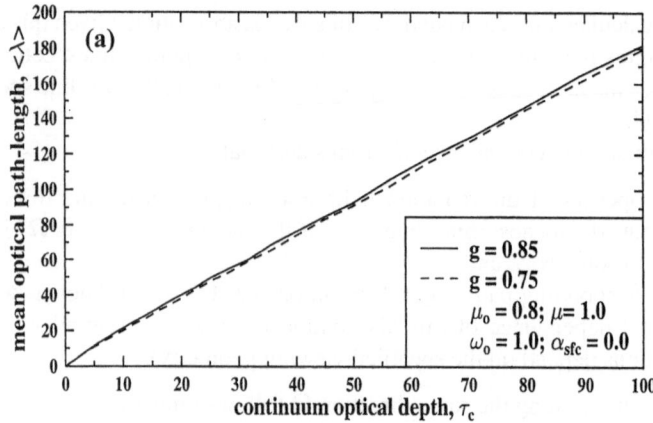

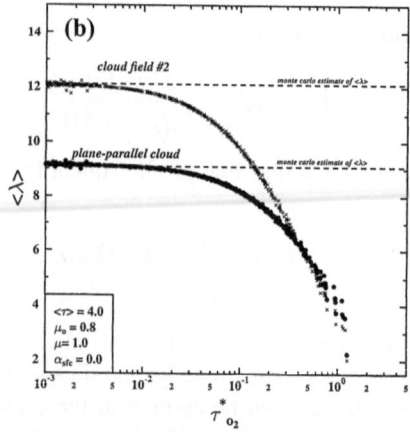

Fig. 13.7. (**a**) Model prediction of the variations of the mean photon path as a function of the continuum optical depth for plane parallel clouds. (**b**) The spectral variation via $\tau_{O_2}(v)$ of the mean photon path with absorption optical depth for two types of cloud fields. The horizontal dashed lines correspond to the values derived from Monte Carlo simulations for a non-absorbing cloud layer. From Stephens and Heidinger (2000), with permission

- For plane parallel clouds, the linear relation between $\langle \lambda \rangle$ and τ_c, as proposed by Stephens and Heidinger (2000), offers an alternative way of estimating optical depth from measurements of absorption lines formed upon reflection (see Notes). This method has one distinct advantage over the more established methods in that the input information required in any inversion are radiance ratios which are less susceptible to errors in absolute calibration. Further examination of this approach for estimating cloud optical depth remains a topic of future research.
- Figure 13.7b also shows the variation of $\langle \lambda \rangle_v$ with $\tau_{O_2}(v)$ for two cases, one of a plane parallel cloud and a second is a 3D varying cloud. The 3D nature of the

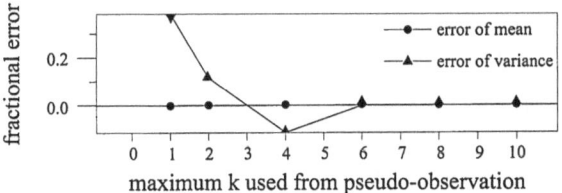

Fig. 13.8. Fractional error of the mean path length and its variance versus maximum observable absorption which is largely established by the spectral resolution properties of a given instrument. From Harrison and Min (1997) with permission. See main text and Notes

medium (described further on) leaves a clear imprint on $\langle \lambda_v \rangle$ compared to the plane parallel counterpart and indicates how the pathlength is enhanced by this 3D structure relative to the plane parallel cloud for the present case of reflection.

13.3.3 Instrumental Constraints on the Estimation of Photon Path Statistics

It is possible in principle to invert (13.15) and determine the distribution $p(\lambda)$. In practice, however, the information about $p(\lambda)$ that might be inferred through such an inversion process using measurements will be greatly influenced by the capability of the instrument providing the measurements. Harrison and Min (1997) discuss the potential for using ground-based spectroscopy for yielding information about photon-pathlengths taking into account specific performance properties typical of existing spectrometer systems, including allowance for measurement noise, finite slit-function width and errors typical of wavelength mis-registration. The essential point of their study is summarized in Fig. 13.8. Shown is the fractional error of the mean photon path and the error in the retrieved variance of the photon path distribution. The abscissa is related to the maximum of the oxygen optical depth resolved by the instrument and serves as a surrogate for instrument spectral resolution. These results were obtained using synthetic spectral data derived from a Monte Carlo simulation of the spectral radiances. The actual photon path distribution derived from the Monte Carlo simulations are also used to check the retrieved moments of the path distributions. Min and Harrison (1999) conclude that mean-pathlength can be derived using low-resolution data and is generally insensitive to many typical instrument errors. Higher-order moments of the distribution, however, require increasingly higher instrument performance and efforts to retrieve more than one or two low-order moments of the distribution are underway at the time of writing; see Min et al. (2004).

13.4 Analyses of Surface Transmission Measurements

The photon path concepts introduced above have, so far, been couched primarily in terms of reflection. It was noted, however, how the same notions also apply to transmission measurements obtained from (say) surface-based zenith-pointing or sun-tracking instruments. The in-band to out-of-band radiance ratio $1 - s_v$ again emerges as the important quantity in analyses of spectral absorption. For surface measurements, however, it is possible to frame the analyses in terms of reference clear-sky spectra and contrast these spectra, obtained at different times, with spectra collected under cloudy skies. For example, Min and Harrison (1999) and Min et al. (2001) derive A-band mean photon pathlengths from data obtained by the Rotating Shadowband Spectrometer (RSS). Although the spectral resolution of the RSS is coarse (approximately 9 nm full-width-half-maximum, FWHM), these data provide reliable estimates of the mean pathlength although but not higher order path distribution statistics.

The in- to out-of-band radiance ratio of the RSS clear sky direct-beam measurements takes the approximate form

$$\frac{I_2}{I_1} \approx \exp[-(\bar{\tau}_2 - \bar{\tau}_1)m] \tag{13.18}$$

where $m = 1/\cos\theta_0$ is the (solar) airmass factor and $\bar{\tau}$ is the optical depth of the absorbing gas averaged over a given spectral channel band-pass of the instrument. Note that this airmass factor for direct beam transmission differs from the airmass factor introduced for reflection and used in (13.1) above. The optical depths of (13.18) can in principle be inferred from the clear-sky direct-beam measurements using simple Langley analyses and then subsequently applied to diffuse sky radiance measurements made by the RSS under cloudy skies.

The ratio of cloudy sky measurements can be expressed as

$$\frac{I_2}{I_1} \approx \exp[-(\bar{\tau}_2 - \bar{\tau}_1)\langle\lambda\rangle] \tag{13.19}$$

where $\bar{\tau}_2$ and $\bar{\tau}_1$ are the previously determined clear-sky direct beam values and $\langle\lambda\rangle$ is the pathlength factor required to account for increased absorption of the diffuse radiation. Therefore the factor $\langle\lambda\rangle - m$ is a measure of the increase in pathlength, relative to that of the direct beam path, due to multiple scattering (see Notes).

Figure 13.9a is an example of RSS data analyzed in this way for the case of the thin cloud layer identified in Fig. 13.9b observed on Dec. 8, 1997. The time series of mean photon pathlength deduced from (13.19) is presented in this figure along with the reference solar airmass and cloud optical depth determined using other channels of the RSS. Figure 13.9b shows the sequence of mm-wavelength cloud radar reflectivity profiles indicating the existence of a low-lying single cloud layer that is relatively homogeneous. Figure 13.9c presents the difference quantity $\langle\lambda\rangle - m$ as a function of cloud optical depth. The results displayed in this way indicate a distinctly linear relationship between τ_c and $\langle\lambda\rangle$ (see Notes). Figures 13.9d-e show results in the same format but for a cloud system with multiple layers. In this case, the relationship between the optical depth of the multi-layered cloud system and the pathlength is no longer linear and is now non-unique. The pathlengths derived from data at times when the multiple layered cloud system was overhead are larger than those at times of a single layer cloud due to enhancements with multiple reflections between the layers (see Notes).

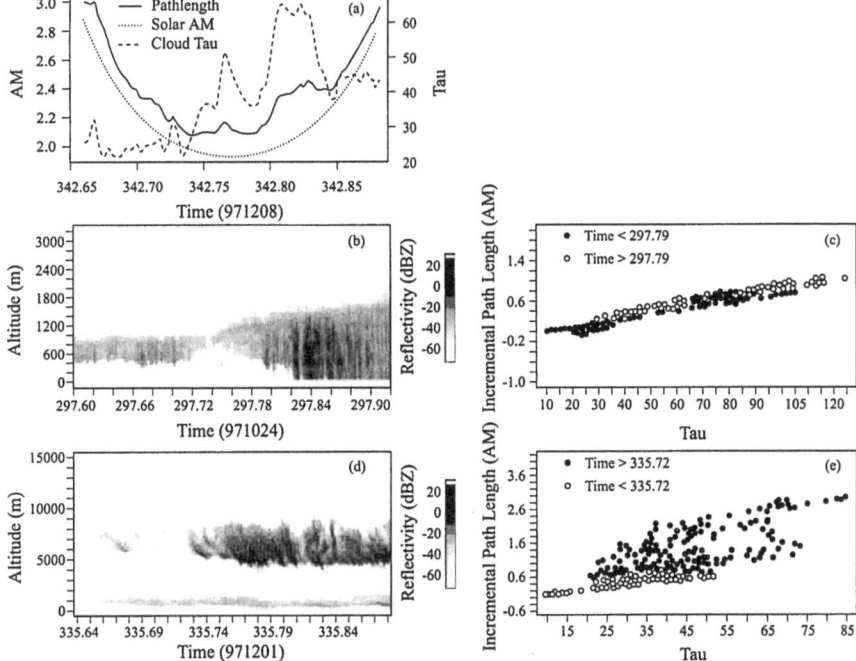

Fig. 13.9. (**a**) An example of the analyses presented by Min and Harrison (1999) of the path-length inferred from diffuse-horizontal irradiances obtained by the RSS for the case of Dec. 8, 1997 at the ARM SGP site. The pathlength (expressed as air mass) is contrasted to the solar airmass and to the optical depth determined from RSS and other measurement types available at the ARM site. (**b**) The radar reflectivity from the ARM cloud radar for the same case as that (a). (**c**) The relationship between the incremental pathlength (described in text) and cloud optical depth for the Oct. 24 case. (**d**) Same as (b) but for the case of Dec. 1, 1997. (**e**) As in (c) but for the case of Dec. 1, 1997. Panels (b)–(e) are from Min et al. (2001) with permission

Another example of this type of research is presented in the work of Portmann et al. (2001). This study provides analyses of measurements in the spectral region of the B-band obtained using a crossed Czerny–Turner fixed grating spectrograph. They use these measurements to infer mean photon paths adopting an analysis procedure similar to that of Min and Harrison (1999). Unlike the RSS, the spectrograph employed in this study had a narrow field of view and a slit function of width 0.8 nm (FWHM). As in the Min and Harrison study, Portmann et al. use the clear-sky direct beam measurements as a reference and derive the pathlength under cloudy skies from nadir radiances obtained with the instrument directed to the zenith.

Figures 13.10a and b show examples of the Portmann et al. analyses. Figure 13.10a refers to observations collected under nearly continuous cloud cover during the observing period on April 6, 1998. The upper panel is the pathlength determined from the measurements compared to a pathlength derived from modeled zenith radiances (second panel) that were obtained with the optical depth (fourth

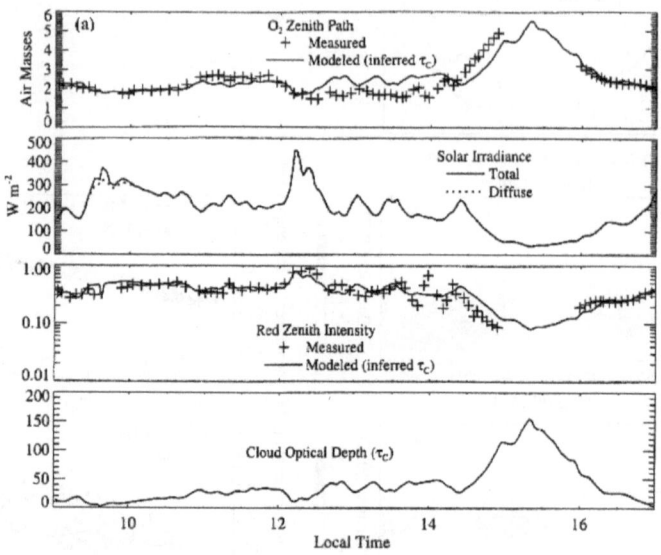

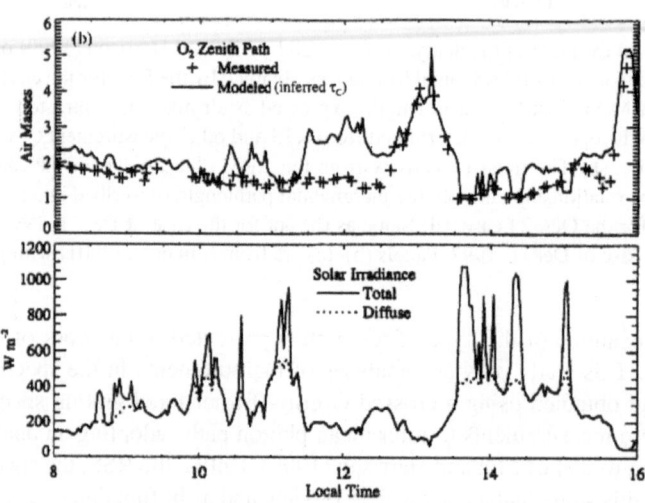

Fig. 13.10. (**a**) O_2 zenith pathlength, solar irradiance, red zenith intensity and optical depth versus local time on April 6, 1998. Periods of continuous cloud cover are inferred from the equivalence of the total and diffuse irradiances. (**b**) O_2 zenith pathlength and solar irradiance versus local time on July 22, 1998. Periods of broken cloud cover are inferred from the lack of equivalence of the total and diffuse irradiances and the high variability of the fluxes. The model results for both (a) and (b) are derived from a plane-parallel radiative transfer model constrained with the optical depths inferred from the irradiance measurements. From Portmann et al. (2001) with permission

panel) inferred from the measured broad band fluxes (third panel) as input. Figure 13.10b presents analyses of data collected on July 22, 1998 and in contrast to the case of Fig. 13.10a, corresponds to a day of broken cloudiness which is evident in comparison of the total to diffuse broadband fluxes (lower panel). Broken clouds were present during the periods between 0800–1130 and 1330–1530 LST during which time the measured path lengths varied from the 1 to 2 air masses which are typical of clear sky conditions.

It has been shown how spatial variations in the optical properties of clouds affect pathlength distributions in ways that are discernible in the absorption spectra of oxygen. It has also been shown that the mean photon path of a homogenous cloud is related in a simple way to the cloud optical depth, a result supported in the analyses of the ground-based data presented by Min et al. (2001). These observations also show how the simple relationship between $\langle \lambda \rangle$ and τ_c breaks down for multi-layered clouds (see Fig. 13.9e). It would also be useful to understand how horizontal variability influences this relationship and whether this influence could be quantified in some way in a remote sensing application (see Notes).

13.5 Spatial Heterogeneity Effects on Radiances and Photon Paths

Heidinger and Stephens (2002) examine the effect of three-dimensional cloud geometry on nadir radiances and the photon path length distributions derived from these radiances. That study was based entirely on model simulations and considered only two cloud cases for illustration. As such, the findings are preliminary and not meant to be comprehensive. A three-dimensional backward Monte Carlo model (see Chap. 4) was used for the purpose of simulating reflected radiances in the nadir. Input into this model were the 3 dimensional geometries of two specific cloud fields. Optical depth distributions representing these cloud fields were derived from two Landsat cloud images representing differing amounts of cloud variability. The details of the cloud fields associated with these images are described elsewhere (Harshvardhan et al., 1994; Barker et al., 1996; Heidinger and Stephens, 2002). The clouds observed were assumed to be confined entirely within a layer 500 m thick. The spatial resolution of the data is 28.5 m over a 58 by 58 km^2 region (2048 by 2048 pixels). One of the images is of a relatively homogeneous cloud field (Fig. 13.11, upper two panels; cloud field #1) and the other (lower two panels; cloud field #2) is a more complex field containing regions of broken cloudiness.

The simulated A-band reflected radiances were obtained for an underlying dark surface, a model atmosphere and the a solar zenith angle, $\mu_0 = 0.8$. The clouds are assumed to be conservatively scattering across the A-band with a scattering phase function specified as a Henyey-Greenstein phase function with an asymmetry parameter of 0.85. The Monte Carlo simulations provided spectral radiances I_c and line profiles s_v (left panels of Fig. 13.12) for cloud field #2 and the path distribution (Fig. 13.13) for cloud fields #1 and #2. The quantity s_v shown in Fig. 13.12 was derived as the ratio of radiances defined at a wavelength for which $\tau_{O_2} = 2.0$

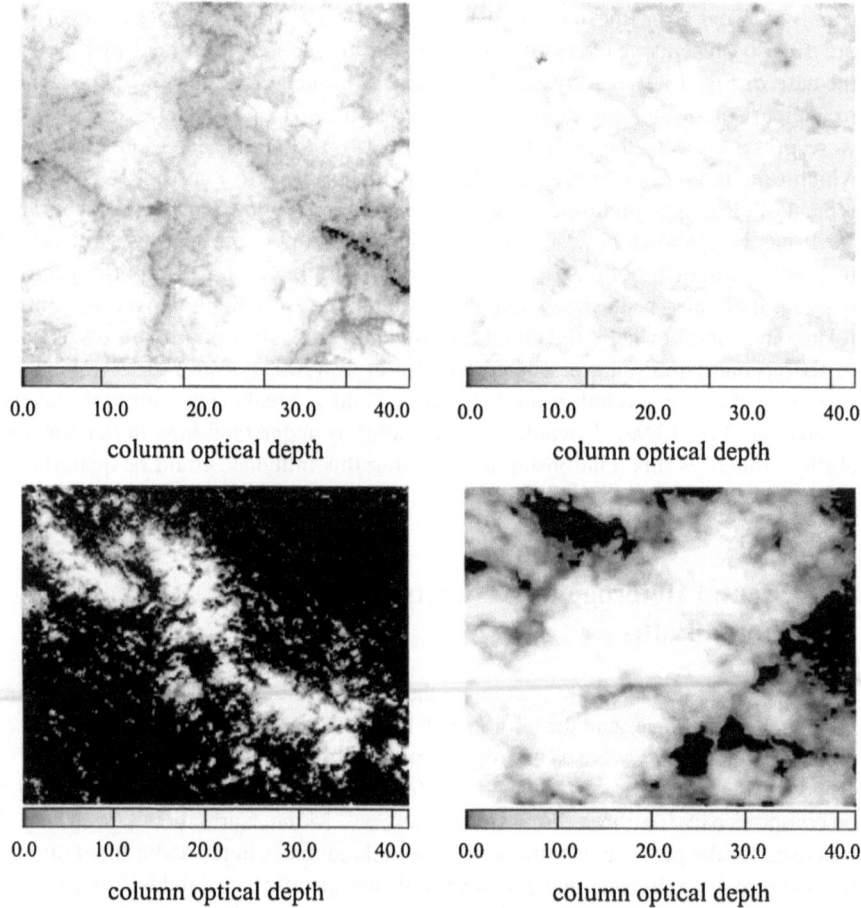

Fig. 13.11. *Upper left* panel shows a 58 × 58 km² Landsat-derived field of optical depth and a 4 × 4 km² sub-section of the same cloud field (*upper right*, cloud field #1). The *lower left panel* shows a same-sized optical depth field derived from a different Landsat scene and a sub-section of the scene is similarly extracted (*lower right*, cloud field #2)

to the continuum radiance I_c. Results obtained using the radiance distributions applied in the Independent Pixel Approximation (IPA) are also shown for comparison. These comparisons are presented in the form of differences between the fully resolved 3D simulations of I_c and s_v and the IPA equivalent of these quantities. These differences, presented in the right panels of Fig. 13.12, show how the IPA tends to over-estimate the continuum radiances in optically thick regions and underestimate the radiances in optically thinner regions. This is a consequence of the inability of the IPA to represent horizontal transport. Under the plane-parallel assumption used to produce the IPA results there is no net horizontal transport of photons whereas in three-dimensional clouds-photons tend to migrate from denser regions of clouds to

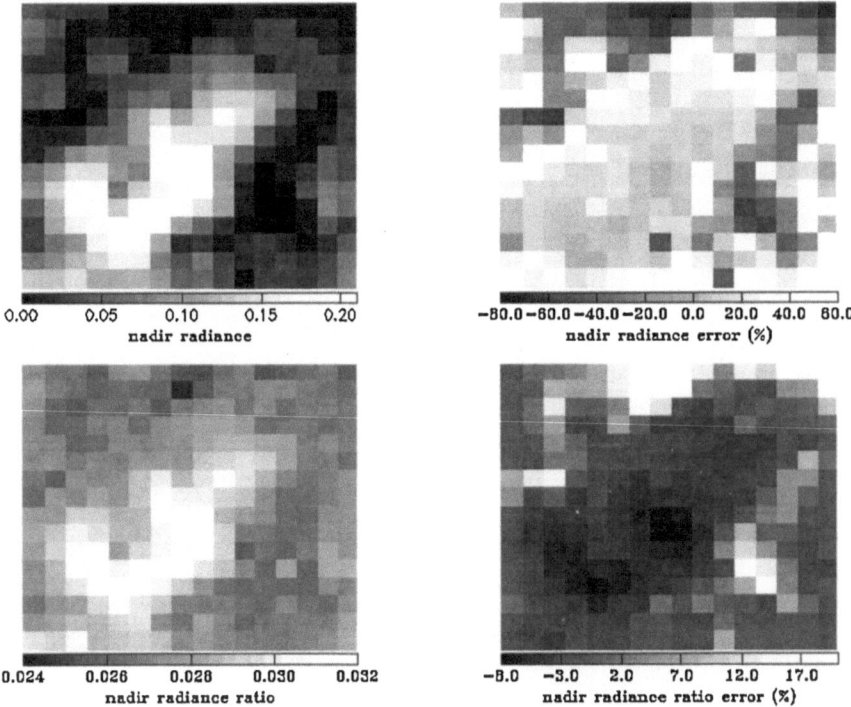

0.00   0.05   0.10   0.15   0.20
nadir radiance

−80.0 −60.0 −40.0 −20.0 0.0 20.0 40.0 60.0
nadir radiance error (%)

0.024   0.026   0.028   0.030   0.032
nadir radiance ratio

−8.0   −3.0   2.0   7.0   12.0   17.0
nadir radiance ratio error (%)

Fig. 13.12. The variation of the continuum nadir radiance field (*upper left*) and radiance ratio (*lower left*) for $\tau_{O_2} = 2.0$ for cloud field #2 of Fig. 13.11. *Right* panels are the respective differences in the quantities between the radiances simulated with a fully-resolved 3D model and radiances derived assuming the IPA. Images are from Heidinger and Stephens (2002), with permission

more tenuous regions thereby smoothing the radiance fields relative to the IPA counterparts (we refer to Chap. 12 for a more detailed description of radiative smoothing). Conversely, the reverse is true for s_v (lower right panel). At wavelengths characteristic of strong absorption, the effect of geometry on the path distribution is relatively weak as those photons that appear in the nadir field do not penetrate far into the medium and therefore do not experience the effect of spatial variations in the photon paths. The negative biases in of the IPA-derived s_v are largely in the opposite direction from the positive biases associated with I_c.

The pathlength distributions for nadir reflectance obtained from the Monte Carlo model for the two cloud fields of Fig. 13.11 are presented in Fig. 13.13. These diagrams present the distributions of optical, λ, and geometrical, L, pathlengths. The geometric pathlengths are normalized by the layer geometric thickness, H, and the optical pathlengths are normalized by the domain mean optical depth, $\langle \tau_c \rangle$, which is the same for both cloud fields. Displaying the pathlength distributions in this manner allows them to be plotted on the same scale. Also shown for comparison are the pathlength distributions for plane-parallel clouds of equal optical depth. The geometric

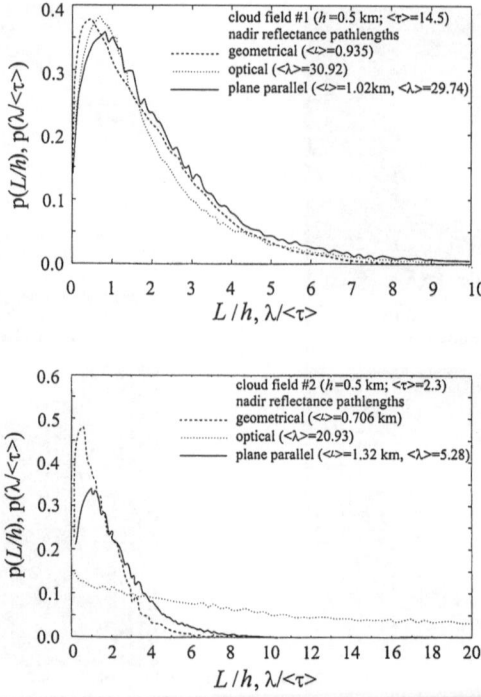

Fig. 13.13. Comparison of pathlength distributions for nadir reflectance for cloud field #1 (*upper panel*) and cloud field #2 (*lower panel*). See text and Notes for discussion of important difference between optical and geometrical paths

and optical pathlength distributions, normalized in the manner described, are identical for the plane-parallel clouds.

As mentioned earlier, spatial variability causes differences in the optical and geometric pathlengths compared to equivalent plane-parallel clouds. Due to the larger degree of spatial heterogeneity in cloud field #2 the difference between the optical and geometric pathlength distributions is much larger than in cloud field #1 where both are close to the equivalent plane-parallel distribution. For cloud field #2, the photons that comprise the nadir reflectance on average travel 1.4 times the layer thickness compared to 2.2 times the layer thickness in the plane-parallel cloud (see Notes).

13.6 Implications for the Remote Sensing of Clouds

The theoretical results presented in the previous section demonstrate how spatial variability affects radiances and radiance ratios in seemingly opposite ways. Furthermore, the pathlength information, encoded in the spectrum of s_v, is also discernibly influenced by this variability. The question then remains as to what extent the effects of spatial variability on I_c differ from the effects of spatial variability on s_v and

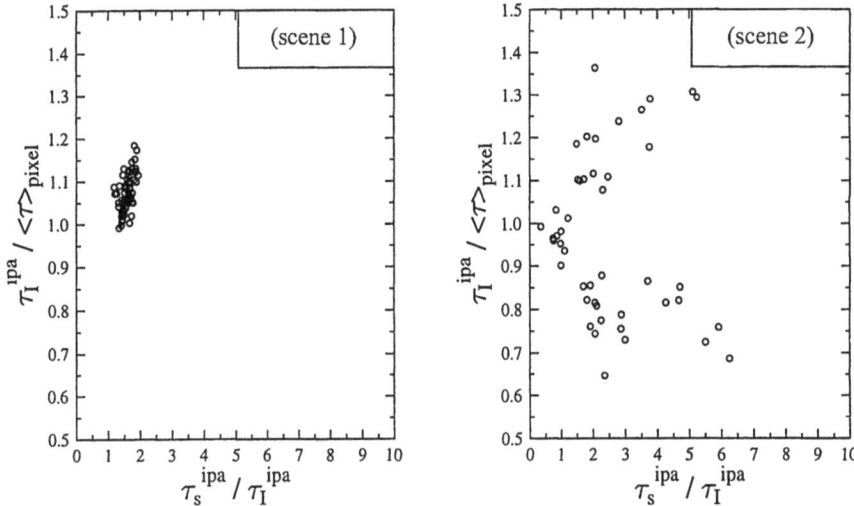

Fig. 13.14. Variation of the optical depth retrieval biases using continuum radiances, I_c, and the radiance ratios, s_v for cloud fields #1 and #2. These retrieved optical depths are denoted respectively τ_s^{ipa} and τ_I^{ipa} while the actual optical depth, averaged over the pixel is $\langle \tau \rangle_{pixel}$

to what extent these different influences might be used to benefit interpretation of cloudy sky radiances in terms of cloud optical properties. In this context, Stephens and Heidinger (2000) showed how the heterogeneity causes the averaged optical depth, inferred directly from radiances, to be underestimated and conversely how the averaged optical depth inferred directly from radiance ratio data is overestimated.

Figure 13.14 shows the effects of the biases on these two different estimates of optical depth retrieved on the pixel level data of cloud fields #1 and #2. The abscissa of Fig. 13.14 is the ratio of the optical depth retrieved using the radiance ratio, s_v, to the optical depth retrieved using the radiance, I_c. The ordinate is the ratio of the retrieved radiance-derived optical depth to the true value of optical depth obtained by averaging the Landsat data over the given pixel. The departure of the value of this ratio from unity thus represents the errors incurred by the conventional radiance-based retrievals such as performed using the AVHRR, GOES and MODIS visible channels.

For plane-parallel cloud fields with no spatial variability, the radiance and radiance ratio derived optical depths will be the same as the true optical depth with all points centering on the point (1.0,1.0) in Fig. 13.14. The values for cloud field # 1, being the more uniform of the two cloud fields cluster tightly around this expected plane-parallel value. The more spatially heterogeneous cloud field #2 is indicated by the points that deviate significantly from the plane-parallel case. Thus the spread of the values from this hypothetical plane-parallel value is a clear indication of the amount of variability in the cloud field. We propose that the value of the $\tau_s^{ipa}/\tau_I^{ipa}$ thus provides a method of diagnosis of when the effect of cloud heterogeneity is too great to permit a meaningful and accurate retrieval of optical depth using a plane-parallel model.

The two branches of points derived for cloud field # 2 in Fig. 13.14 indicates how the optical depths retrieved from radiance data alone can be less than or greater than the actual true value. The physical reason for this behavior relates back to the radiance smoothing that occurs under 3D variability. Under the influence of this smoothing, optically thick regions appear less reflective compared to a plane-parallel cloud of equivalent optical depth and, conversely, the optically thinner portions appear more reflective than the equivalent plane-parallel cloud.

13.7 Summary and Concluding Comments

This chapter describes a systematic attempt to demonstrate the value of spectroscopic measurements of the reflection and transmission by clouds and the insights these measurements provide in understanding radiative transfer through the cloudy atmosphere. We arbitrarily introduce the terminology ultra-fine resolution in reference to reflection and transmission measurements of resolution $1\,\mathrm{cm}^{-1}$ or less and proceed to demonstrate how ultra-fine measurements of absorption, formed by reflection and transmission in clouds, can be interpreted in terms of photon pathlength information. Such information is a fundamental property of the radiative transfer and, in particular, is germane to 3D transport problems for at least two reasons:

i. the characteristics of 3D radiative transport, at least as they effects measurable radiation fields, are uniquely defined by these path distributions; and
ii. the path distribution information can also be extracted from observations made under conditions of 3D geometry, thereby providing a quantitative measurable of this 3D transport.

The concepts for using absorption spectroscopy to infer photon path length are introduced in a simple way at first using the airborne spectrograph data of O'Brien et al. (1999). The analyses presented by these authors underscores the real value of highly resolved spectral measurements in separating effects of scattering from effects of absorption processes in clouds. These authors also demonstrate how ultra-fine spectral measurements offer an observable way of differentiating the effects of different processes that create diffuse sunlight. The effects of these processes are evident in the variations of the pathlengths inferred from measured reflection or transmission spectra. Thus it is proposed that measurements of these spectra contain relevant information about the scattering properties that defines the multiple scattering process.

The concept of photon path is further developed via the equivalence theorem and it is demonstrated how this theorem can be practically used in the context of Monte Carlo models to derive spectrally resolved solar fluxes through a complex 3D cloud system. We highlighted the simple relationship between the mean photon path $\langle\lambda\rangle$ and cloud optical depth τ_c for horizontally uniform clouds. We also demonstrated the influence of the 3D nature of the medium on $\langle\lambda\rangle$. These simulation results are supported by the observational studies of Min and Harrison (1999) and Portmann et al. (2001) based on analyses of surface spectrometer data. These studies indeed confirm the known relationship between $\langle \lambda \rangle$ and cloud optical depth τ_c for uniform clouds. The observations also confirm how this simple relationship between $\langle \lambda \rangle$ and τ_c relationship breaks down for layered cloud systems or under conditions of patchy cloudiness, thereby confirming the influence of 3D geometry on photon paths.

The effects of 3D geometry on photon path properties was explored theoretically. The theoretical results presented demonstrate how spatial variability affects reflected radiances and radiance ratios in seemingly reciprocal ways. Furthermore, the pathlength information, encoded in the spectra of radiance ratios, was shown to be discernibly influenced by this variability. It was then demonstrated how the effects of spatial variability on reflected continuum radiance I_c and optical depth inferred from it differ in opposite ways from the optical depth deduced from radiance ratios and pathlength relationships. The possible significance of these results to the topic of cloud optical property remote sensing was briefly addressed.

Notes and Further Reading

Section 13.1

- Use of reflected sunlight to estimate optical depth of clouds and other column properties is an approach adopted in many studies. Two examples can be found in the work of Nakajima and King (1990) and Minnis et al. (1993).
- The Global Ozone Monitoring Experiment (GOME) is an instrument that flies on the European Space Agency's (ESA) second European Remote Sensing Satellite (ERS-2) launched in 1995. GOME measures in the spectral region between 290-790 nm at a resolution varying between 0.2 and 0.4 nm. GOME views Earth in the nadir with a 960 km swath and a spatial resolution of 40×320 km^2.
- The studies of Mitchell and O'Brien (1987) and later O'Brien and Mitchell (1992) detail many of the issues involved in extracting information about gaseous concentration from absorption spectra. The interests of those authors was in retrieving surface pressure based on estimating the column path of oxygen by inverting measurements in the O$_2$ A-band (see also Stephens (1994) for a discussion on the basis of this measurement approach). The challenge is to retrieve oxygen very accurately requiring some correction for those photons that do not pass through the entire atmospheric column as a consequence of scattering. These authors showed how even small effects of multiple scattering by optically thin aerosol could be removed with sufficient spectral resolution. The same ideas, although applied at a much coarser spectral resolution, have been explored over a number of years to estimate cloud top pressure (e.g., Fischer and Grassl, 1991). See Rozanov et al. (2004) for operational retrievals of cloud properties from O$_2$ A-band using GOME and an overview of the relevant literature.

Section 13.3

- The Equivalence Theorem introduced by Irvine (1964) is also described by van de Hulst (1980). The power of this principle is not often realized despite its widespread use in a variety of atmospheric radiative transfer problems.
- In the presence of scattering, the optical path λ increments by unity with each scattering event whereas the geometric counterpart L increments by one mean-free-path. In optically thick *homogeneous* media, there is obviously not much difference between λ and L since there is, on average and by definition, one mean-free-path (i.e., $1/\sigma_e$) between each scattering. However, in *variable* media the mean-free-path is no longer the only moment of interest in the distribution of free paths (Davis and Marshak, 2004), so the one-to-one connection between λ and L is broken. Note that, while observational access to L is via well-mixed gases such as oxygen or water vapor, for λ we should use absorption features of the (liquid or ice) cloud particles.
- It is well-known that the mean order-of-scattering of reflected photons is proportional to the optical depth. See Chap. 12 (Sect. 12.3, and references therein) for diffusion theoretical arguments leading to this linear connection between $\langle \lambda \rangle$, which is essentially the mean number of scatterings, and τ_c (cf. Fig. 13.7a).
- To reinforce the argument that τ_c can be derived from $\langle \lambda \rangle$, we note that cloud geometrical thickness h has already been successfully derived from the distribution of multiple-scattering geometrical paths L observed directly using "off-beam" cloud lidar technology from ground (Love et al., 2001) and from space (Davis et al., 2001). It is shown in those same studies that τ_c has a rather weak impact on $\langle L \rangle$ for reflected photons; this is in sharp contrast with transmitted photons (cf. Sect. 13.4).
- In Fig. 13.8, the spectral resolution of the instrument is expressed in terms of an equivalent absorption coefficient which differs only from the optical depth by a constant factor. Stephens and Heidinger (2000) introduced the notion of expressing the instrument spectral resolution in terms of an effective maximum optical depth resolved by the instrument. In reality, the optical depth of oxygen (defined monochromatically) varies across the A-Band by 7 orders of magnitude for total column paths through the atmosphere and our ability to extract information from measurements actually depends sensitively on the range of optical depths that can be sampled. The finite resolution capability of instruments degrade the effective maximum and minimum optical depths observed. An instrument with a nominal 1 cm^{-1} resolution resolves maximum oxygen optical depths that are between 2 and 3 orders of magnitude below the monochromatic upper limit (cf. Fig. 1 of Stephens and Heidinger (2000)).

Section 13.4

- The availability of spectrally resolved data collected from surface instruments has increased over the past few years (e.g., Pfeilsticker et al. (1998), Min and Harrison (1999), Min et al. (2001), Portmann et al. (2001), and Min et al. (2004), among others). These data provide an opportunity to study the path properties inferred from oxygen absorption measurements as described in this chapter. Plans continue to develop an airborne spectrometer. Two such instruments have flown on aircraft in the past.

- The Rotating Shadowband Spectroradiometer (RSS) provides spectral direct, diffuse, and total horizontal irradiances over the wavelength range from 350 to 1075 nm. The RSS implements an automated shadowbanding technique used by other instruments and operates quasi-routinely at the U.S. Department of Energy Atmospheric measurement site located at Oklahoma as part of the Atmospheric Radiation Measurement (ARM) program (see http://www.arm.gov). A description of the Rotating Shadowband Spectroradiometer and its operation is described in Harrison et al. (1999) and further details about the instrument can be obtained at http://www.arm.gov/docs/instruments/static/rss.html/.

- Harrison and Min (1997) proposed a quite different technique for inferring pathlength information from A-band spectroscopy than sketched in (13.18)–(13.19) where we can assume $\bar{\tau}_1 \approx 0$ in the continuum. Rather than the exponential law in (13.19) for mean optical pathlength $\langle\lambda\rangle$, they assume an expression with a power-law tail in mean geometrical path $\langle L\rangle$ that is controlled by the the variance of L. Specifically, they assume a Gamma-law in L appropriately weighted by the pressure difference through the cloudy region and then normalized to one air mass and, invoking the Equivalence Theorem, take its Laplace transform. This is the technique implemented by Pfeilsticker et al. (1998), Min and Harrison (1999), Min et al. (2001, 2004).

- As noted above, we have $\langle\lambda\rangle$ ($\approx$ mean number of scatterings) proportional to τ_c for *reflection* from a homogeneous slab. For *transmission* however, we have for the mean number of scatterings (again $\approx \langle\lambda\rangle$) a law in $(1-g)\tau_c^2$ where g is the asymmetry factor; see Chap. 12 (Sect. 12.3) for a simple derivation in the diffusion limit. This translates to $\langle L\rangle/h \propto (1-g)\tau_c$ where h is the geometrical thickness of the (single) cloud layer, and essentially what Min et al. (2001) measure in Figs. 13.9a and c.

- As demonstrated empirically by Pfeilsticker (1999), the anomalous/Lévy diffusion transport model of Davis and Marshak (1997) explains both the linear relation between $\langle\lambda\rangle$ and τ_c for single cloud layers (a standard diffusion result) as well as its observed breakdown in more complex situations, specifically, when the photon diffusion becomes anomalous (see Fig. 13.9c and e).

Section 13.5

- To understand the relationship between optical depth and photon path statistics in more detail and how this relationship is affected by cloud structure, refer to Stephens and Heidinger (2000) and further by Heidinger and Stephens (2002), and references therein.

- In Chap. 12, Fig. 12.11 illustrates the systematic effect of cloud horizontal variability on mean geometrical path $\langle L\rangle$. Similar conclusions are drawn as here. Lévy

transport theory generalizes diffusion (photon random walk) theory to highly variable media and thus captures 3D effects on path.

Section 13.6

- Using ultra-fine O_2 spectroscopy, Pfeilsticker (1999) performs a ground-based retrieval of the variability parameter introduced in Lévy transport theory.

References

Barker, H.W., B.A. Wielicki, and L. Parker (1996). A parameterization for computing grid-averaged solar fluxes for inhomogeneous marine boundary layer clouds -
Part 2, Validation using satellite data. *J. Atmos. Sci.*, **53**, 2304–2316.

Burrows, J.P., M. Weber, M. Buchwitz, V.V Rozanov, A. Ladstaetter-Weisenmeyer,
A. Richter, R. de Beek, R. Hoogen, K. Bramstadt, K.U. Eichmann, M Eisinger, and
D. Perner (1999). The Global Ozone Monitoring Experiment (GOME): Mission
concept and first scientific results. *J. Atmos. Sci.*, **56**, 151–175.

Chamberlain, J.W. and D.M. Hunten (1987). *Theory of Planetary Atmospheres*.
Academic Press, San Diego (CA).

Davis, A. and A. Marshak (1997). Lévy kinetics in slab geometry: Scaling of transmission probability. In *Fractal Frontiers*. M.M. Novak and T.G. Dewey (eds.).
World Scientific, Singapore, pp. 63–72.

Davis, A.B. and A. Marshak (2004). Photon propagation in heterogeneous optical media with spatial correlations: Enhanced mean-free-paths and wider-than-
exponential free-path distributions. *J. Quant. Spectrosc. Radiat. Transfer*, **84**, 3–
34.

Davis, A.B., D.M. Winker, and M.A. Vaughan (2001). First retrievals of dense cloud
properties from off-beam/multiple-scattering lidar data collected in space. In *Laser
Remote Sensing of the Atmosphere, Selected Papers from the 20th International
Conference on Laser Radar*. A. Dabas and J. Pelon (eds.). Vichy (France), pp. 35–
38.

Fischer, J. and H. Grassl (1991). Detection of cloud top height from backscattered
radiances within the oxygen A-band. Part I: Theoretical study. *J. Appl. Meteor.*,
30, 1245–1259.

Harrison, L.C. and Q. Min (1997). Photon path distributions from O_2 A-Band ab-
sorption. In *IRS'96 Current Problems in Atmospheric Radiation (Proc. Intl. Rad.
Symposium, Fairbanks AK)*. W.L. Smith and K. Stamnes (eds.). A. Deepak Press,
Hampton (VA), pp. 594–598.

Harrison, L.C., M. Beauharnois, J. Berndt, P. Kierdrom, J. Michalsky, and Q-L. Min
(1999). The rotating shadowband radiometer (RSS) at the Southern Great Plains
(SGP). *Geophys. Res. Lett.*, **26**, 1715–1718.

Harshvardhan, B. Wielicki, and K.M. Ginger (1994). The interpretation of remotely
sensed cloud properties from a model paramerization perspective. *J. Climate*, **7**,
1987–1998.

Heidinger, A. and G.L. Stephens (2000). Molecular line absorption in a scattering atmosphere. II: Application to remote sensing in the O_2 A-band. *J. Atmos. Sci.*, **57**, 1615–1634.

Heidinger, A. and G.L. Stephens (2002). Molecular line absorption in a scattering atmosphere. III: Path length characteristics and effects of spatially heterogeneous clouds. *J. Atmos. Sci.*, **59**, 1641–1654.

Intergovernmental Panel on Climate Change (2001). Synthesis Report; Stand-alone edition, R.T. Watson et al. (eds.). Intergovernmental Panel on Climate Change, Geneva (Switzerland).

Irvine, W.M. (1964). The formation of absorption bands and the distribution of photon optical paths in a scattering atmosphere. *Bull. Astron. Inst. Neth.*, **17**, 226–279.

Love, S.P., A.B. Davis, C. Ho, and C.A. Rohde (2001). Remote sensing of cloud thickness and liquid water content with Wide-Angle Imaging Lidar (WAIL). *Atmos. Res.*, **59-60**, 295–312.

Min, Q.-L. and L.C. Harrison (1999). Joint statistics of photon pathlength and cloud optical depth. *Geophys. Res. Lett.*, **26**, 1425–1428.

Min, Q.-L., L.C. Harrison, and E.E. Clothiaux (2001). Joint statistics of photon pathlength and cloud optical depth: Case studies. *J. Geophys. Res.*, **106**, 7375–7385.

Min, Q.-L., L.C. Harrison, P. Kiedron, J. Berndt, and E. Joseph (2004). A high-resolution oxygen A-band and water vapor band spectrometer. *J. Geophys. Res.*, **109**, D02202, doi:10.1029/2003JD003540.

Minnis, P., P.W. Heck, D.F. Young, and B.J. Snider (1993). Stratocumulus cloud properties derived from simultaneous satellite and island-based instruments during FIRE. *J. Atmos. Sci.*, **54**, 1525–1532.

Mitchell, R.M. and D.M. O'Brien (1987). Error estimates for passive satellite measurement of surface pressure using absorption in the A-band of oxygen. *J. Atmos. Sci.*, **44**, 1981–1990.

Nakajima, T. and M.D. King (1990). Determination of optical thickness and effective radius of clouds from reflected solar radiation measurements: Part I: Theory. *J. Atmos. Sci.*, **47**, 1878–1893.

O'Brien, D. and R.M. Mitchell (1992). Error estimates for the retrieval of cloud top pressure using absorption in the A-band of Oxygen. *J. Appl. Meteor.*, **31**, 1179–1192.

O'Brien, D., R.M. Mitchell, S.A. English, and G.A. Da Costa (1999). Airborne measurements of air mass from O_2 A-band absorption spectra. *J. Atmos. Oceanic Tech.*, **15**, 1272–1286.

Partain, P., A. Heidinger, and G.L. Stephens (2000). Spectral resolution atmospheric radiative transfer: Application of equivalence theorem. *J. Geophys. Res.*, **105**, 2163–2177.

Pfeilsticker, K. (1999). First geometrical pathlengths probability density function derivation of the skylight from spectroscopically highly resolving oxygen A-band observations. 2. Derivation of the Lévy-index for the skylight transmitted by mid-latitude clouds. *J. Geophys. Res.*, **104**, 4101–4116.

Pfeilsticker, K., F. Erle, O. Funk, H. Veitel, and U. Platt (1998). First geometrical pathlengths probability density function derivation of the skylight from spectroscopically highly resolving oxygen A-band observations: 1. Measurement technique, atmospheric observations, and model calculations. *J. Geophys. Res.*, **103**, 11,483–11,504.

Portmann, R.W., S. Solomon, R.W. Sanders, J.S. Daniel, and E. Dutton (2001). Cloud modulation of zenith sky oxygen path lengths over Boulder, Colorado: Measurement versus model. *J. Geophys. Res.*, **106**, 1139–1155.

Rozanov, V.V., A.A. Kokhanovsky, and J.P. Burrows (2004). The determination of cloud altitudes using GOME reflectance spectra: Multilayered cloud systems. *IEEE Trans. Geosci. and Remote Sens.*, **42**, 1009–1017.

Stephens, G.L. (1994). *Remote Sensing of the Lower Atmosphere: An Introduction.* Oxford University Press, New York (NY).

Stephens, G.L. (1999). Radiative effects of clouds and water vapor. In *Global Energy and Water Cycles.* Browning and Gurney (eds.). Cambridge University Press, New York (NY), pp. 71–90.

Stephens, G.L. and T.J. Greenwald (1990). The Earth's radiation budget and its relation to atmospheric hydrology: 2. Observations of cloud effects. *J. Geophys. Res.*, **96**, 15,325–15,340.

Stephens, G.L. and A. Heidinger (2000). Line absorption in a scattering atmosphere. I: Theory. *J. Atmos. Sci.*, **57**, 1599–1614.

Stephens, G.L., D. Vane, R. Boain, G. Mace, K. Sassen, Z. Wang, A. Illingworth, E. O'Connor, W. Rossow, S. Durden, S. Miller, R. Austin, A. Benedetti, C. Mitrescu, and CloudSat Science Team (2002). The CloudSat mission and the A-Train: A new dimension of space-based observations of clouds and precipitation. *Bull. Amer. Metereol. Soc.*, **83**, 1771–1790.

van de Hulst, H.C. (1980). *Multiple Light Scattering: Tables, Formulae and Applications.* Academic Press, San Diego (CA).

Yamamoto, G.A. and D.Q. Wark (1961). Discussion of the letter by R.A. Hanel, Determination of cloud altitude from satellite. *J. Geophys. Res.*, **66**, 3596.

14

3D Radiative Transfer in Vegetation Canopies and Cloud-Vegetation Interaction

Y. Knyazikhin, A. Marshak and R.B. Myneni

14.1 Introduction 617

14.2 Vegetation Canopy Structure and Optics 620

14.3 Radiative Transfer in Vegetation Canopies 627

14.4 Canopy Spectral Invariants 631

14.5 Vegetation Canopy as a Boundary Condition to Atmospheric Radiative Transfer 639

14.6 Summary 644

References 645

Suggested Reading 649

14.1 Introduction

Interaction of photons with a host medium is described by a linear transport equation. This equation has a very simple physical interpretation; it is a mathematical statement of the energy conservation law. In spite of the different physics behind radiation transfer in clouds and vegetation, these media have certain macro-and micro-scale features in common. First, both are characterized by strong horizontal and vertical variations, and thus their three-dimensionality is important to correctly describe the photon transport. Second, the radiation regime is substantially influenced by the sizes of scattering centers that constitute the medium. Drop and leaf size distribution functions are the most important variables characterizing the micro-scale structure of clouds and vegetation canopies, respectively. Third, the independent (or incoherent) scattering concept underlies the derivation of the extinction coefficient and scattering phase function in both theories (van de Hulst, 1980, pp. 4–5; Ross, 1981, p. 144). This allows the transport equation to relate micro-scale properties of the medium to the photon distribution in the entire medium. From a mathematical point of view, these three features determine common properties of radiative transfer in clouds and vegetation.

However, the governing radiative transfer equation for leaf canopies, in both three-dimensional (3D) and one-dimensional (1D) geometries, has certain unique features. The extinction coefficient is a function of the direction of photon travel. Also, the differential scattering cross section is not, as a rule, rotationally invariant, i.e., it generally depends on the absolute directions of photon travel Ω and Ω', before and after scattering, respectively, and not just the scattering angle $\cos^{-1}(\Omega \cdot \Omega')$. Finally, the single-scattering albedo is also a function of spatial and directional variables. These properties make solving of the radiative transfer equation more complicated; for example, the expansion of the differential scattering cross section in spherical harmonics (see Chap. 4) cannot be used.

In contrast to radiative transfer in clouds, the extinction coefficient in vegetation canopies introduced by Ross (1981) is wavelength independent, considering the size of scattering elements (leaves, branches, twigs, etc.) relative to the wavelength of solar radiation. Although the scattering and absorption processes are different at different wavelengths, the optical distance between two arbitrary points within the vegetation canopy does not depend on the wavelength. This spectral invariance results in various unique relationships which, to some degree, compensate for difficulties in solving the radiative transfer equation due to the above-mentioned features of the extinction and the differential scattering cross sections.

We idealize a vegetation canopy as a medium filled with small planar elements of negligible thickness. We ignore all organs other than green leaves in this chapter. In addition, we neglect the finite size of vegetation canopy elements. Thus, the vegetation canopy is treated as a gas with nondimensional planar scattering centers, i.e., a turbid medium. In other words, one cuts leaves residing in an elementary volume into "dimensionless pieces" and uniformly distributes them within the elementary volume. Two variables, the leaf area density distribution function and the leaf normal distribution, are used in the theory of radiative transfer in vegetation canopies to convey "information" about the total leaf area and leaf orientations in the elementary volume before "converting the leaves into the gas."

It should be noted that the turbid medium assumption is a mathematical idealization of canopy structure, which ignores finite size of leaves. In reality, finite size scatterers can cast shadows. This causes a very sharp peak in reflected radiation about the retro-solar direction. This phenomenon is referred to as the "hot spot" effect (Fig. 14.1). It is clear that point scatters cannot cast shadows and thus the turbid medium concept in its original formulation (Ross, 1981) fails to predict or duplicate experimental observation of exiting radiation about the retro-illumination direction (Kuusk, 1985; Gerstl and Simmer, 1986; Marshak, 1989; Verstrate et al., 1990). Recently, Zhang et al. (2002) showed that if the solution to the radiative transfer equation is treated as a Schwartz distribution, then an additional term must be added to the solution of the radiative transfer equation. This term describes the hot spot effect. This result justifies the use of the transport equation as the basis to model canopy-radiation regime. Here we will follow classical radiative transfer theory in vegetation canopies proposed by Ross (1981) with an emphasis on canopy spectral response to

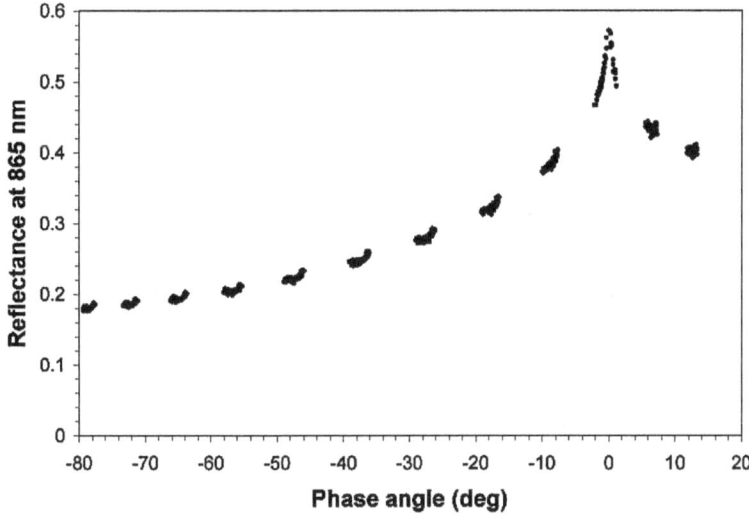

Fig. 14.1. Mean reflectance of evergreen broadleaf forest at near-infrared (865-nm) wavelength as a function of phase angle (after Bréon et al., 2001). Data were acquired by the Polarization and Directionality of the Earth's Reflectances (POLDER) multi-angle spaceborne instrument (Deschamps et al., 1994). Phase angles are shown with a negative sign if the POLDER view azimuth was greater than 180°. Canopy reflectance exhibits a sharp peak about the retro-illumination direction (the "hot spot" effect), which classical transport equation can not predict

the incident radiation. For the mathematical theory of Schwartz distributions applicable to the transport equation, the reader is referred to Germogenova (1986), Choulli and Stefanov (1996) and Antyufeev (1996).

Finally, what are our motivations to include a chapter on radiative transfer in vegetation canopies in a volume on atmospheric radiative transfer? Why is the vegetation canopy a special type of surface? First of all, vegetated surfaces play an important role in the Earth's energy balance and have a significant impact on the global carbon cycle. The problem of accurately evaluating the exchange of carbon between the atmosphere and terrestrial vegetation has received scientific (Intergovernmental Panel on Climate Change, 1995) and political (Steffen et al., 1998) attention. The next motivation is both the similarity and the unique features of the radiative transfer equations that govern radiative transfer processes in these neighboring media. Because of their radiative interactions, the vegetation canopy and the atmosphere are coupled together; each serves as a boundary condition to the radiative transfer equations in the adjacent medium. To better understand radiative processes in these media we need an accurate description of their interactions. This chapter therefore complements the rest of the book and mainly deals with radiative transfer in vegetation canopies. The last section outlines a technique needed to describe canopy–cloud interaction.

14.2 Vegetation Canopy Structure and Optics

Solar radiation scattered from a vegetation canopy results from the interaction of photons traversing through the foliage medium, bounded at the bottom by a radiatively participating surface. Therefore, to estimate the canopy radiation regime, three important features must be carefully formulated. They are (1) the architecture of individual plants and of the entire canopy; (2) optical properties of vegetation elements (leaves, stems) and soil; the former depends on physiological conditions (water status, pigment concentration); and (3) atmospheric conditions, which determine the incident radiation field (Ross, 1981). For radiative transfer in clouds, they correspond respectively to cloud micro- (e.g., distribution of cloud drop sizes) and macro- (e.g., cloud type and geometry) structures, cloud drop optical properties, and boundary (illumination) conditions. Note that radiation incident on the top of the atmosphere is a monodirectional solar beam while the vegetation canopies are illuminated both by a monodirectional beam attenuated by the atmosphere and radiation scattered by the atmosphere (diffuse radiation). Photon transport theory aims at deriving the solar radiation regime, both within the vegetation canopy and cloudy atmosphere, using the above-mentioned attributes as input data. For vegetation canopy, the leaf area density distribution, u_L, leaf normal orientation distribution, g_L, leaf scattering phase function, γ_L, and boundary conditions specify these input (Ross, 1981; Myneni et al., 1990; Knyazikhin, 1991; Pinty and Verstraete, 1997). We will start with definitions of these variables.

14.2.1 Vegetation Canopy Structure

At the very least, two important wavelength-independent structural attributes – leaf area density and leaf normal orientation distribution functions – need to be defined in order to quantify vegetation-photon interactions.

The one-sided green leaf area per unit volume in the vegetation canopy at location $x = (x, y, z)^T$ is defined as the leaf area density distribution $u_L(x)$ (in m²/m³ or simply m⁻¹). Realistic modeling of $u_L(x)$ is a challenge for it requires simulated vegetation canopies with computer graphics and tedious field measurements. The dimensionless quantity

$$L(x, y) = \int_0^H u_L(x, y, z)\mathrm{d}z \,, \tag{14.1}$$

is called the *leaf area index*, one-sided green leaf area per unit ground area at (x, y). Here H is the depth of the vegetation canopy.

Figures 14.2 and 14.3 demonstrate a computer-generated Norway spruce stand and corresponding leaf area index $L(x, y)$ at a spatial resolution of 50 cm (i.e., distribution of the mean leaf area index $L(x, y)$ taken over each of 50 by 50 cm ground cells). Leaf area index is the key variable in most ecosystem productivity models, and in global models of climate, hydrology, biogeochemistry and ecology that attempt to describe the exchange of fluxes of energy, mass (e.g., water and CO_2), and

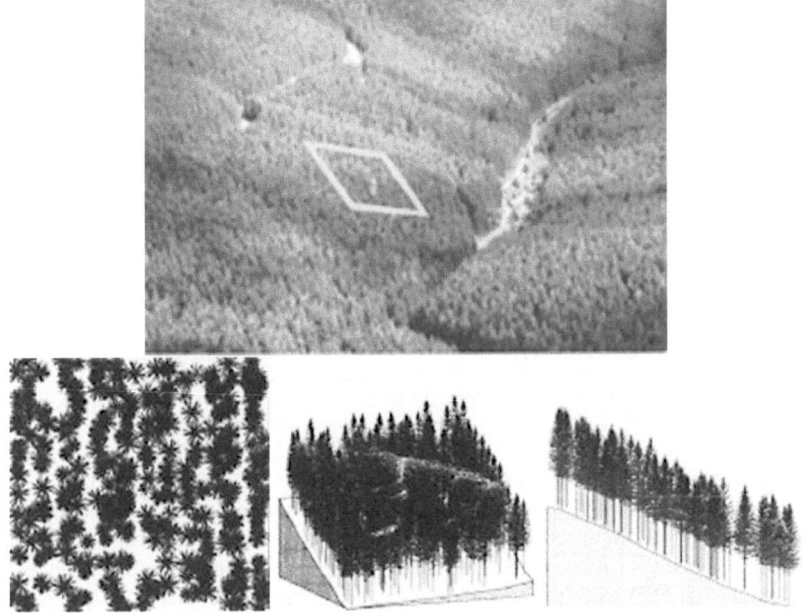

Fig. 14.2. Photo shows a Norway spruce stand about 50 km east of Göttingen in the Harz mountains. The forest is about 45 years old and situated on the south slope. For the sake of detailed examination of the canopy structure, a site covering an area of approximately $40 \times 40\,\mathrm{m}^2$ was chosen (shown as a square) and taken as representative for the whole stand. The canopy space is limited by the slope and a plane parallel to the slope at the height of the tallest tree of 12.5 m. There are in total 297 trees in the sample stand. The tree trunk diameters varied from 6 to 28 cm. The stand is rather dense but with some localized gaps. For the needs of modeling, the trees were divided into five groups with respect to the tree trunk diameter. A model of a Norway spruce based on fractal theory was then used to build a representative of each group (Kranigk and Gravenhorst, 1993; Kranigk et al., 1994; Knyazikhin et al., 1996). Given the distribution of tree trunks in the stand and the diameter of each tree, the entire sample site was generated. Bottom panels demonstrate the computer-generated Norway spruce stand shown from different directions: crown map (*left* panel), front view (*middle* panel), and cross section (*right* panel). Figures in the *lower* panels reprinted from Knyazikhin et al. (1996) with permission from Elsevier

momentum between the surface and the atmosphere. In order to quantitatively and accurately model global dynamics of these processes, differentiate short-term from long-term trends as well as to distinguish regional from global phenomena, this parameter must often be collected for a long period of time and should represent every region of the Earth land surface. The leaf area index has been operationally produced from data provided by two instruments, the moderate resolution imaging spectroradiometer (MODIS) and multiangle imaging spectroradiometer (MISR), during the Earth Observing System (EOS) Terra mission (Myneni et al., 2002; Knyazikhin et al., 1998a,b). A global map of MODIS leaf area index at 1-km resolution is shown in

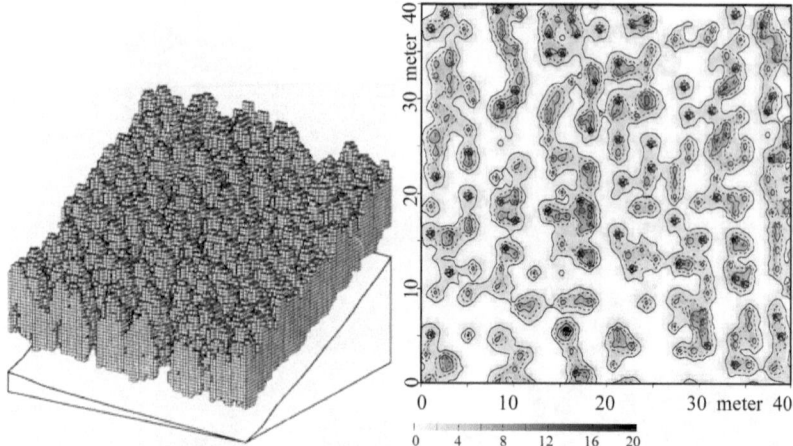

Fig. 14.3. A fine spatial mesh of the resolution of $50 \times 50 \times 50\,\text{cm}$ was imposed on the computer-generated sample site shown in Fig. 14.2 and the leaf area density $u_L(x)$ in each of the fine cells was evaluated. The canopy space contains 160,000 fine cells. The *left* panel shows three-dimensional distribution of foliated cells. This distribution is described by an indicator function $\chi(x)$ whose value is 1 if $u_L(x) \neq 0$, and 0 otherwise. The leaf area index $L(x, y)$ (14.1) derived from the computer-generated leaf area density $u_L(x)$ is shown in the *right* panel. Reprinted from Knyazikhin et al. (1996) with permission from Elsevier

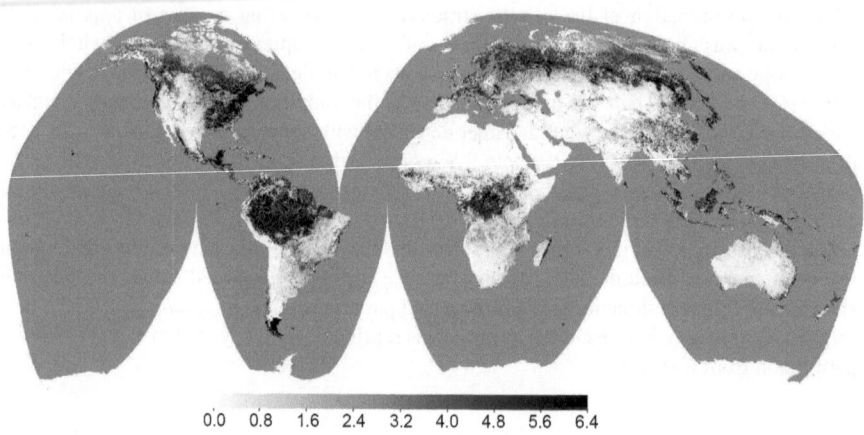

Fig. 14.4. Grey-coded map of the leaf area index at 1-km resolution for July 20–27, 2001, derived from data acquired by MODIS on board the EOS Terra platform

Fig. 14.4. Note that the theory of radiative transfer in vegetation canopies presented in this chapter underlies the retrieval technique for producing global leaf area index from satellite data (Knyazikhin et al., 1998a,b).

Let Ω_L be the upward normal to the leaf element. The following function characterizes the leaf normal distribution: $g_L(\Omega_L)/2\pi$ is the probability density of the leaf normal distribution with respect to the upward hemisphere (Ξ_+)

$$\frac{1}{2\pi} \int_{\Xi_+} g_L(\mathbf{\Omega}_L)\, d\mathbf{\Omega}_L = 1 \,. \tag{14.2}$$

If the polar angle θ_L and azimuth ϕ_L of the normal $\mathbf{\Omega}_L$ are assumed independent, then

$$\frac{1}{2\pi} g_L(\mathbf{\Omega}_L) = \bar{g}_L(\theta_L) \frac{1}{2\pi} h_L(\phi_L) \,, \tag{14.3}$$

where $\bar{g}_L(\theta_L)$ and $h_L(\phi_L)/2\pi$ are the *probability density functions of leaf normal inclination* and *azimuth*, respectively, and

$$\int_0^{\pi/2} \bar{g}_L(\theta_L) \sin\theta_L d\theta_L = 1, \quad \frac{1}{2\pi} \int_0^{2\pi} h_L(\phi_L)\, d\phi_L = 1 \,. \tag{14.4}$$

The functions $g_L(\mathbf{\Omega}_L)$, $\bar{g}_L(\theta_L)$ and $h_L(\phi_L)$ depend on the location x in the vegetation canopy but this is implicit in the remainder of the text.

The following example model distribution functions for leaf normal inclination were proposed by DeWit (1965): *planophile* – mostly horizontal leaves; *erectophile* – mostly erect leaves; *plagiophile* – mostly leaves at 45 degrees; *extremophile* – mostly horizontal and vertical leaves; and *uniform* – all possible inclinations. These distributions can be expressed as (cf. Bunnik, 1978),

$$\text{Planophile:} \qquad \bar{g}_L(\theta_L) = \frac{2}{\pi}\left(\frac{1 + \cos 2\theta_L}{\sin \theta_L}\right), \tag{14.5a}$$

$$\text{Erectophile:} \qquad \bar{g}_L(\theta_L) = \frac{2}{\pi}\left(\frac{1 - \cos 2\theta_L}{\sin \theta_L}\right), \tag{14.5b}$$

$$\text{Plagiophile:} \qquad \bar{g}_L(\theta_L) = \frac{2}{\pi}\left(\frac{1 - \cos 4\theta_L}{\sin \theta_L}\right), \tag{14.5c}$$

$$\text{Extremophile:} \qquad \bar{g}_L(\theta_L) = \frac{2}{\pi}\left(\frac{1 + \cos 4\theta_L}{\sin \theta_L}\right), \tag{14.5d}$$

$$\text{Uniform:} \qquad \bar{g}_L(\theta_L) = 1 \,. \tag{14.5e}$$

Example distributions of the leaf normal inclination are shown in Fig. 14.5.

Certain plants, such as soybeans and sunflowers, exhibit heliotropism, where the leaf azimuths have a preferred orientation with respect to the solar azimuth. A simple model for h_L in such canopies is $h_L(\phi_L, \phi_0) = 2\cos^2(\phi_0 - \phi_L - \eta)$, where the parameter η is the difference between the azimuth of the maximum of the distribution function h_L and the fixed azimuth of the incident photon ϕ_0 (cf. Verstraete, 1987). In the case of diaheliotropic distributions, which tend to maximize the projected leaf area to the incident stream, $\eta = 0$. On the other hand, paraheliotropic distributions tend to minimize the leaf area projected to the incident stream and give $\eta = \pi/2$. A more general model for the leaf normal orientations is the β-distribution, the parameters of which can be obtained from fits to field measurements of the leaf normal orientation (Goel and Strebel, 1984).

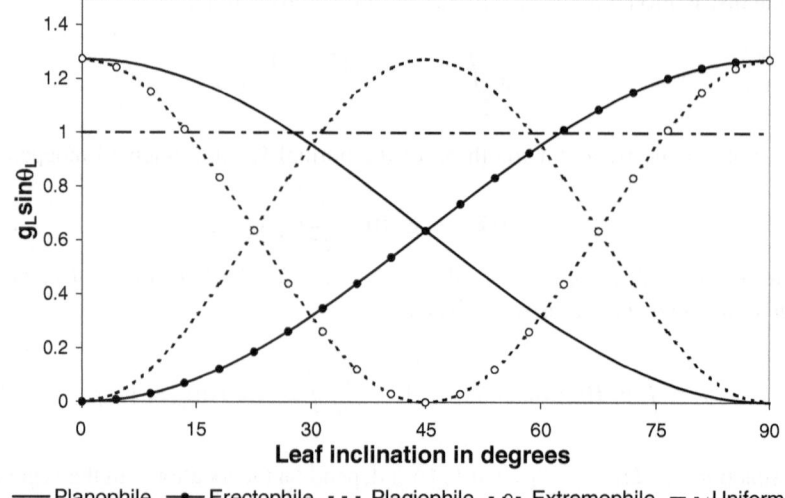

Fig. 14.5. The probability density functions for planophile (14.5a), erectrophile (14.5b), plagiophile (14.5c), extremophile (14.5d) and uniform (14.5e) distributions. Note that the vertical axis shows values of $g_L(\theta_L) \sin \theta_L$

14.2.2 Vegetation Canopy Optics

A photon incident on a leaf surface can either be absorbed or scattered. If the scattered photon emerges from the same side of the leaf as the incident photon, the event is termed reflection. Likewise, if the scattered photon exits the leaf from the opposite side, the event is termed transmission. Scattering of solar radiation by green leaves does not involve frequency-shifting interactions, but is dependent on the wavelength.

The angular distribution of radiant energy scattered by a leaf element is a key variable, and is specified by the leaf element scattering phase function $\gamma_{L,\lambda}(\Omega' \rightarrow \Omega, \Omega_L)$. For a leaf with upward normal Ω_L, this phase function is the fraction of intercepted energy from the photon initially traveling in Ω' that is scattered into an element of solid angle $d\Omega$ about Ω.

The radiant energy may be incident on the upper or the lower sides of the leaf element and the scattering event may be either reflection or transmission. Integration of the leaf scattering phase function over the appropriate solid angles gives the *hemispherical leaf reflectance* $\rho_{L,\lambda}(\Omega', \Omega_L)$ *and transmittance* $\tau_{L,\lambda}(\Omega', \Omega_L)$ coefficients. The leaf scattering phase function, when integrated over all scattered photon directions (Ξ), yields the leaf albedo

$$\omega_{L,\lambda}(\Omega', \Omega_L) = \int_{\Xi} \gamma_{L,\lambda}(\Omega' \rightarrow \Omega, \Omega_L) \, d\Omega \, , \qquad (14.6)$$

where Ξ is the unit sphere. The leaf albedo $\omega_{L,\lambda}(\Omega', \Omega_L)$ is simply the sum of $\rho_{L,\lambda}(\Omega', \Omega_L)$ and $\tau_{L,\lambda}(\Omega', \Omega_L)$. The reflectance and transmittance of an individual

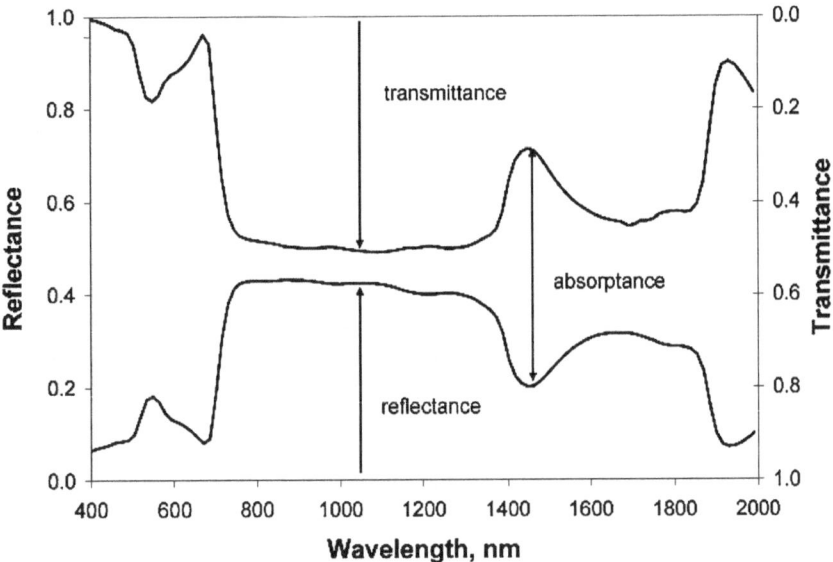

Fig. 14.6. Typical reflectance (*left axis*) and transmittance (*right axis*) spectra of an individual plant leaf from 400 to 2000 nm for normal incidence

leaf depends on wavelength, tree species, growth conditions, leaf age and its location in the canopy.

A photon incident on a leaf element can either be specularly reflected from the surface depending on its roughness or emerge diffused from interactions in the leaf interior. Some leaves can be quite smooth from a coat of wax-like material, while other leaves can have hairs making the surface rough. Light reflected from the leaf surface can be polarized as well. Photons that do not suffer leaf surface reflection enter the interior of the leaf, where they are either absorbed or refracted because of the many refractive index discontinuities between the cell walls and intervening air cavities. Photons that are not absorbed in the interior of the leaf emerge on both sides, generally diffused in all directions. Figure 14.6 shows typical diffuse hemispherical reflectance, $\rho_{L,\lambda}(-\Omega_L, \Omega_L)$, and transmittance, $\tau_{L,\lambda}(-\Omega_L, \Omega_L)$, spectra of an individual leaf under normal illumination. For more details on diffuse leaf scattering and specular reflection from the leaf surface, the reader is referred to Walter-Shea and Norman (1991), and Vanderbilt et al. (1991).

14.2.3 Extinction, Scattering and Absorption Events

To derive an expression for the extinction coefficient, consider photons at x traveling along Ω. The total extinction coefficient is the probability, per unit pathlength of photon travel, that the photon encounters a leaf element,

$$\sigma(x, \Omega) = u_L(x) G(x, \Omega) , \tag{14.7}$$

where $G(x, \Omega)$ is the *geometry factor* first proposed by Ross (1981), defined as the projection of unit leaf area at x onto a plane perpendicular to the direction of photon travel Ω, i.e.,

$$G(x, \Omega) = \frac{1}{2\pi} \int_{\Xi_+} g_L(x, \Omega_L) \, | \Omega \cdot \Omega_L | \, d\Omega_L \, . \qquad (14.8)$$

The geometry factor G satisfies $(2\pi)^{-1} \int_{\Xi_+} G(x, \Omega) d\Omega = 1/2$. Upon collision with a leaf element, a photon can be either absorbed or scattered. So the extinction coefficient can be broken down into its scattering and absorption components, $\sigma = \sigma_{s,\lambda} + \sigma_{a,\lambda}$. It is important to note that the geometry factor is an explicit function of the direction of photon travel Ω in the general case of non-uniformly distributed leaf normals. This imbues directional dependence to the extinction coefficient in the case of vegetation canopies. Only in the case of uniformly distributed leaf normals ($g_L \equiv 1$) its dependence on Ω disappears and $G \equiv 1/2$. A noteworthy point is the *wavelength independence* of σ (and thus the photon's mean-free-path), that is, the extinction probabilities for photons in vegetation media are determined by the structure of the canopy rather than photon frequency or the optics of the canopy.

Consider photons impinging on leaf elements of area density u_L at location x along Ω'. The probability density, per unit pathlength, that these photons would be intercepted and then scattered into the direction Ω is given by the differential scattering coefficient

$$\sigma'_{s,\lambda}(x, \Omega' \to \Omega) = u_L(x) \frac{1}{\pi} \Gamma_\lambda(x, \Omega' \to \Omega)$$
$$= u_L(x) \frac{1}{2\pi} \int_{\Xi_+} g_L(x, \Omega_L) \, | \Omega' \cdot \Omega_L |$$
$$\times \gamma_{L,\lambda}(\Omega' \to \Omega, \Omega_L) \, d\Omega_L \, , \qquad (14.9)$$

where Γ_λ / π is the *area* scattering phase function first proposed by Ross (1981). The scattering phase function combines diffuse scattering from the interior of a leaf and specular reflection from the leaf surface. For more details on the diffuse and specular components of the area scattering phase function, the reader is referred to Ross (1981), Marshak (1989), Myneni (1991), and Knyazikhin and Marshak (1991).

It is important to note that the differential scattering coefficient is an explicit function of the polar coordinates of Ω' and Ω, that is, non-rotationally invariant. Only in some limited cases can it be reduced to the rotationally invariant form, $\sigma'_{s,\lambda}(x, \Omega \to \Omega') \equiv \sigma'_{s,\lambda}(x, \Omega \cdot \Omega')$. This property precludes the use of Legendre polynomial expansions and of the addition theorem often used in transport theory for handling the scattering integral (see Chap. 4).

Integration of the differential scattering coefficient (14.9) over all scattered photon directions results in the scattering coefficient

$$\sigma_{s,\lambda}(x, \Omega) = u_L(x) \frac{1}{2\pi} \int_{\Xi_+} g_L(x, \Omega_L) \, | \Omega \cdot \Omega_L | \, \omega_{L,\lambda}(\Omega, \Omega_L) \, d\Omega_L \, . \qquad (14.10)$$

The absorption coefficient can be specified as

$$\sigma_{a,\lambda}(x,\Omega) = \sigma(x,\Omega) - \sigma_{s,\lambda}(x,\Omega)$$

$$= u_L(x)\frac{1}{2\pi}\int_{\Xi_+} g_L(x,\Omega_L)\,|\,\Omega\cdot\Omega_L\,|$$

$$\times \left[1 - \omega_{L,\lambda}(\Omega,\Omega_L)\right]\mathrm{d}\Omega_L \ . \tag{14.11}$$

The magnitude of scattering by leaves in an elementary volume is described using the single-scattering albedo defined as the ratio of energy scattered by an elementary volume to energy intercepted by this volume

$$\varpi_{0,\lambda}(x,\Omega) = \frac{\sigma_{s,\lambda}(x,\Omega)}{\sigma_{s,\lambda}(x,\Omega) + \sigma_{a,\lambda}(x,\Omega)}$$

$$= \frac{\int_{\Xi_+} g(x,\Omega_L)\,|\,\Omega\cdot\Omega_L\,|\,\omega_{L,\lambda}(\Omega,\Omega_L)\,\mathrm{d}\Omega_L}{\int_{\Xi_+} g(x,\Omega_L)\,|\,\Omega\cdot\Omega_L\,|\,\mathrm{d}\Omega_L} \ .$$

Given the single-scattering albedo, the scattering and absorption coefficients can be expressed via the extinction coefficient as

$$\sigma_{s,\lambda}(x,\Omega) = \varpi_{0,\lambda}(x,\Omega)\,\sigma(x,\Omega), \tag{14.12a}$$

$$\sigma_{a,\lambda}(x,\Omega) = [1 - \varpi_{0,\lambda}(x,\Omega)]\,\sigma(x,\Omega) \ . \tag{14.12b}$$

Note that only if the leaf albedo $\omega_{L,\lambda}$ defined by (14.6) does not depend on the leaf normal Ω_L, will the single-scattering albedo coincide with the leaf albedo, i.e., $\varpi_{0,\lambda}(x,\Omega) \equiv \omega_{L,\lambda}(x,\Omega)$.

14.3 Radiative Transfer in Vegetation Canopies

Let the domain V in which a vegetation canopy is located, be a parallelepiped of horizontal dimensions X_S, Y_S, and height Z_S. The top ∂V_t, bottom ∂V_b, and lateral ∂V_l surfaces of the parallelepiped form the canopy boundary $\partial V = \partial V_t \cup \partial V_b \cup \partial V_l$. Note the boundary ∂V is excluded from the definition of V. The function characterizing the radiative field in V is the monochromatic intensity $I_\lambda(x,\Omega)$ depending on wavelength, λ, location x and direction Ω. Assuming no polarization and emission within the canopy, the monochromatic intensity distribution function is given by the steady-state radiative transfer equation (Ross, 1981; Myneni et al., 1990; Myneni, 1991):

$$\Omega\cdot\nabla I_\lambda(x,\Omega) + G(x,\Omega)\,u_L(x)\,I_\lambda(x,\Omega)$$

$$= \frac{u_L(x)}{\pi}\int_{\Xi} \Gamma_\lambda(x,\Omega'\to\Omega)\,I_\lambda(x,\Omega')\,\mathrm{d}\Omega' \ . \tag{14.13}$$

14.3.1 Boundary Value Problem for the Radiative Transfer Equation in the Vegetation Canopy

Equation (14.13) alone does not provide a full description of the radiative transfer process. It is necessary to specify the incident radiation at the canopy boundary, ∂V, i.e., specification of the boundary conditions. Because the canopy is adjacent to the atmosphere, a neighboring canopy, and the soil, all of which have different reflection properties, the following boundary conditions are used to describe the incoming radiation (Kranigk et al., 1994; Knyazikhin et al., 1997):

$$I_\lambda(x_t, \Omega) = I_{0,d,\lambda}(x_t, \Omega)$$
$$+ I_{0,\lambda}(x_t)\,\delta(\Omega - \Omega_0), \quad x_t \in \partial V_t, \Omega \bullet n_t < 0, \tag{14.14a}$$

$$I_\lambda(x_l, \Omega) = \frac{1}{\pi} \int_{\Omega' \bullet n_l > 0} \rho_{\lambda,l}(\Omega', \Omega)\, I_\lambda(x_l, \Omega')\, |\,\Omega' \bullet n_l\,|\, d\Omega' + L_{d,\lambda}(x_l, \Omega)$$
$$+ L_{m,\lambda}(x_l)\,\delta(\Omega - \Omega_0), \quad x_l \in \partial V_l, \Omega \bullet n_l < 0, \tag{14.14b}$$

$$I_\lambda(x_b, \Omega) = \frac{1}{\pi} \int_{\Omega' \bullet n_b > 0} \rho_{\lambda,b}(\Omega', \Omega)$$
$$\times I_\lambda(x_b, \Omega')\, |\,\Omega' \bullet n_b\,|\, d\Omega', \quad x_b \in \partial V_b, \Omega \bullet n_b < 0, \tag{14.14c}$$

where $I_{0,d,\lambda}$ and $I_{0,\lambda}$ are intensities of the diffuse and monodirectional components of solar radiation incident on the top of the canopy boundary, ∂V_t; Ω_0 denotes direction of the monodirectional solar component; δ is the Dirac delta-function; $L_{m,\lambda}(x_t)$ is the intensity of the monodirectional solar radiation arriving at a point $x_l \in \partial V_l$ along Ω_0 without experiencing an interaction with the neighboring canopies; $L_{d,\lambda}$ is the diffuse radiation penetrating through the lateral surface in the stand; $\rho_{\lambda,l}$ and $\rho_{\lambda,b}$ are the bi-directional reflectance factors of the lateral and the bottom surfaces; n_t, n_l and n_b are the outward normals at points $x_t \in \partial V_t$, $x_l \in \partial V_l$, and $x_b \in \partial V_b$, respectively.

A neighboring environment, as well as the fraction of the monodirectional solar radiation, in the total incident radiation, influences the radiative regime in the vegetation canopy. In order to demonstrate the range of this influence we simulate two extreme situations for a 40 by 40 m sample stand shown in Fig. 14.2. In the first one, we "cut" the forest surrounding the sample plot. The incoming solar radiation can reach the sides of the sample stand without experiencing a collision in this case. The boundary condition (14.14c) with $\rho_{\lambda,l} = 0$, $L_{d,\lambda} = I_{0,d,\lambda}$, and $L_{m,\lambda} = I_{0,\lambda}$ was used to describe photons penetrating into the canopy through the lateral surface. In the second situation, we "plant" a forest of an extremely high density around the sample stand so that no solar radiation can penetrate into the stand through the lateral boundary ∂V_l. The lateral boundary condition (14.14c) takes the form $I(x_l, \Omega) = 0$. The radiative regimes in a real stand usually vary between these extreme situations. For each situation, the boundary value problem { (14.13), (14.14) } was solved and a vertical profile of mean downward radiation flux density was evaluated. Figure 14.7 demonstrates downward fluxes normalized by the incident flux at noon on both a cloudy and clear sunny day. A downward radiation flux density evaluated by averaging the extinction coefficient (14.7) and area scattering phase function (14.9) over the 40 by 40 m area first and then solving a one-dimensional radiative transfer equation is also plotted in this figure. One can see that the radiative regime in the sample stand is more sensitive to the lateral boundary conditions during cloudy days. In both cases, a 3D medium transmits more radiation than those predicted by the 1D transport equation. This is consistent with radiation transmitted through inhomogeneous clouds (see Chap. 12) and with Jensen's inequality as applied to transmission $[T((L_1 + L_2)/2) \leq (T(L_1) + T(L_2))/2$ where T is concave ($T'' \geq 0$) and L is the leaf area index].

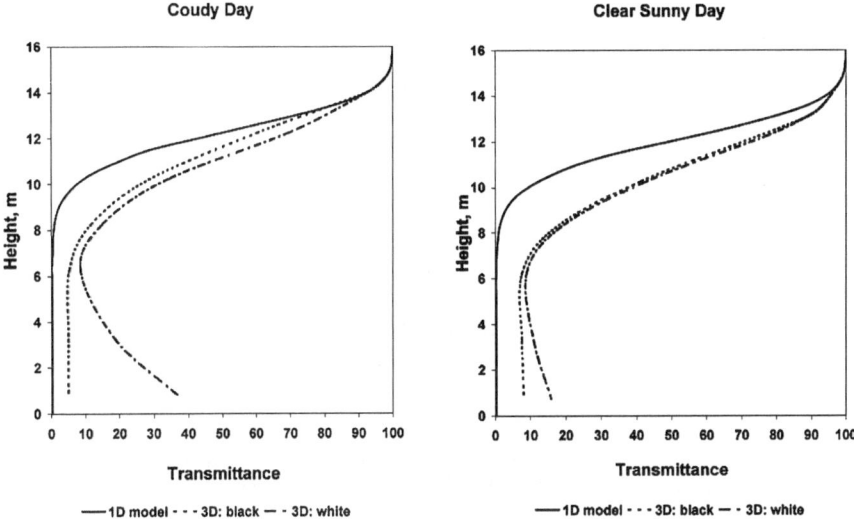

Fig. 14.7. Vertical profile of the downward radiation flux normalized by the incident flux derived from the one-dimensional (1D model) and three-dimensional (*3D: black* and *3D: white*) models on a cloudy day and on a clear sunny day. Curves *3D: black* correspond to a forest stand surrounded by the optically black lateral boundary, and curves *3D: white* to an isolated forest stand of the same size and structure. Reprinted from Knyazikhin et al. (1997) with permission from Elsevier

The radiation penetrating through the lateral sides of the canopy depends on the neighboring environment. Its influence on the radiation field within the canopy is especially pronounced near the lateral canopy boundary. Therefore inaccuracies in the lateral boundary conditions may cause distortions in the simulated radiation field within the domain V. These features should be taken into account when 3D radiation distribution in a vegetation canopy of a small area is investigated. The problem of photon transport in such canopies arises, for example, in the context of optimal planting and cutting of industrial wood (Knyazikhin et al., 1994), land surface climatology, and plant physiology. The lateral side effects, however, decrease with distance from this boundary toward the center of the domain. The size of the "distorted area" depends on the adjoining vegetation and atmospheric conditions (Kranigk, 1996). In particular, it has been shown that these lateral effects can be neglected when the radiation regime is analyzed in a rather extended canopy (see Chap. 12 for comparison with clouds). A "zero" boundary condition for the lateral surface can then be used to simulate canopy radiation regime.

14.3.2 Reflection of Vegetated Surfaces

Solution of the boundary value problem {(14.13), (14.14)} describes the radiative regime in a vegetation canopy and, as a consequence, reflectance properties of the vegetation canopy. As demonstrated in Fig. 14.7, the canopy-radiation regime is sensitive to the partition between the mono-directional and diffuse components of the incoming radiation. The ratio $f_{\text{dir},\lambda}$ of the mono-directional to the total radiation flux incident on the canopy is used to parameterize canopy reflectance properties; that is,

$$f_{\text{dir},\lambda} = \frac{\int_{\partial V_t} I_{0,\lambda}(x_t) \,|\, \mu_0 \,|\, dx_t}{\int_{\partial V_t} dx_t \int_{n_t \,\bullet\, \Omega' < 0} I_\lambda(x_t, \Omega') \,|\, \Omega' \bullet n_t \,|\, d\Omega'} \,. \qquad (14.15)$$

The hemispherical-directional reflectance factor (HDRF, dimensionless) for nonisotropic incident radiation is defined as the ratio of the mean radiance leaving the top of the plant canopy to the mean radiance reflected from an ideal Lambertian target into the same beam geometry and illuminated under identical atmospheric conditions (Martonchik et al., 2000); this can be expressed by the solution of (14.13) and (14.14) as

$$r_\lambda(\Omega) = \frac{\int_{\partial V_t} I_\lambda(x_t, \Omega) \, dx_t}{\frac{1}{\pi} \int_{\partial V_t} dx_t \int_{n_t \,\bullet\, \Omega' < 0} I_\lambda(x_t, \Omega') \,|\, \Omega' \bullet n_t \,|\, d\Omega'}, \Omega \bullet n_t > 0 \,. \qquad (14.16)$$

The bi-hemispherical reflectance for nonisotropic incident radiation (BHR, dimensionless) is defined as the ratio of the exiting flux to the incident flux (Martonchik et al., 2000); that is,

$$R(\lambda) = \frac{\int_{\partial V_t} dx_t \int_{n_t \,\bullet\, \Omega > 0} I_\lambda(x_t, \Omega) \,|\, \Omega \bullet n_t \,|\, d\Omega}{\int_{\partial V_t} dx_t \int_{n_t \,\bullet\, \Omega < 0} I_\lambda(x_t, \Omega) \,|\, \Omega \bullet n_t \,|\, d\Omega} \,. \qquad (14.17)$$

The HDRF and BHR depend on the ratio $f_{\text{dir},\lambda}$ of mono-directional irradiance on the top of the vegetation canopy to the total incident irradiance. If $f_{\text{dir},\lambda} = 1$, the HDRF and BHR become the bi-directional reflectance factor (BRF, dimensionless) and the directional hemispherical reflectance (DHR, dimensionless), respectively. These variables can be derived from data acquired by satellite-borne sensors (Diner et al., 1999) which, in turn, are input to various techniques for retrieval of biophysical parameters from space. It should be noted that the HDRF and BHR, in general, depend on the direction Ω_0 of solar beam. However, HDRF = BRF = DHR = BHR for Lambertian surfaces.

In remote sensing, the dimension of the upper boundary ∂V_t often coincides with a footprint of the imagery. Taking the size of ∂V_t to zero results in a BRF value defined at a given spatial point x_t. Such a BRF is used to describe the lower boundary condition for the radiative transfer in the atmosphere (Sect. 14.5).

14.4 Canopy Spectral Invariants

The extinction coefficient in vegetation canopies is treated as wavelength independent considering the size of the scattering elements (leaves, branches, twigs, etc.)
relative to the wavelength of solar radiation. This results in canopy spectral invariants stating that some simple algebraic combinations of the single-scattering albedo
and canopy spectral transmittances and reflectances eliminate their dependencies
on wavelength through the specification of canopy-specific wavelength independent
variables. These variables specify an accurate relationship between the spectral response of a vegetation canopy to incident solar radiation at the leaf and the canopy
scale. In terms of these variables, the partitioning of the incident solar radiation into
canopy transmission and interception can be described by expressions that relate
canopy transmittance and interception at an arbitrary wavelength to transmittances
and interceptions at all other wavelengths in the solar spectrum (Knyazikhin et al.,
1998b; Panferov et al., 2001; Shabanov et al., 2003). Furthermore, the canopy spectral invariants allow us to separate a small set of independent variables that fully
describes the law of conservation in vegetation canopies at any wavelength in the
visible and near-infrared parts of the solar spectrum (Wang et al., 2003). These characteristic features of radiative transfer in vegetation canopies are discussed in this
section.

14.4.1 Canopy Spectral Behavior in the Case of an Absorbing Ground

Consider a vegetation canopy confined to $0 < z < H$. The plane surfaces $z = 0$
and $z = H$ constitute its upper and lower boundaries, respectively. In other words, a
"horizontally extended" area is considered; thus no lateral illumination is assumed or
needed. The spectral composition of the incident radiation is altered by interactions
with phytoelements. The magnitude of scattering by the foliage elements is characterized by the hemispherical leaf reflectance and transmittance introduced earlier.
The reflectance and transmittance of an individual leaf depends on wavelength, tree
species, growth conditions, leaf age and its location in the canopy. We start with the
simplest case where reflectance of the ground below the vegetation is zero. Results
presented in this subsection are required to extend our analysis to the general case of
a reflecting ground below the vegetation.

Let a parallel beam of unit intensity be incident on the upper boundary in the
direction Ω_0. Equation (14.13) describes the radiative transfer process within the
vegetation canopy. In this case, the boundary conditions are simpler than in (14.14):

$$I_\lambda(x_t, \Omega) = \delta(\Omega - \Omega_0), \qquad x_t \in \partial V_t, \Omega \cdot n_t < 0, \tag{14.18a}$$

$$I_\lambda(x_b, \Omega) = 0, \qquad x_b \in \partial V_b, \Omega \cdot n_b < 0. \tag{14.18b}$$

Indeed because we examine the radiative regime in a horizontally extended area
the boundary condition (14.14c) for the lateral surface ∂V_l is irrelevant. We investigate canopy spectral properties using operator theory (Vladimirov, 1963; Richtmyer,
1978) by introducing the differential, L, and integral, S_λ, operators,

$$LI_\lambda = \mathbf{\Omega} \cdot \nabla I_\lambda + G(\mathbf{x}, \mathbf{\Omega}) u_L(\mathbf{x}) I_\lambda(\mathbf{x}, \mathbf{\Omega}) \; ;$$

$$S_\lambda I_\lambda = \frac{u_L(\mathbf{x})}{\pi} \int_\Xi \Gamma_\lambda(\mathbf{x}, \mathbf{\Omega}' \to \mathbf{\Omega}) I_\lambda(\mathbf{x}, \mathbf{\Omega}') \, d\mathbf{\Omega}' \; . \tag{14.19}$$

The solution I_λ of the boundary value problem $\{(14.13), (14.18)\}$ can be represented as the sum of two components, viz. $I_\lambda = Q + \varphi_\lambda$. Here, a *wavelength-independent* function Q is the probability density that a photon in the incident radiation will arrive at $\mathbf{x}$ along $\mathbf{\Omega}_0$ without suffering a collision. In other words, the function Q describes the radiation field generated by *uncollided* photons. It satisfies the equation $LQ = 0$ and the boundary condition (14.18). Because the uncollided radiation field excludes the scattering event, its solution takes zero values in upward directions. It should be emphasized that the differential operator L does not depend on wavelength and thus Q is *wavelength independent*.

The second term, φ_λ, describes a collided, or diffuse, radiation field; that is, radiation field generated by photons scattered one or more times by phytoelements. It satisfies $L\varphi_\lambda = S_\lambda \varphi_\lambda + S_\lambda Q$ and zero boundary conditions. By letting $K_\lambda = L^{-1} S_\lambda$, the latter can be transformed to

$$\varphi_\lambda = K_\lambda \varphi_\lambda + K_\lambda Q \; . \tag{14.20}$$

Substituting $\varphi_\lambda = I_\lambda - Q$ into this equation results in an integral equation for I_λ (Vladimirov, 1963; Bell and Glasstone, 1970; see also Chap. 3, Sect. 3.8),

$$I_\lambda = K_\lambda I_\lambda + Q \; . \tag{14.21}$$

It follows from (14.21) that $I_\lambda - K_\lambda I_\lambda$ does not depend on λ, and involves the validity of the following relationship

$$I_\lambda - K_\lambda I_\lambda = I_{\lambda'} - K_{\lambda'} I_{\lambda'} = Q \; , \tag{14.22}$$

where I_λ and $I_{\lambda'}$ are solutions of the boundary value problem $\{(14.13), (14.18)\}$ at wavelengths λ and λ', respectively. Equation (14.22) originally derived by Zhang et al. (2002) expresses the canopy spectral invariant in a general form. It also follows from (14.20) and (14.21) that $\varphi_\lambda = K_\lambda I_\lambda$. This simple relationship will be used in further derivations (e.g., (14.28)).

To quantify the spectral invariant, we introduce two coefficients defined as

$$\gamma_0(\lambda) = \frac{\int_V d\mathbf{x} \int_\Xi \sigma(\mathbf{x}, \mathbf{\Omega}) \, \varphi_\lambda(\mathbf{x}, \mathbf{\Omega}) \, d\mathbf{\Omega}}{\int_V d\mathbf{x} \int_\Xi \sigma(\mathbf{x}, \mathbf{\Omega}) I_\lambda(\mathbf{x}, \mathbf{\Omega}) \, d\mathbf{\Omega}} \; , \tag{14.23}$$

$$\omega_0(\lambda) = \frac{\int_V d\mathbf{x} \int_\Xi \sigma_{s,\lambda}(\mathbf{x}, \mathbf{\Omega}) I_\lambda(\mathbf{x}, \mathbf{\Omega}) \, d\mathbf{\Omega}}{\int_V d\mathbf{x} \int_\Xi \sigma(\mathbf{x}, \mathbf{\Omega}) I_\lambda(\mathbf{x}, \mathbf{\Omega}) \, d\mathbf{\Omega}} \; . \tag{14.24}$$

Here σ and $\sigma_{s,\lambda}$ are the extinction and scattering coefficients; the spatial integration is performed over a sufficiently extended domain V for which the lateral side effects can be neglected. The first coefficient, $\gamma_0(\lambda)$, is the portion of collided radiation in the total radiation field intercepted by a vegetation canopy while the second coefficient, $\omega_0(\lambda)$, is the probability that a photon will be scattered as a result of an interaction within the domain V. We can think of $\omega_0(\lambda)$ as a *mean* single-scattering albedo. In general, it can depend on canopy structure, domain V and radiation incident on the vegetation canopy. However, if the leaf albedo (14.6) does not vary with spatial and directional variables, $\omega_0(\lambda)$ coincides with the single-scattering albedo $\varpi_{0,\lambda}$.

The ratio

$$p_0 = \gamma_0(\lambda)/\omega_0(\lambda) \qquad (14.25)$$

is the probability that a *scattered* photon will hit a leaf in the domain V since $\int_V d\mathbf{x} \int_\Xi \sigma(\mathbf{x}, \mathbf{\Omega}) \varphi_\lambda(\mathbf{x}, \mathbf{\Omega}) d\Omega$ is the number of *scattered photons intercepted by leaves* and $\int_V d\mathbf{x} \int_\Xi \sigma_{s,\lambda}(\mathbf{x}, \mathbf{\Omega}) I_\lambda(\mathbf{x}, \mathbf{\Omega}) d\Omega$ is the *total number of scattered photons*. This event is determined solely by the structural properties of vegetation which, in turn, are described by the leaf area density and leaf normal orientation distribution functions, each being a wavelength-independent variable. This suggests that the probability p_0 should be a wavelength-independent variable, too. As it follows from the analysis below, this property is indeed the case!

Let $i(\lambda)$ and $A(\lambda)$ be canopy interception and absorptance defined as

$$i(\lambda) = \frac{\int_V d\mathbf{x} \int_\Xi \sigma(\mathbf{x}, \mathbf{\Omega}) I_\lambda(\mathbf{x}, \mathbf{\Omega}) d\Omega}{\int_{\partial V_t} d\mathbf{x}_t \int_{\mathbf{n}_t \cdot \mathbf{\Omega} < 0} I_\lambda(\mathbf{x}_t, \mathbf{\Omega}) |\mathbf{\Omega} \cdot \mathbf{n}_t| d\Omega},$$

$$A(\lambda) = \frac{\int_V d\mathbf{x} \int_\Xi \sigma_{a,\lambda}(\mathbf{x}, \mathbf{\Omega}) I_\lambda(\mathbf{x}, \mathbf{\Omega}) d\Omega}{\int_{\partial V_t} d\mathbf{x}_t \int_{\mathbf{n}_t \cdot \mathbf{\Omega} < 0} I_\lambda(\mathbf{x}_t, \mathbf{\Omega}) |\mathbf{\Omega} \cdot \mathbf{n}_t| d\Omega}. \qquad (14.26)$$

Here σ and $\sigma_{a,\lambda}$ are the extinction and absorption coefficients, respectively. For a vegetation canopy bounded at the bottom by a black surface, $i(\lambda)$ is the average number of photon interactions with the leaves in V before either being absorbed or exiting the domain V. It follows from (14.24), (14.26) and (14.12) that

$$\frac{A(\lambda)}{i(\lambda)} = 1 - \omega_0(\lambda). \qquad (14.27)$$

This equation has a simple physical interpretation: the capacity of a vegetation canopy for absorption results from the absorption $1 - \omega_0(\lambda)$ of an average leaf multiplied by the average number $i(\lambda)$ of photon interactions with leaves.

Multiplying (14.22) by the extinction coefficient σ and integrating over V and all directions $\mathbf{\Omega}$ and normalizing by the incident radiation flux one obtains

$$i(\lambda) - \gamma_0(\lambda) i(\lambda) = i(\lambda') - \gamma_0(\lambda') i(\lambda') = q. \qquad (14.28)$$

Here q is the probability that a photon entering V will hit a leaf while $1 - q$ is the probability that a photon in the incident radiation will arrive at the canopy bottom without experiencing an interaction with leaves. This probability can be derived from both model calculations and/or field measurements. A question then arises of how $\gamma_0(\lambda)$ can be evaluated.

An eigenvalue of the radiative transfer equation is a number $\tilde{\gamma}(\lambda)$ such that there exists a function $\psi_\lambda(x, \Omega)$ that satisfies $K_\lambda \psi_\lambda = \tilde{\gamma}(\lambda)\psi_\lambda$. Under some general conditions (Vladimirov, 1963), the set of eigenvalues $\tilde{\gamma}_k$, $k = 0, 1, 2, \ldots$ and eigenvectors $\psi_{\lambda,k}(x, \Omega)$, $k = 0, 1, 2, \ldots$ is a discrete set; the eigenvectors satisfy a condition of orthogonality. The radiative transfer equation has a unique positive eigenvalue that corresponds to a unique positive eigenvector (Vladimirov, 1963). This eigenvalue, say $\tilde{\gamma}_0(\lambda)$, is greater than the absolute magnitudes of all the remaining eigenvalues. This means that only one eigenvector, $\psi_{\lambda,0}(x, \Omega)$, takes on positive values for any x and Ω. If this positive eigenvector is treated as a source within a vegetation canopy (i.e., uncollided radiation), then $K_\lambda \psi_{\lambda,0}$ describes the collided radiation field. This suggests that the portion, $\gamma_0(\lambda)$, of collided radiation in the total radiation field intercepted by a vegetation canopy can be approximated by the maximum eigenvalue of the radiative transfer equation, i.e., $\gamma_0(\lambda) \approx \tilde{\gamma}_0(\lambda)$. The unique positive eigenvalue $\tilde{\gamma}_0(\lambda)$ can be represented as a product of the mean single-scattering albedo $\omega_0(\lambda)$ and a wavelength-independent factor $[1 - \exp(-k)]$, i.e., (Knyazikhin and Marshak, 1991)

$$\gamma_0(\lambda) \approx \tilde{\gamma}_0(\lambda) = \omega_0(\lambda)[1 - \exp(-k)] . \tag{14.29}$$

Here k is a coefficient that depends on canopy structure but not on wavelength. Thus, if our hypothesis about the use of the maximum eigenvalue is correct, then the ratio $p_0 = \gamma_0(\lambda)/\omega_0(\lambda)$ does not depend on λ. Substituting $\gamma_0(\lambda) = p_0\omega_0(\lambda)$ into (14.28) and solving for p_0 yields

$$p_0 = \frac{i(\lambda) - i(\lambda')}{\omega_0(\lambda)i(\lambda) - \omega_0(\lambda')i(\lambda')} . \tag{14.30}$$

This equation expresses the canopy spectral invariant for canopy interception; that is, the difference between numbers of photons intercepted by the vegetation canopy at two arbitrary wavelengths is proportional to the difference between numbers of scattered photons at the same wavelengths. A similar property is valid for canopy transmittance (Panferov et al., 2001); that is, the quantity

$$p_t = \frac{T(\lambda) - T(\lambda')}{\omega_0(\lambda)T(\lambda) - \omega_0(\lambda')T(\lambda')} \tag{14.31}$$

does not depend on leaf optical properties. Here, the hemispherical canopy transmittance, $T(\lambda)$, for non-isotropic incident radiation is the ratio of the downward radiation flux density at the canopy bottom to the incident radiant,

$$T(\lambda) = \frac{\int_{\partial V_b} dx_b \int_{n_b \cdot \Omega > 0} I_\lambda(x_b, \Omega) |\Omega \cdot n_b| d\Omega}{\int_{\partial V_t} dx_t \int_{n_t \cdot \Omega < 0} I_\lambda(x_t, \Omega) |\Omega \cdot n_t| d\Omega} . \tag{14.32}$$

Solving (14.31) for $T(\lambda)$ we get

$$T(\lambda) = \frac{1 - p_t\omega_0(\lambda')}{1 - p_t\omega_0(\lambda)}T(\lambda') . \tag{14.33}$$

Similarly, solving (14.30) for $i(\lambda)$ and taking into account (14.27), we have

$$A\left(\lambda\right) = \frac{1 - \omega_0\left(\lambda\right)}{1 - \omega_0\left(\lambda'\right)} \frac{1 - p_0\omega_0\left(\lambda'\right)}{1 - p_0\omega_0\left(\lambda\right)} A\left(\lambda'\right) . \tag{14.34}$$

For a vegetation canopy with a non-reflecting surface the determination of canopy transmittance $T(\lambda)$ and absorptance $A(\lambda)$ as a function of wavelength λ follows from these equations in terms of their values at the chosen wavelength λ'. The canopy reflectance $R(\lambda)$ (14.17) is determined via the energy conservation law as (compare with (12.14) from Chap. 12)

$$T\left(\lambda\right) + R\left(\lambda\right) + A\left(\lambda\right) = 1 . \tag{14.35}$$

Panferov et al. (2001) measured spectral hemispherical reflectances and transmittances of individual leaves and the entire canopy to see if the spectral invariances of (14.30) and (14.31) hold true. Spectra of $T(\lambda)$ and $R(\lambda)$ in the region from 400 to 1100 nm, at 1-nm resolution, were sampled at two sites representative of equatorial rainforests and temperate coniferous forests with a dark ground. A number of leaves from different parts of tree crowns were cut and their spectral transmittances and reflectances (from 400 to 1100 nm, at 1-nm resolution) were measured in a laboratory. Mean spectral leaf albedo over collected data were taken as the mean single-scattering albedo $\omega_0(\lambda)$. Given measured $T(\lambda)$, $R(\lambda)$, and $\omega_0(\lambda)$ at 700 different wavelengths, the canopy interception $i(\lambda)$ was evaluated using (14.27) and (14.35); that is,

$$i\left(\lambda\right) = \frac{1 - T\left(\lambda\right) - R\left(\lambda\right)}{1 - \omega_0\left(\lambda\right)} . \tag{14.36}$$

Figure 14.8a shows the cumulative distribution function of p_0 derived from measured values of the right-hand side of (14.30) corresponding to all available combinations of λ and λ' for which $\lambda > \lambda'$. One can see that this function is very close to the Heaviside function, with a sharp jump from 0 to 1 at $p_0 \approx 0.94$. Its density distribution function behaves as the Dirac delta-function. It means that with a very high probability, p_0 is wavelength independent. Solving (14.30) for $i(\lambda)$ one obtains

$$i\left(\lambda\right) = \frac{1 - p_0\omega_0\left(\lambda'\right)}{1 - p_0\omega_0\left(\lambda\right)} i\left(\lambda'\right) . \tag{14.37}$$

Thus, given canopy interception at an arbitrarily chosen wavelength λ', one can evaluate this variable at any other wavelength. Figure 14.8b shows the correlation between canopy interceptions derived from field data using (14.36) and evaluated with (14.37) using $p_0 = 0.94$ for many wavelength pairs. One can see that field data follow regularities predicted by (14.30) and, therefore, supports our hypothesis regarding the maximum eigenvalue of the radiative transfer equation. For details of the mathematical derivation of the canopy spectral invariant for canopy interception and transmittance, the reader is referred to Knyazikhin et al. (1998a) and Panferov et al. (2001).

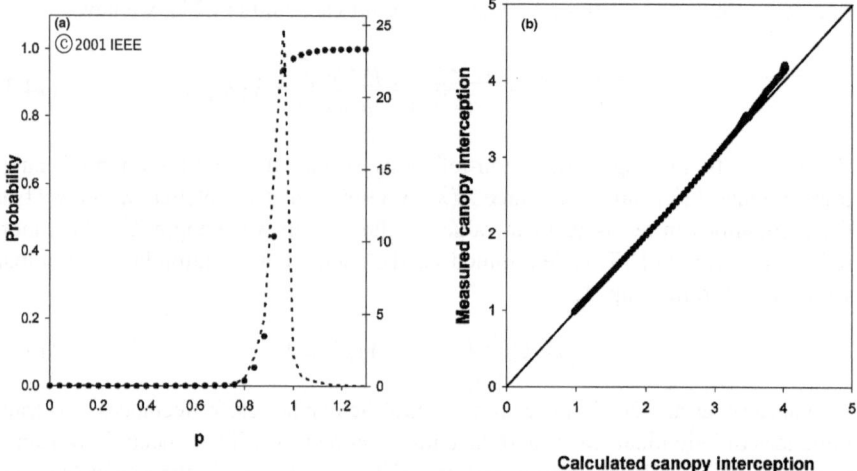

Fig. 14.8. (**a**) Cumulative distribution function (left axis, *dots*) and density distribution function (right axis, *dashed line*) of p_0 derived from field data. (**b**) Correlation between canopy interceptions derived from measurements using (14.36) and from calculations using (14.37). Reprinted from Panferov et al. (2001) with permission from IEEE

14.4.2 Canopy Spectral Behavior in the Case of a Reflective Ground

The canopy spectral invariants discussed in the previous section and the Green function approach (Sect. 3.10) allow us to fully describe the canopy spectral response to incident solar radiation in terms of a small set of independent variables (Wang et al., 2003). As an example, consider a vegetation canopy layer $0 < z < H$ bounded from below by a Lambertian surface with albedo $\rho_{\lambda,b}$. The plane surfaces $z = 0$ and $z = H$ constitute its upper, ∂V_t, and lower, ∂V_b, boundaries, respectively. Let a parallel beam of intensity $I_{0,\lambda}$ be incident on the upper boundary. The intensity $I_\lambda(\boldsymbol{x}, \boldsymbol{\Omega})$ of radiation at the wavelength λ, at a spatial point $\boldsymbol{x}$ and in direction $\boldsymbol{\Omega}$ normalized by $I_{0,\lambda}$ satisfies (14.13) and boundary conditions

$$I_\lambda(\boldsymbol{x}_t, \boldsymbol{\Omega}) = \delta(\boldsymbol{\Omega} - \boldsymbol{\Omega}_0), \qquad \boldsymbol{x}_t \in \partial V_t, \mu < 0, \tag{14.38a}$$

$$I_\lambda(\boldsymbol{x}_b, \boldsymbol{\Omega}) = \frac{1}{\pi}\rho_{\lambda,b} \int_{\Xi_-} I_\lambda(\boldsymbol{x}_b, \boldsymbol{\Omega}') \,|\mu'|\, d\boldsymbol{\Omega}', \qquad \boldsymbol{x}_b \in \partial V_b, \mu > 0, \tag{14.38b}$$

where μ and μ' are cosines of zenith angles of $\boldsymbol{\Omega}$ and $\boldsymbol{\Omega}'$, respectively and Ξ_- is the downward hemisphere.

The three-dimensional radiation field can be represented as a sum of two components: the radiation calculated for a non-reflecting ("black") surface, $I_{\mathrm{blk},\lambda}(\boldsymbol{x}, \boldsymbol{\Omega})$, and the remaining radiation, $I_{\mathrm{rem},\lambda}(\boldsymbol{x}, \boldsymbol{\Omega})$; that is,

$$I_\lambda(\boldsymbol{x}, \boldsymbol{\Omega}) = I_{\mathrm{blk},\lambda}(\boldsymbol{x}, \boldsymbol{\Omega}) + I_{\mathrm{rem},\lambda}(\boldsymbol{x}, \boldsymbol{\Omega}). \tag{14.39}$$

The second component, $I_{\text{rem},\lambda}(x, \Omega)$, accounts for the radiation field due to surface–vegetation multiple interactions and can be expressed as (see Chap. 3, Sect. 3.10)

$$I_{\text{rem},\lambda}(x, \Omega) = \rho_{\lambda,b} \int\limits_{x_b \in \partial V_b} F_\lambda(x_b)\, J_\lambda(x, \Omega; x_b)\, dS(x_b)\ , \qquad (14.40)$$

where

$$J_\lambda(x, \Omega; x_b) = \frac{1}{\pi} \int\limits_{\Xi_+} G_{S,\lambda}(x, \Omega; x_b, \Omega')\, d\Omega' \qquad (14.41)$$

describes the radiation field (in $\text{sr}^{-1}\text{m}^{-2}$) in the canopy layer generated by an isotropic source $\pi^{-1}\delta(x - x_b)$ (in $\text{sr}^{-1}\text{m}^{-2}$) located at a point $x_b \in \partial V_b$, and $G_{S,\lambda}$ is the surface Green function (see Sect. 3.10). The downward radiation flux density at the canopy bottom satisfies the following integral equation (see Chap. 3, Sect. 3.8)

$$F_\lambda(x_b) = T_{\text{blk},\lambda}(x_b)$$
$$+ \rho_{\lambda,b} \int\limits_{x_b' \in \partial V_b} R_\lambda^*(x_b, x_b')\, F_\lambda(x_b')\, dS(x_b')\ , \qquad x_b \in \partial V_b\ . \quad (14.42)$$

Here $T_{\text{blk},\lambda}(x_b)$ is the downward flux density at the canopy bottom calculated for the black surface, and $R^*(x_b, x_b')$ is downward flux density at $x_b \in \partial V_b$ generated by the point isotropic source $\pi^{-1}\delta(x - x_b')$ located at $x_b' \in \partial V_b$.

The use of the Green function allows us to split the radiative transfer problem into two subproblems with purely absorbing boundaries. They are 3D radiation fields generated (a) by the radiation penetrating into the canopy through the upper boundary and (b) by a point isotropic source located at the canopy bottom. The canopy spectral invariant can be applied to each of them. Given solutions of these subproblems, the downward radiation flux density $F_\lambda(x)$ and radiation field $I_{\text{rem},\lambda}(x, \Omega)$ due to the surface–vegetation multiple interactions can be specified via (14.40) and (14.42).

In the case of horizontally homogeneous vegetation canopy, fluxes F_λ and $F_{\text{blk},\lambda}$ and the probability $R_\lambda^* = \int_{x_b' \in \partial V_b} dS(x_b') R_\lambda^*(x_b, x_b')$ that a photon entering through the lower canopy boundary will be reflected by the vegetation, do not depend on horizontal coordinates and thus a solution to (14.42) can by given in an explicit form; that is,

$$F_\lambda = \frac{T_{\text{blk},\lambda}}{1 - \rho_{\lambda,b} R_\lambda^*}\ . \qquad (14.43)$$

Here, $T_{\text{blk},\lambda}$ coincides with the canopy transmittance defined by (14.32). It follows from (14.39), (14.40) and (14.43) that the solution, $I_\lambda(z, \Omega)$, to the 1D radiative transfer equation can be written as

$$I_\lambda(z, \Omega) = I_{\text{blk},\lambda}(z, \Omega) + \frac{\rho_{\lambda,b}}{1 - \rho_{\lambda,b} R_\lambda^*} T_{\text{blk},\lambda} J_\lambda^*(z, \Omega)\ . \qquad (14.44)$$

Here

$$J_\lambda^* (z, \mathbf{\Omega}) = \int\limits_{x_b \in \partial V_b} J_\lambda (z, \mathbf{\Omega}; x_b) \, \mathrm{d}S(x_b) \tag{14.45}$$

is the radiation field generated by isotropic sources uniformly distributed over surface underneath the vegetation canopy. This decomposition of the radiation field allows the separation of three independent variables responsible for the distribution of solar radiation in vegetation canopies: $\rho_{\lambda,b}$, $I_{\mathrm{blk},\lambda}$ and J_λ^*. The reflectance $\rho_{\lambda,b}$ of the underlying surface does not depend on the vegetation canopy; $I_{\mathrm{blk},\lambda}$ and J_λ^* are surface-independent quantities since there is no interaction between the medium and the underlying surface. These variables have intrinsic canopy information.

Let $A_{\mathrm{blk},\lambda}$ and $R_{\mathrm{blk},\lambda}$, defined by (14.26) and (14.17), be the canopy absorptance and reflectance calculated for the case of the black surface underneath the vegetation canopy. We denote the probabilities that photons from isotropic sources on the canopy bottom will escape the canopy through the upper boundary and be absorbed by the vegetation layer by T_λ^* and A_λ^*, respectively. These variables can be expressed as

$$T_\lambda^* = \int\limits_{\Xi_+} J_\lambda^* (H, \mathbf{\Omega}) \mu \mathrm{d}\mathbf{\Omega}, \qquad A_\lambda^* = \int\limits_0^H \mathrm{d}z \int\limits_\Xi \sigma_{a,\lambda} J_\lambda^* (z, \mathbf{\Omega}) \, \mathrm{d}\mathbf{\Omega} \,. \tag{14.46}$$

It follows from here that the canopy reflectance, R_λ, transmittance, T_λ, and absorptance, A_λ, can be expressed as

$$R_\lambda = R_{\mathrm{blk},\lambda} + \frac{\rho_{\lambda,b}}{1 - \rho_{\lambda,b} R_\lambda^*} T_{\mathrm{blk},\lambda} T_\lambda^* \,, \tag{14.47a}$$

$$T_\lambda = T_{\mathrm{blk},\lambda} + \frac{\rho_{\lambda,b}}{1 - \rho_{\lambda,b} R_\lambda^*} T_{\mathrm{blk},\lambda} R_\lambda^* = \frac{T_{\mathrm{blk},\lambda}}{1 - \rho_{\lambda,b} R_\lambda^*} \,, \tag{14.47b}$$

$$A_\lambda = A_{\mathrm{blk},\lambda} + \frac{\rho_{\lambda,b}}{1 - \rho_{\lambda,b} R_\lambda^*} T_{\mathrm{blk},\lambda} A_\lambda^* \,. \tag{14.47c}$$

The canopy spectral invariants can be applied to $A_{\mathrm{blk},\lambda}$, $T_{\mathrm{blk},\lambda}$, T_λ^* and A_λ^*. The corresponding reflectances $R_{\mathrm{blk},\lambda}$ and R_λ^* can be obtained via the conservation law; that is,

$$R_{\mathrm{blk},\lambda} + T_{\mathrm{blk},\lambda} + A_{\mathrm{blk},\lambda} = 1, \qquad R_\lambda^* + T_\lambda^* + A_\lambda^* = 1 \,. \tag{14.48}$$

Thus, a small set of independent variables generally seems to suffice when attempting to describe the spectral response of a vegetation canopy to incident solar radiation. This set includes the soil reflectance $\rho_{\lambda,b}$, the mean single-scattering albedo, $\omega_0(\lambda)$, canopy transmittance and absorptance at an arbitrary wavelength, and two wavelength-independent parameters p_0 and p_t (see (14.30) and (14.31)) calculated for the black soil problem and for the same canopy illuminated from below by isotropic sources. All of these are measurable parameters. In terms of these variables, solar radiation reflected, transmitted and absorbed by the vegetation canopy at any given wavelength at the solar spectrum can be expressed via (14.47) and spectral invariant relations described in Sect. 14.5. Note that a similar statement holds true for non-Lambertian surfaces. For more details on canopy spectral behavior in general case, the reader is referred to Knyazikhin et al. (1998a,b).

14.5 Vegetation Canopy as a Boundary Condition to Atmospheric Radiative Transfer

A decomposition similar to (14.39)–(14.42) is also valid for the atmosphere (Chap. 3, Sect. 3.10) where intensities $I_{\text{blk},\lambda}(x, \Omega)$ and $I_{\text{rem},\lambda}(x, \Omega)$ are determined by the atmospheric radiative transfer equation in which the canopy bi-directional reflectance factor (BRF, Sect. 14.3.2) appears in the lower boundary condition. The BRF can be parameterized in terms of variables introduced earlier in this chapter. As a result, we have a set of surface and atmosphere variables required to quantitatively describe canopy–cloud interaction. As an example, consider a 3D cloudy layer bounded from below by a horizontally inhomogeneous vegetation canopy, V (Fig. 14.9). We idealize the vegetation canopy as a horizontally inhomogeneous Lambertian surface, i.e., the canopy BRF at a given spatial point x_t on the canopy top is independent of directions of incident and reflected radiation. For ease of the analysis, we assume that photons can interact only with cloud drops. Keeping this in mind and combining (14.39)–(14.41) results in

$$I_\lambda (x, \Omega) = I_{\text{blk},\lambda} (x, \Omega) + \int_{x_t \in \partial V_t} F_\lambda^{\text{up}} (x_t) \, J_\lambda (x, \Omega; x_t) \, \mathrm{d}S (x_t) \qquad (14.49)$$

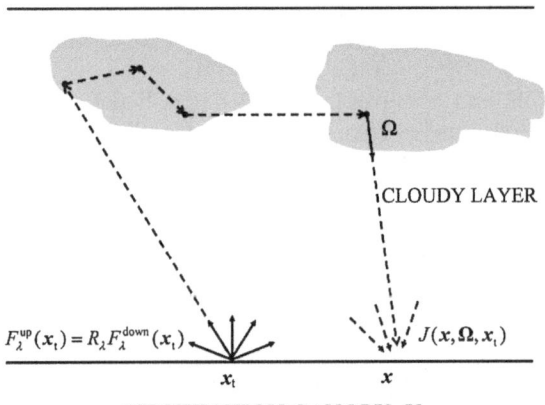

VEGETATION CANOPY, V

Fig. 14.9. A three-dimensional cloudy layer bounded from below by a horizontally inhomogeneous vegetation canopy, V. Each point on the canopy upper boundary is assumed to reradiate the incident radiation isotropically, i.e., radiant intensity of reflected radiation is independent of direction. The function $J(x, \Omega; x_t)$ is the probability that a photon from an isotropic source $1/\pi$ located at x_t arrives at the point x along direction Ω as a result of photon–cloud interactions

for any point x in the cloudy layer or on its lower boundary, which is ∂V_t (Fig. 14.9). Here $F_\lambda^{\text{up}}(x_t)$ is the upward flux density at x_t, $I_{\text{blk},\lambda}$ is the radiation field in the cloudy layer calculated for a black canopy. In (14.49), I and F_λ^{up} refer to intensities and flux density normalized by the intensity of the solar beam hitting the top of the atmosphere in the corresponding spectral interval. The term J_λ is the probability that a photon from an isotropic source $1/\pi$ located at x_t arrives at the point x along direction Ω as a result of photon–cloud interaction. Under the above assumptions, J_λ is a surface-independent variable. Following (14.41) (see also Chap. 3, Sect. 3.10),

$$J_\lambda (x, \Omega; x_t) = \frac{1}{\pi} \int_{\Xi_-} G_{S,\lambda} (x, \Omega; x_t, \Omega') \, d\Omega' \tag{14.50}$$

where the surface Green function $G_{S,\lambda}$ describes the cloud radiative response to the point mono-directional source $\delta(x - x_t)\delta(\Omega - \Omega')$ located at the top of the canopy. If the canopy is idealized as a horizontally homogeneous surface, then $F_\lambda^{\text{up}}(x_t) = R_\lambda F_\lambda^{\text{dn}}(x_t)$, where the canopy reflectance (canopy albedo), R_λ, is defined by (14.47a).

In contrast to a vegetation canopy, variations in cloud optical properties (single-scattering albedo, extinction coefficient and asymmetry factor) in the spectral region between 500 and 900 nm are small and, as a first approximation, can be assumed to be wavelength independent. It follows from this assumption that J_λ is also wavelength independent and $I_{\text{blk},\lambda_1} = I_{\text{blk},\lambda_2}$ for any wavelength in this spectral region. Thus,

$$\begin{aligned} I_{\lambda_1} (x, \Omega) &- I_{\lambda_2} (x, \Omega) \\ &= \int_{x_t \in \partial V_t} \left[R_{\lambda_1} F_{\lambda_1}^{\text{dn}} (x_t) - R_{\lambda_2} F_{\lambda_2}^{\text{dn}} (x_t) \right] J (x, \Omega; x_t) \, dS (x_t) . \end{aligned} \tag{14.51}$$

Below we will limit ourselves by considering only points x located at the top of the canopy, i.e., both x and x_t belong to ∂V_t. We first define a bottom-of-atmosphere reflectance (BOAR), $s_\lambda(x, \Omega)$, for isotropic incident radiation as

$$s_\lambda (x, \Omega) = \frac{\int_{x_t \in \partial V_t} F_\lambda^{\text{up}} (x_t) J (x, \Omega; x_t) \, dS (x_t)}{F_\lambda^{\text{up}} (x)} . \tag{14.52}$$

This variable describes cloud–surface interaction as if the cloudy atmosphere were illuminated from below by horizontally heterogeneous isotropic sources of intensity $F_\lambda^{\text{up}}(x)$. For a horizontally homogeneous cloud layer, the upward radiation flux density F_λ^{up} does not depend on position and thus the BOAR coincides with the probability that isotropically illuminated clouds reflect the radiation into direction Ω.

For horizontally inhomogeneous clouds, the downward radiation flux and thus the upward flux density F_λ^{up} can vary significantly. However, it does not necessarily involve large variation in $s_\lambda(x, \Omega)$. A theoretical explanation of this result can be found in the linear operator analysis (Krein, 1967) and, specifically, in its applications to radiative transfer theory (Knyazikhin, 1991; Kaufmann et al., 2000; Zhang et al., 2002; Lyapustin and Knyazikhin, 2002). One of the theorems of the operator theory states that for a continuous positive linear operator B, minimum, m_n, and maximum, M_n, values of the function $\eta_n = \sqrt[n]{B^n u}$ converge to the maximum eigenvalue, $\rho(B)$, of the operator B from below and above for any arbitrarily chosen positive function u, i.e., $m_n \leq \rho(B) \leq M_n$ and $M_n - m_n$ tends to zero as n tends to infinity. For the problem of atmospheric radiative transfer over common land–surface types, including vegetation, soil sand, and snow, the proximity of m_n and M_n to a high accuracy holds at $n \geq 2$ (Lyapustin and Knyazikhin, 2002). Here for a given direction Ω, the numerator in (14.52) can be treated as a positive integral operator B with a wavelength-independent kernel $J(x, \Omega; x_t)$. Its maximum eigenvalue, $\rho(J, \Omega)$, therefore, is also wavelength independent and can be approximated by $s_\lambda(x, \Omega)$, i.e., $s_\lambda(x, \Omega) \approx \rho(J, \Omega)$. As a result, (14.51) can be rewritten as

$$\frac{I_{\lambda_1}(x, \Omega) - I_{\lambda_2}(x, \Omega)}{R_{\lambda_1} F_{\lambda_1}^{dn}(x) - R_{\lambda_2} F_{\lambda_2}^{dn}(x)} \approx \rho(J, \Omega) . \tag{14.53}$$

Thus, a simple algebraic combination of radiance, downward flux density, and canopy albedo eliminate their dependencies on wavelength through the specification of a cloud-structure wavelength-independent variable. The right-hand side of (14.53) is related to a wavelength-independent cloud optical depth above x. Indeed, it follows from the above-mentioned theorem that $\rho(J, \Omega)$ can be estimated from below and above by the minimum and maximum values of the function $\eta_1 = Bu |_{u \equiv 1}$ calculated for $u \equiv 1$. We take this variable as the first approximation to $\rho(J, \Omega)$, i.e.,

$$\rho(J, \Omega) \approx \eta_1 = \int_{x_t \in \partial V_t} J(x, \Omega; x_t) \, dx_t . \tag{14.54}$$

For plane-parallel geometry, x-independent $\rho(J, \Omega)$ is simply cloud reflection in the direction Ω from isotropic illumination, a monotonic function of cloud optical depth.

Assuming that surface measurements of downward radiance and flux densities are available at two wavelengths with the strongest surface contrast (say, RED, $\lambda_1 = 0.66 \, \mu m$ and NIR, $\lambda_2 = 0.86 \, \mu m$), Barker and Marshak (2001) and Barker et al. (2002) following Marshak et al. (2000) used the ratio (14.53) to retrieve ρ and thus cloud optical depth above x for horizontally inhomogeneous and even broken clouds. The relationship between ρ and cloud optical depth was approximated using 1D radiative transfer. A modified function J that depends only on the horizontal distance between points where a photon enters and exits the cloud is related to the cloud radiative transfer Green function. Based on the diffusion approximation for slab geometry, Chap. 12 shows an analytic relationship between given cloud optical and geometrical thicknesses and the horizontal distance of photon travel.

Another approach to retrieve cloud optical properties above vegetated surfaces is the use of the normalized difference cloud index (NDCI). This index is defined as the ratio between the difference and the sum of two normalized zenith radiances measured at $\lambda_1 = $ RED and $\lambda_2 = $ NIR narrow spectral bands (Marshak et al., 2000),

$$\mathrm{NDCI}(x) = \frac{I_{\mathrm{NIR}}(x) - I_{\mathrm{RED}}(x)}{I_{\mathrm{NIR}}(x) + I_{\mathrm{RED}}(x)} . \tag{14.55}$$

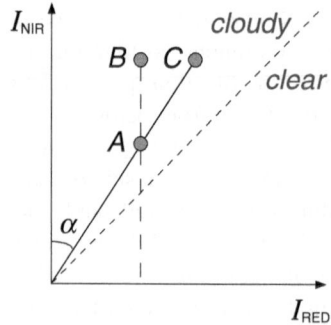

Fig. 14.10. A schematic illustration of cloud–vegetation interaction on the spectral plane. Each measured pair $(I_{\text{RED}}, I_{\text{NIR}})$ of downwelling radiances can be depicted as a point on the RED versus NIR plane. Its location is specified by a radius vector $\eta = \sqrt{I_{\text{RED}}^2 + I_{\text{NIR}}^2}$ and a polar angle $\alpha = \arctan(I_{\text{RED}}/I_{\text{NIR}})$. Points below the diagonal correspond to a clear sky while points above the diagonal correspond to a cloudy sky. For discussion on the location of points A, B and C, see the text

Compared to a two-valued optical depth versus zenith radiance relationship, the NDCI is a monotonic function with respect to optical depth and is less sensitive to 3D cloud structure (Marshak et al., 2000). However, it reduces the spectral information to one number. In other words, instead of *two* spectral values of zenith radiances in RED and NIR, only *one*, NDCI, is used. Indeed, each measurement can be depicted as a point on the RED versus NIR plane, which has two coordinates (Fig. 14.10):

$$\eta = \sqrt{I_{\text{RED}}^2 + I_{\text{NIR}}^2} \tag{14.56}$$

$$\alpha = \tan^{-1}(I_{\text{RED}}/I_{\text{NIR}}) . \tag{14.57}$$

Both coordinates depend on the cloud optical depth. The NDCI is a function of α only and, as was shown by Marshak et al. (2004), cloud optical depth can vary considerably with α unchanged. The first coordinate, η, is required to specify the cloud optical depth.

Consider the RED vs. NIR plane in Fig. 14.10. In the RED spectral region, the chlorophyll in a green leaf absorbs 90–95% of solar radiation (see Fig. 14.6); thus, the vegetation albedo R_{RED} is very low and, as it follows from (14.49), $I_{\text{RED}} \approx I_{\text{blk,RED}}$. In contrast, in the NIR region, a green leaf reflects about 90% of incident radiation resulting in a high value of R_{NIR} (Fig. 14.6). In this spectral region, the green vegetation acts as a powerful reflector that "illuminates" clouds from below. Because of cloud-to-surface photon reflection (integral term in (14.49)), points below the diagonal correspond to clear sky while points above the diagonal correspond to cloudy sky. This also follows from (14.51). Indeed, ignoring aerosols and Rayleigh scattering for a cloud-free atmosphere, the probability J that a photon will be scattered by clouds is equal to zero; thus, $I_{\text{NIR}} = I_{\text{RED}}$. However, due to Rayleigh scattering (that is higher in RED than in NIR), $I_{\text{blk,NIR}} < I_{\text{blk,RED}}$ and, consequently, $I_{\text{NIR}} < I_{\text{RED}}$. By contrast, for cloudy conditions, the probability J takes a non-zero value, and the difference between $I_{blk,NIR}$ and $I_{blk,RED}$ is much smaller than the integral term in (14.51). This results in $I_{NIR} > I_{RED}$.

Consider the cloudy part. Points A and B with equal RED radiances are assumed to have the same cloud optical depth and do not "feel" the surface. Their abscissas are mainly determined by radiation, $I_{blk,RED}$, transmitted through clouds. However, points A and B have different ordinates, NIR radiances, because they are determined not only by radiation, $I_{blk,NIR}$, transmitted through clouds, but also radiation reflected both by the vegetation canopy and clouds (see integral term in (14.49)). In other words, there are more cloud–surface interactions at point B than at point A; hence more photons reach the surface at B, i.e., B corresponds to a smaller cloud fraction. As a result, location on the RED vs. NIR plane gives us information not only about *overhead* cloud optical depth but also about the *radiatively effective cloud fraction* in the whole sky. Thus, we consider both radiances I_{NIR} and I_{RED} to be functions of optical depth, τ, and cloud fraction, A_c:

$$I_{RED} = I_{RED}(\tau, A_c) \tag{14.58a}$$
$$I_{NIR} = I_{NIR}(\tau, A_c) . \tag{14.58b}$$

Note that points A and C have the same NDCI ratio but obviously different τ and A_c.

The above technique has been applied to data from the Atmospheric Radiation Measurement (ARM) Southern Great Plains (SGP) site Cimel radiometer. Cimel is a multichannel sunphotometer with a narrow field of view of $1.2°$ and four filters at 0.44, 0.67, 0.87 and $1.02\,\mu m$ that are designed to retrieve aerosol properties in clear-sky conditions. Cimel sunphotometers are the main part of the Aerosol Robotic Network (AERONET, http://aeronet.gsfc.nasa.gov/) – a ground-based network for monitoring aerosol optical properties (Holben et al., 1998). The vegetation reflectances R_{RED} and R_{NIR} needed to estimate (I_{RED}, I_{NIR}) are available globally from satellite standard surface products (e.g., EOS Terra MODIS and MISR data) at moderate spatial resolutions with temporal frequencies of 8–16 days (Schaaf et al., 2002; Martonchik et al., 1998). Figure 14.11 shows a DISORT-calculated (Stamnes et al., 1988) set of curves for various τ and A_c (on a plane modified from Fig. 14.10 to spread the curves out better), and three Cimel-measured groups of 10 data-points each. The data-point groups, while being located at different positions on the plane, have almost the same NDCI (the straight line); hence, if retrieved using NDCI alone, all three groups would have the same optical depth τ (80 for $A_c = 1.0$). However, as follows from the plot, these groups correspond to different pairs ($A_c = 0.9; \tau = 28$), ($A_c = 0.8; \tau = 22$) and ($A_c = 0.4; \tau = 12$) with different optical depths. Note that cloud fraction here is not a visual cloud fraction but a "radiatively effective" one that also compensates for cloud horizontal inhomogeneity not accounted for by 1D radiative transfer (Marshak et al., 2004).

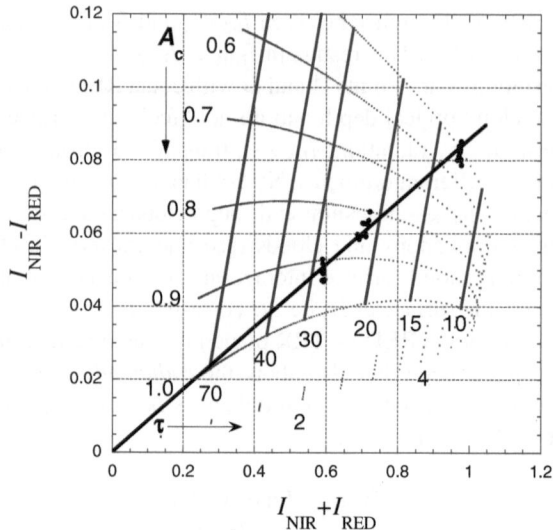

Fig. 14.11. DISORT calculated values of $I_{NIR} + I_{RED}$ and $I_{NIR} - I_{RED}$ for a wide range of optical depth τ and cloud fraction A_c for SZA=62° ± 3°, surface albedos $R_{RED} = 0.092$ and $R_{NIR} = 0.289$. When A_c is constant and τ is varying, the set of calculated values define a cloud fraction isoline. When τ is constant and A_c is varying, the set of calculated values define an optical depth isoline. Values $I_{NIR} + I_{RED}$ and $I_{NIR} - I_{RED}$ are from Cimel measurements at the ARM site on July 28, 2002. Measurements were taken around 13:45, 13:58 and 14:11 UT, respectively (towards decreasing $I_{NIR} + I_{RED}$). A straight line through $(0,0)$ corresponds to the NDCI ≈ 0.08

14.6 Summary

Solar radiation scattered from a vegetation canopy results from the interaction of photons traversing the foliage medium, bounded at the bottom by a radiatively participating surface. To estimate the canopy radiation regime, four important features must be carefully formulated. They are (1) the architecture of individual plants and the entire canopy; (2) optical properties of vegetation elements (leaves, stems), (3) reflective properties of the ground underneath the canopy, and (4) atmospheric conditions. Photon transport theory aims at deriving the solar radiation regime within the vegetation canopy using the above-mentioned attributes as input. The first two attributes are accounted for in the extinction and differential scattering coefficients, which appear in the radiative transfer equation. Their specification requires models for the leaf area density function, the probability density of leaf normal orientation, and the leaf scattering phase function (Sect. 14.2). Reflective properties of the ground and atmospheric conditions determine the boundary conditions for the radiative transfer equation required to describe incoming radiation and radiation reflected by the ground. A solution of the boundary value problem describes the radiative regime in a vegetation canopy and, as a consequence, its reflective properties (Sect. 14.3). The latter, for example, are required to specify boundary conditions for the transfer of radiation in a cloudy atmosphere adjoining a vegetation canopy (Sect. 14.5).

In contrast to radiative transfer in clouds, the extinction coefficient in vegetation canopies does not depend on wavelength. This feature results in canopy spectral invariants; that is, some simple algebraic combinations of the single-scattering albedo and canopy spectral transmittances and reflectances eliminate their dependencies on wavelength through the specification of canopy-structure-dependent variables. These variables are related to the maximum positive eigenvalues of linear operators describing the canopy transmittance and radiation field in the vegetation canopy. The eigenvalues have a simple physical interpretation. They represent the collided radiation portions of the total transmitted and intercepted radiation. In terms of the maximum eigenvalue, the partitioning of the incident solar radiation into canopy transmission and absorption can be described by expressions that relate canopy transmittance and interception at an arbitrarily chosen wavelength to transmittances and interceptions at any other wavelength in the solar spectrum. Furthermore, the canopy spectral invariant allows us to separate a small set of independent variables that fully describe the law of conservation in vegetation canopies at any given wavelength in the solar spectrum (Sect 14.4). This feature of the solution of the canopy radiative transfer equation provides an accurate parameterization of the boundary value problem for the radiative transfer in a cloudy atmosphere adjoining the vegetation canopy (Sect.14.5). Finally we showed that interactions between vegetation and horizontally inhomogeneous clouds can be exploited to retrieve cloud optical depth and "effective" cloud fraction from ground-based radiance measurements.

References

Antyufeev, V.S. (1996). Solution of the generalized transport equation with a peak-shaped indicatrix by the Monte Carlo method. *Russ. J. Numer. Anal. and Modeling*, **11**, 113–137.

Barker, H.W. and A. Marshak (2001). Inferring optical depth of broken clouds above green vegetation using surface solar radiometric measurements. *J. Atmos. Sci.*, **58**, 2989–3006.

Barker, H.W., A. Marshak, W. Szyrmer, A. Trishchenko, J.-P. Blanchet, and Z. Li (2002). Inference of cloud optical properties from aircraft-based solar radiometric measurements. *J. Atmos. Sci.*, **59**, 2093–2111.

Bell, G.I. and S. Glasstone (1970). *Nuclear Reactor Theory*. Van Nostrand Reinholt, New York (NY).

Bréon, F.-M., F. Maignan, M. Leroy, and I. Grant (2001). A statistical analysis of hot spot directional signatures measured from space. In *Proceedings of 8th International Symposium Physical Measurements and Signatures in Remote Sensing*, Centre Nationale d'Études Spatiales (CNES), Toulouse (France), 343–353.

Bunnik, N.J.J. (1978). *The Multispectral Reflectance of Shortwave Radiation by Agricultural Crops in Relation with Their Morphological and Optical Properties*. Pudoc Publications, Wageningen (The Netherlands).

Choulli, M. and P. Stefanov (1996). Reconstruction of the coefficient of the stationary transport equation from boundary measurements. *Inverse Problems*, **12**, L19–L23.

Deschamps, P.Y., F.M. Bréon, M. Leroy, A. Podaire, A. Bricaud, J.C. Buriez, and G. Sèze (1994). The POLDER mission: Instrument characteristics and scientific objectives. *IEEE Trans. Geosci. and Remote Sens.*, **32**, 598–615.

DeWit, C.T. (1965). *Photosynthesis of Leaf Canopies*. Technical Report (Agric. Res. Report 663), Pudoc Publ., Wageningen (The Netherlands).

Diner, D.J., G.P. Asner, R. Davies, Yu. Knyazikhin, J.-P. Muller, A.W. Nolin, B. Pinty, C.B. Schaaf, and J. Stroeve (1999). New directions in Earth observing: Scientific application of multi-angle remote sensing. *Bull. Amer. Meteor. Soc.*, **80**, 2209–2228.

Germogenova, T.A. (1986). *The Local Properties of the Solution of the Transport Equation* (in Russian). Nauka, Moscow (Russia).

Gerstl, S.A. and C. Simmer (1986). Radiation physics and modeling for off-nadir satellite-sensing of non-Lambertian surfaces. *Remote Sens. Environ.*, **20**, 1–29.

Goel, N.S. and D.E. Strebel (1984). Simple Beta distribution representation of leaf orientation in vegetation canopies. *Agron. J.*, **76**, 800–802.

Holben, B.N., T.F. Eck, I. Slutsker, D. Tanre, J.P. Buis, A. Setzer, E. Vermote, J.A. Reagan, Y.J. Kaufman, T. Nakajima, F. Lavenu, I. Jankowiak, and A. Smirnov (1998). AERONET - A federated instrument network and data archive for aerosol characterization. *Remote Sens. Environ.*, **66**, 1–16.

Intergovernmental Panel on Climate Change (1995). Climate Change: Report of the Intergovernmental Panel on Climate Change (IPCC), J.T. Houghton et al. (eds.). Cambridge University Press, New York (NY).

Kaufmann, R.K., L. Zhou, Yu. Knyazikhin, N.V. Shabanov, R.B. Myneni, and C.J. Tucker (2000). Effect of orbital drift and sensor changes on the time series of AVHRR vegetation index data. *IEEE Trans. Geosci. and Remote Sens.*, **38**, 2584–2597.

Knyazikhin, Yu. (1991). On the solvability of plane-parallel problems in the theory of radiation transport. *USSR Computer Maths. and Math. Phys. (English version)*, **30**, 145–154.

Knyazikhin, Yu. and A. Marshak (1991). Fundamental equations of radiative transfer in leaf canopies and iterative methods for their solution. In *Photon-Vegetation Interactions: Applications in Plant Physiology and Optical Remote Sensing*. R.B. Myneni and J. Ross (eds.). Springer-Verlag, New York (NY), pp. 9–43.

Knyazikhin, Yu., A. Marshak, D. Schulze, R. Myneni, and G. Gravenhorst (1994). Optimisation of solar radiation input in forest canopy as a tool for planting patterns of trees. *Transport Theory and Statist. Phys.*, **23**, 671–700.

Knyazikhin, Yu., J. Kranigk, G. Miessen, O. Panfyorov, N. Vygodskaya, and G. Gravenhorst (1996). Modelling three-dimensional distribution of photosynthetically active radiation in sloping coniferous stands. *Biomass and Bioenergy*, **11**, 189–200.

Knyazikhin, Yu., G. Miessen, O. Panfyorov, and G. Gravenhorst (1997). Small-Scale study of three-dimensional distribution of photosynthetically active radiation in a forest. *Agric. For. Meteorol.*, **88**, 215–239.

Knyazikhin, Yu., J.V. Martonchik, D.J. Diner, R.B. Myneni, M.M. Verstraete, B. Pinty, and N. Gobron (1998a). Estimation of vegetation canopy leaf area index and fraction of absorbed photosynthetically active radiation from atmosphere corrected MISR data. *J. Geophys. Res.*, **103**, 32,239–32,256.

Knyazikhin, Yu., J.V. Martonchik, R.B. Myneni, D.J. Diner, and S.W. Running (1998b). Synergistic algorithm for estimating vegetation canopy leaf area index and fraction of absorbed photosynthetically active radiation from MODIS and MISR data. *J. Geophys. Res.*, **103**, 32,257–32,276.

Kranigk, J. (1996). *Ein Model für den Strahlungstransport in Fichtenbeständen.* Cuvillier, Göttingen (Germany).

Kranigk, J. and G. Gravenhorst (1993). Ein dreidimensionales modell für fichtenkronen. *Allg. Forst Jagdztg.*, **164**, 146–149.

Kranigk, J., F. Gruber, J. Heimann, and A. Thorwest (1994). Ein Model für die Kronenraumstruktur und die räumliche Verteilung der Nadeloberfläche in einem Fichtenbestand. *Allg. Forst Jagdztg.*, **165**, 193–197.

Krein, S.G. (1967). *Functional Analysis.* Foreign Technol. Div., Wright-Patterson Air Force Base (Ohio).

Kuusk, A. (1985). The hot spot effect of a uniform vegetative cover. *Sov. J. Remote Sens.*, **3**, 645–658.

Lyapustin, A.I. and Yu. Knyazikhin (2002). Green's function method for the radiative transfer problem. 2. Spatially heterogeneous anisotropic surface. *Applied Optics*, **41**, 5600–5606.

Marshak, A. (1989). Effect of the hot spot on the transport equation in plant canopies. *J. Quant. Spectrosc. Radiat. Transfer*, **42**, 615–630.

Marshak, A., Yu. Knyazikhin, A.B. Davis, W.J. Wiscombe, and P. Pilewskie (2000). Cloud - vegetation interaction: Use of normalized difference cloud index for estimation of cloud optical thickness. *Geophys. Res. Lett.*, **27**, 1695–1698.

Marshak, A., Yu. Knyazikhin, K.D. Evans, and W.J. Wiscombe (2004). The "RED versus NIR" plane to retrieve broken-cloud optical depth from ground-based measurements. *J. Atmos. Sci.*, **61**, 1911–1925.

Martonchik, J.V., D.J. Diner, B. Pinty, M.M. Verstaete, R.B. Myneni, Yu. Knyazikhin, and H.R. Gordon (1998). Determination of land and ocean reflective, radiative and biophysical properties using multi-angle imaging. *IEEE Trans. Geosci. Remote Sens.*, **36**, 1266–1281.

Martonchik, J.V., C.J. Bruegge, and A. Strahler (2000). A review of reflectance nomenclature used in remote sensing. *Remote Sens. Rev.*, **19**, 9–20.

Myneni, R.B. (1991). Modeling radiative transfer and photosynthesis in three-dimensional vegetation canopies. *Agric. For. Meteorol.*, **55**, 323–344.

Myneni, R.B., G. Asrar, and S.A.W. Gerstl (1990). Radiative transfer in three dimensional leaf canopies. *Transp. Theory and Stat. Phys.*, **19**, 205–250.

Myneni, R.B., S. Hoffman, Yu. Knyazikhin, J.L. Privette, J. Glassy, Y. Tian, Y. Wang, X. Song, Y. Zhang, G.R. Smith, A. Lotsch, M. Friedl, J.T. Morisette, P. Votava, R.R. Nemani, and S.W. Running (2002). Global products of vegetation leaf area and fraction absorbed par from year one of MODIS data. *Remote Sens. Environ.*, **83**, 214–231.

Panferov, O., Yu. Knyazikhin, R.B. Myneni, J. Szarzynski, S. Engwald, K.G. Schnitzler, and G. Gravenhorst (2001). The role of canopy structure in the spectral
variation of transmission and absorption of solar radiation in vegetation canopies.
IEEE Trans. Geosci. and Remote Sens., **39**, 241–253.

Pinty, B. and M.M. Verstraete (1997). Modeling the scattering of light by vegetation
in optical remote sensing. *J. Atmos. Sci.*, **55**, 137–150.

Richtmyer, R.D. (1978). *Principles of Advanced Mathematical Physics*, volume 1.
Springer-Verlag, New York (NY).

Ross, J. (1981). *The Radiation Regime and Architecture of Plant Stands*. Dr. W.
Junk, Norwell (MA).

Schaaf, C.B., F. Gao, A.H. Strahler, W. Lucht, X. Li, T. Tsang, N. Strugnell,
Z. Xiaoyang, Y. Jin, J.-P. Muller, P. Lewis, M.J. Barnsley, P.H. Hobson, M.I.
Disney, G. Roberts, M. Dunderdale, C. Doll., R.P. D'Entremont, B. Hu, S. Liang,
J.L. Privette, and D. Roy (2002). First operational BRDF, albedo nadir reflectance
products from MODIS. *Remote Sens. Environ.*, **83**, 135–148.

Shabanov, N.V., Y. Wang, W. Buerman, J. Dong, S. Hoffman, G.R. Smith, Y. Tian,
Y. Knyazikhin, and R.B. Myneni (2003). Effect of foliage spatial heterogeneity
on the MODIS LAI and FPAR algorithm over broadleaf forests. *Remote Sens.
Environ.*, **85**, 410–423.

Stamnes, K., S.-C. Tsay, W.J. Wiscombe, and K. Jayaweera (1988). Numerically stable algorithm for discrete-ordinate-method radiative transfer in multiple scattering
and emitting layered media. *Appl. Opt.*, **27**, 2502–2509.

Steffen, W., I. Noble, J. Canadell, M. Apps, E.D. Schulze, P.G. Jarvis, D. Baldocchi,
P. Ciais, W. Cramer, J. Ehleringer, G. Farquhar, C.B. Field, A. Ghazi, R. Gif-
ford, M. Heimann, R. Houghton, P. Kabat, C. Korner, E. Lambin, S. Linder, H.A.
Mooney, D. Murdiyarso, W.M. Post, I.C. Prentice, M.R. Raupach, D.S. Schimel,
A. Shvidenko, and R. Valentini (1998). The terrestrial carbon cycle: Implications
for the Kyoto Protocol. *Science*, **280**, 1393–1397.

van de Hulst, H.C. (1980). *Multiple Light Scattering: Tables, Formulae and Applications*, volume I. Academic Press, San Diego (CA).

Vanderbilt, V.C., L. Grant, and S.L. Ustin (1991). Polarization of light by vegetation. In *Photon-Vegetation Interactions*. R.B. Myneni and J. Ross (eds.). Springer-
Verlag, New York (NY), pp. 191–228.

Verstraete, M.M. (1987). Radiation transfer in plant canopies: Transmission of direct
solar radiation and the role of leaf orientation. *J. Geophys. Res.*, **92**, 10,985–
10,995.

Verstrate, M.M., B. Pinty, and R.E. Dickinson (1990). A physical model of the
bidirectional reflectance of vegetation canopies. 1. Theory. *J. Geophys. Res.*, **95**,
11,755–11,765.

Vladimirov, V.S. (1963). Mathematical problems in the one-velocity theory of particle transport. Technical Report AECL-1661, Atomic Energy of Canada Ltd.,
Ottawa (Canada).

Walter-Shea, E.A. and J.M. Norman (1991). Leaf optical properties. In *Photon-
Vegetation Interactions: Applications in Plant Physiology and Optical Remote*

Sensing. R.B. Myneni and J. Ross (eds.).   Springer-Verlag, New York (NY), pp. 229–251.

Wang, Y., W. Buermann, P. Stenberg, P. Voipio, H. Smolander, T. Häme, Y. Tian, J. Hu, Yu. Knyazikhin, and R.B. Myneni (2003). A new parameterization of canopy spectral response to incident solar radiation: Case study with hyperspectral data from pine dominant forest. *Remote Sens. Environ.*, **85**, 304–315.

Zhang, Y., N. Shabanov, Yu. Knyazikhin, and R.B. Myneni (2002). Assessing the information content of multiangle satellite data for mapping biomes. II. Theory. *Remote Sens. Environ.*, **80**, 435–446.

Suggested Reading

Ross, J. (1981). *The Radiation Regime and Architecture of Plant Stands.* 391pp, Dr. W. Junk, Norwell (MA).

This is a classical monograph in which Juhan Ross originally formulated radiative transfer in vegetation canopies as a problem in mathematical physics as well as parameterized the architecture of the vegetation canopy and optical properties of vegetation elements in terms of the leaf area density function, the probability density of leaf normal orientation, and the leaf scattering phase function. He also provided a solution of the radiative transfer for several special cases.

———————

Myneni, R.B. and Ross, J. (eds.) (1991). *Photon-Vegetation Interactions: Applications in Plant Physiology and Optical Remote Sensing.* 565pp, Springer-Verlag, New York (NY).

This edited volume summarizes progress in the radiative transfer theory for plant canopies since the publication of Ross' book. This book consists of 17 chapters that present new approaches to the boundary value problem for radiative transfer equation in vegetation canopies, new numerical and analytical methods of its solution. A special emphasis is given to the application of the theory in the plant ecology and optical remote sensing.

———————

Kuusk, A. (1985). The hot spot effect of a uniform vegetative cover. *Sov. J. Remote Sens.*, **3**, 645–658.

Simmer, C. and S.A. Gerstl (1985). Remote sensing of angular characteristics of canopy reflectances. *IEEE Trans. Geosci. and Remote Sens.*, **23**, 648–658.

Marshak, A. (1989). Effect of the hot spot on the transport equation in plant canopies. *J. Quant. Spectrosc. Radiat. Transfer*, **42**, 615–630.

Verstraete, M.M., B. Pinty and R.E. Dickenson (1990). A physical model of the bidirectional reflectance of vegetation canopies. 1. Theory. *J. Geophys. Res.*, **95**, 11,765–11,775.

Myneni, R.B. and G. Asrar (1991). Photon interaction cross sections for aggregations of finite dimensional leaves. *Remote Sens. Environ.*, **37**, 219–224.

Pinty, B. and M.M. Verstraete (1997). Modeling the scattering of light by vegetation in optical remote sensing. *J. Atmos. Sci.*, **55**, 137–150.

The hot spot effect (Fig. 14.1) is a result of cross-shading between finite dimensional leaves in the canopy leading to a peak in reflectance in the retro-illumination direction. The radiative transfer equation in its original formulation (Ross, 1981) fails to predict or duplicate this effect. These papers present models of the canopy bi-directional reflectance factor, which account for the hot spot effect.

Knyazikhin, Y., J.V. Martonchik, D.J. Diner, R.B. Myneni, M.M. Verstraete, B. Pinty and N. Gobron (1998). Estimation of vegetation canopy leaf area index and fraction of absorbed photosynthetically active radiation from atmosphere corrected MISR data. *J. Geophys. Res.*, **103**, 32,239–32,256.

Knyazikhin, Y., J.V. Martonchik, R.B. Myneni, D.J. Diner and S.W. Running (1998). Synergistic algorithm for estimating vegetation canopy leaf area index and fraction of absorbed photosynthetically active radiation from MODIS and MISR data. *J. Geophys. Res.*, **103**, 32,257–32,276.

These two papers describe the operational algorithm of global leaf area index (LAI) and fraction of photosynthetically active radiation, (FPAR) 400–700 nm absorbed by vegetation for the MODerate resolution Imaging Spectroradiometer (MODIS) and Multiangle Imaging Spectro-Radiometer (MISR) instruments of the Earth Observing System Terra mission and, in the case of MODIS, its Aqua counterpart. The theory of radiative transfer in vegetation canopies presented in this chapter underlies this algorithm. It is also shown in the second paper that a rather wide family of canopy radiation models designed to account for the hot spot effect conflict with the law of energy conservation. This law was taken as a basic tool to constrain the MODIS/MISR LAI/FPAR retrieval techniques. A theoretical derivation of the canopy spectral invariants was originally presented in these papers.

Zhang, Y., Y. Tian, R.B. Myneni, Y. Knyazikhin and C.E. Woodcock (2002). Assessing the information content of multiangle satellite data for mapping biomes. I. Statistical analysis. *Remote Sens. Environ*, **80**, 418–434.

Zhang, Y., N. Shabanov, Y. Knyazikhin and R.B. Myneni (2002). Assessing the information content of multiangle satellite data for mapping biomes. II. Theory, *Remote Sens. Environ.*, **80**, 435–446.

Knyazikhin, Yu., A. Marshak, W.J. Wiscombe, J. Martonchik and R.B. Myneni (2002). A missing solution to the transport equation and its effect on stimation of cloud absorptive properties. *J. Atmos. Sci.*, **59**, 3572–5385.

It is shown in the first two papers that if the solution to the radiative transfer equation is treated as a Schwartz distribution, then an additional term must be added to the solution of the radiative transfer equation. This term describes the hot spot effect. A similar result takes place in cloud radiative transfer; this is documented in the third paper.

Panferov, O., Y. Knyazikhin, R.B. Myneni, J. Szarzynski, S. Engwald, K.G. Schnitzler and G. Gravenhorst (2001). The role of canopy structure in the spectral variation of transmission and absorption of solar radiation in vegetation canopies. *IEEE Trans. Geosci. Remote Sens.*, **39**, 241–253.

Wang, Y., W. Buermann, P. Stenberg, P. Voipio, H. Smolander, T. Häme, Y. Tian, J. Hu, Y. Knyazikhin and R.B. Myneni (2003). A new parameterization of canopy spectral response to incident solar radiation: Case study with hyperspectral data from pine dominant forest. *Remote Sens. Environ.*, **85**, 304–315.

Shabanov, N.V., Y. Wang, W. Buerman, J. Dong, S. Hoffman, G.R. Smith, Y. Tian, Y. Knyazikhin and R.B. Myneni (2003). Effect of foliage heterogeneity on the MODIS LAI and FPAR over broadleaf forests. *Remote Sens. Environ.*, **85**, 410–423.

An experimental derivation of the canopy spectral invariants as well as their use in parameterization of canopy spectral response to incident solar radiation is discussed in these papers.

Marshak, A., Yu. Knyazikhin, A.B. Davis, W.J. Wiscombe and P. Pilewskie (2000). Cloud – Vegetation interaction: Use of normalized difference cloud index for estimation of cloud optical thickness. *Geoph. Res. Lett.*, **27**,1695–1698.

Knyazikhin, Yu. and A. Marshak (2000). Mathematical aspects of BRDF modeling: Adjoint problem and Green's function. *Remote Sens. Review*, **18**, 263–280.

Barker, H.W. and A. Marshak (2001). Inferring optical depth of broken clouds above green vegetation using surface solar radiometric measurements. *J. Atmos. Sci.*, **58**, 2989–3006.

Marshak, A., Yu. Knyazikhin, K.D. Evans and W.J. Wiscombe (2004). The "RED versus NIR" plane to retrieve broken-cloud optical depth from ground-based measurements. *J. Atmos. Sci.*, **61**, 1911–1925.

These papers discuss cloud–vegetation interactions and use Green function as a cloud radiative response to the illumination by a point mono-directional source located on the vegetated surface. They also propose a method for inferring cloud optical depth for inhomogeneous clouds using surface-based radiometric observations at two wavelengths with the strongest surface contrast and similar cloud optical properties. A special emphasis is given to the case of broken clouds.

Appendix:
Scale-by-Scale Analysis and Fractal Cloud Models

A. Marshak and A.B. Davis

A.1 Scale-by-Scale Analysis 653

A.2 Fractal Models 658

References 662

A.1 Scale-by-Scale Analysis

Geophysical systems in general, and clouds in particular, exhibit structure over a wide range of scales and have high levels of variability. Thus, to better understand cloud structure, we compile statistical information on a scale-by-scale basis and seek simple connections that relate properties at one scale to another. Based on the turbulent nature of clouds and following a well-established tradition in turbulence study, we seek power laws in the scale parameter, r. The physical meaning of a statistical power law in r is that the same physical processes dominate over a wide range of scales. Smaller parts of the system therefore look like scaled-down versions of larger parts, and vice versa. The most important quantity is then the exponent of the power law. A power-law statistic r^α is invariant under a change of scale $r \to \lambda r$ in the sense that only the scale ratio λ and the exponent α are required to predict the new value. Observation of a power law therefore reflects a statistical invariance under change of scale, called "scale-invariance" or just "scaling." In practice, this means linear relations in log(statistic) – log(scale) plots.

A.1.1 Wavenumber Spectrum and Autocorrelation Function

Let us assume that we have a stochastic process $\phi(x), 0 \leq x \leq L$ and let $\widetilde{\phi}(k), -\infty < k < \infty$ be its Fourier transform. The wavenumber (or energy, or power) spectrum (or spectral density) $E(k)$ of ϕ is defined as

$$E(k) = \frac{1}{L}\langle |\widetilde{\phi}(k)|^2 + |\widetilde{\phi}(-k)|^2 \rangle, \; k > 0 \,, \tag{A.1}$$

where $\langle \cdot \rangle$ designates ensemble averaging, i.e., over all possible realizations of ϕ. For scale-invariant processes, the wavenumber spectrum follows a power law

$$E(k) \propto k^{-\beta} \tag{A.2}$$

over the large range of wavenumbers $k = 1/r$. The spectral exponent β contains valuable information on the variability in ϕ.

For a real stationary process, a power spectrum can be obtained from a cosine transform of its autocorrelation (e.g., Papoulis, 1965, p. 338),

$$E(k) = 2 \int_0^\infty \cos(2\pi k r)\, G(r)\, dr \tag{A.3}$$

where

$$G(r) = \langle \phi(x+r)\, \phi(x) \rangle \tag{A.4}$$

is the autocorrelation function (assuming $\langle \phi(x) \rangle = 0$). The stationarity assumption translates here in finding no dependence of $\langle \phi(x+r)\, \phi(x) \rangle$ on x upon ensemble averaging. Conversely, we have

$$G(r) = 2 \int_0^\infty \cos(2\pi k r)\, E(k)\, dk \,. \tag{A.5}$$

Note that

$$\langle |\phi^2(x)| \rangle = G(0) = 2 \int_0^\infty E(k)\, dk \,. \tag{A.6}$$

Finally, it follows from (A.4) that

$$\langle [\phi(x+r) - \phi(x)]^2 \rangle = 2[G(0) - G(r)] \geq 0 \,. \tag{A.7}$$

So if the autocorrelation function $G(r)$ is continuous at $r = 0$, process $\phi(x)$ is stochastically continuous, meaning that (A.7) goes to 0 with r.

For real measurements, we generally have only a small number of "realizations" with a finite spatial sampling,

$$\phi_i = \phi(x_i), x_i = il \; (i = 1, 2, \ldots, N) \,, \tag{A.8}$$

where $N = L/l$ is the total number of points, L the spatial length of record, and l the step size. Let us for simplicity assume that N is a power of 2, and denote the discrete Fourier transform of the data in (A.8) as

$$\widetilde{\phi}_{\pm j} = \widetilde{\phi}(\pm k_j), \; k_j = \frac{j}{L}\left(j = 0, 1, \ldots, \frac{N}{2}\right) \,. \tag{A.9}$$

The associated energy density, as a discrete counterpart of (A.1) is

$$E(k_j) = \frac{2}{L}|\tilde{\phi}_j|^2, \left(j = 1, 2, \ldots, \frac{N}{2} - 1\right) \tag{A.10}$$

since $\tilde{\phi}_{-j} = \tilde{\phi}_{+j}^*$ for real data in (A.8).

Now plotting $N/2 - 1$ values of $E(k_j)$ versus k_j in a log-log plot gives us an estimate of a slope β. However, without a judicious weighting scheme, a least square fit to the power-law in form (A.2) on a log E versus log k plot will be dominated by the smallest scales r (largest wavenumbers k). The contribution of large scales to the exponent β becomes virtually nil. To make all scales contribute equally and simultaneously yield log-log plots that are easy to interpret visually, we average $E(k)$ by octaves, that is, a factor of 2 in k,

$$\overline{E_m} = \frac{1}{2^{m-1}} \sum_{j=2^{m-1}}^{2^m - 1} E(k_j), \quad m = 1, \ldots, \log_2 N - 1. \tag{A.11}$$

This corresponds to average wavenumber

$$\overline{k_m} = \frac{1}{2^{m-1}} \sum_{j=2^{m-1}}^{2^m - 1} k_j = \frac{3}{2}2^{m-1} - \frac{1}{2}, \quad m = 1, \ldots, \log_2 N - 1 \tag{A.12}$$

which are equally spaced on a log scale only in the limit $m \gg 1$. We thus obtain exactly $\log_2 N - 1$ estimates for $E(k)$.

A.1.2 Structure Functions

For a stochastic process $\phi(x)$, let us define the absolute increments across scale r,

$$\Delta\phi(r; x) = |\phi(x + r) - \phi(x)|, \quad 0 \le r \le L, 0 \le x \le L - r \tag{A.13}$$

and consider their statistical moments. We assume that the statistical properties of $\Delta\phi(r; x)$ are independent of position x. This is weaker than the assumption of stationary increments. Then, because of the scale-invariance, we expect

$$S_q(r) = \langle \Delta\phi(r; x)^q \rangle \equiv \langle \Delta\phi(r)^q \rangle \propto r^{\zeta(q)}, \tag{A.14}$$

where $S_q(r)$ is called the structure function of order $q \ge 0$. The family of exponents $\zeta(q)$, as a function of q, has the following properties:

(a) it is normalized,

$$\zeta(0) = 0 ; \tag{A.15a}$$

(b) it is a convex function (Frisch and Parisi, 1985), i.e.,

$$\zeta''(q) \le 0 ; \tag{A.15b}$$

(c) finally, if the increments in (A.13) are bounded, $\zeta(q)$ is nondecreasing (Frisch, 1991; Marshak et al., 1994), i.e.

$$\zeta'(q) \geq 0 . \tag{A.15c}$$

These last two inequality properties are contingent on the fact that the q-dependence of the prefactors (proportionality constants) in (A.14) is weak enough to neglect.

Two low-order exponents are well known. For $q = 1$, there is

$$0 \leq \zeta(1) = H_1 \leq 1 , \tag{A.16}$$

which is called the roughness or Hurst exponent. It characterizes the smoothness of the signal: the larger H_1 the smoother the signal. Referring back to (A.14), the limit $H_1 \rightarrow 1$ corresponds to almost everywhere differentiable signals. The opposite limit $H_1 \rightarrow 0$ leads to a signal with scale-independent increments. White noise is the most famous example of a signal with $H_1 = 0$, but any scale-invariant stationary process yields the same answer.

The second order ($q = 2$) structure function

$$S_2(r) = \langle |\phi(x + r) - \phi(x)|^2 \rangle \propto r^{\zeta(2)} \tag{A.17}$$

is related to a wavenumber spectrum through the Wiener-Khinchine theorem. This theorem generalized to nonstationary processes with stationary increments (Monin and Yaglom, 1975) reads as

$$1 \leq \beta = \zeta(2) + 1 \leq 3 . \tag{A.18}$$

We will use these moments in the examples below.

A.1.3 Examples with Cloud Liquid Water Data

We illustrate the above concepts on cloud liquid water content (LWC) data measured during the Atlantic Stratocumulus Transition Experiment (ASTEX) from an aircraft in June 1992 (Albrecht et al., 1995). Figure A.1a shows a 16384-point data stream sampled approximately every 8 m for an overall length of about 130 km (Davis et al., 1994). So, in the notations of (A.8), we have $N = 2^{14} = 16384$, $l = 8$ m, and $L = 130$ km. The wavenumber spectrum of this data set (panel A.1b) follows a power-law behavior (A.2) with spectral exponent $\beta \approx 1.5$ over a quite large range of scales from tens of meters to tens of kilometers. In addition to $N/2 - 1$ wavenumber points, as in (A.10), we also plotted $\log_2 N - 1 = 13$ octave averaged dots as in (A.11)–(A.12). Note that because of fewer contributions from small scales, the octave averaged log-log plot yields slightly smaller spectral exponent ($\beta \approx 1.45$). A detailed discussion of different estimates of spectral exponents for wavenumber spectra can be found in Davis et al. (1996).

The first five integer moments of the absolute increments over scale r, (A.13), for the LWC data in panel A.1a are shown in panel A.1c. Again we see a remarkably

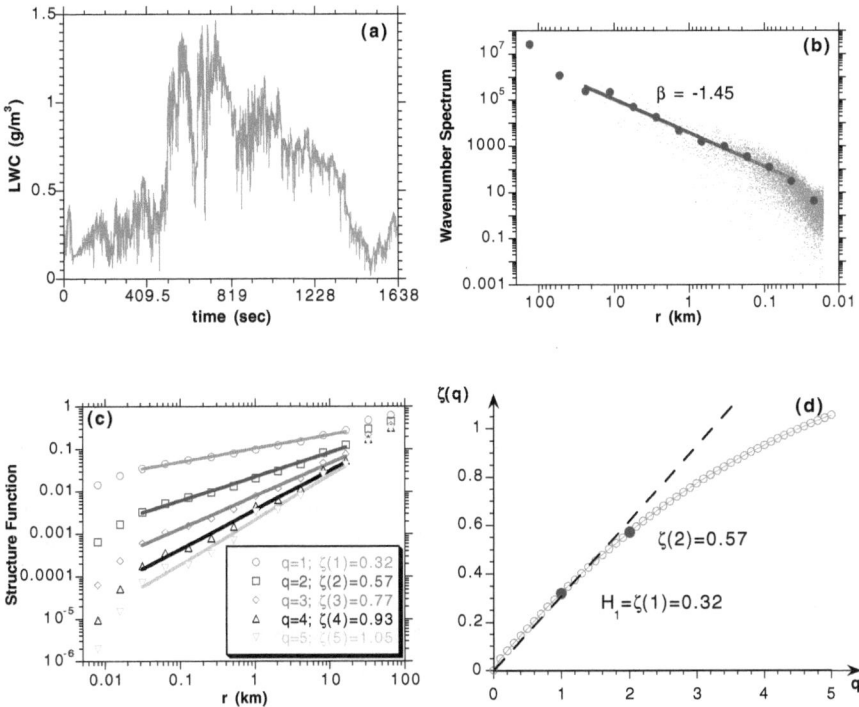

Fig. A.1. Wavenumber spectrum and structure functions of a real-world data set. (**a**) A horizontal transect of cloud liquid water content (LWC) sampled at 10 Hz, 8 m at aircraft speed 80 m/s. The data were collected with the PVM-100 (Gerber, 1991) during the ASTEX experiment in June 1992. (**b**) Wavenumber spectrum for the data on panel (**a**). Circles correspond to the octave-averaged data described in (A.11)–(A.12). (**c**) Structure functions of order 1 to 5 plotted versus scale r. Scaling is indicated from several tens of meters to several tens of kilometers. (**d**) The resulting structure function exponents versus q. The linear relation corresponds to fractional Brownian motion with the same Hurst exponent

good scaling (with correlation coefficient of 0.99) over three orders of magnitude for all five moments. The exponents $\zeta(q)$ of structure function $S_q(r)$ are plotted in panel A.1d. We can easily see that $\zeta(q)$ is a convex nondecreasing function with $\zeta(0) = 0$ as stated in (A.15a)–(A.15c). The Hurst exponent $H_1 = 0.32$ which is typical for marine Sc (Davis et al., 1994, 1999; Marshak et al., 1997). Note that $\zeta(2)+1 = 1.57$ that is close to spectral exponent β in panel A.1b and thus consistent with the Wiener-Khinchine relation in (A.18).

The straight dashed line in panel Fig. A.1d would correspond to fractional Brownian motion (fBm) (Mandelbrot, 1977) with the same Hurst exponent $H = H_1$; i.e.,

$$\zeta(q) = qH \qquad (A.19)$$

is a linear function. This linearity is the hallmark of "monoscaling" and, statistically, it corresponds to relatively narrow distributions of the increments across all scales, e.g., Gaussian distributions which are determined entirely by their variance (and fBm is Gaussian by definition).

The observed nonlinearity or "multiscaling" of $\zeta(q)$ for LWC data indicates a level of intermittency in the data that follows from the non-Gaussian nature of the turbulent signal. Notice that $\zeta(2)$ is visibly lower than the monoscaling prediction $2H_1$. More analysis results and references to data analysis of LWC or other kinds of cloud data can be found, for instance, in Lovejoy and Schertzer (1990), Tessier et al. (1993), Ivanova and Ackerman (1999), and Davis et al. (1999).

A.2 Fractal Models

In this section we describe stochastic models that simulate fluctuations of cloud liquid water. As we saw in the previous section, LWC fluctuations inside marine Sc obey power-law statistics over at least three orders of magnitude in scale. Hence, the main feature we seek in a stochastic model is scale-invariance. In turbulence studies the most popular scale-invariant models are multiplicative cascade models. The construction of a mass-conserving cascade model is as follows (Fig. A.2). Start with a homogeneous slab of length L. Divide into 2 parts and then transfer a fraction f_1 of the mass from one half to the other in a randomly chosen direction. This is equivalent to multiplying the originally uniform density field on either side by factors $W_1^{(\pm)} = 1 \pm f_1$. The same procedure is repeated recursively at ever smaller scales using fractions $f_i(i = 2, 3, \ldots)$ on segments of length r_{i+1} where $r_{i+1} = L/2^i$.

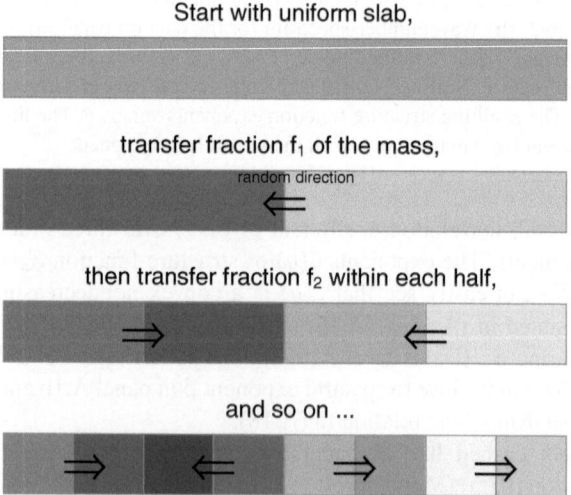

Start with uniform slab,

transfer fraction f₁ of the mass,

random direction

then transfer fraction f₂ within each half,

and so on ...

Fig. A.2. Schematic construction of a mass-conservative multiplicative cascade model

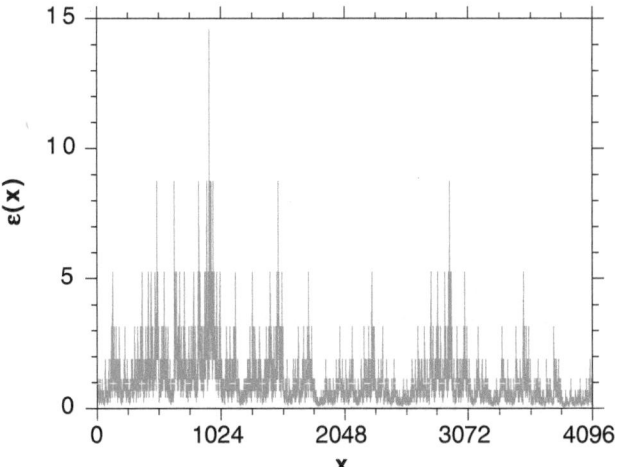

Fig. A.3. A multiplicative 12 cascades "*p*-model" with unit mean and $p = 0.375$, thus from (A.21) $\beta_\varepsilon(p) = 0.91$

A.2.1 Singular Cascades

We now parameterize the multiplicative weights as

$$W_i^{(\pm)} = 1 \pm f_i = 1 \pm (1 - 2p) = \begin{cases} 2p, \text{ or} \\ 2(1-p) \end{cases}, \quad 0 \le p < \frac{1}{2}, \tag{A.20}$$

independently of $i = 1, 2, \ldots$, with 50/50 probability for the signs. This leads to a singular (multi)fractal model $\varepsilon(x)$ called the "*p*-model" (Meneveau and Sreenivasan, 1987); in this model, parameter p controls the degree of mass (energy) redistribution at each cascade step. The *p*-model (illustrated in Fig. A.3) is scale-invariant; its wavenumber spectrum exhibits a power law, $E_\varepsilon(k) \propto k^{-\beta_\varepsilon}, k > 0$, with spectral exponent

$$0 \le \beta_\varepsilon(p) = 1 - \log_2[1 + (1 - 2p)^2] < 1. \tag{A.21}$$

Singular cascade models $\varepsilon(x)$ have interesting intermittency properties but their spectra with $\beta_\varepsilon < 1$ do not scale as observed cloud liquid water fields (that have $\beta > 1$) and therefore they do not show stochastic continuity.

A.2.2 Bounded Cascades

A simple way to obtain $\beta > 1$ is to reduce the variance of the multiplicative weights in (A.20) at each cascade step. Taking

$$W_i^{(\pm)} = 1 \pm (1 - 2p)2^{-H(i-1)}, 0 \le p < 1/2, H > 0, i = 1, 2, \ldots \tag{A.22}$$

leads to "bounded" cascade models (Cahalan, 1994). The limit $H \to \infty$ yields a single jump (Heaviside step) from $2p$ to $2(1 - p)$ at $x = L/2$.

By reducing the size of the jumps as the scale decreases, we are effectively introducing a degree of continuity into the model. One can show that in this case, its autocorrelation function is a continuous function at $r = 0$ and a generated field $\phi(x)$ is stochastically continuous (A.7). As a result, the spectral exponent has moved into the range

$$1 < \beta_\phi(H) = \min\{2H, 1\} + 1 \leq 2 , \tag{A.23}$$

independently of p. Figure A.4 (top panel) shows a realization of a 14-step bounded cascade model with $H = 1/3$ and $p = 0.375$.

In the limit of an infinite number of cascade-steps, the structure function exponents $\zeta(q)$ have a nonlinear form (Marshak et al., 1994),

$$\zeta(q) = \min\{qH, 1\} = \begin{cases} qH, & 0 \leq q \leq 1/H \\ 1, & 1/H \leq q < \infty \end{cases} ; \tag{A.24}$$

so (A.23) follows from (A.18) and (A.24). Since the spectral exponents of a mono-scaling fBm is a linear function, $\zeta(q) = qH$, the bounded cascade model cannot be distinguished from fBm for moments smaller than $q = 1/H$. This is clearly seen in Fig. A.5 which shows theoretical structure function exponents for both bounded models (A.24) and fBm (A.19).

To summarize, the bounded cascade model is a good tutorial model for cloud horizontal inhomogeneity. To a first approximation, it reproduces lower-order statistical moments of cloud liquid water distribution. However, as follows from (A.24), its $\zeta(q) = 1$ for $q \geq 1/H$ whereas the higher-order moments of LWC fluctuations have exponents that substantially exceed unity and show strong curvature even for low values of q. In the next subsection we describe another model, fractionally integrated cascades (Schertzer and Lovejoy, 1987) that overcomes these limitations.

A.2.3 Fractional Integration

Another way of transforming singular cascades with $\beta_\varepsilon < 1$ into a more realistic one with $\beta_\phi > 1$ is power-law filtering in Fourier space (Schertzer and Lovejoy, 1987); this will bring the spectral exponent to any prescribed value. In particular, we have

$$\beta_\phi(p, H^*) = \beta_\varepsilon(p) + 2H^* \tag{A.25}$$

where $0 < H^* < 1$ describes the low-pass filter in k^{-H^*}. Mathematically, this operation – also known as "fractional integration" (FI) – is a convolution with a weakly singular kernel:

$$\phi(x) = \int \varepsilon(y)|x - y|^{H^*-1}dy . \tag{A.26}$$

This is called FI since for $H^* = 1$ it corresponds to ordinary integration. Here again, thanks to the FI term in (A.25), field $\phi(x)$ is stochastically continuous. As an example, Fig. A.4 shows a realization of the FI cascade model with $p = 0.375$ (thus

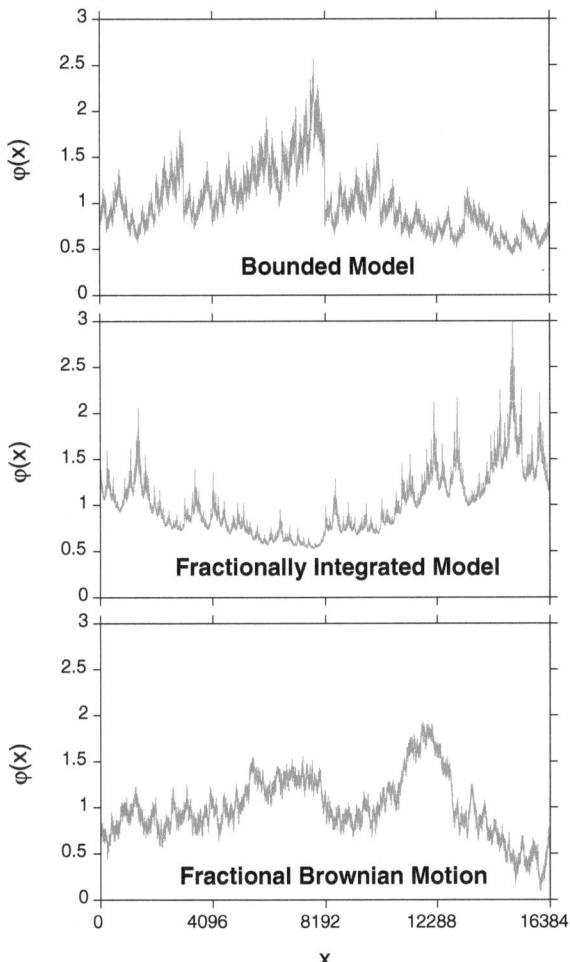

Fig. A.4. Three fractal models with the same spectral exponent $-5/3$ and overall mean and variance. From *top* to *bottom*: bounded (Bo) cascade model with $H = 1/3$ and $p = 0.375$; fractionally integrated (FI) cascade model with $p = 0.375$ and $H^* = 0.38$; fractional Brownian motion (fBm) with $H = 1/3$. Bo and FI are generated with 14 cascade steps; fBm is generated using mid-point displacement method (e.g., Peitgen and Saupe, 1988) also with 14 steps. As a result, all models have $2^{14} = 16384$ pixels. In all three models average $\langle \phi \rangle = 1$ and standard deviation $= 1/3$. Note that, by construction, both Bo and FI multiplicative cascade models have only positive values while $\phi(x)$ for fBm can be either positive or negative since its probability density function is Gaussian

$\beta_\varepsilon(p) = 0.91$) and $H^* = 0.38$ (thus $\beta_\phi(p, H^*) = 5/3$). In contrast to the bounded model, we have only approximate formulas for the structure function exponents for FI model. They are exact for $q = 0$ and $q = 2$ and quite accurate for all low-order moments and, moreover, numerical results are always available.

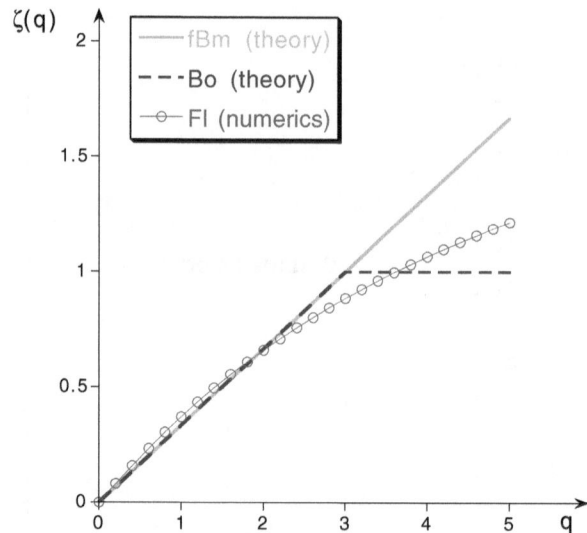

Fig. A.5. Comparison of exponent functions $\zeta(q)$ for three fractal models. Fractional Brownian motion (fBm) with $H = 1/3$, bounded cascades (Bo) with $H = 1/3$ and fractionally integrated (FI) cascades with $p = 0.375$ and $H^* = 0.38$. Note that while for fBm and Bo models we give theoretical curves, FI shows the results of numerical calculations with 15 cascade steps averaged over 100 realizations

To compare the three models (monoscaling fBm and two multiscaling models, bounded and FI cascades), we plotted their structure function exponents $\zeta(q)$ in Fig. A.5 for q going from 0 to 5. As we see, the bounded model is the most intermittent among them since its $\zeta(q)$ is the most nonlinear. However, with at least one more tunable parameter than bounded cascades, FI cascades are better candidates for simulating observed liquid water fluctuations.

References

Albrecht, B.A., C.S. Bretherton, D. Johnson, W.H. Schubert, and A.S. Frisch (1995). The Atlantic Stratocumulus Transition Experiment – ASTEX. *Bull. Amer. Meteor. Soc.*, **76**, 889–904.

Cahalan, R.F. (1994). Bounded cascade clouds: Albedo and effective thickness. *Nonlinear Proc. Geophys.*, **1**, 156–167.

Davis, A., A. Marshak, W.J. Wiscombe, and R.F. Cahalan (1994). Multifractal characterizations of nonstationarity and intermittency in geophysical fields: Observed, retrieved, or simulated. *J. Geophys. Res.*, **99**, 8055–8072.

Davis, A., A. Marshak, W.J. Wiscombe, and R.F. Cahalan (1996). Scale-invariance in liquid water distributions in marine stratocumulus, Part I, Spectral properties and stationarity issues. *J. Atmos. Sci.*, **53**, 1538–1558.

Davis, A.B., A. Marshak, H. Gerber, and W.J. Wiscombe (1999). Horizontal structure of marine boundary-layer clouds from cm- to km-scales. *J. Geophys. Res.*, **104**, 6123–6144.

Frisch, U. (1991). From global scaling, à la Kolmogorov, to local multifractal scaling in fully developed turbulence. *Proc. Roy. Soc. London. A*, **434**, 89–99.

Frisch, U. and G. Parisi (1985). A multifractal model of intermittency. In *Turbulence and Predictability in Geophysical Fluid Dynamics*. M. Ghil, R. Benzi and G. Parisi (eds.). North Holland, Amsterdam, The Netherlands, pp. 84–88.

Gerber, H. (1991). Direct measurement of suspended particulate volume concentration and far-infrared extinction coefficient with a laser-diffraction instrument. *Atmos. Res.*, **30**, 4824–4831.

Ivanova, K. and T.P. Ackerman (1999). Multifractal characterization of liquid water in clouds. *Phys. Rev. E*, **59**, 2778–2782.

Lovejoy, S. and D. Schertzer (1990). Multifractals, universality classes and satellite and radar measurements of cloud and rain fields. *J. Geophys. Res.*, **95**, 2021–2034.

Mandelbrot, B.B. (1977). *Fractals: Form, Chance, and Dimension*. W.H. Freeman, San Francisco (CA).

Marshak, A., A. Davis, R.F. Cahalan, and W.J. Wiscombe (1994). Bounded cascade models as non-stationary multifractals. *Phys. Rev. E*, **49**, 55–69.

Marshak, A., A. Davis, W.J. Wiscombe, and R.F. Cahalan (1997). Scale-invariance of liquid water distributions in marine stratocumulus, Part 2 – Multifractal properties and intermittency issues. *J. Atmos. Sci.*, **54**, 1423–1444.

Meneveau, C. and K.R. Sreenivasan (1987). Simple multifractal cascade model for fully developed turbulence. *Phys. Rev. Lett.*, **59**, 1424–1427.

Monin, A.S. and A.M. Yaglom (1975). *Statistical Fluid Mechanics*, volume 2. MIT Press, Boston (MA).

Papoulis, A. (1965). *Probability, Random Variables, and Stochastic Processes*. McGraw-Hill, New York (NY).

Peitgen, H.-O. and D. Saupe (1988). *The Science of Fractal Images*. Springer-Verlag, New York (NY).

Schertzer, D. and S. Lovejoy (1987). Physical modeling and analysis of rain and clouds by anisotropic scaling multiplicative processes. *J. Geophys. Res.*, **92**, 9693–9714.

Tessier, Y., S. Lovejoy, and D. Schertzer (1993). Universal multifractals: Theory and observations for rain and clouds. *J. Appl. Meteor.*, **32**, 223–250.

Epilogue: What Happens Next?

Radiative Transfer Problems

Let's make the circumstantially reasonable assumption that we have a tough radiative transfer (RT) problem to solve. Maybe clouds are in the picture. For simplicity, let's assume it is a "forward" RT problem where we know the optical properties of the atmosphere and surface. If we are faced with an "inverse" problem, as is often the case in remote sensing, then we first need to establish that we can solve the forward problem anyway. We are thus asked to compute some radiance values $I(x, \Omega)$ for a given wavelength where x is position in three-dimensional (3D) space and Ω is a direction of propagation on the unit sphere.

In remote sensing, x is likely to be the position of a sensor and $-\Omega$ will be the direction it is looking into, defined (say) by one of many pixels at the focal plane.[1] In radiation energy budget modeling, we are only interested in angular integrals of $I(x, \Omega)$, weighted or not with $|\Omega_z|$ (the vertical direction cosine). At any rate, we now have to somehow find a solution of the integro-differential RT equation that looks like

$$\Omega \cdot \nabla I = -\sigma(x)[I(x, \Omega) - \varpi_0 \int_{4\pi} p(\Omega' \cdot \Omega)I(x, \Omega')d\Omega'] + Q(x, \Omega)$$

in the relatively simple case where scattering properties, single-scattering albedo ϖ_0 and phase function p, are assumed uniform but the extinction coefficient $\sigma(x)$ varies spatially. $Q(x, \Omega)$ is a given volume source term, e.g., thermal emission. There will also be boundary conditions to satisfy and, for simplicity, we will assume these boundaries are flat, at a constant z-coordinate: the upper (top-of-atmosphere, TOA) condition is typically an incoming uniform collimated solar beam; the lower (surface) condition can be quite complex, with spatially and angularly varying reflection and/or emission. If there is an interest in polarization, then I is a 4-vector and p is a

---

[1] From large stand-off distances, x can also be viewed as a variable point on the upper boundary of the atmosphere-surface medium and Ω is then the constant (or x-dependent, depending on distance) direction of the sensor.

4×4 matrix – a relatively minor complication in view of all the spatial and angular variables in the balance.

The above stated 3D RT problem includes the majority of current remote sensing and climate needs. But we still have to find a way to solve the scalar or vector RT equation.

Classic Solutions, and Their Limitations

There are only two situations where exact solutions exist, clearly a desirable scenario and especially in remote sensing:

1. $\varpi_0 = 0$ (no scattering) and a non-reflecting surface, hence no coupling at all between beams in direction or in polarization. In this case, $\sigma(x)$ and $Q(x, \Omega)$ can be arbitrarily complicated, and the same is true for the surface source distribution. Here, we can even write an analytical expression for $I(x, \Omega)$: a one-dimensional integral along the beam $\{x, \Omega\}$ using the boundary source, Q, and transmission functions (negative exponentials of line-integrals of σ under the present monochromatic assumption).

2. $\sigma(x) \equiv \sigma(z)$ and $Q(x, \Omega) \equiv Q(z, \Omega)$, hence $I(x, \Omega) \equiv I(z, \Omega)$. This is the famous horizontally uniform atmosphere/surface plane-parallel system of one-dimensional (1D) RT. Under the present assumptions of vertically uniform scattering properties, $0 < \varpi_0 \leq 1$ and $p(\cdot)$, this multiple-scattering problem is amenable, via invariant embedding and superposition, to Chandrasekhar's H-, or X- and Y-functions using the dimensionless coordinate $d\tau = \sigma(z)dz$.

If, in problem #1, the surface becomes partially reflective then there will be at the most one reflection off the (flat) ground. Thus, if a surface point x_S is in view, an angular integral over the down-welling hemisphere (weighted by surface reflectivity and $|\Omega_z|$) of the closed-form solutions at x_S is required to adjust the boundary source term. Then the exact 1D solution along the beam going from the sensor to x_S can be applied. Early releases of the remote-sensing workhorse code known as MODTRAN were designed specifically to solve this problem for a 1D (i.e., stratified) atmosphere.[2] If, in problem #2, the scattering properties ϖ_0 and p become dependent on the sole spatial coordinate $\tau(z)$ of the 1D RT problem, then we know how to obtain accurate numerical solutions using popular codes such as DISORT or, for spectral details, a recent release of MODTRAN (which will in fact call DISORT for the multiple scattering).

---

[2] The main problem that MODTRAN designers, and contributing molecular spectroscopy experts, address is the weighted (sensor-response) integration over potentially complicated and fast spectral variability of absorbing atmospheric gases computed on the fly for a given atmospheric composition. The MODTRAN solution to the spectral variability problem generally requires averaging over many values of the extinction/absorption coefficient σ, even at the smallest allowable wavelength interval. This spectral averaging in turn leads to transmission functions that are not exponential in the amounts of absorbers. MODTRAN developers have also enabled spherical 1D (stratified) geometry, including refractivity.

That's it. Every other solution of the 3D RT equation will call for some sophisticated mathematical analysis, an intricate time-consuming numerical implementation, and then probably too many computer cycles ... so only research-mode utilization is contemplated. At least that is the conventional wisdom. It is therefore not surprising that virtually all operational remote sensing and radiation energy budget estimations are done with one of the above pair of text-book solutions, too often regardless of their applicability.

Take, for instance, atmospheric sounding – extracting (say) temperature and water vapor profiles, via $\sigma(z)$ and $Q(z)$, from measured I(ground, zenith) or I(TOA, nadir) at a number of wavelengths. This application uses the "no-scattering" solution #1, often in the microwave region where its strong assumptions are considered reasonably valid. But are they really? Apparently not in the presence of heavy precipitation.

In radiation energy budget problems, two "streams" of radiation (one up and one down) are usually considered sufficient. Schuster, arguably the founding father of atmospheric RT, showed – one hundred years ago at the time of writing – that the "scattering/absorption" problem #2 then becomes analytically tractable in closed-form. This outcome is always a plus in applications such as GCM-based climate simulations where very many RT computations need to be performed (one per layer and per cell and at as many time-steps as possible). But can we be confident in century-old transport physics for such a critical aspect of the the climate problem?

A New Perspective on Atmospheric Radiative Transfer

The above formulation of 3D RT in the atmosphere/surface system is not just a challenging problem in computational physics. It should be viewed as an accurate description of the photon flow that is actually unfolding in Nature. It is a codified representation of the reality we are dealing with, hardly a "problem." So the preferred *mathematically exact* solutions #1 and #2 are in fact just *physical approximations*. Indeed, a purely emitting/absorbing medium is an abstraction ... as we know from practical thermodynamics that it is impossible to make a perfect black-body (every material will reflect or scatter at least a little).[3] And of course, an exactly uniform plane-parallel cloud layer has never materialized since cloud structure and evolution are part of normal tropospheric dynamics. These fluid dynamics (a.k.a. "weather") are clearly stratified (lapse rates, boundary layers, inversion layers, etc.) but they are also very 3D in nature at all scales (general circulation, fronts, waves, jets, convection, precipitation, instability, turbulence, and so on).

We can make an immediate damage assessment here. Any volume scattering perturbs solution #1 and horizontal gradients in anything perturb solution #2 for scattering media. In cloud remote sensing at optical wavelengths, blind faith in solution #2 is to believe that satellite pixels are radiatively independent. That would be reasonable only if they were huge. But then again, if we approximated that situation by

---

[3] For the more theoretically minded, this is traceable to the Kramers-Kronig causality relations, a precursor to the linear response formalism for transport coefficient estimation.

aggregating enough pixels deemed too-small, would it be reasonable to assume the pixels are internally uniform?

Much of the 3D RT literature so far has used the uniform plane-parallel cloud model as a benchmark, estimating the bias associated with that 1D model simply because it is so widely accepted. We propose that for the next century of atmospheric RT, the plane-parallel model be considered what it really is: an approximation that, like any other, needs validation or at least a convincing justification. It should not be a convenient assumption we can make ... just because most everyone else does, just because that is what MODTRAN delivers.

For instance, letting $\Omega = (\sin\theta\cos\phi, \sin\theta\sin\phi, \cos\theta)^T$ in the usual polar representation and looking back at the *left-hand side* of the above 3D RT equation, a modeler intent on using 1D RT should explain to the end-user why

$$\sin\theta\left|\cos\phi\frac{\partial I}{\partial x} + \sin\phi\frac{\partial I}{\partial y}\right| \ll \left|\cos\theta\frac{\partial I}{\partial z}\right|.$$

This case for weak "3D-ness" can be made by stating the (preferably low) value of ϖ_0 and/or by noting that $\sigma(x)$ is a stronger function of z than of $\vec{x} = (x, y)^T$, by making a solid theoretical argument based on a conceptual model, by using a computational 3D model as a benchmark, by analyzing real-world data which is 3D-compliant by definition, or any combination of these approaches. At least this way we collectively reinforce the awareness that 1D RT is an approximate solution of the real 3D RT problem at hand.

Assuming the case for using a 1D model has been made, what will be done with the residual horizontal variability ($\vec{x}$-dependence) of $\sigma(x)$ and $Q(x, \Omega)$ on the *right-hand side* of the transport equation? Will it be averaged out? Will it be treated parametetrically? That is, extinction in the 1D transport solver will be processed as $\sigma(z; \vec{x})$ where the semi-colon emphasizes the difference between variables and parameters. This is known as the local Independent Column Approximation (ICA). Will we horizontally average the outcome of the local ICA result, say, for $I(\text{TOA,nadir})$, over all $\vec{x}$ values? Will an "effective" extinction be used that attempts to capture 3D effects in some statistical sense? And what will we do about $Q(x, \Omega)$? At any rate, we can no longer say: *What else than 1D RT can be done?* This monograph demonstrates that much more can be done, and not necessarily at a large computational cost.

The above questions collide head-on with a new and interesting quantity that arises in 3D RT modeling. What is the "scale" r of interest? & What are the RT "dynamics" of $\bar{I}_r(x, \Omega)$, the radiance field coarse-grained to scale r? The instinctive answer will likely be that r is the resolution of the instrument (the pixel-size) or the mesh spacing in the climate- or cloud-dynamics model (the grid-constant). These are important but *artificial scales*. First, sensors as well as computational meshes are becoming ever-more "adaptive," so pixel- or grid-scales are moving targets. Second, there are equally if not more important *physical scales* and dimensionless ratios thereof to consider: the photon mean-free-path, the cloud-layer thickness, the photon diffusion scale, cloud aspect ratio, dominant variability scales, and so on.

Where observational or computational resolution plays an essential role, at least from the 3D RT modeling perspective, is in telling us whether or not the variability of concern is resolved. What comes next?

- In the case of resolved variability, we have to worry about (the divergence of net) radiative fluxes that cross boundaries of pixels or grid-cells. These fluxes will invalidate a preemptive 1D treatment of the RT. Is there a simple way to estimate them and remove the bias from the 1D model?
- In the case of unresolved variability, we have to make statistical assumptions about it and try to estimate its likely impact on the resolved scales. Can we formulate an "effective medium" theory where modified optical properties will incorporate the bulk effects of small-scale fluctuations?

In 3D reality of course, both processes are at work to some extent. At any rate, we must learn how to manage all the identified radiative processes at various scales and make informed decisions about their importance based on experience and on programmatic priorities. This is by no means easy. All too frequently, the "ignorance is bliss" modus-operandi prevails.

Time: The New Frontier in Atmospheric Remote Sensing

Time-dependence is a dimension of RT that we have only started to explore in any depth for real – that is, 3D – atmospheric RT. So now we consider $I(t, x, \Omega)$, typically with $t \geq 0$, that obeys a RT equation in "3 + 1" dimensions: the volume- and/or boundary-sources become t-dependent and, most notably, the monokinetic advection operator $\Omega \cdot \nabla$ becomes[4]

$$\frac{1}{c} \frac{\partial}{\partial t} + \Omega \cdot \nabla .$$

Even for steady-state photon transport in 1D geometry, the dominant number of scattering events or mean photon time-of-flight for a given outcome (e.g., reflection versus transmission) are profoundly informative quantities. There is already a core international community invested in differential absorption spectroscopy in the oxygen A-band as a means of accessing solar photon path length statistics. These new observables have proven to be highly sensitive to the degree of spatial complexity in cloudiness, e.g., single/unbroken layers versus multiple and/or broken layers. Pulsed (laser) sources or rapidly time-varying sources (e.g., lightning flashes) can be traced along convoluted multiple-scattering paths through dense clouds. Thanks to recent technological advances, we can now observe the resulting waveforms, and these can be analyzed in terms of cloud or source properties using time-dependent RT, in 3D as needed.

---

[4] Somewhat miraculously, this is not so new as it appears at first glance, at least for pulsed sources. A temporal Laplace transform indeed converts the t-derivative into what looks like a spatially uniform (hence gaseous) absorption term with a coefficient varying from 0 to ∞ while the δ-in-time source term becomes a constant. The converse statement – if we know the detailed t-dependence, then we can compute the effect of any level of gaseous absorption – is known as the "Equivalence Theorem."

Last but not least, temporal (i.e., pulse-stretching) scales and some of the above-mentioned spatial scales are so tightly connected through the fractal nature of photon random walks in the Earth's cloudy atmosphere that they are really just two sides of the same coin. This insight has far-reaching ramifications that, in particular, have yet to be exploited for the teaching of radiative transfer through geometry and probability rather than through calculus.

In summary, we feel that future generations of atmospheric and remote-sensing scientists, whether or not they become "RT" experts, need to be exposed to the rich phenomenology of 3D radiative transfer and encouraged to explore it to their young heart's content.

Los Alamos, New Mexico   *Anthony Davis*
January, 2005   *Alexander Marshak*

Notations

In this volume, we have adopted standard notations for radiance/intensity $I_\lambda(x, \Omega)$, irradiance/flux $F_\lambda(x)$, transmittance T, optical depth τ, extinction σ, and so on, from the astrophysical and transport-theoretical literatures because they are also very well known in the geophysics community. However, some readers more familiar with remote-sensing textbooks will recognize respectively L_λ, E_λ, τ, δ, β, etc.

Square brackets designate an [alternate] name or unit for some quantity. Parentheses identify an (optional qualifier) of some quantity's name, or units that can be omitted by proper normalization and/or spectral integration. A dagger[†] designates a quantity that can be "spectral," and therefore carry a subscript λ or ν as in Planck function; in this case, (/μm) or (/cm^{-1}) appears in their units. In solar problems, where one can set $\mu_0 F_0$ to unity, these quantities can be non-dimensionalized altogether (hence the optional units of W/m^2). Many other optical quantities, e.g., σ or τ, can depend *parameterically* on λ or ν but this does not affect their units.

1 Scalar and Vector Quantities, about Units

In the following table, four different SI-derived units of length are used – sometimes in combinations – and each has its dedicated purpose in accounting for cloud optics, microphysics, radiation sources/sinks and transport:

- the "μm" is assigned to wavelengths and particle sizes (hence to spectral and/or single-scattering considerations in cloud/aerosol optics);
- the "cm" is assigned to droplet densities (hence to cloud microphysical considerations);
- the "m" is assigned to flux/irradiance units (hence to radiation budget considerations);
- the "km" is assigned to photon transport coefficients and scales of variability in clouds, including outer scales such as physical thickness (hence to all radiative transfer considerations).

In the volume, yet another derived SI unit is used occasionally, following established tradition: the "mm," for vertically-integrated liquid- and/or ice-water in cloud layers, while the "cm" is often used for total precipitable water in the atmosphere (always dominated by the vapor phase).

1. SCALAR AND VECTOR QUANTITIES, BEGIN

Symbol	Name	Relation to Others	Units (·: none)
A	absorptance		–
A_c	cloud fraction		–
$B_v(T)$	Planck('s) function	$2h\nu^3/c^2(e^{h\nu/kT}-1)$	$W/m^2/cm^{-1}/sr$
$B_\lambda(T)$	Planck('s) function, by wavelength	$B_{v(\lambda)}(T)\,\lvert d\nu/d\lambda\rvert(\lambda)$	$W/m^2/\mu m/sr$
BRF	(boundary) radiance as a Bidirectional Reflection Function	$\pi I/\mu_0 F_0$	–
d	distance		km
D	(radiative [photon]) diffusivity	$c\ell_t/3$	km^2/s
D'	diffusivity, for steady-state	$D/c = \ell_t/3$	km
$\mathbf{F}$	(radiant energy) flux vector[†]	$\int \mathbf{\Omega} I(\cdot, \mathbf{\Omega})\mathrm{d}\Omega$	$(W/m^2(\mu m))$
$\mathbf{F}'$	photon flux [current density] vector[†]	$\mathbf{F}/h\nu$	$(m^{-2}\,s^{-1}(\mu m^{-1}))$
$F_u, F_\uparrow$	upward (hemispherical) flux [irradiance][†]	$\int_{\mu>0}\mu I(\cdot, \mathbf{\Omega})\mathrm{d}\Omega$	$(W/m^2(\mu m))$
$F_d, F_\downarrow$	downward (hemispherical) flux [irradiance][†]	$\int_{\mu<0}\lvert\mu\rvert I(\cdot, \mathbf{\Omega})\mathrm{d}\Omega$	$(W/m^2(\mu m))$
$F_0, F_\odot$	solar constant [flux] [irradiance][†]		$(W/m^2(\mu m))$
g	asymmetry factor (of phase function)	$2\pi \int \mu_s P(\mu_s)\mathrm{d}\mu_s$	–
h	cloud geometrical [physical] thickness		km
H	(vertical integral of) horizontal flux (divergence)	$1 - (A + R + T)$	–
$I(\mathbf{x}, \mathbf{\Omega})$	radiance [(specific) intensity][†]		$(W/m^2(\mu m))/sr$
$I_0(\mathbf{x})$	directly transmitted [un-collided] radiance[†]		$(W/m^2(\mu m))/sr$
J	scalar [spherical, or actinic] flux[†]	$\int I(\cdot, \mathbf{\Omega})\mathrm{d}\Omega$	$(W/m^2(\mu m))$
J'	scalar [spherical, or actinic] photon flux[†]	$J/h\nu$	$(m^{-2}\,s^{-1}(\mu m^{-1}))$
k	wavenumber [Fourier-conjugate of position]	$(2\pi\times)1/r$	$(rad)km^{-1}$
k_s, k_v, k_λ	(gaseous) absorption coefficient		km^{-1}
$K, \mathcal{K}$	kernel of an integral RT equation		$km^{-3}sr^{-1}$

1. SCALAR AND VECTOR QUANTITIES, CONTINUED

Symbol	Name	Relation to Others	Units (-: none)
$L, \mathcal{L}$	integro-differential operator in RT equation		km^{-1}
L	(cumulative [total]) (geometrical) (photon) path((-)length)	ct	km
L_d	diffusion length	$\ell/\sqrt{3(1-\varpi_0)(1-\varpi_0 g)}$	km
L_D	horizontal domain size [outer scale]		km
$\mathcal{L}, L_{wc}$	liquid water content		g/m^3
ℓ	(local) (photon) mean-free-path, or "MFP"	$1/\sigma$	km
ℓ_t	transport mean-free-path	$\ell/(1-\varpi_0 g)$	km
n	order [number] of scattering[s]		–
$n(r)$	(cloud) droplet [particle] size-distribution [-spectrum]		$cm^{-3}/\mu m$
N	(cloud) droplet [particle] density	$\int n(r)dr$	cm^{-3}
$p(\Omega \to \Omega')$	(scattering) phase function	$\int_{4\pi} p(\Omega \to \Omega')d\Omega = 1$	1/sr
$p(\cos\theta_s)$	(azimuthally-symmetric) phase function	$2\pi\int_{-1}^{+1} p(\cos\theta_s)d\cos\theta_s = 1$	1/sr
$P(\theta_s)$	(alternately normalized) phase function	$\int_{-\pi}^{+\pi} P(\theta_s)\sin\theta_s d\theta_s = 2$	–
$Q(x,\Omega)$	source term† in RT equation		$(W/m^2(/\mu m))/sr/km$
r	distance [scale (in spatial statistics)]	$[(2\pi\times)1/k]$	km
r	droplet [particle] radius		μm
$\langle r^q \rangle$	qth moment of droplet [particle] radius	$\int r^q n(r)dr/N$	μm^q
r_e	effective (droplet) radius	$\langle r^3\rangle/\langle r^2\rangle$	μm
R	reflectance [albedo]		–
s	cross-section (per particle)		m^2
s	(photon) free path [step] (between two collisions)		m
s	Laplace-conjugate (variable) of time t		s^{-1}
S	similarity factor	ℓ_t/L_d	–
$S(x,\Omega)$	source function†	$\sigma_s(x)\int p(\Omega' \to \Omega)I(x,\Omega')d\Omega'$	$(W/m^2(/\mu m))/sr/km$
t	time		s

1. SCALAR AND VECTOR QUANTITIES, CONTINUED

Symbol	Name	Relation to Others	Units (-: none)
T	temperature		K [°C]
T	(total) transmittance	$T_{0[\text{dir}]} + T_{\text{d(if)}}$	–
T_0, T_{dir}, T	direct [un-collided] transmittance		–
$T_{\text{d}}, T_{\text{dif}}$	diffuse transmittance		
U	photon [radiant energy] density	J/c	m^{-3} [J/m^3]
W	liquid water path	$\int \mathcal{L} dz$	g/m^2
W'	liquid water path	W/ρ_w	cm [mm]
x, y, z	(Cartesian) coordinates, with z increasing upwards		km
$\mathbf{x}$	position (in 3D space)	$(x, y, z)^{\mathrm{T}}$	km
$\vec{x}$	position in the horizontal plane	$(x, y)^{\mathrm{T}}$	km
$\mathbf{x}_{\text{S}}$	position on the 2D boundary of a 3D volume		km
α	(planar) albedo		–
α_{S}	surface albedo		–
χ	(numerical) extrapolation length factor	$O(1)$	–
$\chi \ell_{\text{t}}$	extrapolation length		km
ε	emissivity		–
λ	wavelength	$10^4 [c]/\nu$	µm [m]
λ	(dimensionless) optical pathlength [cumulated extinction]	σL	–
μ	cosine of zenith [polar] angle	$\cos \theta$	–
$\mu_0 > 0$	cosine of solar (zenith) angle	$\cos \theta_0$	–
μ_{s}	cosine of scattering angle	$\cos \theta_{\text{s}}$	–
ν	frequency [wavenumber] (of photon [E-M wave])	$10^4 [c]/\lambda$	cm^{-1} [Hz]

1. SCALAR AND VECTOR QUANTITIES, END

Symbol	Name	Relation to Others	Units (-: none)
ϕ	azimuthal angle		°[rad]
ϕ_0	solar azimuthal angle		°[rad]
ϕ_s	scattering azimuthal angle		°[rad]
ϖ_0	single-scattering albedo	σ_s/σ	-
ρ	horizontal radius [displacement]	$\sqrt{x^2 + y^2}$	km
ρ	(ground) reflectivity [albedo]		-
ρ, n, N	(particle [mass]) density		m^{-3} (kg/m^3)
σ, σ_e	extinction (coefficient)	$\sigma_a + \sigma_s$	km^{-1}
σ_a	absorption coefficient	$s_a N$	km^{-1}
σ_s	scattering coefficient	$s_s N$	km^{-1}
τ	optical distance	$\int \sigma ds$	-
τ, τ_c	(cloud) optical depth [thickness]	$h/\ell = \sigma h$	-
τ_t	transport [rescaled] optical depth [thickness]	$h/\ell_t = (1 - \varpi_0 g)\tau$	-
θ	zenith [polar] angle, with $\theta = 0$ meaning "up"		°[rad]
$\theta_0 < \pi/2$	solar (zenith) [incidence] angle		°[rad]
θ_s	scattering angle	$\cos^{-1}(\mathbf{\Omega} \cdot \mathbf{\Omega'})$	°[rad]
$\mathbf{\Omega}(\mu, \phi)$	direction of propagation [viewing direction]		-
Ω_x	direction-cosine in x	$\sin\theta\cos\phi$	-
Ω_y	direction-cosine in y	$\sin\theta\sin\phi$	-
Ω_z	direction-cosine in z	$\cos\theta$	-
$\mathbf{\Omega}_0$	direction of incoming (collimated) (solar) radiation	$\mathbf{\Omega}(-\mu_0, \pi - \phi_0)$	-
$d\mathbf{\Omega}$	element of solid angle	$d\mu d\phi$	sr
$ds/d\Omega$	differential cross-section	$\sigma_s p(\cdot)/N$	m^2/sr

2. PHYSICAL CONSTANTS

Symbol	Name	Value	Units
c	velocity of light (in vacuum)	2.99825×10^8	m/s
h	Planck's constant	6.62618×10^{34}	J s
k	Boltzmann's constant	1.38066×10^{23}	J/K
σ_B	Stefan-Boltzmann constant, $\pi \int B_\nu(T) d\nu / T^4 = 2\pi^5 k^4 / 15 c^2 h^3$	5.67032×10^{-8}	W/m^2/K^4
$T\lambda_{max}$	Wien's wavelength displacement constant, $0.20141 \cdots \times hc/k$	0.28978	cm K
Tc/ν_{max}	Wien's wavenumber displacement constant, $0.35443 \cdots \times hc/k$	0.50995	cm K
T_s	triple-point of water	273.16	K
ρ_w	density of liquid water (at S.T.P.)	10^3 [1]	kg/m^3 [g/cm^3]

3. SOLAR, TERRESTRIAL AND ATMOSPHERIC CONSTANTS

Symbol	Name	Value	Units
$T_\odot$	Sun's effective emission temperature	≈ 5775	K
$R_\odot$	Sun's (equatorial) radius	6.960×10^5	km
AU	(mean) Sun-Earth distance [Astronomical Unit]	1.4960×10^8	km
$2R_\odot/\text{AU}$	angular diameter of Sun from Earth [31.9876 arc min]	9.3048	mrad
$\delta\Omega_\odot$	solid angle subtended by Sun from Earth, $\pi R_\odot^2/\text{AU}^2$	6.799×10^{-5}	sr
F_0	(mean) solar constant, $(\sigma T_\odot^4/\pi)\delta\Omega_\odot = \sigma T_\odot^4 (R_\odot/\text{AU})^2$	$\approx 1.365 \times 10^3$	W/m^2
$R_\oplus$	Earth's (equatorial) radius	6.378388×10^3	km
g	standard surface gravity	9.80665	m/s^2
p_s	standard surface pressure (N.B. 10^5 Pa = 10^3 hPa = 10^3 mbar)	1.01325	10^5 Pa [J/m^3]
R_a	specific gas constant of dry air	2.8704×10^2	J/K/kg
n_s	density of dry air (at S.T.P.)	2.686754×10^{25}	m^{-3}
ρ_s	mass density of dry air (at S.T.P.), $R_a T_s/p_s$	1.2925	kg/m^3
C_v	constant-volume specific heat of dry air (at S.T.P.)	0.718×10^3	J/K/kg
C_p	constant-pressure specific heat of dry air (at S.T.P.), $\approx (7/5)C_v$	1.005×10^3	J/K/kg
H	(nominal) scale-height of atmosphere, $R_a T_s/g$	7.994	km

4. SETS AND TOPOLOGY

Symbol	Name	Definition, Relation to Others
$\mathbb{N}$	all (non-negative) integers	$\{0, 1, 2, \ldots\}$
$\mathbb{Z}$	all (signed) integers	$\{\ldots, -2, -1, 0, 1, 2, \ldots\}$
$\mathbb{R}$	all real numbers	$(-\infty, +\infty)$
$\mathbb{R}^+$	non-negative real numbers	$[0, +\infty)$
$[a, b]$	closed interval of $\mathbb{R}$	$\{x \in \mathbb{R} : a \leq x \leq b\}$
(a, b)	open interval of $\mathbb{R}$	$\{x \in \mathbb{R} : a < x < b\}$
$[a, b)$	a semi-open interval of $\mathbb{R}$	$\{x \in \mathbb{R} : a \leq x < b\}$
$(a, b]$	another semi-open interval	$\{x \in \mathbb{R} : a < x \leq b\}$
Ξ	unit sphere (of direction vectors)	$\{\boldsymbol{\Omega} \in \mathbb{R}^3 : \|\boldsymbol{\Omega}\| = 1\}$
Ξ_+	upward hemisphere (of directions)	$\{\boldsymbol{\Omega}(\mu, \phi) \in \Xi : \mu \geq 0\}$
Ξ_-	downward hemisphere (of directions)	$\{\boldsymbol{\Omega}(\mu, \phi) \in \Xi : \mu \leq 0\}$
$S_1 \times S_2$	(Cartesian) product of 2 sets, e.g., $(\boldsymbol{x}, \boldsymbol{\Omega}) \in \mathbb{R}^3 \times \Xi$	
$1_S(\boldsymbol{x})$	indicator function of set S	$= 1$ if $\boldsymbol{x} \in S$, $= 0$ otherwise
$\underline{S}$	closure of a point-set S [itself *and* limits of all its infinite sequences]	$\supseteq S$, $= S$ if it is closed
M	((open) convex) (optical) medium	$\subseteq \mathbb{R}^3$
∂M	boundary of a(n optical) medium	

5. MATHEMATICAL ANALYSIS

Symbol	Name	Definition, Relation to Others
$\delta q, \Delta q$	a small amount [perturbation] of quantity q	
$f_a(x), f(a;x)$	function f of (fixed) parameter a and variable x	
$\hat{f}(s), L[f](s)$	Laplace transform of $f(t)$	$\int f(t) e^{-st} dt$
$\overrightarrow{x}$	2D (horizontal position) vector	
$\tilde{f}(\mathbf{k}), F[f](\mathbf{k})$	2D Fourier transform of $f(\overrightarrow{x})$	$\int f(\overrightarrow{x}) e^{\pm(2\pi)\mathrm{i}\mathbf{k}\bullet\overrightarrow{x}} \mathrm{d}\overrightarrow{x}$
$f(\overrightarrow{x}), F^{-1}[\tilde{f}](\overrightarrow{x})$	inverse Fourier transform of $\tilde{f}(\mathbf{k})$ in 2D	$\int \tilde{f}(\mathbf{k}) e^{\mp(2\pi)\mathrm{i}\mathbf{k}\bullet\overrightarrow{x}} \mathrm{d}\mathbf{k}(/(2\pi)^2)$
a	an arbitrary 3D vector	
ab	tensor [outer] product of vectors a and b	$\mathrm{Trace}[ab]$
$a \bullet b$	scalar [inner] product of vectors a and b	$(a\bullet a)^{1/2}$
$a, \|a\|$	(Euclidian) norm of vector a	
$< f, g >, (f, g)$	scalar [inner] product of functions f and g	$\int f(\overrightarrow{x}) g(\overrightarrow{x}) \mathrm{d}x$
$\|f\|^2$	(L_2) norm of function f, squared	$< f, f >$
$\partial/\partial\xi, \partial_\xi$	partial derivative with respect to $\xi = t, x, y, z$ or $\ldots$	
∇	gradient [nabla] (operator)	$(\partial_x, \partial_y, \partial_z)^{\mathrm{T}}$
∇^2	Laplacian (operator)	$\nabla \bullet \nabla = \partial_x^2 + \partial_y^2 + \partial_z^2$
$\delta(\cdot)$	(Dirac) delta-function, has $1/(\cdot)^d$-units in d dimensions	integral of $\delta(\cdot)$ in $d=1$
$H(\cdot), \Theta(\cdot)$	(Heaviside) step-function	
$\Gamma(a)$	(Euler's) Gamma-function $\int_0^\infty x^{a-1} e^{-x} \mathrm{d}x$	$\Gamma(n+1) = n!, n \in \mathbb{N}$
$n(x)$	(outward) (unit) normal vector (to boundary ∂M) (at x)	

6. PROBABILITY AND STATISTICS

Symbol	Name	Definition, Relation to Others
$\mathrm{Pr}\{\cdot\}$	probability of event defined in argument	
p_i	probability of discrete event with index "i"	
$P(\cdot)$	(cumulative) probability distribution function,	a definite integral of $p(\cdot)$
$p(\cdot)$	probability density function, has $1/(\cdot)$-units	$\pm dP/d(\cdot)$
$\mathcal{E}(\cdot)$	(mathematical) expectation of argument	
$\mathcal{D}(\cdot)$	variance of argument	$\mathcal{E}(([\cdot] - \mathcal{E}(\cdot))^2) = \mathcal{E}([\cdot]^2) - \mathcal{E}(\cdot)^2$
m_f	mean of quantity f	an alternative for $\mathcal{E}(f)$
σ_f, s_f	standard deviation of quantity f	an alternative for $\mathcal{D}(f)^{1/2}$
$\overline{(\cdots)}$	spatial average of $(\cdots)$ over a domain S	$\int_S (\cdots) d\vec{x} / \int_S d\vec{x}$
$\langle \cdots \rangle$	spatial or ensemble average (or both)	$\int (\cdots) dP(\cdots)$
$E_f(\mathbf{k})$	(wavenumber [energy] [power]) spectrum of $f(\vec{x})$	$\propto |\tilde{f}(\mathbf{k})|^2$
$E_f(k)$	(1D (wavenumber [etc.]) spectrum of $f(x)$	$\int E_f(\mathbf{k}')\delta(\|\mathbf{k}'\| - k)d\mathbf{k}'$
β	spectral exponent	$E(k) \propto k^{-\beta}$

Index

3D radiative transfer   465, 488, 523, 576

absorptance   165, 191, 474, 633
absorption   115, 116, 170, 551, 597
   coefficient   112, 176, 456, 493, 546, 627
addition theorem   244, 248, 626
Aerosol Robotic Network (AERONET)
      643
aerosol(s)   20, 131, 174, 195, 216, 371, 461
air mass   594
albedo   185, 346, 453, 524, 529, 531
   bias   428, 434, 532
   leaf   624, 635
   planetary   452
   single-scattering   170
   spherical   185
   surface   215, 454, 590
anomalous diffusion   366
asymmetry factor   174, 265, 303
Atmospheric Radiation Measurement
      (ARM)   VII, 9, 18, 26, 41, 44, 95,
      134, 142, 386, 417, 455, 479, 506
AVHRR   4, 529, 589
azimuthal angle   245, 247

backscattering   177
backward Monte Carlo   272, 273, 509, 549
band model   492
Bccr Law   168, 366
bidirectional reflectance distribution function
      (BRDF)   184, 185, 188, 222
bidirectional reflectance factor (BRF)   163,
      185, 630, 639, 650
bihemispherical reflectance (BHR)   630

blackbody   114, 118, 512
boundary conditions   194, 247–250, 252,
      255, 457, 488, 556, 628, 636
boundary value problem   243, 555, 630

CALIPSO   16, 591
cirrus   6, 38, 142, 490, 525
climate   317, 386, 425, 449, 479
   model   8, 9, 25, 95, 284, 456, 491
cloud droplet   8, 100, 102, 459
cloud fraction   76, 95, 97, 346, 372, 428,
      453, 462, 498
cloud overlap   463, 467, 475
cloud particles   99, 100, 104, 106, 137
cloud radiative forcing   439, 450
CloudSat   15, 21, 591
continuous random number   262
cooling rate   190, 462, 497
cross-section   110–112, 115, 117, 121, 123,
      168, 172
   absorption   176
   differential   172, 176
   extinction   169, 170
   maximum   261, 274
cumulonimbus   226, 525
cumulus   142, 276, 320, 417, 471, 493, 525

diffuse field   159, 595
diffuse hemispherical reflectance   625
diffuse irradiance   134, 144, 145
diffuse radiation   182, 244, 628, 630
diffuse source   161
diffusion equation   287, 306, 386
diffusivity   289, 560

Dirac δ-function   213, 217, 269, 628, 635
directional reciprocity   223, 530
Discrete Ordinate Method (DOM)   252, 253
discrete ordinates   37, 244–247, 252–259
discrete random number   263
DISORT   21, 246, 549
doubling-adding method   246, 599

effective cloud fraction   331, 501, 643
effective radius   41, 46, 123, 169, 370
eigenmatrix method   246
eigenvalue   246, 304, 634, 641, 645
eigenvector   304, 634
elementary volume   28, 189, 618, 627
emission   113, 182, 184, 508
emissivity   184, 186, 196, 495, 505
energy balance   546
energy budget   164
equivalence theorem   597
escape   178, 198, 204, 559
extinction   48, 49, 121, 123, 166–171, 625
    coefficient   110, 166, 170, 262, 596, 625, 627
extrapolation length   207, 555

flux   158, 159
    actinic   164, 285
    hemispherical   162, 163, 303
    horizontal   162, 543, 547, 548
    net   164
    scalar   164, 304, 546
    vector   164, 304, 322
    vertical   162
forward Monte Carlo   267, 509
Fourier series   244, 250
Fourier transform   250, 259, 298, 653, 654
free path   179–181, 366, 612
frequency   167

Global Ozone Monitoring Experiment (GOME)   589, 611
Green function   213, 298, 555, 578
    surface   214, 315, 637, 640
    volume   213
grid cells   95, 259

heating rate   14, 190, 296, 473, 477, 515
Hemispherical-Directional Reflectance Factor (HDRF)   630

Henyey–Greenstein phase function   174, 460
homogeneous clouds   34, 533
horizontal flux divergence   315, 547
horizontal transport   546, 568, 606
hot spot   618

ice particle   102, 105
ice water content   103
Independent Column/Pixel Approximation (ICA/IPA)   324, 427, 433, 607
    accuracy   548–552, 568, 569
    bias   330, 433
integral radiative transfer equation (integral RTE)   202, 267
integro-differential
    adjoint   218
    equation   189
    operator   189, 309
intensity   159, 510
interception   182, 631, 633–635
irradiance   188
    broadband (BB)   450, 459
    longwave   135, 138, 350
    shortwave   134, 135, 144
    spectral   454

k-distribution   24, 455, 493
kernel
    smoothing   578
    transport   202, 214, 267
    weakly singular   660

Lambertian
    emittance   163
    reflection   186
    surface   185, 187
LandSat   96, 274, 543, 572, 606
Legendre
    coefficients   176, 190, 244
    functions   245, 248
    polynomials   173, 244
    series   244
lidar   44, 46–48, 134, 136, 142, 195, 218, 302, 321
line profile   605
line-by-line (LBL) radiative transfer model (RTM)   115, 454, 492, 510
linear transport equation   617

liquid water content (LWC)   100–102, 145,
     370, 471, 656–658
liquid water path (LWP)   433
longwave radiation   109, 124, 452, 488
Lorenz-Mie theory   175

mean photon geometric path   597
mean photon pathlength   555, 602
mean-free-path   179, 262, 289, 319, 626
microwave   27, 29, 34, 40, 124, 135, 142
MISR   125–129, 132, 526, 533, 535
MODIS   16, 45, 46, 125–129, 131, 132,
     143, 527, 622
monochromatic   244, 454, 492
monodirectional   213, 628
Monte Carlo   28, 35, 37, 48, 243, 261, 272,
     374, 458, 508
MTI   210
multi-angle remote sensing   524
multiple scattering   52, 183, 591

net flux   162, 323, 450, 547
Neumann series   268, 272
Nonlocal Independent Pixel Approximation
     (NIPA)   314, 576–582
Normalized Difference Cloud Index (NDCI)
     641
number of scatterings   263, 555

optical depth   45, 46, 49, 170, 263, 427,
     529
   absorption   468, 506
   aerosol   590
   continuum   599
   mean   351
   oxygen   601, 612
   transport   332
   water vapor   467
optical distance   167, 201, 268, 618
optical medium   154
optical thickness   205
   absorption   117
   effective   427
   horizontal   430
   scattering   117
   total   120
   vertical   429
order-of-scattering   252, 261, 368, 559, 612
oxygen

A-band   369, 456, 593
   absorption   117, 589, 613
   lines   321, 594

particle density   169
particle size   105, 120, 489
particle size distribution   94
pathlength   321, 412, 596, 602, 625
photon mean-free-path   see mean-free-path
photon pathlength   see pathlength
Planck function   114, 184, 488
plane-parallel albedo bias   428, 436, 531
plane-parallel medium   195, 204, 310
polarization   160
POLDER   185, 619
probability of a clear line of sight (PCL)
     499

quadrature   245, 253

radar   34, 39, 47, 121, 133, 134
radiance   159–163, 285, 302, 309
radiant energy   158, 161
radiative
   forcing   10, 450
   roughening   570, 572
   smoothing   302, 315, 543, 570, 572, 582
radiative transfer equation
   adjoint   218
   integral   202, 267
   integro-differential   189
   plane-parallel   244
   stochastic   412
radiosity   306, 308
random walk   290, 559
Rayleigh scattering   175
   cross-section   175
   phase function   177
reciprocity   221, 530
reflectance   204
   bidirectional   163, 184
   bottom of atmosphere   640
   canopy   638
   leaf   624, 625
   local   547
   top of atmosphere   163
reflection   186, 188, 535, 630
remote sensing   523, 608
   clouds   319

Rotating Shadowband Spectrometer (RSS)
    602

scattering   170–178, 290
    angle   109, 174, 244
    center   617, 618
    coefficient   171, 626
    cross-section   116
    phase function   171, 244, 289, 304, 626
shortwave radiation   419
single-scattering albedo   170, 263, 627
solar constant   182
solar zenith angle   529
solid angle   108, 159, 162
source function   183, 201, 257
source term   181, 184, 555
spectrum
    electromagnetic (EM)   345
    longwave (thermal)   156, 165, 186, 488
    shortwave (solar)   165, 458
    wavenumber (power)   430, 568, 653
spherical harmonics   173, 244, 248, 311
Spherical Harmonics Discrete Ordinate
        Method (SHDOM)   257–261
stochastic   358, 386–465
stratocumulus   8, 318, 352, 371, 505, 543
structure function   655

surface reflection   181, 186

top-of-atmosphere (TOA)   163
total path   368, 561
transmission   110, 168, 179, 298
transmittance   204, 455
    canopy   634
    diffuse   463
    leaf   624
    local   547
transport mean-free-path   555
two-stream approximation   347

water vapor   117, 126, 598
wavelength
    centimeter (cm)   34
    infrared   6, 39
    microwave   39, 137, 142
    millimeter (mm)   15, 133, 602
    near-infrared   131
    solar   6, 39, 458
    sub-millimeter   31
wavenumber   488
    spectrum   565, 653, 655
window   488

zenith angle   185